时代教育·国外高校优秀教材精选

# 静 力 学

影印版·原书第 12 版

STATICS TWELFTH EDITION

［美］ R. C. 希伯勒（R. C. Hibbeler） 编著

机 械 工 业 出 版 社

本套书的目的是清晰、全面地向学生介绍理论力学的原理和应用。全套书分为两册：静力学与动力学。

本册为静力学分册，共 8 章，包括基本原理、力矢量、质点的平衡、力系的简化、刚体的平衡、结构的平衡、摩擦，以及虚功原理。

本套书可作为普通高校工科各专业理论力学课程双语教学用书，也可供相关专业的技术人员参考。

Authorized reprint from the Singapore edition of the original united English language edition, Engineering Mechanics: Statics, twelfth edition, by HIBBELER, RUSSELL C., published by Pearson Education, Inc., publishing as Prentice Hall. Copyright © 2010 by R. C. Hibbeler. Singapore adapted edition, Engineering Mechanics: Statics, twelfth Edition in SI units, adapted by S. C. Fan, published by PEARSON EDUCATION SOUTH ASIA PTE LTD., publishing as Prentice Hall. Copyright © R. C. Hibbeler.

All rights reserved. No part of this book may be reproduced or transmitted in any form or by any means, electronic or mechanical, including photocopying, recording or by any information storage retrieval system, without permission from Pearson Education, Inc.

China edition published by PEARSON EDUCATION ASIA LTD., and CHINA MACHINE PRESS. Copyright © 2013.

Authorized for sale and distribution in the Chinese mainland (excluding Hong Kong SAR, Macau SAR and Taiwan).

仅限于中国大陆地区（不包括香港、澳门特别行政区和台湾地区）销售发行。

本书英文影印版由培生教育出版公司授权机械工业出版社合作出版，未经出版者书面许可，不得以任何形式复制或抄袭本书的任何部分。

本书封面贴有 Pearson Education（培生教育出版集团）激光防伪标签。无标签者不得销售。

北京市版权局著作权登记号：01-2013-1429。

## 图书在版编目（CIP）数据

静力学：第 12 版：英文/（美）希伯勒（Hibbeler, R. C.）编著. —影印本. —北京：机械工业出版社，2013.11（2023.8 重印）
（时代教育：国外高校优秀教材精选）
ISBN 978-7-111-44734-4

Ⅰ. ①静… Ⅱ. ①希… Ⅲ. ①静力学—高等学校—教材—英文 Ⅳ. ①O312

中国版本图书馆 CIP 数据核字（2013）第 267739 号

机械工业出版社（北京市百万庄大街 22 号　邮政编码 100037）
策划编辑：姜　凤　　责任编辑：姜　凤
责任校对：陈立辉　　封面设计：张　静
责任印刷：单爱军
北京虎彩文化传播有限公司印刷
2023 年 8 月第 1 版第 5 次印刷
205mm×235mm · 29.75 印张 · 834 千字
标准书号：ISBN 978-7-111-44734-4
定价：89.00 元

| 电话服务 | 网络服务 |
| --- | --- |
| 客服电话：010-88361066 | 机　工　官　网：www.cmpbook.com |
| 　　　　　010-88379833 | 机　工　官　博：weibo.com/cmp1952 |
| 　　　　　010-68326294 | 金　书　网：www.golden-book.com |
| **封底无防伪标均为盗版** | 机工教育服务网：www.cmpedu.com |

# 国外高校优秀教材审定委员会

**主任委员：**

　　杨叔子

**委员（按姓氏笔画为序）：**

| 王先逵 | 王大康 | 白峰杉 | 史荣昌 | 朱孝禄 |
| 陆启韶 | 张润琦 | 张　策 | 张三慧 | 张福润 |
| 张延华 | 吴宗泽 | 吴　麒 | 宋心琦 | 李俊峰 |
| 佘远斌 | 陈文楷 | 陈立周 | 单辉祖 | 俞正光 |
| 赵汝嘉 | 郭可谦 | 翁海珊 | 龚光鲁 | 章栋恩 |
| 黄永畅 | 谭泽光 |

# 出 版 说 明

随着我国加入 WTO，国际间的竞争越来越激烈，而国际间的竞争实际上也就是人才的竞争、教育的竞争。为了加快培养具有国际竞争力的高水平技术人才，加快我国教育改革的步伐，国家教育部出台了一系列倡导高校开展双语教学、引进原版教材的政策。以此为契机，机械工业出版社推出了一系列国外影印版教材，其内容涉及高等学校公共基础课，以及机、电、信息领域的专业基础课和专业课。

引进国外优秀原版教材，在有条件的学校推动开展英语授课或双语教学，自然也引进了先进的教学思想和教学方法，这对提高我国自编教材的水平，加强学生的英语实际应用能力，使我国的高等教育尽快与国际接轨，必将起到积极的推动作用。

为了做好教材的引进工作，机械工业出版社特别成立了由著名专家组成的国外高校优秀教材审定委员会。这些专家对实施双语教学做了深入细致的调查研究，对引进原版教材提出许多建设性意见，并慎重地对每一本将要引进的原版教材一审再审，精选再精选，确认教材本身的质量水平，以及权威性和先进性，以期所引进的原版教材能适应我国学生的外语水平和学习特点。在引进工作中，审定委员会还结合我国高校教学课程体系的设置和要求，对原版教材的教学思想和方法的先进性、科学性严格把关，同时尽量考虑原版教材的系统性和经济性。

这套教材出版后，我们将根据各高校的双语教学计划，及时地将其推荐给各高校选用。希望高校师生在使用教材后及时反馈意见和建议，使我们更好地为教学改革服务。

机械工业出版社<br>高等教育分社

# 前　言

本书的目的是清晰、全面地向学生介绍理论力学的原理和应用。为实现这个目标，本书在编写过程中汲取了许多从事教学工作的专家和作者本人的学生的批评及意见。本书为第12版，与前一版相比多处作了较大改进，希望能为教师和学生带来更大帮助。

### 新特点

**基础题**　在每章例题之后都安排了基础题。通过这些题，可以培养学生对一些基本概念的简单应用能力，使学生掌握处理简单问题的基本技巧，为后续求解一般问题打下良好基础。由于附录已经给出这些问题的全部计算结果和部分解题过程，因此也可将这些题作为延伸性例题。另外，从考试角度，基础题为学生提供了一种非常高效的学习方式。同时，在理论力学考试准备前期，可以通过这些练习复习基础知识。

**内容修订**　本版对书中每一部分内容都进行了仔细核查。为更好地解释概念，对许多章节的材料进行了重新编写。为更加突出一些重要概念的应用，增加和更换了部分例题。

**概念题**　每章结尾处通常安排一些概念题，它们与该章中的力学原理的应用相关。安排这些分析和设计型问题的目的，是引导学生通过照片所描述的实际生活场景进行思考。在学生做过一些相关类型的练习后，可将概念题留为作业或练习。

**附加的照片**　整本书更新或增加近60幅照片，它们反映了本书讲授的知识在实际生活中的实用性。这些照片一般用于解释如何在现实世界中应用力学原理处理问题。在某些章节，也通过照片展示工程师处理实际问题的基本过程：首先必须建立一个便于分析的理想模型，然后画受力图，最后应用力学原理求解。

**新习题**　在本版中，新增或更新近800道习题，约占习题总量的50%。它们涉及航空航天、石油工程和生物力学等领域。新版的习题总量也比上一版增加近17%。

### 其他特点

除了上面提到的新特点外，现将本书正文的一些其他显著特点叙述如下：

**内容组织**　本版对章节的每一部分都进行了精心组织和安排，包括特定主题的解释、说明性例题和课外习题。每节的主题作为该节副标题，以黑体字形式表示。这样做的目的是为引入每个新定义或概念提供一种结构化方法，同时也方便以后的复习和参考。

**章节内容**　每章开篇都以生活或工程中常见的实例为基础引入和论证一些应用广泛的力学原理。每章章首部分的粗体圆点清单概括了本章的主要内容。

**突出受力图**　画受力图是求解力学问题的关键所在。因此，本书从始至终一直强调"画受力图"的重要性，并且通过特定章节和具体例题详细讲解"画受力图"的步骤和注意事项。同时，通过相关课外习题的训练，使学生熟练掌握"画受力图"的方法步骤。

**分析过程** 本书第 1 章 1.6 节给出力学问题的一般分析过程。它适用于书中所有类型的习题。在后续章节学习中，这一特点给初学者提供了一种在应用理论时可遵循的合理方法。为阐明这种方法的应用，本书的例题也采用上述方法求解。实际上，随着学生对相关原理理解的不断深入和自信心的日渐增强，他们将形成自己的解题过程。

**重点** 这部分主要总结和回顾每节中最重要的概念，并强调这些概念对应用原理求解问题的重要性。

**概念的理解** 本书对各章照片中的力学问题都进行了简化处理，然后应用力学原理求解这些问题。这样更便于阐明原理中一些更重要的概念，解释方程中所涉及术语的物理意义。通过这些简化应用，不仅增强学生对本学科的兴趣和对例题的理解，同时也为学生进一步求解习题奠定了良好基础。

**课外习题** 除基础题和概念性习题外，本书还包括如下类型题：

- **画受力图的习题** 本书的部分章节包含一些介绍性问题，对于问题中的一些特例仅需画受力图便可求解出来。这些习题可使学生清楚"正确画出受力图"对求解任何平衡问题都至关重要。

- **一般分析和设计题** 本书的大部分习题取材于工程实际问题。一些习题来自于工业应用的实际产品。希望通过这些实例，激发学生对理论力学的兴趣，培养将实际物理描述简化为理论模型或符号表示（这些描述更便于应用力学原理）的能力。

- **计算机的问题** 本书精心设计了一些习题，这些习题必须通过一定的数值计算才能得出结论，数值求解过程可在台式计算机或可编程便携计算器上完成。这样做的好处是既能扩展其他方面的数理分析能力，又能有更多精力关注力学原理应用本身。这种类型习题的题号前加了符号"■"作为标记。

新版中的课外习题非常多，具体可分为三类：一类是比较简单的习题，书后附录给出参考答案，题号前没有任何标记；第二类习题的题号前加上圆点（●）作为标记，书后附录给出了建议、关键公式和计算结果；最后一类习题的题号前加上星号（＊）作为标记，书中没有给出参考答案。

**精确性** 与前一版相比，新版对书中所有文字和习题解答都进行了校核。除作者以外，还有如下人员参与了校核工作：弗吉尼亚理工大学的 Scott Hendricks、南佛罗里达大学的 Karim Nohra、劳雷尔技术学院综合出版服务部的 Kurt Norlin。此外，工程师 Kai Beng 不仅对本书作出精确评论，同时还给出许多内容改进方面的建议。

**内容** 全套书共分为 19 章。每章内容安排循序渐进，应用力学原理先处理简单情况，再处理更复杂的情况。一般地，每个原理首先应用于质点，然后应用于受平面力系作用的刚体，最后应用于受空间力系作用的刚体。

本册为静力学分册，共 8 章。第 1 章，介绍力学概况，讨论单位制问题。第 2 章介绍矢量的概念和汇交力系的特点。然后，在第 3 章中，将上述理论应用于质点的平衡问题。第 4 章对集中力系和分布力系进行了一般性讨论，并给出这些力系的简化方法。第 5 章建立刚体平衡原理。接着第 6 章应用这些原理求解涉及桁架、框架和机构的平衡等特殊问题。第 7 章讨论涉及摩擦力的应用问题。本书中标有星号（＊）的章节包含更深一些的内容，如果时间允许可以讲授。这些内容大多包含在第 8 章（虚功原理）中。注意，在更深的课程中讨论基本原理时，这些材料是非常适合的参考资料。最后，附录 A 提供书中求解问题所必需的数学公式列表。

**讲授次序** 书中有些章节，在不影响本书连续性前提下，教师可自行安排不同顺序。例如，可先学习第 2 章和 4.2 节（矢量积），之后再引入力的概念并讲解所有必要的矢量分析方法，然后在介绍第 4 章（力系的简化）的其余部分以后，可以讨论第 3 章和第 5 章的平衡问题。

---

### 致 谢

我力求写好本书，以引起教师和学生的兴趣。在本书这些年的编写过程中，得到许多人的帮助。我非常感谢他们给予的许多宝贵建议和评述。在这里，我还要特别感谢下面这些人，在本书第 12 版的准备过程中，他们提供了许多中肯的建议。

Yesh P. Singh，得克萨斯大学圣安东尼奥分校

Manoj Chopra，中佛罗里达大学

Kathryn McWilliams，萨省大学

Daniel Linzell，宾夕法尼亚州立大学

Larry Banta，西弗吉尼亚大学

Manohar L. Arora，科罗拉多矿业大学

Robert Rennaker，俄克拉荷马大学

Ahmad M. Itani，内华达大学

另外还有一些人有理由获得特殊赞誉。Vince O'Brien（项目组管理主任）和 Rose Kernan（制作编辑）多年来给予我许多鼓励和支持。坦诚地讲，如果没有他们的帮助，这本书的全面修订不可能完成。还有，Kai Beng Yap，我的多年好友和同事，在全书手稿的审核及习题解答的准备过程中给予我极大的帮助。在这份致谢中，我还要特别感谢劳雷尔技术学院综合出版服务部的 Kurt Norlin。在本书出版过程中，我也要感谢我的夫人 Conny 和女儿 Mary Ann 给予的帮助，她们承担了出版底稿准备中的大量校对和打印工作。

最后，我还要感谢我的全体学生和教育同仁们，他们利用大量的休息时间通过电子邮件提供了许多宝贵的建议和意见。由于篇幅所限不能一一列举，在此谨致以诚挚的谢意。

无论何时收到您关于本版的任何评论、建议或问题，我都将不胜感激。

**RUSSELL CHARLES HIBBELER**
hibbeler@bellsouth.net

# 目 录

前言
静力学基本公式
SI 词头和换算系数

**第 1 章　基本原理** ………………… 3
　本章目标 ……………………………… 3
　1.1　力学 ……………………………… 3
　1.2　基本概念 ………………………… 4
　1.3　计量单位 ………………………… 8
　1.4　国际单位制 ……………………… 9
　1.5　数值计算 ………………………… 10
　1.6　一般分析过程 …………………… 12

**第 2 章　力矢量** …………………… 17
　本章目标 ……………………………… 17
　2.1　标量与矢量 ……………………… 17
　2.2　矢量运算 ………………………… 18
　2.3　力的矢量合成 …………………… 20
　2.4　平面力系的合成 ………………… 32
　2.5　笛卡儿矢量 ……………………… 43
　2.6　笛卡儿矢量的相加 ……………… 46
　2.7　位置矢量 ………………………… 56
　2.8　沿直线方向的力矢量 …………… 59
　2.9　标量积 …………………………… 69

**第 3 章　质点的平衡** ……………… 85
　本章目标 ……………………………… 85
　3.1　质点的平衡条件 ………………… 85
　3.2　受力图 …………………………… 86
　3.3　共面力系 ………………………… 89
　3.4　空间共点力系 …………………… 103

**第 4 章　力系的简化** ……………… 117
　本章目标 ……………………………… 117
　4.1　力矩——标量公式 ……………… 117
　4.2　矢量积 …………………………… 121
　4.3　力矩——矢量公式 ……………… 124
　4.4　合力矩定理 ……………………… 128
　4.5　力对轴的矩 ……………………… 139
　4.6　力偶矩 …………………………… 148
　4.7　力系的简化 ……………………… 160
　4.8　力系的最简结果 ………………… 170
　4.9　简单分布载荷的简化 …………… 183

**第 5 章　刚体的平衡** ……………… 199
　本章目标 ……………………………… 199
　5.1　刚体平衡的条件 ………………… 199
　5.2　受力分析：平面力系 …………… 201
　5.3　平衡方程：平面力系 …………… 214
　5.4　二力构件以及三力构件 ………… 224
　5.5　受力分析：空间力系 …………… 237
　5.6　平衡方程：空间力系 …………… 242
　5.7　约束及静定问题 ………………… 243

**第 6 章　结构的平衡** ……………… 263
　本章目标 ……………………………… 263
　6.1　简单桁架 ………………………… 263
　6.2　节点法 …………………………… 266
　6.3　零力杆 …………………………… 272
　6.4　截面法 …………………………… 280
　*6.5　空间桁架 ………………………… 290
　6.6　构架与机具 ……………………… 294

## 第 7 章　摩擦 ······ 329
　本章目标 ······ 329
　7.1　干摩擦的性质 ······ 329
　7.2　含干摩擦的问题 ······ 334
　7.3　楔块 ······ 354
　7.4　螺钉的摩擦力 ······ 356
　7.5　传动带的摩擦力 ······ 363
　*7.6　环形推力轴承、枢轴承和圆盘的摩擦力 ······ 371
　7.7　滑动轴承的摩擦力 ······ 374
　*7.8　滚动摩阻 ······ 376

## 第 8 章　虚功原理 ······ 389
　本章目标 ······ 389
　8.1　功的定义 ······ 389
　8.2　虚功原理 ······ 391
　*8.3　虚功原理在连接刚体系统中的应用 ······ 393
　*8.4　保守力 ······ 405
　*8.5　势能 ······ 406
　*8.6　平衡的势能判据 ······ 408
　*8.7　平衡的稳定性 ······ 409

## 附录 ······ 424
　附录 A　数学复习和公式 ······ 424

## 基础题的部分解答和答案 ······ 428

## 习题答案 ······ 440

# CONTENTS

## 1
### General Principles 3

Chapter Objectives 3
1.1 Mechanics 3
1.2 Fundamental Concepts 4
1.3 Units of Measurement 8
1.4 The International System of Units 9
1.5 Numerical Calculations 10
1.6 General Procedure for Analysis 12

## 2
### Force Vectors 17

Chapter Objectives 17
2.1 Scalars and Vectors 17
2.2 Vector Operations 18
2.3 Vector Addition of Forces 20
2.4 Addition of a System of Coplanar Forces 32
2.5 Cartesian Vectors 43
2.6 Addition of Cartesian Vectors 46
2.7 Position Vectors 56
2.8 Force Vector Directed Along a Line 59
2.9 Dot Product 69

## 3
### Equilibrium of a Particle 85

Chapter Objectives 85
3.1 Condition for the Equilibrium of a Particle 85
3.2 The Free-Body Diagram 86
3.3 Coplanar Force Systems 89
3.4 Three-Dimensional Force Systems 103

## 4
### Force System Resultants 117

Chapter Objectives 117
4.1 Moment of a Force—Scalar Formulation 117
4.2 Cross Product 121
4.3 Moment of a Force—Vector Formulation 124
4.4 Principle of Moments 128
4.5 Moment of a Force about a Specified Axis 139
4.6 Moment of a Couple 148
4.7 Simplification of a Force and Couple System 160
4.8 Further Simplification of a Force and Couple System 170
4.9 Reduction of a Simple Distributed Loading 183

# 5
## Equilibrium of a Rigid Body 199

Chapter Objectives 199
- **5.1** Conditions for Rigid-Body Equilibrium 199
- **5.2** Free-Body Diagrams 201
- **5.3** Equations of Equilibrium 214
- **5.4** Two- and Three-Force Members 224
- **5.5** Free-Body Diagrams 237
- **5.6** Equations of Equilibrium 242
- **5.7** Constraints and Statical Determinacy 243

# 6
## Structural Analysis 263

Chapter Objectives 263
- **6.1** Simple Trusses 263
- **6.2** The Method of Joints 266
- **6.3** Zero-Force Members 272
- **6.4** The Method of Sections 280
- \***6.5** Space Trusses 290
- **6.6** Frames and Machines 294

# 7
## Friction 329

Chapter Objectives 329
- **7.1** Characteristics of Dry Friction 329
- **7.2** Problems Involving Dry Friction 334
- **7.3** Wedges 354
- **7.4** Frictional Forces on Screws 356
- **7.5** Frictional Forces on Flat Belts 363
- \***7.6** Frictional Forces on Collar Bearings, Pivot Bearings, and Disks 371
- **7.7** Frictional Forces on Journal Bearings 374
- \***7.8** Rolling Resistance 376

# 8
## Virtual Work 389

Chapter Objectives 389
- **8.1** Definition of Work 389
- **8.2** Principle of Virtual Work 391
- \***8.3** Principle of Virtual Work for a System of Connected Rigid Bodies 393
- \***8.4** Conservative Forces 405
- \***8.5** Potential Energy 406
- \***8.6** Potential-Energy Criterion for Equilibrium 408
- \***8.7** Stability of Equilibrium Configuration 409

# Appendix

A. Mathematical Review and Expressions 424

# Fundamental Problems Partial Solutions And Answers 428

# Answers to Selected Problems 440

# Fundamental Equations of Statics

## Cartesian Vector

$$\mathbf{A} = A_x\mathbf{i} + A_y\mathbf{j} + A_z\mathbf{k}$$

### Magnitude

$$A = \sqrt{A_x^2 + A_y^2 + A_z^2}$$

### Directions

$$\mathbf{u}_A = \frac{\mathbf{A}}{A} = \frac{A_x}{A}\mathbf{i} + \frac{A_y}{A}\mathbf{j} + \frac{A_z}{A}\mathbf{k}$$
$$= \cos\alpha\,\mathbf{i} + \cos\beta\,\mathbf{j} + \cos\gamma\,\mathbf{k}$$
$$\cos^2\alpha + \cos^2\beta + \cos^2\gamma = 1$$

## Dot Product

$$\mathbf{A}\cdot\mathbf{B} = AB\cos\theta$$
$$= A_xB_x + A_yB_y + A_zB_z$$

## Cross Product

$$\mathbf{C} = \mathbf{A}\times\mathbf{B} = \begin{vmatrix} \mathbf{i} & \mathbf{j} & \mathbf{k} \\ A_x & A_y & A_z \\ B_x & B_y & B_z \end{vmatrix}$$

## Cartesian Position Vector

$$\mathbf{r} = (x_2 - x_1)\mathbf{i} + (y_2 - y_1)\mathbf{j} + (z_2 - z_1)\mathbf{k}$$

## Cartesian Force Vector

$$\mathbf{F} = F\mathbf{u} = F\left(\frac{\mathbf{r}}{r}\right)$$

## Moment of a Force

$$M_o = Fd$$
$$\mathbf{M}_o = \mathbf{r}\times\mathbf{F} = \begin{vmatrix} \mathbf{i} & \mathbf{j} & \mathbf{k} \\ r_x & r_y & r_z \\ F_x & F_y & F_z \end{vmatrix}$$

## Moment of a Force About a Specified Axis

$$M_a = \mathbf{u}\cdot\mathbf{r}\times\mathbf{F} = \begin{vmatrix} u_x & u_y & u_z \\ r_x & r_y & r_z \\ F_x & F_y & F_z \end{vmatrix}$$

## Simplification of a Force and Couple System

$$\mathbf{F}_R = \Sigma\mathbf{F}$$
$$(\mathbf{M}_R)_O = \Sigma\mathbf{M} + \Sigma\mathbf{M}_O$$

## Equilibrium

### Particle

$$\Sigma F_x = 0,\ \Sigma F_y = 0,\ \Sigma F_z = 0$$

### Rigid Body-Two Dimensions

$$\Sigma F_x = 0,\ \Sigma F_y = 0,\ \Sigma M_O = 0$$

### Rigid Body-Three Dimensions

$$\Sigma F_x = 0,\ \Sigma F_y = 0,\ \Sigma F_z = 0$$
$$\Sigma M_{x'} = 0,\ \Sigma M_{y'} = 0,\ \Sigma M_{z'} = 0$$

## Friction

Static (maximum)  $F_s = \mu_s N$
Kinetic  $F_k = \mu_k N$

## Center of Gravity

### Particles or Discrete Parts

$$\bar{r} = \frac{\Sigma\tilde{r}W}{\Sigma W}$$

### Body

$$\bar{r} = \frac{\int\tilde{r}\,dW}{\int dW}$$

## Area and Mass Moments of Inertia

$$I = \int r^2\,dA \qquad I = \int r^2\,dm$$

### Parallel-Axis Theorem

$$I = \bar{I} + Ad^2 \qquad I = \bar{I} + md^2$$

### Radius of Gyration

$$k = \sqrt{\frac{I}{A}} \qquad k = \sqrt{\frac{I}{m}}$$

## Virtual Work

$$\delta U = 0$$

## SI Prefixes

| Multiple | Exponential Form | Prefix | SI Symbol |
|---|---|---|---|
| 1 000 000 000 | $10^9$ | giga | G |
| 1 000 000 | $10^6$ | mega | M |
| 1 000 | $10^3$ | kilo | k |

| Submultiple | | | |
|---|---|---|---|
| 0.001 | $10^{-3}$ | milli | m |
| 0.000 001 | $10^{-6}$ | micro | $\mu$ |
| 0.000 000 001 | $10^{-9}$ | nano | n |

## Conversion Factors (FPS) to (SI)

| Quantity | Unit of Measurement (FPS) | Equals | Unit of Measurement (SI) |
|---|---|---|---|
| Force | lb | | 4.4482 N |
| Mass | slug | | 14.5938 kg |
| Length | ft | | 0.3048 m |

## Conversion Factors (FPS)

1 ft = 12 in. (inches)
1 mi. (mile) = 5280 ft
1 kip (kilopound) = 1000 lb
1 ton = 2000 lb

# ENGINEERING MECHANICS

# STATICS

**TWELFTH EDITION**

The design of this rocket and gantry structure requires a basic knowledge of both statics and dynamics, which form the subject matter of engineering mechanics.

# General Principles

**CHAPTER OBJECTIVES**

- To provide an introduction to the basic quantities and idealizations of mechanics.
- To give a statement of Newton's Laws of Motion and Gravitation.
- To review the principles for applying the SI system of units.
- To examine the standard procedures for performing numerical calculations.
- To present a general guide for solving problems.

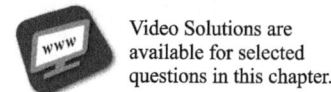

Video Solutions are available for selected questions in this chapter.

## 1.1 Mechanics

*Mechanics* is a branch of the physical sciences that is concerned with the state of rest or motion of bodies that are subjected to the action of forces. In general, this subject can be subdivided into three branches: *rigid-body mechanics, deformable-body mechanics*, and *fluid mechanics*. In this book, we will study rigid-body mechanics since it is a basic requirement for the study of the mechanics of deformable bodies and the mechanics of fluids. Furthermore, rigid-body mechanics is essential for the design and analysis of many types of structural members, mechanical components, or electrical devices encountered in engineering.

Rigid-body mechanics is divided into two areas: statics and dynamics. *Statics* deals with the equilibrium of bodies, that is, those that are either at rest or move with a constant velocity; whereas *dynamics* is concerned with the accelerated motion of bodies. We can consider statics as a special case of dynamics, in which the acceleration is zero; however, statics deserves separate treatment in engineering education since many objects are designed with the intention that they remain in equilibrium.

**Historical Development.** The subject of statics developed very early in history because its principles can be formulated simply from measurements of geometry and force. For example, the writings of Archimedes (287–212 B.C.) deal with the principle of the lever. Studies of the pulley, inclined plane, and wrench are also recorded in ancient writings—at times when the requirements for engineering were limited primarily to building construction.

Since the principles of dynamics depend on an accurate measurement of time, this subject developed much later. Galileo Galilei (1564–1642) was one of the first major contributors to this field. His work consisted of experiments using pendulums and falling bodies. The most significant contributions in dynamics, however, were made by Isaac Newton (1642–1727), who is noted for his formulation of the three fundamental laws of motion and the law of universal gravitational attraction. Shortly after these laws were postulated, important techniques for their application were developed by such notables as Euler, D'Alembert, Lagrange, and others.

## 1.2 Fundamental Concepts

Before we begin our study of engineering mechanics, it is important to understand the meaning of certain fundamental concepts and principles.

**Basic Quantities.** The following four quantities are used throughout mechanics.

**Length.** *Length* is used to locate the position of a point in space and thereby describe the size of a physical system. Once a standard unit of length is defined, one can then use it to define distances and geometric properties of a body as multiples of this unit.

**Time.** *Time* is conceived as a succession of events. Although the principles of statics are time independent, this quantity plays an important role in the study of dynamics.

**Mass.** *Mass* is a measure of a quantity of matter that is used to compare the action of one body with that of another. This property manifests itself as a gravitational attraction between two bodies and provides a measure of the resistance of matter to a change in velocity.

**Force.** In general, *force* is considered as a "push" or "pull" exerted by one body on another. This interaction can occur when there is direct contact between the bodies, such as a person pushing on a wall, or it can occur through a distance when the bodies are physically separated. Examples of the latter type include gravitational, electrical, and magnetic forces. In any case, a force is completely characterized by its magnitude, direction, and point of application.

## 1.2 Fundamental Concepts

**Idealizations.** Models or idealizations are used in mechanics in order to simplify application of the theory. Here, we will consider three important idealizations.

**Particle.** A *particle* has a mass, but a size that can be neglected. For example, the size of the earth is insignificant compared to the size of its orbit, and therefore the earth can be modeled as a particle when studying its orbital motion. When a body is idealized as a particle, the principles of mechanics reduce to a rather simplified form since the geometry of the body *will not be involved* in the analysis of the problem.

**Rigid Body.** A *rigid body* can be considered as a combination of a large number of particles in which all the particles remain at a fixed distance from one another, both before and after applying a load. This model is important because the material properties of any body that is assumed to be rigid will not have to be considered when studying the effects of forces acting on the body. In most cases, the actual deformations occurring in structures, machines, mechanisms, and the like are relatively small, and the rigid-body assumption is suitable for analysis.

**Concentrated Force.** A *concentrated force* represents the effect of a loading which is assumed to act at a point on a body. We can represent a load by a concentrated force, provided the area over which the load is applied is very small compared to the overall size of the body. An example would be the contact force between a wheel and the ground.

Three forces act on the hook at $A$. Since these forces all meet at a point, then for any force analysis, we can assume the hook to be represented as a particle.

Steel is a common engineering material that does not deform very much under load. Therefore, we can consider this railroad wheel to be a rigid body acted upon by the concentrated force of the rail.

### Newton's Three Laws of Motion.
Engineering mechanics is formulated on the basis of Newton's three laws of motion, the validity of which is based on experimental observation. These laws apply to the motion of a particle as measured from a *nonaccelerating* reference frame. They may be briefly stated as follows.

**First Law.** A particle originally at rest, or moving in a straight line with constant velocity, tends to remain in this state provided the particle is *not* subjected to an unbalanced force, Fig. 1–1a.

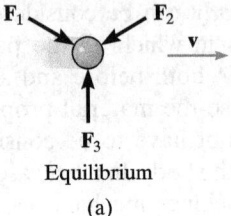

Equilibrium
(a)

**Second Law.** A particle acted upon by an *unbalanced force* **F** experiences an acceleration **a** that has the same direction as the force and a magnitude that is directly proportional to the force, Fig. 1–1b.* If **F** is applied to a particle of mass $m$, this law may be expressed mathematically as

$$\mathbf{F} = m\mathbf{a} \qquad (1\text{–}1)$$

Accelerated motion
(b)

**Third Law.** The mutual forces of action and reaction between two particles are equal, opposite, and collinear, Fig. 1–1c.

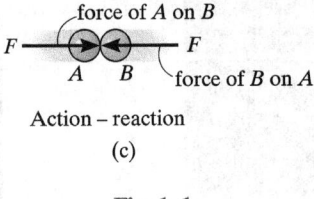

Action – reaction
(c)

**Fig. 1–1**

---

*Stated another way, the unbalanced force acting on the particle is proportional to the time rate of change of the particle's linear momentum.

**Newton's Law of Gravitational Attraction.** Shortly after formulating his three laws of motion, Newton postulated a law governing the gravitational attraction between any two particles. Stated mathematically,

$$F = G\frac{m_1 m_2}{r^2} \qquad (1\text{--}2)$$

where

$F$ = force of gravitation between the two particles

$G$ = universal constant of gravitation; according to experimental evidence, $G = 66.73(10^{-12})$ m³/(kg·s²)

$m_1, m_2$ = mass of each of the two particles

$r$ = distance between the two particles

**Weight.** According to Eq. 1–2, any two particles or bodies have a mutual attractive (gravitational) force acting between them. In the case of a particle located at or near the surface of the earth, however, the only gravitational force having any sizable magnitude is that between the earth and the particle. Consequently, this force, termed the *weight*, will be the only gravitational force considered in our study of mechanics.

From Eq. 1–2, we can develop an approximate expression for finding the weight $W$ of a particle having a mass $m_1 = m$. If we assume the earth to be a nonrotating sphere of constant density and having a mass $m_2 = M_e$, then if $r$ is the distance between the earth's center and the particle, we have

$$W = G\frac{mM_e}{r^2}$$

Letting $g = GM_e/r^2$ yields

$$\boxed{W = mg} \qquad (1\text{--}3)$$

By comparison with $\mathbf{F} = m\mathbf{a}$, we can see that $g$ is the acceleration due to gravity. Since it depends on $r$, then the weight of a body is *not* an absolute quantity. Instead, its magnitude is determined from where the measurement was made. For most engineering calculations, however, $g$ is determined at sea level and at a latitude of 45°, which is considered the "standard location."

The astronaut is weightless, for all practical purposes, since she is far removed from the gravitational field of the earth.

## 1.3 Units of Measurement

The four basic quantities—length, time, mass, and force—are not all independent from one another; in fact, they are *related* by Newton's second law of motion, $\mathbf{F} = m\mathbf{a}$. Because of this, the *units* used to measure these quantities cannot *all* be selected arbitrarily. The equality $\mathbf{F} = m\mathbf{a}$ is maintained only if three of the four units, called *base units*, are *defined* and the fourth unit is then *derived* from the equation.

**SI Units.** The International System of units, abbreviated SI after the French "Système International d'Unités," is a modern version of the metric system which has received worldwide recognition. As shown in Table 1–1, the SI system defines length in meters (m), time in seconds (s), and mass in kilograms (kg). The unit of force, called a newton (N), is *derived* from $\mathbf{F} = m\mathbf{a}$. Thus, 1 newton is equal to a force required to give 1 kilogram of mass an acceleration of 1 m/s² ($N = kg \cdot m/s^2$).

If the weight of a body located at the "standard location" is to be determined in newtons, then Eq. 1–3 must be applied. Here measurements give $g = 9.806\ 65$ m/s²; however, for calculations, the value $g = 9.81$ m/s² will be used. Thus,

$$W = mg \quad (g = 9.81 \text{ m/s}^2) \quad (1\text{–}4)$$

Therefore, a body of mass 1 kg has a weight of 9.81 N, a 2-kg body weighs 19.62 N, and so on, Fig. 1–2.

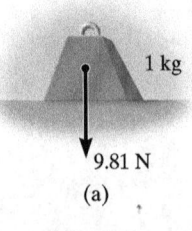

9.81 N
(a)

Fig. 1–2

| TABLE 1–1 | Systems of Units | | | |
|---|---|---|---|---|
| Name | Length | Time | Mass | Force |
| International System of Units<br>SI | meter<br><br>m | second<br><br>s | kilogram<br><br>kg | newton*<br>$N$<br>$\left(\dfrac{kg \cdot m}{s^2}\right)$ |

## 1.4 The International System of Units

The SI system of units is used extensively in this book since it is intended to become the worldwide standard for measurement. Therefore, we will now present some of the rules for its use and some of its terminology relevant to engineering mechanics.

**Prefixes.** When a numerical quantity is either very large or very small, the units used to define its size may be modified by using a prefix. Some of the prefixes used in the SI system are shown in Table 1–2. Each represents a multiple or submultiple of a unit which, if applied successively, moves the decimal point of a numerical quantity to every third place.* For example, 4 000 000 N = 4 000 kN (kilo-newton) = 4 MN (mega-newton), or 0.005 m = 5 mm (milli-meter). Notice that the SI system does not include the multiple deca (10) or the submultiple centi (0.01), which form part of the metric system. Except for some volume and area measurements, the use of these prefixes is to be avoided in science and engineering.

### TABLE 1–2  Prefixes

|  | Exponential Form | Prefix | SI Symbol |
|---|---|---|---|
| *Multiple* | | | |
| 1 000 000 000 | $10^9$ | giga | G |
| 1 000 000 | $10^6$ | mega | M |
| 1 000 | $10^3$ | kilo | k |
| *Submultiple* | | | |
| 0.001 | $10^{-3}$ | milli | m |
| 0.000 001 | $10^{-6}$ | micro | $\mu$ |
| 0.000 000 001 | $10^{-9}$ | nano | n |

* The kilogram is the only base unit that is defined with a prefix.

**Rules for Use.** Here are a few of the important rules that describe the proper use of the various SI symbols:

- Quantities defined by several units which are multiples of one another are separated by a *dot* to avoid confusion with prefix notation, as indicated by $N = kg \cdot m/s^2 = kg \cdot m \cdot s^{-2}$. Also, m · s (meter-second), whereas ms (milli-second).
- The exponential power on a unit having a prefix refers to both the unit *and* its prefix. For example, $\mu N^2 = (\mu N)^2 = \mu N \cdot \mu N$. Likewise, $mm^2$ represents $(mm)^2 = mm \cdot mm$.
- With the exception of the base unit the kilogram, in general avoid the use of a prefix in the denominator of composite units. For example, do not write N/mm, but rather kN/m; also, m/mg should be written as Mm/kg.
- When performing calculations, represent the numbers in terms of their *base or derived units* by converting all prefixes to powers of 10. The final result should then be expressed using a *single prefix*. Also, after calculation, it is best to keep numerical values between 0.1 and 1000; otherwise, a suitable prefix should be chosen. For example,

$$(50 \text{ kN})(60 \text{ nm}) = [50(10^3) \text{ N}][60(10^{-9}) \text{ m}]$$
$$= 3000(10^{-6}) \text{ N} \cdot \text{m} = 3(10^{-3}) \text{ N} \cdot \text{m} = 3 \text{ mN} \cdot \text{m}$$

## 1.5 Numerical Calculations

Numerical work in engineering practice is most often performed by using handheld calculators and computers. It is important, however, that the answers to any problem be reported with both justifiable accuracy and appropriate significant figures. In this section, we will discuss these topics together with some other important aspects involved in all engineering calculations.

Computers are often used in engineering for advanced design and analysis.

**Dimensional Homogeneity.** The terms of any equation used to describe a physical process must be *dimensionally homogeneous;* that is, each term must be expressed in the same units. Provided this is the case, all the terms of an equation can then be combined if numerical values are substituted for the variables. Consider, for example, the equation $s = vt + \frac{1}{2}at^2$, where, in SI units, $s$ is the position in meters, m, $t$ is time in seconds, s, $v$ is velocity in m/s and $a$ is acceleration in m/s². Regardless of how this equation is evaluated, it maintains its dimensional homogeneity. In the form stated, each of the three terms is expressed in meters $[m, (m/\cancel{s})\cancel{s}, (m/\cancel{s^2})\cancel{s^2},]$ or solving for $a$, $a = 2s/t^2 - 2v/t$, the terms are each expressed in units of m/s² [m/s², m/s², (m/s)/s].

Keep in mind that problems in mechanics always involve the solution of dimensionally homogeneous equations; so, this fact can then be used as a partial check for algebraic manipulations of an equation.

**Significant Figures.** The number of significant figures contained in any number determines the accuracy of the number. For instance, the number 4981 contains four significant figures. However, if zeros occur at the end of a whole number, it may be unclear as to how many significant figures the number represents. For example, 23 400 might have three (234), four (2340), or five (23 400) significant figures. To avoid these ambiguities, we will use *engineering notation* to report a result. This requires that numbers be rounded off to the appropriate number of significant digits and then expressed in multiples of $(10^3)$, such as $(10^3)$, $(10^6)$, or $(10^{-9})$. For instance, if 23 400 has five significant figures, it is written as $23.400(10^3)$, but if it has only three significant figures, it is written as $23.4(10^3)$.

If zeros occur at the beginning of a number that is less than one, then the zeros are not significant. For example, 0.00821 has three significant figures. Using engineering notation, this number is expressed as $8.21(10^{-3})$. Likewise, 0.000582 can be expressed as $0.582(10^{-3})$ or $582(10^{-6})$.

**Rounding Off Numbers.** Rounding off a number is necessary so that the accuracy of the result will be the same as that of the problem data. As a general rule, any numerical figure ending in five or greater is rounded up and a number less than five is rounded down. The rules for rounding off numbers are best illustrated by examples. Suppose the number 3.5587 is to be rounded off to *three* significant figures. Because the fourth digit (8) is *greater than* 5, the third number is rounded up to 3.56. Likewise 0.5896 becomes 0.590 and 9.3866 becomes 9.39. If we round off 1.341 to three significant figures, because the fourth digit (1) is *less than* 5, then we get 1.34. Likewise, 0.3762 becomes 0.376 and 9.871 becomes 9.87. There is a special case for any number that has a 5 with zeroes following it. As a general rule, if the digit preceding the 5 is an *even number*, then this digit is *not* rounded up. If the digit preceding the 5 is an *odd number*, then it is rounded up. For example, 75.25 rounded off to three significant digits becomes 75.2, 0.1275 becomes 0.128, and 0.2555 becomes 0.256.

**Calculations.** When a sequence of calculations is performed, it is best to store the intermediate results in the calculator. In other words, do not round off calculations until expressing the final result. This procedure maintains precision throughout the series of steps to the final solution. In this text, we will generally round off the answers to three significant figures since most of the data in engineering mechanics, such as geometry and loads, may be reliably measured to this accuracy.

## 1.6 General Procedure for Analysis

The most effective way of learning the principles of engineering mechanics is to *solve problems*. To be successful at this, it is important to always present the work in a *logical* and *orderly manner*, as suggested by the following sequence of steps:

- Read the problem carefully and try to correlate the actual physical situation with the theory studied.
- Tabulate the problem data and draw any necessary diagrams.
- Apply the relevant principles, generally in mathematical form. When writing any equations, be sure they are dimensionally homogeneous.
- Solve the necessary equations, and report the answer with no more than three significant figures.
- Study the answer with technical judgment and common sense to determine whether or not it seems reasonable.

When solving problems, do the work as neatly as possible. Being neat will stimulate clear and orderly thinking, and vice versa.

### Important Points

- Statics is the study of bodies that are at rest or move with constant velocity.
- A particle has a mass but a size that can be neglected.
- A rigid body does not deform under load.
- Concentrated forces are assumed to act at a point on a body.
- Newton's three laws of motion should be memorized.
- Mass is measure of a quantity of matter that does not change from one location to another.
- Weight refers to the gravitational attraction of the earth on a body or quantity of mass. Its magnitude depends upon the elevation at which the mass is located.
- In the SI system, the unit of force, the newton, is a derived unit. The meter, second, and kilogram are base units.
- Prefixes G, M, k, m, $\mu$, and n are used to represent large and small numerical quantities. Their exponential size should be known, along with the rules for using the SI units.
- Perform numerical calculations with several significant figures, and then report the final answer to three significant figures.
- Algebraic manipulations of an equation can be checked in part by verifying that the equation remains dimensionally homogeneous.
- Know the rules for rounding off numbers.

## 1.6 GENERAL PROCEDURE FOR ANALYSIS

### EXAMPLE 1.1

Convert 2 km/h to m/s.

**SOLUTION**
Since 1 km = 1000 m and 1 h = 3600 s, the factors of conversion are arranged in the following order, so that a cancellation of the units can be applied:

$$2 \text{ km/h} = \frac{2 \cancel{\text{km}}}{\cancel{\text{h}}} \left(\frac{1000 \text{ m}}{\cancel{\text{km}}}\right)\left(\frac{1 \cancel{\text{h}}}{3600 \text{ s}}\right)$$

$$= \frac{2000 \text{ m}}{3600 \text{ s}} = 0.556 \text{ m/s} \qquad Ans.$$

**NOTE:** Remember to round off the final answer to three significant figures.

### EXAMPLE 1.2

Evaluate each of the following and express with SI units having an appropriate prefix: (a) (50 mN)(6 GN), (b) (400 mm)(0.6 MN)$^2$, (c) 45 MN$^3$/900 Gg.

**SOLUTION**
First convert each number to base units, perform the indicated operations, then choose an appropriate prefix.

**Part (a)**

$$(50 \text{ mN})(6 \text{ GN}) = [50(10^{-3}) \text{ N}][6(10^9) \text{ N}]$$

$$= 300(10^6) \text{ N}^2$$

$$= 300(10^6) \cancel{\text{N}^2} \left(\frac{1 \text{ kN}}{10^3 \cancel{\text{N}}}\right)\left(\frac{1 \text{ kN}}{10^3 \cancel{\text{N}}}\right)$$

$$= 300 \text{ kN}^2 \qquad Ans.$$

**NOTE:** Keep in mind the convention kN$^2$ = (kN)$^2$ = $10^6$ N$^2$.

# EXAMPLE 1.2 (Continued)

**Part (b)**

$$(400 \text{ mm})(0.6 \text{ MN})^2 = [400(10^{-3}) \text{ m}][0.6(10^6) \text{ N}]^2$$
$$= [400(10^{-3}) \text{ m}][0.36(10^{12}) \text{ N}^2]$$
$$= 144(10^9) \text{ m} \cdot \text{N}^2$$
$$= 144 \text{ Gm} \cdot \text{N}^2 \qquad Ans.$$

We can also write

$$144(10^9) \text{ m} \cdot \text{N}^2 = 144(10^9) \text{ m} \cdot \cancel{\text{N}}^2 \left(\frac{1 \text{ MN}}{10^6 \cancel{\text{N}}}\right)\left(\frac{1 \text{ MN}}{10^6 \cancel{\text{N}}}\right)$$
$$= 0.144 \text{ m} \cdot \text{MN}^2 \qquad Ans.$$

**Part (c)**

$$\frac{45 \text{ MN}^3}{900 \text{ Gg}} = \frac{45(10^6 \text{ N})^3}{900(10^6) \text{ kg}}$$
$$= 50(10^9) \cancel{\text{N}}^3/\text{kg}$$
$$= 50(10^9) \cancel{\text{N}}^3 \left(\frac{1 \text{ kN}}{10^3 \cancel{\text{N}}}\right)^3 \frac{1}{\text{kg}}$$
$$= 50 \text{ kN}^3/\text{kg} \qquad Ans.$$

# PROBLEMS

**1–1.** Round off the following numbers to three significant figures: (a) 4.65735 m, (b) 55.578 s, (c) 4555 N, and (d) 2768 kg.

**1–2.** Represent each of the following combinations of units in the correct SI form using an appropriate prefix: (a) $\mu$MN, (b) N/$\mu$m, (c) MN/ks$^2$, and (d) kN/ms.

**1–3.** Represent each of the following quantities in the correct SI form using an appropriate prefix: (a) 0.000431 kg, (b) 35.3($10^3$) N, and (c) 0.00532 km.

**\*1–4.** Represent each of the following combinations of units in the correct SI form: (a) Mg/ms, (b) N/mm, and (c) mN/(kg·$\mu$s).

**1–5.** Represent each of the following combinations of units in the correct SI form using an appropriate prefix: (a) kN/$\mu$s, (b) Mg/mN, and (c) MN/(kg·ms).

**1–6.** Represent each of the following to three significant figures and express each answer in SI units using an appropriate prefix: (a) 45 320 kN, (b) 568($10^5$) mm, and (c) 0.005 63 mg.

**1–7.** A rocket has a mass of 3.65($10^6$) kg on earth. Specify its weight in SI units. If the rocket is on the moon, where the acceleration due to gravity is $g_m$ = 1.62 m/s$^2$, determine to three significant figures its weight in SI units and its mass in SI units.

**\*1–8.** If a car is traveling at 88 km/h, determine its speed in meters per second.

**1–9.** The *pascal* (Pa) is actually a very small unit of pressure. Given (1 Pa = 1 N/m$^2$) and atmospheric pressure at sea level is 101.325 kN/m$^2$. How many pascals is this?

**1–10.** What is the weight in newtons of an object that has a mass of: (a) 10 kg, (b) 0.5 g, and (c) 4.50 Mg? Express the result to three significant figures. Use an appropriate prefix.

**1–11.** Evaluate each of the following to three significant figures and express each answer in SI units using an appropriate prefix: (a) 354 mg(45 km)/(0.0356 kN), (b) (0.004 53 Mg)(201 ms), and (c) 435 MN/23.2 mm.

**\*1–12.** The specific weight (wt./vol.) of brass is 85 kN/m$^3$. Determine its density (mass/vol.) in SI units. Use an appropriate prefix.

**\*1–13.** Two particles have a mass of 8 kg and 12 kg, respectively. If they are 800 mm apart, determine the force of gravity acting between them. Compare this result with the weight of each particle.

**1–14.** Determine the mass in kilograms of an object that has a weight of (a) 20 mN, (b) 150 kN, and (c) 60 MN. Express the answer to three significant figures.

**1–15.** Evaluate each of the following to three significant figures and express each answer in SI units using an appropriate prefix: (a) (200 kN)$^2$, (b) (0.005 mm)$^2$, and (c) (400 m)$^3$.

**1–16.** Using the base units of the SI system, show that Eq. 1–2 is a dimensionally homogeneous equation which gives $F$ in newtons. Determine to three significant figures the gravitational force acting between two spheres that are touching each other. The mass of each sphere is 200 kg and the radius is 300 mm.

**\*1–17.** Evaluate each of the following to three significant figures and express each answer in SI units using an appropriate prefix: (a) (0.631 Mm)/(8.60 kg)$^2$, and (b) (35 mm)$^2$(48 kg)$^3$.

**1–18.** Evaluate (204 mm)(0.00457 kg)/(34.6 N) to three significant figures and express the answer in SI units using an appropriate prefix.

This bridge tower is stabilized by cables that exert forces at the points of connection. In this chapter we will show how to express these forces as Cartesian vectors and then determine the resultant force.

# Force Vectors

## CHAPTER OBJECTIVES

- To show how to add forces and resolve them into components using the Parallelogram Law.
- To express force and position in Cartesian vector form and explain how to determine the vector's magnitude and direction.
- To introduce the dot product in order to determine the angle between two vectors or the projection of one vector onto another.

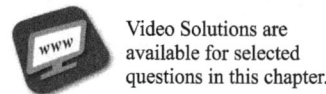

Video Solutions are available for selected questions in this chapter.

## 2.1 Scalars and Vectors

All physical quantities in engineering mechanics are measured using either scalars or vectors.

**Scalar.** A *scalar* is any positive or negative physical quantity that can be completely specified by its *magnitude*. Examples of scalar quantities include length, mass, and time.

**Vector.** A *vector* is any physical quantity that requires both a *magnitude* and a *direction* for its complete description. Examples of vectors encountered in statics are force, position, and moment. A vector is shown graphically by an arrow. The length of the arrow represents the *magnitude* of the vector, and the angle $\theta$ between the vector and a fixed axis defines the *direction of its line of action*. The head or tip of the arrow indicates the *sense of direction* of the vector, Fig. 2–1.

In print, vector quantities are represented by bold face letters such as **A**, and its magnitude of the vector is italicized, $A$. For handwritten work, it is often convenient to denote a vector quantity by simply drawing an arrow on top of it, $\vec{A}$.

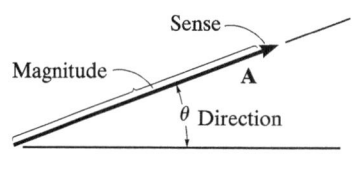

Fig. 2–1

## 2.2 Vector Operations

**Multiplication and Division of a Vector by a Scalar.** If a vector is multiplied by a positive scalar, its magnitude is increased by that amount. When multiplied by a negative scalar it will also change the directional sense of the vector. Graphic examples of these operations are shown in Fig. 2–2.

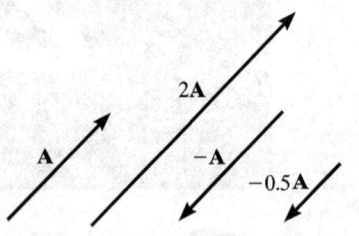

Scalar multiplication and division

Fig. 2–2

**Vector Addition.** All vector quantities obey the *parallelogram law of addition*. To illustrate, the two "*component*" vectors **A** and **B** in Fig. 2–3a are added to form a "*resultant*" vector **R** = **A** + **B** using the following procedure:

- First join the tails of the components at a point so that it makes them concurrent, Fig. 2–3b.
- From the head of **B**, draw a line parallel to **A**. Draw another line from the head of **A** that is parallel to **B**. These two lines intersect at point *P* to form the adjacent sides of a parallelogram.
- The diagonal of this parallelogram that extends to *P* forms **R**, which then represents the resultant vector **R** = **A** + **B**, Fig. 2–3c.

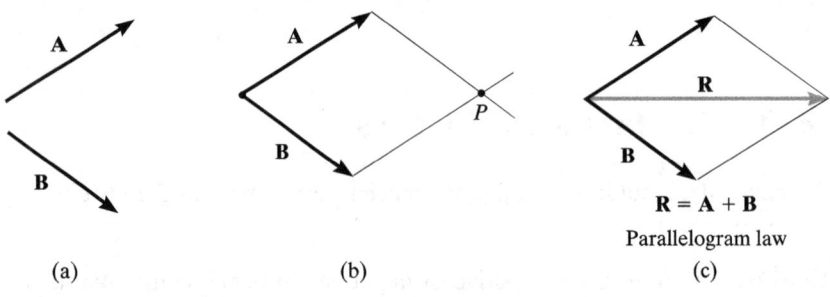

(a)        (b)        **R** = **A** + **B**
Parallelogram law
(c)

Fig. 2–3

We can also add **B** to **A**, Fig. 2–4a, using the *triangle rule*, which is a special case of the parallelogram law, whereby vector **B** is added to vector **A** in a "head-to-tail" fashion, i.e., by connecting the head of **A** to the tail of **B**, Fig. 2–4b. The resultant **R** extends from the tail of **A** to the head of **B**. In a similar manner, **R** can also be obtained by adding **A** to **B**, Fig. 2–4c. By comparison, it is seen that vector addition is commutative; in other words, the vectors can be added in either order, i.e., **R** = **A** + **B** = **B** + **A**.

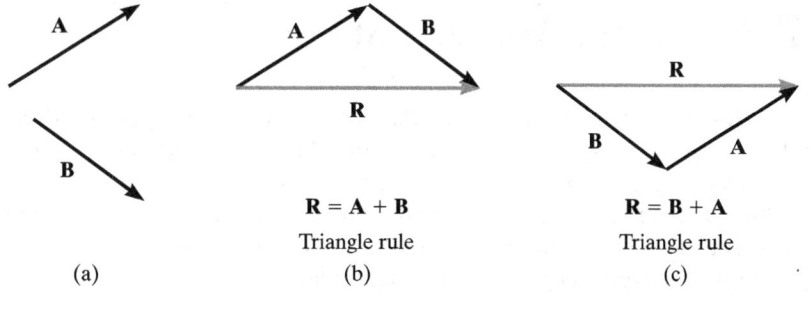

**Fig. 2–4**

As a special case, if the two vectors **A** and **B** are *collinear*, i.e., both have the same line of action, the parallelogram law reduces to an *algebraic* or *scalar addition* $R = A + B$, as shown in Fig. 2–5.

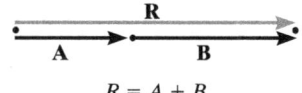

Addition of collinear vectors

**Fig. 2–5**

## Vector Subtraction.

The resultant of the *difference* between two vectors **A** and **B** of the same type may be expressed as

$$\mathbf{R}' = \mathbf{A} - \mathbf{B} = \mathbf{A} + (-\mathbf{B})$$

This vector sum is shown graphically in Fig. 2–6. Subtraction is therefore defined as a special case of addition, so the rules of vector addition also apply to vector subtraction.

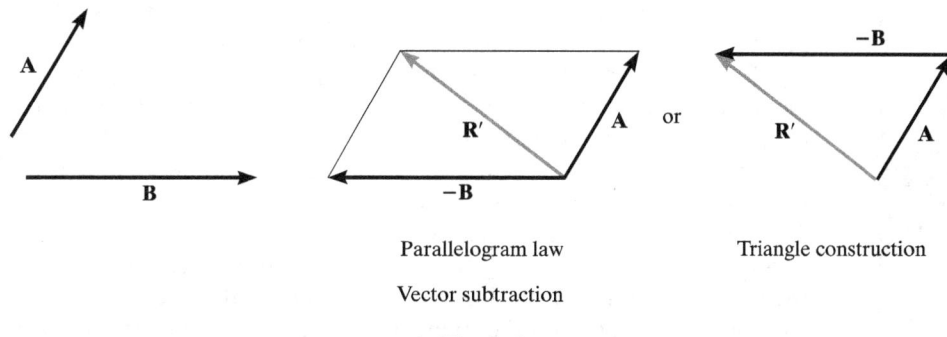

Parallelogram law    Triangle construction

Vector subtraction

**Fig. 2–6**

## 2.3 Vector Addition of Forces

Experimental evidence has shown that a force is a vector quantity since it has a specified magnitude, direction, and sense and it adds according to the parallelogram law. Two common problems in statics involve either finding the resultant force, knowing its components, or resolving a known force into two components. We will now describe how each of these problems is solved using the parallelogram law.

The parallelogram law must be used to determine the resultant of the two forces acting on the hook.

**Finding a Resultant Force.** The two component forces $F_1$ and $F_2$ acting on the pin in Fig. 2–7a can be added together to form the resultant force $F_R = F_1 + F_2$, as shown in Fig. 2–7b. From this construction, or using the triangle rule, Fig. 2–7c, we can apply the law of cosines or the law of sines to the triangle in order to obtain the magnitude of the resultant force and its direction.

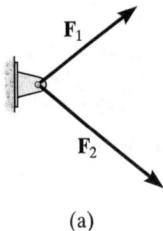
(a)

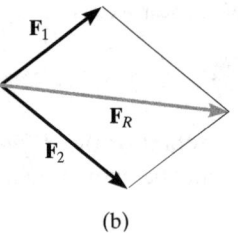
(b)

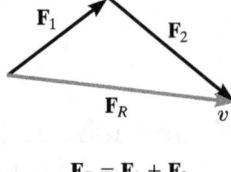
$F_R = F_1 + F_2$
(c)

**Fig. 2–7**

Using the parallelogram law force **F** caused by the vertical member can be resolved into components acting along the suspension cables $u$ and $v$.

**Finding the Components of a Force.** Sometimes it is necessary to resolve a force into two *components* in order to study its pulling or pushing effect in two specific directions. For example, in Fig. 2–8a, **F** is to be resolved into two components along the two members, defined by the $u$ and $v$ axes. In order to determine the magnitude of each component, a parallelogram is constructed first, by drawing lines starting from the tip of **F**, one line parallel to $u$, and the other line parallel to $v$. These lines then intersect with the $v$ and $u$ axes, forming a parallelogram. The force components $F_u$ and $F_v$ are then established by simply joining the tail of **F** to the intersection points on the $u$ and $v$ axes, Fig. 2–8b. This parallelogram can then be reduced to a triangle, which represents the triangle rule, Fig. 2–8c. From this, the law of sines can then be applied to determine the unknown magnitudes of the components.

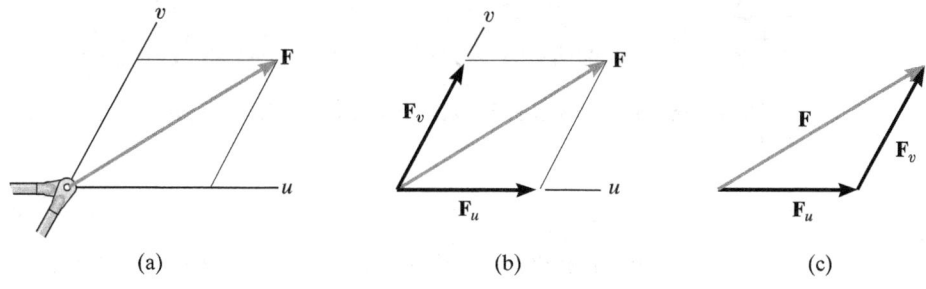

(a)　　　　　　　　　(b)　　　　　　　　　(c)

Fig. 2–8

**Addition of Several Forces.** If more than two forces are to be added, successive applications of the parallelogram law can be carried out in order to obtain the resultant force. For example, if three forces $\mathbf{F}_1$, $\mathbf{F}_2$, $\mathbf{F}_3$ act at a point $O$, Fig. 2–9, the resultant of any two of the forces is found, say, $\mathbf{F}_1 + \mathbf{F}_2$—and then this resultant is added to the third force, yielding the resultant of all three forces; i.e., $\mathbf{F}_R = (\mathbf{F}_1 + \mathbf{F}_2) + \mathbf{F}_3$. Using the parallelogram law to add more than two forces, as shown here, often requires extensive geometric and trigonometric calculation to determine the numerical values for the magnitude and direction of the resultant. Instead, problems of this type are easily solved by using the "rectangular-component method," which is explained in Sec. 2.4.

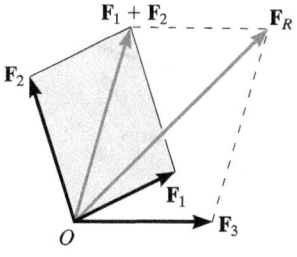

Fig. 2–9

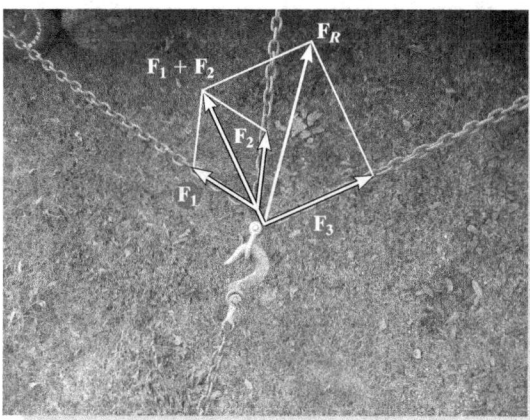

The resultant force $\mathbf{F}_R$ on the hook requires the addition of $\mathbf{F}_1 + \mathbf{F}_2$, then this resultant is added to $\mathbf{F}_3$.

## 22     CHAPTER 2   FORCE VECTORS

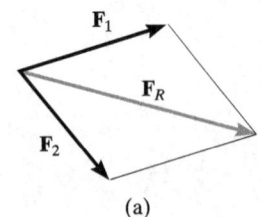

(a)

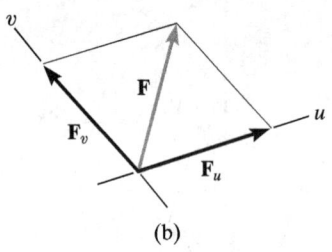

(b)

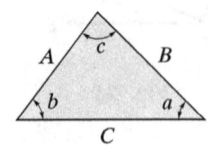

Cosine law:
$C = \sqrt{A^2 + B^2 - 2AB \cos c}$
Sine law:
$\dfrac{A}{\sin a} = \dfrac{B}{\sin b} = \dfrac{C}{\sin c}$

(c)

**Fig. 2–10**

## Procedure for Analysis

Problems that involve the addition of two forces can be solved as follows:

### Parallelogram Law.

- Two "component" forces $\mathbf{F}_1$ and $\mathbf{F}_2$ in Fig. 2–10a add according to the parallelogram law, yielding a *resultant* force $\mathbf{F}_R$ that forms the diagonal of the parallelogram.

- If a force $\mathbf{F}$ is to be resolved into *components* along two axes $u$ and $v$, Fig. 2–10b, then start at the head of force $\mathbf{F}$ and construct lines parallel to the axes, thereby forming the parallelogram. The sides of the parallelogram represent the components, $\mathbf{F}_u$ and $\mathbf{F}_v$.

- Label all the known and unknown force magnitudes and the angles on the sketch and identify the two unknowns as the magnitude and direction of $\mathbf{F}_R$, or the magnitudes of its components.

### Trigonometry.

- Redraw a half portion of the parallelogram to illustrate the triangular head-to-tail addition of the components.

- From this triangle, the magnitude of the resultant force can be determined using the law of cosines, and its direction is determined from the law of sines. The magnitudes of two force components are determined from the law of sines. The formulas are given in Fig. 2–10c.

## Important Points

- A scalar is a positive or negative number.

- A vector is a quantity that has a magnitude, direction, and sense.

- Multiplication or division of a vector by a scalar will change the magnitude of the vector. The sense of the vector will change if the scalar is negative.

- As a special case, if the vectors are collinear, the resultant is formed by an algebraic or scalar addition.

## EXAMPLE 2.1

The screw eye in Fig. 2–11a is subjected to two forces, $\mathbf{F}_1$ and $\mathbf{F}_2$. Determine the magnitude and direction of the resultant force.

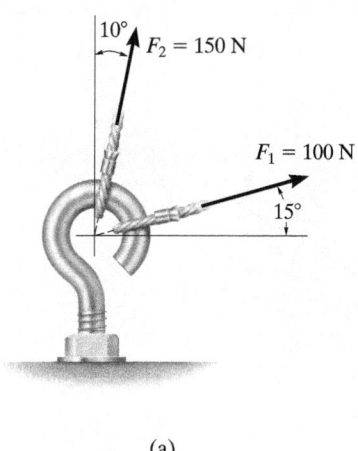

(a)

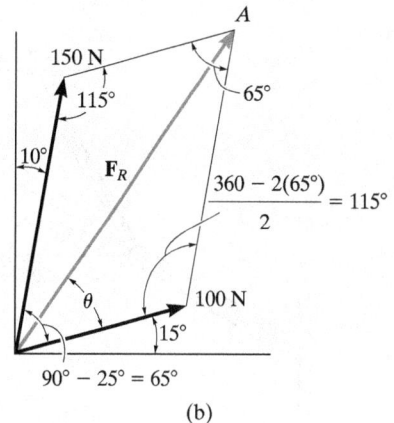

(b)

### SOLUTION

**Parallelogram Law.** The parallelogram is formed by drawing a line from the head of $\mathbf{F}_1$ that is parallel to $\mathbf{F}_2$, and another line from the head of $\mathbf{F}_2$ that is parallel to $\mathbf{F}_1$. The resultant force $\mathbf{F}_R$ extends to where these lines intersect at point $A$, Fig. 2–11b. The two unknowns are the magnitude of $\mathbf{F}_R$ and the angle $\theta$ (theta).

**Trigonometry.** From the parallelogram, the vector triangle is constructed, Fig. 2–11c. Using the law of cosines

$$F_R = \sqrt{(100\ \text{N})^2 + (150\ \text{N})^2 - 2(100\ \text{N})(150\ \text{N}) \cos 115°}$$
$$= \sqrt{10\,000 + 22\,500 - 30\,000(-0.4226)} = 212.6\ \text{N}$$
$$= 213\ \text{N} \qquad Ans.$$

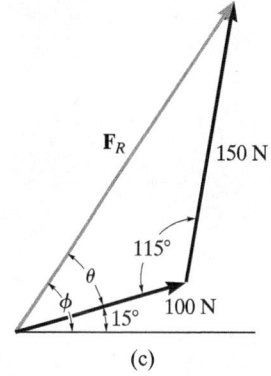

(c)

**Fig. 2–11**

Applying the law of sines to determine $\theta$,

$$\frac{150\ \text{N}}{\sin \theta} = \frac{212.6\ \text{N}}{\sin 115°} \qquad \sin \theta = \frac{150\ \text{N}}{212.6\ \text{N}} (\sin 115°)$$

$$\theta = 39.8°$$

Thus, the direction $\phi$ (phi) of $\mathbf{F}_R$, measured from the horizontal, is

$$\phi = 39.8° + 15.0° = 54.8° \qquad Ans.$$

**NOTE:** The results seem reasonable, since Fig. 2–11b shows $F_R$ to have a magnitude larger than its components and a direction that is between them.

## EXAMPLE 2.2

Resolve the horizontal 600-N force in Fig. 2–12a into components acting along the $u$ and $v$ axes and determine the magnitudes of these components.

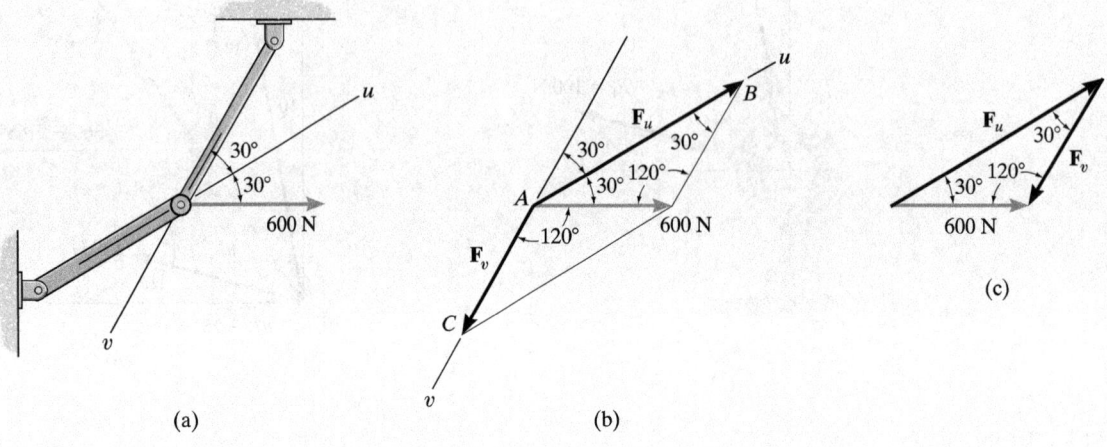

Fig. 2–12

### SOLUTION

The parallelogram is constructed by extending a line from the *head* of the 600-N force parallel to the $v$ axis until it intersects the $u$ axis at point $B$, Fig. 2–12b. The arrow from $A$ to $B$ represents $\mathbf{F}_u$. Similarly, the line extended from the head of the 600-N force drawn parallel to the $u$ axis intersects the $v$ axis at point $C$, which gives $\mathbf{F}_v$.

The vector addition using the triangle rule is shown in Fig. 2–12c. The two unknowns are the magnitudes of $\mathbf{F}_u$ and $\mathbf{F}_v$. Applying the law of sines,

$$\frac{F_u}{\sin 120°} = \frac{600 \text{ N}}{\sin 30°}$$

$$F_u = 1039 \text{ N} \qquad \textit{Ans.}$$

$$\frac{F_v}{\sin 30°} = \frac{600 \text{ N}}{\sin 30°}$$

$$F_v = 600 \text{ N} \qquad \textit{Ans.}$$

**NOTE:** The result for $F_u$ shows that sometimes a component can have a greater magnitude than the resultant.

## EXAMPLE 2.3

Determine the magnitude of the component force **F** in Fig. 2–13a and the magnitude of the resultant force **F**$_R$ if **F**$_R$ is directed along the positive $y$ axis.

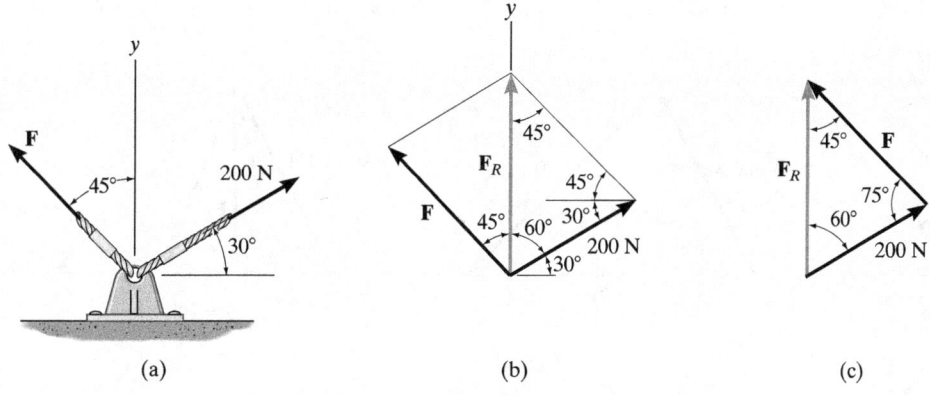

Fig. 2–13

### SOLUTION

The parallelogram law of addition is shown in Fig. 2–13b, and the triangle rule is shown in Fig. 2–13c. The magnitudes of **F**$_R$ and **F** are the two unknowns. They can be determined by applying the law of sines.

$$\frac{F}{\sin 60°} = \frac{200 \text{ N}}{\sin 45°}$$

$$F = 245 \text{ N} \qquad \qquad Ans.$$

$$\frac{F_R}{\sin 75°} = \frac{200 \text{ N}}{\sin 45°}$$

$$F_R = 273 \text{ N} \qquad \qquad Ans.$$

# EXAMPLE 2.4

It is required that the resultant force acting on the eyebolt in Fig. 2–14a be directed along the positive x axis and that $\mathbf{F}_2$ have a *minimum* magnitude. Determine this magnitude, the angle $\theta$, and the corresponding resultant force.

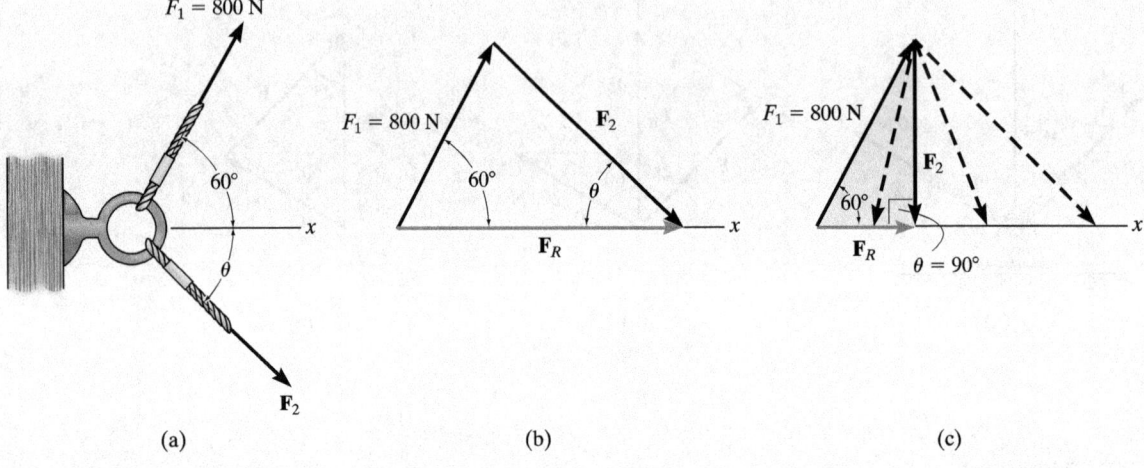

(a) (b) (c)

Fig. 2–14

## SOLUTION

The triangle rule for $\mathbf{F}_R = \mathbf{F}_1 + \mathbf{F}_2$ is shown in Fig. 2–14b. Since the magnitudes (lengths) of $\mathbf{F}_R$ and $\mathbf{F}_2$ are not specified, then $\mathbf{F}_2$ can actually be any vector that has its head touching the line of action of $\mathbf{F}_R$, Fig. 2–14c. However, as shown, the magnitude of $\mathbf{F}_2$ is a *minimum* or the shortest length when its line of action is *perpendicular* to the line of action of $\mathbf{F}_R$, that is, when

$$\theta = 90° \qquad \text{Ans.}$$

Since the vector addition now forms a right triangle, the two unknown magnitudes can be obtained by trigonometry.

$$F_R = (800 \text{ N})\cos 60° = 400 \text{ N} \qquad \text{Ans.}$$

$$F_2 = (800 \text{ N})\sin 60° = 693 \text{ N} \qquad \text{Ans.}$$

## FUNDAMENTAL PROBLEMS*

**F2–1.** Determine the magnitude of the resultant force acting on the screw eye and its direction measured clockwise from the x axis.

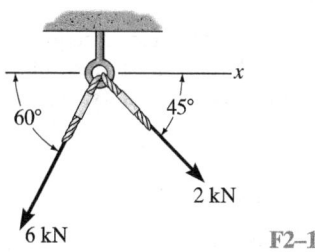

F2–1

**F2–2.** Two forces act on the hook. Determine the magnitude of the resultant force.

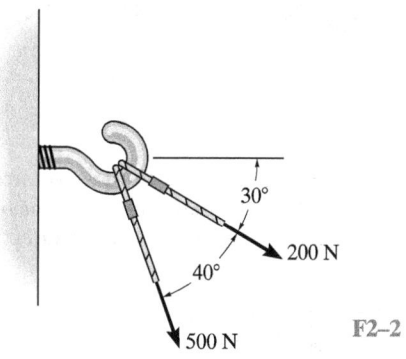

F2–2

**F2–3.** Determine the magnitude of the resultant force and its direction measured counterclockwise from the positive x axis.

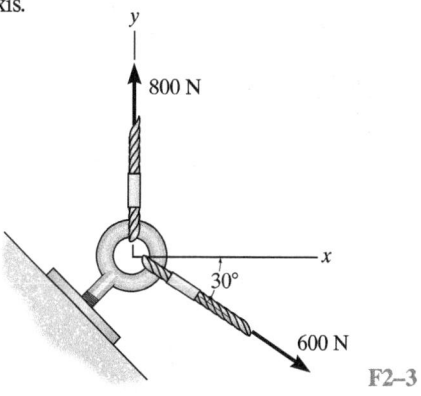

F2–3

**F2–4.** Resolve the 300-N force into components along the u and v axes, and determine the magnitude of each of these components.

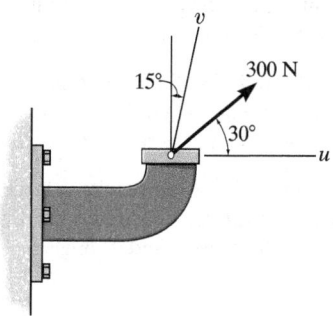

F2–4

**F2–5.** The force $F = 900$ N acts on the frame. Resolve this force into components acting along members $AB$ and $AC$, and determine the magnitude of each component.

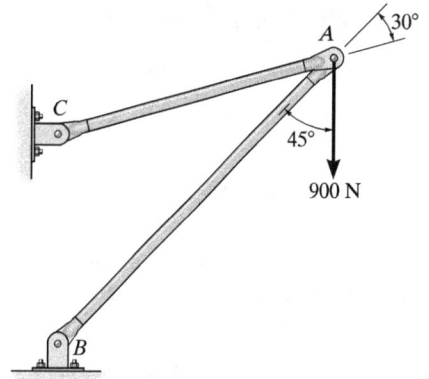

F2–5

**F2–6.** If force **F** is to have a component along the $u$ axis of $F_u = 6$ kN, determine the magnitude of **F** and the magnitude of its component $\mathbf{F}_v$ along the $v$ axis.

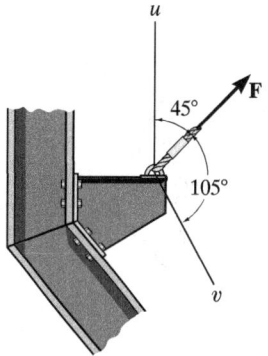

F2–6

*Partial solutions and answers to all Fundamental Problems are given in the back of the book.

## PROBLEMS

**•2–1.** If $\theta = 30°$ and $T = 6$ kN, determine the magnitude of the resultant force acting on the eyebolt and its direction measured clockwise from the positive $x$ axis.

**2–2.** If $\theta = 60°$ and $T = 5$ kN, determine the magnitude of the resultant force acting on the eyebolt and its direction measured clockwise from the positive $x$ axis.

**2–3.** If the magnitude of the resultant force is to be 9 kN directed along the positive $x$ axis, determine the magnitude of force **T** acting on the eyebolt and its angle $\theta$.

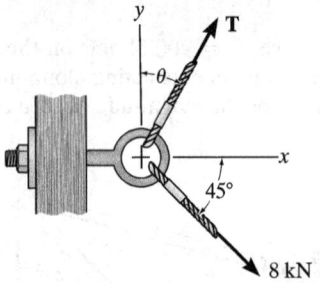

Probs. 2–1/2/3

***2–4.** Determine the magnitude of the resultant force acting on the bracket and its direction measured counterclockwise from the positive $u$ axis.

**•2–5.** Resolve $\mathbf{F}_1$ into components along the $u$ and $v$ axes, and determine the magnitudes of these components.

**2–6.** Resolve $\mathbf{F}_2$ into components along the $u$ and $v$ axes, and determine the magnitudes of these components.

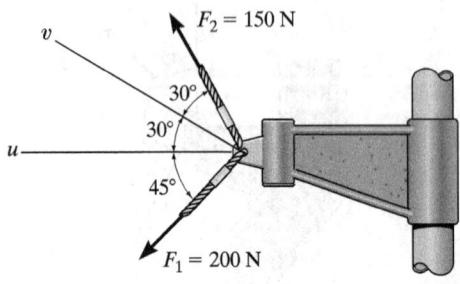

Probs. 2–4/5/6

**2–7.** If $F_B = 2$ kN and the resultant force acts along the positive $u$ axis, determine the magnitude of the resultant force and the angle $\theta$.

***2–8.** If the resultant force is required to act along the positive $u$ axis and have a magnitude of 5 kN, determine the required magnitude of $\mathbf{F}_B$ and its direction $\theta$.

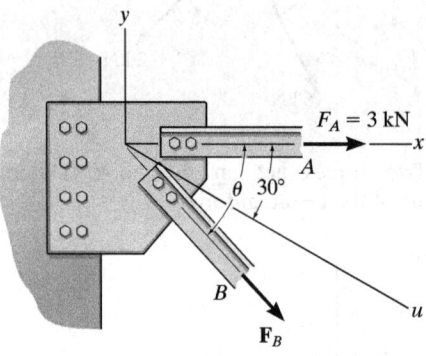

Probs. 2–7/8

**•2–9.** The plate is subjected to the two forces at $A$ and $B$ as shown. If $\theta = 60°$, determine the magnitude of the resultant of these two forces and its direction measured clockwise from the horizontal.

**2–10.** Determine the angle of $\theta$ for connecting member $A$ to the plate so that the resultant force of $\mathbf{F}_A$ and $\mathbf{F}_B$ is directed horizontally to the right. Also, what is the magnitude of the resultant force?

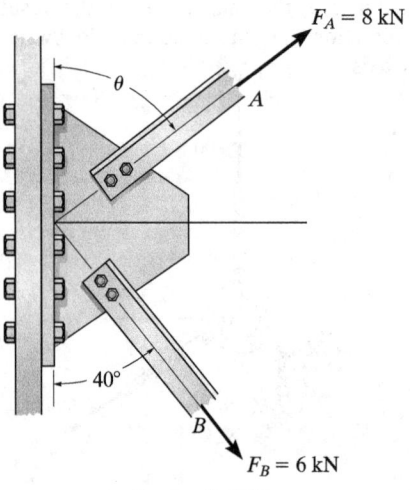

Probs. 2–9/10

**2–11.** If the tension in the cable is 400 N, determine the magnitude and direction of the resultant force acting on the pulley. This angle is the same angle $\theta$ of line $AB$ on the tailboard block.

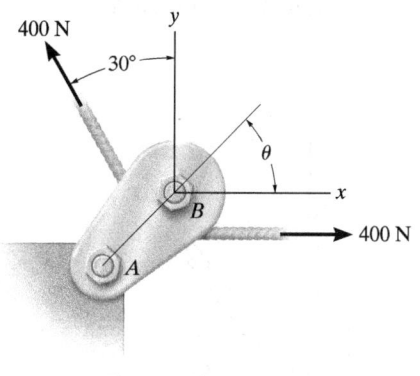

Prob. 2–11

**\*2–12.** The device is used for surgical replacement of the knee joint. If the force acting along the leg is 360 N, determine its components along the $x$ and $y'$ axes.

**•2–13.** The device is used for surgical replacement of the knee joint. If the force acting along the leg is 360 N, determine its components along the $x'$ and $y$ axes.

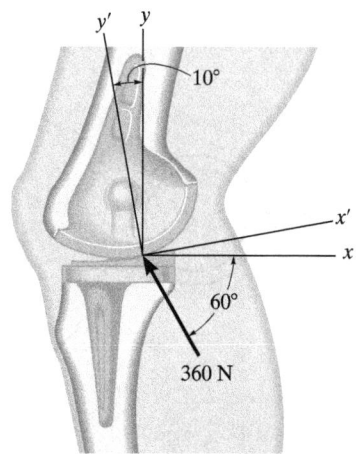

Probs. 2–12/13

**2–14.** Determine the design angle $\theta$ ($0° \leq \theta \leq 90°$) for strut $AB$ so that the 800-N horizontal force has a component of 1000 N directed from $A$ towards $C$. What is the component of force acting along member $AB$? Take $\phi = 40°$.

**2–15.** Determine the design angle $\phi$ ($0° \leq \phi \leq 90°$) between struts $AB$ and $AC$ so that the 800-N horizontal force has a component of 1200 N which acts up to the left, in the same direction as from $B$ towards $A$. Take $\theta = 30°$.

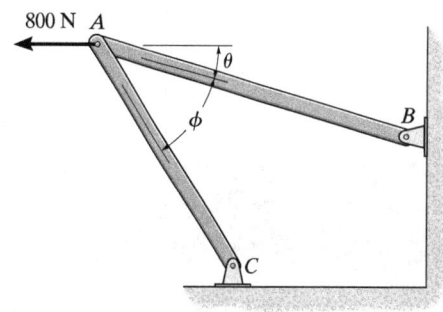

Probs. 2–14/15

**\*2–16.** Resolve $\mathbf{F}_1$ into components along the $u$ and $v$ axes and determine the magnitudes of these components.

**•2–17.** Resolve $\mathbf{F}_2$ into components along the $u$ and $v$ axes and determine the magnitudes of these components.

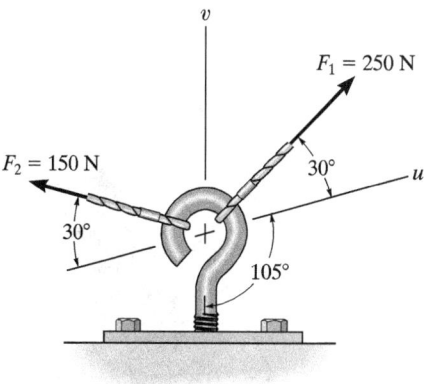

Probs. 2–16/17

**2–18.** The truck is to be towed using two ropes. Determine the magnitudes of forces $\mathbf{F}_A$ and $\mathbf{F}_B$ acting on each rope in order to develop a resultant force of 950 N directed along the positive $x$ axis. Set $\theta = 50°$.

**2–19.** The truck is to be towed using two ropes. If the resultant force is to be 950 N, directed along the positive $x$ axis, determine the magnitudes of forces $\mathbf{F}_A$ and $\mathbf{F}_B$ acting on each rope and the angle $\theta$ of $\mathbf{F}_B$ so that the magnitude of $F_B$ is a *minimum*. $\mathbf{F}_A$ acts at 20° from the $x$ axis as shown.

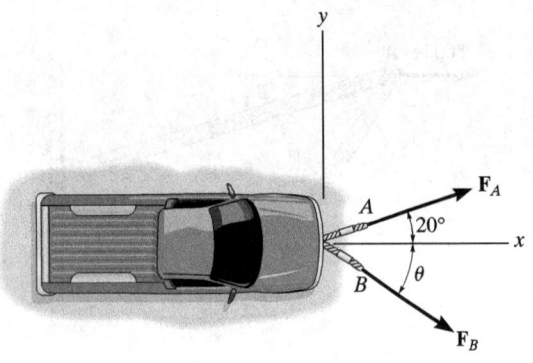

Prob. 2–18/19

**\*2–20.** If $\phi = 45°$, $F_1 = 5$ kN, and the resultant force is 6 kN directed along the positive $y$ axis, determine the required magnitude of $\mathbf{F}_2$ and its direction $\theta$.

**•2–21.** If $\phi = 30°$ and the resultant force is to be 6 kN directed along the positive $y$ axis, determine the magnitudes of $\mathbf{F}_1$ and $\mathbf{F}_2$ and the angle $\theta$ if $F_2$ is required to be a minimum.

**2–22.** If $\phi = 30°$, $F_1 = 5$ kN, and the resultant force is to be directed along the positive $y$ axis, determine the magnitude of the resultant force if $F_2$ is to be a minimum. Also, what is $F_2$ and the angle $\theta$?

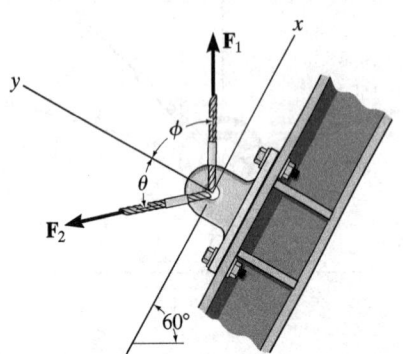

Probs. 2–20/21/22

**2–23.** If $\theta = 30°$ and $F_2 = 6$ kN, determine the magnitude of the resultant force acting on the plate and its direction measured clockwise from the positive $x$ axis.

**\*2–24.** If the resultant force $\mathbf{F}_R$ is directed along a line measured 75° clockwise from the positive $x$ axis and the magnitude of $\mathbf{F}_2$ is to be a minimum, determine the magnitudes of $\mathbf{F}_R$ and $\mathbf{F}_2$ and the angle $\theta \leq 90°$.

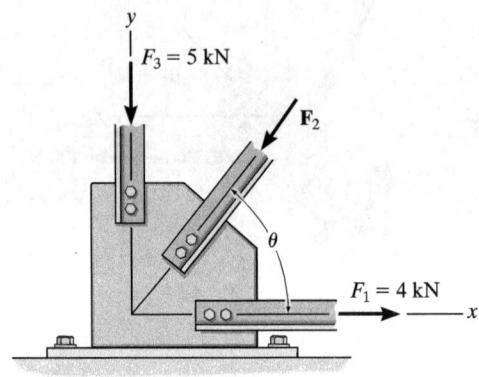

Probs. 2–23/24

**•2–25.** Two forces $\mathbf{F}_1$ and $\mathbf{F}_2$ act on the screw eye. If their lines of action are at an angle $\theta$ apart and the magnitude of each force is $F_1 = F_2 = F$, determine the magnitude of the resultant force $\mathbf{F}_R$ and the angle between $\mathbf{F}_R$ and $\mathbf{F}_1$.

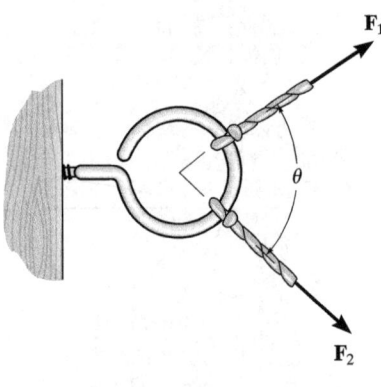

Prob. 2–25

**2–26.** The log is being towed by two tractors $A$ and $B$. Determine the magnitudes of the two towing forces $\mathbf{F}_A$ and $\mathbf{F}_B$ if it is required that the resultant force have a magnitude $F_R = 10$ kN and be directed along the $x$ axis. Set $\theta = 15°$.

**2–27.** The resultant $\mathbf{F}_R$ of the two forces acting on the log is to be directed along the positive $x$ axis and have a magnitude of 10 kN, determine the angle $\theta$ of the cable, attached to $B$ such that the magnitude of force $\mathbf{F}_B$ in this cable is a minimum. What is the magnitude of the force in each cable for this situation?

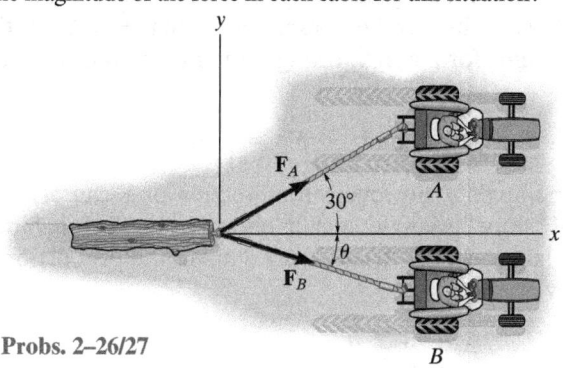

Probs. 2–26/27

**\*2–28.** The beam is to be hoisted using two chains. Determine the magnitudes of forces $\mathbf{F}_A$ and $\mathbf{F}_B$ acting on each chain in order to develop a resultant force of 600 N directed along the positive $y$ axis. Set $\theta = 45°$.

**•2–29.** The beam is to be hoisted using two chains. If the resultant force is to be 600 N directed along the positive $y$ axis, determine the magnitudes of forces $\mathbf{F}_A$ and $\mathbf{F}_B$ acting on each chain and the angle $\theta$ of $\mathbf{F}_B$ so that the magnitude of $\mathbf{F}_B$ is a *minimum*. $\mathbf{F}_A$ acts at 30° from the $y$ axis, as shown.

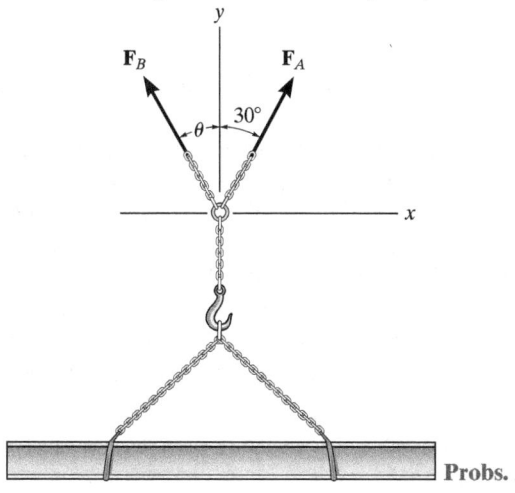

Probs. 2–28/29

**2–30.** Three chains act on the bracket such that they create a resultant force having a magnitude of 1000 N. If two of the chains are subjected to known forces, as shown, determine the angle $\theta$ of the third chain measured clockwise from the positive $x$ axis, so that the magnitude of force $\mathbf{F}$ in this chain is a *minimum*. All forces lie in the $x$–$y$ plane. What is the magnitude of $\mathbf{F}$? *Hint*: First find the resultant of the two known forces. Force $\mathbf{F}$ acts in this direction.

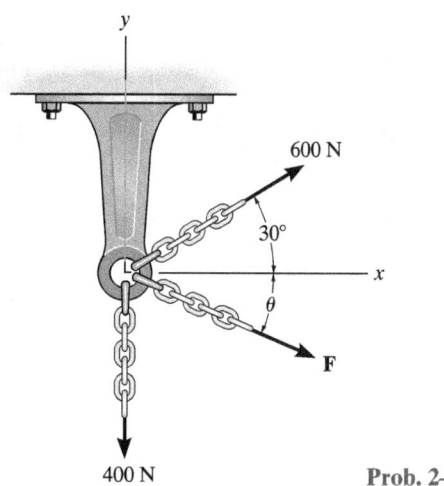

Prob. 2–30

**2–31.** Three cables pull on the pipe such that they create a resultant force having a magnitude of 1800 N. If two of the cables are subjected to known forces, as shown in the figure, determine the angle $\theta$ of the third cable so that the magnitude of force $\mathbf{F}$ in this cable is a *minimum*. All forces lie in the $x$–$y$ plane. What is the magnitude of $\mathbf{F}$? *Hint*: First find the resultant of the two known forces.

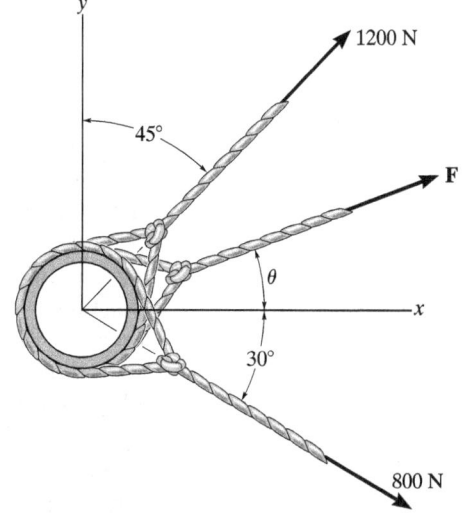

Prob. 2–31

## 2.4 Addition of a System of Coplanar Forces

When a force is resolved into two components along the $x$ and $y$ axes, the components are then called *rectangular components*. For analytical work we can represent these components in one of two ways, using either scalar notation or Cartesian vector notation.

**Scalar Notation.** The rectangular components of force $\mathbf{F}$ shown in Fig. 2–15a are found using the parallelogram law, so that $\mathbf{F} = \mathbf{F}_x + \mathbf{F}_y$. Because these components form a right triangle, their magnitudes can be determined from

$$F_x = F \cos \theta \quad \text{and} \quad F_y = F \sin \theta$$

Instead of using the angle $\theta$, however, the direction of $\mathbf{F}$ can also be defined using a small "slope" triangle, such as shown in Fig. 2–15b. Since this triangle and the larger shaded triangle are similar, the proportional length of the sides gives

$$\frac{F_x}{F} = \frac{a}{c}$$

or

$$F_x = F\left(\frac{a}{c}\right)$$

and

$$\frac{F_y}{F} = \frac{b}{c}$$

or

$$F_y = -F\left(\frac{b}{c}\right)$$

Here the $y$ component is a negative scalar since $\mathbf{F}_y$ is directed along the negative $y$ axis.

It is important to keep in mind that this positive and negative scalar notation is to be used only for computational purposes, not for graphical representations in figures. Throughout the book, the *head of a vector arrow* in any figure indicates the sense of the vector *graphically*; algebraic signs are not used for this purpose. Thus, the vectors in Figs. 2–15a and 2–15b are designated by using boldface (vector) notation.* Whenever italic symbols are written near vector arrows in figures, they indicate the *magnitude* of the vector, which is *always* a *positive* quantity.

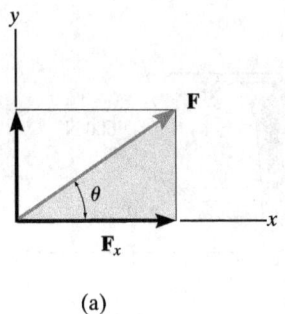

(a)

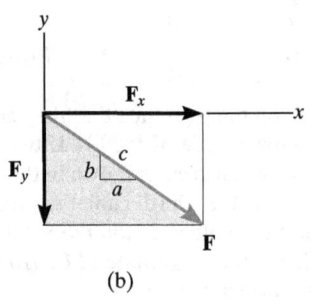

(b)

Fig. 2–15

---

*Negative signs are used only in figures with boldface notation when showing equal but opposite pairs of vectors, as in Fig. 2–2.

## 2.4 ADDITION OF A SYSTEM OF COPLANAR FORCES

**Cartesian Vector Notation.** It is also possible to represent the $x$ and $y$ components of a force in terms of Cartesian unit vectors $\mathbf{i}$ and $\mathbf{j}$. Each of these unit vectors has a dimensionless magnitude of one, and so they can be used to designate the *directions* of the $x$ and $y$ axes, respectively, Fig. 2–16.*

Since the *magnitude* of each component of $\mathbf{F}$ is *always a positive quantity*, which is represented by the (positive) scalars $F_x$ and $F_y$, then we can express $\mathbf{F}$ as a *Cartesian vector*,

$$\mathbf{F} = F_x \mathbf{i} + F_y \mathbf{j}$$

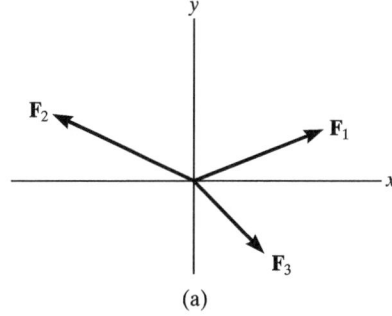

Fig. 2–16

**Coplanar Force Resultants.** We can use either of the two methods just described to determine the resultant of several *coplanar forces*. To do this, each force is first resolved into its $x$ and $y$ components, and then the respective components are added using *scalar algebra* since they are collinear. The resultant force is then formed by adding the resultant components using the parallelogram law. For example, consider the three concurrent forces in Fig. 2–17a, which have $x$ and $y$ components shown in Fig. 2–17b. Using *Cartesian vector notation*, each force is first represented as a Cartesian vector, i.e.,

$$\mathbf{F}_1 = F_{1x}\mathbf{i} + F_{1y}\mathbf{j}$$
$$\mathbf{F}_2 = -F_{2x}\mathbf{i} + F_{2y}\mathbf{j}$$
$$\mathbf{F}_3 = F_{3x}\mathbf{i} - F_{3y}\mathbf{j}$$

The vector resultant is therefore

$$\mathbf{F}_R = \mathbf{F}_1 + \mathbf{F}_2 + \mathbf{F}_3$$
$$= F_{1x}\mathbf{i} + F_{1y}\mathbf{j} - F_{2x}\mathbf{i} + F_{2y}\mathbf{j} + F_{3x}\mathbf{i} - F_{3y}\mathbf{j}$$
$$= (F_{1x} - F_{2x} + F_{3x})\mathbf{i} + (F_{1y} + F_{2y} - F_{3y})\mathbf{j}$$
$$= (F_{Rx})\mathbf{i} + (F_{Ry})\mathbf{j}$$

If *scalar notation* is used, then we have

$$(\xrightarrow{+}) \quad F_{Rx} = F_{1x} - F_{2x} + F_{3x}$$
$$(+\uparrow) \quad F_{Ry} = F_{1y} + F_{2y} - F_{3y}$$

These are the *same* results as the $\mathbf{i}$ and $\mathbf{j}$ components of $\mathbf{F}_R$ determined above.

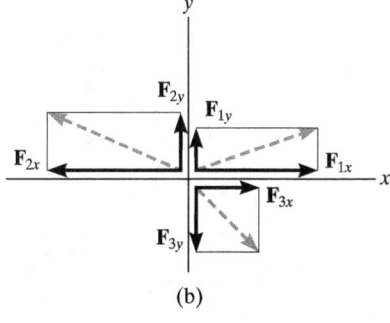

(a)

(b)

Fig. 2–17

---

*For handwritten work, unit vectors are usually indicated using a circumflex, e.g., $\hat{i}$ and $\hat{j}$. These vectors have a dimensionless magnitude of unity, and their sense (or arrowhead) will be described analytically by a plus or minus sign, depending on whether they are pointing along the positive or negative $x$ or $y$ axis.

We can represent the components of the resultant force of any number of coplanar forces symbolically by the algebraic sum of the $x$ and $y$ components of all the forces, i.e.,

$$F_{Rx} = \Sigma F_x$$
$$F_{Ry} = \Sigma F_y \qquad (2\text{–}1)$$

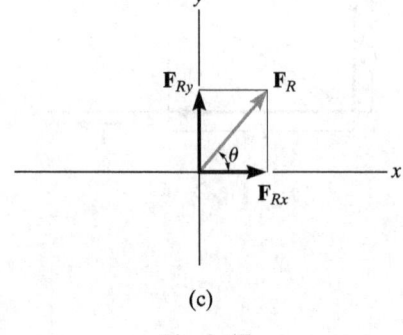

(c)

**Fig. 2–17**

Once these components are determined, they may be sketched along the $x$ and $y$ axes with their proper sense of direction, and the resultant force can be determined from vector addition, as shown in Fig. 2–17. From this sketch, the magnitude of $\mathbf{F}_R$ is then found from the Pythagorean theorem; that is,

$$F_R = \sqrt{F_{Rx}^2 + F_{Ry}^2}$$

Also, the angle $\theta$, which specifies the direction of the resultant force, is determined from trigonometry:

$$\theta = \tan^{-1}\left|\frac{F_{Ry}}{F_{Rx}}\right|$$

The above concepts are illustrated numerically in the examples which follow.

### Important Points

- The resultant of several coplanar forces can easily be determined if an $x$, $y$ coordinate system is established and the forces are resolved along the axes.

- The direction of each force is specified by the angle its line of action makes with one of the axes, or by a sloped triangle.

- The orientation of the $x$ and $y$ axes is arbitrary, and their positive direction can be specified by the Cartesian unit vectors $\mathbf{i}$ and $\mathbf{j}$.

- The $x$ and $y$ components of the *resultant force* are simply the algebraic addition of the components of all the coplanar forces.

- The magnitude of the resultant force is determined from the Pythagorean theorem, and when the components are sketched on the $x$ and $y$ axes, the direction can be determined from trigonometry.

The resultant force of the four cable forces acting on the supporting bracket can be determined by adding algebraically the separate x and y components of each cable force. This resultant FR produces the same pulling effect on the bracket as all four cables.

## EXAMPLE 2.5

Determine the $x$ and $y$ components of $\mathbf{F}_1$ and $\mathbf{F}_2$ acting on the boom shown in Fig. 2–18a. Express each force as a Cartesian vector.

### SOLUTION

**Scalar Notation.** By the parallelogram law, $\mathbf{F}_1$ is resolved into $x$ and $y$ components, Fig. 2–18b. Since $\mathbf{F}_{1x}$ acts in the $-x$ direction, and $\mathbf{F}_{1y}$ acts in the $+y$ direction, we have

$$F_{1x} = -200 \sin 30° \text{ N} = -100 \text{ N} = 100 \text{ N} \leftarrow \qquad Ans.$$

$$F_{1y} = 200 \cos 30° \text{ N} = 173 \text{ N} = 173 \text{ N} \uparrow \qquad Ans.$$

The force $\mathbf{F}_2$ is resolved into its $x$ and $y$ components as shown in Fig. 2–18c. Here the *slope* of the line of action for the force is indicated. From this "slope triangle" we could obtain the angle $\theta$, e.g., $\theta = \tan^{-1}(\frac{5}{12})$, and then proceed to determine the magnitudes of the components in the same manner as for $\mathbf{F}_1$. The easier method, however, consists of using proportional parts of similar triangles, i.e.,

$$\frac{F_{2x}}{260 \text{ N}} = \frac{12}{13} \qquad F_{2x} = 260 \text{ N}\left(\frac{12}{13}\right) = 240 \text{ N}$$

Similarly,

$$F_{2y} = 260 \text{ N}\left(\frac{5}{13}\right) = 100 \text{ N}$$

Notice how the magnitude of the *horizontal component*, $F_{2x}$, was obtained by multiplying the force magnitude by the ratio of the *horizontal leg* of the slope triangle divided by the hypotenuse; whereas the magnitude of the *vertical component*, $F_{2y}$, was obtained by multiplying the force magnitude by the ratio of the *vertical leg* divided by the hypotenuse. Hence,

$$F_{2x} = 240 \text{ N} = 240 \text{ N} \rightarrow \qquad Ans.$$

$$F_{2y} = -100 \text{ N} = 100 \text{ N} \downarrow \qquad Ans.$$

**Cartesian Vector Notation.** Having determined the magnitudes and directions of the components of each force, we can express each force as a Cartesian vector.

$$\mathbf{F}_1 = \{-100\mathbf{i} + 173\mathbf{j}\} \text{ N} \qquad Ans.$$

$$\mathbf{F}_2 = \{240\mathbf{i} - 100\mathbf{j}\} \text{ N} \qquad Ans.$$

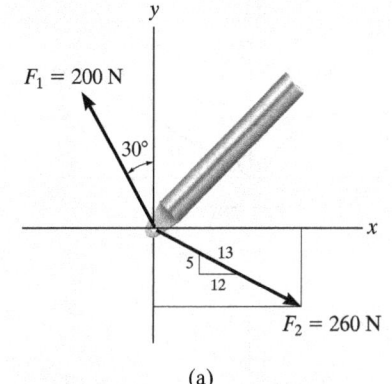

(a)

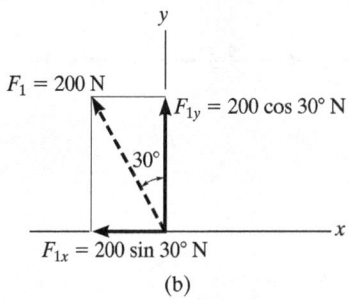

(b)

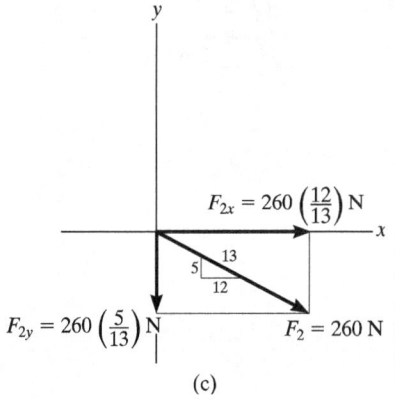

(c)

Fig. 2–18

## EXAMPLE 2.6

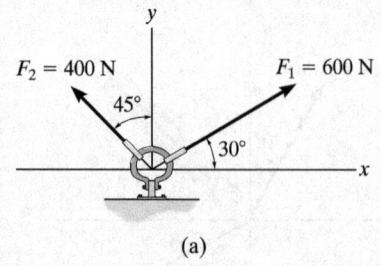

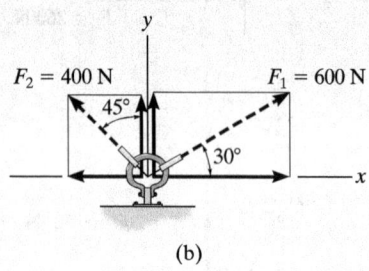

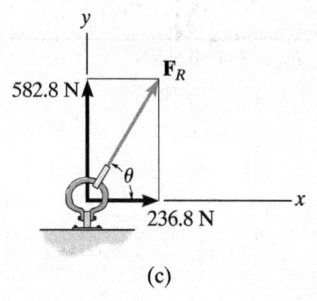

Fig. 2–19

The link in Fig. 2–19a is subjected to two forces $\mathbf{F}_1$ and $\mathbf{F}_2$. Determine the magnitude and direction of the resultant force.

### SOLUTION I

**Scalar Notation.** First we resolve each force into its $x$ and $y$ components, Fig. 2–19b, then we sum these components algebraically.

$\stackrel{+}{\rightarrow} F_{Rx} = \Sigma F_x;$ $\qquad F_{Rx} = 600 \cos 30°\text{ N} - 400 \sin 45°\text{ N}$
$\qquad\qquad\qquad\qquad = 236.8\text{ N} \rightarrow$

$+\uparrow F_{Ry} = \Sigma F_y;$ $\qquad F_{Ry} = 600 \sin 30°\text{ N} + 400 \cos 45°\text{ N}$
$\qquad\qquad\qquad\qquad = 582.8\text{ N} \uparrow$

The resultant force, shown in Fig. 2–18c, has a *magnitude* of

$$F_R = \sqrt{(236.8\text{ N})^2 + (582.8\text{ N})^2}$$
$$= 629\text{ N} \qquad\qquad Ans.$$

From the vector addition,

$$\theta = \tan^{-1}\left(\frac{582.8\text{ N}}{236.8\text{ N}}\right) = 67.9° \qquad Ans.$$

### SOLUTION II

**Cartesian Vector Notation.** From Fig. 2–19b, each force is first expressed as a Cartesian vector.

$$\mathbf{F}_1 = \{600 \cos 30°\mathbf{i} + 600 \sin 30°\mathbf{j}\}\text{ N}$$
$$\mathbf{F}_2 = \{-400 \sin 45°\mathbf{i} + 400 \cos 45°\mathbf{j}\}\text{ N}$$

Then,

$$\mathbf{F}_R = \mathbf{F}_1 + \mathbf{F}_2 = (600 \cos 30°\text{ N} - 400 \sin 45°\text{ N})\mathbf{i}$$
$$+ (600 \sin 30°\text{ N} + 400 \cos 45°\text{ N})\mathbf{j}$$
$$= \{236.8\mathbf{i} + 582.8\mathbf{j}\}\text{ N}$$

The magnitude and direction of $\mathbf{F}_R$ are determined in the same manner as before.

**NOTE:** Comparing the two methods of solution, notice that the use of scalar notation is more efficient since the components can be found *directly*, without first having to express each force as a Cartesian vector before adding the components. Later, however, we will show that Cartesian vector analysis is very beneficial for solving three-dimensional problems.

# EXAMPLE 2.7

The end of the boom $O$ in Fig. 2–20a is subjected to three concurrent and coplanar forces. Determine the magnitude and direction of the resultant force.

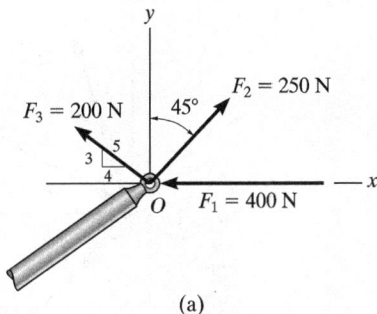

(a)

## SOLUTION

Each force is resolved into its $x$ and $y$ components, Fig. 2–20b. Summing the $x$ components, we have

$$\xrightarrow{+} F_{Rx} = \Sigma F_x; \quad F_{Rx} = -400 \text{ N} + 250 \sin 45° \text{ N} - 200\left(\tfrac{4}{5}\right) \text{N}$$

$$= -383.2 \text{ N} = 383.2 \text{ N} \leftarrow$$

The negative sign indicates that $F_{Rx}$ acts to the left, i.e., in the negative $x$ direction, as noted by the small arrow. Obviously, this occurs because $F_1$ and $F_3$ in Fig. 2–20b contribute a greater pull to the left than $F_2$ which pulls to the right. Summing the $y$ components yields

$$+\uparrow F_{Ry} = \Sigma F_y; \quad F_{Ry} = 250 \cos 45° \text{ N} + 200\left(\tfrac{3}{5}\right) \text{N}$$

$$= 296.8 \text{ N} \uparrow$$

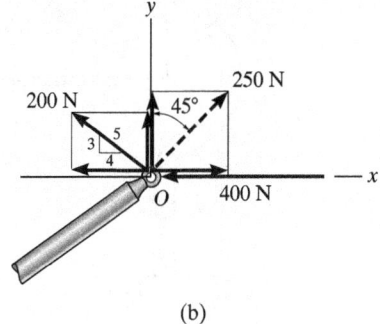

(b)

The resultant force, shown in Fig. 2–20c, has a *magnitude* of

$$F_R = \sqrt{(-383.2 \text{ N})^2 + (296.8 \text{ N})^2}$$

$$= 485 \text{ N} \qquad \qquad \textit{Ans.}$$

From the vector addition in Fig. 2–20c, the direction angle $\theta$ is

$$\theta = \tan^{-1}\left(\frac{296.8}{383.2}\right) = 37.8° \qquad \textit{Ans.}$$

**NOTE:** Application of this method is more convenient, compared to using two applications of the parallelogram law, first to add $\mathbf{F}_1$ and $\mathbf{F}_2$ then adding $\mathbf{F}_3$ to this resultant.

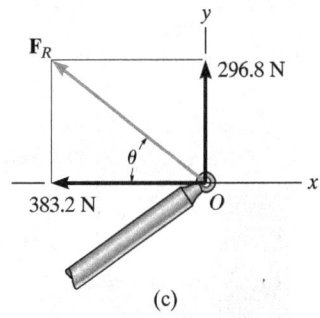

(c)

Fig. 2–20

## FUNDAMENTAL PROBLEMS

**F2–7.** Resolve each force acting on the post into its $x$ and $y$ components.

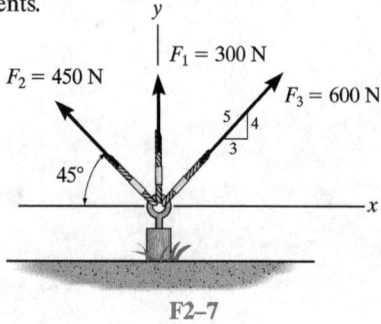

F2–7

**F2–8.** Determine the magnitude and direction of the resultant force.

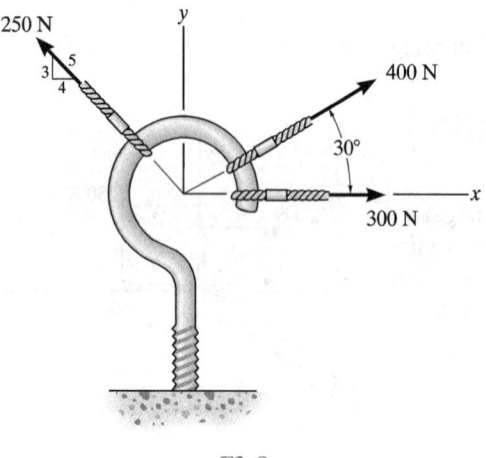

F2–8

**F2–9.** Determine the magnitude of the resultant force acting on the corbel and its direction $\theta$ measured counterclockwise from the $x$ axis.

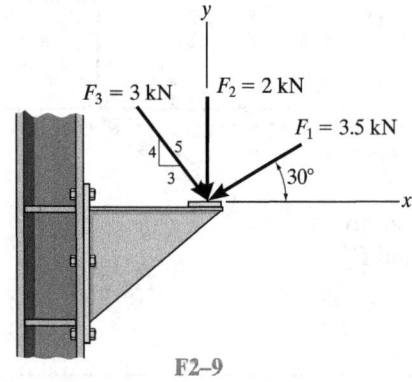

F2–9

**F2–10.** If the resultant force acting on the bracket is to be 750 N directed along the positive $x$ axis, determine the magnitude of F and its direction $\theta$.

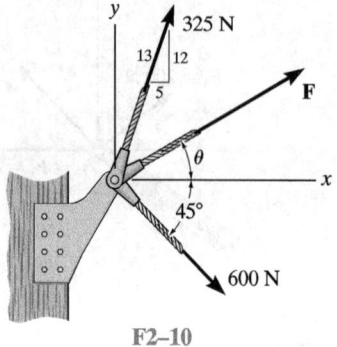

F2–10

**F2–11.** If the magnitude of the resultant force acting on the bracket is to be 400 N directed along the $u$ axis, determine the magnitude of F and its direction $\theta$.

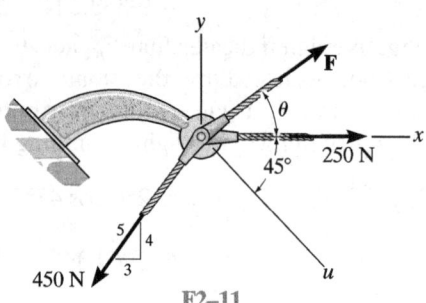

F2–11

**F2–12.** Determine the magnitude of the resultant force and its direction $\theta$ measured counterclockwise from the positive $x$ axis.

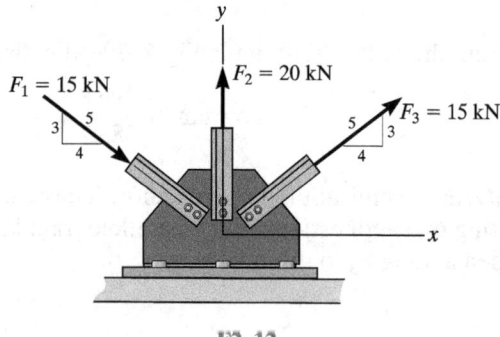

F2–12

## PROBLEMS

**\*2–32.** Determine the magnitude of the resultant force acting on the pin and its direction measured clockwise from the positive $x$ axis.

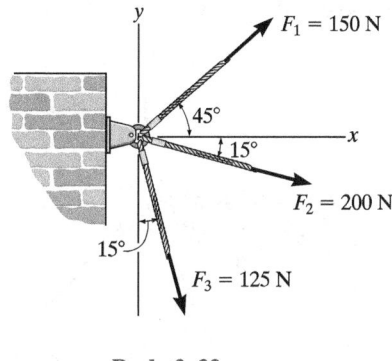

Prob. 2–32

**2–35.** The contact point between the femur and tibia bones of the leg is at $A$. If a vertical force of 875 N is applied at this point, determine the components along the $x$ and $y$ axes. Note that the $y$ component represents the normal force on the load-bearing region of the bones. Both the $x$ and $y$ components of this force cause synovial fluid to be squeezed out of the bearing space.

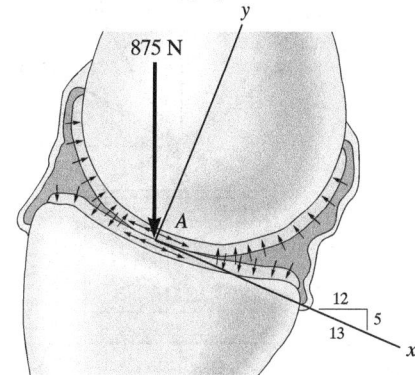

Prob. 2–35

**•2–33.** If $F_1 = 600$ N and $\phi = 30°$, determine the magnitude of the resultant force acting on the eyebolt and its direction measured clockwise from the positive $x$ axis.

**2–34.** If the magnitude of the resultant force acting on the eyebolt is 600 N and its direction measured clockwise from the positive $x$ axis is $\theta = 30°$, determine the magnitude of $\mathbf{F}_1$ and the angle $\phi$.

**\*2–36.** If $\phi = 30°$ and $F_2 = 3$ kN, determine the magnitude of the resultant force acting on the plate and its direction $\theta$ measured clockwise from the positive $x$ axis.

**•2–37.** If the magnitude for the resultant force acting on the plate is required to be 6 kN and its direction measured clockwise from the positive $x$ axis is $\theta = 30°$, determine the magnitude of $\mathbf{F}_2$ and its direction $\phi$.

**2–38.** If $\phi = 30°$ and the resultant force acting on the gusset plate is directed along the positive $x$ axis, determine the magnitudes of $\mathbf{F}_2$ and the resultant force.

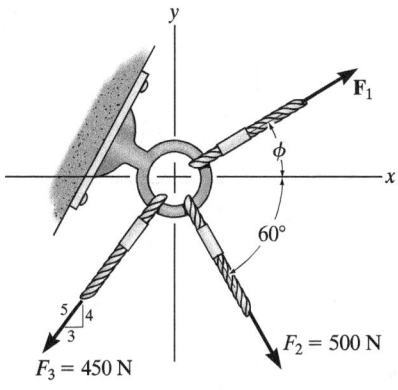

Probs. 2–33/34

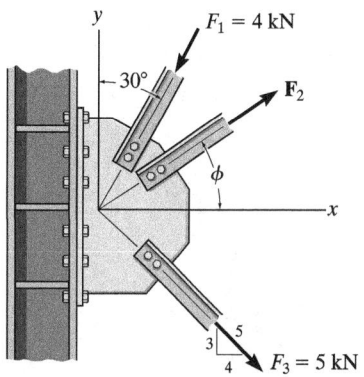

Probs. 2–36/37/38

**2–39.** Determine the magnitude of $\mathbf{F}_1$ and its direction $\theta$ so that the resultant force is directed vertically upward and has a magnitude of 800 N.

**\*2–40.** Determine the magnitude and direction measured counterclockwise from the positive $x$ axis of the resultant force of the three forces acting on the ring $A$. Take $F_1 = 500$ N and $\theta = 20°$.

**2–43.** If $\phi = 30°$ and $F_1 = 1.25$ kN, determine the magnitude of the resultant force acting on the bracket and its direction measured clockwise from the positive $x$ axis.

**\*2–44.** If the magnitude of the resultant force acting on the bracket is 2 kN directed along the positive $x$ axis, determine the magnitude of $\mathbf{F}_1$ and its direction $\phi$.

**•2–45.** If the resultant force acting on the bracket is to be directed along the positive $x$ axis and the magnitude of $\mathbf{F}_1$ is required to be a minimum, determine the magnitudes of the resultant force and $\mathbf{F}_1$.

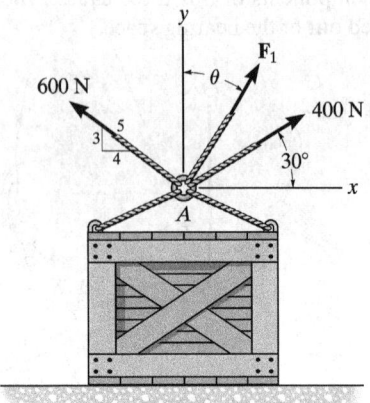

Probs. 2–39/40

**•2–41.** Determine the magnitude and direction $\theta$ of $\mathbf{F}_B$ so that the resultant force is directed along the positive $y$ axis and has a magnitude of 1500 N.

**2–42.** Determine the magnitude and angle measured counterclockwise from the positive $y$ axis of the resultant force acting on the bracket if $F_B = 600$ N and $\theta = 20°$.

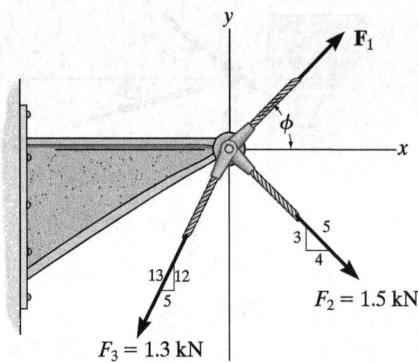

Probs. 2–43/44/45

**2–46.** The three concurrent forces acting on the screw eye produce a resultant force $\mathbf{F}_R = 0$. If $F_2 = \frac{2}{3} F_1$ and $\mathbf{F}_1$ is to be 90° from $\mathbf{F}_2$ as shown, determine the required magnitude of $\mathbf{F}_3$ expressed in terms of $F_1$ and the angle $\theta$.

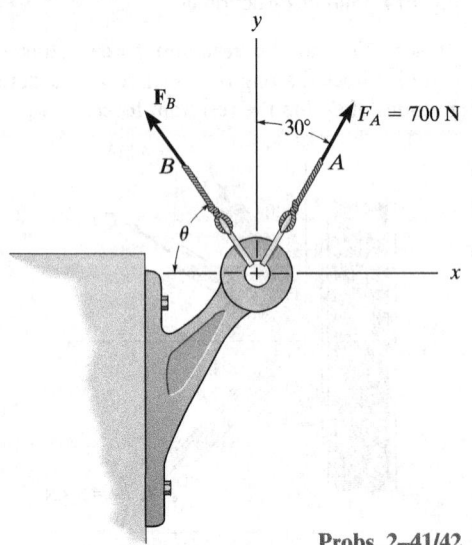

Probs. 2–41/42

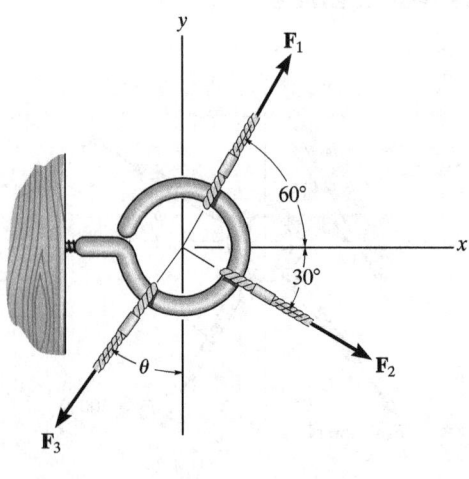

Prob. 2–46

**2–47.** Determine the magnitude of $\mathbf{F}_A$ and its direction $\theta$ so that the resultant force is directed along the positive $x$ axis and has a magnitude of 1250 N.

**\*2–48.** Determine the magnitude and direction measured counterclockwise from the positive $x$ axis of the resultant force acting on the ring at $O$ if $F_A = 750$ N and $\theta = 45°$.

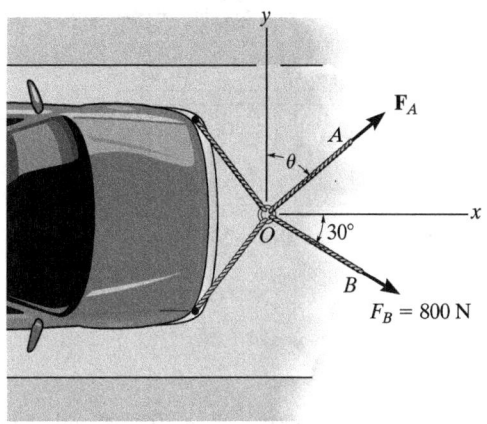

Probs. 2–47/48

**•2–49.** Determine the magnitude of the resultant force and its direction measured counterclockwise from the positive $x$ axis.

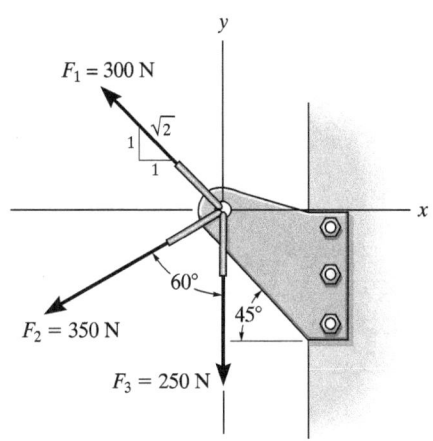

Prob. 2–49

**2–50.** The three forces are applied to the bracket. Determine the range of values for the magnitude of force $\mathbf{P}$ so that the resultant of the three forces does not exceed 2400 N.

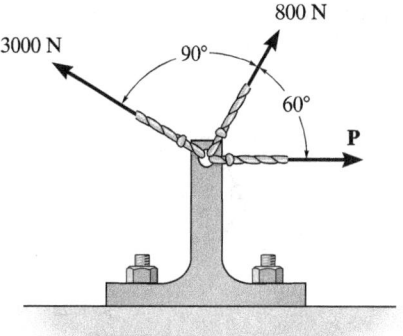

Prob. 2–50

**2–51.** If $F_1 = 150$ N and $\phi = 30°$, determine the magnitude of the resultant force acting on the bracket and its direction measured clockwise from the positive $x$ axis.

**\*2–52.** If the magnitude of the resultant force acting on the bracket is to be 450 N directed along the positive $u$ axis, determine the magnitude of $\mathbf{F}_1$ and its direction $\phi$.

**•2–53.** If the resultant force acting on the bracket is required to be a minimum, determine the magnitudes of $\mathbf{F}_1$ and the resultant force. Set $\phi = 30°$.

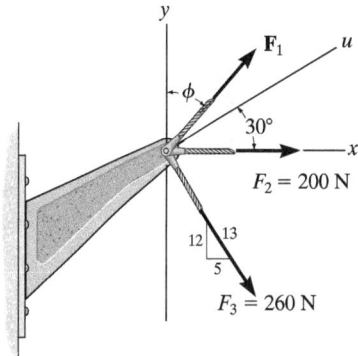

Probs. 2–51/52/53

**2–54.** Three forces act on the bracket. Determine the magnitude and direction $\theta$ of $\mathbf{F}_2$ so that the resultant force is directed along the positive $u$ axis and has a magnitude of 250 N.

**2–55.** If $F_2 = 750$ N and $\theta = 55°$, determine the magnitude and direction measured clockwise from the positive $x$ axis of the resultant force of the three forces acting on the bracket.

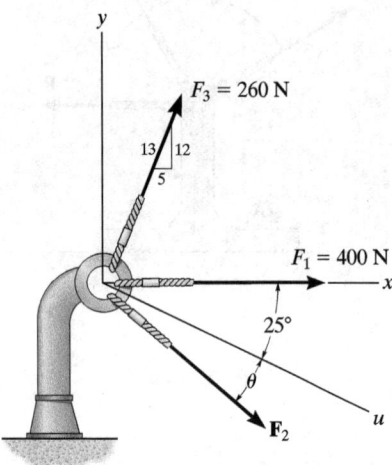

Probs. 2–54/55

**\*2–56.** The three concurrent forces acting on the post produce a resultant force $\mathbf{F}_R = \mathbf{0}$. If $F_2 = \frac{1}{2}F_1$, and $\mathbf{F}_1$ is to be 90° from $\mathbf{F}_2$ as shown, determine the required magnitude of $F_3$ expressed in terms of $F_1$ and the angle $\theta$.

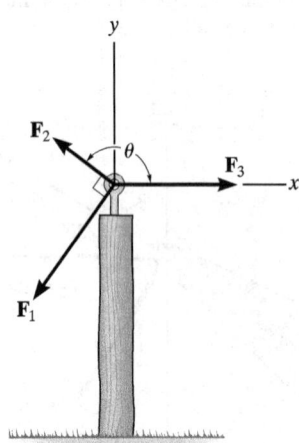

Prob. 2–56

**•2–57.** Determine the magnitude of force $\mathbf{F}$ so that the resultant force of the three forces is as small as possible. What is the magnitude of this smallest resultant force?

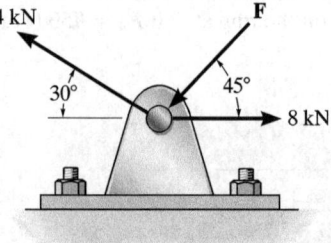

Prob. 2–57

**2–58.** Express each of the three forces acting on the bracket in Cartesian vector form with respect to the $x$ and $y$ axes. Determine the magnitude and direction $\theta$ of $\mathbf{F}_1$ so that the resultant force is directed along the positive $x'$ axis and has a magnitude of $F_R = 600$ N.

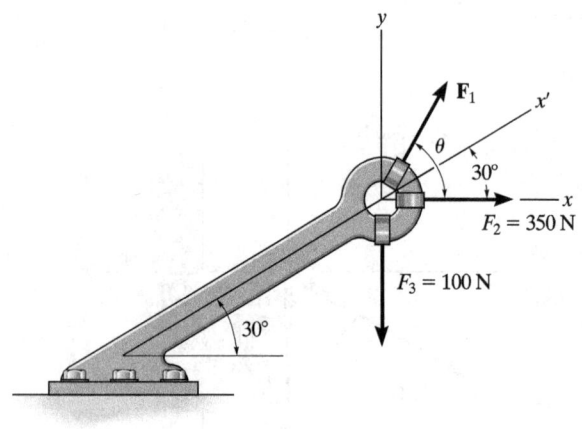

Prob. 2–58

## 2.5 Cartesian Vectors

The operations of vector algebra, when applied to solving problems in *three dimensions*, are greatly simplified if the vectors are first represented in Cartesian vector form. In this section we will present a general method for doing this; then in the next section we will use this method for finding the resultant force of a system of concurrent forces.

**Right-Handed Coordinate System.** We will use a right-handed coordinate system to develop the theory of vector algebra that follows. A rectangular coordinate system is said to be *right-handed* if the thumb of the right hand points in the direction of the positive $z$ axis when the right-hand fingers are curled about this axis and directed from the positive $x$ towards the positive $y$ axis, Fig. 2–21.

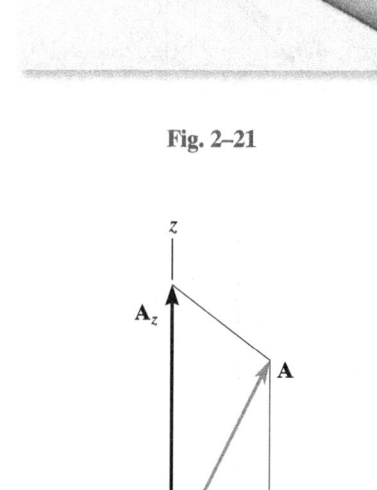

Fig. 2–21

**Rectangular Components of a Vector.** A vector **A** may have one, two, or three rectangular components along the $x$, $y$, $z$ coordinate axes, depending on how the vector is oriented relative to the axes. In general, though, when **A** is directed within an octant of the $x$, $y$, $z$ frame, Fig. 2–22, then by two successive applications of the parallelogram law, we may resolve the vector into components as $\mathbf{A} = \mathbf{A}' + \mathbf{A}_z$ and then $\mathbf{A}' = \mathbf{A}_x + \mathbf{A}_y$. Combining these equations, to eliminate $\mathbf{A}'$, **A** is represented by the vector sum of its *three* rectangular components,

$$\mathbf{A} = \mathbf{A}_x + \mathbf{A}_y + \mathbf{A}_z \tag{2-2}$$

**Cartesian Unit Vectors.** In three dimensions, the set of Cartesian unit vectors, **i**, **j**, **k**, is used to designate the directions of the $x$, $y$, $z$ axes, respectively. As stated in Sec. 2.4, the *sense* (or arrowhead) of these vectors will be represented analytically by a plus or minus sign, depending on whether they are directed along the positive or negative $x$, $y$, or $z$ axes. The positive Cartesian unit vectors are shown in Fig. 2–23.

Fig. 2–22

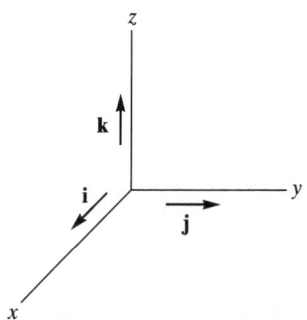

Fig. 2–23

**Cartesian Vector Representation.** Since the three components of **A** in Eq. 2–2 act in the positive **i**, **j**, and **k** directions, Fig. 2–24, we can write **A** in Cartesian vector form as

$$\mathbf{A} = A_x\mathbf{i} + A_y\mathbf{j} + A_z\mathbf{k} \quad (2\text{–}3)$$

There is a distinct advantage to writing vectors in this manner. Separating the *magnitude* and *direction* of each *component vector* will simplify the operations of vector algebra, particularly in three dimensions.

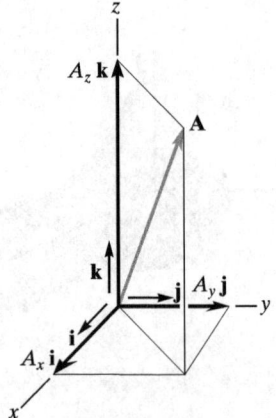

Fig. 2–24

**Magnitude of a Cartesian Vector.** It is always possible to obtain the magnitude of **A** provided it is expressed in Cartesian vector form. As shown in Fig. 2–25, from the blue right triangle, $A = \sqrt{A'^2 + A_z^2}$, and from the gray right triangle, $A' = \sqrt{A_x^2 + A_y^2}$. Combining these equations to eliminate $A'$, yields

$$A = \sqrt{A_x^2 + A_y^2 + A_z^2} \quad (2\text{–}4)$$

Hence, the magnitude of **A** is equal to the positive square root of the sum of the squares of its components.

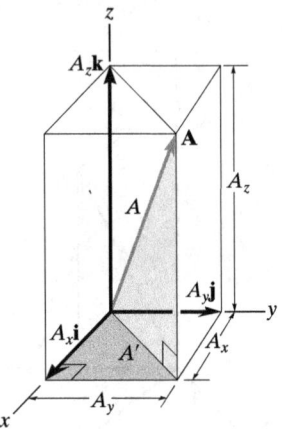

Fig. 2–25

**Direction of a Cartesian Vector.** We will define the *direction* of **A** by the *coordinate direction angles* $\alpha$ (alpha), $\beta$ (beta), and $\gamma$ (gamma), measured between the *tail* of **A** and the *positive x, y, z axes* provided they are located at the tail of **A**, Fig. 2–26. Note that regardless of where **A** is directed, each of these angles will be between 0° and 180°.

To determine $\alpha$, $\beta$, and $\gamma$, consider the projection of **A** onto the *x, y, z* axes, Fig. 2–27. Referring to the blue colored right triangles shown in each figure, we have

$$\cos\alpha = \frac{A_x}{A} \quad \cos\beta = \frac{A_y}{A} \quad \cos\gamma = \frac{A_z}{A} \quad (2\text{–}5)$$

These numbers are known as the *direction cosines* of **A**. Once they have been obtained, the coordinate direction angles $\alpha$, $\beta$, $\gamma$ can then be determined from the inverse cosines.

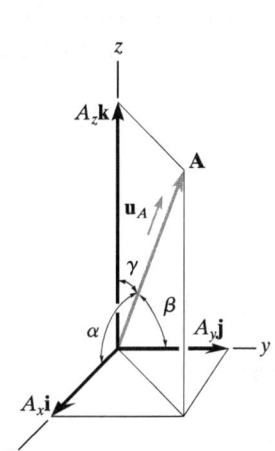 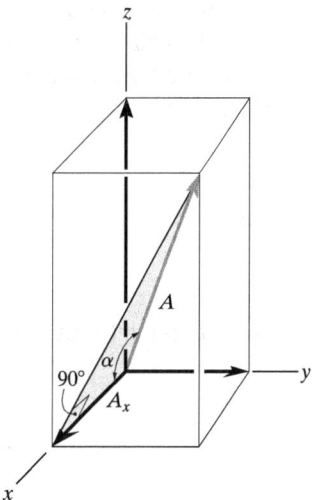

Fig. 2–26

An easy way of obtaining these direction cosines is to form a unit vector $\mathbf{u}_A$ in the direction of $\mathbf{A}$, Fig. 2–26. If $\mathbf{A}$ is expressed in Cartesian vector form, $\mathbf{A} = A_x\mathbf{i} + A_y\mathbf{j} + A_z\mathbf{k}$, then $\mathbf{u}_A$ will have a magnitude of one and be dimensionless provided $\mathbf{A}$ is divided by its magnitude, i.e.,

$$\mathbf{u}_A = \frac{\mathbf{A}}{A} = \frac{A_x}{A}\mathbf{i} + \frac{A_y}{A}\mathbf{j} + \frac{A_z}{A}\mathbf{k} \qquad (2\text{--}6)$$

where $A = \sqrt{A_x^2 + A_y^2 + A_z^2}$. By comparison with Eqs. 2–7, it is seen that *the $\mathbf{i},\mathbf{j},\mathbf{k}$ components of $\mathbf{u}_A$ represent the direction cosines of $\mathbf{A}$*, i.e.,

$$\mathbf{u}_A = \cos\alpha\,\mathbf{i} + \cos\beta\,\mathbf{j} + \cos\gamma\,\mathbf{k} \qquad (2\text{--}7)$$

Since the magnitude of a vector is equal to the positive square root of the sum of the squares of the magnitudes of its components, and $\mathbf{u}_A$ has a magnitude of one, then from the above equation an important relation between the direction cosines can be formulated as

$$\cos^2\alpha + \cos^2\beta + \cos^2\gamma = 1 \qquad (2\text{--}8)$$

Here we can see that if only *two* of the coordinate angles are known, the third angle can be found using this equation.

Finally, if the magnitude and coordinate direction angles of $\mathbf{A}$ are known, then $\mathbf{A}$ may be expressed in Cartesian vector form as

$$\begin{aligned}\mathbf{A} &= A\mathbf{u}_A \\ &= A\cos\alpha\,\mathbf{i} + A\cos\beta\,\mathbf{j} + A\cos\gamma\,\mathbf{k} \\ &= A_x\mathbf{i} + A_y\mathbf{j} + A_z\mathbf{k}\end{aligned} \qquad (2\text{--}9)$$

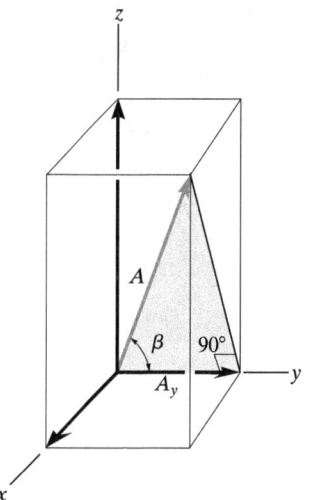
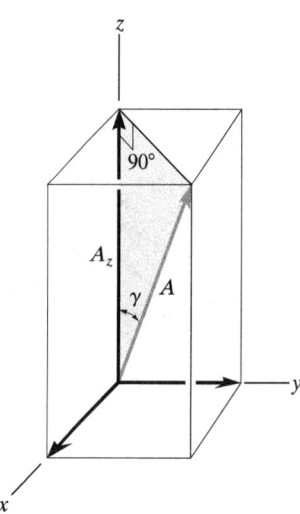

Fig 2–27

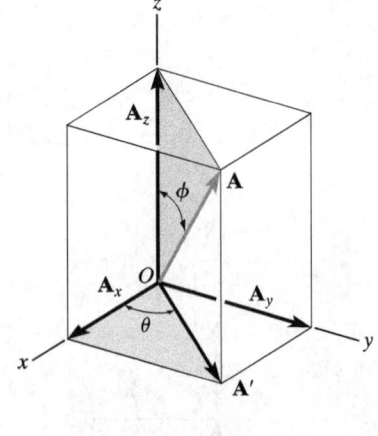

Fig. 2–28

Sometimes, the direction of **A** can be specified using two angles, $\theta$ and $\phi$ (phi), such as shown in Fig. 2–28. The components of **A** can then be determined by applying trigonometry first to the blue right triangle, which yields

$$A_z = A \cos \phi$$

and

$$A' = A \sin \phi$$

Now applying trigonometry to the other shaded right triangle,

$$A_x = A' \cos \theta = A \sin \phi \cos \theta$$

$$A_y = A' \sin \theta = A \sin \phi \sin \theta$$

Therefore **A** written in Cartesian vector form becomes

$$\mathbf{A} = A \sin \phi \cos \theta \, \mathbf{i} + A \sin \phi \sin \theta \, \mathbf{j} + A \cos \phi \, \mathbf{k}$$

You should not memorize this equation, rather it is important to understand how the components were determined using trigonometry.

## 2.6 Addition of Cartesian Vectors

The addition (or subtraction) of two or more vectors are greatly simplified if the vectors are expressed in terms of their Cartesian components. For example, if $\mathbf{A} = A_x\mathbf{i} + A_y\mathbf{j} + A_z\mathbf{k}$ and $\mathbf{B} = B_x\mathbf{i} + B_y\mathbf{j} + B_z\mathbf{k}$, Fig. 2–29, then the resultant vector, **R**, has components which are the scalar sums of the **i, j, k** components of **A** and **B**, i.e.,

$$\mathbf{R} = \mathbf{A} + \mathbf{B} = (A_x + B_x)\mathbf{i} + (A_y + B_y)\mathbf{j} + (A_z + B_z)\mathbf{k}$$

If this is generalized and applied to a system of several concurrent forces, then the force resultant is the vector sum of all the forces in the system and can be written as

$$\boxed{\mathbf{F}_R = \Sigma\mathbf{F} = \Sigma F_x\mathbf{i} + \Sigma F_y\mathbf{j} + \Sigma F_z\mathbf{k}} \qquad (2\text{–}10)$$

Here $\Sigma F_x$, $\Sigma F_y$, and $\Sigma F_z$ represent the algebraic sums of the respective $x$, $y$, $z$ or **i, j, k** components of each force in the system.

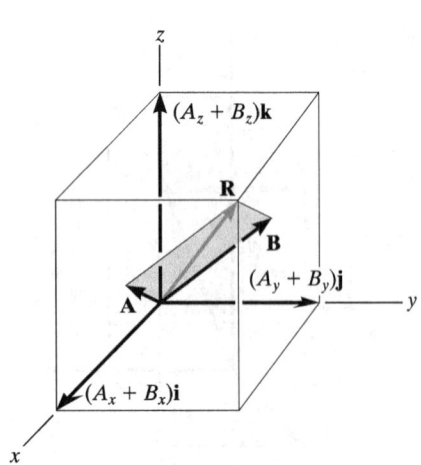

Fig. 2–29

## Important Points

- Cartesian vector analysis is often used to solve problems in three dimensions.
- The positive directions of the $x$, $y$, $z$ axes are defined by the Cartesian unit vectors $\mathbf{i}, \mathbf{j}, \mathbf{k}$, respectively.
- The *magnitude* of a Cartesian vector is $A = \sqrt{A_x^2 + A_y^2 + A_z^2}$.
- The *direction* of a Cartesian vector is specified using coordinate direction angles $\alpha$, $\beta$, $\gamma$ which the tail of the vector makes with the positive $x$, $y$, $z$ axes, respectively. The components of the unit vector $\mathbf{u}_A = \mathbf{A}/A$ represent the direction cosines of $\alpha$, $\beta$, $\gamma$. Only two of the angles $\alpha$, $\beta$, $\gamma$ have to be specified. The third angle is determined from the relationship $\cos^2 \alpha + \cos^2 \beta + \cos^2 \gamma = 1$.
- Sometimes the direction of a vector is defined using the two angles $\theta$ and $\phi$ as in Fig. 2–28. In this case, the vector components are obtained by vector resolution using trigonometry.
- To find the *resultant* of a concurrent force system, express each force as a Cartesian vector and add the $\mathbf{i}, \mathbf{j}, \mathbf{k}$ components of all the forces in the system.

The resultant force acting on the bow the ship can be determined by first representing each rope force as a Cartesian vector and then summing the $\mathbf{i}, \mathbf{j}$, and $\mathbf{k}$ components.

## EXAMPLE 2.8

Express the force $\mathbf{F}$ shown in Fig. 2–30 as a Cartesian vector.

### SOLUTION

Since only two coordinate direction angles are specified, the third angle $\alpha$ must be determined from Eq. 2–8; i.e.,

$$\cos^2 \alpha + \cos^2 \beta + \cos^2 \gamma = 1$$
$$\cos^2 \alpha + \cos^2 60° + \cos^2 45° = 1$$
$$\cos \alpha = \sqrt{1 - (0.5)^2 - (0.707)^2} = \pm 0.5$$

Hence, two possibilities exist, namely,

$$\alpha = \cos^{-1}(0.5) = 60° \quad \text{or} \quad \alpha = \cos^{-1}(-0.5) = 120°$$

By inspection, it is necessary that $\alpha = 60°$, since $\mathbf{F}_x$ must be in the $+x$ direction.

Using Eq. 2–9, with $F = 200$ N, we have

$$\mathbf{F} = F \cos \alpha \mathbf{i} + F \cos \beta \mathbf{j} + F \cos \gamma \mathbf{k}$$
$$= (200 \cos 60° \text{ N})\mathbf{i} + (200 \cos 60° \text{ N})\mathbf{j} + (200 \cos 45° \text{ N})\mathbf{k}$$
$$= \{100.0\mathbf{i} + 100.0\mathbf{j} + 141.4\mathbf{k}\} \text{ N} \qquad Ans.$$

Show that indeed the magnitude of $F = 200$ N.

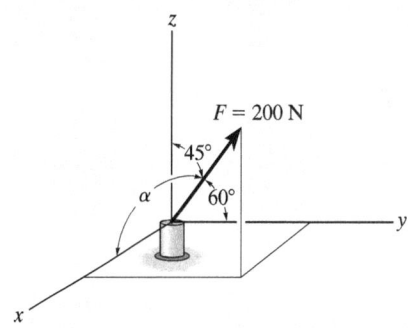

Fig. 2–30

## EXAMPLE 2.9

Determine the magnitude and the coordinate direction angles of the resultant force acting on the ring in Fig. 2–31a.

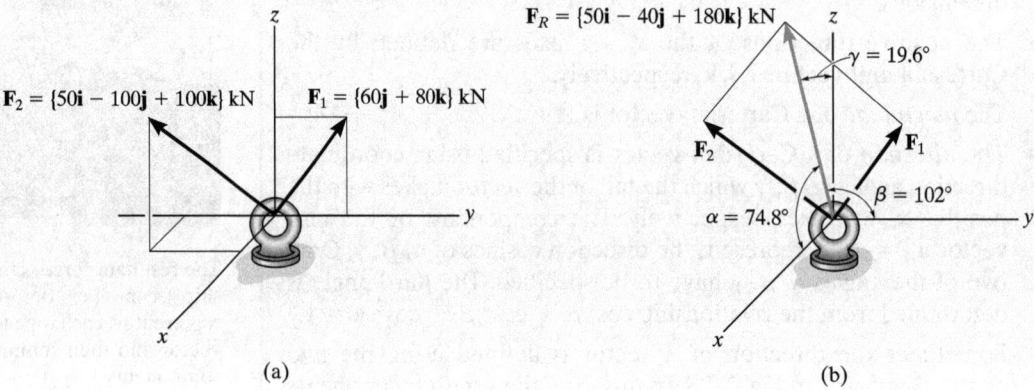

Fig. 2–31

### SOLUTION

Since each force is represented in Cartesian vector form, the resultant force, shown in Fig. 2–31b, is

$$\mathbf{F}_R = \Sigma \mathbf{F} = \mathbf{F}_1 + \mathbf{F}_2 = \{60\mathbf{j} + 80\mathbf{k}\} \text{ kN} + \{50\mathbf{i} - 100\mathbf{j} + 100\mathbf{k}\} \text{ kN}$$
$$= \{50\mathbf{i} - 40\mathbf{j} + 180\mathbf{k}\} \text{ kN}$$

The magnitude of $\mathbf{F}_R$ is

$$F_R = \sqrt{(50 \text{ kN})^2 + (-40 \text{ kN})^2 + (180 \text{ kN})^2} = 191.0 \text{ kN}$$
$$= 191 \text{ kN} \qquad \qquad Ans.$$

The coordinate direction angles $\alpha, \beta, \gamma$ are determined from the components of the unit vector acting in the direction of $\mathbf{F}_R$.

$$\mathbf{u}_{F_R} = \frac{\mathbf{F}_R}{F_R} = \frac{50}{191.0}\mathbf{i} - \frac{40}{191.0}\mathbf{j} + \frac{180}{191.0}\mathbf{k}$$
$$= 0.2617\mathbf{i} - 0.2094\mathbf{j} + 0.9422\mathbf{k}$$

so that

| | | |
|---|---|---|
| $\cos \alpha = 0.2617$ | $\alpha = 74.8°$ | Ans. |
| $\cos \beta = -0.2094$ | $\beta = 102°$ | Ans. |
| $\cos \gamma = 0.9422$ | $\gamma = 19.6°$ | Ans. |

These angles are shown in Fig. 2–31b.

**NOTE:** In particular, notice that $\beta > 90°$ since the **j** component of $\mathbf{u}_{F_R}$ is negative. This seems reasonable considering how $\mathbf{F}_1$ and $\mathbf{F}_2$ add according to the parallelogram law.

## EXAMPLE 2.10

Express the force **F** shown in Fig. 2–32a as a Cartesian vector.

### SOLUTION

The angles of 60° and 45° defining the direction of **F** are *not* coordinate direction angles. Two successive applications of the parallelogram law are needed to resolve **F** into its x, y, z components. First $\mathbf{F} = \mathbf{F}' + \mathbf{F}_z$, then $\mathbf{F}' = \mathbf{F}_x + \mathbf{F}_y$, Fig. 2–32b. By trigonometry, the magnitudes of the components are

$$F_z = 100 \sin 60° \text{ kN} = 86.6 \text{ kN}$$

$$F' = 100 \cos 60° \text{ kN} = 50 \text{ kN}$$

$$F_x = F' \cos 45° = 50 \cos 45° \text{ kN} = 35.4 \text{ kN}$$

$$F_y = F' \sin 45° = 50 \sin 45° \text{ kN} = 35.4 \text{ kN}$$

Realizing that $\mathbf{F}_y$ has a direction defined by $-\mathbf{j}$, we have

$$\mathbf{F} = \{35.4\mathbf{i} - 35.4\mathbf{j} + 86.6\mathbf{k}\} \text{ kN} \qquad Ans.$$

To show that the magnitude of this vector is indeed 100 kN, apply Eq. 2–4,

$$F = \sqrt{F_x^2 + F_y^2 + F_z^2}$$

$$= \sqrt{(35.4)^2 + (-35.4)^2 + (86.6)^2} = 100 \text{ kN}$$

If needed, the coordinate direction angles of **F** can be determined from the components of the unit vector acting in the direction of **F**. Hence,

$$\mathbf{u} = \frac{\mathbf{F}}{F} = \frac{F_x}{F}\mathbf{i} + \frac{F_y}{F}\mathbf{j} + \frac{F_z}{F}\mathbf{k}$$

$$= \frac{35.4}{100}\mathbf{i} - \frac{35.4}{100}\mathbf{j} + \frac{86.6}{100}\mathbf{k}$$

$$= 0.354\mathbf{i} - 0.354\mathbf{j} + 0.866\mathbf{k}$$

so that

$$\alpha = \cos^{-1}(0.354) = 69.3°$$

$$\beta = \cos^{-1}(-0.354) = 111°$$

$$\gamma = \cos^{-1}(0.866) = 30.0°$$

These results are shown in Fig. 2–32c.

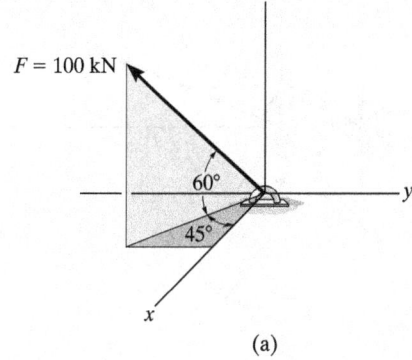
(a)

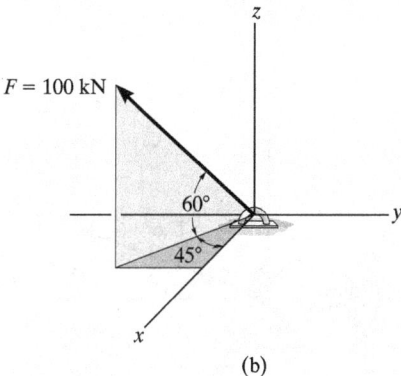
(b)

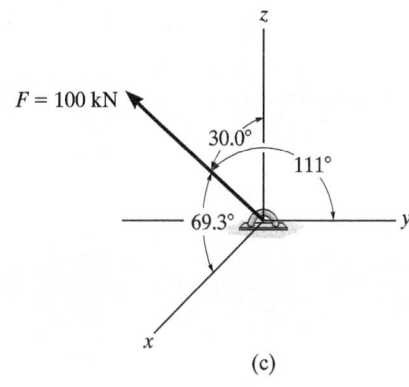
(c)

Fig. 2–32

# EXAMPLE 2.11

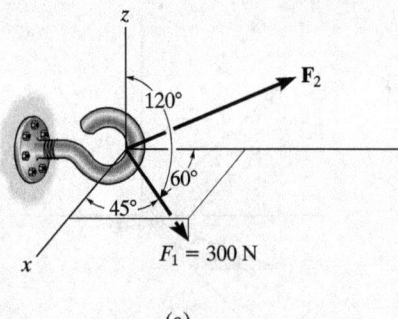

(a)

Fig. 2–33

Two forces act on the hook shown in Fig. 2–33a. Specify the magnitude of $\mathbf{F}_2$ and its coordinate direction angles of $\mathbf{F}_2$ that the resultant force $\mathbf{F}_R$ acts along the positive $y$ axis and has a magnitude of 800 N.

## SOLUTION

To solve this problem, the resultant force $\mathbf{F}_R$ and its two components, $\mathbf{F}_1$ and $\mathbf{F}_2$, will each be expressed in Cartesian vector form. Then, as shown in Fig. 2–33a, it is necessary that $\mathbf{F}_R = \mathbf{F}_1 + \mathbf{F}_2$.

Applying Eq. 2–9,

$$\mathbf{F}_1 = F_1 \cos \alpha_1 \mathbf{i} + F_1 \cos \beta_1 \mathbf{j} + F_1 \cos \gamma_1 \mathbf{k}$$
$$= 300 \cos 45° \mathbf{i} + 300 \cos 60° \mathbf{j} + 300 \cos 120° \mathbf{k}$$
$$= \{212.1\mathbf{i} + 150\mathbf{j} - 150\mathbf{k}\} \text{ N}$$
$$\mathbf{F}_2 = F_{2x}\mathbf{i} + F_{2y}\mathbf{j} + F_{2z}\mathbf{k}$$

Since $\mathbf{F}_R$ has a magnitude of 800 N and acts in the $+\mathbf{j}$ direction,

$$\mathbf{F}_R = (800 \text{ N})(+\mathbf{j}) = \{800\mathbf{j}\} \text{ N}$$

We require

$$\mathbf{F}_R = \mathbf{F}_1 + \mathbf{F}_2$$
$$800\mathbf{j} = 212.1\mathbf{i} + 150\mathbf{j} - 150\mathbf{k} + F_{2x}\mathbf{i} + F_{2y}\mathbf{j} + F_{2z}\mathbf{k}$$
$$800\mathbf{j} = (212.1 + F_{2x})\mathbf{i} + (150 + F_{2y})\mathbf{j} + (-150 + F_{2z})\mathbf{k}$$

To satisfy this equation, the $\mathbf{i}, \mathbf{j}, \mathbf{k}$ components of $\mathbf{F}_R$ must be equal to the corresponding $\mathbf{i}, \mathbf{j}, \mathbf{k}$ components of $(\mathbf{F}_1 + \mathbf{F}_2)$. Hence,

$$0 = 212.1 + F_{2x} \qquad F_{2x} = -212.1 \text{ N}$$
$$800 = 150 + F_{2y} \qquad F_{2y} = 650 \text{ N}$$
$$0 = -150 + F_{2z} \qquad F_{2z} = 150 \text{ N}$$

The magnitude of $\mathbf{F}_2$ is thus

$$F_2 = \sqrt{(-212.1 \text{ N})^2 + (650 \text{ N})^2 + (150 \text{ N})^2}$$
$$= 700 \text{ N} \qquad \text{Ans.}$$

We can use Eq. 2–9 to determine $\alpha_2, \beta_2$, and $\gamma_2$.

$$\cos \alpha_2 = \frac{-212.1}{700}; \qquad \alpha_2 = 108° \qquad \text{Ans.}$$

$$\cos \beta_2 = \frac{650}{700}; \qquad \beta_2 = 21.8° \qquad \text{Ans.}$$

$$\cos \gamma_2 = \frac{150}{700}; \qquad \gamma_2 = 77.6° \qquad \text{Ans.}$$

These results are shown in Fig. 2–32b.

## FUNDAMENTAL PROBLEMS

**F2–13.** Determine the coordinate direction angles of the force.

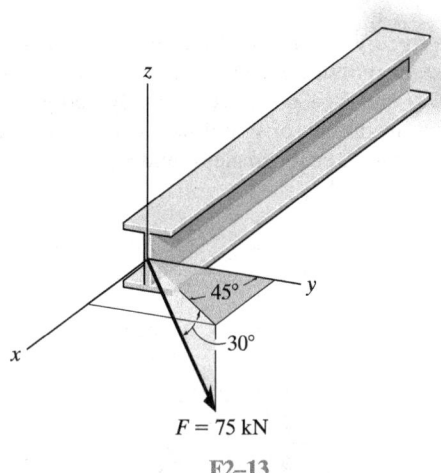

F2–13

**F2–14.** Express the force as a Cartesian vector.

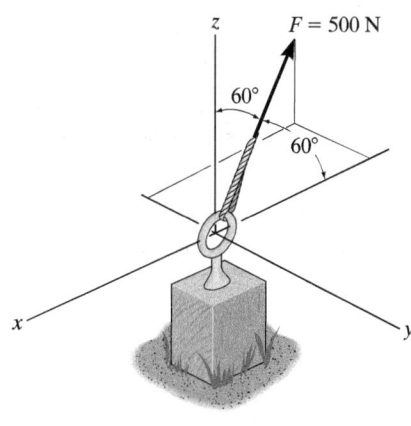

F2–14

**F2–15.** Express the force as a Cartesian vector.

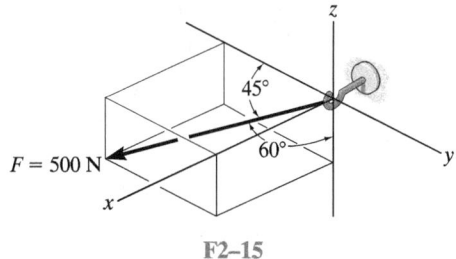

F2–15

**F2–16.** Express the force as a Cartesian vector.

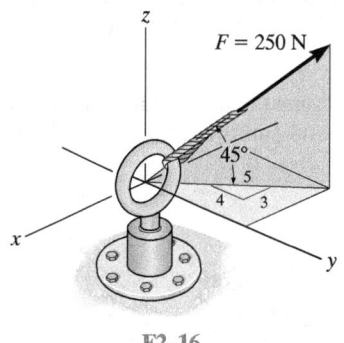

F2–16

**F2–17.** Express the force as a Cartesian vector.

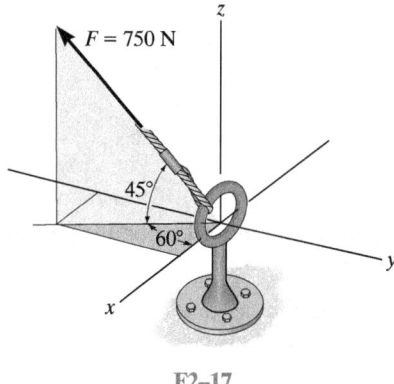

F2–17

**F2–18.** Determine the resultant force acting on the hook.

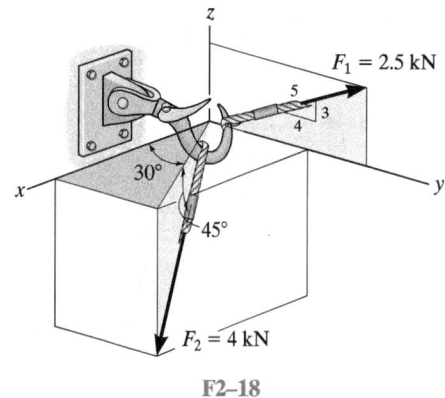

F2–18

## PROBLEMS

**2–59.** Determine the coordinate angle $\gamma$ for $F_2$ and then express each force acting on the bracket as a Cartesian vector.

***2–60.** Determine the magnitude and coordinate direction angles of the resultant force acting on the bracket.

**2–63.** The force $F$ acts on the bracket within the octant shown. If $F = 400$ N, $\beta = 60°$, and $\gamma = 45°$, determine the $x, y, z$ components of $F$.

**\*2–64.** The force $F$ acts on the bracket within the octant shown. If the magnitudes of the $x$ and $z$ components of $F$ are $F_x = 300$ N and $F_z = 600$ N, respectively, and $\beta = 60°$, determine the magnitude of $F$ and its $y$ component. Also, find the coordinate direction angles $\alpha$ and $\gamma$.

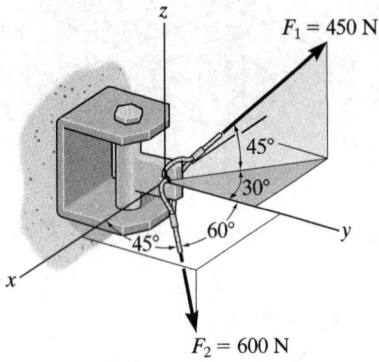

Probs. 2–59/60

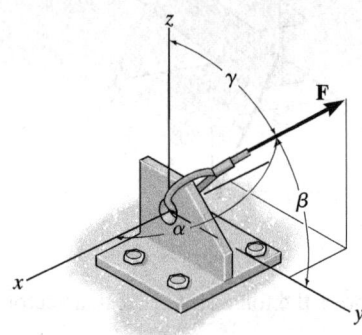

Probs. 2–63/64

**•2–61.** Express each force acting on the pipe assembly in Cartesian vector form.

**2–62.** Determine the magnitude and direction of the resultant force acting on the pipe assembly.

**•2–65.** The two forces $F_1$ and $F_2$ acting at $A$ have a resultant force of $F_R = \{-100\mathbf{k}\}$ N. Determine the magnitude and coordinate direction angles of $F_2$.

**2–66.** Determine the coordinate direction angles of the force $F_1$ and indicate them on the figure.

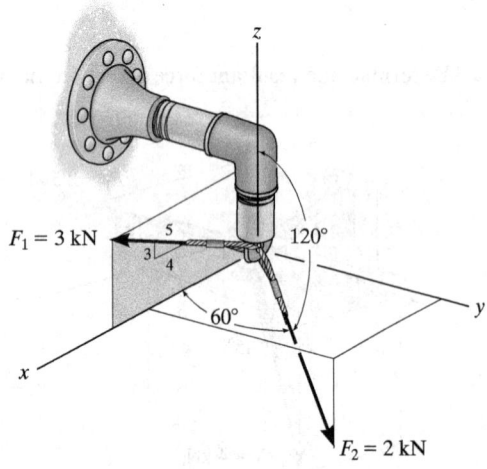

Probs. 2–61/62

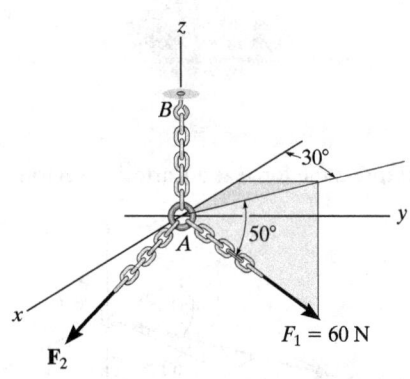

Probs. 2–65/66

**2–67.** The spur gear is subjected to the two forces caused by contact with other gears. Express each force as a Cartesian vector.

**\*2–68.** The spur gear is subjected to the two forces caused by contact with other gears. Determine the resultant of the two forces and express the result as a Cartesian vector.

**2–71.** If $\alpha = 120°$, $\beta < 90°$, $\gamma = 60°$, and $F = 400$ N, determine the magnitude and coordinate direction angles of the resultant force acting on the hook.

**\*2–72.** If the resultant force acting on the hook is $\mathbf{F}_R = \{-200\mathbf{i} + 800\mathbf{j} + 150\mathbf{k}\}$ N, determine the magnitude and coordinate direction angles of $\mathbf{F}$.

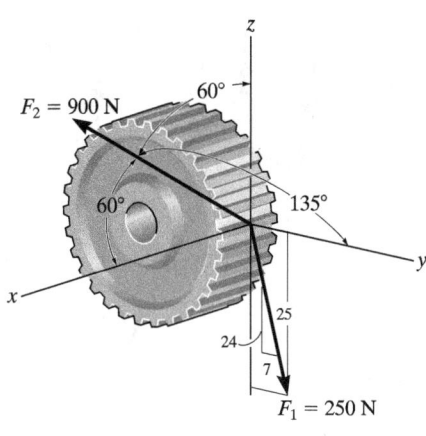

Probs. 2–67/68

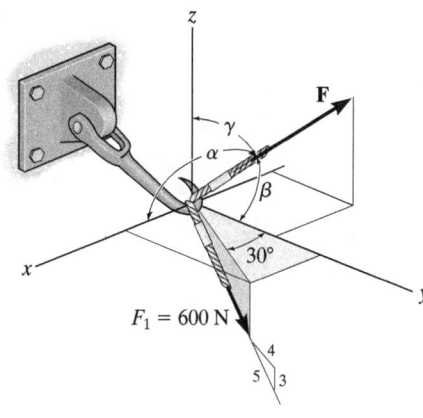

Probs. 2–71/72

**•2–69.** If the resultant force acting on the bracket is $\mathbf{F}_R = \{-300\mathbf{i} + 650\mathbf{j} + 250\mathbf{k}\}$ N, determine the magnitude and coordinate direction angles of $\mathbf{F}$.

**2–70.** If the resultant force acting on the bracket is to be $\mathbf{F}_R = \{800\mathbf{j}\}$ N, determine the magnitude and coordinate direction angles of $\mathbf{F}$.

**•2–73.** The shaft $S$ exerts three force components on the die $D$. Find the magnitude and coordinate direction angles of the resultant force. Force $\mathbf{F}_2$ acts within the octant shown.

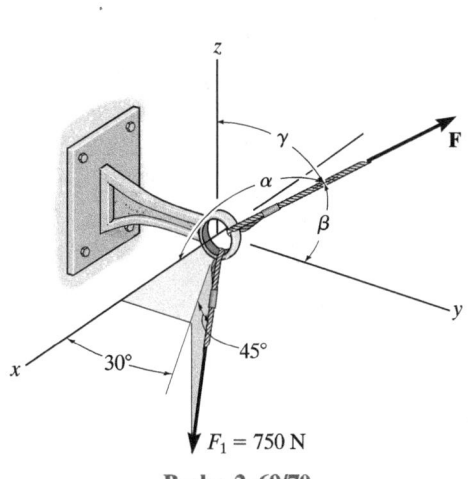

Probs. 2–69/70

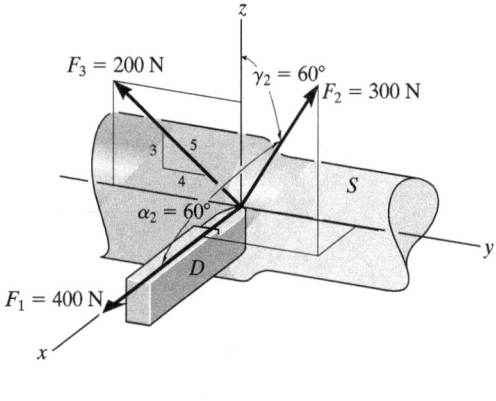

Prob. 2–73

**2–74.** The mast is subjected to the three forces shown. Determine the coordinate direction angles $\alpha_1, \beta_1, \gamma_1$ of $\mathbf{F}_1$ so that the resultant force acting on the mast is $\mathbf{F}_R = \{350\mathbf{i}\}$ N.

**2–75.** The mast is subjected to the three forces shown. Determine the coordinate direction angles $\alpha_1, \beta_1, \gamma_1$ of $\mathbf{F}_1$ so that the resultant force acting on the mast is zero.

**2–78.** If the resultant force acting on the bracket is directed along the positive $y$ axis, determine the magnitude of the resultant force and the coordinate direction angles of $\mathbf{F}$ so that $\beta < 90°$.

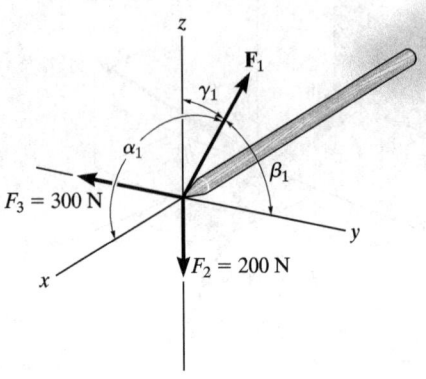

Probs. 2–74/75

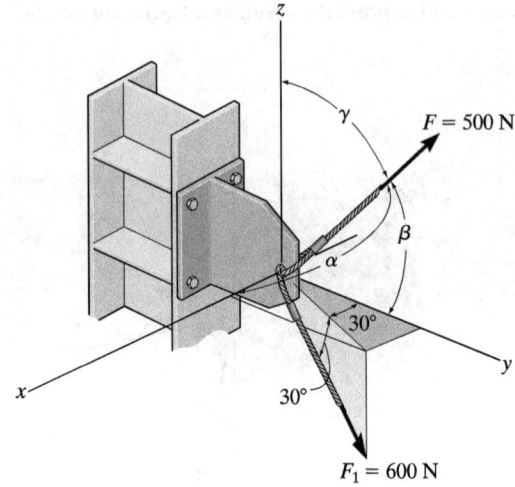

Prob. 2–78

**\*2–76.** Determine the magnitude and coordinate direction angles of $\mathbf{F}_2$ so that the resultant of the two forces acts along the positive $x$ axis and has a magnitude of 500 N.

**•2–77.** Determine the magnitude and coordinate direction angles of $\mathbf{F}_2$ so that the resultant of the two forces is zero.

**2–79.** Specify the magnitude of $\mathbf{F}_3$ and its coordinate direction angles $\alpha_3, \beta_3, \gamma_3$ so that the resultant force $\mathbf{F}_R = \{9\mathbf{j}\}$ kN.

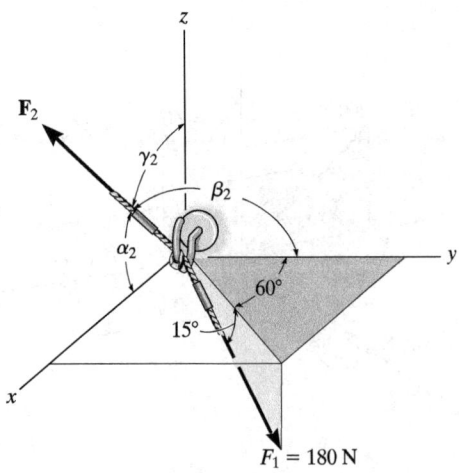

Probs. 2–76/77

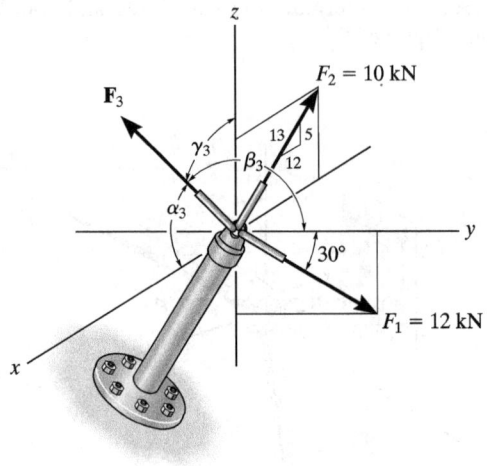

Prob. 2–79

**\*2–80.** If $F_3 = 9$ kN, $\theta = 30°$, and $\phi = 45°$, determine the magnitude and coordinate direction angles of the resultant force acting on the ball-and-socket joint.

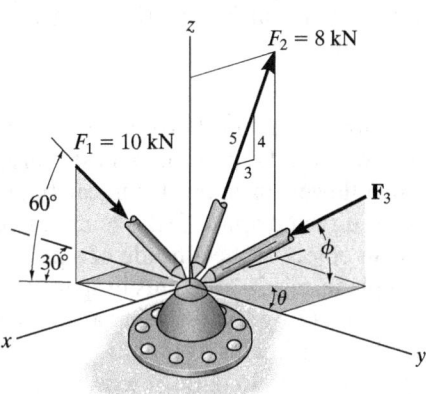

Prob. 2–80

**•2–81.** The pole is subjected to the force **F**, which has components acting along the $x, y, z$ axes as shown. If the magnitude of **F** is 3 kN, $\beta = 30°$, and $\gamma = 75°$, determine the magnitudes of its three components.

**2–82.** The pole is subjected to the force **F** which has components $F_x = 1.5$ kN and $F_z = 1.25$ kN. If $\beta = 75°$, determine the magnitudes of **F** and **F**$_y$.

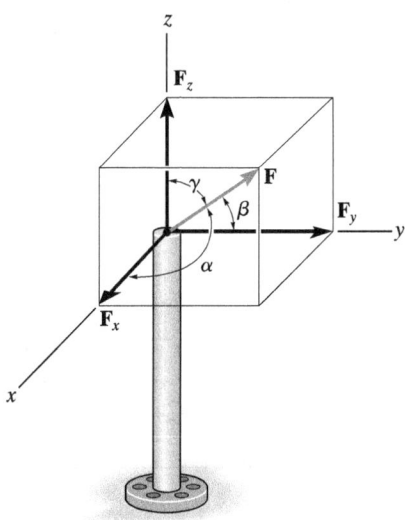

Probs. 2–81/82

**2–83.** Three forces act on the ring. If the resultant force **F**$_R$ has a magnitude and direction as shown, determine the magnitude and the coordinate direction angles of force **F**$_3$.

**\*2–84.** Determine the coordinate direction angles of **F**$_1$ and **F**$_R$.

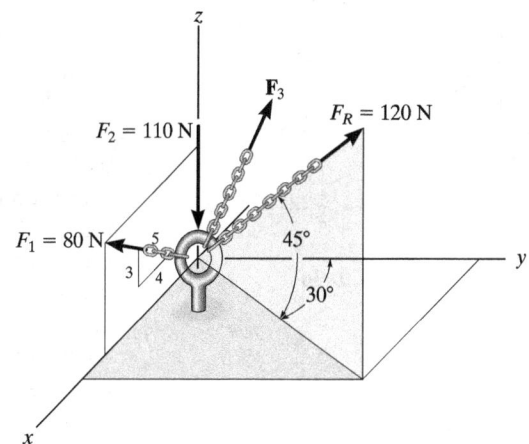

Probs. 2–83/84

**•2–85.** Two forces **F**$_1$ and **F**$_2$ act on the bolt. If the resultant force **F**$_R$ has a magnitude of 50 N and coordinate direction angles $\alpha = 110°$ and $\beta = 80°$, as shown, determine the magnitude of **F**$_2$ and its coordinate direction angles.

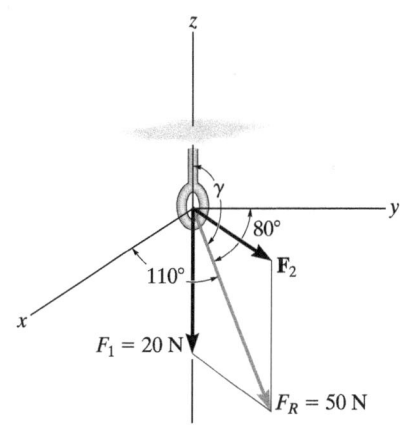

Prob. 2–85

## 2.7 Position Vectors

In this section we will introduce the concept of a position vector. It will be shown that this vector is of importance in formulating a Cartesian force vector directed between two points in space.

**x, y, z Coordinates.** Throughout the book we will use a *right-handed* coordinate system to reference the location of points in space. We will also use the convention followed in many technical books, which requires the positive z axis to be directed *upward* (the zenith direction) so that it measures the height of an object or the altitude of a point. The x, y axes then lie in the horizontal plane, Fig. 2–34. Points in space are located relative to the origin of coordinates, O, by successive measurements along the x, y, z axes. For example, the coordinates of point A are obtained by starting at O and measuring $x_A = +4$ m along the x axis, then $y_A = +2$ m along the y axis, and finally $z_A = -6$ m along the z axis. Thus, $A(4$ m, $2$ m, $-6$ m$)$. In a similar manner, measurements along the x, y, z axes from O to B yield the coordinates of B, i.e., $B(6$ m, $-1$ m, $4$ m$)$.

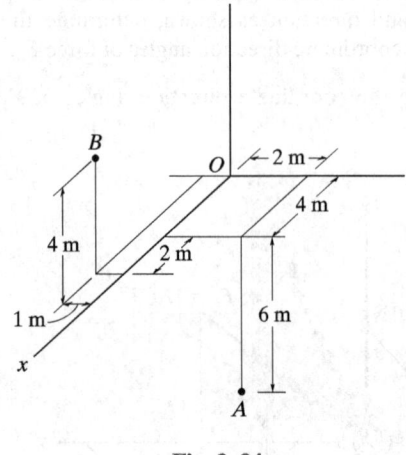

Fig. 2–34

**Position Vector.** A *position vector* **r** is defined as a fixed vector which locates a point in space relative to another point. For example, if **r** extends from the origin of coordinates, O, to point $P(x, y, z)$, Fig. 2–35a, then **r** can be expressed in Cartesian vector form as

$$\mathbf{r} = x\mathbf{i} + y\mathbf{j} + z\mathbf{k}$$

Note how the head-to-tail vector addition of the three components yields vector **r**, Fig. 2–35b. Starting at the origin O, one "travels" x in the +**i** direction, then y in the +**j** direction, and finally z in the +**k** direction to arrive at point $P(x, y, z)$.

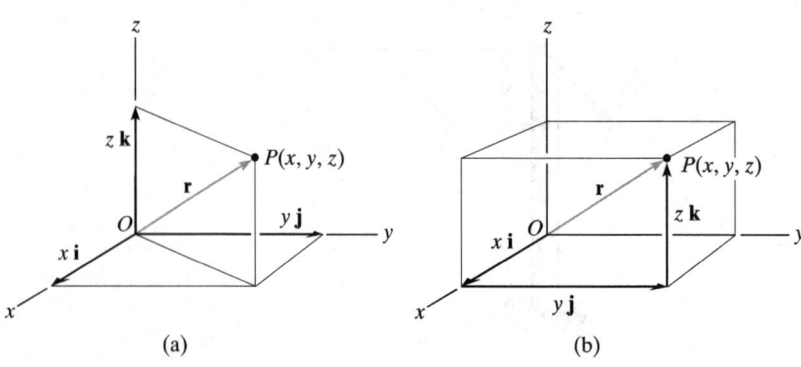

Fig. 2–35

## 2.7 Position Vectors

In the more general case, the position vector may be directed from point $A$ to point $B$ in space, Fig. 2–36a. This vector is also designated by the symbol $\mathbf{r}$. As a matter of convention, we will *sometimes* refer to this vector with *two subscripts* to indicate from and to the point where it is directed. Thus, $\mathbf{r}$ can also be designated as $\mathbf{r}_{AB}$. Also, note that $\mathbf{r}_A$ and $\mathbf{r}_B$ in Fig. 2–36a are referenced with only one subscript since they extend from the origin of coordinates.

From Fig. 2–36a, by the head-to-tail vector addition, using the triangle rule, we require

$$\mathbf{r}_A + \mathbf{r} = \mathbf{r}_B$$

Solving for $\mathbf{r}$ and expressing $\mathbf{r}_A$ and $\mathbf{r}_B$ in Cartesian vector form yield

$$\mathbf{r} = \mathbf{r}_B - \mathbf{r}_A = (x_B\mathbf{i} + y_B\mathbf{j} + z_B\mathbf{k}) - (x_A\mathbf{i} + y_A\mathbf{j} + z_A\mathbf{k})$$

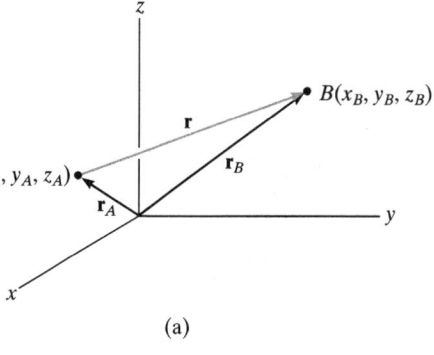

(a)

or

$$\mathbf{r} = (x_B - x_A)\mathbf{i} + (y_B - y_A)\mathbf{j} + (z_B - z_A)\mathbf{k} \qquad (2\text{–}11)$$

Thus, the $\mathbf{i}$, $\mathbf{j}$, $\mathbf{k}$ components of the position vector $\mathbf{r}$ may be formed by taking the coordinates of the tail of the vector $A(x_A, y_A, z_A)$ and subtracting them from the corresponding coordinates of the head $B(x_B, y_B, z_B)$. We can also form these components *directly*, Fig. 2–36b, by starting at $A$ and moving through a distance of $(x_B - x_A)$ along the positive $x$ axis (+$\mathbf{i}$), then $(y_B - y_A)$ along the positive $y$ axis (+$\mathbf{j}$), and finally $(z_B - z_A)$ along the positive $z$ axis (+$\mathbf{k}$) to get to $B$.

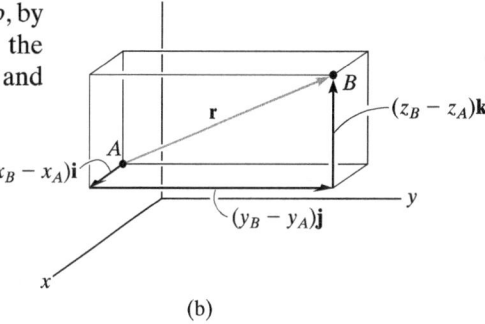

(b)

Fig. 2–36

If an $x, y, z$ coordinate system is established, then the coordinates of points $A$ and $B$ can be determined. From this, the position vector $\mathbf{r}$ acting along the cable can be formulated. Its magnitude represents the length of the cable, and its unit vector, $\mathbf{u} = \mathbf{r}/r$, gives the direction defined by $\alpha, \beta, \gamma$.

## EXAMPLE 2.12

An elastic rubber band is attached to points A and B as shown in Fig. 2–37a. Determine its length and its direction measured from A toward B.

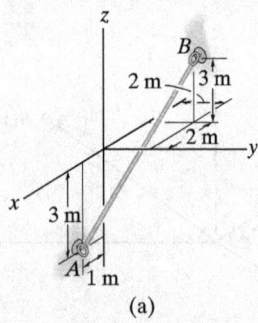

(a)

### SOLUTION

We first establish a position vector from A to B, Fig. 2–37b. In accordance with Eq. 2–11, the coordinates of the tail $A(1\text{ m}, 0, -3\text{ m})$ are subtracted from the coordinates of the head $B(-2\text{ m}, 2\text{ m}, 3\text{ m})$, which yields

$$\mathbf{r} = [-2\text{ m} - 1\text{ m}]\mathbf{i} + [2\text{ m} - 0]\mathbf{j} + [3\text{ m} - (-3\text{ m})]\mathbf{k}$$
$$= \{-3\mathbf{i} + 2\mathbf{j} + 6\mathbf{k}\}\text{ m}$$

These components of $\mathbf{r}$ can also be determined *directly* by realizing that they represent the direction and distance one must travel along each axis in order to move from A to B, i.e., along the x axis $\{-3\mathbf{i}\}$ m, along the y axis $\{2\mathbf{j}\}$ m, and finally along the z axis $\{6\mathbf{k}\}$ m.

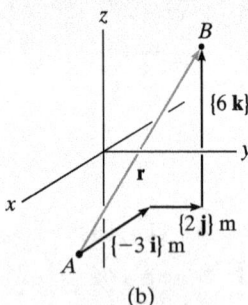

(b)

The length of the rubber band is therefore

$$r = \sqrt{(-3\text{ m})^2 + (2\text{ m})^2 + (6\text{ m})^2} = 7\text{ m} \qquad Ans.$$

Formulating a unit vector in the direction of $\mathbf{r}$, we have

$$\mathbf{u} = \frac{\mathbf{r}}{r} = -\frac{3}{7}\mathbf{i} + \frac{2}{7}\mathbf{j} + \frac{6}{7}\mathbf{k}$$

The components of this unit vector give the coordinate direction angles

$$\alpha = \cos^{-1}\left(-\frac{3}{7}\right) = 115° \qquad Ans.$$

$$\beta = \cos^{-1}\left(\frac{2}{7}\right) = 73.4° \qquad Ans.$$

$$\gamma = \cos^{-1}\left(\frac{6}{7}\right) = 31.0° \qquad Ans.$$

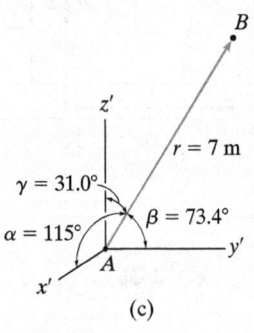

(c)

Fig. 2–37

**NOTE:** These angles are measured from the *positive axes* of a localized coordinate system placed at the tail of $\mathbf{r}$, as shown in Fig. 2–37c.

## 2.8 Force Vector Directed Along a Line

Quite often in three-dimensional statics problems, the direction of a force is specified by two points through which its line of action passes. Such a situation is shown in Fig. 2–38, where the force **F** is directed along the cord $AB$. We can formulate **F** as a Cartesian vector by realizing that it has the *same direction* and *sense* as the position vector **r** directed from point $A$ to point $B$ on the cord. This common direction is specified by the *unit vector* $\mathbf{u} = \mathbf{r}/r$. Hence,

$$\mathbf{F} = F\mathbf{u} = F\left(\frac{\mathbf{r}}{r}\right) = F\left(\frac{(x_B - x_A)\mathbf{i} + (y_B - y_A)\mathbf{j} + (z_B - z_A)\mathbf{k}}{\sqrt{(x_B - x_A)^2 + (y_B - y_A)^2 + (z_B - z_A)^2}}\right)$$

Although we have represented **F** symbolically in Fig. 2–38, note that it has *units of force*, unlike **r**, which has units of length.

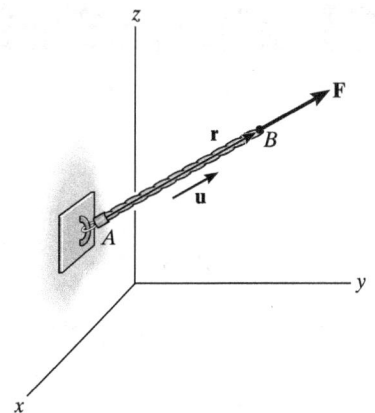

Fig. 2–38

The force **F** acting along the chain can be represented as a Cartesian vector by establishing $x, y, z$ axes and first forming a position vector **r** along the length of the chain. Then, the corresponding unit vector $\mathbf{u} = \mathbf{r}/r$ that defines the direction of both the chain and the force can be determined. Finally, the magnitude of the force is combined with its direction, $\mathbf{F} = F\mathbf{u}$.

### Important Points

- A position vector locates one point in space relative to another point.
- The easiest way to formulate the components of a position vector is to determine the distance and direction that must be traveled along the $x, y, z$ directions—going from the tail to the head of the vector.
- A force **F** acting in the direction of a position vector **r** can be represented in Cartesian form if the unit vector **u** of the position vector is determined and it is multiplied by the magnitude of the force, i.e., $\mathbf{F} = F\mathbf{u} = F(\mathbf{r}/r)$.

## EXAMPLE 2.13

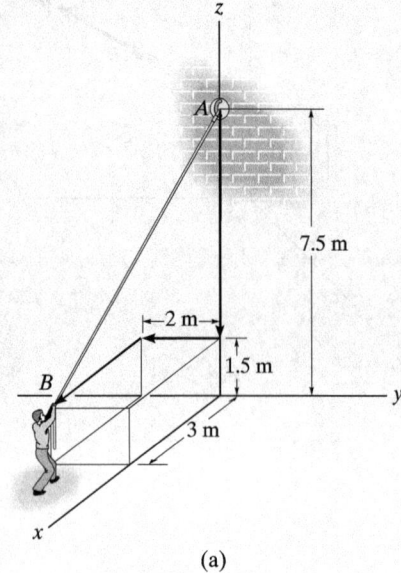

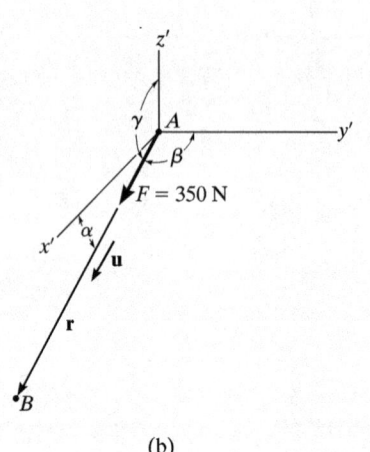

Fig. 2–39

The man shown in Fig. 2–39a pulls on the cord with a force of 350 N. Represent this force acting on the support $A$ as a Cartesian vector and determine its direction.

### SOLUTION

Force **F** is shown in Fig. 2–39b. The *direction* of this vector, **u**, is determined from the position vector **r**, which extends from $A$ to $B$. Rather than using the coordinates of the end points of the cord, **r** can be determined *directly* by noting in Fig. 2–39a that one must travel from $A$ $\{-6\mathbf{k}\}$ m, then $\{-2\mathbf{j}\}$ m, and finally $\{3\mathbf{i}\}$ m to get to $B$. Thus,

$$\mathbf{r} = \{3\mathbf{i} - 2\mathbf{j} - 6\mathbf{k}\} \text{ m}$$

The magnitude of **r**, which represents the *length* of cord $AB$, is

$$r = \sqrt{(3 \text{ m})^2 + (-2 \text{ m})^2 + (-6 \text{ m})^2} = 7 \text{ m}$$

Forming the unit vector that defines the direction and sense of both **r** and **F**, we have

$$\mathbf{u} = \frac{\mathbf{r}}{r} = \frac{3}{7}\mathbf{i} - \frac{2}{7}\mathbf{j} - \frac{6}{7}\mathbf{k}$$

Since **F**, has a *magnitude* of 350 N and a *direction* specified by **u**, then

$$\mathbf{F} = F\mathbf{u} = 350 \text{ N}\left(\frac{3}{7}\mathbf{i} - \frac{2}{7}\mathbf{j} - \frac{6}{7}\mathbf{k}\right)$$
$$= \{150\mathbf{i} - 100\mathbf{j} - 300\mathbf{k}\} \text{ N} \qquad Ans.$$

The coordinate direction angles are measured between **r** (or **F**) and the *positive axes* of a localized coordinate system with origin placed at $A$, Fig. 2–39b. From the components of the unit vector:

$$\alpha = \cos^{-1}\left(\frac{3}{7}\right) = 64.6° \qquad Ans.$$

$$\beta = \cos^{-1}\left(\frac{-2}{7}\right) = 107° \qquad Ans.$$

$$\gamma = \cos^{-1}\left(\frac{-6}{7}\right) = 149° \qquad Ans.$$

**NOTE:** These results make sense when compared with the angles identified in Fig. 2–39b.

## EXAMPLE 2.14

The force in Fig. 2–40a acts on the hook. Express it as a Cartesian vector.

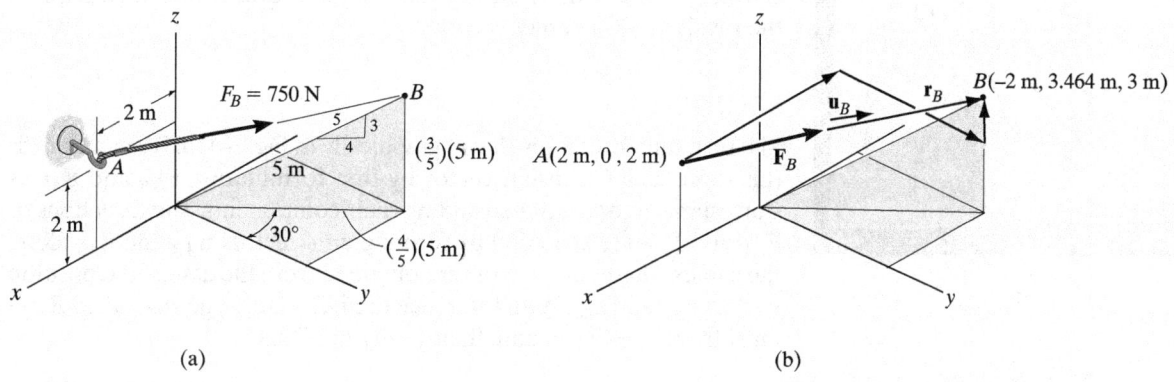

Fig. 2–40

### SOLUTION
As shown in Fig. 2–40b, the coordinates for points $A$ and $B$ are

$$A(2 \text{ m}, 0, 2 \text{ m})$$

and

$$B\left[-\left(\frac{4}{5}\right)5 \sin 30° \text{ m}, \left(\frac{4}{5}\right)5 \cos 30° \text{ m}, \left(\frac{3}{5}\right)5 \text{ m}\right]$$

or

$$B(-2 \text{ m}, 3.464 \text{ m}, 3 \text{ m})$$

Therefore, to go from $A$ to $B$, one must travel $\{4\mathbf{i}\}$ m, then $\{3.464\,\mathbf{j}\}$ m, and finally $\{1\,\mathbf{k}\}$ m. Thus,

$$\mathbf{u}_B = \left(\frac{\mathbf{r}_B}{r_B}\right) = \frac{\{-4\mathbf{i} + 3.464\mathbf{j} + 1\mathbf{k}\} \text{ m}}{\sqrt{(-4 \text{ m})^2 + (3.464 \text{ m})^2 + (1 \text{ m})^2}}$$

$$= -0.7428\mathbf{i} + 0.6433\mathbf{j} + 0.1857\mathbf{k}$$

Force $\mathbf{F}_B$ expressed as a Cartesian vector becomes

$$\mathbf{F}_B = F_B \mathbf{u}_B = (750 \text{ N})(-0.7428\mathbf{i} + 0.6433\mathbf{j} + 0.1857\mathbf{k})$$

$$= \{-557\mathbf{i} + 482\mathbf{j} + 139\mathbf{k}\} \text{ N} \qquad Ans.$$

## EXAMPLE 2.15

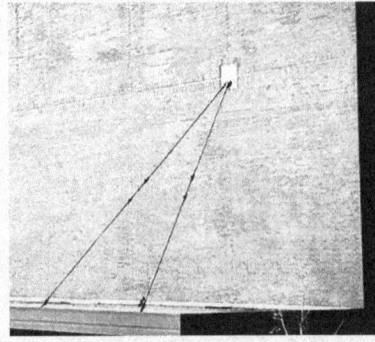

The roof is supported by cables as shown in the photo. If the cables exert forces $F_{AB} = 100$ N and $F_{AC} = 120$ N on the wall hook at $A$ as shown in Fig. 2–41a, determine the resultant force acting at $A$. Express the result as a Cartesian vector.

### SOLUTION

The resultant force $\mathbf{F}_R$ is shown graphically in Fig. 2–41b. We can express this force as a Cartesian vector by first formulating $\mathbf{F}_{AB}$ and $\mathbf{F}_{AC}$ as Cartesian vectors and then adding their components. The directions of $\mathbf{F}_{AB}$ and $\mathbf{F}_{AC}$ are specified by forming unit vectors $\mathbf{u}_{AB}$ and $\mathbf{u}_{AC}$ along the cables. These unit vectors are obtained from the associated position vectors $\mathbf{r}_{AB}$ and $\mathbf{r}_{AC}$. With reference to Fig. 2–41a, to go from $A$ to $B$, we must travel $\{-4\mathbf{k}\}$ m and, then $\{-4\mathbf{i}\}$ m. Thus,

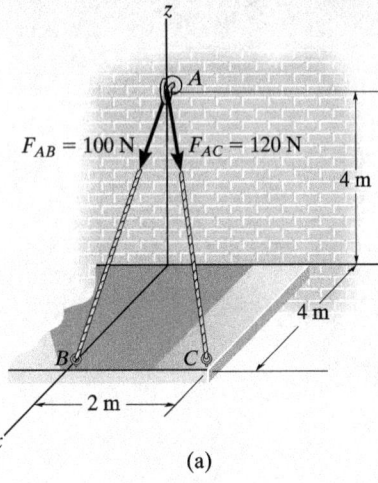

$$\mathbf{r}_{AB} = \{4\mathbf{i} - 4\mathbf{k}\} \text{ m}$$

$$r_{AB} = \sqrt{(4 \text{ m})^2 + (-4 \text{ m})^2} = 5.66 \text{ m}$$

$$\mathbf{F}_{AB} = F_{AB}\left(\frac{\mathbf{r}_{AB}}{r_{AB}}\right) = (100 \text{ N})\left(\frac{4}{5.66}\mathbf{i} - \frac{4}{5.66}\mathbf{k}\right)$$

$$\mathbf{F}_{AB} = \{70.7\mathbf{i} - 70.7\mathbf{k}\} \text{ N}$$

To go from $A$ to $C$, we must travel $\{-4\mathbf{k}\}$ m, then $\{2\mathbf{j}\}$ m, and finally $\{4\mathbf{i}\}$. Thus,

$$\mathbf{r}_{AC} = \{4\mathbf{i} + 2\mathbf{j} - 4\mathbf{k}\} \text{ m}$$

$$r_{AC} = \sqrt{(4 \text{ m})^2 + (2 \text{ m})^2 + (-4 \text{ m})^2} = 6 \text{ m}$$

$$\mathbf{F}_{AC} = F_{AC}\left(\frac{\mathbf{r}_{AC}}{r_{AC}}\right) = (120 \text{ N})\left(\frac{4}{6}\mathbf{i} + \frac{2}{6}\mathbf{j} - \frac{4}{6}\mathbf{k}\right)$$

$$= \{80\mathbf{i} + 40\mathbf{j} - 80\mathbf{k}\} \text{ N}$$

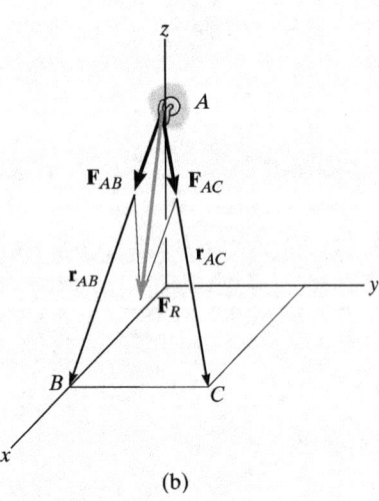

The resultant force is therefore

$$\mathbf{F}_R = \mathbf{F}_{AB} + \mathbf{F}_{AC} = \{70.7\mathbf{i} - 70.7\mathbf{k}\} \text{ N} + \{80\mathbf{i} + 40\mathbf{j} - 80\mathbf{k}\} \text{ N}$$

$$= \{151\mathbf{i} + 40\mathbf{j} - 151\mathbf{k}\} \text{ N} \qquad Ans.$$

Fig. 2–41

## FUNDAMENTAL PROBLEMS

**F2–19.** Express the position vector $\mathbf{r}_{AB}$ in Cartesian vector form, then determine its magnitude and coordinate direction angles.

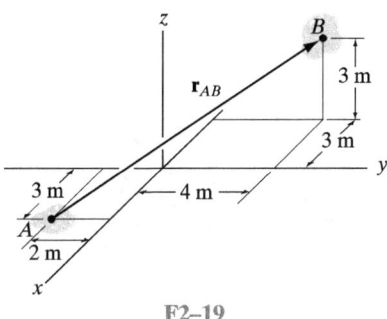

F2–19

**F2–20.** Determine the length of the rod and the position vector directed from $A$ to $B$. What is the angle $\theta$?

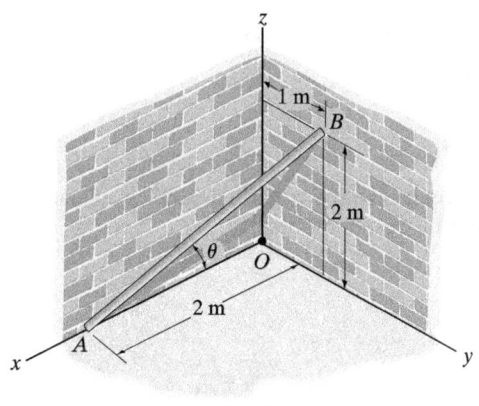

F2–20

**F2–21.** Express the force as a Cartesian vector.

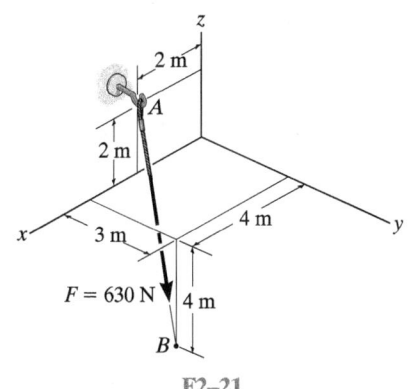

F2–21

**F2–22.** Express the force as a Cartesian vector.

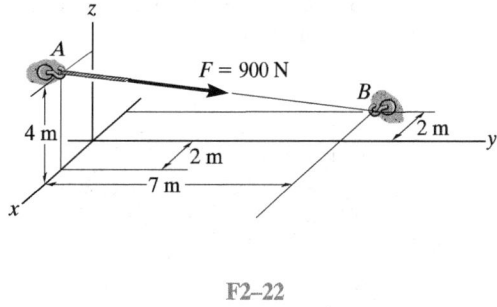

F2–22

**F2–23.** Determine the magnitude of the resultant force at $A$.

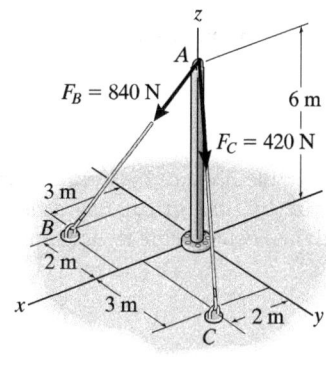

F2–23

**F2–24.** Determine the resultant force at $A$.

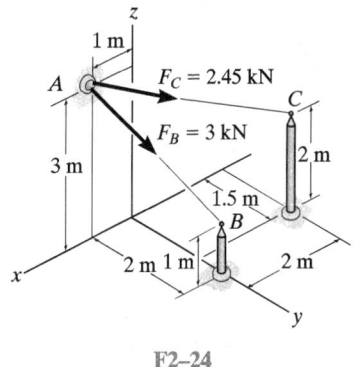

F2–24

## PROBLEMS

**2–86.** Determine the position vector **r** directed from point $A$ to point $B$ and the length of cord $AB$. Take $z = 4$ m.

**2–87.** If the cord $AB$ is 7.5 m long, determine the coordinate position $+z$ of point $B$.

**•2–89.** Determine the magnitude and coordinate direction angles of the resultant force acting at $A$.

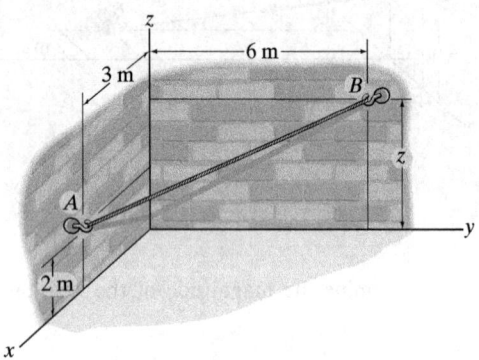

Probs. 2–86/87

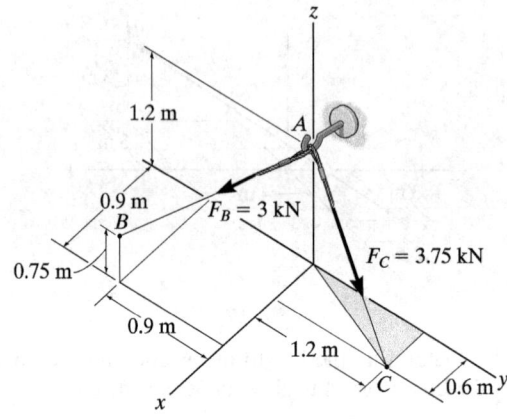

Prob. 2–89

**\*2–88.** Determine the distance between the end points $A$ and $B$ on the wire by first formulating a position vector from $A$ to $B$ and then determining its magnitude.

**2–90.** Determine the magnitude and coordinate direction angles of the resultant force.

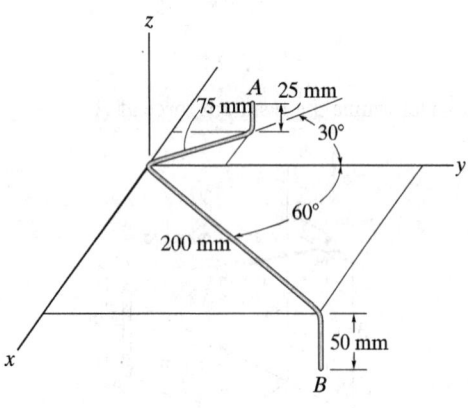

Prob. 2–88

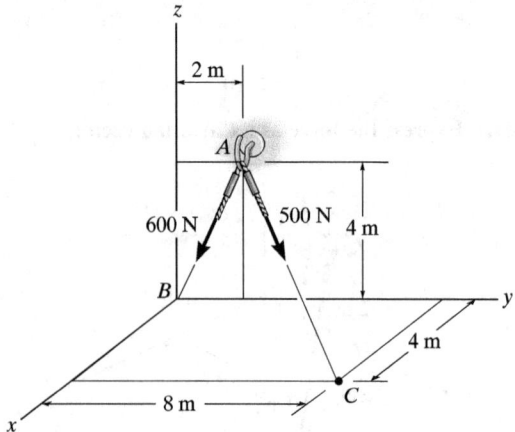

Prob. 2–90

**2–91.** Determine the magnitude and coordinate direction angles of the resultant force acting at $A$.

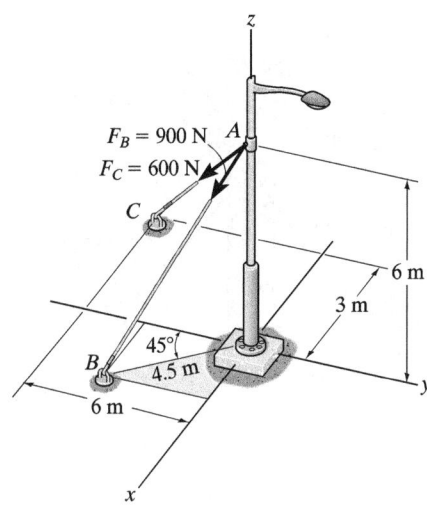

Prob. 2–91

**•2–93.** The chandelier is supported by three chains which are concurrent at point $O$. If the force in each chain has a magnitude of 300 N, express each force as a Cartesian vector and determine the magnitude and coordinate direction angles of the resultant force.

**2–94.** The chandelier is supported by three chains which are concurrent at point $O$. If the resultant force at $O$ has a magnitude of 650 N and is directed along the negative $z$ axis, determine the force in each chain.

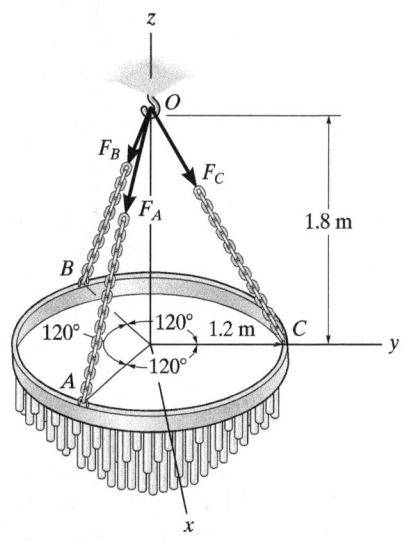

Probs. 2–93/94

**\*2–92.** Determine the magnitude and coordinate direction angles of the resultant force.

**2–95.** Express force **F** as a Cartesian vector; then determine its coordinate direction angles.

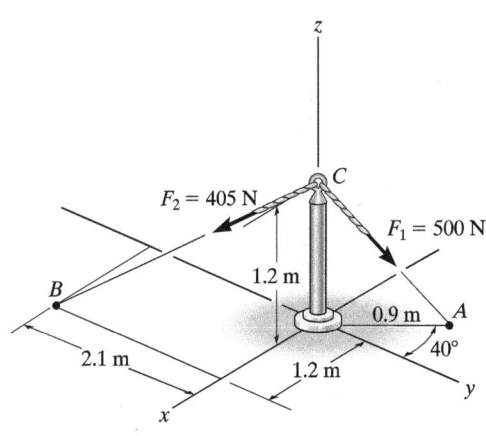

Prob. 2–92

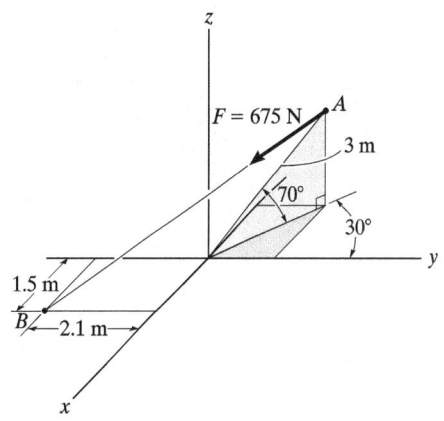

Prob. 2–95

**\*2–96.** The tower is held in place by three cables. If the force of each cable acting on the tower is shown, determine the magnitude and coordinate direction angles $\alpha, \beta, \gamma$ of the resultant force. Take $x = 20$ m, $y = 15$ m.

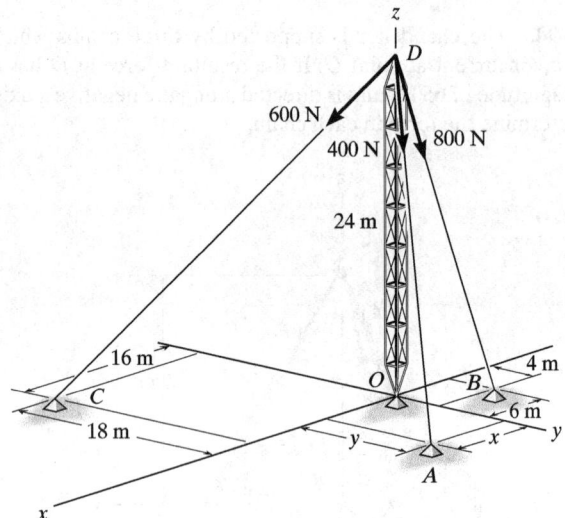

Prob. 2–96

**•2–97.** The door is held opened by means of two chains. If the tension in $AB$ and $CD$ is $F_A = 300$ N and $F_C = 250$ N, respectively, express each of these forces in Cartesian vector form.

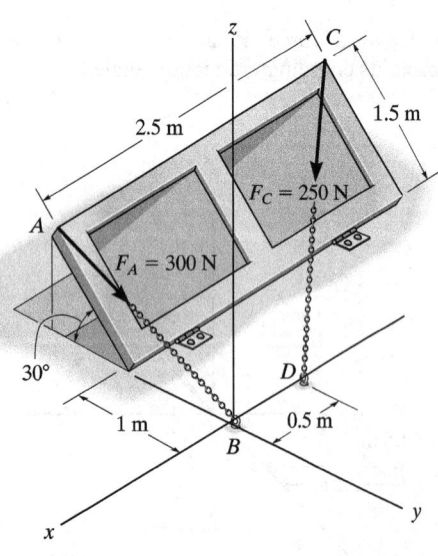

Prob. 2–97

**2–98.** The guy wires are used to support the telephone pole. Represent the force in each wire in Cartesian vector form. Neglect the diameter of the pole.

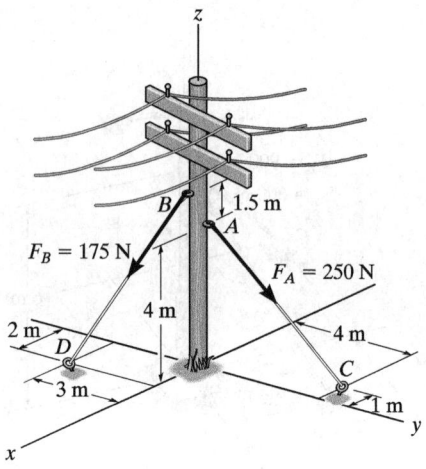

Prob. 2–98

**2–99.** Two cables are used to secure the overhang boom in position and support the 1500-N load. If the resultant force is directed along the boom from point $A$ towards $O$, determine the magnitudes of the resultant force and forces $\mathbf{F}_B$ and $\mathbf{F}_C$. Set $x = 3$ m and $z = 2$ m.

**\*2–100.** Two cables are used to secure the overhang boom in position and support the 1500-N load. If the resultant force is directed along the boom from point $A$ towards $O$, determine the values of $x$ and $z$ for the coordinates of point $C$ and the magnitude of the resultant force. Set $F_B = 1610$ N and $F_C = 2400$ N.

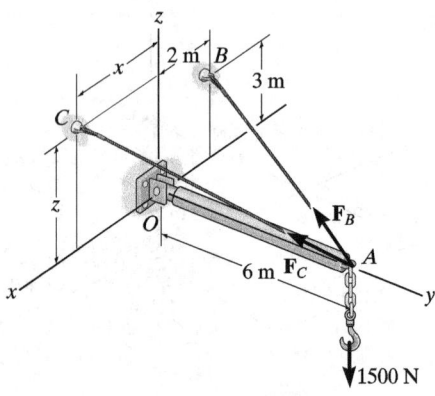

Probs. 2–99/100

•2–101. The cable $AO$ exerts a force on the top of the pole of $\mathbf{F} = \{-120\mathbf{i} - 90\mathbf{j} - 80\mathbf{k}\}$ N. If the cable has a length of 1.02 m, determine the height $z$ of the pole and the location $(x, y)$ of its base.

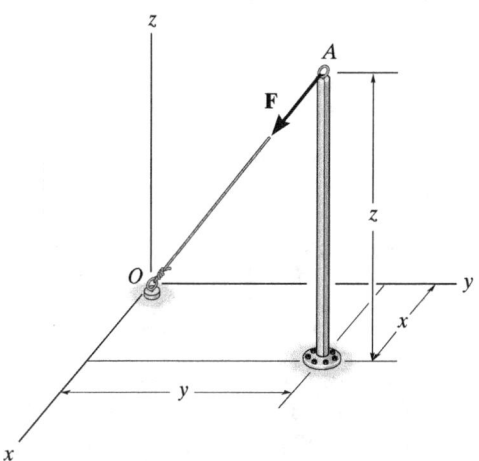

Prob. 2–101

2–102. If the force in each chain has a magnitude of 2.25 kN, determine the magnitude and coordinate direction angles of the resultant force.

2–103. If the resultant of the three forces is $\mathbf{F}_R = \{-4.5\mathbf{k}\}$ kN, determine the magnitude of the force in each chain.

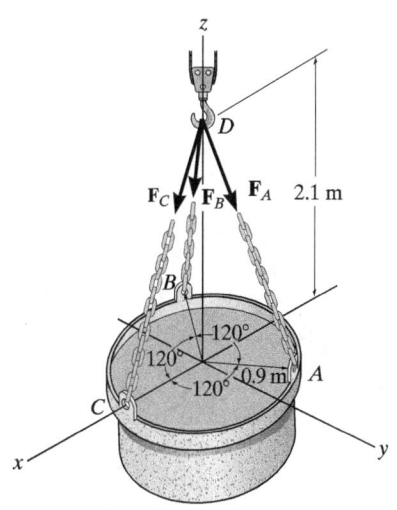

Probs. 2–102/103

*2–104. The antenna tower is supported by three cables. If the forces of these cables acting on the antenna are $F_B = 520$ N, $F_C = 680$ N, and $F_D = 560$ N, determine the magnitude and coordinate direction angles of the resultant force acting at $A$.

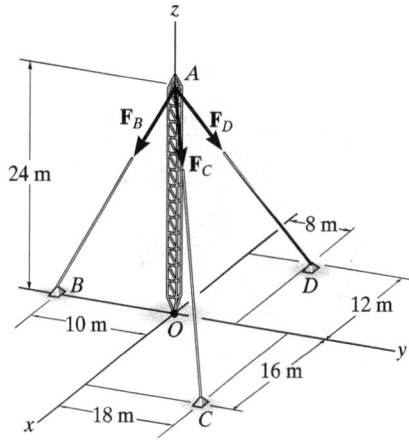

Prob. 2–104

•2–105. If the force in each cable tied to the bin is 350 N, determine the magnitude and coordinate direction angles of the resultant force.

2–106. If the resultant of the four forces is $\mathbf{F}_R = \{-1.8\mathbf{k}\}$ kN, determine the tension developed in each cable. Due to symmetry, the tension in the four cables is the same.

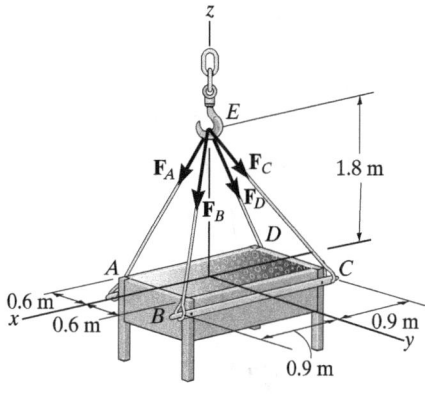

Probs. 2–105/106

**2–107.** The pipe is supported at its end by a cord $AB$. If the cord exerts a force of $F = 60$ N on the pipe at $A$, express this force as a Cartesian vector.

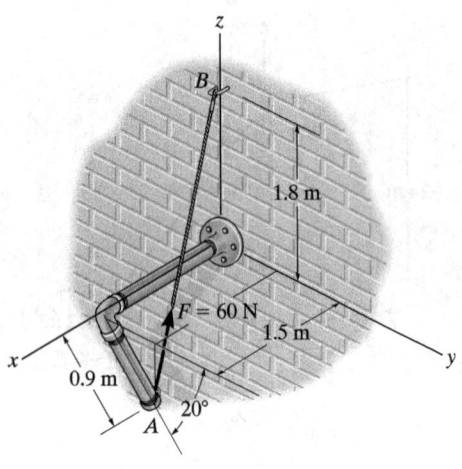

Prob. 2–107

**\*2–108.** The load at $A$ creates a force of 200 N in wire $AB$. Express this force as a Cartesian vector, acting on $A$ and directed towards $B$.

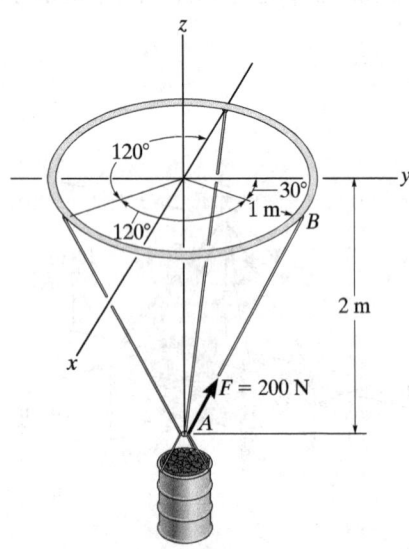

Prob. 2–108

**•2–109.** The cylindrical plate is subjected to the three cable forces which are concurrent at point $D$. Express each force which the cables exert on the plate as a Cartesian vector, and determine the magnitude and coordinate direction angles of the resultant force.

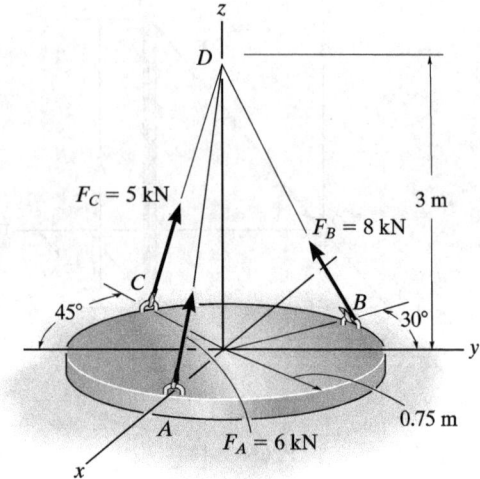

Prob. 2–109

**2–110.** The cable attached to the shear-leg derrick exerts a force on the derrick of $F = 1.75$ kN. Express this force as a Cartesian vector.

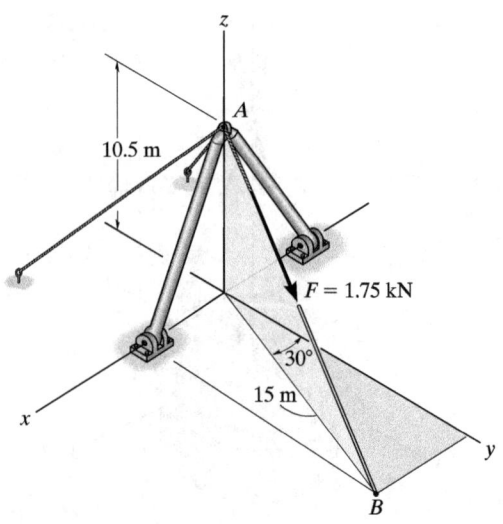

Prob. 2–110

## 2.9 Dot Product

Occasionally in statics one has to find the angle between two lines or the components of a force parallel and perpendicular to a line. In two dimensions, these problems can readily be solved by trigonometry since the geometry is easy to visualize. In three dimensions, however, this is often difficult, and consequently vector methods should be employed for the solution. The dot product, which defines a particular method for "multiplying" two vectors, will be is used to solve the above-mentioned problems.

The *dot product* of vectors **A** and **B**, written **A** · **B**, and read "**A** dot **B**" is defined as the product of the magnitudes of **A** and **B** and the cosine of the angle $\theta$ between their tails, Fig. 2–42. Expressed in equation form,

$$\mathbf{A} \cdot \mathbf{B} = AB \cos \theta \qquad (2\text{--}12)$$

where $0° \leq \theta \leq 180°$. The dot product is often referred to as the *scalar product* of vectors since the result is a *scalar* and not a vector.

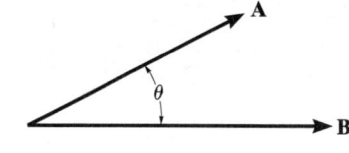

Fig. 2–42

### Laws of Operation.

1. Commutative law: $\mathbf{A} \cdot \mathbf{B} = \mathbf{B} \cdot \mathbf{A}$
2. Multiplication by a scalar: $a(\mathbf{A} \cdot \mathbf{B}) = (a\mathbf{A}) \cdot \mathbf{B} = \mathbf{A} \cdot (a\mathbf{B})$
3. Distributive law: $\mathbf{A} \cdot (\mathbf{B} + \mathbf{D}) = (\mathbf{A} \cdot \mathbf{B}) + (\mathbf{A} \cdot \mathbf{D})$

It is easy to prove the first and second laws by using Eq. 2–12. The proof of the distributive law is left as an exercise (see Prob. 2–111).

### Cartesian Vector Formulation.
Equation 2–12 must be used to find the dot product for any two Cartesian unit vectors. For example, $\mathbf{i} \cdot \mathbf{i} = (1)(1) \cos 0° = 1$ and $\mathbf{i} \cdot \mathbf{j} = (1)(1) \cos 90° = 0$. If we want to find the dot product of two general vectors **A** and **B** that are expressed in Cartesian vector form, then we have

$$\mathbf{A} \cdot \mathbf{B} = (A_x\mathbf{i} + A_y\mathbf{j} + A_z\mathbf{k}) \cdot (B_x\mathbf{i} + B_y\mathbf{j} + B_z\mathbf{k})$$
$$= A_xB_x(\mathbf{i} \cdot \mathbf{i}) + A_xB_y(\mathbf{i} \cdot \mathbf{j}) + A_xB_z(\mathbf{i} \cdot \mathbf{k})$$
$$+ A_yB_x(\mathbf{j} \cdot \mathbf{i}) + (A_yB_y(\mathbf{j} \cdot \mathbf{j}) + A_yB_z(\mathbf{j} \cdot \mathbf{k})$$
$$+ A_zB_x(\mathbf{k} \cdot \mathbf{i}) + A_zB_y(\mathbf{k} \cdot \mathbf{j}) + A_zB_z(\mathbf{k} \cdot \mathbf{k})$$

Carrying out the dot-product operations, the final result becomes

$$\mathbf{A} \cdot \mathbf{B} = A_xB_x + A_yB_y + A_zB_z \qquad (2\text{--}13)$$

Thus, to determine the dot product of two Cartesian vectors, multiply their corresponding x, y, z components and sum these products algebraically. Note that the result will be either a positive or negative *scalar*.

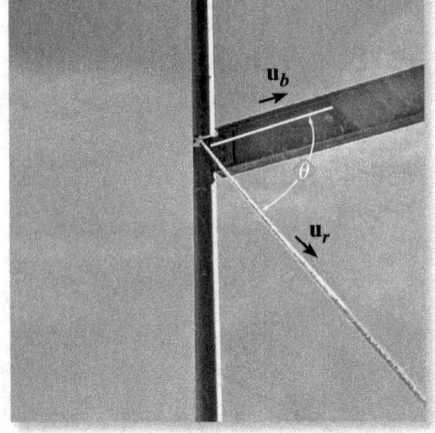

The angle $\theta$ between the rope and the beam can be determined by formulating unit vectors along the beam and rope and then using the dot product $\mathbf{u}_b \cdot \mathbf{u}_r = (1)(1)\cos\theta$.

**Applications.** The dot product has two important applications in mechanics.

- *The angle formed between two vectors or intersecting lines.* The angle $\theta$ between the tails of vectors $\mathbf{A}$ and $\mathbf{B}$ in Fig. 2–42 can be determined from Eq. 2–12 and written as

$$\theta = \cos^{-1}\left(\frac{\mathbf{A} \cdot \mathbf{B}}{AB}\right) \qquad 0° \leq \theta \leq 180°$$

Here $\mathbf{A} \cdot \mathbf{B}$ is found from Eq. 2–13. In particular, notice that if $\mathbf{A} \cdot \mathbf{B} = 0, \theta = \cos^{-1} 0 = 90°$ so that $\mathbf{A}$ will be *perpendicular* to $\mathbf{B}$.

- *The components of a vector parallel and perpendicular to a line.* The component of vector $\mathbf{A}$ parallel to or collinear with the line $aa'$ in Fig. 2–43 is defined by $A_a$ where $A_a = A\cos\theta$. This component is sometimes referred to as the *projection* of $\mathbf{A}$ onto the line, since a *right angle* is formed in the construction. If the *direction* of the line is specified by the unit vector $\mathbf{u}_a$, then since $u_a = 1$, we can determine the magnitude of $\mathbf{A}_a$ directly from the dot product (Eq. 2–12); i.e.,

$$A_a = A\cos\theta = \mathbf{A} \cdot \mathbf{u}_a$$

Hence, the scalar projection of $\mathbf{A}$ along a line is determined from the dot product of $\mathbf{A}$ and the unit vector $\mathbf{u}_a$ which defines the direction of the line. Notice that if this result is positive, then $\mathbf{A}_a$ has a directional sense which is the same as $\mathbf{u}_a$, whereas if $A_a$ is a negative scalar, then $\mathbf{A}_a$ has the opposite sense of direction to $\mathbf{u}_a$.

The component $\mathbf{A}_a$ represented as a *vector* is therefore

$$\mathbf{A}_a = A_a \mathbf{u}_a$$

The component of $\mathbf{A}$ that is perpendicular to line $aa$ can also be obtained, Fig. 2–43. Since $\mathbf{A} = \mathbf{A}_a + \mathbf{A}_\perp$, then $\mathbf{A}_\perp = \mathbf{A} - \mathbf{A}_a$. There are two possible ways of obtaining $A_\perp$. One way would be to determine $\theta$ from the dot product, $\theta = \cos^{-1}(\mathbf{A} \cdot \mathbf{u}_A/A)$, then $A_\perp = A\sin\theta$. Alternatively, if $A_a$ is known, then by Pythagorean's theorem we can also write $A_\perp = \sqrt{A^2 - A_a^2}$.

The projection of the cable force $\mathbf{F}$ along the beam can be determined by first finding the unit vector $\mathbf{u}_b$ that defines this direction. Then apply the dot product, $F_b = \mathbf{F} \cdot \mathbf{u}_b$.

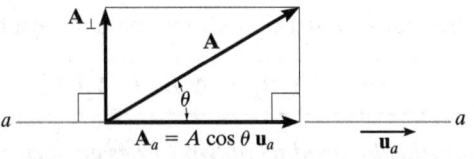

Fig. 2–43

## Important Points

- The dot product is used to determine the angle between two vectors or the projection of a vector in a specified direction.
- If vectors **A** and **B** are expressed in Cartesian vector form, the dot product is determined by multiplying the respective $x, y, z$ scalar components and algebraically adding the results, i.e., $\mathbf{A} \cdot \mathbf{B} = A_x B_x + A_y B_y + A_z B_z$.
- From the definition of the dot product, the angle formed between the tails of vectors **A** and **B** is $\theta = \cos^{-1}(\mathbf{A} \cdot \mathbf{B}/AB)$.
- The magnitude of the projection of vector **A** along a line $aa$ whose direction is specified by $\mathbf{u}_a$ is determined from the dot product $A_a = \mathbf{A} \cdot \mathbf{u}_a$.

### EXAMPLE 2.16

Determine the magnitudes of the projection of the force **F** in Fig. 2–44 onto the $u$ and $v$ axes.

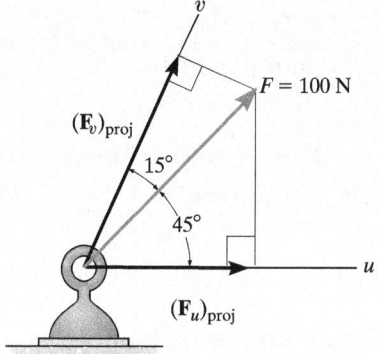

Fig. 2–44

**SOLUTION**

**Projections of Force.** The graphical representation of the *projections* is shown in Fig. 2–44. From this figure, the magnitudes of the projections of **F** onto the $u$ and $v$ axes can be obtained by trigonometry:

$$(F_u)_{\text{proj}} = (100 \text{ N})\cos 45° = 70.7 \text{ N} \qquad \textit{Ans.}$$
$$(F_v)_{\text{proj}} = (100 \text{ N})\cos 15° = 96.6 \text{ N} \qquad \textit{Ans.}$$

**NOTE:** These projections are not equal to the magnitudes of the components of force **F** along the $u$ and $v$ axes found from the parallelogram law. They will only be equal if the $u$ and $v$ axes are *perpendicular* to one another.

## EXAMPLE 2.17

The frame shown in Fig. 2–45a is subjected to a horizontal force $\mathbf{F} = \{300\mathbf{j}\}$. Determine the magnitude of the components of this force parallel and perpendicular to member $AB$.

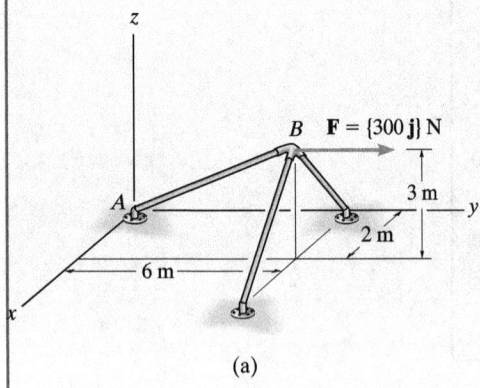

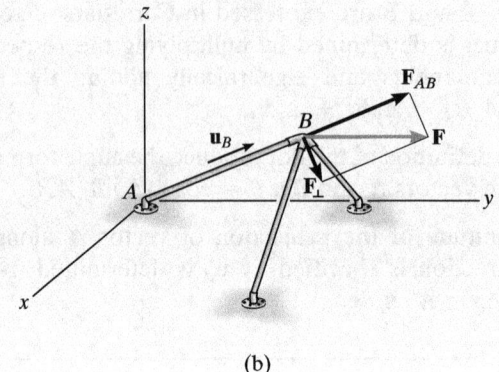

(a)

(b)

Fig 2–45

### SOLUTION

The magnitude of the component of $\mathbf{F}$ along $AB$ is equal to the dot product of $\mathbf{F}$ and the unit vector $\mathbf{u}_B$, which defines the direction of $AB$, Fig. 2–45b. Since

$$\mathbf{u}_B = \frac{\mathbf{r}_B}{r_B} = \frac{2\mathbf{i} + 6\mathbf{j} + 3\mathbf{k}}{\sqrt{(2)^2 + (6)^2 + (3)^2}} = 0.286\mathbf{i} + 0.857\mathbf{j} + 0.429\mathbf{k}$$

then

$$F_{AB} = F \cos\theta = \mathbf{F} \cdot \mathbf{u}_B = (300\mathbf{j}) \cdot (0.286\mathbf{i} + 0.857\mathbf{j} + 0.429\mathbf{k})$$
$$= (0)(0.286) + (300)(0.857) + (0)(0.429)$$
$$= 257.1 \text{ N} \qquad \textit{Ans.}$$

Since the result is a positive scalar, $\mathbf{F}_{AB}$ has the same sense of direction as $\mathbf{u}_B$, Fig. 2–45b.

Expressing $\mathbf{F}_{AB}$ in Cartesian vector form, we have

$$\mathbf{F}_{AB} = F_{AB}\mathbf{u}_B = (257.1 \text{ N})(0.286\mathbf{i} + 0.857\mathbf{j} + 0.429\mathbf{k})$$
$$= \{73.5\mathbf{i} + 220\mathbf{j} + 110\mathbf{k}\}\text{N} \qquad \textit{Ans.}$$

The perpendicular component, Fig. 2–45b, is therefore

$$\mathbf{F}_\perp = \mathbf{F} - \mathbf{F}_{AB} = 300\mathbf{j} - (73.5\mathbf{i} + 220\mathbf{j} + 110\mathbf{k})$$
$$= \{-73.5\mathbf{i} + 80\mathbf{j} - 110\mathbf{k}\} \text{ N}$$

Its magnitude can be determined either from this vector or by using the Pythagorean theorem, Fig. 2–45b:

$$F_\perp = \sqrt{F^2 - F_{AB}^2} = \sqrt{(300 \text{ N})^2 - (257.1 \text{ N})^2}$$
$$= 155 \text{ N} \qquad \textit{Ans.}$$

## EXAMPLE 2.18

The pipe in Fig. 2–46a is subjected to the force of $F = 800$ N. Determine the angle $\theta$ between **F** and the pipe segment $BA$ and the projection of **F** along this segment.

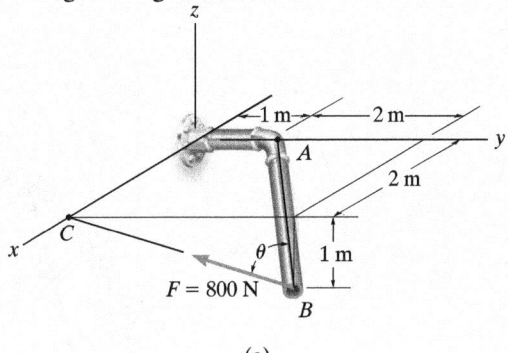

(a)

### SOLUTION

**Angle $\theta$.** First, we will establish position vectors from $B$ to $A$ and $B$ to $C$; Fig. 2–46b. Then, we will determine the angle $\theta$ between the tails of these two vectors.

$$\mathbf{r}_{BA} = \{-2\mathbf{i} - 2\mathbf{j} + 1\mathbf{k}\} \text{ m}, \; r_{BA} = 3 \text{ m}$$
$$\mathbf{r}_{BC} = \{-3\mathbf{j} + 1\mathbf{k}\} \text{ ft}, \; r_{BC} = \sqrt{10} \text{ m}$$

Thus,

$$\cos\theta = \frac{\mathbf{r}_{BA} \cdot \mathbf{r}_{BC}}{r_{BA} r_{BC}} = \frac{(-2)(0) + (-2)(-3) + (1)(1)}{3\sqrt{10}} = 0.7379$$

$$\theta = 42.5° \qquad \text{Ans.}$$

**Components of F.** The component of **F** along $BA$ is shown in Fig. 2–46b. We must first formulate the unit vector along $BA$ and force **F** as Cartesian vectors.

$$\mathbf{u}_{BA} = \frac{\mathbf{r}_{BA}}{r_{BA}} = \frac{(-2\mathbf{i} - 2\mathbf{j} + 1\mathbf{k})}{3} = -\frac{2}{3}\mathbf{i} - \frac{2}{3}\mathbf{j} + \frac{1}{3}\mathbf{k}$$

$$\mathbf{F} = 800 \text{ N}\left(\frac{\mathbf{r}_{BC}}{r_{BC}}\right) = 800\left(\frac{-3\mathbf{j} + 1\mathbf{k}}{\sqrt{10}}\right) = \left(-758.9\mathbf{j} + 253.0\mathbf{k}\right) \text{ N}$$

Thus,

$$F_{BA} = \mathbf{F} \cdot \mathbf{u}_{BA} = (-758.9\mathbf{j} + 253.0\mathbf{k}) \cdot \left(-\frac{2}{3}\mathbf{i} - \frac{2}{3}\mathbf{j} + \frac{1}{3}\mathbf{k}\right)$$

$$= 0\left(-\frac{2}{3}\right) + (-758.9)\left(-\frac{2}{3}\right) + (253.0)\left(\frac{1}{3}\right)$$

$$= 590 \text{ N} \qquad \text{Ans.}$$

**NOTE:** Since $\theta$ is known, then also, $F_{BA} = F \cos\theta = 800 \text{ N} \cos 42.5° = 590 \text{ N}$.

Fig. 2–46

# FUNDAMENTAL PROBLEMS

**F2–25.** Determine the angle $\theta$ between the force and the line $AO$.

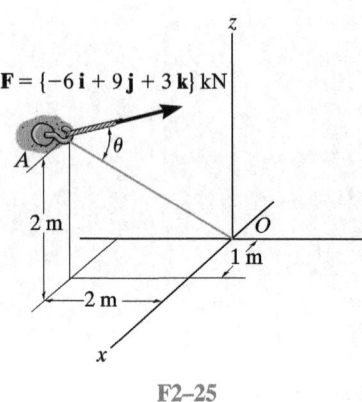

F2–25

**F2–26.** Determine the angle $\theta$ between the force and the line $AB$.

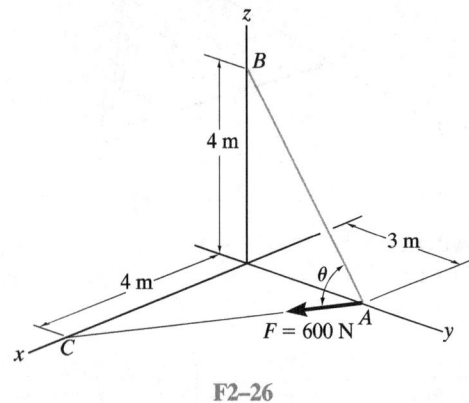

F2–26

**F2–27.** Determine the angle $\theta$ between the force and the line $OA$.

**F2–28.** Determine the component of projection of the force along the line $OA$.

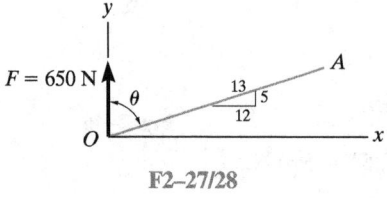

F2–27/28

**F2–29.** Find the magnitude of the projected component of the force along the pipe.

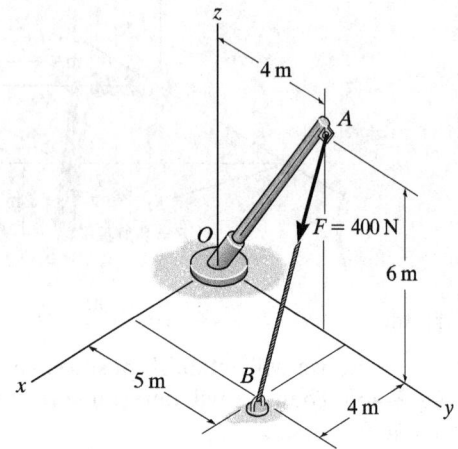

F2–29

**F2–30.** Determine the components of the force acting parallel and perpendicular to the axis of the pole.

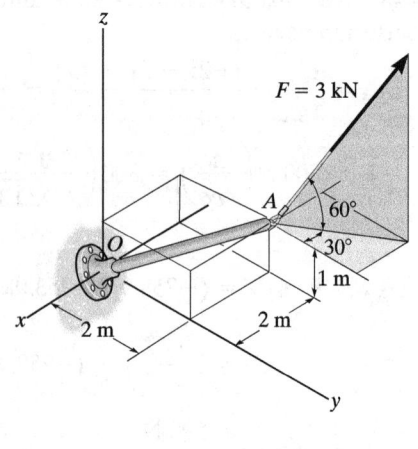

F2–30

## PROBLEMS

**2–111.** Given the three vectors A, B, and D, show that $\mathbf{A} \cdot (\mathbf{B} + \mathbf{D}) = (\mathbf{A} \cdot \mathbf{B}) + (\mathbf{A} \cdot \mathbf{D})$.

**\*2–112.** Determine the projected component of the force $F_{AB} = 560$ N acting along cable AC. Express the result as a Cartesian vector.

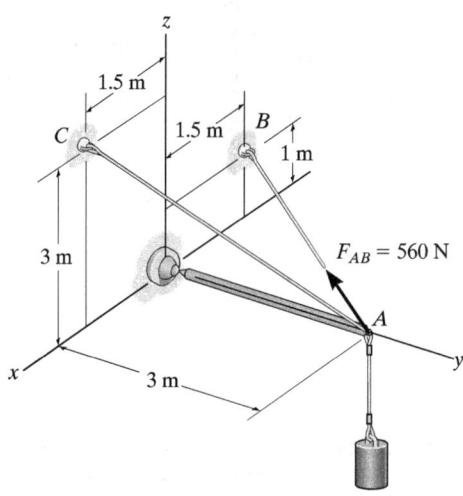

Prob. 2–112

**•2–113.** Determine the magnitudes of the components of force $F = 56$ N acting along and perpendicular to line AO.

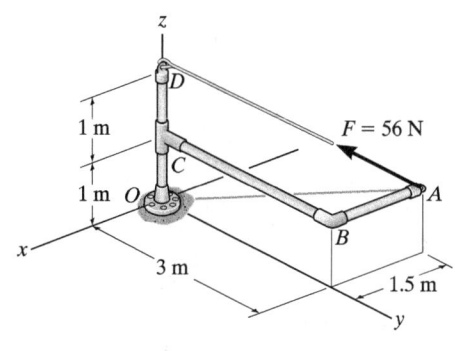

Prob. 2–113

**2–114.** Determine the length of side BC of the triangular plate. Solve the problem by finding the magnitude of $\mathbf{r}_{BC}$; then check the result by first finding $\theta$, $r_{AB}$, and $r_{AC}$ and then using the cosine law.

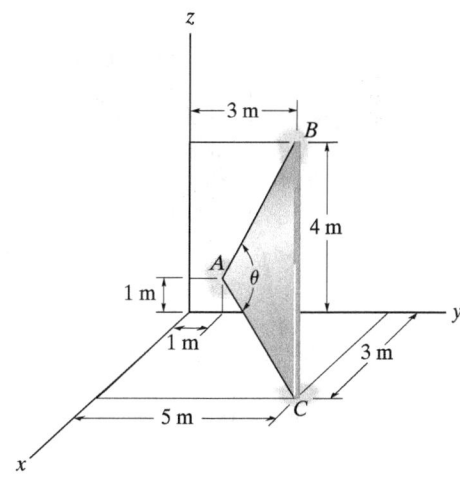

Prob. 2–114

**2–115.** Determine the magnitudes of the components of $F = 600$ N acting along and perpendicular to segment DE of the pipe assembly.

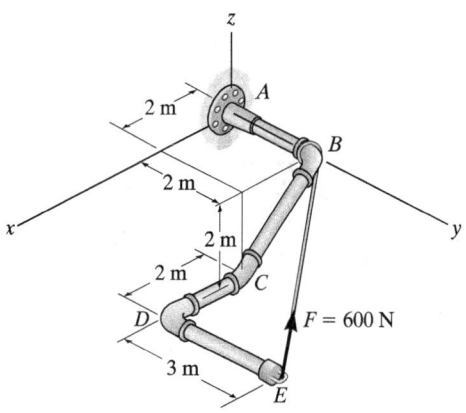

Prob. 2–115

**\*2–116.** Two forces act on the hook. Determine the angle $\theta$ between them. Also, what are the projections of $\mathbf{F}_1$ and $\mathbf{F}_2$ along the y axis?

**•2–117.** Two forces act on the hook. Determine the magnitude of the projection of $\mathbf{F}_2$ along $\mathbf{F}_1$.

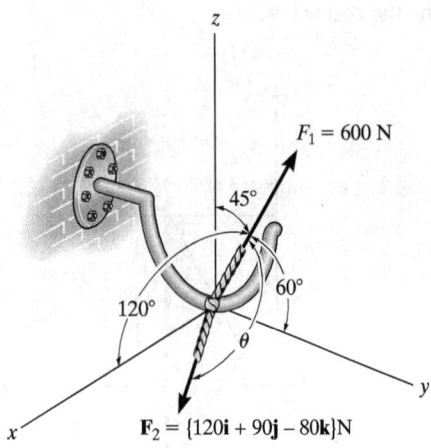

Probs. 2–116/117

**2–118.** Determine the projection of force $F = 80$ N along line $BC$. Express the result as a Cartesian vector.

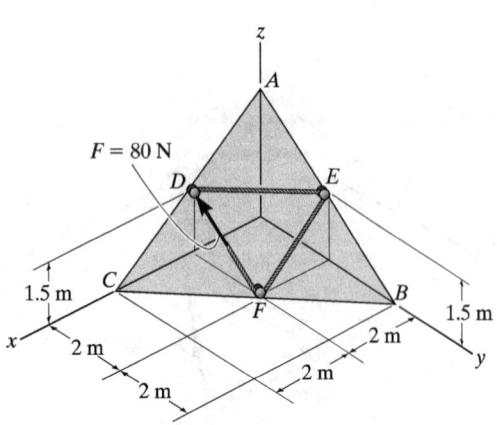

Prob. 2–118

**2–119.** The clamp is used on a jig. If the vertical force acting on the bolt is $\mathbf{F} = \{-500\mathbf{k}\}$ N, determine the magnitudes of its components $\mathbf{F}_1$ and $\mathbf{F}_2$ which act along the $OA$ axis and perpendicular to it.

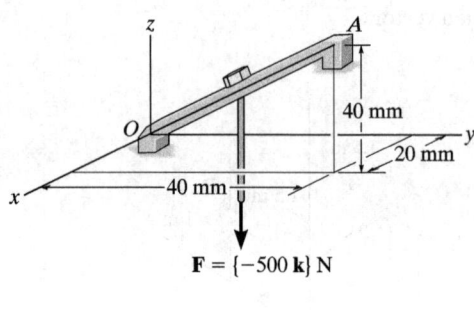

Prob. 2–119

**\*2–120.** Determine the magnitude of the projected component of force $\mathbf{F}_{AB}$ acting along the z axis.

**•2–121.** Determine the magnitude of the projected component of force $\mathbf{F}_{AC}$ acting along the z axis.

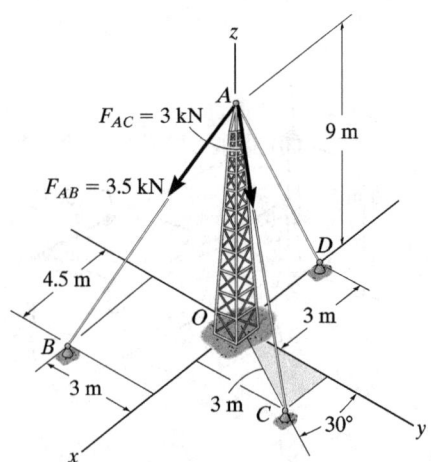

Probs. 2–120/121

**2–122.** Determine the projection of force $F = 400$ N acting along line $AC$ of the pipe assembly. Express the result as a Cartesian vector.

**2–123.** Determine the magnitudes of the components of force $F = 400$ N acting parallel and perpendicular to segment $BC$ of the pipe assembly.

**2–126.** The cables each exert a force of 400 N on the post. Determine the magnitude of the projected component of $\mathbf{F}_1$ along the line of action of $\mathbf{F}_2$.

**2–127.** Determine the angle $\theta$ between the two cables attached to the post.

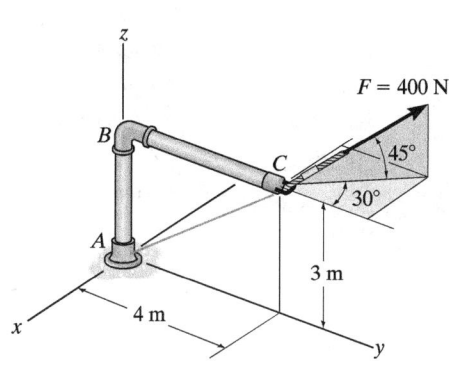

Probs. 2–122/123

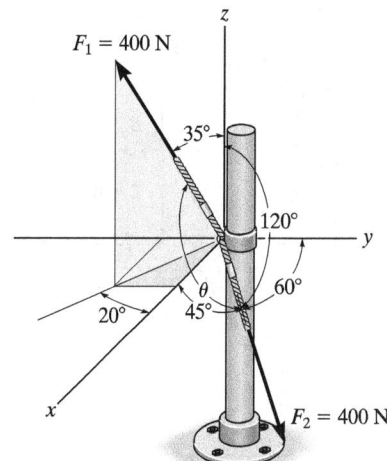

Probs. 2–126/127

**\*2–124.** Cable $OA$ is used to support column $OB$. Determine the angle $\theta$ it makes with beam $OC$.

**•2–125.** Cable $OA$ is used to support column $OB$. Determine the angle $\phi$ it makes with beam $OD$.

**\*2–128.** A force of $F = 80$ N is applied to the handle of the wrench. Determine the angle $\theta$ between the tail of the force and the handle $AB$.

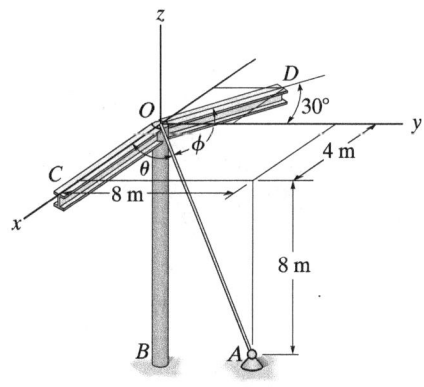

Probs. 2–124/125

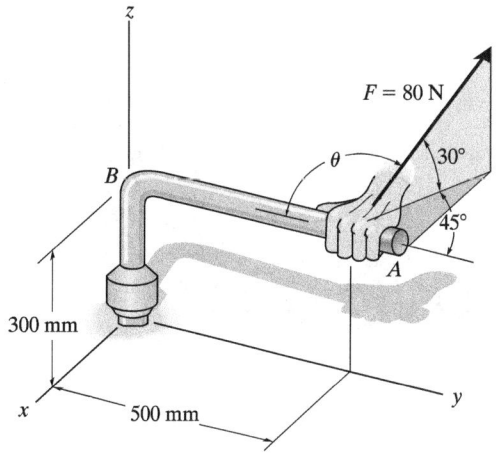

Prob. 2–128

•2–129. Determine the angle θ between cables AB and AC.

2–130. If **F** has a magnitude of 250 N, determine the magnitude of its projected components acting along the x axis and along cable AC.

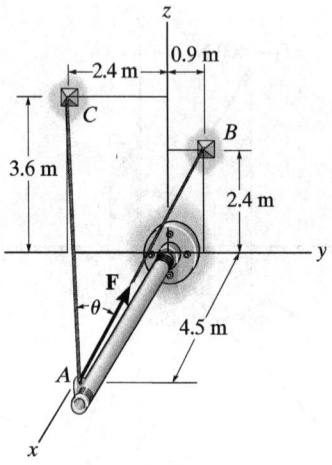

Probs. 2–129/130

*2–132. Determine the magnitude of the projected component of the force F = 300 N acting along line OA.

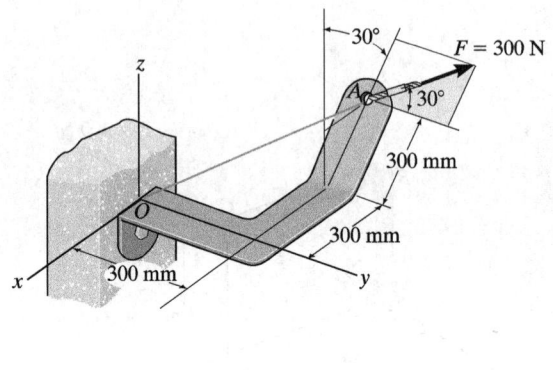

Prob. 2–132

2–131. Determine the magnitudes of the projected components of the force F = 300 N acting along the x and y axes.

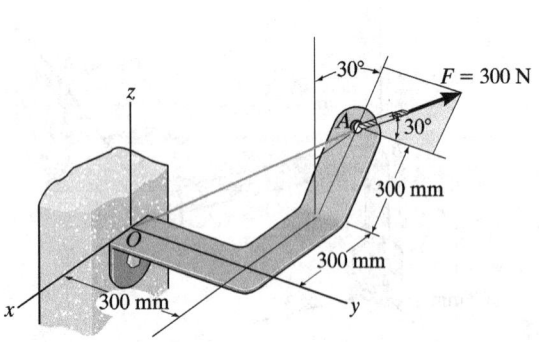

Prob. 2–131

•2–133. Two cables exert forces on the pipe. Determine the magnitude of the projected component of **F**$_1$ along the line of action of **F**$_2$.

2–134. Determine the angle θ between the two cables attached to the pipe.

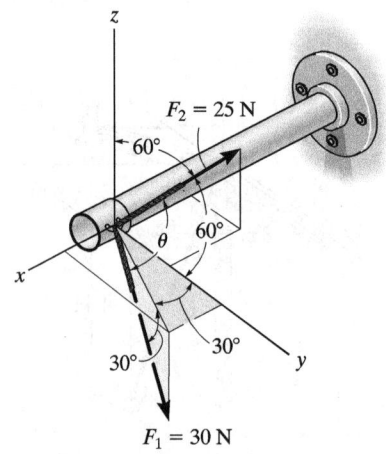

Probs. 2–133/134

# CHAPTER REVIEW

A scalar is a positive or negative number; e.g., mass and temperature.

A vector has a magnitude and direction, where the arrowhead represents the sense of the vector.

---

Multiplication or division of a vector by a scalar will change only the magnitude of the vector. If the scalar is negative, the sense of the vector will change so that it acts in the opposite sense.

---

If vectors are collinear, the resultant is simply the algebraic or scalar addition.

$$R = A + B$$

---

**Parallelogram Law**

Two forces add according to the parallelogram law. The *components* form the sides of the parallelogram and the *resultant* is the diagonal.

To find the components of a force along any two axes, extend lines from the head of the force, parallel to the axes, to form the components.

To obtain the components or the resultant, show how the forces add by tip-to-tail using the triangle rule, and then use the law of cosines and the law of sines to calculate their values.

$$F_R = \sqrt{F_1^2 + F_2^2 - 2 F_1 F_2 \cos \theta_R}$$

$$\frac{F_1}{\sin \theta_1} = \frac{F_2}{\sin \theta_2} = \frac{F_R}{\sin \theta_R}$$

## Rectangular Components: Two Dimensions

Vectors $\mathbf{F}_x$ and $\mathbf{F}_y$ are rectangular components of $\mathbf{F}$.

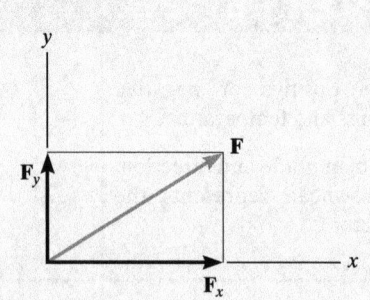

The resultant force is determined from the algebraic sum of its components.

$$F_{Rx} = \Sigma F_x$$

$$F_{Ry} = \Sigma F_y$$

$$F_R = \sqrt{(F_{Rx})^2 + (F_{Ry})^2}$$

$$\theta = \tan^{-1}\left|\frac{F_{Ry}}{F_{Rx}}\right|$$

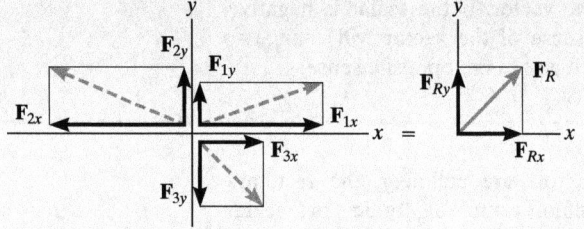

## Cartesian Vectors

The unit vector $\mathbf{u}$ has a length of one, no units, and it points in the direction of the vector $\mathbf{F}$.

$$\mathbf{u} = \frac{\mathbf{F}}{F}$$

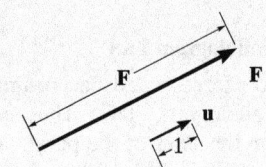

A force can be resolved into its Cartesian components along the $x$, $y$, $z$ axes so that $\mathbf{F} = F_x\mathbf{i} + F_y\mathbf{j} + F_z\mathbf{k}$.

The magnitude of $\mathbf{F}$ is determined from the positive square root of the sum of the squares of its components.

$$F = \sqrt{F_x^2 + F_y^2 + F_z^2}$$

The coordinate direction angles $\alpha, \beta, \gamma$ are determined by formulating a unit vector in the direction of $\mathbf{F}$. The $x$, $y$, $z$ components of $\mathbf{u}$ represent $\cos\alpha, \cos\beta, \cos\gamma$.

$$\mathbf{u} = \frac{\mathbf{F}}{F} = \frac{F_x}{F}\mathbf{i} + \frac{F_y}{F}\mathbf{j} + \frac{F_z}{F}\mathbf{k}$$

$$\mathbf{u} = \cos\alpha\,\mathbf{i} + \cos\beta\,\mathbf{j} + \cos\gamma\,\mathbf{k}$$

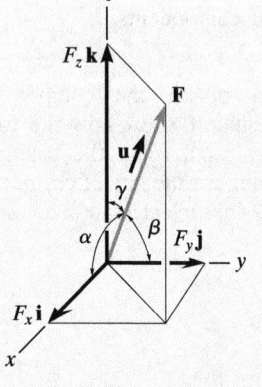

| | | |
|---|---|---|
| The coordinate direction angles are related so that only two of the three angles are independent of one another. | $\cos^2 \alpha + \cos^2 \beta + \cos^2 \gamma = 1$ | |
| To find the resultant of a concurrent force system, express each force as a Cartesian vector and add the **i**, **j**, **k** components of all the forces in the system. | $\mathbf{F}_R = \Sigma \mathbf{F} = \Sigma \mathbf{F}_x \mathbf{i} + \Sigma F_y \mathbf{j} + \Sigma F_z \mathbf{k}$ | |
| **Position and Force Vectors** A position vector locates one point in space relative to another. The easiest way to formulate the components of a position vector is to determine the distance and direction that one must travel along the $x$, $y$, and $z$ directions—going from the tail to the head of the vector. | $\mathbf{r} = (x_B - x_A)\mathbf{i}$ $+ (y_B - y_A)\mathbf{j}$ $+ (z_B - z_A)\mathbf{k}$ | 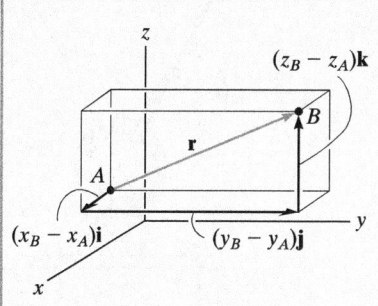 |
| If the line of action of a force passes through points A and B, then the force acts in the same direction as the position vector **r**, which is defined by the unit vector $u$. The force can then be expressed as a Cartesian vector. | $\mathbf{F} = F\mathbf{u} = F\left(\dfrac{\mathbf{r}}{r}\right)$ | 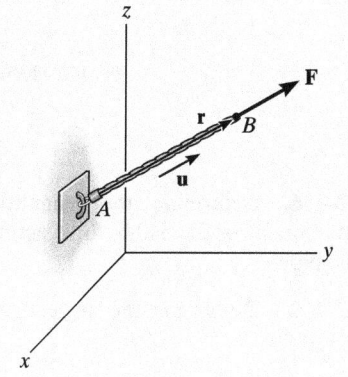 |
| **Dot Product** The dot product between two vectors **A** and **B** yields a scalar. If **A** and **B** are expressed in Cartesian vector form, then the dot product is the sum of the products of their $x, y,$ and $z$ components The dot product can be used to determine the angle between **A** and **B**. The dot product is also used to determine the projected component of a vector **A** onto an axis $aa$ defined by its unit vector $\mathbf{u}_a$. | $\mathbf{A} \cdot \mathbf{B} = AB \cos \theta$ $= A_x B_x + A_y B_y + A_z B_z$ $\theta = \cos^{-1}\left(\dfrac{\mathbf{A} \cdot \mathbf{B}}{AB}\right)$ $\mathbf{A}_a = A \cos \theta\, \mathbf{u}_a = (\mathbf{A} \cdot \mathbf{u}_a)\mathbf{u}_a$ | 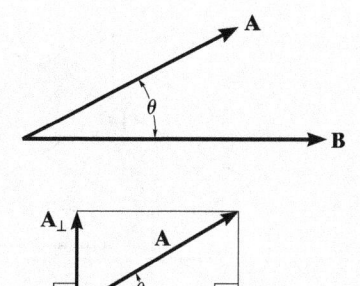 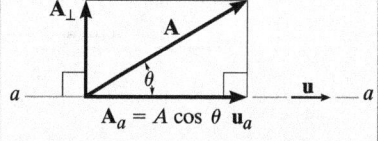 |

# REVIEW PROBLEMS

**2–135.** Determine the $x$ and $y$ components of the 700-N force.

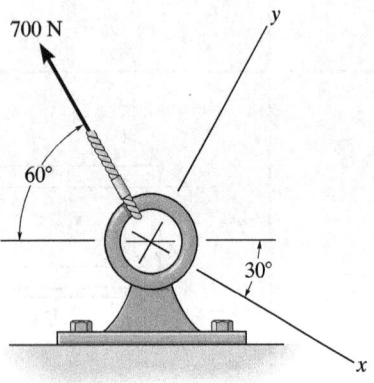

Prob. 2–135

**\*2–136.** Determine the magnitude of the projected component of the 500-N force acting along the axis $BC$ of the pipe.

**•2–137.** Determine the angle $\theta$ between pipe segments $BA$ and $BC$.

**2–138.** Determine the magnitude and direction of the resultant $\mathbf{F}_R = \mathbf{F}_1 + \mathbf{F}_2 + \mathbf{F}_3$ of the three forces by first finding the resultant $\mathbf{F}' = \mathbf{F}_1 + \mathbf{F}_3$ and then forming $\mathbf{F}_R = \mathbf{F}' + \mathbf{F}_2$. Specify its direction measured counterclockwise from the positive $x$ axis.

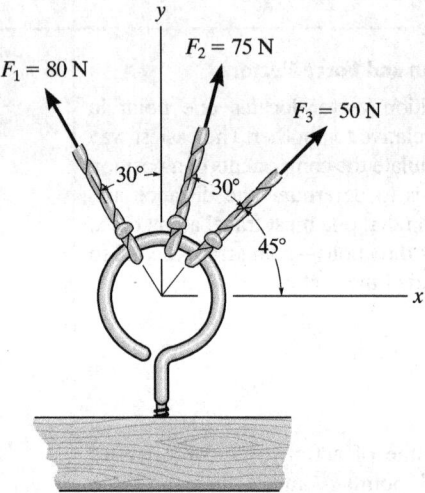

Prob. 2–138

**2–139.** Determine the design angle $\theta$ ($\theta < 90°$) between the two struts so that the 500-N horizontal force has a component of 600 N directed from $A$ toward $C$. What is the component of force acting along member $BA$?

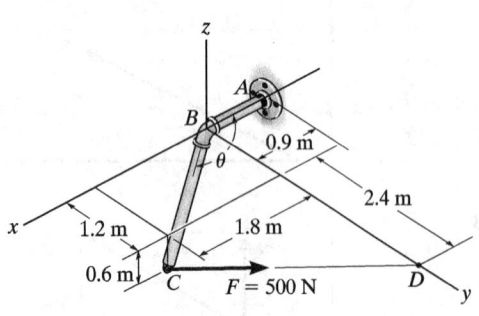

Probs. 2–136/137

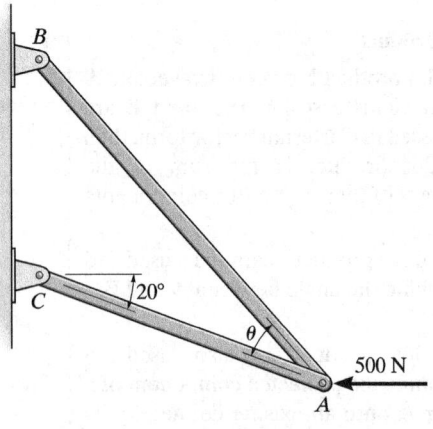

Prob. 2–139

**\*2–140.** Determine the magnitude and direction of the *smallest* force $\mathbf{F}_3$ so that the resultant force of all three forces has a magnitude of 100 N.

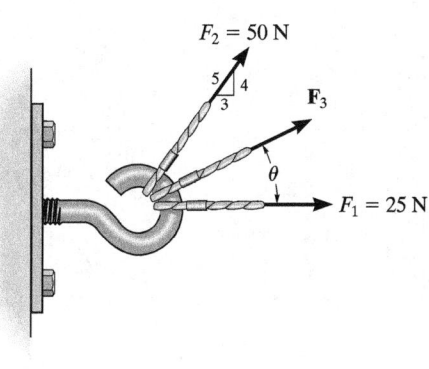

Prob. 2–140

**•2–141.** Resolve the 250-N force into components acting along the $u$ and $v$ axes and determine the magnitudes of these components.

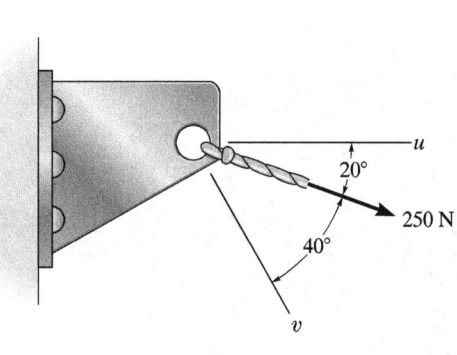

Prob. 2–141

**2–142.** Cable $AB$ exerts a force of 80 N on the end of the 3-m-long boom $OA$. Determine the magnitude of the projection of this force along the boom.

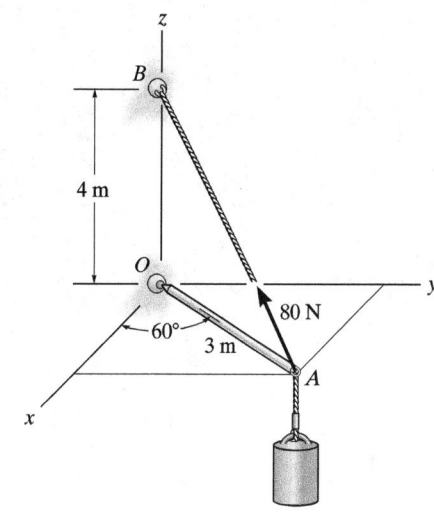

Prob. 2–142

**2–143.** The three supporting cables exert the forces shown on the sign. Represent each force as a Cartesian vector.

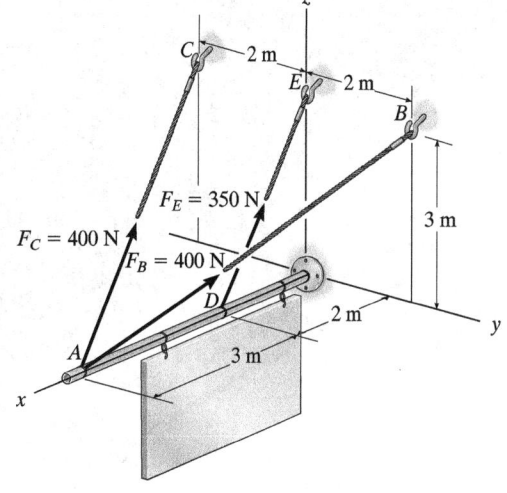

Prob. 2–143

Whenever cables are used for hoisting loads, they must be selected so that they do not fail when they are placed at their points of attachment. In this chapter, we will show how to calculate cable loadings for such cases.

# Equilibrium of a Particle

# 3

## CHAPTER OBJECTIVES

- To introduce the concept of the free-body diagram for a particle.
- To show how to solve particle equilibrium problems using the equations of equilibrium.

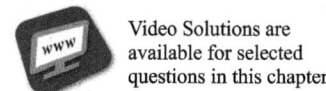

Video Solutions are available for selected questions in this chapter.

## 3.1 Condition for the Equilibrium of a Particle

A particle is said to be in *equilibrium* if it remains at rest if originally at rest, or has a constant velocity if originally in motion. Most often, however, the term "equilibrium" or, more specifically, "static equilibrium" is used to describe an object at rest. To maintain equilibrium, it is *necessary* to satisfy Newton's first law of motion, which requires the *resultant force* acting on a particle to be equal to *zero*. This condition may be stated mathematically as

$$\Sigma \mathbf{F} = \mathbf{0} \tag{3-1}$$

where $\Sigma \mathbf{F}$ is the vector *sum of all the forces* acting on the particle.

Not only is Eq. 3–1 a necessary condition for equilibrium, it is also a *sufficient* condition. This follows from Newton's second law of motion, which can be written as $\Sigma \mathbf{F} = m\mathbf{a}$. Since the force system satisfies Eq. 3–1, then $m\mathbf{a} = \mathbf{0}$, and therefore the particle's acceleration $\mathbf{a} = \mathbf{0}$. Consequently, the particle indeed moves with constant velocity or remains at rest.

## 3.2 The Free-Body Diagram

To apply the equation of equilibrium, we must account for *all* the known and unknown forces ($\Sigma \mathbf{F}$) which act *on* the particle. The best way to do this is to think of the particle as isolated and "free" from its surroundings. A drawing that shows the particle with *all* the forces that act on it is called a *free-body diagram (FBD)*.

Before presenting a formal procedure as to how to draw a free-body diagram, we will first consider two types of connections often encountered in particle equilibrium problems.

**Springs.** If a *linearly elastic spring* (or cord) of undeformed length $l_o$ is used to support a particle, the length of the spring will change in direct proportion to the force $\mathbf{F}$ acting on it, Fig. 3–1. A characteristic that defines the "elasticity" of a spring is the *spring constant* or *stiffness k*.

The magnitude of force exerted on a linearly elastic spring which has a stiffness $k$ and is deformed (elongated or compressed) a distance $s = l - l_o$, measured from its *unloaded* position, is

$$F = ks \tag{3-2}$$

If $s$ is positive, causing an elongation, then $\mathbf{F}$ must pull on the spring; whereas if $s$ is negative, causing a shortening, then $\mathbf{F}$ must push on it. For example, if the spring in Fig. 3–1 has an unstretched length of 0.8 m and a stiffness $k = 500$ N/m and it is stretched to a length of 1 m, so that $s = l - l_o = 1$ m $- 0.8$ m $= 0.2$ m, then a force $F = ks = 500$ N/m$(0.2$ m$) = 100$ N is needed.

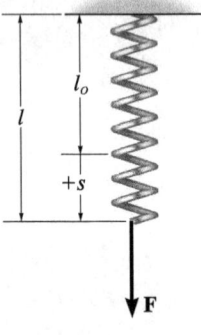

Fig. 3–1

**Cables and Pulleys.** Unless otherwise stated, throughout this book, except in Sec. 7.4, all cables (or cords) will be assumed to have negligible weight and they cannot stretch. Also, a cable can support *only* a tension or "pulling" force, and this force always acts in the direction of the cable. In Chapter 5, it will be shown that the tension force developed in a *continuous cable* which passes over a frictionless pulley must have a *constant* magnitude to keep the cable in equilibrium. Hence, for any angle $\theta$, shown in Fig. 3–2, the cable is subjected to a constant tension $T$ throughout its length.

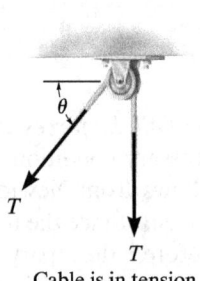

Cable is in tension

Fig. 3–2

## 3.2 THE FREE-BODY DIAGRAM

### Procedure for Drawing a Free-Body Diagram

Since we must account for *all the forces acting on the particle* when applying the equations of equilibrium, the importance of first drawing a free-body diagram cannot be overemphasized. To construct a free-body diagram, the following three steps are necessary.

#### Draw Outlined Shape.

Imagine the particle to be *isolated* or cut "free" from its surroundings by drawing its outlined shape.

#### Show All Forces.

Indicate on this sketch *all* the forces that act *on the particle*. These forces can be *active forces*, which tend to set the particle in motion, or they can be *reactive forces* which are the result of the constraints or supports that tend to prevent motion. To account for all these forces, it may be helpful to trace around the particle's boundary, carefully noting each force acting on it.

#### Identify Each Force.

The forces that are *known* should be labeled with their proper magnitudes and directions. Letters are used to represent the magnitudes and directions of forces that are unknown.

The bucket is held in equilibrium by the cable, and instinctively we know that the force in the cable must equal the weight of the bucket. By drawing a free-body diagram of the bucket we can understand why this is so. This diagram shows that there are only two forces *acting on the bucket*, namely, its weight **W** and the force **T** of the cable. For equilibrium, the resultant of these forces must be equal to zero, and so $T = W$.

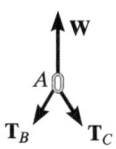

The spool has a weight $W$ and is suspended from the crane boom. If we wish to obtain the forces in cables $AB$ and $AC$, then we should consider the free-body diagram of the ring at $A$. Here the cables $AD$ exert a resultant force of **W** on the ring and the condition of equilibrium is used to obtain $\mathbf{T}_B$ and $\mathbf{T}_C$.

# EXAMPLE 3.1

The sphere in Fig. 3–3a has a mass of 6 kg and is supported as shown. Draw a free-body diagram of the sphere, the cord CE, and the knot at C.

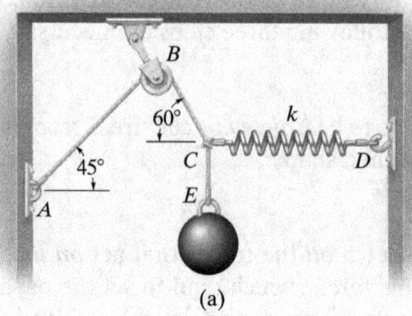

(a)

## SOLUTION

**Sphere.** By inspection, there are only two forces acting on the sphere, namely, its weight, 6 kg (9.81 m/s$^2$) = 58.9 N, and the force of cord CE. The free-body diagram is shown in Fig. 3–3b.

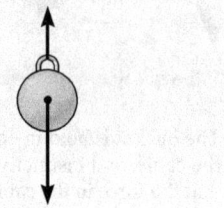

$F_{CE}$ (Force of cord CE acting on sphere)

58.9 N (Weight or gravity acting on sphere)
(b)

**Cord CE.** When the cord CE is isolated from its surroundings, its free-body diagram shows only two forces acting on it, namely, the force of the sphere and the force of the knot, Fig. 3–3c. Notice that $F_{CE}$ shown here is equal but opposite to that shown in Fig. 3–3b, a consequence of Newton's third law of action–reaction. Also, $F_{CE}$ and $F_{EC}$ pull on the cord and keep it in tension so that it doesn't collapse. For equilibrium, $F_{CE} = F_{EC}$.

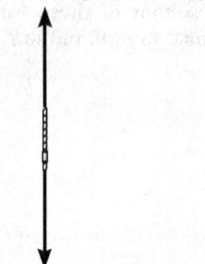

$F_{EC}$ (Force of knot acting on cord CE)

$F_{CE}$ (Force of sphere acting on cord CE)
(c)

**Knot.** The knot at C is subjected to three forces, Fig. 3–3d. They are caused by the cords CBA and CE and the spring CD. As required, the free-body diagram shows all these forces labeled with their magnitudes and directions. It is important to recognize that the weight of the sphere does not directly act on the knot. Instead, the cord CE subjects the knot to this force.

$F_{CBA}$ (Force of cord CBA acting on knot)

$F_{CD}$ (Force of spring acting on knot)

$F_{CE}$ (Force of cord CE acting on knot)
(d)

Fig. 3–3

## 3.3 Coplanar Force Systems

If a particle is subjected to a system of coplanar forces that lie in the $x$–$y$ plane as in Fig. 3–4, then each force can be resolved into its $\mathbf{i}$ and $\mathbf{j}$ components. For equilibrium, these forces must sum to produce a zero force resultant, i.e.,

$$\Sigma \mathbf{F} = \mathbf{0}$$
$$\Sigma F_x \mathbf{i} + \Sigma F_y \mathbf{j} = \mathbf{0}$$

For this vector equation to be satisfied, the force's $x$ and $y$ components must both be equal to zero. Hence,

$$\boxed{\begin{aligned} \Sigma F_x &= 0 \\ \Sigma F_y &= 0 \end{aligned}} \quad (3\text{–}3)$$

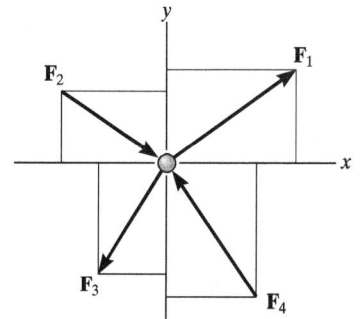

Fig. 3–4

These two equations can be solved for at most two unknowns, generally represented as angles and magnitudes of forces shown on the particle's free-body diagram.

When applying each of the two equations of equilibrium, we must account for the sense of direction of any component by using an *algebraic sign* which corresponds to the arrowhead direction of the component along the $x$ or $y$ axis. It is important to note that if a force has an *unknown magnitude*, then the arrowhead sense of the force on the free-body diagram can be *assumed*. Then if the *solution* yields a *negative scalar*, this indicates that the sense of the force is opposite to that which was assumed.

For example, consider the free-body diagram of the particle subjected to the two forces shown in Fig. 3–5. Here it is *assumed* that the *unknown force* $\mathbf{F}$ acts to the right to maintain equilibrium. Applying the equation of equilibrium along the $x$ axis, we have

$$\xrightarrow{+} \Sigma F_x = 0; \qquad +F + 10\text{ N} = 0$$

Both terms are "positive" since both forces act in the positive $x$ direction. When this equation is solved, $F = -10$ N. Here the *negative sign* indicates that $\mathbf{F}$ must act to the left to hold the particle in equilibrium, Fig. 3–5. Notice that if the $+x$ axis in Fig. 3–5 were directed to the left, both terms in the above equation would be negative, but again, after solving, $F = -10$ N, indicating that $\mathbf{F}$ would be directed to the left.

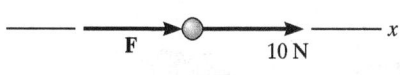

Fig. 3–5

## Procedure for Analysis

Coplanar force equilibrium problems for a particle can be solved using the following procedure.

### Free-Body Diagram.

- Establish the $x$, $y$ axes in any suitable orientation.
- Label all the known and unknown force magnitudes and directions on the diagram.
- The sense of a force having an unknown magnitude can be assumed.

### Equations of Equilibrium.

- Apply the equations of equilibrium, $\Sigma F_x = 0$ and $\Sigma F_y = 0$.
- Components are positive if they are directed along a positive axis, and negative if they are directed along a negative axis.
- If more than two unknowns exist and the problem involves a spring, apply $F = ks$ to relate the spring force to the deformation $s$ of the spring.
- Since the magnitude of a force is always a positive quantity, then if the solution for a force yields a negative result, this indicates its sense is the reverse of that shown on the free-body diagram.

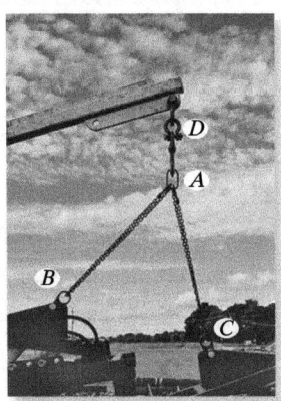
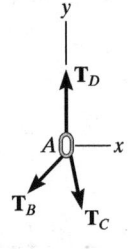

The chains exert three forces on the ring at $A$, as shown on its free-body diagram. The ring will not move, or will move with constant velocity, provided the summation of these forces along the $x$ and along the $y$ axis is zero. If one of the three forces is known, the magnitudes of the other two forces can be obtained from the two equations of equilibrium.

## EXAMPLE 3.2

Determine the tension in cables $BA$ and $BC$ necessary to support the 60-kg cylinder in Fig. 3-6a.

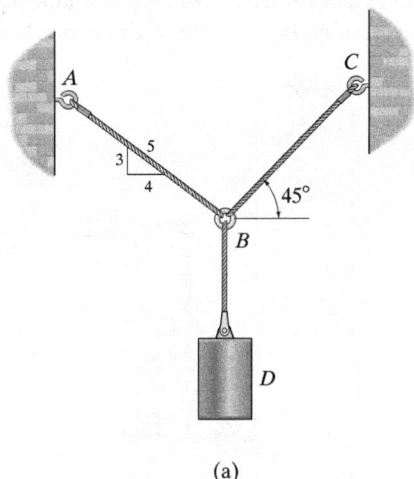

(a)

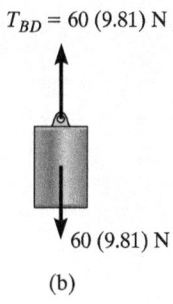

(b)

### SOLUTION
**Free-Body Diagram.** Due to equilibrium, the weight of the cylinder causes the tension in cable $BD$ to be $T_{BD} = 60(9.81)$ N, Fig. 3-6b. The forces in cables $BA$ and $BC$ can be determined by investigating the equilibrium of ring $B$. Its free-body diagram is shown in Fig. 3-6c. The magnitudes of $\mathbf{T}_A$ and $\mathbf{T}_C$ are unknown, but their directions are known.

**Equations of Equilibrium.** Applying the equations of equilibrium along the $x$ and $y$ axes, we have

$\xrightarrow{+} \Sigma F_x = 0;$  $\quad T_C \cos 45° - \left(\frac{4}{5}\right)T_A = 0$  (1)

$+\uparrow \Sigma F_y = 0;$ $\quad T_C \sin 45° + \left(\frac{3}{5}\right)T_A - 60(9.81) \text{ N} = 0$  (2)

Equation (1) can be written as $T_A = 0.8839 T_C$. Substituting this into Eq. (2) yields

$$T_C \sin 45° + \left(\frac{3}{5}\right)(0.8839 T_C) - 60(9.81) \text{ N} = 0$$

So that

$$T_C = 475.66 \text{ N} = 476 \text{ N} \qquad Ans.$$

Substituting this result into either Eq. (1) or Eq. (2), we get

$$T_A = 420 \text{ N} \qquad Ans.$$

**NOTE:** The accuracy of these results, of course, depends on the accuracy of the data, i.e., measurements of geometry and loads. For most engineering work involving a problem such as this, the data as measured to three significant figures would be sufficient.

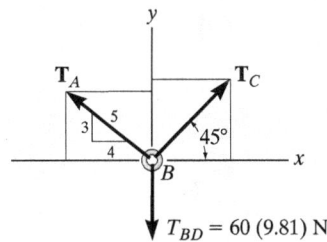

(c)

Fig. 3-6

# EXAMPLE 3.3

The 200-kg crate in Fig. 3-7a is suspended using the ropes $AB$ and $AC$. Each rope can withstand a maximum force of 10 kN before it breaks. If $AB$ always remains horizontal, determine the smallest angle $\theta$ to which the crate can be suspended before one of the ropes breaks.

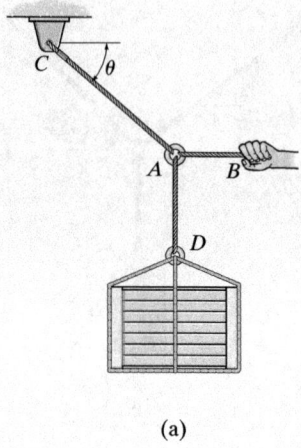

(a)

## SOLUTION

**Free-Body Diagram.** We will study the equilibrium of ring $A$. There are three forces acting on it, Fig. 3-7b. The magnitude of $\mathbf{F}_D$ is equal to the weight of the crate, i.e., $F_D = 200 \,(9.81) \text{ N} = 1962 \text{ N} < 10$ kN.

**Equations of Equilibrium.** Applying the equations of equilibrium along the $x$ and $y$ axes,

$$\xrightarrow{+} \Sigma F_x = 0; \quad -F_C \cos\theta + F_B = 0; \quad F_C = \frac{F_B}{\cos\theta} \quad (1)$$

$$+\uparrow \Sigma F_y = 0; \quad F_C \sin\theta - 1962 \text{ N} = 0 \quad (2)$$

From Eq. (1), $F_C$ is always greater than $F_B$ since $\cos\theta \leq 1$. Therefore, rope $AC$ will reach the maximum tensile force of 10 kN *before* rope $AB$. Substituting $F_C = 10$ kN into Eq. (2), we get

$$[10(10^3)\text{ N}] \sin\theta - 1962 \text{ N} = 0$$

$$\theta = \sin^{-1}(0.1962) = 11.31° \quad \textit{Ans.}$$

The force developed in rope $AB$ can be obtained by substituting the values for $\theta$ and $F_C$ into Eq. (1).

$$10(10^3) \text{ N} = \frac{F_B}{\cos 11.31°}$$

$$F_B = 9.81 \text{ kN}$$

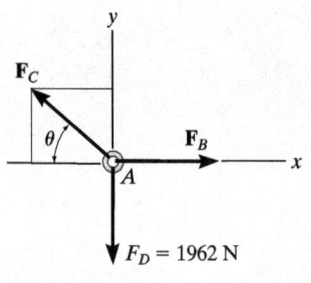

(b)

Fig. 3–7

## EXAMPLE 3.4

Determine the required length of cord $AC$ in Fig. 3–8a so that the 8-kg lamp can be suspended in the position shown. The *undeformed* length of spring $AB$ is $l'_{AB} = 0.4$ m, and the spring has a stiffness of $k_{AB} = 300$ N/m.

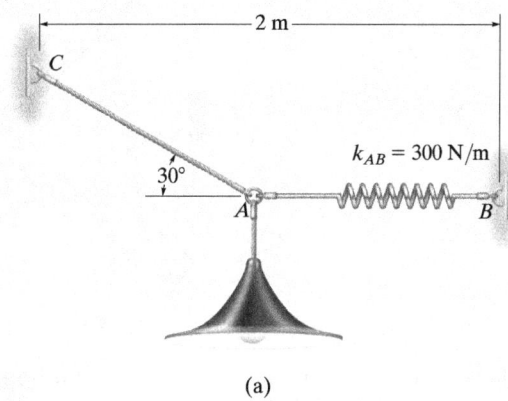
(a)

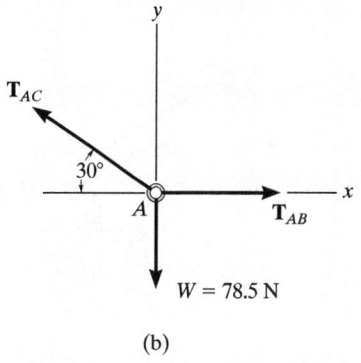
(b)

Fig. 3–8

### SOLUTION

If the force in spring $AB$ is known, the stretch of the spring can be found using $F = ks$. From the problem geometry, it is then possible to calculate the required length of $AC$.

**Free-Body Diagram.** The lamp has a weight $W = 8(9.81) = 78.5$ N and so the free-body diagram of the ring at $A$ is shown in Fig. 3–8b.

**Equations of Equilibrium.** Using the $x$, $y$ axes,

$\xrightarrow{+} \Sigma F_x = 0$;    $T_{AB} - T_{AC} \cos 30° = 0$
$+\uparrow \Sigma F_y = 0$;    $T_{AC} \sin 30° - 78.5 \text{ N} = 0$

Solving, we obtain

$$T_{AC} = 157.0 \text{ N}$$
$$T_{AB} = 135.9 \text{ N}$$

The stretch of spring $AB$ is therefore

$T_{AB} = k_{AB} s_{AB}$;    $135.9 \text{ N} = 300 \text{ N/m}(s_{AB})$

$$s_{AB} = 0.453 \text{ m}$$

so the stretched length is

$$l_{AB} = l'_{AB} + s_{AB}$$
$$l_{AB} = 0.4 \text{ m} + 0.453 \text{ m} = 0.853 \text{ m}$$

The horizontal distance from $C$ to $B$, Fig. 3–8a, requires

$$2 \text{ m} = l_{AC} \cos 30° + 0.853 \text{ m}$$
$$l_{AC} = 1.32 \text{ m} \qquad \text{Ans.}$$

## FUNDAMENTAL PROBLEMS

*All problem solutions must include an FBD.*

**F3–1.** The crate has a weight of 2.75 kN. Determine the force in each supporting cable.

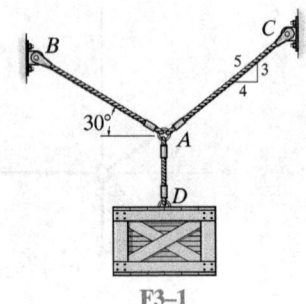

F3–1

**F3–2.** The beam has a weight of 3.5 kN. Determine the shortest cable $ABC$ that can be used to lift it if the maximum force the cable can sustain is 7.5 kN.

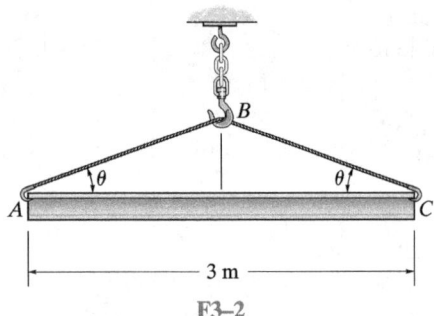

F3–2

**F3–3.** If the 5-kg block is suspended from the pulley $B$ and the sag of the cord is $d = 0.15$ m, determine the force in cord $ABC$. Neglect the size of the pulley.

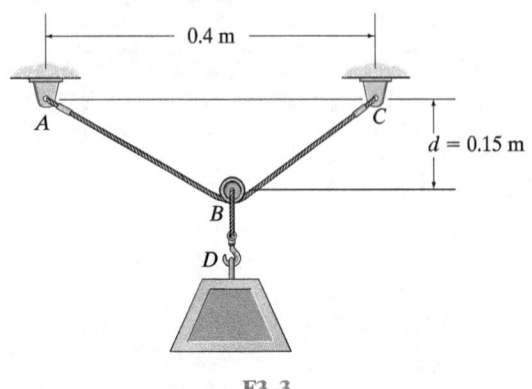

F3–3

**F3–4.** The block has a mass of 5 kg and rests on the smooth plane. Determine the unstretched length of the spring.

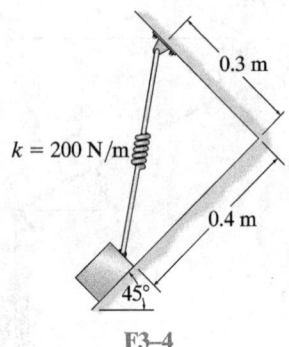

F3–4

**F3–5.** If the mass of cylinder $C$ is 40 kg, determine the mass of cylinder $A$ in order to hold the assembly in the position shown.

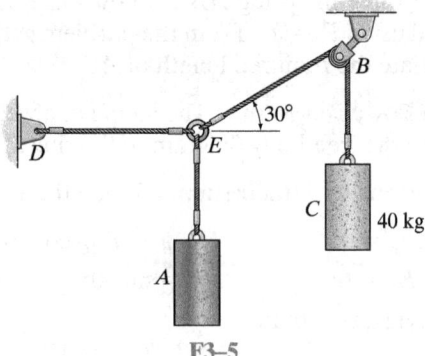

F3–5

**F3–6.** Determine the tension in cables $AB$, $BC$, and $CD$, necessary to support the 10-kg and 15-kg traffic lights at $B$ and $C$, respectively. Also, find the angle $\theta$.

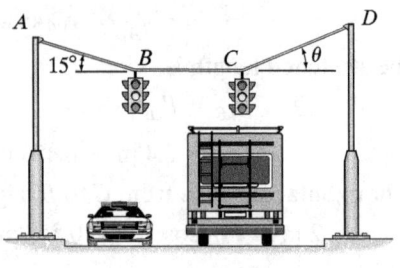

F3–6

## PROBLEMS

*All problem solutions must include an FBD.*

**•3–1.** Determine the force in each cord for equilibrium of the 200-kg crate. Cord $BC$ remains horizontal due to the roller at $C$, and $AB$ has a length of 1.5 m. Set $y = 0.75$ m.

**3–2.** If the 1.5-m-long cord $AB$ can withstand a maximum force of 3500 N, determine the force in cord $BC$ and the distance $y$ so that the 200-kg crate can be supported.

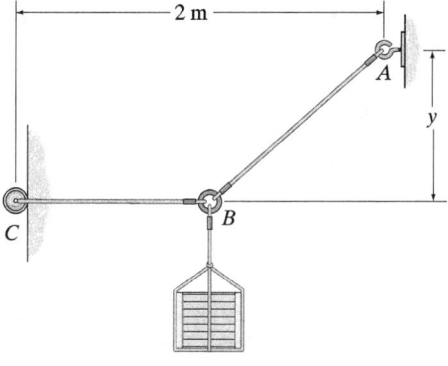

Probs. 3–1/2

**•3–5.** The members of a truss are connected to the gusset plate. If the forces are concurrent at point $O$, determine the magnitudes of **F** and **T** for equilibrium. Take $\theta = 30°$.

**3–6.** The gusset plate is subjected to the forces of four members. Determine the force in member $B$ and its proper orientation $\theta$ for equilibrium. The forces are concurrent at point $O$. Take $F = 12$ kN.

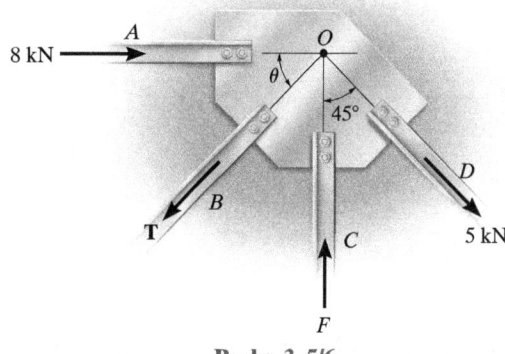

Probs. 3–5/6

**3–3.** If the mass of the girder is 3 Mg and its center of mass is located at point $G$, determine the tension developed in cables $AB$, $BC$, and $BD$ for equilibrium.

**\*3–4.** If cables $BD$ and $BC$ can withstand a maximum tensile force of 20 kN, determine the maximum mass of the girder that can be suspended from cable $AB$ so that neither cable will fail. The center of mass of the girder is located at point $G$.

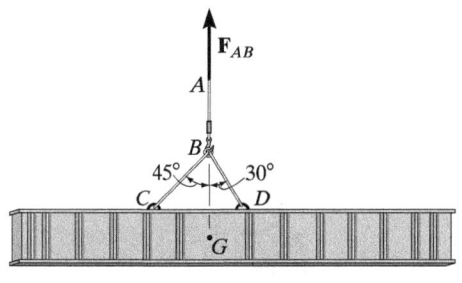

Probs. 3–3/4

**3–7.** The towing pendant $AB$ is subjected to the force of 50 kN exerted by a tugboat. Determine the force in each of the bridles, $BC$ and $BD$, if the ship is moving forward with constant velocity.

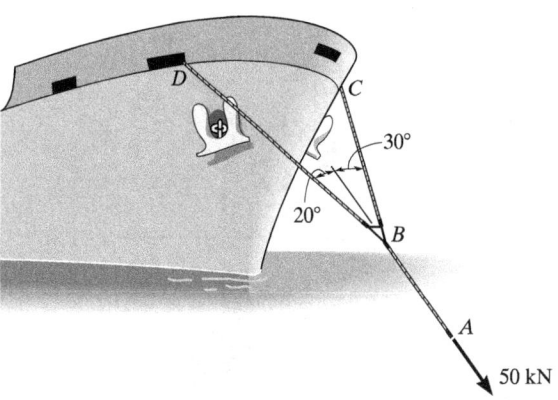

Prob. 3–7

**\*3–8.** Members $AC$ and $AB$ support the 100-kg crate. Determine the tensile force developed in each member.

**•3–9.** If members $AC$ and $AB$ can support a maximum tension of 1500 N and 1250 N, respectively, determine the largest weight of the crate that can be safely supported.

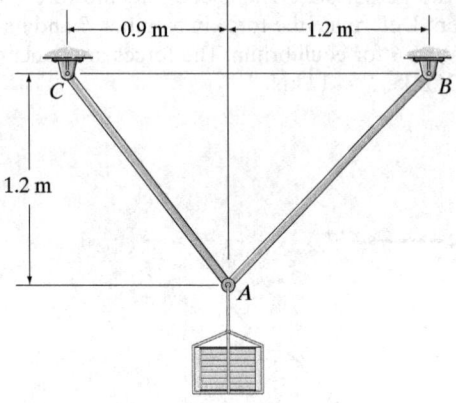

Probs. 3–8/9

**\*3–12.** If block $B$ weighs 1 kN and block $C$ weighs 0.5 kN, determine the required weight of block $D$ and the angle $\theta$ for equilibrium.

**•3–13.** If block $D$ weighs 1.5 kN and block $B$ weighs 1.375 kN, determine the required weight of block $C$ and the angle $\theta$ for equilibrium.

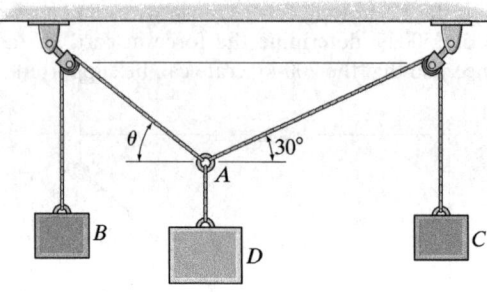

Probs. 3–12/13

**3–10.** The members of a truss are connected to the gusset plate. If the forces are concurrent at point $O$, determine the magnitudes of $\mathbf{F}$ and $\mathbf{T}$ for equilibrium. Take $\theta = 90°$.

**3–11.** The gusset plate is subjected to the forces of three members. Determine the tension force in member $C$ and its angle $\theta$ for equilibrium. The forces are concurrent at point $O$. Take $F = 8$ kN.

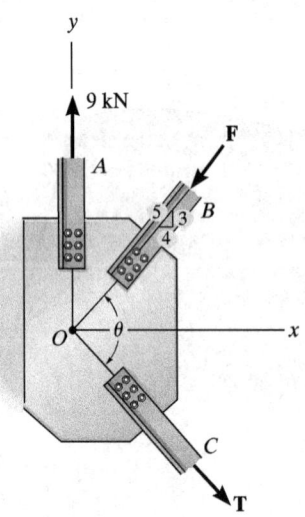

Probs. 3–10/11

**3–14.** Determine the stretch in springs $AC$ and $AB$ for equilibrium of the 2-kg block. The springs are shown in the equilibrium position.

**3–15.** The unstretched length of spring $AB$ is 3 m. If the block is held in the equilibrium position shown, determine the mass of the block at $D$.

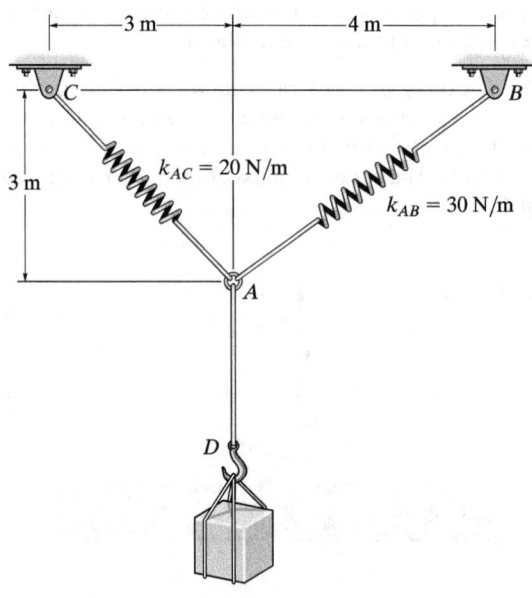

Probs. 3–14/15

**\*3–16.** Determine the tension developed in wires $CA$ and $CB$ required for equilibrium of the 10-kg cylinder. Take $\theta = 40°$.

**•3–17.** If cable $CB$ is subjected to a tension that is twice that of cable $CA$, determine the angle $\theta$ for equilibrium of the 10-kg cylinder. Also, what are the tensions in wires $CA$ and $CB$?

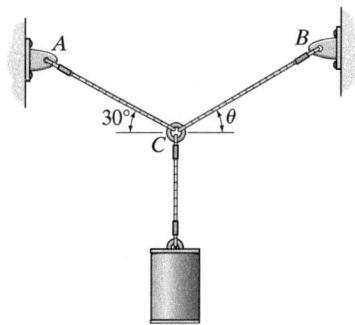

Probs. 3–16/17

**3–18.** Determine the forces in cables $AC$ and $AB$ needed to hold the 20-kg ball $D$ in equilibrium. Take $F = 300$ N and $d = 1$ m.

**3–19.** The ball $D$ has a mass of 20 kg. If a force of $F = 100$ N is applied horizontally to the ring at $A$, determine the dimension $d$ so that the force in cable $AC$ is zero.

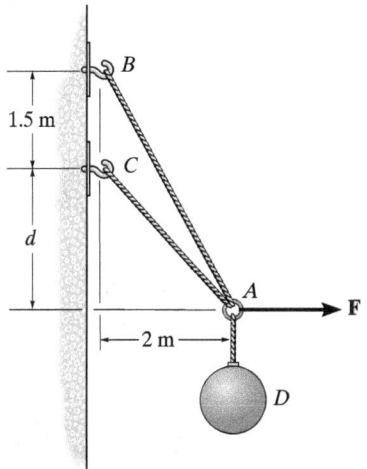

Probs. 3–18/19

**\*3–20.** Determine the tension developed in each wire used to support the 50-kg chandelier.

**•3–21.** If the tension developed in each of the four wires is not allowed to exceed 600 N, determine the maximum mass of the chandelier that can be supported.

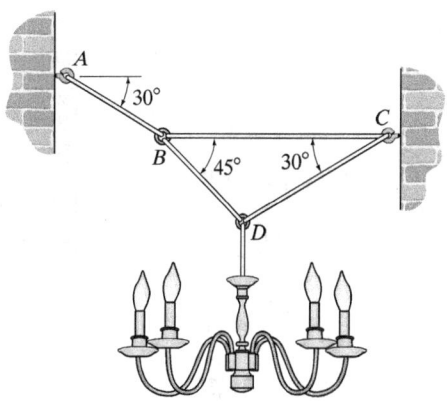

Prob. 3–20/21

**■3–22.** A vertical force $P = 50$ N is applied to the ends of the 0.6 m cord $AB$ and spring $AC$. If the spring has an unstretched length of 0.6 m, determine the angle $\theta$ for equilibrium. Take $k = 250$ N/m.

**3–23.** Determine the unstretched length of spring $AC$ if a force $P = 400$ N causes the angle $\theta = 60°$ for equilibrium. Cord $AB$ is 0.6 m long. Take $k = 850$ N/m.

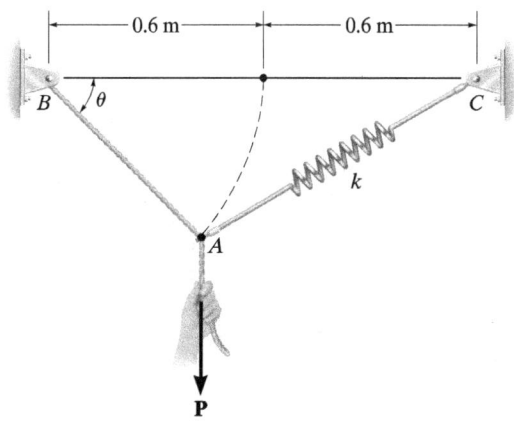

Probs. 3–22/23

**\*3–24.** If the bucket weighs 0.25 kN, determine the tension developed in each of the wires.

**•3–25.** Determine the maximum weight of the bucket that the wire system can support so that no single wire develops a tension exceeding 0.5 kN.

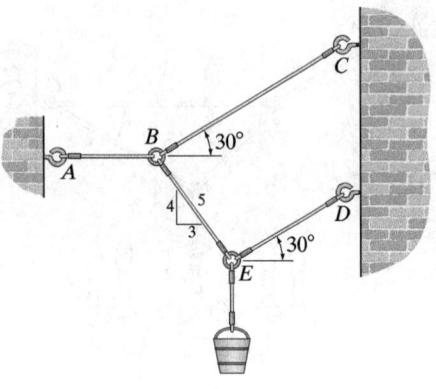

Probs. 3–24/25

**3–26.** Determine the tensions developed in wires $CD, CB,$ and $BA$ and the angle $\theta$ required for equilibrium of the 15-kg cylinder $E$ and the 30-kg cylinder $F$.

**3–27.** If cylinder $E$ weighs 150 N and $\theta = 15°$, determine the weight of cylinder $F$.

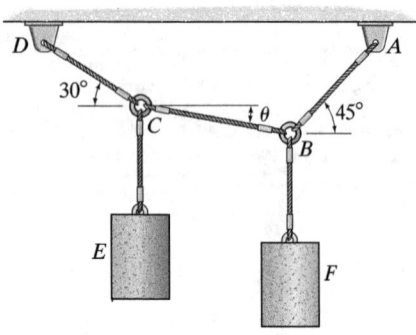

Probs. 3–26/27

**\*3–28.** Two spheres $A$ and $B$ have an equal mass and are electrostatically charged such that the repulsive force acting between them has a magnitude of 20 mN and is directed along line $AB$. Determine the angle $\theta$, the tension in cords $AC$ and $BC$, and the mass $m$ of each sphere.

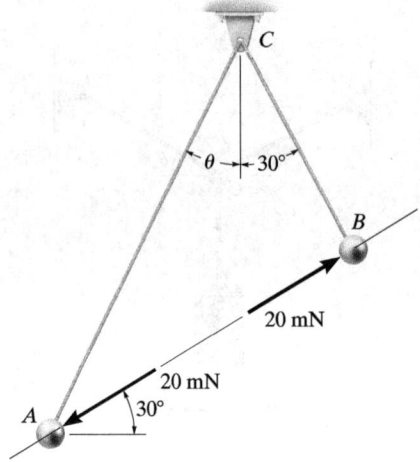

Prob. 3–28

**•3–29.** The cords $BCA$ and $CD$ can each support a maximum load of 0.5 kN. Determine the maximum weight of the crate that can be hoisted at constant velocity and the angle $\theta$ for equilibrium. Neglect the size of the smooth pulley at $C$.

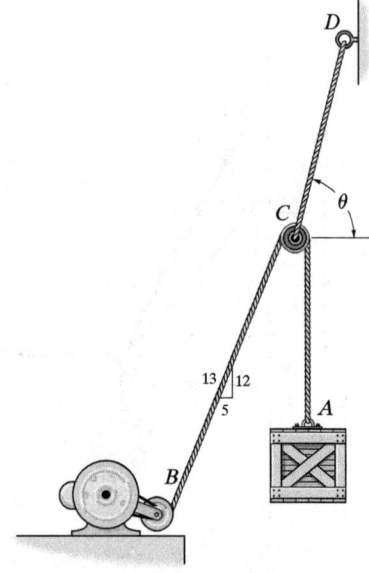

Prob. 3–29

**3–30.** The springs on the rope assembly are originally unstretched when $\theta = 0°$. Determine the tension in each rope when $F = 450$ N. Neglect the size of the pulleys at $B$ and $D$.

**3–31.** The springs on the rope assembly are originally stretched 0.3 m when $\theta = 0°$. Determine the vertical force F that must be applied so that $\theta = 30°$.

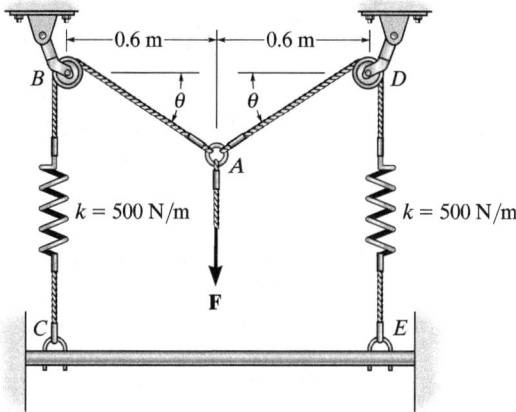

Probs. 3–30/31

*3–32.** Determine the magnitude and direction $\theta$ of the equilibrium force $F_{AB}$ exerted along link $AB$ by the tractive apparatus shown. The suspended mass is 10 kg. Neglect the size of the pulley at $A$.

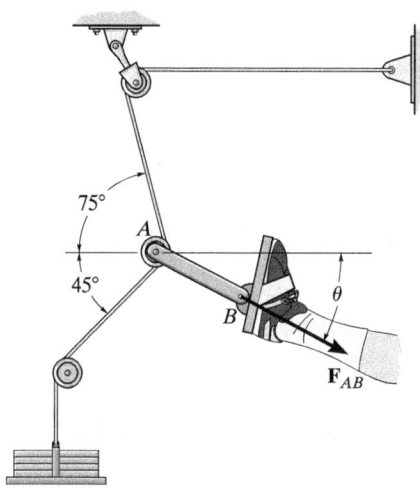

Prob. 3–32

•**3–33.** The wire forms a loop and passes over the small pulleys at $A, B, C,$ and $D$. If its end is subjected to a force of $P = 50$ N, determine the force in the wire and the magnitude of the resultant force that the wire exerts on each of the pulleys.

**3–34.** The wire forms a loop and passes over the small pulleys at $A, B, C,$ and $D$. If the maximum *resultant force* that the wire can exert on each pulley is 120 N, determine the greatest force $P$ that can be applied to the wire as shown.

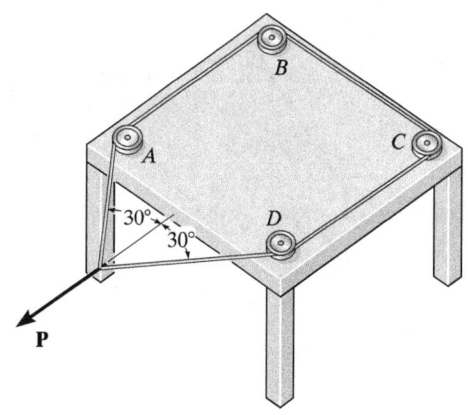

Probs. 3–33/34

**3–35.** The picture has a weight of 50 N and is to be hung over the smooth pin $B$. If a string is attached to the frame at points $A$ and $C$, and the maximum force the string can support is 75 N, determine the shortest string that can be safely used.

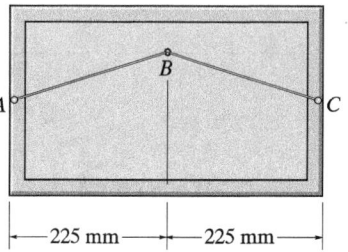

Prob. 3–35

**\*3–36.** The 100-kg uniform tank is suspended by means of a 3-m-long cable, which is attached to the sides of the tank and passes over the small pulley located at O. If the cable can be attached at either points A and B or C and D, determine which attachment produces the least amount of tension in the cable. What is this tension?

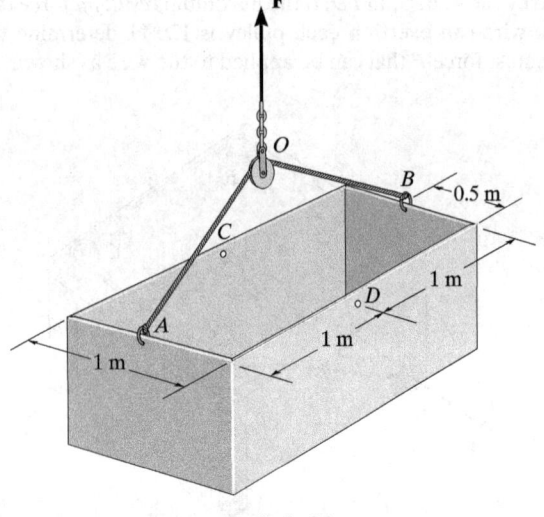

Prob. 3–36

**•3–37.** The 5-kg weight is supported by the cord AC and roller and by the spring that has a stiffness of $k = 2000$ N/m and an unstretched length of 300 mm. Determine the distance d to where the weight is located when it is in equilibrium.

**3–38.** The 5-kg weight is supported by the cord AC and roller and by a spring. If the spring has an unstretched length of 200 mm and the weight is in equilibrium when $d = 100$ mm, determine the stiffness k of the spring.

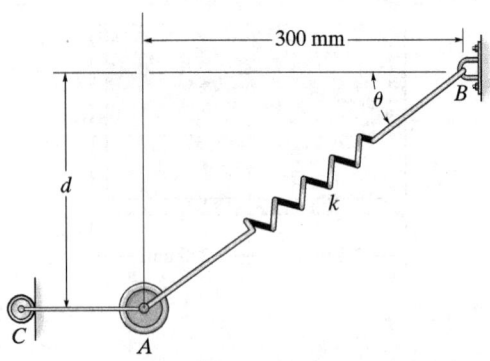

Probs. 3–37/38

**3–39.** A "scale" is constructed with a 1.2-m-long cord and the 5-kg block D. The cord is fixed to a pin at A and passes over two small pulleys at B and C. Determine the weight of the suspended block at B if the system is in equilibrium.

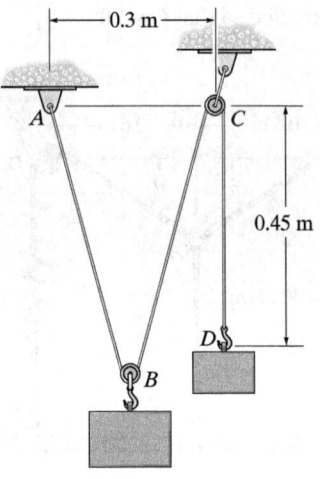

Prob. 3–39

**•\*3–40.** The spring has a stiffness of $k = 800$ N/m and an unstretched length of 200 mm. Determine the force in cables BC and BD when the spring is held in the position shown.

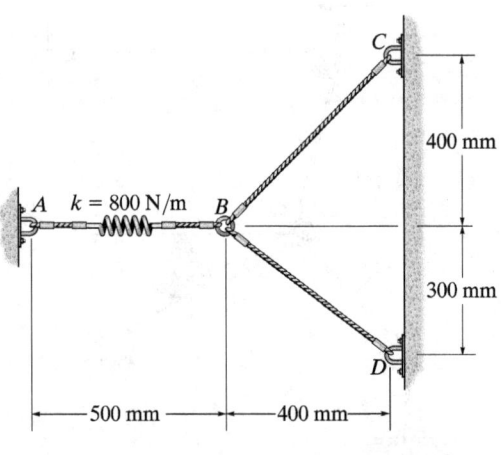

Prob. 3–40

•3–41. A continuous cable of total length 4 m is wrapped around the *small* pulleys at A, B, C, and D. If each spring is stretched 300 mm, determine the mass m of each block. Neglect the weight of the pulleys and cords. The springs are unstretched when d = 2 m.

3–43. The pail and its contents have a mass of 60 kg. If the cable BAC is 15 m long, determine the distance y to the pulley at A for equilibrium. Neglect the size of the pulley.

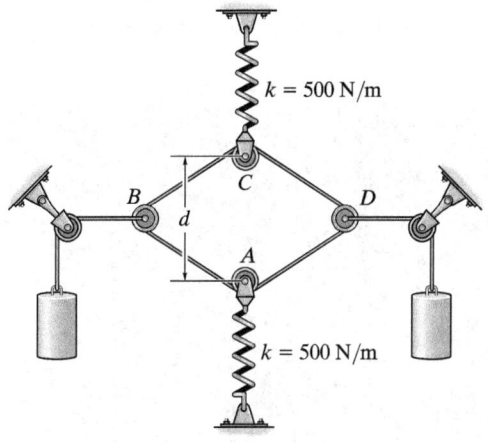

Prob. 3–41

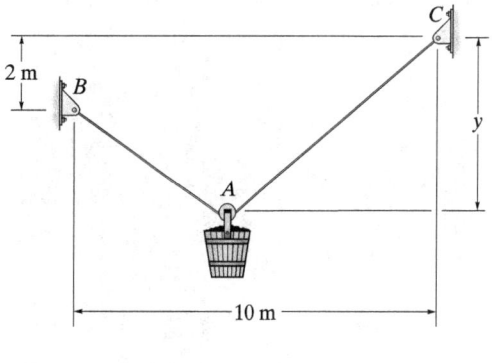

Prob. 3–43

3–42. Determine the mass of each of the two cylinders if they cause a sag of s = 0.5 m when suspended from the rings at A and B. Note that s = 0 when the cylinders are removed.

•*3–44. A scale is constructed using the 10-kg mass, the 2-kg pan P, and the pulley and cord arrangement. Cord BCA is 2 m long. If s = 0.75 m, determine the mass D in the pan. Neglect the size of the pulley.

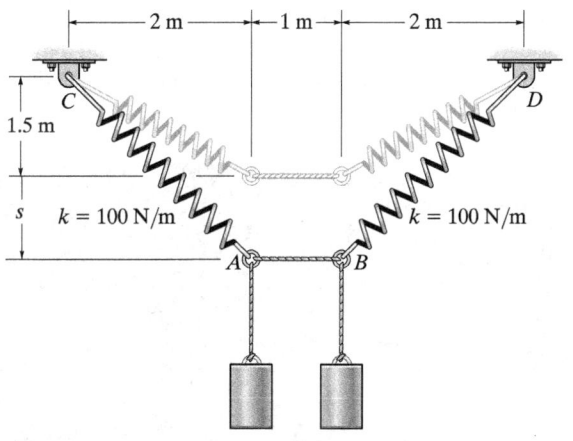

Prob. 3–42

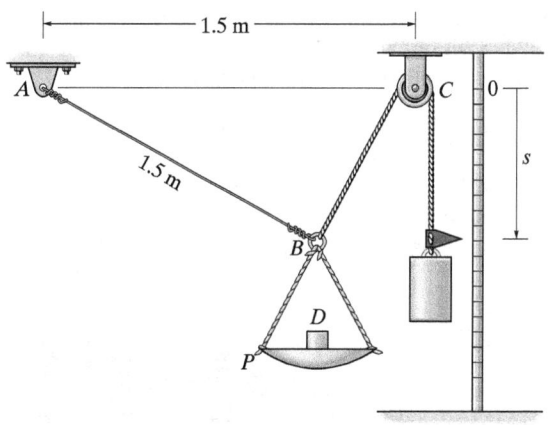

Prob. 3–44

## CONCEPTUAL PROBLEMS

**P3–1.** The concrete wall panel is hoisted into position using the two cables $AB$ and $AC$ of equal length. Establish appropriate dimensions and use an equilibrium analysis to show that the longer the cables the less the force in each cable.

**P3–2.** The truss is hoisted using cable $ABC$ that passes through a very small pulley at $B$. If the truss is placed in a tipped position, show that it will always return to the horizontal position to maintain equilibrium.

**P3–3.** The device $DB$ is used to pull on the chain $ABC$ so as to hold a door closed on the bin. If the angle between $AB$ and the horizontal segment $BC$ is 30°, determine the angle between $DB$ and the horizontal for equilibrium.

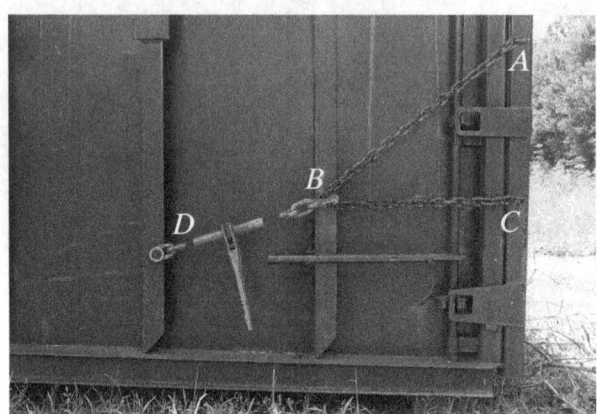

**P3–4.** The two chains $AB$ and $AC$ have equal lengths and are subjected to the vertical force **F**. If $AB$ is replaced by a shorter chain $AB'$, show that this chain would have to support a larger tensile force than $AB$ in order to maintain equilibrium.

## 3.4 Three-Dimensional Force Systems

In Section 3.1, we stated that the necessary and sufficient condition for particle equilibrium is

$$\Sigma \mathbf{F} = \mathbf{0} \quad (3\text{-}4)$$

In the case of a three-dimensional force system, as in Fig. 3–9, we can resolve the forces into their respective $\mathbf{i}$, $\mathbf{j}$, $\mathbf{k}$ components, so that $\Sigma F_x \mathbf{i} + \Sigma F_y \mathbf{j} + \Sigma F_z \mathbf{k} = \mathbf{0}$. To satisfy this equation we require

$$\begin{array}{c} \Sigma F_x = 0 \\ \Sigma F_y = 0 \\ \Sigma F_z = 0 \end{array} \quad (3\text{-}5)$$

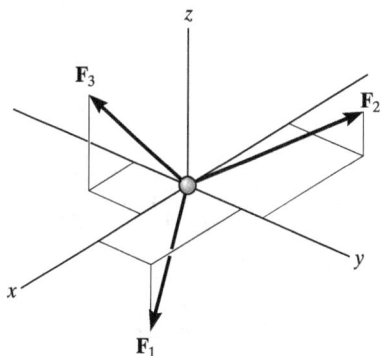

Fig. 3–9

These three equations state that the *algebraic sum* of the components of all the forces acting on the particle along each of the coordinate axes must be zero. Using them we can solve for at most three unknowns, generally represented as coordinate direction angles or magnitudes of forces shown on the particle's free-body diagram.

### Procedure for Analysis

Three-dimensional force equilibrium problems for a particle can be solved using the following procedure.

Free-Body Diagram.

- Establish the $x$, $y$, $z$ axes in any suitable orientation.
- Label all the known and unknown force magnitudes and directions on the diagram.
- The sense of a force having an unknown magnitude can be assumed.

Equations of Equilibrium.

- Use the scalar equations of equilibrium, $\Sigma F_x = 0$, $\Sigma F_y = 0$, $\Sigma F_z = 0$, in cases where it is easy to resolve each force into its $x$, $y$, $z$ components.
- If the three-dimensional geometry appears difficult, then first express each force on the free-body diagram as a Cartesian vector, substitute these vectors into $\Sigma \mathbf{F} = \mathbf{0}$, and then set the $\mathbf{i}$, $\mathbf{j}$, $\mathbf{k}$ components equal to zero.
- If the solution for a force yields a negative result, this indicates that its sense is the reverse of that shown on the free-body diagram.

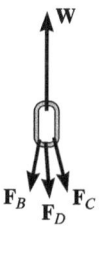

The ring at $A$ is subjected to the force from the hook as well as forces from each of the three chains. If the electromagnet and its load have a weight $W$, then the force at the hook will be $\mathbf{W}$, and the three scalar equations of equilibrium can be applied to the free-body diagram of the ring in order to determine the chain forces, $\mathbf{F}_B$, $\mathbf{F}_C$, and $\mathbf{F}_D$.

## EXAMPLE 3.5

A 90-N load is suspended from the hook shown in Fig. 3–10a. If the load is supported by two cables and a spring having a stiffness $k = 500$ N/m, determine the force in the cables and the stretch of the spring for equilibrium. Cable $AD$ lies in the $x$–$y$ plane and cable $AC$ lies in the $x$–$z$ plane.

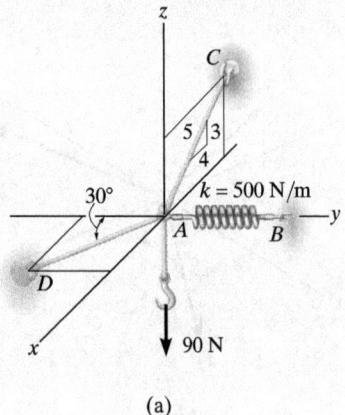

### SOLUTION

The stretch of the spring can be determined once the force in the spring is determined.

**Free-Body Diagram.** The connection at $A$ is chosen for the equilibrium analysis since the cable forces are concurrent at this point. The free-body diagram is shown in Fig. 3–10b.

**Equations of Equilibrium.** By inspection, each force can easily be resolved into its $x$, $y$, $z$ components, and therefore the three scalar equations of equilibrium can be used. Considering components directed along each positive axis as "positive," we have

$$\Sigma F_x = 0; \quad F_D \sin 30° - \left(\tfrac{4}{5}\right)F_C = 0 \quad (1)$$
$$\Sigma F_y = 0; \quad -F_D \cos 30° + F_B = 0 \quad (2)$$
$$\Sigma F_z = 0; \quad \left(\tfrac{3}{5}\right)F_C - 90 \text{ N} = 0 \quad (3)$$

Solving Eq. (3) for $F_C$, then Eq. (1) for $F_D$, and finally Eq. (2) for $F_B$, yields

$$F_C = 150 \text{ N} \quad \text{Ans.}$$
$$F_D = 240 \text{ N} \quad \text{Ans.}$$
$$F_B = 207.8 \text{ N} \quad \text{Ans.}$$

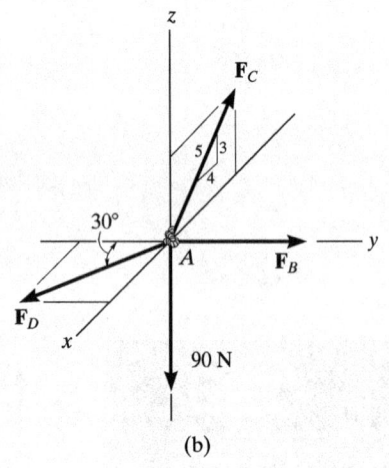

Fig. 3–10

The stretch of the spring is therefore

$$F_B = ks_{AB}$$
$$207.8 \text{ N} = (500 \text{ N/m})(s_{AB})$$
$$s_{AB} = 0.416 \text{ m} \quad \text{Ans.}$$

**NOTE:** Since the results for all the cable forces are positive, each cable is in tension; that is, it pulls on point $A$ as expected, Fig. 3–10b.

## EXAMPLE 3.6

The 10-kg lamp in Fig. 3-11a is suspended from the three equal-length cords. Determine its smallest vertical distance $s$ from the ceiling if the force developed in any cord is not allowed to exceed 50 N.

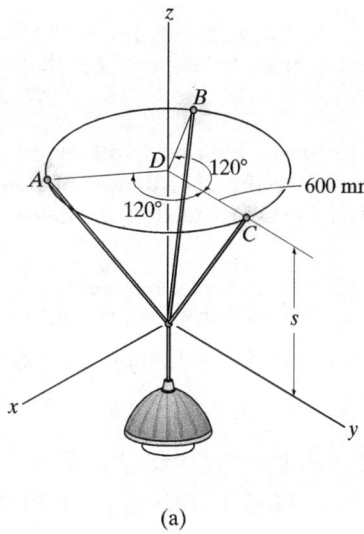

(a)

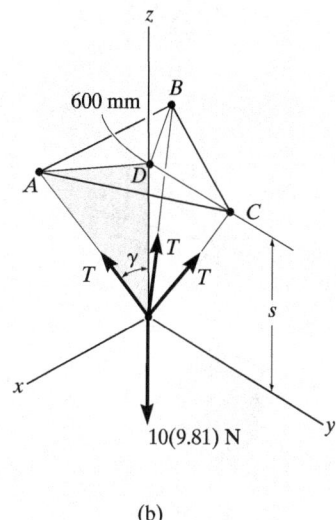

(b)

Fig. 3–11

### SOLUTION

**Free-Body Diagram.** Due to symmetry, Fig. 3-11b, the distance $DA = DB = DC = 600$ mm. It follows that from $\Sigma F_x = 0$ and $\Sigma F_y = 0$, the tension $T$ in each cord will be the same. Also, the angle between each cord and the $z$ axis is $\gamma$.

**Equation of Equilibrium.** Applying the equilibrium equation along the $z$ axis, with $T = 50$ N, we have

$\Sigma F_z = 0;$ $\quad 3[(50 \text{ N}) \cos \gamma] - 10(9.81) \text{ N} = 0$

$$\gamma = \cos^{-1} \frac{98.1}{150} = 49.16°$$

From the shaded triangle shown in Fig. 3-11b,

$$\tan 49.16° = \frac{600 \text{ mm}}{s}$$

$$s = 519 \text{ mm} \qquad \qquad Ans.$$

## EXAMPLE 3.7

Determine the force in each cable used to support the 40-kN ($\approx$ 4000-kg) crate shown in Fig. 3–12a.

### SOLUTION

**Free-Body Diagram.** As shown in Fig. 3–12b, the free-body diagram of point $A$ is considered in order to "expose" the three unknown forces in the cables.

**Equations of Equilibrium.** First we will express each force in Cartesian vector form. Since the coordinates of points $B$ and $C$ are $B(-3\text{ m}, -4\text{ m}, 8\text{ m})$ and $C(-3\text{ m}, 4\text{ m}, 8\text{ m})$, we have

$$\mathbf{F}_B = F_B\left[\frac{-3\mathbf{i} - 4\mathbf{j} + 8\mathbf{k}}{\sqrt{(-3)^2 + (-4)^2 + (8)^2}}\right]$$

$$= -0.318F_B\mathbf{i} - 0.424F_B\mathbf{j} + 0.848F_B\mathbf{k}$$

$$\mathbf{F}_C = F_C\left[\frac{-3\mathbf{i} + 4\mathbf{j} + 8\mathbf{k}}{\sqrt{(-3)^2 + (4)^2 + (8)^2}}\right]$$

$$= -0.318F_C\mathbf{i} + 0.424F_C\mathbf{j} + 0.848F_C\mathbf{k}$$

$$\mathbf{F}_D = F_D\mathbf{i}$$

$$\mathbf{W} = \{-40\mathbf{k}\}\text{ kN}$$

Equilibrium requires

$$\Sigma\mathbf{F} = \mathbf{0}; \qquad \mathbf{F}_B + \mathbf{F}_C + \mathbf{F}_D + \mathbf{W} = \mathbf{0}$$

$$-0.318F_B\mathbf{i} - 0.424F_B\mathbf{j} + 0.848F_B\mathbf{k}$$
$$- 0.318F_C\mathbf{i} + 0.424F_C\mathbf{j} + 0.848F_C\mathbf{k} + F_D\mathbf{i} - 40\mathbf{k} = \mathbf{0}$$

Equating the respective $\mathbf{i}, \mathbf{j}, \mathbf{k}$ components to zero yields

$$\Sigma F_x = 0; \qquad -0.318F_B - 0.318F_C + F_D = 0 \qquad (1)$$

$$\Sigma F_y = 0; \qquad -0.424F_B + 0.424F_C = 0 \qquad (2)$$

$$\Sigma F_z = 0; \qquad 0.848F_B + 0.848F_C - 40 = 0 \qquad (3)$$

Equation (2) states that $F_B = F_C$. Thus, solving Eq. (3) for $F_B$ and $F_C$ and substituting the result into Eq. (1) to obtain $F_D$, we have

$$F_B = F_C = 23.6\text{ kN} \qquad \textit{Ans.}$$

$$F_D = 15.0\text{ kN} \qquad \textit{Ans.}$$

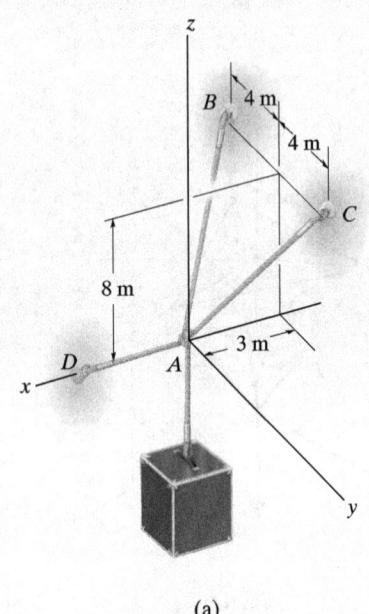

(a)

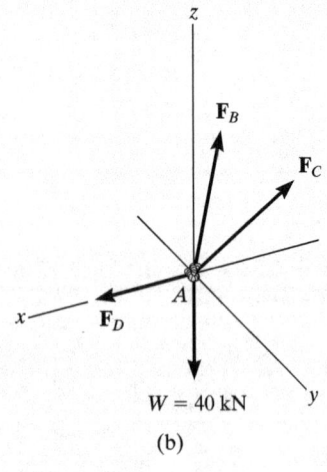

(b)

Fig. 3–12

## EXAMPLE 3.8

Determine the tension in each cord used to support the 100-kg crate shown in Fig. 3–13a.

### SOLUTION

**Free-Body Diagram.** The force in each of the cords can be determined by investigating the equilibrium of point $A$. The free-body diagram is shown in Fig. 3–13b. The weight of the crate is $W = 100(9.81) = 981$ N.

**Equations of Equilibrium.** Each force on the free-body diagram is first expressed in Cartesian vector form. Using Eq. 2–9 for $\mathbf{F}_C$ and noting point $D(-1 \text{ m}, 2 \text{ m}, 2 \text{ m})$ for $\mathbf{F}_D$, we have

$$\mathbf{F}_B = F_B \mathbf{i}$$

$$\mathbf{F}_C = F_C \cos 120°\mathbf{i} + F_C \cos 135°\mathbf{j} + F_C \cos 60°\mathbf{k}$$
$$= -0.5 F_C \mathbf{i} - 0.707 F_C \mathbf{j} + 0.5 F_C \mathbf{k}$$

$$\mathbf{F}_D = F_D \left[ \frac{-1\mathbf{i} + 2\mathbf{j} + 2\mathbf{k}}{\sqrt{(-1)^2 + (2)^2 + (2)^2}} \right]$$
$$= -0.333 F_D \mathbf{i} + 0.667 F_D \mathbf{j} + 0.667 F_D \mathbf{k}$$

$$\mathbf{W} = \{-981\mathbf{k}\} \text{ N}$$

Equilibrium requires

$$\Sigma \mathbf{F} = 0; \qquad \mathbf{F}_B + \mathbf{F}_C + \mathbf{F}_D + \mathbf{W} = 0$$

$$F_B \mathbf{i} - 0.5 F_C \mathbf{i} - 0.707 F_C \mathbf{j} + 0.5 F_C \mathbf{k}$$
$$- 0.333 F_D \mathbf{i} + 0.667 F_D \mathbf{j} + 0.667 F_D \mathbf{k} - 981\mathbf{k} = 0$$

Equating the respective $\mathbf{i}, \mathbf{j}, \mathbf{k}$ components to zero,

$$\Sigma F_x = 0; \qquad F_B - 0.5 F_C - 0.333 F_D = 0 \qquad (1)$$
$$\Sigma F_y = 0; \qquad -0.707 F_C + 0.667 F_D = 0 \qquad (2)$$
$$\Sigma F_z = 0; \qquad 0.5 F_C + 0.667 F_D - 981 = 0 \qquad (3)$$

Solving Eq. (2) for $F_D$ in terms of $F_C$ and substituting this into Eq. (3) yields $F_C$. $F_D$ is then determined from Eq. (2). Finally, substituting the results into Eq. (1) gives $F_B$. Hence,

$$F_C = 813 \text{ N} \qquad \text{Ans.}$$
$$F_D = 862 \text{ N} \qquad \text{Ans.}$$
$$F_B = 694 \text{ N} \qquad \text{Ans.}$$

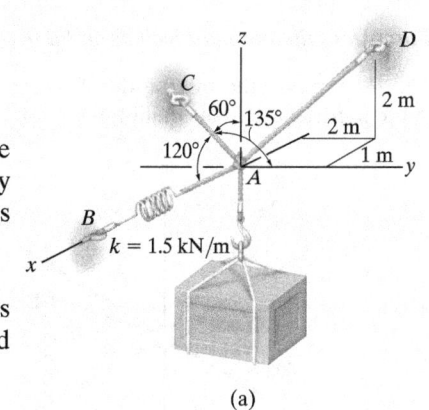

(a)

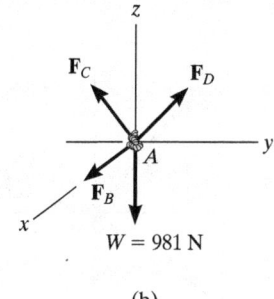

(b)

Fig. 3–13

## FUNDAMENTAL PROBLEMS

*All problem solutions must include an FBD.*

**F3–7.** Determine the magnitude of forces $\mathbf{F}_1$, $\mathbf{F}_2$, $\mathbf{F}_3$, so that the particle is held in equilibrium.

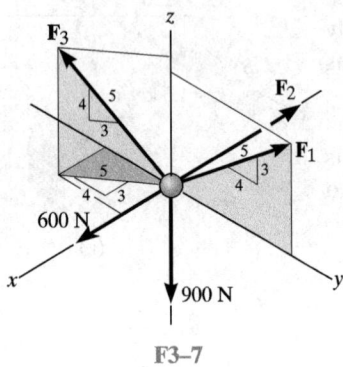

F3–7

**F3–8.** Determine the tension developed in cables $AB$, $AC$, and $AD$.

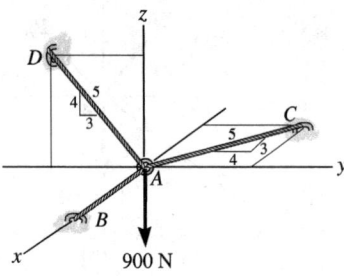

F3–8

**F3–9.** Determine the tension developed in cables $AB$, $AC$, and $AD$.

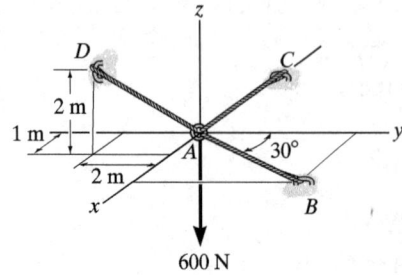

F3–9

**F3–10.** Determine the tension developed in cables $AB$, $AC$, and $AD$.

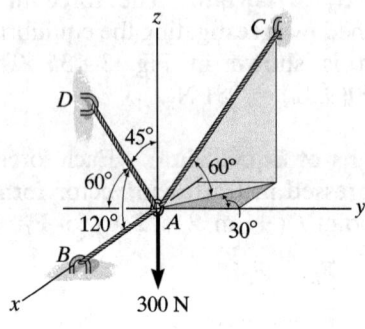

F3–10

**F3–11.** The 75-kg crate is supported by cables $AB$, $AC$, and $AD$. Determine the tension in these wires.

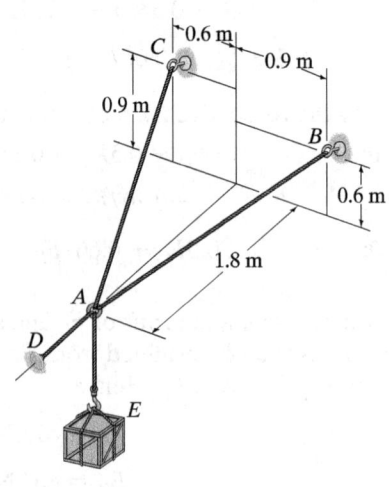

F3–11

## PROBLEMS

*All problem solutions must include an FBD.*

**•3–45.** Determine the tension in the cables in order to support the 100-kg crate in the equilibrium position shown.

**3–46.** Determine the maximum mass of the crate so that the tension developed in any cable does not exceeded 3 kN.

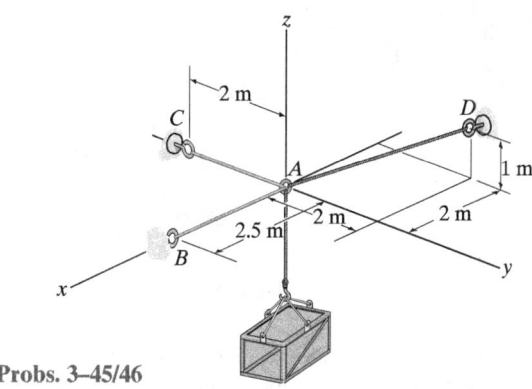

Probs. 3–45/46

**3–47.** The shear leg derrick is used to haul the 200-kg net of fish onto the dock. Determine the compressive force along each of the legs $AB$ and $CB$ and the tension in the winch cable $DB$. Assume the force in each leg acts along its axis.

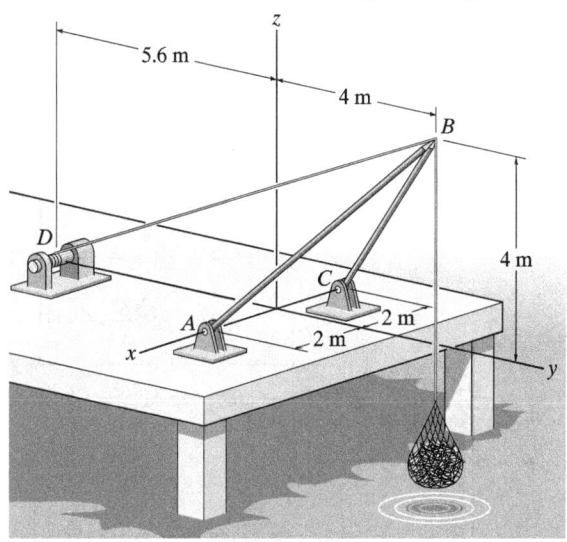

Prob. 3–47

**\*3–48.** Determine the tension developed in cables $AB$, $AC$, and $AD$ required for equilibrium of the 150-kg crate.

**•3–49.** Determine the maximum weight of the crate so that the tension developed in any cable does not exceed 2250 N.

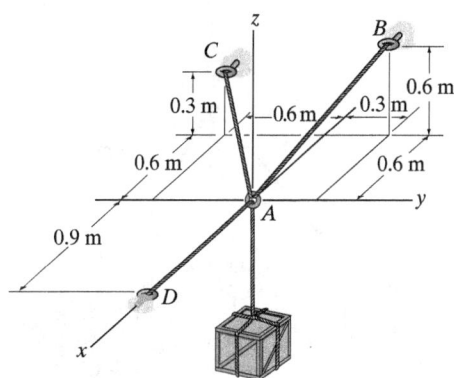

Probs. 3–48/49

**3–50.** Determine the force in each cable needed to support the 17.5-kN ($\approx$ 1750-kg) platform. Set $d = 0.6$ m.

**3–51.** Determine the force in each cable needed to support the 17.5-kN ($\approx$ 1750-kg) platform. Set $d = 1.2$ m.

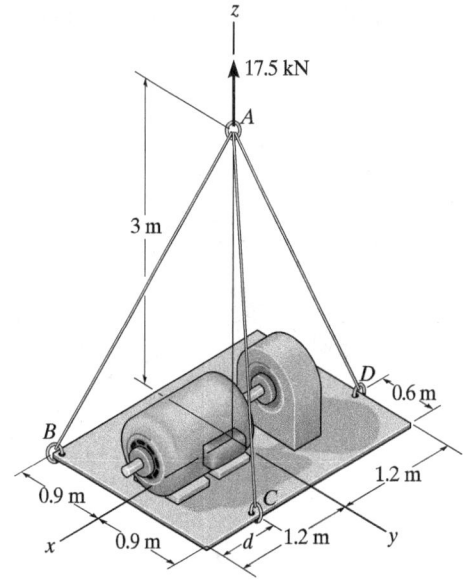

Probs. 3–50/51

**\*3–52.** Determine the force in each of the three cables needed to lift the tractor which has a mass of 8 Mg.

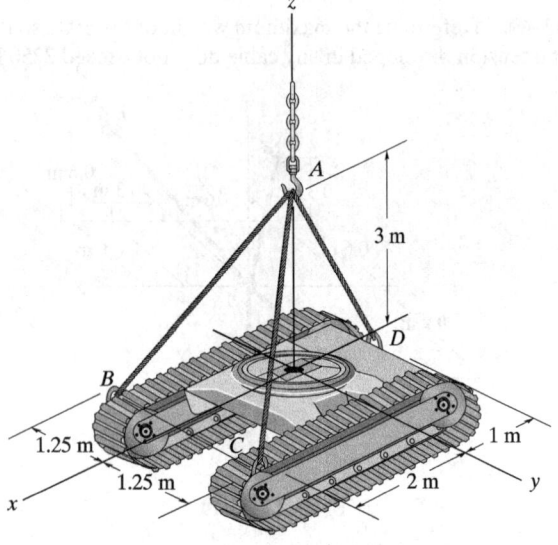

Prob. 3–52

**•3–53.** Determine the force acting along the axis of each of the three struts needed to support the 500-kg block.

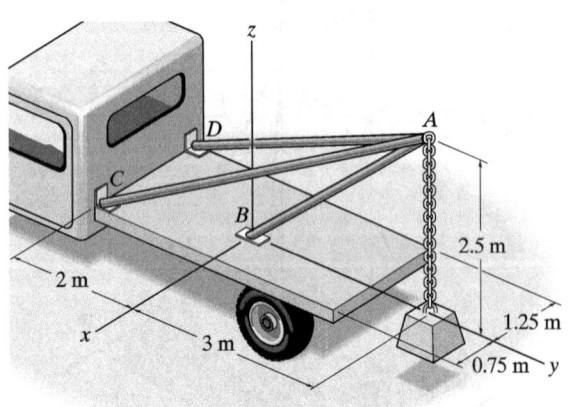

Prob. 3–53

**3–54.** If the mass of the flowerpot is 50 kg, determine the tension developed in each wire for equilibrium. Set $x = 1.5$ m and $z = 2$ m.

**3–55.** If the mass of the flowerpot is 50 kg, determine the tension developed in each wire for equilibrium. Set $x = 2$ m and $z = 1.5$ m.

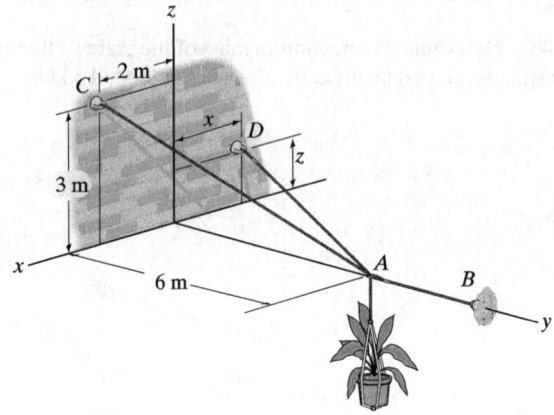

Probs. 3–54/55

**\*3–56.** The ends of the three cables are attached to a ring at $A$ and to the edge of a uniform 150-kg plate. Determine the tension in each of the cables for equilibrium.

**•3–57.** The ends of the three cables are attached to a ring at $A$ and to the edge of the uniform plate. Determine the largest mass the plate can have if each cable can support a maximum tension of 15 kN.

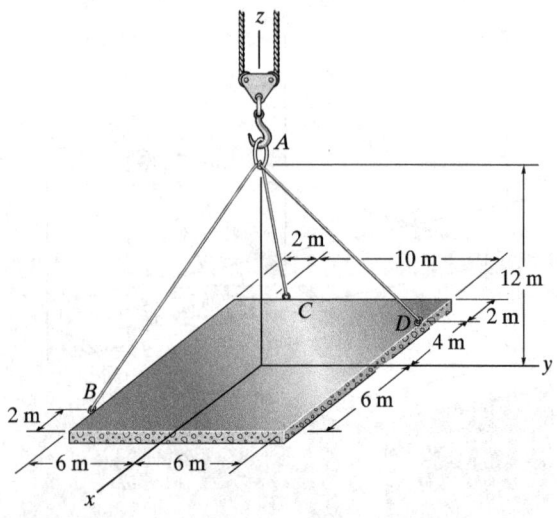

Probs. 3–56/57

## 3.4 Three-Dimensional Force Systems

**3–58.** Determine the tension developed in cables $AB$, $AC$, and $AD$ required for equilibrium of the 75-kg cylinder.

**3–59.** If each cable can withstand a maximum tension of 1000 N, determine the largest mass of the cylinder for equilibrium.

**3–62.** A force of $F = 500$ N holds the 200-kg crate in equilibrium. Determine the coordinates $(0, y, z)$ of point $A$ if the tension in cords $AC$ and $AB$ is 3500 N each.

**3–63.** If the maximum allowable tension in cables $AB$ and $AC$ is 2500 N, determine the maximum height $z$ to which the 100-kg crate can be lifted. What horizontal force $F$ must be applied? Take $y = 2.4$ m.

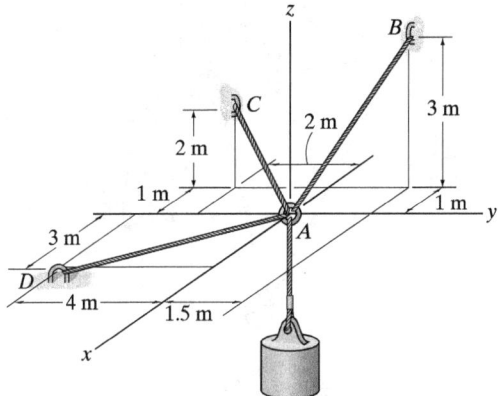

Probs. 3–58/59

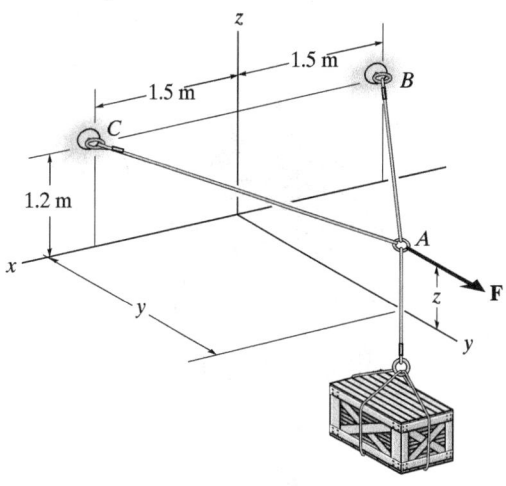

Probs. 3–62/63

*__3–60.__ The 50-kg pot is supported from $A$ by the three cables. Determine the force acting in each cable for equilibrium. Take $d = 2.5$ m.

•**3–61.** Determine the height $d$ of cable $AB$ so that the force in cables $AD$ and $AC$ is one-half as great as the force in cable $AB$. What is the force in each cable for this case? The flower pot has a mass of 50 kg.

*__3–64.__ The thin ring can be adjusted vertically between three equally long cables from which the 100-kg chandelier is suspended. If the ring remains in the horizontal plane and $z = 600$ mm, determine the tension in each cable.

•**3–65.** The thin ring can be adjusted vertically between three equally long cables from which the 100-kg chandelier is suspended. If the ring remains in the horizontal plane and the tension in each cable is not allowed to exceed 1 kN, determine the smallest allowable distance $z$ required for equilibrium.

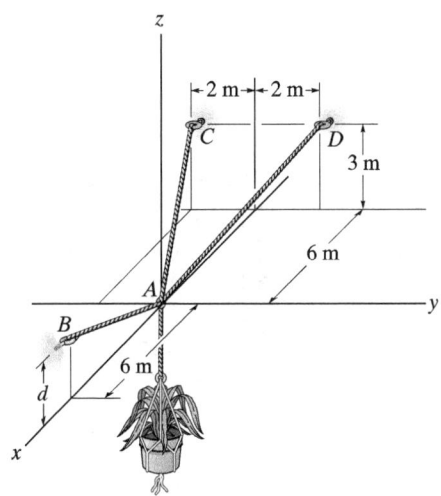

Probs. 3–60/61

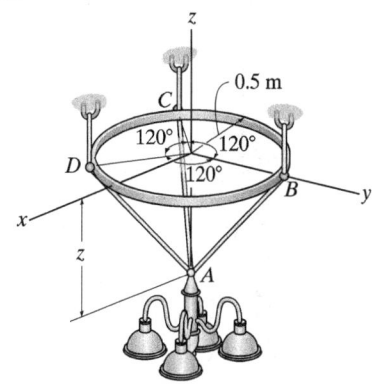

Probs. 3–64/65

**3–66.** The bucket has a weight of 400 N and is being hoisted using three springs, each having an unstretched length of $l_0 = 0.45$ m and stiffness of $k = 800$ N/m. Determine the vertical distance $d$ from the rim to point $A$ for equilibrium.

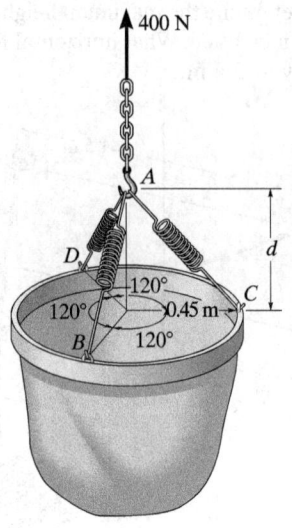

Prob. 3–66

**3–67.** Three cables are used to support a 450-kg ring. Determine the tension in each cable for equilibrium.

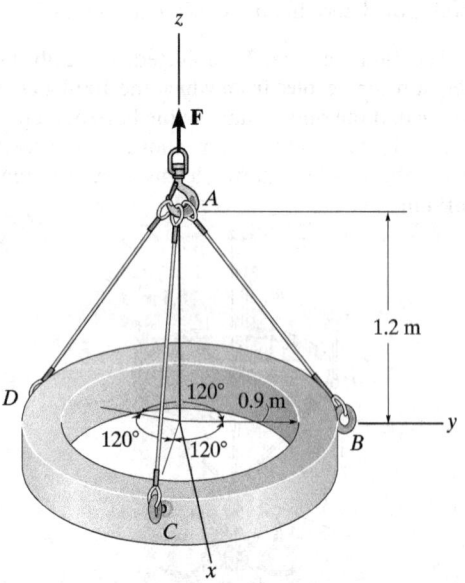

Prob. 3–67

**\*3–68.** The three outer blocks each have a mass of 2 kg, and the central block $E$ has a mass of 3 kg. Determine the sag $s$ for equilibrium of the system.

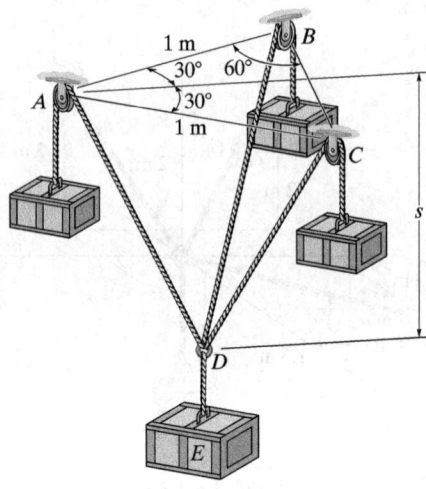

Prob. 3–68

**•3–69.** Determine the angle $\theta$ such that an equal force is developed in legs $OB$ and $OC$. What is the force in each leg if the force is directed along the axis of each leg? The force $F$ lies in the $x$–$y$ plane. The supports at $A$, $B$, $C$ can exert forces in either direction along the attached legs.

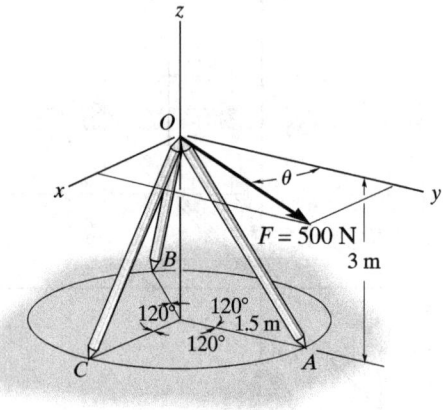

Prob. 3–69

# CHAPTER REVIEW

**Particle Equilibrium**

When a particle is at rest or moves with constant velocity, it is in equilibrium. This requires that all the forces acting on the particle form a zero resultant force.

In order to account for all the forces that act on a particle, it is necessary to draw its free-body diagram. This diagram is an outlined shape of the particle that shows all the forces listed with their known or unknown magnitudes and directions.

$$\mathbf{F}_R = \Sigma \mathbf{F} = \mathbf{0}$$

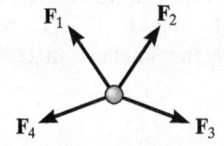

**Two Dimensions**

The two scalar equations of force equilibrium can be applied with reference to an established $x, y$ coordinate system.

$$\Sigma F_x = 0$$
$$\Sigma F_y = 0$$

The tensile force developed in a *continuous cable* that passes over a frictionless pulley must have a *constant* magnitude throughout the cable to keep the cable in equilibrium.

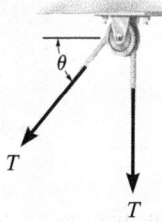

Cable is in tension

If the problem involves a linearly elastic spring, then the stretch or compression $s$ of the spring can be related to the force applied to it.

$$F = ks$$

**Three Dimensions**

If the three-dimensional geometry is difficult to visualize, then the equilibrium equation should be applied using a Cartesian vector analysis. This requires first expressing each force on the free-body diagram as a Cartesian vector. When the forces are summed and set equal to zero, then the **i**, **j**, and **k** components are also zero.

$$\Sigma \mathbf{F} = \mathbf{0}$$

$$\Sigma F_x = 0$$
$$\Sigma F_y = 0$$
$$\Sigma F_z = 0$$

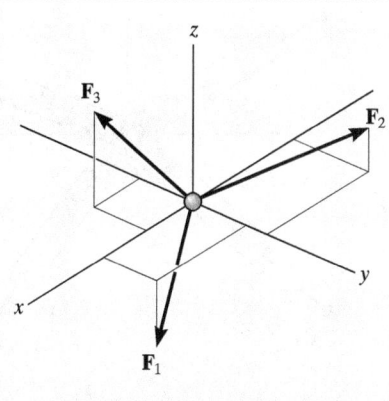

## REVIEW PROBLEMS

**3-70.** The 250-kg crate is hoisted using the ropes $AB$ and $AC$. Each rope can withstand a maximum tension of 12.5 kN before it breaks. If $AB$ always remains horizontal, determine the smallest angle $\theta$ to which the crate can be hoisted.

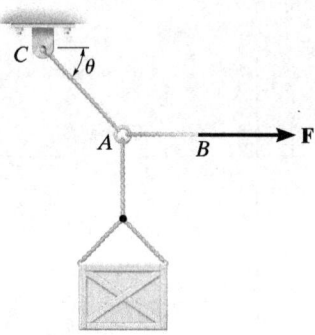

Prob. 3-70

**3-71.** The members of a truss are pin connected at joint $O$. Determine the magnitude of $\mathbf{F}_1$ and its angle $\theta$ for equilibrium. Set $F_2 = 6$ kN.

**\*3-72.** The members of a truss are pin connected at joint $O$. Determine the magnitudes of $\mathbf{F}_1$ and $\mathbf{F}_2$ for equilibrium. Set $\theta = 60°$.

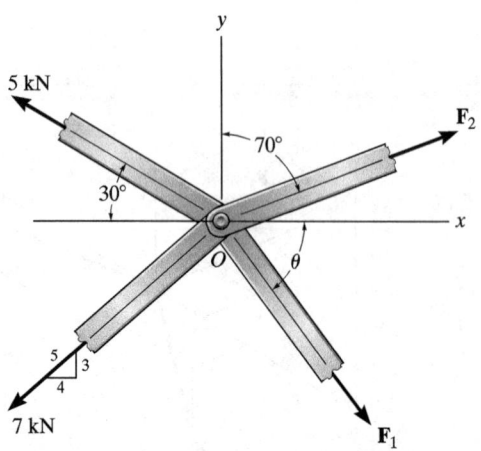

Prob. 3-71/72

**•3-73.** Two electrically charged pith balls, each having a mass of 0.15 g, are suspended from light threads of equal length. Determine the magnitude of the horizontal repulsive force, $F$, acting on each ball if the measured distance between them is $r = 200$ mm.

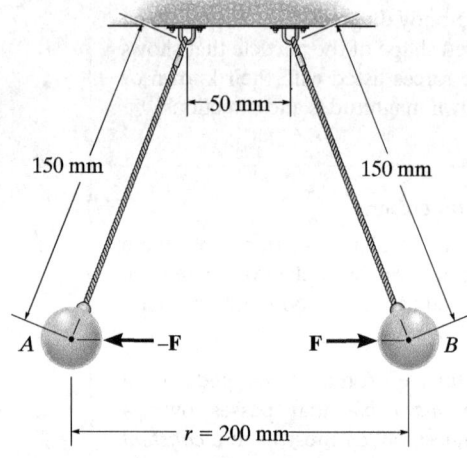

Prob. 3-73

**3-74.** The lamp has a mass of 15 kg and is supported by a pole $AO$ and cables $AB$ and $AC$. If the force in the pole acts along its axis, determine the forces in $AO$, $AB$, and $AC$ for equilibrium.

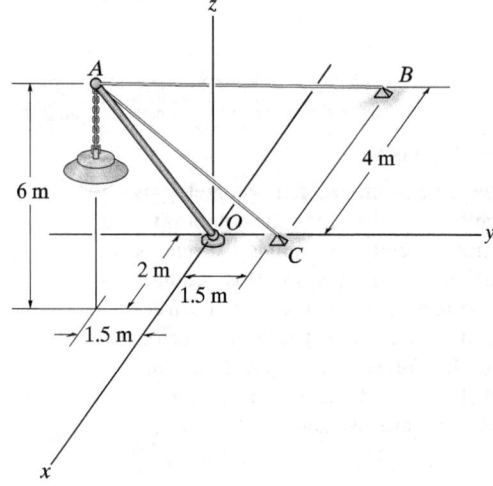

Prob. 3-74

**3–75.** Determine the magnitude of **P** and the coordinate direction angles of $\mathbf{F}_3$ required for equilibrium of the particle. Note that $\mathbf{F}_3$ acts in the octant shown.

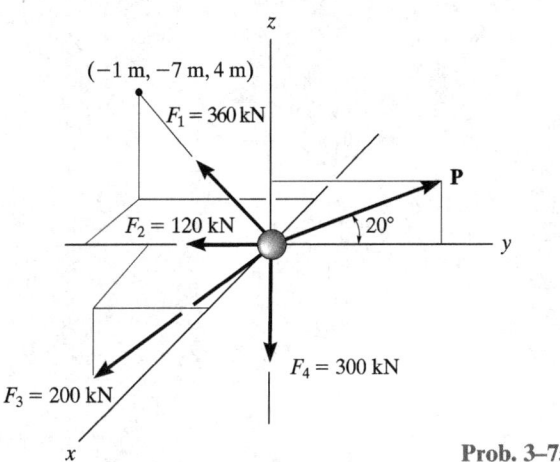

Prob. 3–75

***3–76.** The ring of negligible size is subjected to a vertical force of 1000 N. Determine the longest length $l$ of cord $AC$ such that the tension acting in $AC$ is 800 N. Also, what is the force acting in cord $AB$? *Hint:* Use the equilibrium condition to determine the required angle $\theta$ for attachment, then determine $l$ using trigonometry applied to $\triangle ABC$.

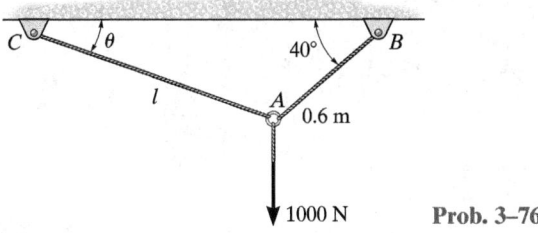

Prob. 3–76

**•3–77.** Determine the magnitudes of $\mathbf{F}_1$, $\mathbf{F}_2$, and $\mathbf{F}_3$ for equilibrium of the particle.

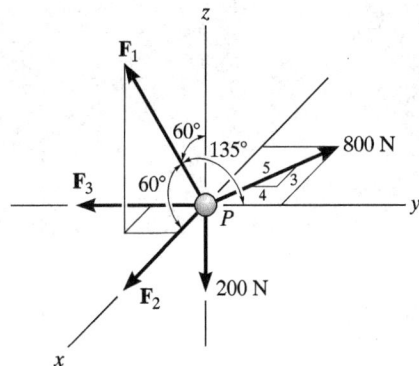

Prob. 3–77

**3–78.** Determine the force in each cable needed to support the 2.5-kN load.

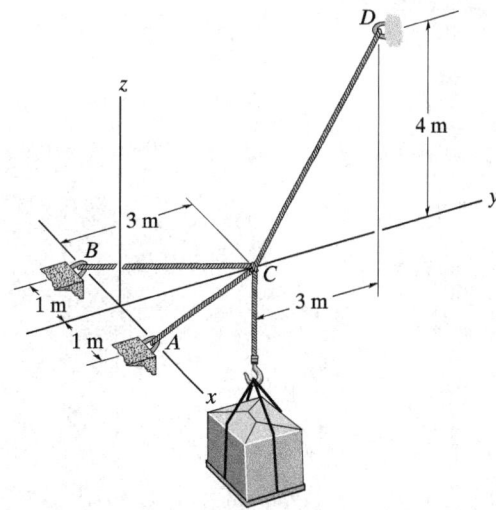

Prob. 3–78

**3–79.** The joint of a space frame is subjected to four member forces. Member $OA$ lies in the $x$–$y$ plane and member $OB$ lies in the $y$–$z$ plane. Determine the forces acting in each of the members required for equilibrium of the joint.

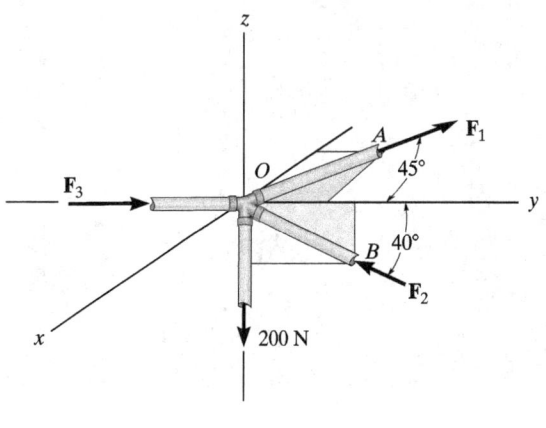

Prob. 3–79

Application of forces to the handles of these wrenches will produce a tendency to rotate each wrench about its end. It is important to know how to calculate this effect and, in some cases, to be able to simplify this system to its resultants.

# Force System Resultants

# 4

## CHAPTER OBJECTIVES

- To discuss the concept of the moment of a force and show how to calculate it in two and three dimensions.
- To provide a method for finding the moment of a force about a specified axis.
- To define the moment of a couple.
- To present methods for determining the resultants of nonconcurrent force systems.
- To indicate how to reduce a simple distributed loading to a resultant force having a specified location.

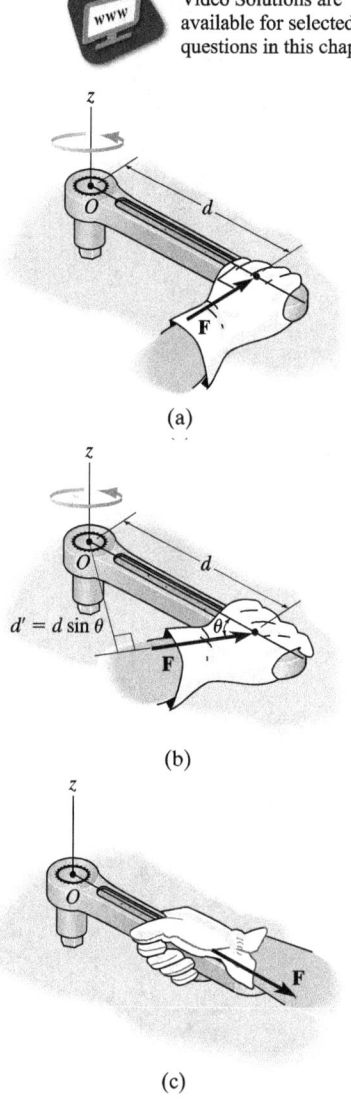

Video Solutions are available for selected questions in this chapter.

## 4.1 Moment of a Force— Scalar Formulation

When a force is applied to a body it will produce a tendency for the body to rotate about a point that is not on the line of action of the force. This tendency to rotate is sometimes called a *torque*, but most often it is called the moment of a force or simply the *moment*. For example, consider a wrench used to unscrew the bolt in Fig. 4–1a. If a force is applied to the handle of the wrench it will tend to turn the bolt about point $O$ (or the $z$ axis). The magnitude of the moment is directly proportional to the magnitude of **F** and the perpendicular distance or *moment arm* $d$. The larger the force or the longer the moment arm, the greater the moment or turning effect. Note that if the force **F** is applied at an angle $\theta \neq 90°$, Fig. 4–1b, then it will be more difficult to turn the bolt since the moment arm $d' = d \sin \theta$ will be smaller than $d$. If **F** is applied along the wrench, Fig. 4–1c, its moment arm will be zero since the line of action of **F** will intersect point $O$ (the $z$ axis). As a result, the moment of **F** about $O$ is also zero and no turning can occur.

Fig. 4–1

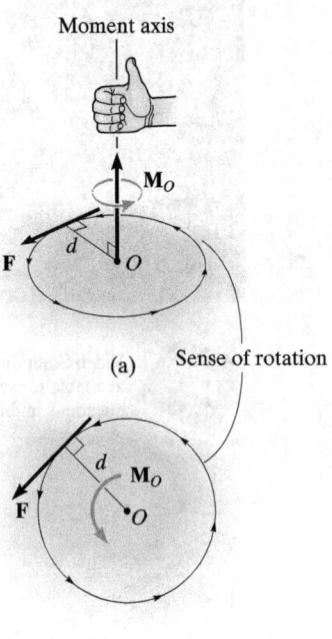

We can generalize the above discussion and consider the force **F** and point $O$ which lie in the shaded plane as shown in Fig. 4–2a. The moment $\mathbf{M}_O$ about point $O$, or about an axis passing through $O$ and perpendicular to the plane, is a *vector quantity* since it has a specified magnitude and direction.

### Magnitude.
The magnitude of $\mathbf{M}_O$ is

$$M_O = Fd \qquad (4\text{–}1)$$

where $d$ is the *moment arm* or *perpendicular distance* from the axis at point $O$ to the line of action of the force. Units of moment magnitude consist of force times distance, e.g., N · m or lb · ft.

### Direction.
The direction of $\mathbf{M}_O$ is defined by its *moment axis*, which is perpendicular to the plane that contains the force **F** and its moment arm $d$. The right-hand rule is used to establish the sense of direction of $\mathbf{M}_O$. According to this rule, the natural curl of the fingers of the right hand, as they are drawn towards the palm, represent the tendency for rotation caused by the moment. As this action is performed, the thumb of the right hand will give the directional sense of $\mathbf{M}_O$, Fig. 4–2a. Notice that the moment vector is represented three-dimensionally by a curl around an arrow. In two dimensions this vector is represented only by the curl as in Fig. 4–2b. Since in this case the moment will tend to cause a counterclockwise rotation, the moment vector is actually directed out of the page.

**Fig. 4–2**

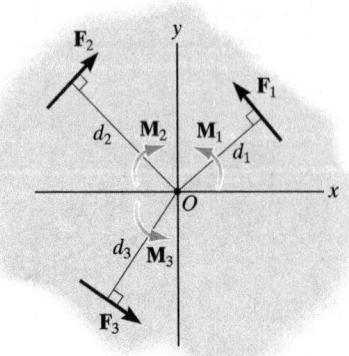

Fig. 4–3

### Resultant Moment.
For two-dimensional problems, where all the forces lie within the $x$–$y$ plane, Fig. 4–3, the resultant moment $(\mathbf{M}_R)_O$ about point $O$ (the $z$ axis) can be determined by *finding the algebraic sum of the moments caused by all the forces in the system*. As a convention, we will generally consider *positive moments* as *counterclockwise* since they are directed along the positive $z$ axis (out of the page). *Clockwise moments will be negative.* Doing this, the directional sense of each moment can be represented by a *plus or minus* sign. Using this sign convention, the resultant moment in Fig. 4–3 is therefore

$$\zeta + (M_R)_O = \Sigma Fd; \qquad (M_R)_O = F_1 d_1 - F_2 d_2 + F_3 d_3$$

If the numerical result of this sum is a positive scalar, $(\mathbf{M}_R)_O$ will be a counterclockwise moment (out of the page); and if the result is negative, $(\mathbf{M}_R)_O$ will be a clockwise moment (into the page).

## EXAMPLE 4.1

For each case illustrated in Fig. 4–4, determine the moment of the force about point $O$.

### SOLUTION (SCALAR ANALYSIS)
The line of action of each force is extended as a dashed line in order to establish the moment arm $d$. Also illustrated is the tendency of rotation of the member as caused by the force. Furthermore, the orbit of the force about $O$ is shown as a colored curl. Thus,

Fig. 4–4a    $M_O = (100\text{ N})(2\text{ m}) = 200\text{ N} \cdot \text{m} \,\downarrow\!\circlearrowright$    Ans.

Fig. 4–4b    $M_O = (50\text{ N})(0.75\text{ m}) = 37.5\text{ N} \cdot \text{m} \,\downarrow\!\circlearrowright$    Ans.

Fig. 4–4c    $M_O = (40\text{ kN})(4\text{ m} + 2\cos 30°\text{ m}) = 229\text{ kN} \cdot \text{m} \,\downarrow\!\circlearrowright$    Ans.

Fig. 4–4d    $M_O = (60\text{ kN})(1\sin 45°\text{ m}) = 42.4\text{ kN} \cdot \text{m} \,\circlearrowleft$    Ans.

Fig. 4–4e    $M_O = (7\text{ kN})(4\text{ m} - 1\text{ m}) = 21.0\text{ kN} \cdot \text{m} \,\circlearrowleft$    Ans.

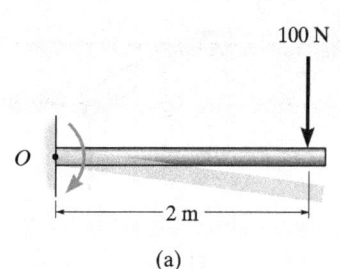

(a)

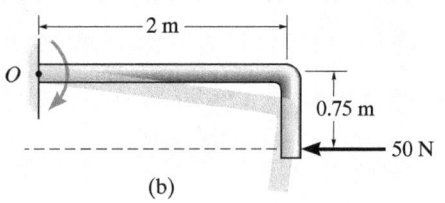

(b)

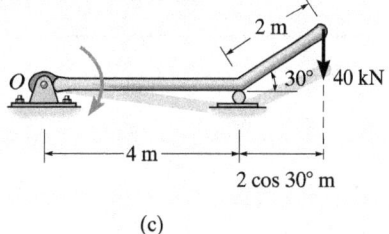

(c)

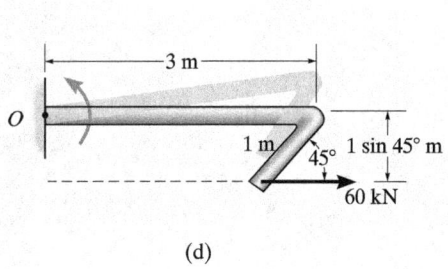

(d)

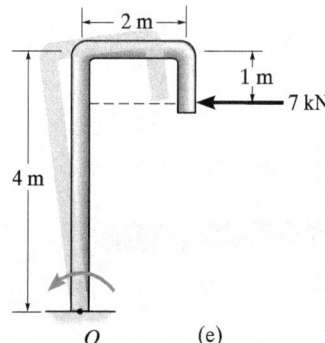

(e)

Fig. 4–4

## EXAMPLE 4.2

Determine the resultant moment of the four forces acting on the rod shown in Fig. 4–5 about point $O$.

### SOLUTION

Assuming that positive moments act in the $+\mathbf{k}$ direction, i.e., counterclockwise, we have

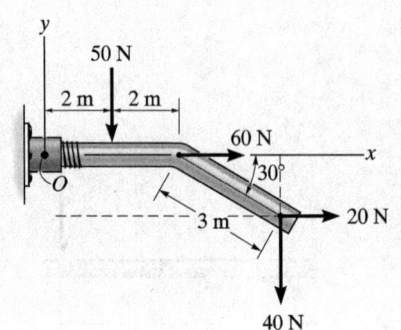

Fig. 4–5

$$\zeta + M_{R_O} = \Sigma Fd;$$
$$M_{R_O} = -50 \text{ N}(2 \text{ m}) + 60 \text{ N}(0) + 20 \text{ N}(3 \sin 30° \text{ m})$$
$$\quad -40 \text{ N}(4 \text{ m} + 3 \cos 30° \text{ m})$$
$$M_{R_O} = -334 \text{ N} \cdot \text{m} = 334 \text{ N} \cdot \text{m} \,\rangle \qquad Ans.$$

For this calculation, note how the moment-arm distances for the 20-N and 40-N forces are established from the extended (dashed) lines of action of each of these forces.

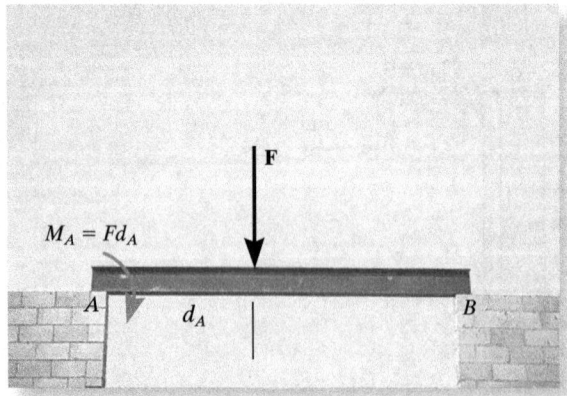

As illustrated by the example problems, the moment of a force does not always cause a rotation. For example, the force **F** tends to rotate the beam clockwise about its support at $A$ with a moment $M_A = Fd_A$. The actual rotation would occur if the support at $B$ were removed.

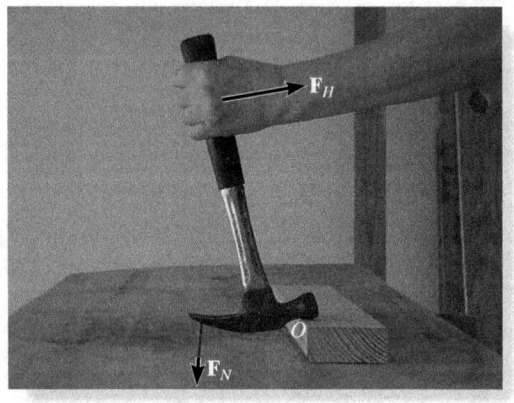

The ability to remove the nail will require the moment of $\mathbf{F}_H$ about point $O$ to be larger than the moment of the force $\mathbf{F}_N$ about $O$ that is needed to pull the nail out.

## 4.2 Cross Product

The moment of a force will be formulated using Cartesian vectors in the next section. Before doing this, however, it is first necessary to expand our knowledge of vector algebra and introduce the cross-product method of vector multiplication.

The *cross product* of two vectors **A** and **B** yields the vector **C**, which is written

$$\mathbf{C} = \mathbf{A} \times \mathbf{B} \qquad (4\text{--}2)$$

and is read "**C** equals **A** cross **B**."

**Magnitude.** The *magnitude* of **C** is defined as the product of the magnitudes of **A** and **B** and the sine of the angle $\theta$ between their tails ($0° \leq \theta \leq 180°$). Thus, $C = AB \sin \theta$.

**Direction.** Vector **C** has a *direction* that is perpendicular to the plane containing **A** and **B** such that **C** is specified by the right-hand rule; i.e., curling the fingers of the right hand from vector **A** (cross) to vector **B**, the thumb points in the direction of **C**, as shown in Fig. 4–6.

Knowing both the magnitude and direction of **C**, we can write

$$\mathbf{C} = \mathbf{A} \times \mathbf{B} = (AB \sin \theta)\mathbf{u}_C \qquad (4\text{--}3)$$

where the scalar $AB \sin \theta$ defines the *magnitude* of **C** and the unit vector $\mathbf{u}_C$ defines the *direction* of **C**. The terms of Eq. 4–3 are illustrated graphically in Fig. 4–6.

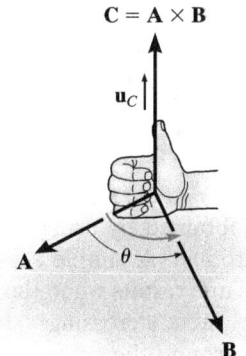

Fig. 4–6

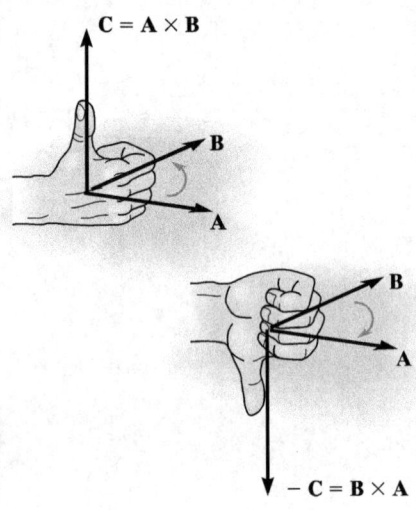

Fig. 4–7

## Laws of Operation.

- The commutative law is *not* valid; i.e., $\mathbf{A} \times \mathbf{B} \neq \mathbf{B} \times \mathbf{A}$. Rather,

$$\mathbf{A} \times \mathbf{B} = -\mathbf{B} \times \mathbf{A}$$

This is shown in Fig. 4–7 by using the right-hand rule. The cross product $\mathbf{B} \times \mathbf{A}$ yields a vector that has the same magnitude but acts in the opposite direction to $\mathbf{C}$; i.e., $\mathbf{B} \times \mathbf{A} = -\mathbf{C}$.

- If the cross product is multiplied by a scalar $a$, it obeys the associative law;

$$a(\mathbf{A} \times \mathbf{B}) = (a\mathbf{A}) \times \mathbf{B} = \mathbf{A} \times (a\mathbf{B}) = (\mathbf{A} \times \mathbf{B})a$$

This property is easily shown since the magnitude of the resultant vector ($|a|AB \sin \theta$) and its direction are the same in each case.

- The vector cross product also obeys the distributive law of addition,

$$\mathbf{A} \times (\mathbf{B} + \mathbf{D}) = (\mathbf{A} \times \mathbf{B}) + (\mathbf{A} \times \mathbf{D})$$

- The proof of this identity is left as an exercise (see Prob. 4–1). It is important to note that *proper order* of the cross products must be maintained, since they are not commutative.

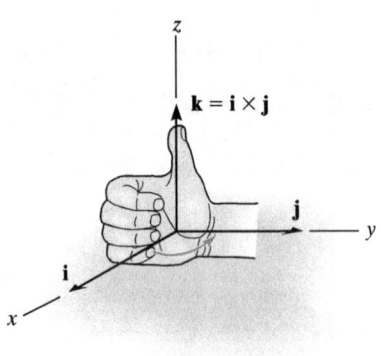

Fig. 4–8

**Cartesian Vector Formulation.** Equation 4–3 may be used to find the cross product of any pair of Cartesian unit vectors. For example, to find $\mathbf{i} \times \mathbf{j}$, the magnitude of the resultant vector is $(i)(j)(\sin 90°) = (1)(1)(1) = 1$, and its direction is determined using the right-hand rule. As shown in Fig. 4–8, the resultant vector points in the $+\mathbf{k}$ direction. Thus, $\mathbf{i} \times \mathbf{j} = (1)\mathbf{k}$. In a similar manner,

$$\mathbf{i} \times \mathbf{j} = \mathbf{k} \quad \mathbf{i} \times \mathbf{k} = -\mathbf{j} \quad \mathbf{i} \times \mathbf{i} = 0$$
$$\mathbf{j} \times \mathbf{k} = \mathbf{i} \quad \mathbf{j} \times \mathbf{i} = -\mathbf{k} \quad \mathbf{j} \times \mathbf{j} = 0$$
$$\mathbf{k} \times \mathbf{i} = \mathbf{j} \quad \mathbf{k} \times \mathbf{j} = -\mathbf{i} \quad \mathbf{k} \times \mathbf{k} = 0$$

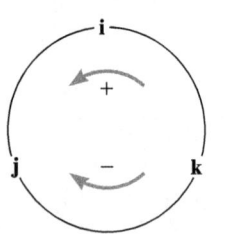

Fig. 4–9

These results should *not* be memorized; rather, it should be clearly understood how each is obtained by using the right-hand rule and the definition of the cross product. A simple scheme shown in Fig. 4–9 is helpful for obtaining the same results when the need arises. If the circle is constructed as shown, then "crossing" two unit vectors in a *counterclockwise* fashion around the circle yields the *positive* third unit vector; e.g., $\mathbf{k} \times \mathbf{i} = \mathbf{j}$. "Crossing" *clockwise*, a *negative* unit vector is obtained; e.g., $\mathbf{i} \times \mathbf{k} = -\mathbf{j}$.

Let us now consider the cross product of two general vectors **A** and **B** which are expressed in Cartesian vector form. We have

$$\mathbf{A} \times \mathbf{B} = (A_x\mathbf{i} + A_y\mathbf{j} + A_z\mathbf{k}) \times (B_x\mathbf{i} + B_y\mathbf{j} + B_z\mathbf{k})$$
$$= A_xB_x(\mathbf{i} \times \mathbf{i}) + A_xB_y(\mathbf{i} \times \mathbf{j}) + A_xB_z(\mathbf{i} \times \mathbf{k})$$
$$+ A_yB_x(\mathbf{j} \times \mathbf{i}) + A_yB_y(\mathbf{j} \times \mathbf{j}) + A_yB_z(\mathbf{j} \times \mathbf{k})$$
$$+ A_zB_x(\mathbf{k} \times \mathbf{i}) + A_zB_y(\mathbf{k} \times \mathbf{j}) + A_zB_z(\mathbf{k} \times \mathbf{k})$$

Carrying out the cross-product operations and combining terms yields

$$\mathbf{A} \times \mathbf{B} = (A_yB_z - A_zB_y)\mathbf{i} - (A_xB_z - A_zB_x)\mathbf{j} + (A_xB_y - A_yB_x)\mathbf{k} \quad (4\text{-}4)$$

This equation may also be written in a more compact determinant form as

$$\mathbf{A} \times \mathbf{B} = \begin{vmatrix} \mathbf{i} & \mathbf{j} & \mathbf{k} \\ A_x & A_y & A_z \\ B_x & B_y & B_z \end{vmatrix} \quad (4\text{-}5)$$

Thus, to find the cross product of any two Cartesian vectors **A** and **B**, it is necessary to expand a determinant whose first row of elements consists of the unit vectors **i**, **j**, and **k** and whose second and third rows represent the *x, y, z* components of the two vectors **A** and **B**, respectively.*

---

*A determinant having three rows and three columns can be expanded using three minors, each of which is multiplied by one of the three terms in the first row. There are four elements in each minor, for example,

$$\begin{vmatrix} A_{11} & A_{12} \\ A_{21} & A_{22} \end{vmatrix}$$

By *definition*, this determinant notation represents the terms $(A_{11}A_{22} - A_{12}A_{21})$, which is simply the product of the two elements intersected by the arrow slanting downward to the right $(A_{11}A_{22})$ *minus* the product of the two elements intersected by the arrow slanting downward to the left $(A_{12}A_{21})$. For a $3 \times 3$ determinant, such as Eq. 4–5, the three minors can be generated in accordance with the following scheme:

For element **i**: $\begin{vmatrix} \mathbf{i} & \mathbf{j} & \mathbf{k} \\ A_x & A_y & A_z \\ B_x & B_y & B_z \end{vmatrix} = \mathbf{i}(A_yB_z - A_zB_y)$

Remember the negative sign

For element **j**: $\begin{vmatrix} \mathbf{i} & \mathbf{j} & \mathbf{k} \\ A_x & A_y & A_z \\ B_x & B_y & B_z \end{vmatrix} = -\mathbf{j}(A_xB_z - A_zB_x)$

For element **k**: $\begin{vmatrix} \mathbf{i} & \mathbf{j} & \mathbf{k} \\ A_x & A_y & A_z \\ B_x & B_y & B_z \end{vmatrix} = \mathbf{k}(A_xB_y - A_yB_x)$

Adding the results and noting that the **j** element *must include the minus sign* yields the expanded form of $\mathbf{A} \times \mathbf{B}$ given by Eq. 4–4.

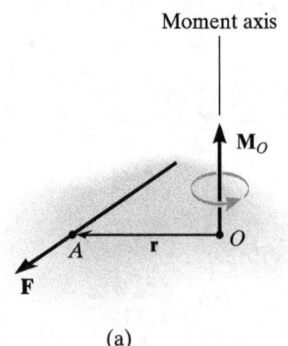

(a)

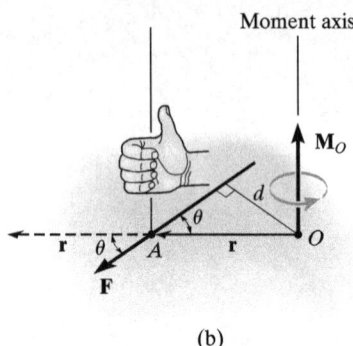

(b)

**Fig. 4–10**

## 4.3 Moment of a Force—Vector Formulation

The moment of a force **F** about point $O$, or actually about the moment axis passing through $O$ and perpendicular to the plane containing $O$ and **F**, Fig. 4–10a, can be expressed using the vector cross product, namely,

$$\mathbf{M}_O = \mathbf{r} \times \mathbf{F} \qquad (4\text{–}6)$$

Here **r** represents a position vector directed *from O to any point* on the line of action of **F**. We will now show that indeed the moment $\mathbf{M}_O$, when determined by this cross product, has the proper magnitude and direction.

**Magnitude.** The magnitude of the cross product is defined from Eq. 4–3 as $M_O = rF \sin \theta$, where the angle $\theta$ is measured between the *tails* of **r** and **F**. To establish this angle, **r** must be treated as a sliding vector so that $\theta$ can be constructed properly, Fig. 4–10b. Since the moment arm $d = r \sin \theta$, then

$$M_O = rF \sin \theta = F(r \sin \theta) = Fd$$

which agrees with Eq. 4–1.

**Direction.** The direction and sense of $\mathbf{M}_O$ in Eq. 4–6 are determined by the right-hand rule as it applies to the cross product. Thus, sliding **r** to the dashed position and curling the right-hand fingers from **r** toward **F**, "**r** cross **F**," the thumb is directed upward or perpendicular to the plane containing **r** and **F** and this is in the *same direction* as $\mathbf{M}_O$, the moment of the force about point $O$, Fig. 4–10b. Note that the "curl" of the fingers, like the curl around the moment vector, indicates the sense of rotation caused by the force. Since the cross product does not obey the commutative law, the order of $\mathbf{r} \times \mathbf{F}$ must be maintained to produce the correct sense of direction for $\mathbf{M}_O$.

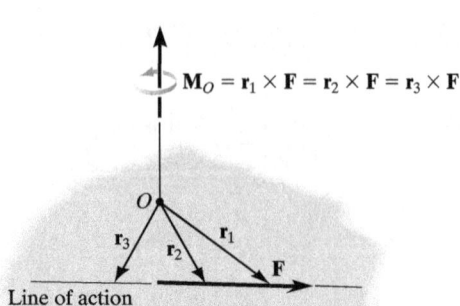

Line of action

**Fig. 4–11**

**Principle of Transmissibility.** The cross product operation is often used in three dimensions since the perpendicular distance or moment arm from point $O$ to the line of action of the force is not needed. In other words, we can use any position vector **r** measured from point $O$ to any point on the line of action of the force **F**, Fig. 4–11. Thus,

$$\mathbf{M}_O = \mathbf{r}_1 \times \mathbf{F} = \mathbf{r}_2 \times \mathbf{F} = \mathbf{r}_3 \times \mathbf{F}$$

Since **F** can be applied at any point along its line of action and still create this *same moment* about point $O$, then **F** can be considered a *sliding vector*. This property is called the *principle of transmissibility* of a force.

## 4.3 Moment of a Force—Vector Formulation

**Cartesian Vector Formulation.** If we establish $x, y, z$ coordinate axes, then the position vector **r** and force **F** can be expressed as Cartesian vectors, Fig. 4–12a. Applying Eq. 4–5 we have

$$\mathbf{M}_O = \mathbf{r} \times \mathbf{F} = \begin{vmatrix} \mathbf{i} & \mathbf{j} & \mathbf{k} \\ r_x & r_y & r_z \\ F_x & F_y & F_z \end{vmatrix} \quad (4\text{–}7)$$

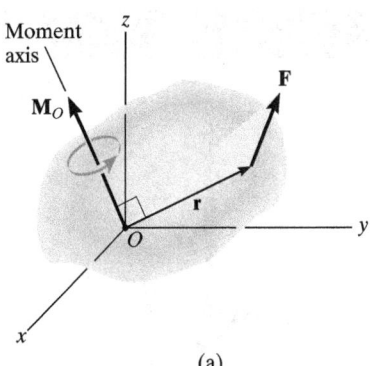

(a)

where

$r_x, r_y, r_z$ represent the $x, y, z$ components of the position vector drawn from point $O$ to *any point* on the line of action of the force

$F_x, F_y, F_z$ represent the $x, y, z$ components of the force vector

If the determinant is expanded, then like Eq. 4–4 we have

$$\mathbf{M}_O = (r_y F_z - r_z F_y)\mathbf{i} - (r_x F_z - r_z F_x)\mathbf{j} + (r_x F_y - r_y F_x)\mathbf{k} \quad (4\text{–}8)$$

The physical meaning of these three moment components becomes evident by studying Fig. 4–12b. For example, the **i** component of $\mathbf{M}_O$ can be determined from the moments of $\mathbf{F}_x$, $\mathbf{F}_y$, and $\mathbf{F}_z$ about the $x$ axis. The component $\mathbf{F}_x$ does *not* create a moment or tendency to cause turning about the $x$ axis since this force is *parallel* to the $x$ axis. The line of action of $\mathbf{F}_y$ passes through point $B$, and so the magnitude of the moment of $\mathbf{F}_y$ about point $A$ on the $x$ axis is $r_z F_y$. By the right-hand rule this component acts in the *negative* **i** direction. Likewise, $\mathbf{F}_z$ passes through point $C$ and so it contributes a moment component of $r_y F_z \mathbf{i}$ about the axis. Thus, $(M_O)_x = (r_y F_z - r_z F_y)$ as shown in Eq. 4–8. As an exercise, establish the **j** and **k** components of $\mathbf{M}_O$ in this manner and show that indeed the expanded form of the determinant, Eq. 4–8, represents the moment of **F** about point $O$. Once $\mathbf{M}_O$ is determined, realize that it will always be *perpendicular* to the shaded plane containing vectors **r** and **F**, Fig. 4–12a.

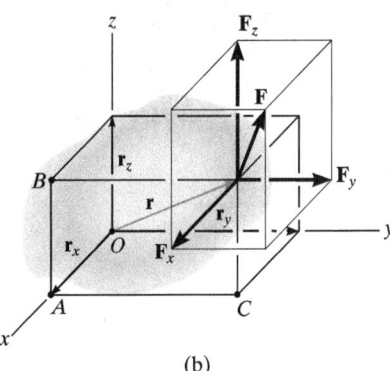

(b)

Fig. 4–12

**Resultant Moment of a System of Forces.** If a body is acted upon by a system of forces, Fig. 4–13, the resultant moment of the forces about point O can be determined by vector addition of the moment of each force. This resultant can be written symbolically as

$$\mathbf{M}_{R_O} = \Sigma(\mathbf{r} \times \mathbf{F}) \quad (4\text{–}9)$$

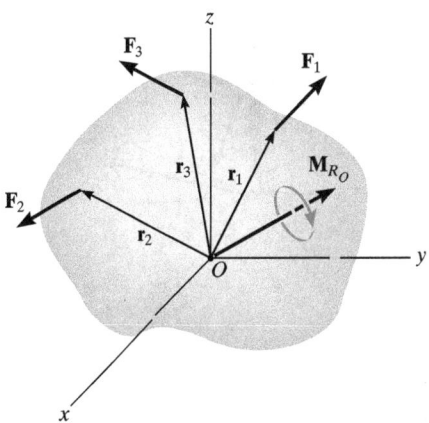

Fig. 4–13

## EXAMPLE 4.3

Determine the moment produced by the force **F** in Fig. 4–14a about point O. Express the result as a Cartesian vector.

### SOLUTION

As shown in Fig. 4–14a, either $\mathbf{r}_A$ or $\mathbf{r}_B$ can be used to determine the moment about point O. These position vectors are

$$\mathbf{r}_A = \{12\mathbf{k}\} \text{ m} \quad \text{and} \quad \mathbf{r}_B = \{4\mathbf{i} + 12\mathbf{j}\} \text{ m}$$

Force **F** expressed as a Cartesian vector is

$$\mathbf{F} = F\mathbf{u}_{AB} = 2 \text{ kN} \left[ \frac{\{4\mathbf{i} + 12\mathbf{j} - 12\mathbf{k}\} \text{ m}}{\sqrt{(4 \text{ m})^2 + (12 \text{ m})^2 + (-12 \text{ m})^2}} \right]$$

$$= \{0.4588\mathbf{i} + 1.376\mathbf{j} - 1.376\mathbf{k}\} \text{ kN}$$

Thus

$$\mathbf{M}_O = \mathbf{r}_A \times \mathbf{F} = \begin{vmatrix} \mathbf{i} & \mathbf{j} & \mathbf{k} \\ 0 & 0 & 12 \\ 0.4588 & 1.376 & -1.376 \end{vmatrix}$$

$$= [0(-1.376) - 12(1.376)]\mathbf{i} - [0(-1.376) - 12(0.4588)]\mathbf{j}$$
$$+ [0(1.376) - 0(0.4588)]\mathbf{k}$$

$$= \{-16.5\mathbf{i} + 5.51\mathbf{j}\} \text{ kN} \cdot \text{m} \qquad \textit{Ans.}$$

or

$$\mathbf{M}_O = \mathbf{r}_B \times \mathbf{F} = \begin{vmatrix} \mathbf{i} & \mathbf{j} & \mathbf{k} \\ 4 & 12 & 0 \\ 0.4588 & 1.376 & -1.376 \end{vmatrix}$$

$$= [12(-1.376) - 0(1.376)]\mathbf{i} - [4(-1.376) - 0(0.4588)]\mathbf{j}$$
$$+ [4(1.376) - 12(0.4588)]\mathbf{k}$$

$$= \{-16.5\mathbf{i} + 5.51\mathbf{j}\} \text{ kN} \cdot \text{m} \qquad \textit{Ans.}$$

**NOTE:** As shown in Fig. 4–14b, $\mathbf{M}_O$ acts perpendicular to the plane that contains **F**, $\mathbf{r}_A$, and $\mathbf{r}_B$. Had this problem been worked using $M_O = Fd$, notice the difficulty that would arise in obtaining the moment arm $d$.

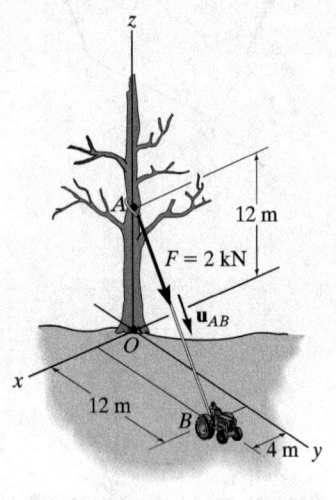

(a)

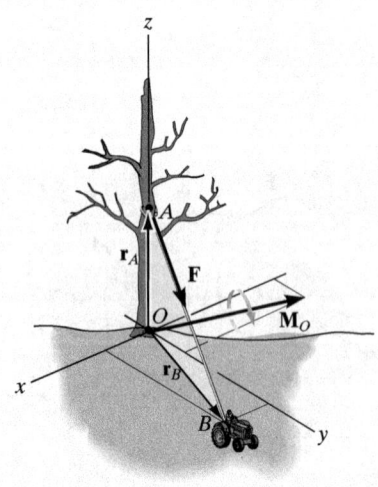

(b)

Fig. 4–14

## EXAMPLE 4.4

Two forces act on the rod shown in Fig. 4–15a. Determine the resultant moment they create about the flange at $O$. Express the result as a Cartesian vector.

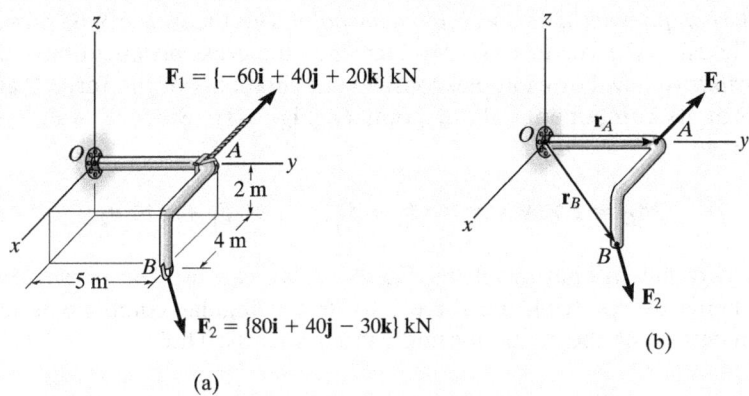

(a)

(b)

### SOLUTION

Position vectors are directed from point $O$ to each force as shown in Fig. 4–15b. These vectors are

$$\mathbf{r}_A = \{5\mathbf{j}\} \text{ m}$$

$$\mathbf{r}_B = \{4\mathbf{i} + 5\mathbf{j} - 2\mathbf{k}\} \text{ m}$$

The resultant moment about $O$ is therefore

$$\mathbf{M}_{R_O} = \Sigma(\mathbf{r} \times \mathbf{F})$$

$$= \mathbf{r}_A \times \mathbf{F}_1 + \mathbf{r}_B \times \mathbf{F}_3$$

$$= \begin{vmatrix} \mathbf{i} & \mathbf{j} & \mathbf{k} \\ 0 & 5 & 0 \\ -60 & 40 & 20 \end{vmatrix} + \begin{vmatrix} \mathbf{i} & \mathbf{j} & \mathbf{k} \\ 4 & 5 & -2 \\ 80 & 40 & -30 \end{vmatrix}$$

$$= [5(20) - 0(40)]\mathbf{i} - [0]\mathbf{j} + [0(40) - (5)(-60)]\mathbf{k}$$

$$+ [5(-30) - (-2)(40)]\mathbf{i} - [4(-30) - (-2)(80)]\mathbf{j} + [4(40) - 5(80)]\mathbf{k}$$

$$= \{30\mathbf{i} - 40\mathbf{j} + 60\mathbf{k}\} \text{ kN} \cdot \text{m} \qquad Ans.$$

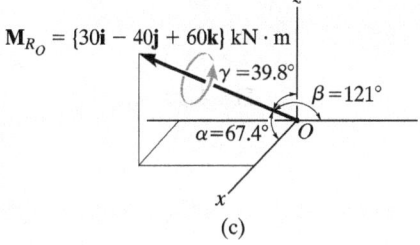

(c)

Fig. 4–15

**NOTE:** This result is shown in Fig. 4–15c. The coordinate direction angles were determined from the unit vector for $\mathbf{M}_{R_O}$. Realize that the two forces tend to cause the rod to rotate about the moment axis in the manner shown by the curl indicated on the moment vector.

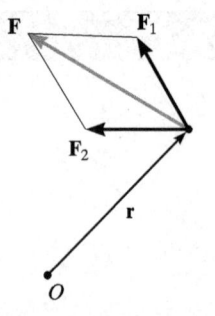

Fig 4–16

## 4.4 Principle of Moments

A concept often used in mechanics is the *principle of moments*, which is sometimes referred to as *Varignon's theorem* since it was originally developed by the French mathematician Varignon (1654–1722). It states that *the moment of a force about a point is equal to the sum of the moments of the components of the force about the point*. This theorem can be proven easily using the vector cross product since the cross product obeys the *distributive law*. For example, consider the moments of the force **F** and two of its components about point $O$. Fig. 4–16. Since $\mathbf{F} = \mathbf{F}_1 + \mathbf{F}_2$ we have

$$\mathbf{M}_O = \mathbf{r} \times \mathbf{F} = \mathbf{r} \times (\mathbf{F}_1 + \mathbf{F}_2) = \mathbf{r} \times \mathbf{F}_1 + \mathbf{r} \times \mathbf{F}_2$$

For two-dimensional problems, Fig. 4–17, we can use the principle of moments by resolving the force into its rectangular components and then determine the moment using a scalar analysis. Thus,

$$M_O = F_x y - F_y x$$

This method is generally easier than finding the same moment using $M_O = Fd$.

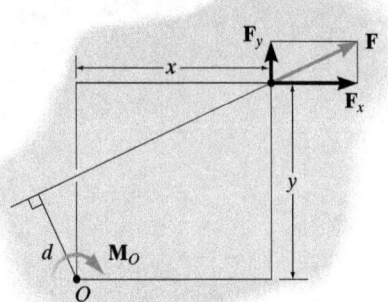

Fig. 4–17

### Important Points

- The moment of a force creates the tendency of a body to turn about an axis passing through a specific point $O$.

- Using the right-hand rule, the sense of rotation is indicated by the curl of the fingers, and the thumb is directed along the moment axis, or line of action of the moment.

- The magnitude of the moment is determined from $M_O = Fd$, where $d$ is called the moment arm, which represents the perpendicular or shortest distance from point $O$ to the line of action of the force.

- In three dimensions the vector cross product is used to determine the moment, i.e., $\mathbf{M}_O = \mathbf{r} \times \mathbf{F}$. Remember that **r** is directed *from point O to any point* on the line of action of **F**.

- The principle of moments states that the moment of a force about a point is equal to the sum of the moments of the force's components about the point. This is a very convenient method to use in two dimensions.

The moment of the applied force **F** about point O is easy to determine if we use the principle of moments. It is simply $M_O = F_x d$.

## EXAMPLE 4.5

Determine the moment of the force in Fig. 4–18a about point O.

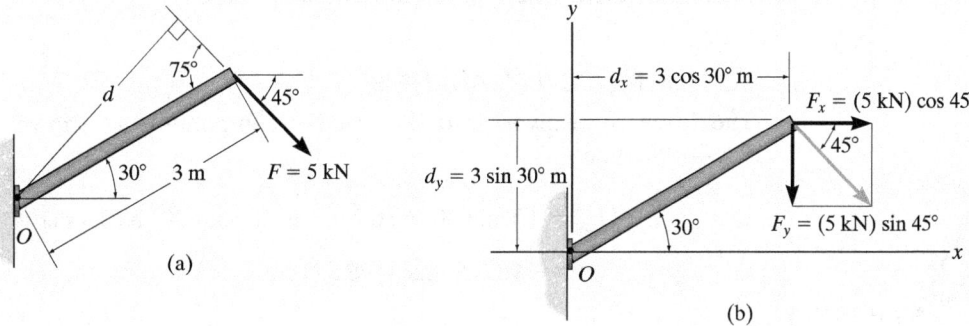

### SOLUTION I
The moment arm $d$ in Fig. 4–18a can be found from trigonometry.

$$d = (3 \text{ m}) \sin 75° = 2.898 \text{ m}$$

Thus,

$$M_O = Fd = (5\text{kN})(2.898 \text{ m}) = 14.5 \text{ kN} \cdot \text{m} \;\downarrow\!\!\!\curvearrowright \quad \text{Ans.}$$

Since the force tends to rotate or orbit clockwise about point $O$, the moment is directed into the page.

### SOLUTION II
The $x$ and $y$ components of the force are indicated in Fig. 4–18b. Considering counterclockwise moments as positive, and applying the principle of moments, we have

$$\zeta + M_O = -F_x d_y - F_y d_x$$
$$= -(5 \cos 45° \text{ kN})(3 \sin 30° \text{ m}) - (5 \sin 45° \text{ kN})(3 \cos 30° \text{ m})$$
$$= -14.5 \text{ kN} \cdot \text{m} = 14.5 \text{ kN} \cdot \text{m} \;\downarrow\!\!\!\curvearrowright \quad \text{Ans.}$$

### SOLUTION III
The $x$ and $y$ axes can be set parallel and perpendicular to the rod's axis as shown in Fig. 4-18c. Here $\mathbf{F}_x$ produces no moment about point $O$ since its line of action passes through this point. Therefore,

$$\zeta + M_O = -F_y d_x$$
$$= -(5 \sin 75° \text{ kN})(3 \text{ m})$$
$$= -14.5 \text{ kN} \cdot \text{m} = 14.5 \text{ kN} \cdot \text{m} \;\downarrow\!\!\!\curvearrowright \quad \text{Ans.}$$

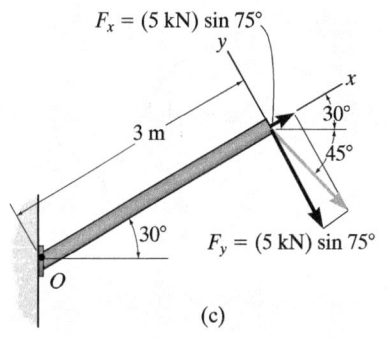

Fig. 4–18

## EXAMPLE 4.6

Force **F** acts at the end of the angle bracket shown in Fig. 4–19a. Determine the moment of the force about point $O$.

### SOLUTION I (SCALAR ANALYSIS)

The force is resolved into its $x$ and $y$ components as shown in Fig. 4–19b, then

$$\zeta + M_O = 400 \sin 30° \text{ N}(0.2 \text{ m}) - 400 \cos 30° \text{ N}(0.4 \text{ m})$$

$$= -98.6 \text{ N} \cdot \text{m} = 98.6 \text{ N} \cdot \text{m} \;\;\rangle$$

or

$$\mathbf{M}_O = \{-98.6\mathbf{k}\} \text{ N} \cdot \text{m} \qquad Ans.$$

### SOLUTION II (VECTOR ANALYSIS)

Using a Cartesian vector approach, the force and position vectors shown in Fig. 4–19c are

$$\mathbf{r} = \{0.4\mathbf{i} - 0.2\mathbf{j}\} \text{ m}$$

$$\mathbf{F} = \{400 \sin 30°\mathbf{i} - 400 \cos 30°\mathbf{j}\} \text{ N}$$

$$= \{200.0\mathbf{i} - 346.4\mathbf{j}\} \text{ N}$$

The moment is therefore

$$\mathbf{M}_O = \mathbf{r} \times \mathbf{F} = \begin{vmatrix} \mathbf{i} & \mathbf{j} & \mathbf{k} \\ 0.4 & -0.2 & 0 \\ 200.0 & -346.4 & 0 \end{vmatrix}$$

$$= 0\mathbf{i} - 0\mathbf{j} + [0.4(-346.4) - (-0.2)(200.0)]\mathbf{k}$$

$$= \{-98.6\mathbf{k}\} \text{ N} \cdot \text{m} \qquad Ans.$$

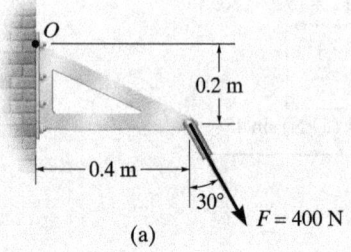

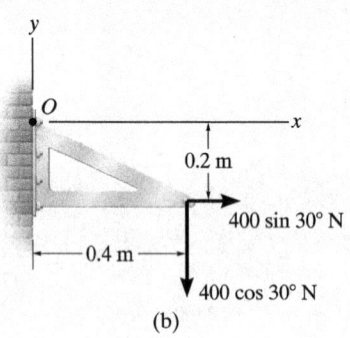

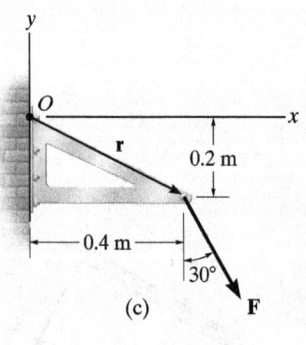

Fig. 4–19

**NOTE:** It is seen that the scalar analysis (Solution I) provides a more *convenient method* for analysis than Solution II since the direction of the moment and the moment arm for each component force are easy to establish. Hence, this method is generally recommended for solving problems displayed in two dimensions, whereas a Cartesian vector analysis is generally recommended only for solving three-dimensional problems.

## FUNDAMENTAL PROBLEMS

**F4–1.** Determine the moment of the force about point $O$.

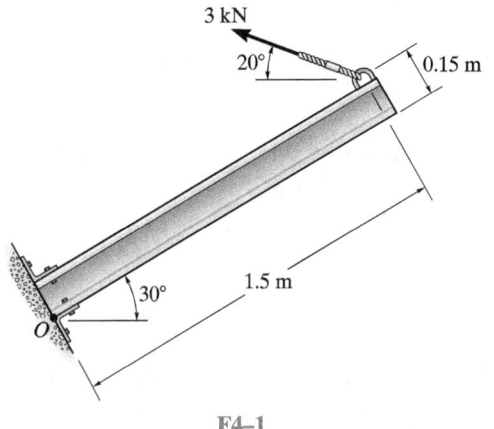

F4–1

**F4–2.** Determine the moment of the force about point $O$.

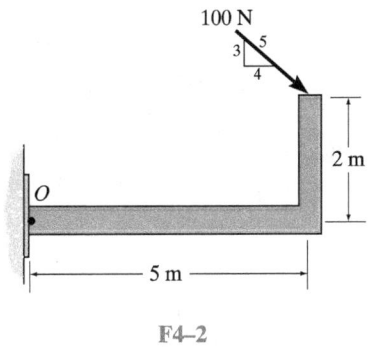

F4–2

**F4–3.** Determine the moment of the force about point $O$.

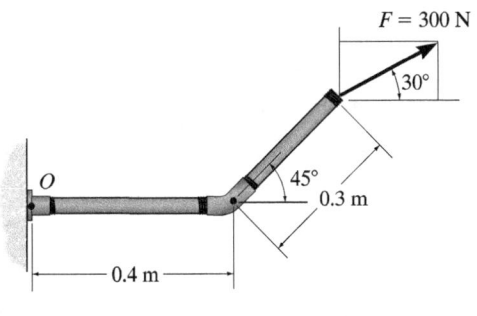

F4–3

**F4–4.** Determine the moment of the force about point $O$.

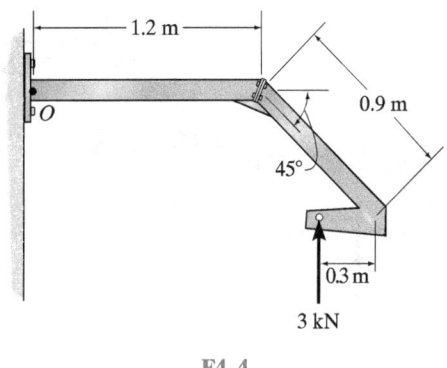

F4–4

**F4–5.** Determine the moment of the force about point $O$. Neglect the thickness of the member.

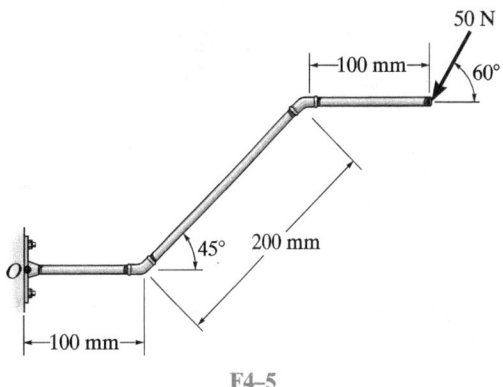

F4–5

**F4–6.** Determine the moment of the force about point $O$.

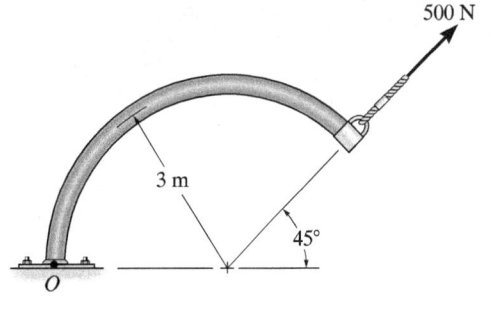

F4–6

**F4–7.** Determine the resultant moment produced by the forces about point $O$.

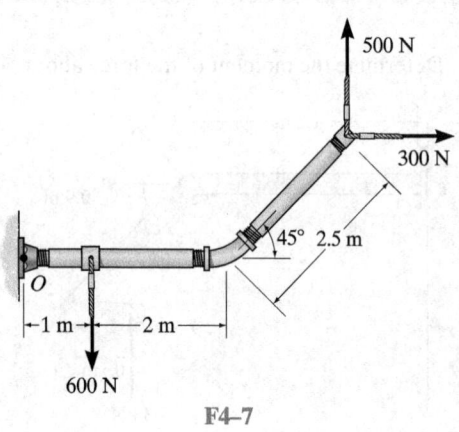

F4–7

**F4–8.** Determine the resultant moment produced by the forces about point $O$.

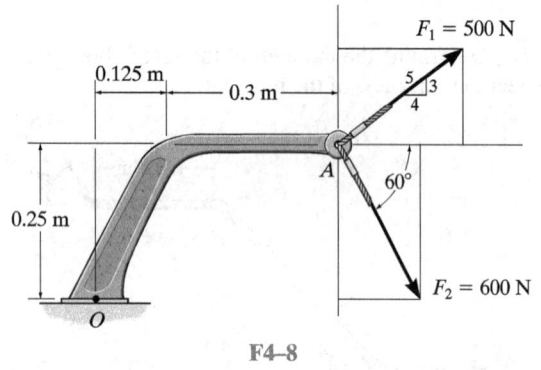

F4–8

**F4–9.** Determine the resultant moment produced by the forces about point $O$.

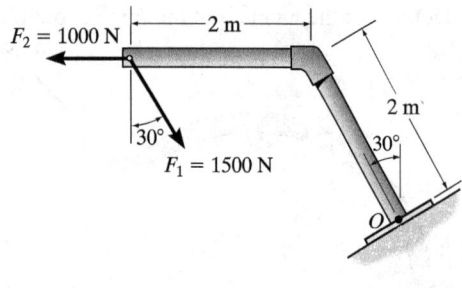

F4–9

**F4–10.** Determine the moment of force $\mathbf{F}$ about point $O$. Express the result as a Cartesian vector.

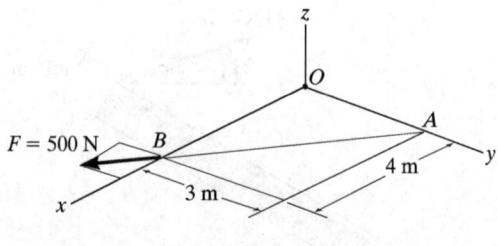

F 4–10

**F4–11.** Determine the moment of force $\mathbf{F}$ about point $O$. Express the result as a Cartesian vector.

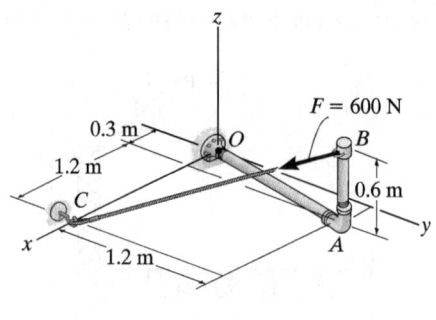

F4–11

**F4–12.** If $\mathbf{F}_1 = \{100\mathbf{i} - 120\mathbf{j} + 75\mathbf{k}\}$ N and $\mathbf{F}_2 = \{-200\mathbf{i} + 250\mathbf{j} + 100\mathbf{k}\}$ N, determine the resultant moment produced by these forces about point $O$. Express the result as a Cartesian vector.

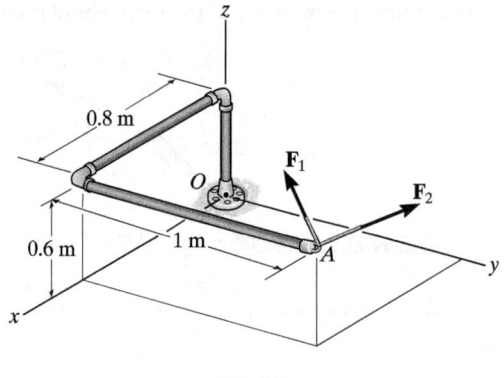

F4–12

## PROBLEMS

**•4–1.** If **A**, **B**, and **D** are given vectors, prove the distributive law for the vector cross product, i.e., **A** × (**B** + **D**) = (**A** × **B**) + (**A** × **D**).

**4–2.** Prove the triple scalar product identity **A** · **B** × **C** = **A** × **B** · **C**.

**4–3.** Given the three nonzero vectors **A**, **B**, and **C**, show that if **A** · (**B** × **C**) = 0, the three vectors *must* lie in the same plane.

**\*4–4.** Two men exert forces of $F = 400$ N and $P = 250$ N on the ropes. Determine the moment of each force about $A$. Which way will the pole rotate, clockwise or counterclockwise?

**•4–5.** If the man at $B$ exerts a force of $P = 150$ N on his rope, determine the magnitude of the force **F** the man at $C$ must exert to prevent the pole from rotating, i.e., so the resultant moment about $A$ of both forces is zero.

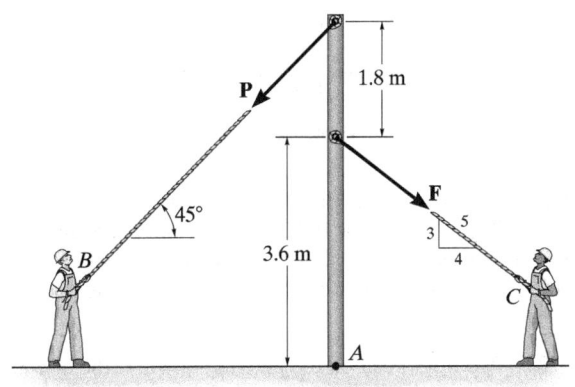

Probs. 4–4/5

**4–6.** If $\theta = 45°$, determine the moment produced by the 4-kN force about point $A$.

**4–7.** If the moment produced by the 4-kN force about point $A$ is 10 kN·m clockwise, determine the angle $\theta$, where $0° \le \theta \le 90°$.

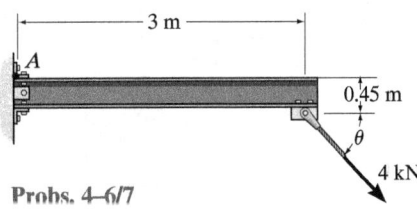

Probs. 4–6/7

**\*4–8.** The handle of the hammer is subjected to the force of $F = 100$ N. Determine the moment of this force about the point $A$.

**•4–9.** In order to pull out the nail at $B$, the force **F** exerted on the handle of the hammer must produce a clockwise moment of 60 N·m about point $A$. Determine the required magnitude of force **F**.

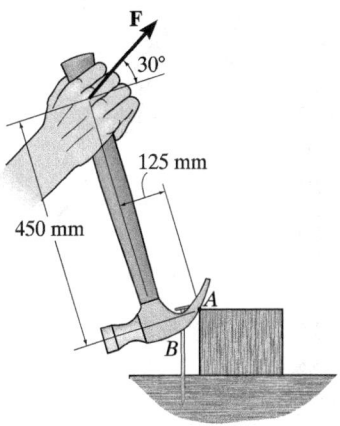

Probs. 4–8/9

**4–10.** The hub of the wheel can be attached to the axle either with negative offset (left) or with positive offset (right). If the tire is subjected to both a normal and radial load as shown, determine the resultant moment of these loads about point $O$ on the axle for both cases.

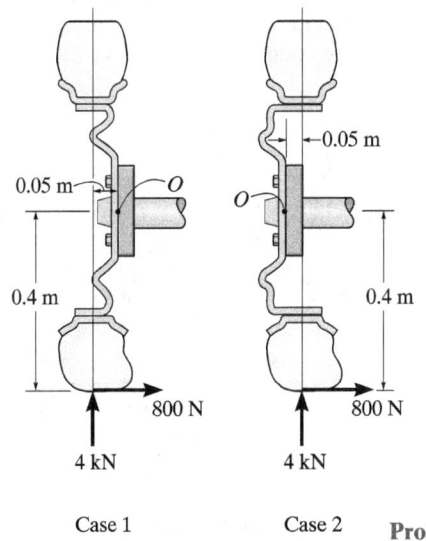

Prob. 4–10

**4–11.** The member is subjected to a force of $F = 6$ kN. If $\theta = 45°$, determine the moment produced by **F** about point $A$.

**\*4–12.** Determine the angle $\theta$ ($0° \leq \theta \leq 180°$) of the force **F** so that it produces a maximum moment and a minimum moment about point $A$. Also, what are the magnitudes of these maximum and minimum moments?

**•4–13.** Determine the moment produced by the force **F** about point $A$ in terms of the angle $\theta$. Plot the graph of $M_A$ versus $\theta$, where $0° \leq \theta \leq 180°$.

**4–15.** The Achilles tendon force of $F_t = 650$ N is mobilized when the man tries to stand on his toes. As this is done, each of his feet is subjected to a reactive force of $N_f = 400$ N. Determine the resultant moment of $\mathbf{F}_t$ and $\mathbf{N}_f$ about the ankle joint $A$.

**\*4–16.** The Achilles tendon force $\mathbf{F}_t$ is mobilized when the man tries to stand on his toes. As this is done, each of his feet is subjected to a reactive force of $N_t = 400$ N. If the resultant moment produced by forces $\mathbf{F}_t$ and $\mathbf{N}_t$ about the ankle joint $A$ is required to be zero, determine the magnitude of $\mathbf{F}_t$.

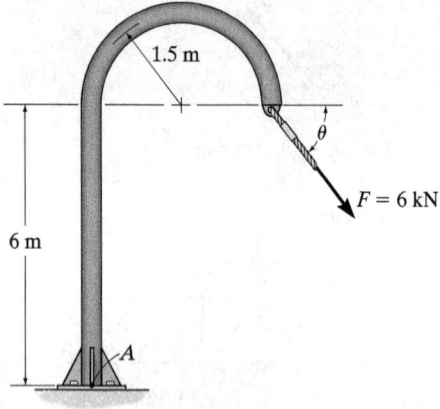

Probs. 4–11/12/13

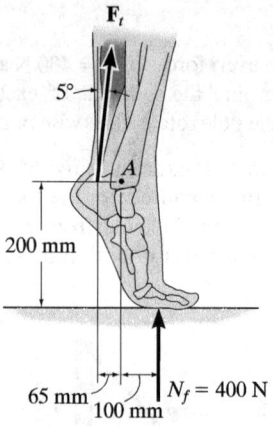

Probs. 4–15/16

**4–14.** Serious neck injuries can occur when a football player is struck in the face guard of his helmet in the manner shown, giving rise to a guillotine mechanism. Determine the moment of the knee force $P = 250$ N about point $A$. What would be the magnitude of the neck force **F** so that it gives the counterbalancing moment about $A$?

**•4–17.** The two boys push on the gate with forces of $F_B = 250$ N and $F_A = 150$ N as shown. Determine the moment of each force about $C$. Which way will the gate rotate, clockwise or counterclockwise? Neglect the thickness of the gate.

**4–18.** Two boys push on the gate as shown. If the boy at $B$ exerts a force of $F_B = 150$ N, determine the magnitude of the force $\mathbf{F}_A$ the boy at $A$ must exert in order to prevent the gate from turning. Neglect the thickness of the gate.

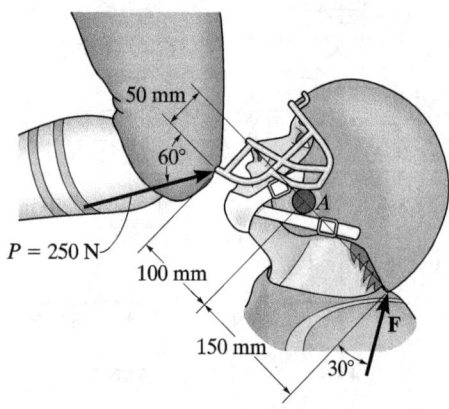

Probs. 4–14

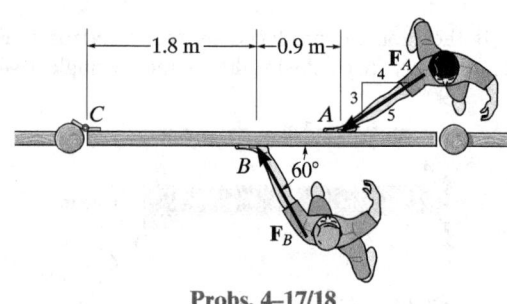

Probs. 4–17/18

**4–19.** The tongs are used to grip the ends of the drilling pipe $P$. Determine the torque (moment) $M_P$ that the applied force $F = 750$ N exerts on the pipe about point $P$ as a function of $\theta$. Plot this moment $M_P$ versus $\theta$ for $0 \le \theta \le 90°$.

**\*4–20.** The tongs are used to grip the ends of the drilling pipe $P$. If a torque (moment) of $M_P = 1200$ N·m is needed at $P$ to turn the pipe, determine the cable force $F$ that must be applied to the tongs. Set $\theta = 30°$.

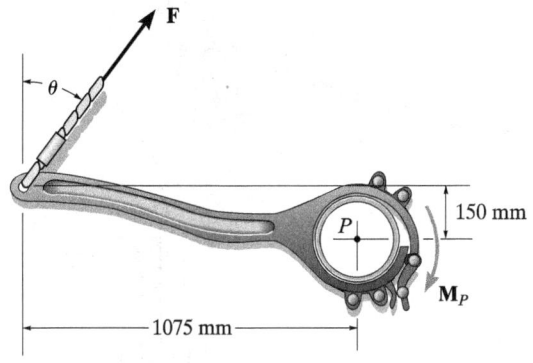

Probs. 4–19/20

**•4–21.** Determine the direction $\theta$ for $0° \le \theta \le 180°$ of the force $\mathbf{F}$ so that it produces the maximum moment about point $A$. Calculate this moment.

**4–22.** Determine the moment of the force $\mathbf{F}$ about point $A$ as a function of $\theta$. Plot the results of $M$ (ordinate) versus $\theta$ (abscissa) for $0° \le \theta \le 180°$.

**4–23.** Determine the minimum moment produced by the force $\mathbf{F}$ about point $A$. Specify the angle $\theta$ ($0° \le \theta \le 180°$).

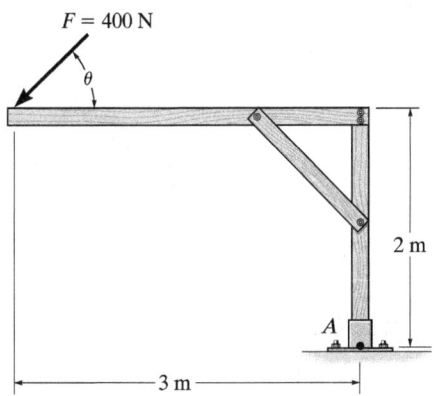

Probs. 4–21/22/23

**\*4–24.** In order to raise the lamp post from the position shown, force $\mathbf{F}$ is applied to the cable. If $F = 1000$ N, determine the moment produced by $\mathbf{F}$ about point $A$.

**•4–25.** In order to raise the lamp post from the position shown, the force $\mathbf{F}$ on the cable must create a counterclockwise moment of $2250$ N·m about point $A$. Determine the magnitude of $\mathbf{F}$ that must be applied to the cable.

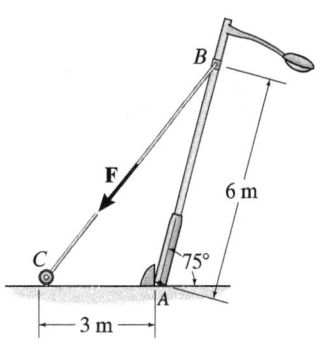

Probs. 4–24/25

**4–26.** The foot segment is subjected to the pull of the two plantarflexor muscles. Determine the moment of each force about the point of contact $A$ on the ground.

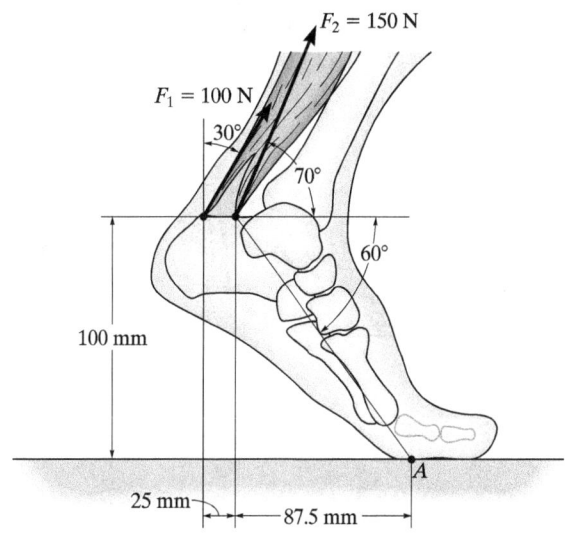

Prob. 4–26

**4–27.** The 70-N force acts on the end of the pipe at B. Determine (a) the moment of this force about point A, and (b) the magnitude and direction of a horizontal force, applied at C, which produces the same moment. Take $\theta = 60°$.

**\*4–28.** The 70-N force acts on the end of the pipe at B. Determine the angles $\theta$ ($0° \leq \theta \leq 180°$) of the force that will produce maximum and minimum moments about point A. What are the magnitudes of these moments?

**4–31.** The rod on the power control mechanism for a business jet is subjected to a force of 80 N. Determine the moment of this force about the bearing at A.

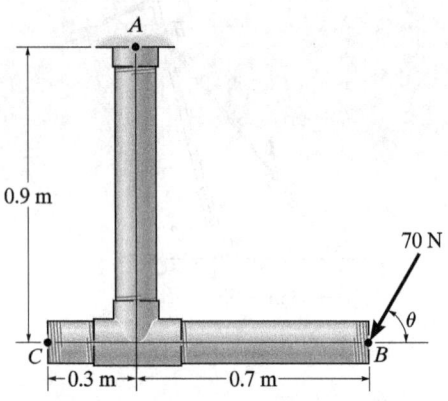

Probs. 4–27/28

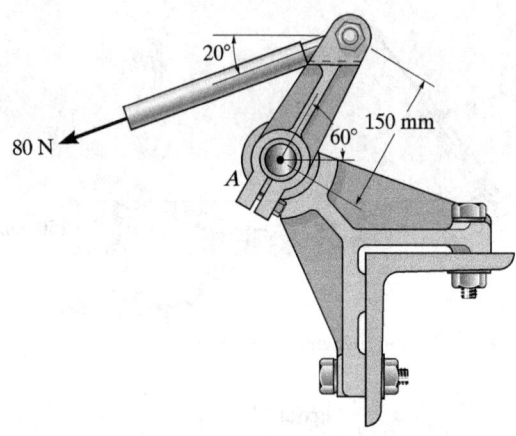

Prob. 4–31

**•4–29.** Determine the moment of each force about the bolt located at A. Take $F_B = 200$ N, $F_C = 250$ N.

**4–30.** If $F_B = 150$ N and $F_C = 225$ N, determine the resultant moment about the bolt located at A.

**\*4–32.** The towline exerts a force of $P = 4$ kN at the end of the 20-m-long crane boom. If $\theta = 30°$, determine the placement $x$ of the hook at A so that this force creates a maximum moment about point O. What is this moment?

**•4–33.** The towline exerts a force of $P = 4$ kN at the end of the 20-m-long crane boom. If $x = 25$ m, determine the position $\theta$ of the boom so that this force creates a maximum moment about point O. What is this moment?

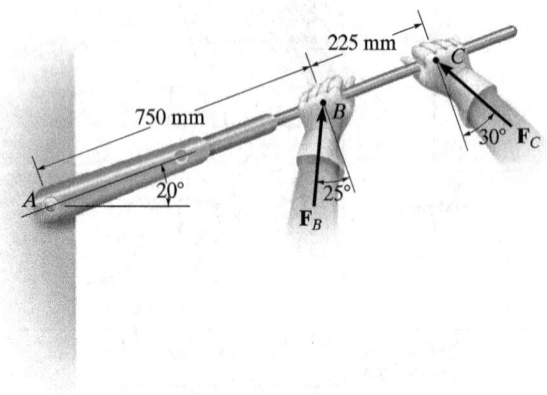

Probs. 4–29/30

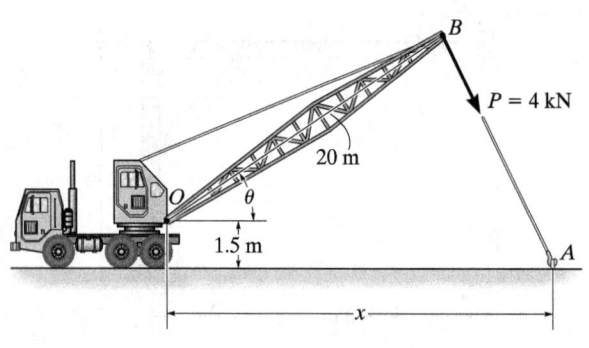

Probs. 4–32/33

**4–34.** In order to hold the wheelbarrow in the position shown, force **F** must produce a counterclockwise moment of 200 N·m about the axle at $A$. Determine the required magnitude of force **F**.

**4–35.** The wheelbarrow and its contents have a mass of 50 kg and a center of mass at $G$. If the resultant moment produced by force **F** and the weight about point $A$ is to be zero, determine the required magnitude of force **F**.

**\*4–36.** The wheelbarrow and its contents have a center of mass at $G$. If $F = 100$ N and the resultant moment produced by force **F** and the weight about the axle at $A$ is zero, determine the mass of the wheelbarrow and its contents.

**\*4–40.** Determine the moment produced by force $\mathbf{F}_B$ about point $O$. Express the result as a Cartesian vector.

**•4–41.** Determine the moment produced by force $\mathbf{F}_C$ about point $O$. Express the result as a Cartesian vector.

**4–42.** Determine the resultant moment produced by forces $\mathbf{F}_B$ and $\mathbf{F}_C$ about point $O$. Express the result as a Cartesian vector.

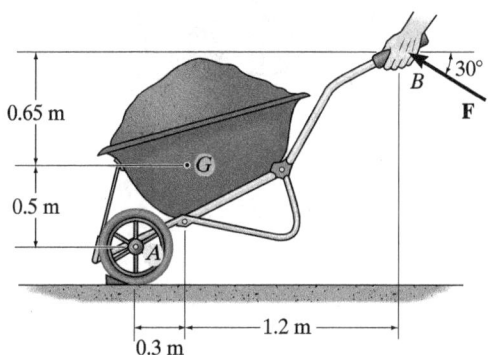

Prob. 4–34/35/36

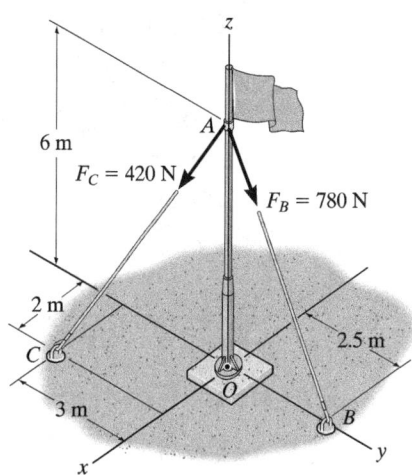

Probs. 4–40/41/42

**•4–37.** Determine the moment produced by $\mathbf{F}_1$ about point $O$. Express the result as a Cartesian vector.

**4–38.** Determine the moment produced by $\mathbf{F}_2$ about point $O$. Express the result as a Cartesian vector.

**4–39.** Determine the resultant moment produced by the two forces about point $O$. Express the result as a Cartesian vector.

**4–43.** Determine the moment produced by each force about point $O$ located on the drill bit. Express the results as Cartesian vectors.

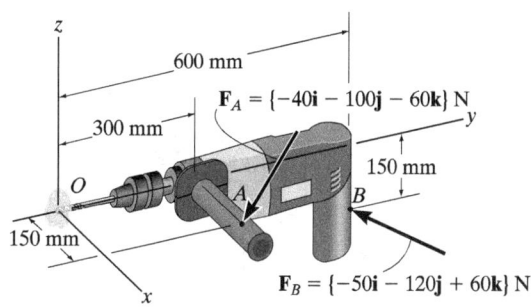

Prob. 4–43

**\*4–44.** A force of $\mathbf{F} = \{6\mathbf{i} - 2\mathbf{j} + 1\mathbf{k}\}$ kN produces a moment of $\mathbf{M}_O = \{4\mathbf{i} + 5\mathbf{j} - 14\mathbf{k}\}$ kN·m about the origin of coordinates, point $O$. If the force acts at a point having an $x$ coordinate of $x = 1$ m, determine the $y$ and $z$ coordinates.

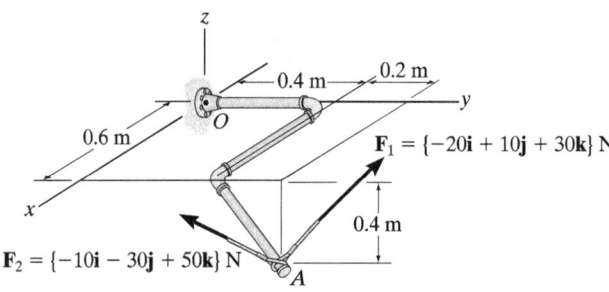

Probs. 4–37/38/39

•4–45. The pipe assembly is subjected to the 80-N force. Determine the moment of this force about point A.

4–46. The pipe assembly is subjected to the 80-N force. Determine the moment of this force about point B.

*4–48. Force **F** acts perpendicular to the inclined plane. Determine the moment produced by **F** about point A. Express the result as a Cartesian vector.

•4–49. Force **F** acts perpendicular to the inclined plane. Determine the moment produced by **F** about point B. Express the result as a Cartesian vector.

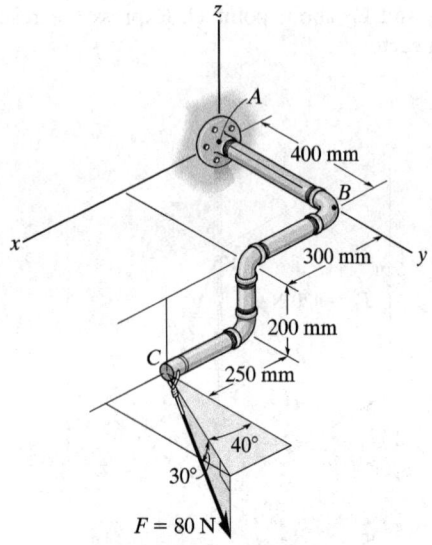

Probs. 4–45/46

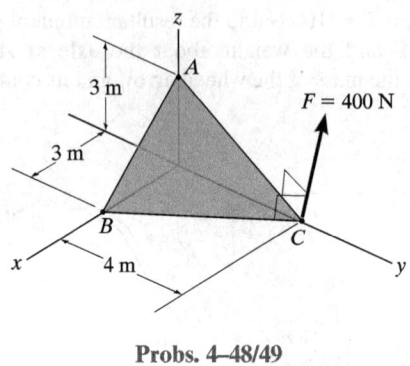

Probs. 4–48/49

4–47. The force $\mathbf{F} = \{6\mathbf{i} + 8\mathbf{j} + 10\mathbf{k}\}$ N creates a moment about point O of $\mathbf{M}_O = \{-14\mathbf{i} + 8\mathbf{j} + 2\mathbf{k}\}$ N·m. If the force passes through a point having an x coordinate of 1 m, determine the y and z coordinates of the point. Also, realizing that $M_O = Fd$, determine the perpendicular distance d from point O to the line of action of **F**.

4–50. A 20-N horizontal force is applied perpendicular to the handle of the socket wrench. Determine the magnitude and the coordinate direction angles of the moment created by this force about point O.

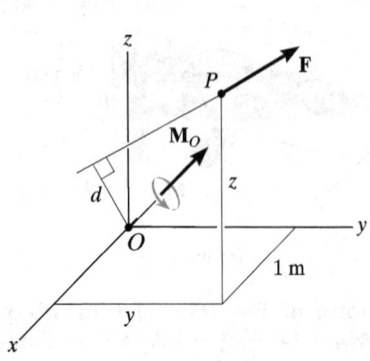

Prob. 4–47

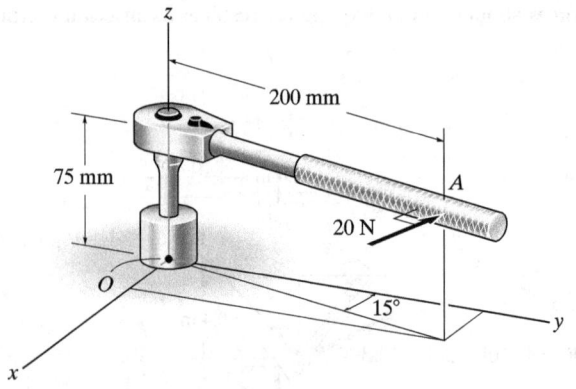

Prob. 4–50

## 4.5 Moment of a Force about a Specified Axis

Sometimes, the moment produced by a force about a *specified axis* must be determined. For example, suppose the lug nut at $O$ on the car tire in Fig. 4–20a needs to be loosened. The force applied to the wrench will create a tendency for the wrench and the nut to rotate about the *moment axis* passing through $O$; however, the nut can only rotate about the $y$ axis. Therefore, to determine the turning effect, only the $y$ component of the moment is needed, and the total moment produced is not important. To determine this component, we can use either a scalar or vector analysis.

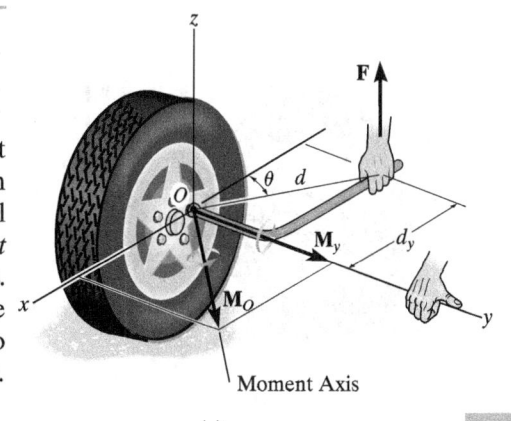

(a)

Fig. 4–20

**Scalar Analysis.** To use a scalar analysis in the case of the lug nut in Fig. 4–20a, the moment arm perpendicular distance from the axis to the line of action of the force is $d_y = d \cos \theta$. Thus, the moment of **F** about the $y$ axis is $M_y = F d_y = F(d \cos \theta)$. According to the right-hand rule, $\mathbf{M}_y$ is directed along the positive $y$ axis as shown in the figure. In general, for any axis $a$, the moment is

$$M_a = F d_a \qquad (4\text{–}10)$$

If large enough, the cable force **F** on the boom of this crane can cause the crane to topple over. To investigate this, the moment of the force must be calculated about an axis passing through the base of the legs at $A$ and $B$.

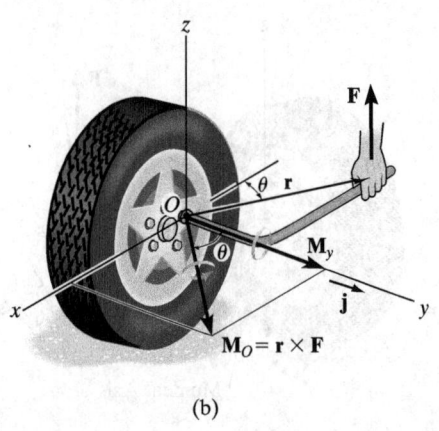

Fig. 4–20

**Vector Analysis.** To find the moment of force **F** in Fig. 4–20b about the y axis using a vector analysis, we must first determine the moment of the force about *any point O* on the y axis by applying Eq. 4–7, $\mathbf{M}_O = \mathbf{r} \times \mathbf{F}$. The component $\mathbf{M}_y$ along the y axis is the *projection* of $\mathbf{M}_O$ onto the y axis. It can be found using the *dot product* discussed in Chapter 2, so that $M_y = \mathbf{j} \cdot \mathbf{M}_O = \mathbf{j} \cdot (\mathbf{r} \times \mathbf{F})$, where **j** is the unit vector for the y axis.

We can generalize this approach by letting $\mathbf{u}_a$ be the unit vector that specifies the direction of the *a* axis shown in Fig. 4–21. Then the moment of **F** about the axis is $M_a = \mathbf{u}_a \cdot (\mathbf{r} \times \mathbf{F})$. This combination is referred to as the *scalar triple product*. If the vectors are written in Cartesian form, we have

$$M_a = [u_{a_x}\mathbf{i} + u_{a_y}\mathbf{j} + u_{a_z}\mathbf{k}] \cdot \begin{vmatrix} \mathbf{i} & \mathbf{j} & \mathbf{k} \\ r_x & r_y & r_z \\ F_x & F_y & F_z \end{vmatrix}$$

$$= u_{a_x}(r_y F_z - r_z F_y) - u_{a_y}(r_x F_z - r_z F_x) + u_{a_z}(r_x F_y - r_y F_x)$$

This result can also be written in the form of a determinant, making it easier to memorize.*

$$M_a = \mathbf{u}_a \cdot (\mathbf{r} \times \mathbf{F}) = \begin{vmatrix} u_{a_x} & u_{a_y} & u_{a_z} \\ r_x & r_y & r_z \\ F_x & F_y & F_z \end{vmatrix} \qquad (4\text{–}11)$$

where

$u_{a_x}, u_{a_y}, u_{a_z}$    represent the x, y, z components of the unit vector defining the direction of the *a* axis

$r_x, r_y, r_z$    represent the x, y, z components of the position vector extended from *any point O* on the *a* axis to *any point A* on the line of action of the force

$F_x, F_y, F_z$    represent the x, y, z components of the force vector.

When $M_a$ is evaluated from Eq. 4–11, it will yield a positive or negative scalar. The sign of this scalar indicates the sense of direction of $\mathbf{M}_a$ along the *a* axis. If it is positive, then $\mathbf{M}_a$ will have the same sense as $\mathbf{u}_a$, whereas if it is negative, then $\mathbf{M}_a$ will act opposite to $\mathbf{u}_a$.

Once $M_a$ is determined, we can then express $\mathbf{M}_a$ as a Cartesian vector, namely,

$$\mathbf{M}_a = M_a \mathbf{u}_a \qquad (4\text{–}12)$$

The examples which follow illustrate numerical applications of the above concepts.

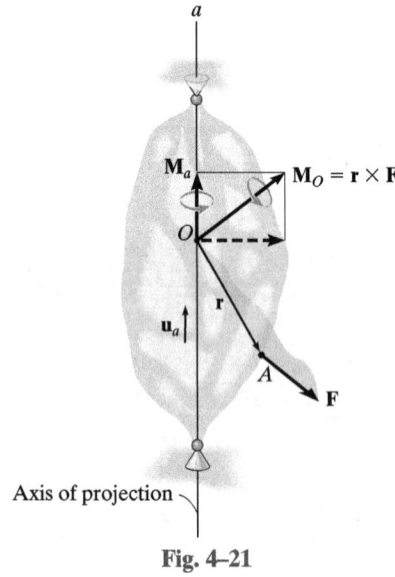

Fig. 4–21

---

*Take a moment to expand this determinant, to show that it will yield the above result.

## 4.5 MOMENT OF A FORCE ABOUT A SPECIFIED AXIS

### Important Points

- The moment of a force about a specified axis can be determined provided the perpendicular distance $d_a$ from the force line of action to the axis can be determined. $M_a = F d_a$.

- If vector analysis is used, $M_a = \mathbf{u}_a \cdot (\mathbf{r} \times \mathbf{F})$, where $\mathbf{u}_a$ defines the direction of the axis and $\mathbf{r}$ is extended from *any point* on the axis to *any point* on the line of action of the force.

- If $M_a$ is calculated as a negative scalar, then the sense of direction of $\mathbf{M}_a$ is opposite to $\mathbf{u}_a$.

- The moment $\mathbf{M}_a$ expressed as a Cartesian vector is determined from $\mathbf{M}_a = M_a \mathbf{u}_a$.

### EXAMPLE 4.7

Determine the resultant moment of the three forces in Fig. 4–22 about the *x* axis, the *y* axis, and the *z* axis.

#### SOLUTION

A force that is *parallel* to a coordinate axis or has a line of action that passes through the axis does *not* produce any moment or tendency for turning about that axis. Therefore, defining the positive direction of the moment of a force according to the right-hand rule, as shown in the figure, we have

$$M_x = (600 \text{ N})(0.2 \text{ m}) + (500 \text{ N})(0.2 \text{ m}) + 0 = 220 \text{ N} \cdot \text{m} \quad Ans.$$

$$M_y = 0 - (500 \text{ N})(0.3 \text{ m}) - (400 \text{ N})(0.2 \text{ m}) = -230 \text{ N} \cdot \text{m} \quad Ans.$$

$$M_z = 0 + 0 - (400 \text{ N})(0.2 \text{ m}) = -80 \text{ N} \cdot \text{m} \quad Ans.$$

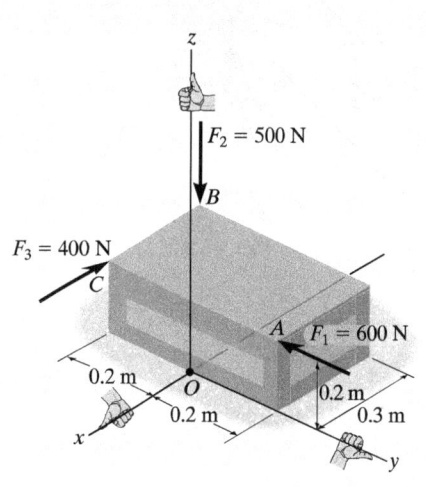

Fig. 4–22

The negative signs indicate that $\mathbf{M}_y$ and $\mathbf{M}_z$ act in the $-y$ and $-z$ directions, respectively.

## EXAMPLE 4.8

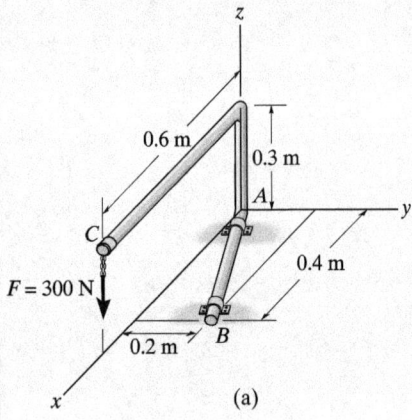

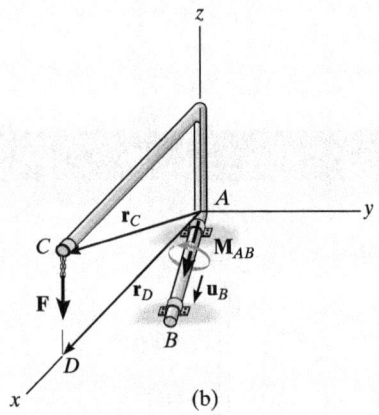

Fig. 4–23

Determine the moment $M_{AB}$ produced by the force $\mathbf{F}$ in Fig. 4–23a, which tends to rotate the rod about the $AB$ axis.

### SOLUTION

A vector analysis using $M_{AB} = \mathbf{u}_B \cdot (\mathbf{r} \times \mathbf{F})$ will be considered for the solution rather than trying to find the moment arm or perpendicular distance from the line of action of $\mathbf{F}$ to the $AB$ axis. Each of the terms in the equation will now be identified.

Unit vector $\mathbf{u}_B$ defines the direction of the $AB$ axis of the rod, Fig. 4–23b, where

$$\mathbf{u}_B = \frac{\mathbf{r}_B}{r_B} = \frac{\{0.4\mathbf{i} + 0.2\mathbf{j}\} \text{ m}}{\sqrt{(0.4 \text{ m})^2 + (0.2 \text{ m})^2}} = 0.8944\mathbf{i} + 0.4472\mathbf{j}$$

Vector $\mathbf{r}$ is directed from *any point* on the $AB$ axis to *any point* on the line of action of the force. For example, position vectors $\mathbf{r}_C$ and $\mathbf{r}_D$ are suitable, Fig. 4–23b. (Although not shown, $\mathbf{r}_{BC}$ or $\mathbf{r}_{BD}$ can also be used.) For simplicity, we choose $\mathbf{r}_D$, where

$$\mathbf{r}_D = \{0.6\mathbf{i}\} \text{ m}$$

The force is

$$\mathbf{F} = \{-300\mathbf{k}\} \text{ N}$$

Substituting these vectors into the determinant form and expanding, we have

$$M_{AB} = \mathbf{u}_B \cdot (\mathbf{r}_D \times \mathbf{F}) = \begin{vmatrix} 0.8944 & 0.4472 & 0 \\ 0.6 & 0 & 0 \\ 0 & 0 & -300 \end{vmatrix}$$

$$= 0.8944[0(-300) - 0(0)] - 0.4472[0.6(-300) - 0(0)]$$
$$\qquad + 0[0.6(0) - 0(0)]$$

$$= 80.50 \text{ N} \cdot \text{m}$$

This positive result indicates that the sense of $\mathbf{M}_{AB}$ is in the same direction as $\mathbf{u}_B$.

Expressing $\mathbf{M}_{AB}$ as a Cartesian vector yields

$$\mathbf{M}_{AB} = M_{AB}\mathbf{u}_B = (80.50 \text{ N} \cdot \text{m})(0.8944\mathbf{i} + 0.4472\mathbf{j})$$
$$= \{72.0\mathbf{i} + 36.0\mathbf{j}\} \text{ N} \cdot \text{m} \qquad \textit{Ans.}$$

The result is shown in Fig. 4–23b.

**NOTE:** If axis $AB$ is defined using a unit vector directed from B toward A, then in the above formulation $-\mathbf{u}_B$ would have to be used. This would lead to $M_{AB} = -80.50$ N·m. Consequently, $\mathbf{M}_{AB} = M_{AB}(-\mathbf{u}_B)$, and the same result would be obtained.

## EXAMPLE 4.9

Determine the magnitude of the moment of force **F** about segment $OA$ of the pipe assembly in Fig. 4–24a.

### SOLUTION

The moment of **F** about the $OA$ axis is determined from $M_{OA} = \mathbf{u}_{OA} \cdot (\mathbf{r} \times \mathbf{F})$, where **r** is a position vector extending from any point on the $OA$ axis to any point on the line of action of **F**. As indicated in Fig. 4–24b, either $\mathbf{r}_{OD}, \mathbf{r}_{OC}, \mathbf{r}_{AD}$, or $\mathbf{r}_{AC}$ can be used; however, $\mathbf{r}_{OD}$ will be considered since it will simplify the calculation.

The unit vector $\mathbf{u}_{OA}$, which specifies the direction of the $OA$ axis, is

$$\mathbf{u}_{OA} = \frac{\mathbf{r}_{OA}}{r_{OA}} = \frac{\{0.3\mathbf{i} + 0.4\mathbf{j}\} \text{ m}}{\sqrt{(0.3 \text{ m})^2 + (0.4 \text{ m})^2}} = 0.6\mathbf{i} + 0.8\mathbf{j}$$

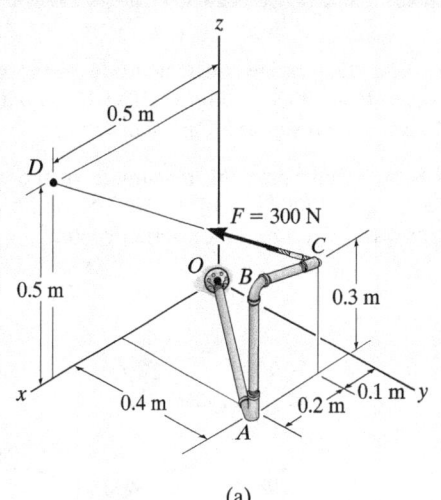

(a)

and the position vector $\mathbf{r}_{OD}$ is

$$\mathbf{r}_{OD} = \{0.5\mathbf{i} + 0.5\mathbf{k}\} \text{ m}$$

The force **F** expressed as a Cartesian vector is

$$\mathbf{F} = F\left(\frac{\mathbf{r}_{CD}}{r_{CD}}\right)$$

$$= (300 \text{ N})\left[\frac{\{0.4\mathbf{i} - 0.4\mathbf{j} + 0.2\mathbf{k}\} \text{ m}}{\sqrt{(0.4 \text{ m})^2 + (-0.4 \text{ m})^2 + (0.2 \text{ m})^2}}\right]$$

$$= \{200\mathbf{i} - 200\mathbf{j} + 100\mathbf{k}\} \text{ N}$$

Therefore,

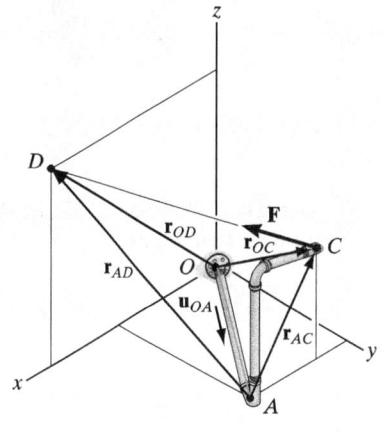

(b)

Fig. 4–24

$$M_{OA} = \mathbf{u}_{OA} \cdot (\mathbf{r}_{OD} \times \mathbf{F})$$

$$= \begin{vmatrix} 0.6 & 0.8 & 0 \\ 0.5 & 0 & 0.5 \\ 200 & -200 & 100 \end{vmatrix}$$

$$= 0.6[0(100) - (0.5)(-200)] - 0.8[0.5(100) - (0.5)(200)] + 0$$

$$= 100 \text{ N·m} \qquad\qquad Ans.$$

## FUNDAMENTAL PROBLEMS

**F4–13.** Determine the magnitude of the moment of the force $\mathbf{F} = \{300\mathbf{i} - 200\mathbf{j} + 150\mathbf{k}\}$ N about the $x$ axis. Express the result as a Cartesian vector.

**F4–14.** Determine the magnitude of the moment of the force $\mathbf{F} = \{300\mathbf{i} - 200\mathbf{j} + 150\mathbf{k}\}$ N about the $OA$ axis. Express the result as a Cartesian vector.

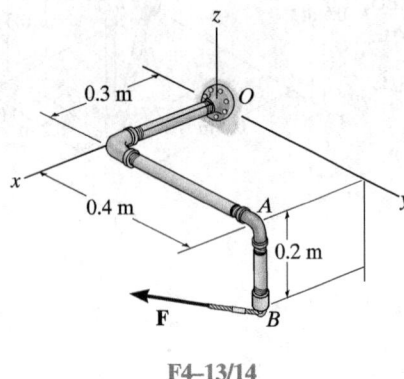

F4–13/14

**F4–15.** Determine the magnitude of the moment of the 200-N force about the $x$ axis.

**F4–16.** Determine the magnitude of the moment of the force about the $y$ axis.
$\mathbf{F} = \{30\mathbf{i} - 20\mathbf{j} + 50\mathbf{k}\}$ N

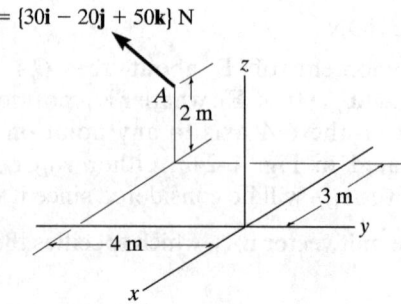

F4–16

**F4–17.** Determine the moment of the force $\mathbf{F} = \{50\mathbf{i} - 40\mathbf{j} + 20\mathbf{k}\}$ N about the $AB$ axis. Express the result as a Cartesian vector.

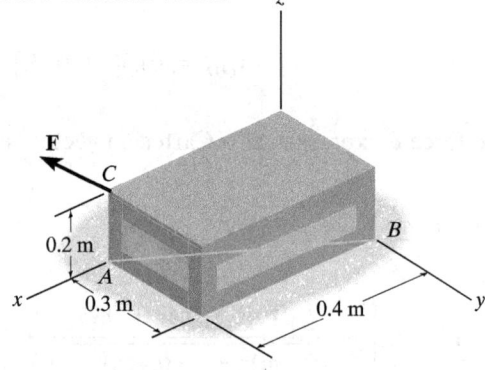

F4–17

**F4–18.** Determine the moment of force $\mathbf{F}$ about the $x$, the $y$, and the $z$ axes. Use a scalar analysis.

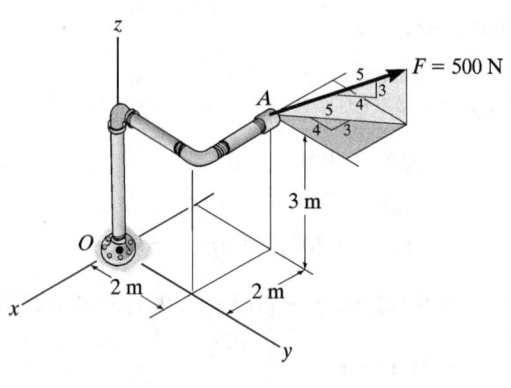

F4–18

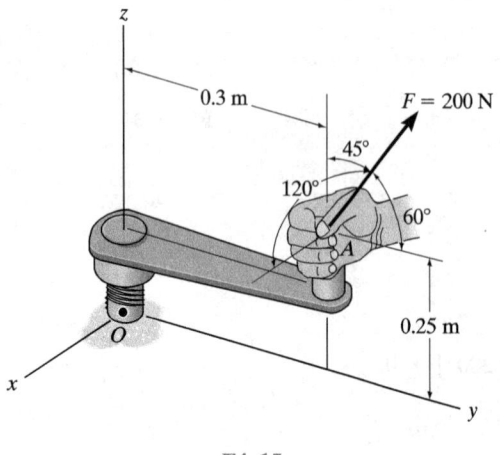

F4–15

## PROBLEMS

**4–51.** Determine the moment produced by force **F** about the diagonal *AF* of the rectangular block. Express the result as a Cartesian vector.

**\*4–52.** Determine the moment produced by force **F** about the diagonal *OD* of the rectangular block. Express the result as a Cartesian vector.

**4–54.** Determine the magnitude of the moments of the force **F** about the *x*, *y*, and *z* axes. Solve the problem (a) using a Cartesian vector approach and (b) using a scalar approach.

**4–55.** Determine the moment of the force **F** about an axis extending between *A* and *C*. Express the result as a Cartesian vector.

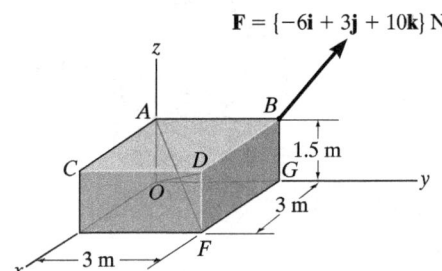

Probs. 4–51/52

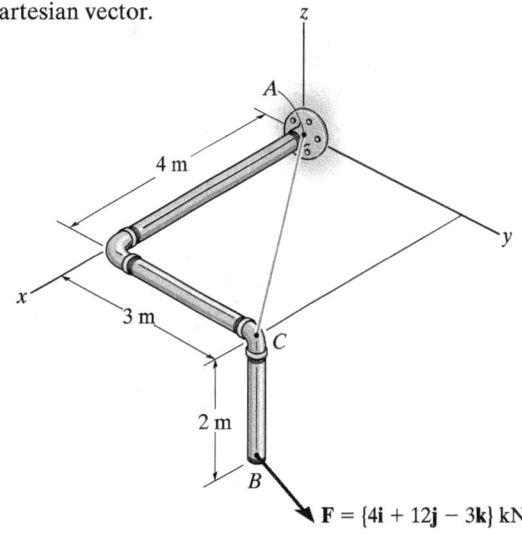

Probs. 4–54/55

**•4–53.** The tool is used to shut off gas valves that are difficult to access. If the force **F** is applied to the handle, determine the component of the moment created about the *z* axis of the valve.

**\*4–56.** Determine the moment produced by force **F** about segment *AB* of the pipe assembly. Express the result as a Cartesian vector.

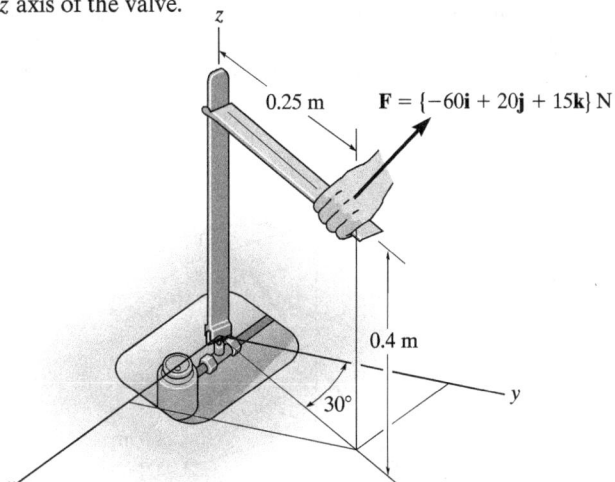

Prob. 4–53

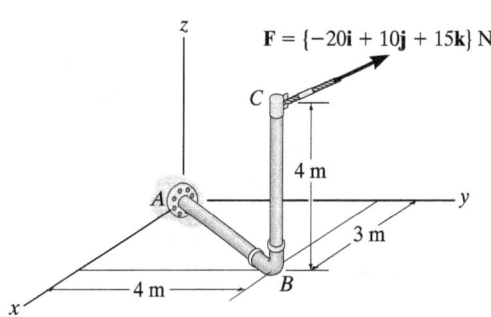

Prob. 4–56

**•4–57.** Determine the magnitude of the moment that the force **F** exerts about the y axis of the shaft. Solve the problem using a Cartesian vector approach and using a scalar approach.

**\*4–60.** Determine the magnitude of the moment produced by the force of $F = 200$ N about the hinged axis (the x axis) of the door.

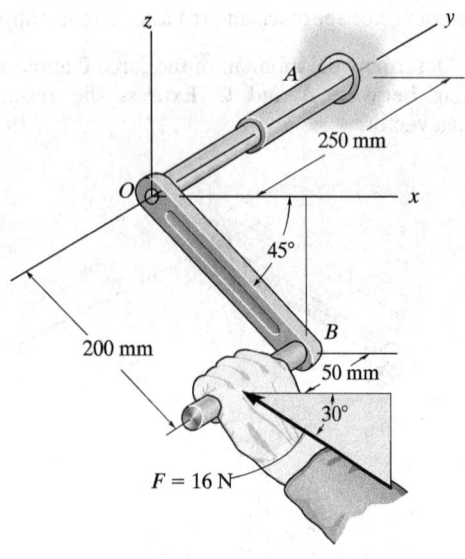

Prob. 4–57

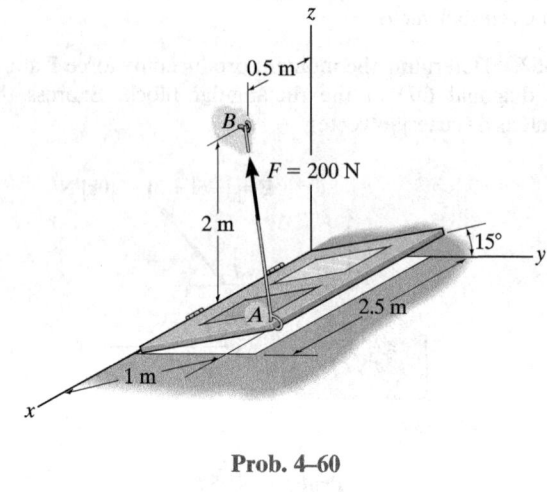

Prob. 4–60

**4–58.** If $F = 450$ N, determine the magnitude of the moment produced by this force about the x axis.

**4–59.** The friction at sleeve A can provide a maximum resisting moment of 125 N · m about the x axis. Determine the largest magnitude of force **F** that can be applied to the bracket so that the bracket will not turn.

**•4–61.** If the tension in the cable is $F = 700$ N, determine the magnitude of the moment produced by this force about the hinged axis, CD, of the panel.

**4–62.** Determine the magnitude of force **F** in cable AB in order to produce a moment of 750 N · m about the hinged axis CD, which is needed to hold the panel in the position shown.

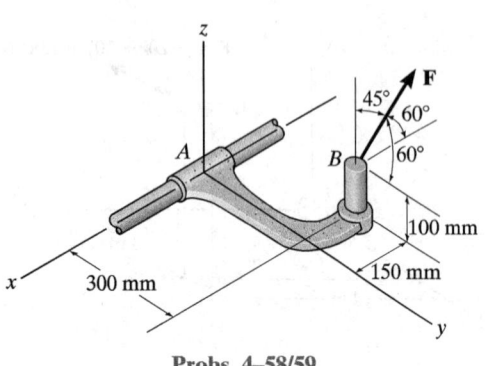

Probs. 4–58/59

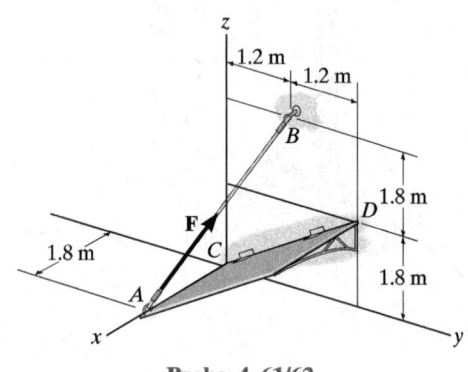

Probs. 4–61/62

**4–63.** The A-frame is being hoisted into an upright position by the vertical force of $F = 400$ N. Determine the moment of this force about the $y'$ axis passing through points $A$ and $B$ when the frame is in the position shown.

**\*4–64.** The A-frame is being hoisted into an upright position by the vertical force of $F = 400$ N. Determine the moment of this force about the $x$ axis when the frame is in the position shown.

**•4–65.** The A-frame is being hoisted into an upright position by the vertical force of $F = 400$ N. Determine the moment of this force about the $y$ axis when the frame is in the position shown.

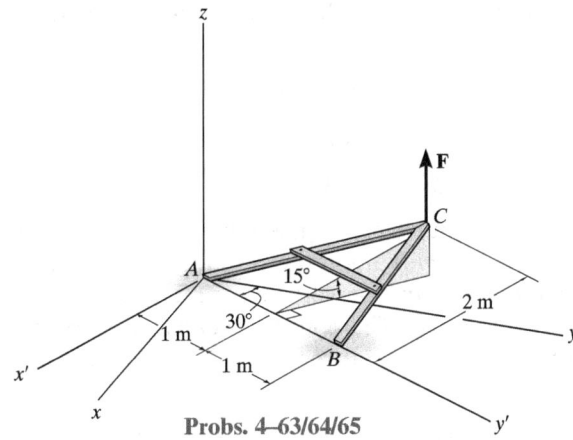

Probs. 4–63/64/65

**4–66.** The flex-headed ratchet wrench is subjected to a force of $P = 80$ N, applied perpendicular to the handle as shown. Determine the moment or torque this imparts along the vertical axis of the bolt at $A$.

**4–67.** If a torque or moment of $10$ N·m is required to loosen the bolt at $A$, determine the force $P$ that must be applied perpendicular to the handle of the flex-headed ratchet wrench.

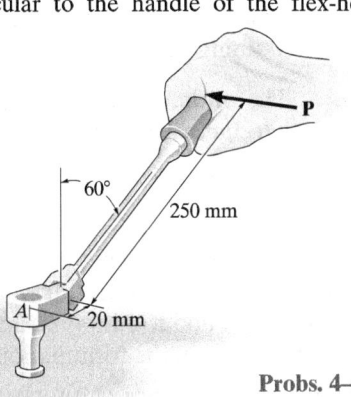

Probs. 4–66/67

**\*4–68.** The pipe assembly is secured on the wall by the two brackets. If the flower pot has a weight of 250 N, determine the magnitude of the moment produced by the weight about the $OA$ axis.

**•4–69.** The pipe assembly is secured on the wall by the two brackets. If the frictional force of both brackets can resist a maximum moment of $225$ N·m, determine the largest weight of the flower pot that can be supported by the assembly without causing it to rotate about the $OA$ axis.

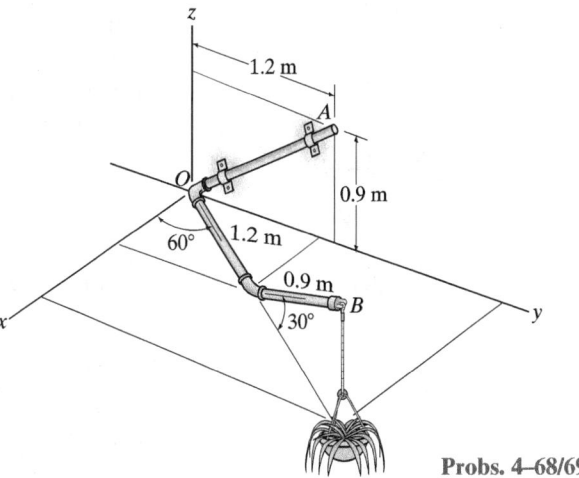

Probs. 4–68/69

**4–70.** A vertical force of $F = 60$ N is applied to the handle of the pipe wrench. Determine the moment that this force exerts along the axis $AB$ ($x$ axis) of the pipe assembly. Both the wrench and pipe assembly $ABC$ lie in the $x$–$y$ plane. *Suggestion:* Use a scalar analysis.

**4–71.** Determine the magnitude of the vertical force **F** acting on the handle of the wrench so that this force produces a component of moment along the $AB$ axis ($x$ axis) of the pipe assembly of $(M_A)_x = \{-5\mathbf{i}\}$ N·m. Both the pipe assembly $ABC$ and the wrench lie in the $x$–$y$ plane. *Suggestion:* Use a scalar analysis.

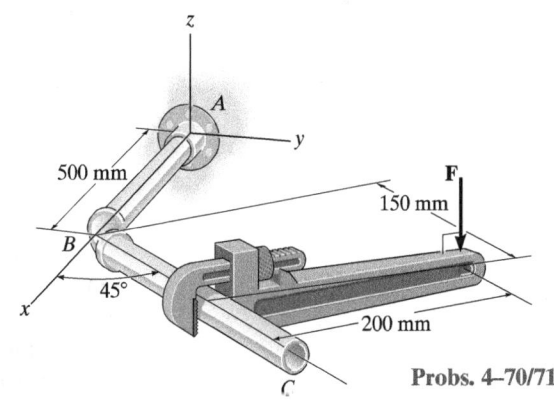

Probs. 4–70/71

## 4.6 Moment of a Couple

A *couple* is defined as two parallel forces that have the same magnitude, but opposite directions, and are separated by a perpendicular distance $d$, Fig. 4–25. Since the resultant force is zero, the only effect of a couple is to produce a rotation or tendency of rotation in a specified direction. For example, imagine that you are driving a car with both hands on the steering wheel and you are making a turn. One hand will push up on the wheel while the other hand pulls down, which causes the steering wheel to rotate.

The moment produced by a couple is called a *couple moment*. We can determine its value by finding the sum of the moments of both couple forces about *any* arbitrary point. For example, in Fig. 4–26, position vectors $\mathbf{r}_A$ and $\mathbf{r}_B$ are directed from point $O$ to points $A$ and $B$ lying on the line of action of $-\mathbf{F}$ and $\mathbf{F}$. The couple moment determined about $O$ is therefore

$$\mathbf{M} = \mathbf{r}_B \times \mathbf{F} + \mathbf{r}_A \times -\mathbf{F} = (\mathbf{r}_B - \mathbf{r}_A) \times \mathbf{F}$$

However $\mathbf{r}_B = \mathbf{r}_A + \mathbf{r}$ or $\mathbf{r} = \mathbf{r}_B - \mathbf{r}_A$, so that

$$\boxed{\mathbf{M} = \mathbf{r} \times \mathbf{F}} \tag{4–13}$$

This result indicates that a couple moment is a *free vector*, i.e., it can act at *any point* since $\mathbf{M}$ depends *only* upon the position vector $\mathbf{r}$ directed *between* the forces and *not* the position vectors $\mathbf{r}_A$ and $\mathbf{r}_B$, directed from the arbitrary point $O$ to the forces. This concept is unlike the moment of a force, which requires a definite point (or axis) about which moments are determined.

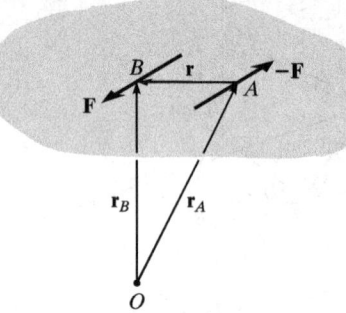

Fig. 4–26

**Scalar Formulation.** The moment of a couple, $\mathbf{M}$, Fig. 4–27, is defined as having a *magnitude* of

$$\boxed{M = Fd} \tag{4–14}$$

where $F$ is the magnitude of one of the forces and $d$ is the perpendicular distance or moment arm between the forces. The *direction* and sense of the couple moment are determined by the right-hand rule, where the thumb indicates this direction when the fingers are curled with the sense of rotation caused by the couple forces. In all cases, $\mathbf{M}$ will act perpendicular to the plane containing these forces.

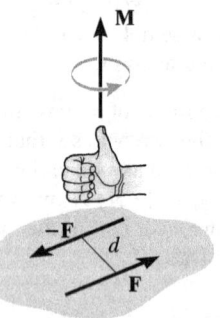

Fig. 4–27

**Vector Formulation.** The moment of a couple can also be expressed by the vector cross product using Eq. 4–13, i.e.,

$$\boxed{\mathbf{M} = \mathbf{r} \times \mathbf{F}} \tag{4–15}$$

Application of this equation is easily remembered if one thinks of taking the moments of both forces about a point lying on the line of action of one of the forces. For example, if moments are taken about point $A$ in Fig. 4–26, the moment of $-\mathbf{F}$ is *zero* about this point, and the moment of $\mathbf{F}$ is defined from Eq. 4–15. Therefore, in the formulation $\mathbf{r}$ is crossed with the force $\mathbf{F}$ to which it is directed.

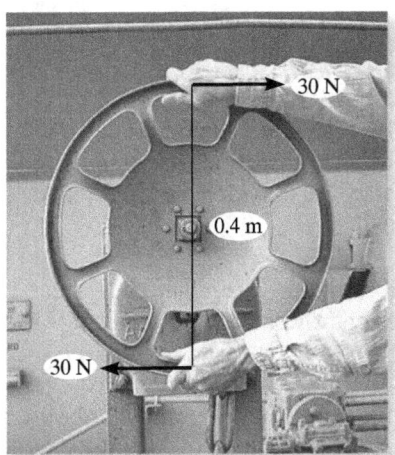 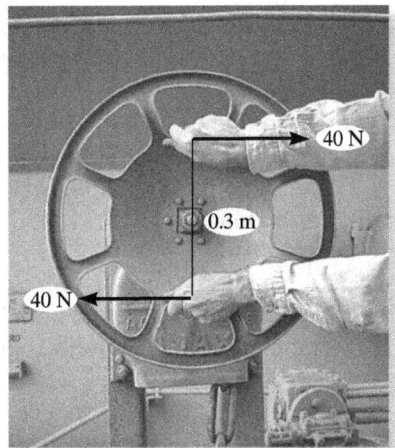

Fig. 4–28

## Equivalent Couples.
If two couples produce a moment with the *same magnitude and direction*, then these two couples are *equivalent*. For example, the two couples shown in Fig. 4–28 are *equivalent* because each couple moment has a magnitude of $M = 30 \text{ N}(0.4 \text{ m}) = 40 \text{ N}(0.3 \text{ m}) = 12 \text{ N} \cdot \text{m}$, and each is directed into the plane of the page. Notice that larger forces are required in the second case to create the same turning effect because the hands are placed closer together. Also, if the wheel was connected to the shaft at a point other than at its center, then the wheel would still turn when each couple is applied since the $12 \text{ N} \cdot \text{m}$ couple is a free vector.

## Resultant Couple Moment.
Since couple moments are vectors, their resultant can be determined by vector addition. For example, consider the couple moments $\mathbf{M}_1$ and $\mathbf{M}_2$ acting on the pipe in Fig. 4–29a. Since each couple moment is a free vector, we can join their tails at any arbitrary point and find the resultant couple moment, $\mathbf{M}_R = \mathbf{M}_1 + \mathbf{M}_2$ as shown in Fig. 4–29b.

If more than two couple moments act on the body, we may generalize this concept and write the vector resultant as

$$\mathbf{M}_R = \Sigma(\mathbf{r} \times \mathbf{F}) \quad (4\text{--}16)$$

These concepts are illustrated numerically in the examples that follow. In general, problems projected in two dimensions should be solved using a scalar analysis since the moment arms and force components are easy to determine.

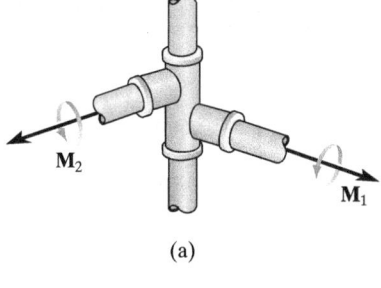

(a)

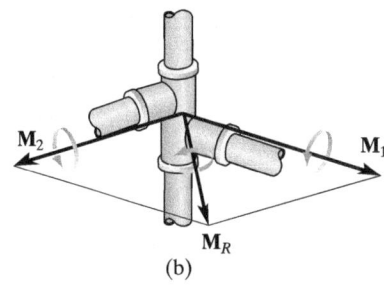

(b)

Fig. 4–29

Steering wheels on vehicles have been made smaller than on older vehicles because power steering does not require the driver to apply a large couple moment to the rim of the wheel.

## Important Points

- A couple moment is produced by two noncollinear forces that are equal in magnitude but opposite in direction. Its effect is to produce pure rotation, or tendency for rotation in a specified direction.

- A couple moment is a free vector, and as a result it causes the same rotational effect on a body regardless of where the couple moment is applied to the body.

- The moment of the two couple forces can be determined about *any point*. For convenience, this point is often chosen on the line of action of one of the forces in order to eliminate the moment of this force about the point.

- In three dimensions the couple moment is often determined using the vector formulation, $\mathbf{M} = \mathbf{r} \times \mathbf{F}$, where $\mathbf{r}$ is directed from *any point* on the line of action of one of the forces to *any point* on the line of action of the other force $\mathbf{F}$.

- A resultant couple moment is simply the vector sum of all the couple moments of the system.

## EXAMPLE 4.10

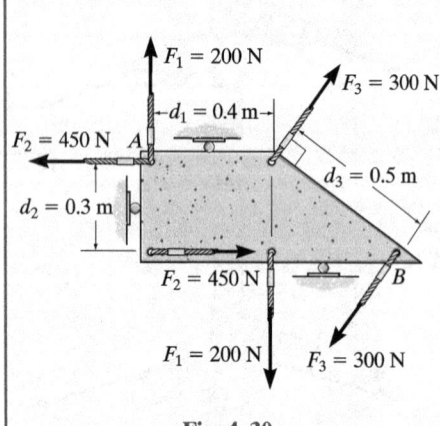

Fig. 4–30

Determine the resultant couple moment of the three couples acting on the plate in Fig. 4–30.

### SOLUTION

As shown the perpendicular distances between each pair of couple forces are $d_1 = 0.4$ m, $d_2 = 0.3$ m, and $d_3 = 0.5$ m. Considering counterclockwise couple moments as positive, we have

$\zeta + M_R = \Sigma M;\quad M_R = -F_1 d_1 + F_2 d_2 - F_3 d_3$

$\quad = (-200 \text{ N})(0.4 \text{ m}) + (450 \text{ N})(0.3 \text{ m}) - (300 \text{ N})(0.5 \text{ m})$

$\quad = -95 \text{ N} \cdot \text{m} = 95 \text{ N} \cdot \text{m} \;\downarrow$ *Ans.*

The negative sign indicates that $\mathbf{M}_R$ has a clockwise rotational sense.

## EXAMPLE 4.11

Determine the magnitude and direction of the couple moment acting on the gear in Fig. 4–31a.

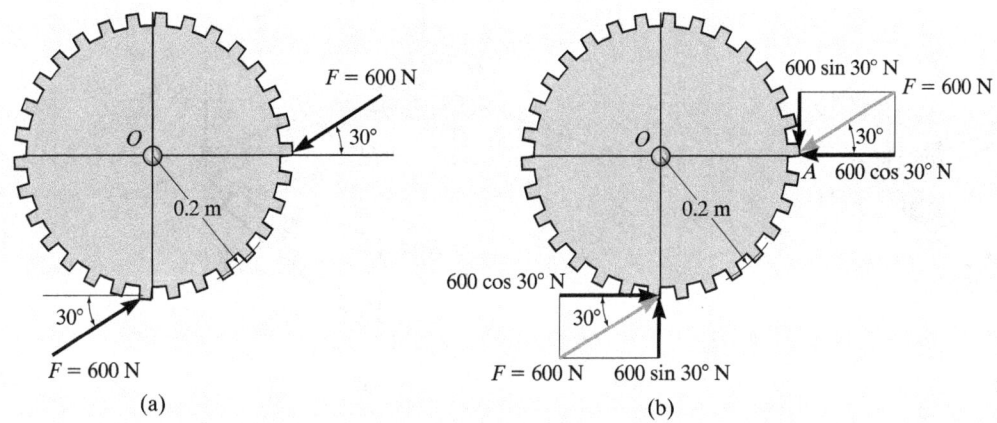

### SOLUTION

The easiest solution requires resolving each force into its components as shown in Fig. 4–31b. The couple moment can be determined by summing the moments of these force components about any point, for example, the center $O$ of the gear or point $A$. If we consider counterclockwise moments as positive, we have

$\zeta + M = \Sigma M_O;$  $M = (600 \cos 30° \text{ N})(0.2 \text{ m}) - (600 \sin 30° \text{ N})(0.2 \text{ m})$
$= 43.9 \text{ N·m} \circlearrowleft$  *Ans.*

or

$\zeta + M = \Sigma M_A;$  $M = (600 \cos 30° \text{ N})(0.2 \text{ m}) - (600 \sin 30° \text{ N})(0.2 \text{ m})$
$= 43.9 \text{ N·m} \circlearrowleft$  *Ans.*

This positive result indicates that **M** has a counterclockwise rotational sense, so it is directed outward, perpendicular to the page.

**NOTE:** The same result can also be obtained using $M = Fd$, where $d$ is the perpendicular distance between the lines of action of the couple forces, Fig. 4–31c. However, the computation for $d$ is more involved. Realize that the couple moment is a free vector and can act at any point on the gear and produce the same turning effect about point $O$.

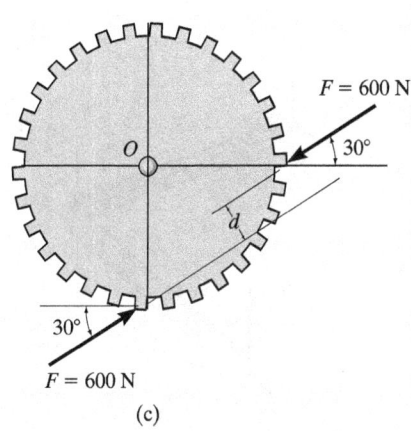

Fig. 4–31

## EXAMPLE 4.12

Determine the couple moment acting on the pipe shown in Fig. 4–32a. Segment $AB$ is directed 30° below the $x$–$y$ plane.

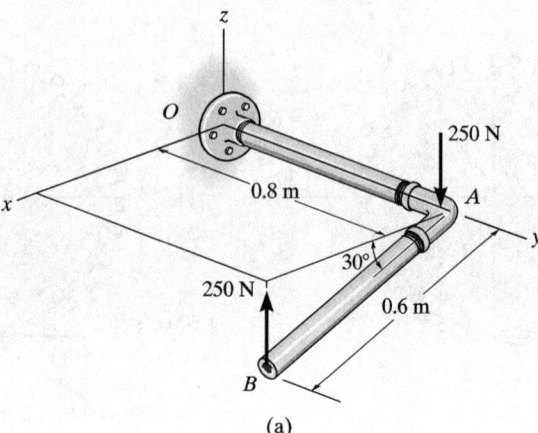

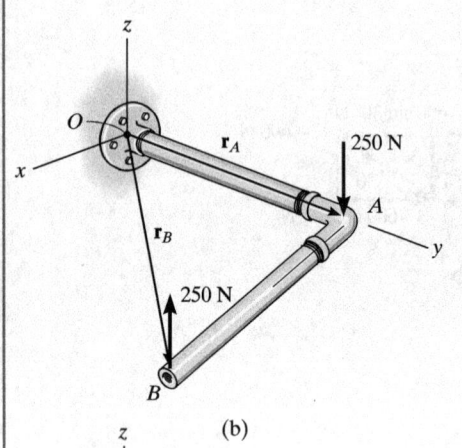

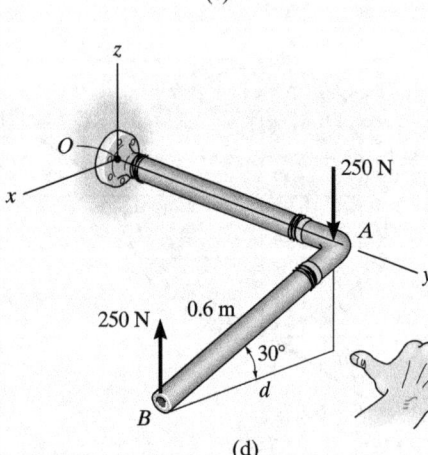

Fig. 4–32

### SOLUTION I (VECTOR ANALYSIS)

The moment of the two couple forces can be found about *any point*. If point $O$ is considered, Fig. 4–32b, we have

$$\mathbf{M} = \mathbf{r}_A \times (-250\mathbf{k}) + \mathbf{r}_B \times (250\mathbf{k})$$
$$= (0.8\mathbf{j}) \times (-250\mathbf{k}) + (0.6\cos 30°\mathbf{i} + 0.8\mathbf{j} - 0.6\sin 30°\mathbf{k}) \times (250\mathbf{k})$$
$$= -200\mathbf{i} - 129.9\mathbf{j} + 200\mathbf{i}$$
$$= \{-130\mathbf{j}\} \text{ N} \cdot \text{m} \qquad Ans.$$

It is *easier* to take moments of the couple forces about a point lying on the line of action of one of the forces, e.g., point $A$, Fig. 4–32c. In this case the moment of the force at $A$ is zero, so that

$$\mathbf{M} = \mathbf{r}_{AB} \times (250\mathbf{k})$$
$$= (0.6\cos 30°\mathbf{i} - 0.6\sin 30°\mathbf{k}) \times (250\mathbf{k})$$
$$= \{-130\mathbf{j}\} \text{ N} \cdot \text{m} \qquad Ans.$$

### SOLUTION II (SCALAR ANALYSIS)

Although this problem is shown in three dimensions, the geometry is simple enough to use the scalar equation $M = Fd$. The perpendicular distance between the lines of action of the couple forces is $d = 0.6\cos 30° = 0.5196$ m, Fig. 4–32d. Hence, taking moments of the forces about either point $A$ or point $B$ yields

$$M = Fd = 250 \text{ N} (0.5196 \text{ m}) = 129.9 \text{ N} \cdot \text{m}$$

Applying the right-hand rule, $\mathbf{M}$ acts in the $-\mathbf{j}$ direction. Thus,

$$\mathbf{M} = \{-130\mathbf{j}\} \text{ N} \cdot \text{m} \qquad Ans.$$

## EXAMPLE 4.13

Replace the two couples acting on the pipe column in Fig. 4–33a by a resultant couple moment.

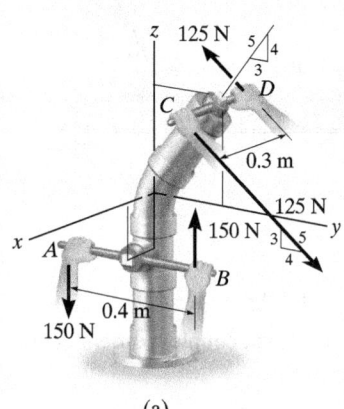

(a)

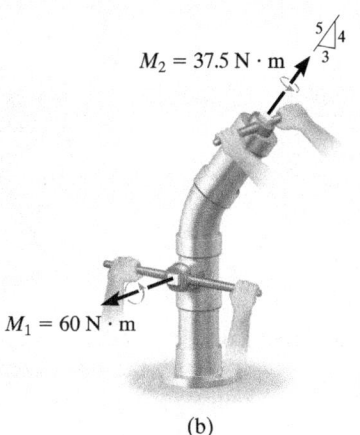

(b)

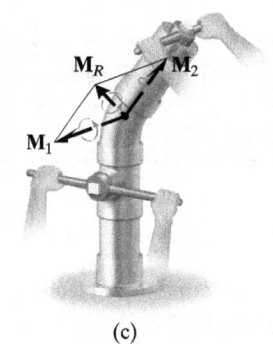

(c)

Fig. 4–33

### SOLUTION (VECTOR ANALYSIS)

The couple moment $\mathbf{M}_1$, developed by the forces at $A$ and $B$, can easily be determined from a scalar formulation.

$$M_1 = Fd = 150 \text{ N}(0.4 \text{ m}) = 60 \text{ N} \cdot \text{m}$$

By the right-hand rule, $\mathbf{M}_1$ acts in the $+\mathbf{i}$ direction, Fig. 4–33b. Hence,

$$\mathbf{M}_1 = \{60\mathbf{i}\} \text{ N} \cdot \text{m}$$

Vector analysis will be used to determine $\mathbf{M}_2$, caused by forces at $C$ and $D$. If moments are computed about point $D$, Fig. 4–33a, $\mathbf{M}_2 = \mathbf{r}_{DC} \times \mathbf{F}_C$, then

$$\mathbf{M}_2 = \mathbf{r}_{DC} \times \mathbf{F}_C = (0.3\mathbf{i}) \times \left[125\left(\tfrac{4}{5}\right)\mathbf{j} - 125\left(\tfrac{3}{5}\right)\mathbf{k}\right]$$
$$= (0.3\mathbf{i}) \times [100\mathbf{j} - 75\mathbf{k}] = 30(\mathbf{i} \times \mathbf{j}) - 22.5(\mathbf{i} \times \mathbf{k})$$
$$= \{22.5\mathbf{j} + 30\mathbf{k}\} \text{ N} \cdot \text{m}$$

Since $\mathbf{M}_1$ and $\mathbf{M}_2$ are free vectors, they may be moved to some arbitrary point and added vectorially, Fig. 4–33c. The resultant couple moment becomes

$$\mathbf{M}_R = \mathbf{M}_1 + \mathbf{M}_2 = \{60\mathbf{i} + 22.5\mathbf{j} + 30\mathbf{k}\} \text{ N} \cdot \text{m} \qquad Ans.$$

## FUNDAMENTAL PROBLEMS

**F4–19.** Determine the resultant couple moment acting on the beam.

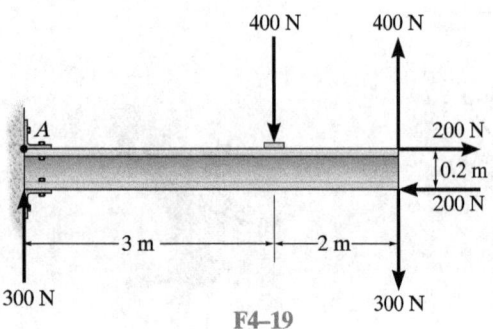

F4–19

**F4–20.** Determine the resultant couple moment acting on the triangular plate.

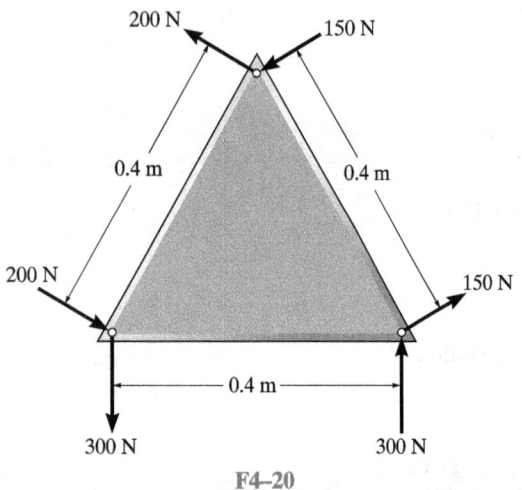

F4–20

**F4–21.** Determine the magnitude of **F** so that the resultant couple moment acting on the beam is 1.5 kN·m clockwise.

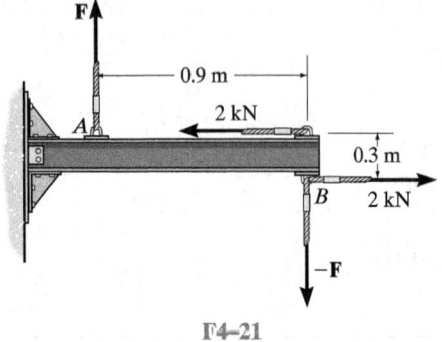

F4–21

**F4–22.** Determine the couple moment acting on the beam.

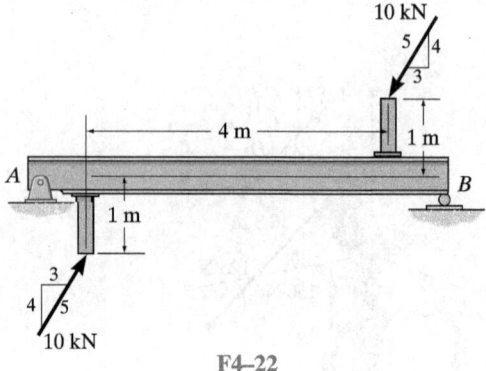

F4–22

**F4–23.** Determine the resultant couple moment acting on the pipe assembly.

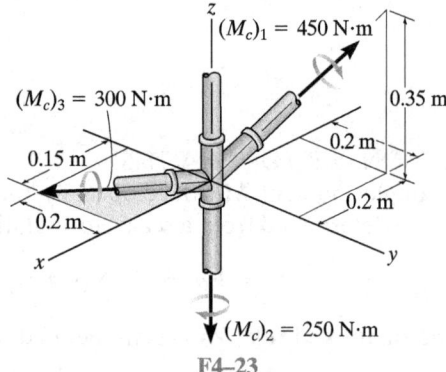

F4–23

**F4–24.** Determine the couple moment acting on the pipe assembly and express the result as a Cartesian vector.

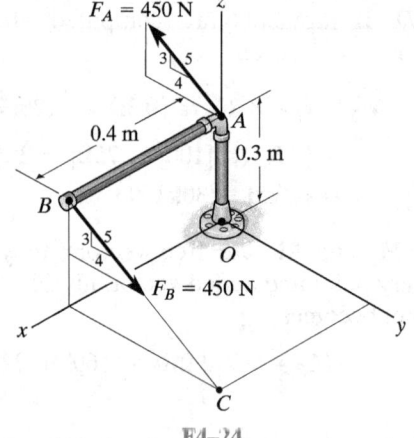

F4–24

## PROBLEMS

**\*4–72.** The frictional effects of the air on the blades of the standing fan creates a couple moment of $M_O = 6\,\text{N}\cdot\text{m}$ on the blades. Determine the magnitude of the couple forces at the base of the fan so that the resultant couple moment on the fan is zero.

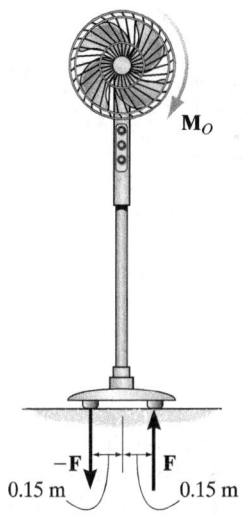

Prob. 4–72

**•4–73.** Determine the required magnitude of the couple moments $\mathbf{M}_2$ and $\mathbf{M}_3$ so that the resultant couple moment is zero.

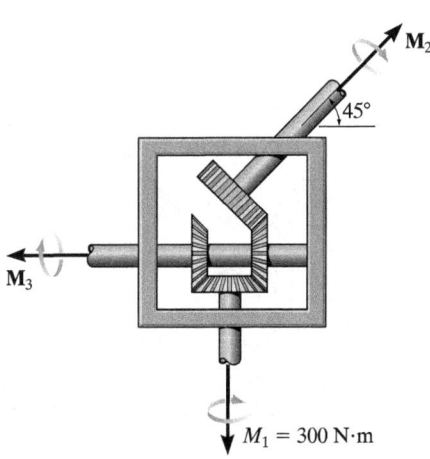

Prob. 4–73

**4–74.** The caster wheel is subjected to the two couples. Determine the forces $F$ that the bearings exert on the shaft so that the resultant couple moment on the caster is zero.

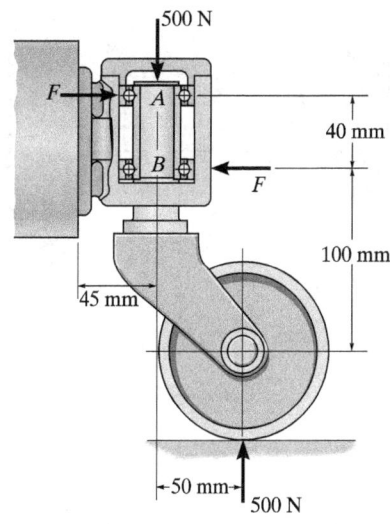

Prob. 4–74

**4–75.** If $F = 2000\,\text{N}$, determine the resultant couple moment.

**\*4–76.** Determine the required magnitude of force $\mathbf{F}$ if the resultant couple moment on the frame is $200\,\text{N}\cdot\text{m}$, clockwise.

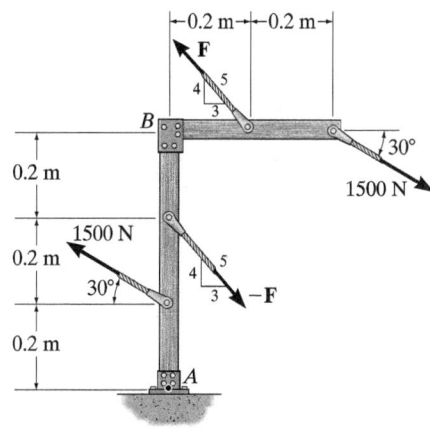

Probs. 4–75/76

**•4–77.** The floor causes a couple moment of $M_A = 40$ N·m and $M_B = 30$ N·m on the brushes of the polishing machine. Determine the magnitude of the couple forces that must be developed by the operator on the handles so that the resultant couple moment on the polisher is zero. What is the magnitude of these forces if the brush at $B$ suddenly stops so that $M_B = 0$?

***4–80.** Two couples act on the beam. Determine the magnitude of **F** so that the resultant couple moment is 450 N·m counterclockwise. Where on the beam does the resultant couple moment act?

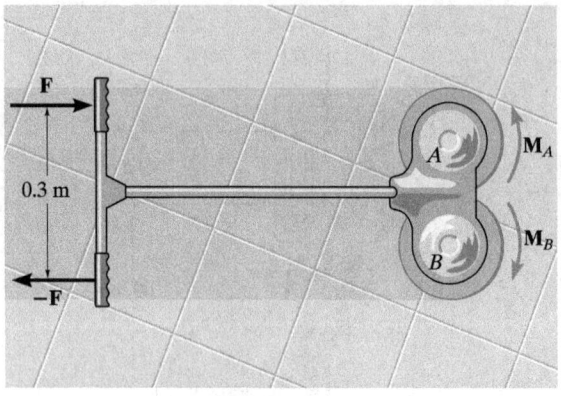

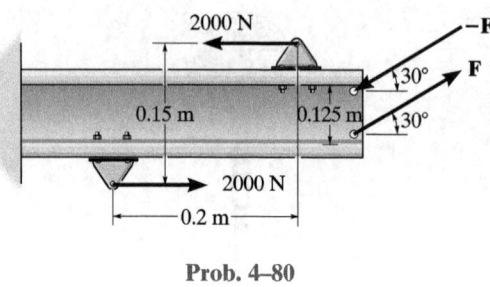

Prob. 4–80

Prob. 4–77

**4–78.** If $\theta = 30°$, determine the magnitude of force **F** so that the resultant couple moment is 100 N·m, clockwise.

**4–79.** If $F = 200$ N, determine the required angle $\theta$ so that the resultant couple moment is zero.

**•4–81.** The cord passing over the two small pegs $A$ and $B$ of the square board is subjected to a tension of 100 N. Determine the required tension $P$ acting on the cord that passes over pegs $C$ and $D$ so that the resultant couple produced by the two couples is 15 N·m acting clockwise. Take $\theta = 15°$.

**4–82.** The cord passing over the two small pegs $A$ and $B$ of the board is subjected to a tension of 100 N. Determine the *minimum* tension $P$ and the orientation $\theta$ of the cord passing over pegs $C$ and $D$, so that the resultant couple moment produced by the two cords is 20 N·m, clockwise.

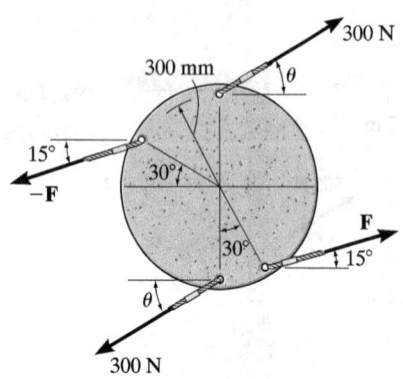

Probs. 4–78/79

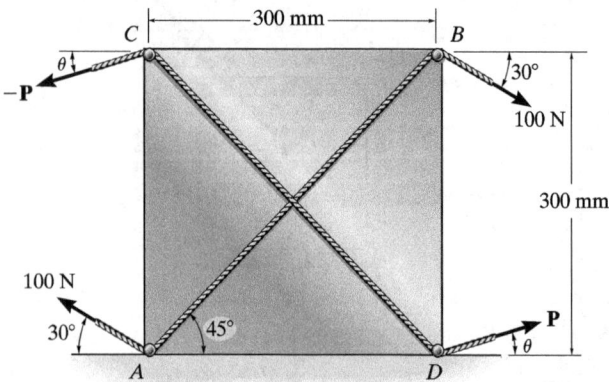

Probs. 4–81/82

**4–83.** A device called a rolamite is used in various ways to replace slipping motion with rolling motion. If the belt, which wraps between the rollers, is subjected to a tension of 15 N, determine the reactive forces N of the top and bottom plates on the rollers so that the resultant couple acting on the rollers is equal to zero.

**•4–85.** Determine the resultant couple moment acting on the beam. Solve the problem two ways: (a) sum moments about point O; and (b) sum moments about point A.

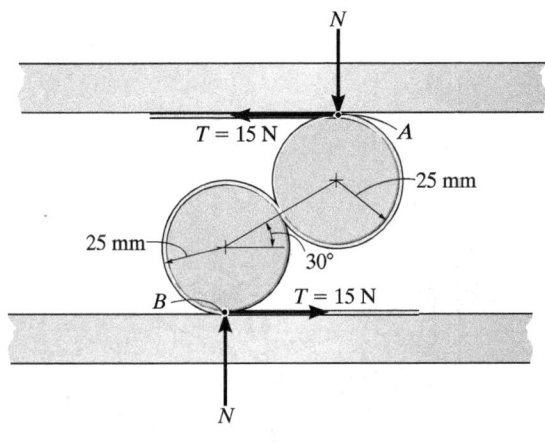

Prob. 4–83

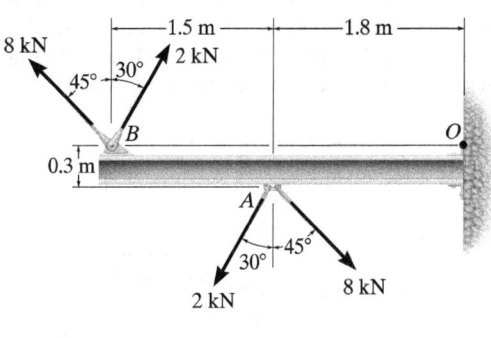

Prob. 4–85

**\*4–84.** Two couples act on the beam as shown. Determine the magnitude of **F** so that the resultant couple moment is 300 N·m counterclockwise. Where on the beam does the resultant couple act?

**4–86.** Two couples act on the cantilever beam. If $F = 6$ kN, determine the resultant couple moment.

**4–87.** Determine the required magnitude of force **F**, if the resultant couple moment on the beam is to be zero.

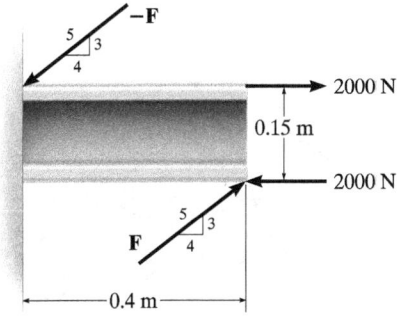

Prob. 4–84

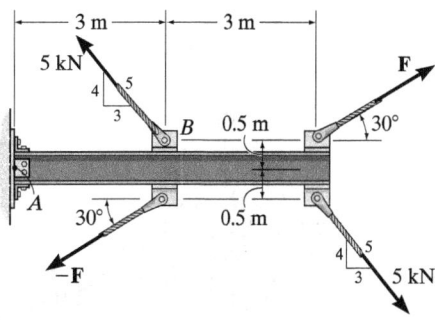

Probs. 4–86/87

**\*4–88.** Two couples act on the frame. If the resultant couple moment is to be zero, determine the distance $d$ between the 200-N couple forces.

**•4–89.** Two couples act on the frame. If $d = 1.2$ m, determine the resultant couple moment. Compute the result by resolving each force into $x$ and $y$ components and (a) finding the moment of each couple (Eq. 4–13) and (b) summing the moments of all the force components about point $A$.

**4–90.** Two couples act on the frame. If $d = 1.2$ m, determine the resultant couple moment. Compute the result by resolving each force into $x$ and $y$ components and (a) finding the moment of each couple (Eq. 4–13) and (b) summing the moments of all the force components about point $B$.

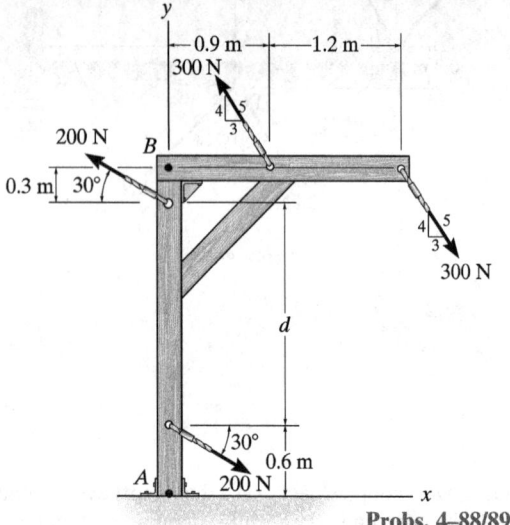

Probs. 4–88/89/90

**4–91.** If $M_1 = 500$ N·m, $M_2 = 600$ N·m, and $M_3 = 450$ N·m, determine the magnitude and coordinate direction angles of the resultant couple moment.

**\*4–92.** Determine the required magnitude of couple moments $M_1$, $M_2$, and $M_3$ so that the resultant couple moment is $M_R = \{-300\mathbf{i} + 450\mathbf{j} - 600\mathbf{k}\}$ N·m.

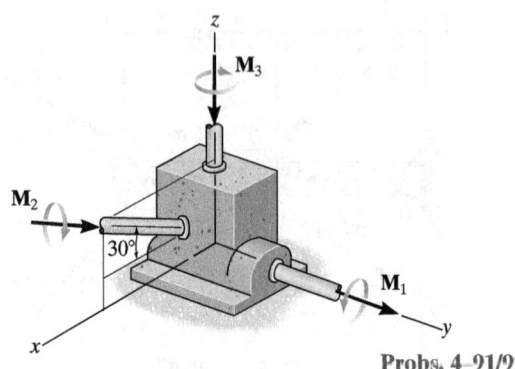

Probs. 4–91/92

**•4–93.** If $F = 80$ N, determine the magnitude and coordinate direction angles of the couple moment. The pipe assembly lies in the $x$–$y$ plane.

**4–94.** If the magnitude of the couple moment acting on the pipe assembly is 50 N·m, determine the magnitude of the couple forces applied to each wrench. The pipe assembly lies in the $x$–$y$ plane.

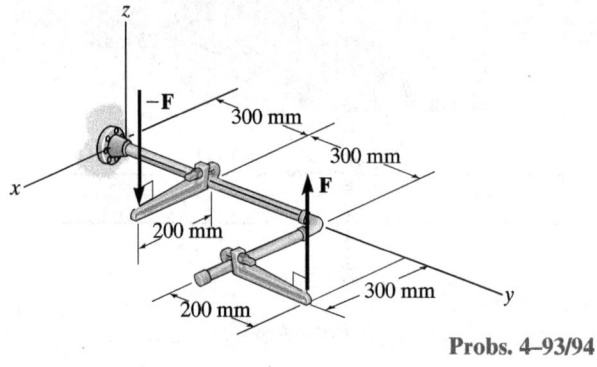

Probs. 4–93/94

**4–95.** From load calculations it is determined that the wing is subjected to couple moments $M_x = 25.5$ kN·m and $M_y = 37.5$ kN·m. Determine the resultant couple moments created about the $x'$ and $y'$ axes. The axes all lie in the same horizontal plane.

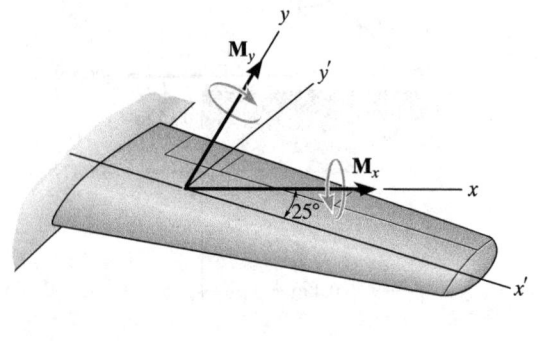

Prob. 4–95

**\*4–96.** Express the moment of the couple acting on the frame in Cartesian vector form. The forces are applied perpendicular to the frame. What is the magnitude of the couple moment? Take $F = 50$ N.

**•4–97.** In order to turn over the frame, a couple moment is applied as shown. If the component of this couple moment along the $x$ axis is $\mathbf{M}_x = \{-20\mathbf{i}\}$ N·m, determine the magnitude $F$ of the couple forces.

**\*4–100.** If $M_1 = 270$ N·m, $M_2 = 135$ N·m, and $M_3 = 180$ N·m, determine the magnitude and coordinate direction angles of the resultant couple moment.

**•4–101.** Determine the magnitudes of couple moments $\mathbf{M}_1$, $\mathbf{M}_2$, and $\mathbf{M}_3$ so that the resultant couple moment is zero.

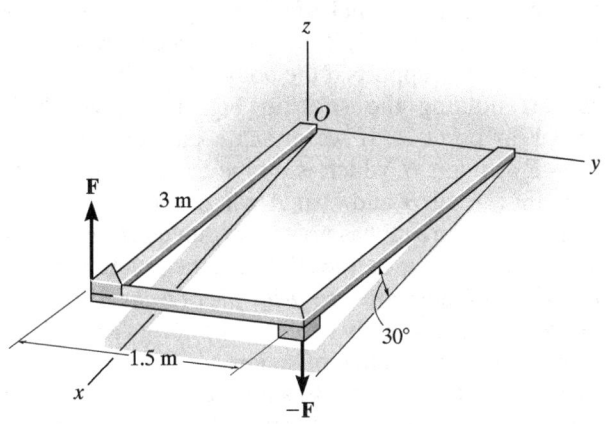

Probs. 4–96/97

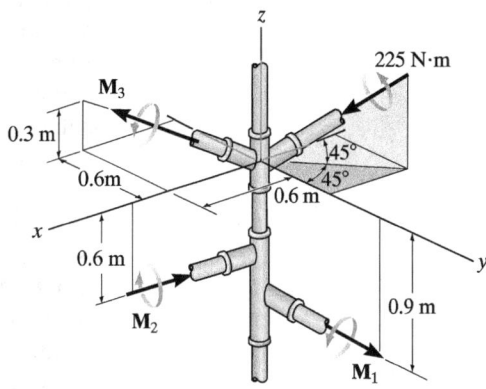

Probs. 4–100/101

**4–98.** Determine the resultant couple moment of the two couples that act on the pipe assembly. The distance from $A$ to $B$ is $d = 400$ mm. Express the result as a Cartesian vector.

**4–99.** Determine the distance $d$ between $A$ and $B$ so that the resultant couple moment has a magnitude of $M_R = 20$ N·m.

**4–102.** If $F_1 = 500$ N and $F_2 = 1000$ N, determine the magnitude and coordinate direction angles of the resultant couple moment.

**4–103.** Determine the magnitude of couple forces $\mathbf{F}_1$ and $\mathbf{F}_2$ so that the resultant couple moment acting on the block is zero.

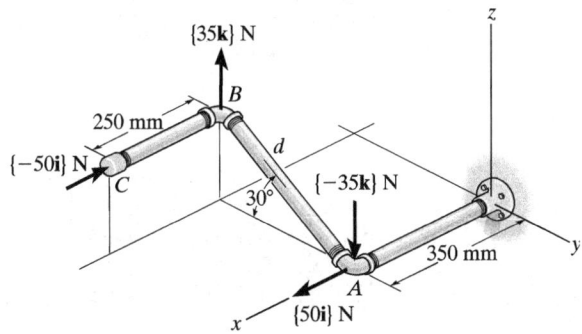

Probs. 4–98/99

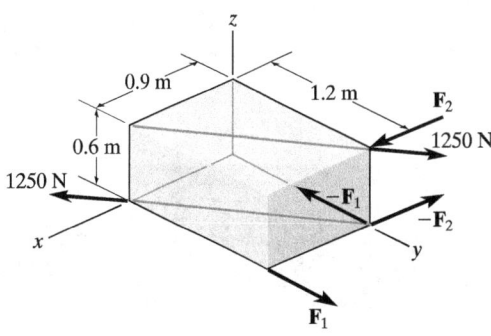

Probs. 4–102/103

## 4.7 Simplification of a Force and Couple System

Sometimes it is convenient to reduce a system of forces and couple moments acting on a body to a simpler form by replacing it with an *equivalent system*, consisting of a single resultant force acting at a specific point and a resultant couple moment. A system is equivalent if the *external effects* it produces on a body are the same as those caused by the original force and couple moment system. In this context, the external effects of a system refer to the *translating and rotating motion* of the body if the body is free to move, or it refers to the *reactive forces* at the supports if the body is held fixed.

For example, consider holding the stick in Fig. 4–34a, which is subjected to the force **F** at point A. If we attach a pair of equal but opposite forces **F** and **–F** at point B, which is *on the line of action* of **F**, Fig. 4–34b, we observe that **–F** at B and **F** at A will cancel each other, leaving only **F** at B, Fig. 4–34c. Force **F** has now been moved from A to B without modifying its *external effects* on the stick; i.e., the reaction at the grip remains the same. This demonstrates the *principle of transmissibility*, which states that a force acting on a body (stick) is a *sliding vector* since it can be applied at any point along its line of action.

We can also use the above procedure to move a force to a point that is *not* on the line of action of the force. If **F** is applied perpendicular to the stick, as in Fig. 4–35a, then we can attach a pair of equal but opposite forces **F** and **–F** to B, Fig. 4–35b. Force **F** is now applied at B, and the other two forces, **F** at A and **–F** at B, form a couple that produces the couple moment $M = Fd$, Fig. 4–35c. Therefore, the force **F** can be moved from A to B provided a couple moment **M** is added to maintain an equivalent system. This couple moment is determined by taking the moment of **F** about B. Since **M** is actually a *free vector*, it can act at any point on the stick. In both cases the systems are equivalent which causes a downward force **F** and clockwise couple moment $M = Fd$ to be felt at the grip.

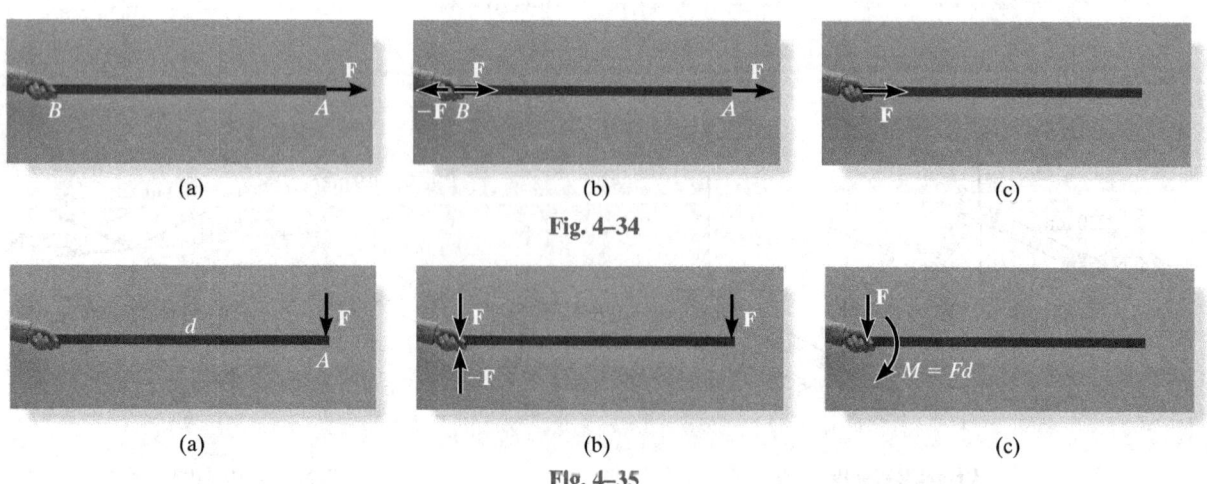

Fig. 4–34

Fig. 4–35

## 4.7 SIMPLIFICATION OF A FORCE AND COUPLE SYSTEM

**System of Forces and Couple Moments.** Using the above method, a system of several forces and couple moments acting on a body can be reduced to an equivalent single resultant force acting at a point $O$ and a resultant couple moment. For example, in Fig. 4–36a, $O$ is not on the line of action of $\mathbf{F}_1$, and so this force can be moved to point $O$ provided a couple moment $\mathbf{M}_1 = \mathbf{r}_1 \times \mathbf{F}$ is added to the body. Similarly, the couple moment $\mathbf{M}_2 = \mathbf{r}_2 \times \mathbf{F}_2$ should be added to the body when we move $\mathbf{F}_2$ to point $O$. Finally, since the couple moment $\mathbf{M}$ is a free vector, it can just be moved to point $O$. By doing this, we obtain the equivalent system shown in Fig. 4–36b, which produces the same external effects (support reactions) on the body as that of the force and couple system shown in Fig. 4–36a. If we sum the forces and couple moments, we obtain the resultant force $\mathbf{F}_R = \mathbf{F}_1 + \mathbf{F}_2$ and the resultant couple moment $(\mathbf{M}_R)_O = \mathbf{M} + \mathbf{M}_1 + \mathbf{M}_2$, Fig. 4–36c.

Notice that $\mathbf{F}_R$ is independent of the location of point $O$; however, $(\mathbf{M}_R)_O$ depends upon this location since the moments $\mathbf{M}_1$ and $\mathbf{M}_2$ are determined using the position vectors $\mathbf{r}_1$ and $\mathbf{r}_2$. Also note that $(\mathbf{M}_R)_O$ is a free vector and can act at *any point* on the body, although point $O$ is generally chosen as its point of application.

We can generalize the above method of reducing a force and couple system to an equivalent resultant force $\mathbf{F}_R$ acting at point $O$ and a resultant couple moment $(\mathbf{M}_R)_O$ by using the following two equations.

$$\mathbf{F}_R = \Sigma \mathbf{F}$$
$$(\mathbf{M}_R)_O = \Sigma \mathbf{M}_O + \Sigma \mathbf{M} \qquad (4\text{–}17)$$

The first equation states that the resultant force of the system is equivalent to the sum of all the forces; and the second equation states that the resultant couple moment of the system is equivalent to the sum of all the couple moments $\Sigma \mathbf{M}$ plus the moments of all the forces $\Sigma \mathbf{M}_O$ about point $O$. If the force system lies in the $x$–$y$ plane and any couple moments are perpendicular to this plane, then the above equations reduce to the following three scalar equations.

$$(F_R)_x = \Sigma F_x$$
$$(F_R)_y = \Sigma F_y \qquad (4\text{–}18)$$
$$(M_R)_O = \Sigma M_O + \Sigma M$$

Here the resultant force is determined from the vector sum of its two components $(F_R)_x$ and $(F_R)_y$.

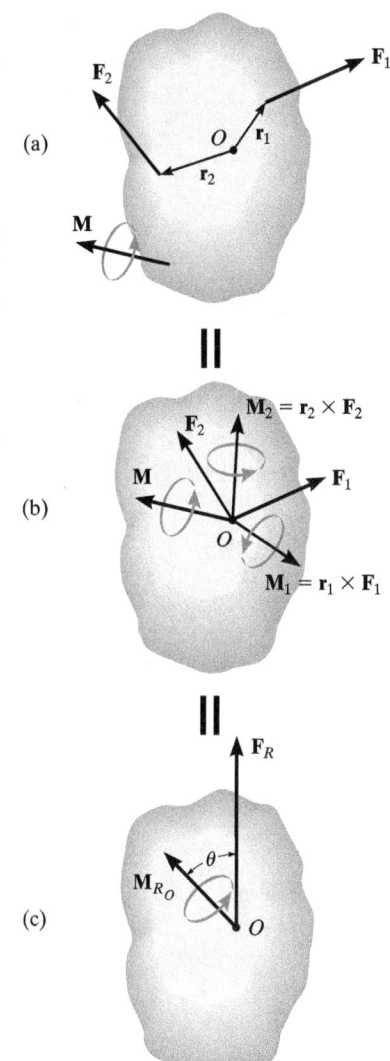

(a)

(b)

(c)

Fig. 4–36

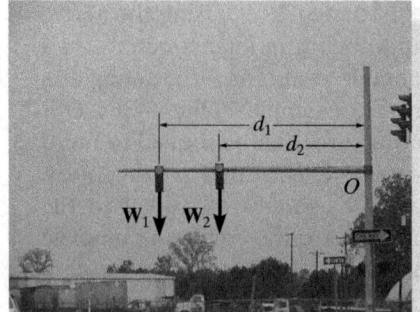 

The weights of these traffic lights can be replaced by their equivalent resultant force $W_R = W_1 + W_2$ and a couple moment $(M_R)_O = W_1 d_1 + W_2 d_2$ at the support, O. In both cases the support must provide the same resistance to translation and rotation in order to keep the member in the horizontal position.

## Procedure for Analysis

The following points should be kept in mind when simplifying a force and couple moment system to an equivalent resultant force and couple system.

- Establish the coordinate axes with the origin located at point $O$ and the axes having a selected orientation.

Force Summation.

- If the force system is *coplanar*, resolve each force into its $x$ and $y$ components. If a component is directed along the positive $x$ or $y$ axis, it represents a positive scalar; whereas if it is directed along the negative $x$ or $y$ axis, it is a negative scalar.

- In three dimensions, represent each force as a Cartesian vector before summing the forces.

Moment Summation.

- When determining the moments of a *coplanar* force system about point $O$, it is generally advantageous to use the principle of moments, i.e., determine the moments of the components of each force, rather than the moment of the force itself.

- In three dimensions use the vector cross product to determine the moment of each force about point $O$. Here the position vectors extend from $O$ to any point on the line of action of each force.

## EXAMPLE 4.14

Replace the force and couple system shown in Fig. 4–37a by an equivalent resultant force and couple moment acting at point $O$.

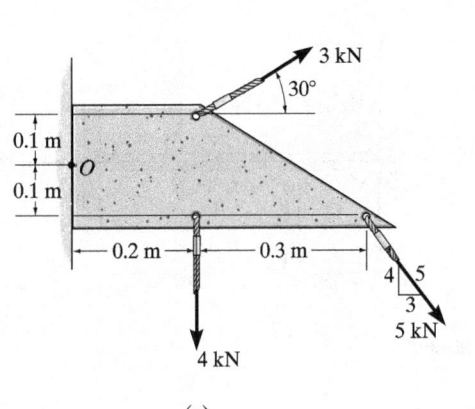
(a)

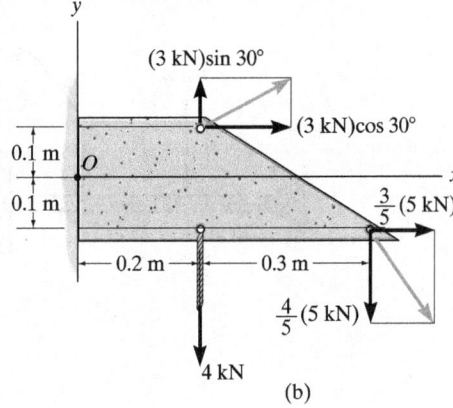
(b)

Fig. 4–37

### SOLUTION

**Force Summation.** The 3 kN and 5 kN forces are resolved into their $x$ and $y$ components as shown in Fig. 4–37b. We have

$$\xrightarrow{+} (F_R)_x = \Sigma F_x; (F_R)_x = (3 \text{ kN})\cos 30° + \left(\tfrac{3}{5}\right)(5 \text{ kN}) = 5.598 \text{ kN} \rightarrow$$

$$+\uparrow (F_R)_y = \Sigma F_y; (F_R)_y = (3 \text{ kN})\sin 30° - \left(\tfrac{4}{5}\right)(5 \text{ kN}) - 4 \text{ kN} = -6.50 \text{ kN} = 6.50 \text{ kN}\downarrow$$

Using the Pythagorean theorem, Fig. 4–37c, the magnitude of $\mathbf{F}_R$ is

$$F_R = \sqrt{(F_{Rx})^2 + (F_{Ry})^2} = \sqrt{(5.598 \text{ kN}^2) + (6.50 \text{ kN}^2)} = 8.58 \text{ kN} \quad Ans.$$

Its direction $\theta$ is

$$\theta = \tan^{-1}\left(\frac{(F_R)_y}{(F_R)_x}\right) = \tan^{-1}\left(\frac{6.50 \text{ kN}}{5.598 \text{ kN}}\right) = 49.3° \quad Ans.$$

**Moment Summation.** The moments of 3 kN and 5 kN about point $O$ will be determined using their $x$ and $y$ components. Referring to Fig. 4–37b, we have
$\zeta + (M_R)_O = \Sigma M_O;$

$$(M_R)_O = (3 \text{ kN})\sin 30°(0.2 \text{ m}) - (3 \text{ kN})\cos 30°(0.1 \text{ m}) + \left(\tfrac{3}{5}\right)(5 \text{ kN})(0.1 \text{ m})$$
$$- \left(\tfrac{4}{5}\right)(5 \text{ kN})(0.5 \text{ m}) - (4 \text{ kN})(0.2 \text{ m})$$
$$= -2.46 \text{ kN} \cdot \text{m} = 2.46 \text{ kN} \cdot \text{m} \,\rotatebox \quad Ans.$$

This clockwise moment is shown in Fig. 4–37c.

**NOTE:** Realize that the resultant force and couple moment in Fig. 4–37c will produce the same external effects or reactions at the supports as those produced by the force system, Fig 4–37a.

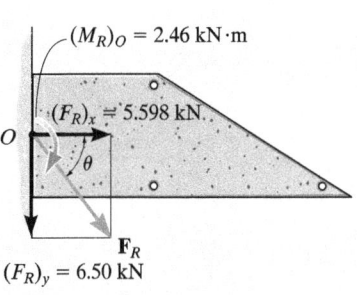
(c)

# EXAMPLE 4.15

Replace the force and couple system acting on the member in Fig. 4–38a by an equivalent resultant force and couple moment acting at point O.

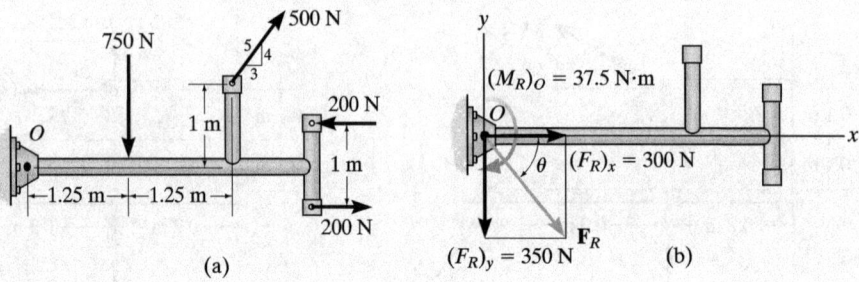

Fig. 4–38

## SOLUTION

**Force Summation.** Since the couple forces of 200 N are equal but opposite, they produce a zero resultant force, and so it is not necessary to consider them in the force summation. The 500-N force is resolved into its $x$ and $y$ components, thus,

$$\xrightarrow{+} (F_R)_x = \Sigma F_x; \quad (F_R)_x = \left(\tfrac{3}{5}\right)(500 \text{ N}) = 300 \text{ N} \rightarrow$$

$$+\uparrow (F_R)_y = \Sigma F_y; \quad (F_R)_y = (500 \text{ N})\left(\tfrac{4}{5}\right) - 750 \text{ N} = -350 \text{ N} = 350 \text{ N} \downarrow$$

From Fig. 4–15b, the magnitude of $\mathbf{F}_R$ is

$$F_R = \sqrt{(F_R)_x^2 + (F_R)_y^2}$$

$$= \sqrt{(300 \text{ N})^2 + (350 \text{ N})^2} = 461 \text{ N} \qquad Ans.$$

And the angle $\theta$ is

$$\theta = \tan^{-1}\left(\frac{(F_R)_y}{(F_R)_x}\right) = \tan^{-1}\left(\frac{350 \text{ N}}{300 \text{ N}}\right) = 49.4° \qquad Ans.$$

**Moment Summation.** Since the couple moment is a free vector, it can act at any point on the member. Referring to Fig. 4–38a, we have

$$\zeta + (M_R)_O = \Sigma M_O + \Sigma M_c;$$

$$(M_R)_O = (500 \text{ N})\left(\tfrac{4}{5}\right)(2.5 \text{ m}) - (500 \text{ N})\left(\tfrac{3}{5}\right)(1 \text{ m})$$

$$- (750 \text{ N})(1.25 \text{ m}) + 200 \text{ N·m}$$

$$= -37.5 \text{ N·m} = 37.5 \text{ N·m} \downarrow \qquad Ans.$$

This clockwise moment is shown in Fig. 4–38b.

## EXAMPLE 4.16

The structural member is subjected to a couple moment **M** and forces $\mathbf{F}_1$ and $\mathbf{F}_2$ in Fig. 4–39a. Replace this system by an equivalent resultant force and couple moment acting at its base, point $O$.

### SOLUTION *(VECTOR ANALYSIS)*
The three-dimensional aspects of the problem can be simplified by using a Cartesian vector analysis. Expressing the forces and couple moment as Cartesian vectors, we have

$$\mathbf{F}_1 = \{-800\mathbf{k}\} \text{ N}$$

$$\mathbf{F}_2 = (300 \text{ N})\mathbf{u}_{CB}$$

$$= (300 \text{ N})\left(\frac{\mathbf{r}_{CB}}{r_{CB}}\right)$$

$$= 300 \text{ N}\left[\frac{\{-0.15\mathbf{i} + 0.1\mathbf{j}\} \text{ m}}{\sqrt{(-0.15 \text{ m})^2 + (0.1 \text{ m})^2}}\right] = \{-249.6\mathbf{i} + 166.4\mathbf{j}\} \text{ N}$$

$$\mathbf{M} = -500\left(\tfrac{4}{5}\right)\mathbf{j} + 500\left(\tfrac{3}{5}\right)\mathbf{k} = \{-400\mathbf{j} + 300\mathbf{k}\} \text{ N} \cdot \text{m}$$

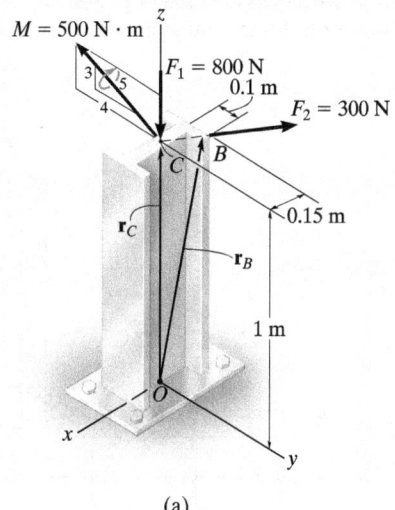

**Force Summation.**

$$\mathbf{F}_R = \Sigma\mathbf{F}; \qquad \mathbf{F}_R = \mathbf{F}_1 + \mathbf{F}_2 = -800\mathbf{k} - 249.6\mathbf{i} + 166.4\mathbf{j}$$

$$= \{-250\mathbf{i} + 166\mathbf{j} - 800\mathbf{k}\} \text{ N} \qquad Ans.$$

**Moment Summation.**

$$\mathbf{M}_{R_O} = \Sigma\mathbf{M} + \Sigma\mathbf{M}_O$$

$$\mathbf{M}_{R_O} = \mathbf{M} + \mathbf{r}_C \times \mathbf{F}_1 + \mathbf{r}_B \times \mathbf{F}_2$$

$$\mathbf{M}_{R_O} = (-400\mathbf{j} + 300\mathbf{k}) + (1\mathbf{k}) \times (-800\mathbf{k}) + \begin{vmatrix} \mathbf{i} & \mathbf{j} & \mathbf{k} \\ -0.15 & 0.1 & 1 \\ -249.6 & 166.4 & 0 \end{vmatrix}$$

$$= (-400\mathbf{j} + 300\mathbf{k}) + (\mathbf{0}) + (-166.4\mathbf{i} - 249.6\mathbf{j})$$

$$= \{-166\mathbf{i} - 650\mathbf{j} + 300\mathbf{k}\} \text{ N} \cdot \text{m} \qquad Ans.$$

The results are shown in Fig. 4–39b.

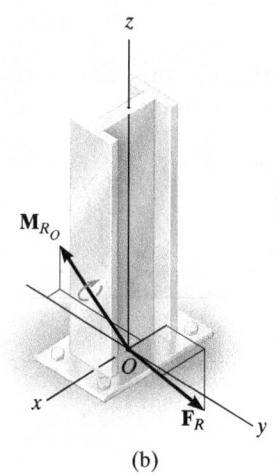

Fig. 4–39

## FUNDAMENTAL PROBLEMS

**F4–25.** Replace the loading system by an equivalent resultant force and couple moment acting at point $A$.

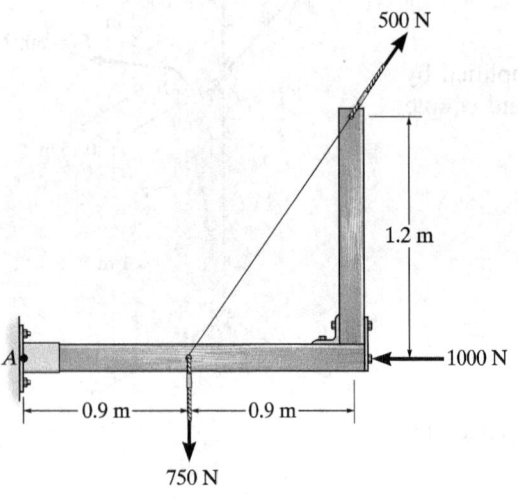

F4–25

**F4–26.** Replace the loading system by an equivalent resultant force and couple moment acting at point $A$.

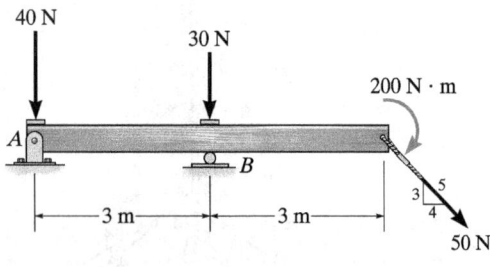

F4–26

**F4–27.** Replace the loading system by an equivalent resultant force and couple moment acting at point $A$.

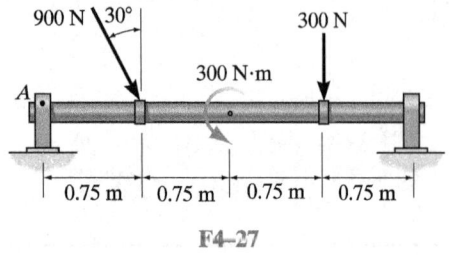

F4–27

**F4–28.** Replace the loading system by an equivalent resultant force and couple moment acting at point $A$.

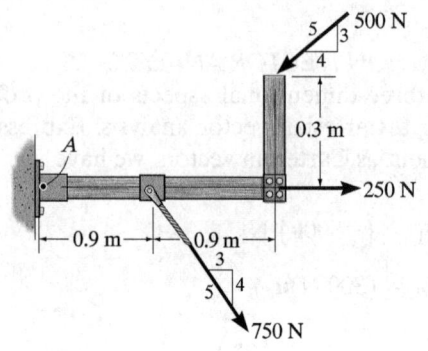

F4–28

**F4–29.** Replace the loading system by an equivalent resultant force and couple moment acting at point $O$.

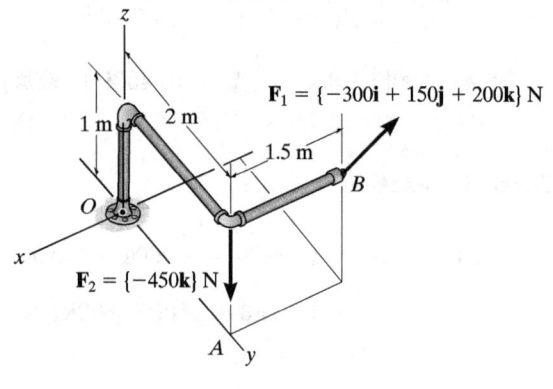

F4–29

**F4–30.** Replace the loading system by an equivalent resultant force and couple moment acting at point $O$.

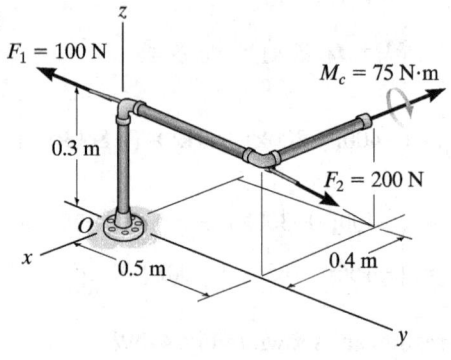

F4–30

## PROBLEMS

**\*4–104.** Replace the force system acting on the truss by a resultant force and couple moment at point $C$.

**4–107.** Replace the two forces by an equivalent resultant force and couple moment at point $O$. Set $F = 100$ N.

**\*4–108.** Replace the two forces by an equivalent resultant force and couple moment at point $O$. Set $F = 75$ N.

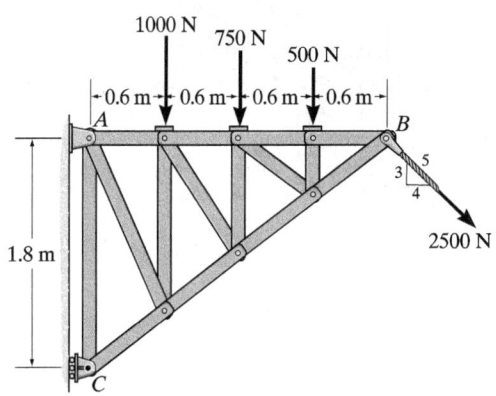

Prob. 4–104

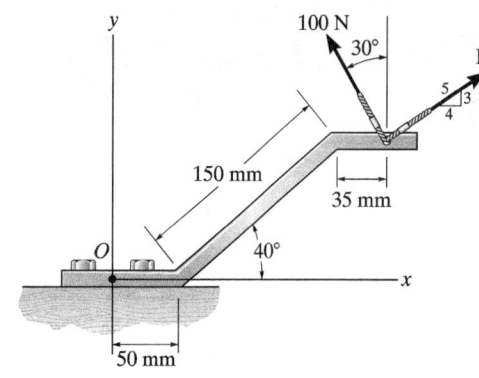

Probs. 4–107/108

**•4–105.** Replace the force system acting on the beam by an equivalent force and couple moment at point $A$.

**4–106.** Replace the force system acting on the beam by an equivalent force and couple moment at point $B$.

**•4–109.** Replace the force system acting on the post by a resultant force and couple moment at point $A$.

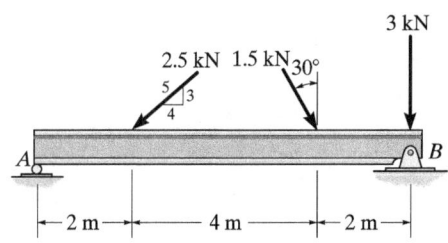

Probs. 4–105/106

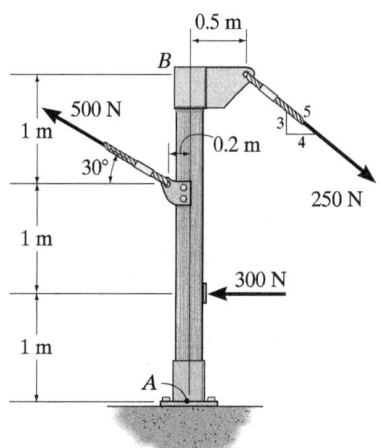

Prob. 4–109

**4–110.** Replace the force and couple moment system acting on the overhang beam by a resultant force and couple moment at point $A$.

***4–112.** Replace the two forces acting on the grinder by a resultant force and couple moment at point $O$. Express the results in Cartesian vector form.

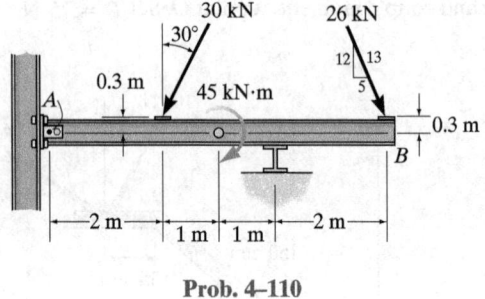

Prob. 4–110

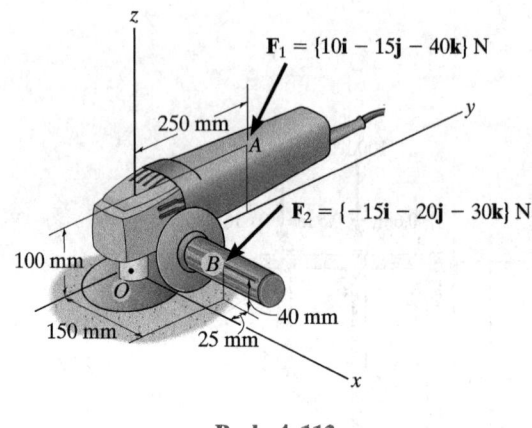

Prob. 4–112

**4–111.** Replace the force system by a resultant force and couple moment at point $O$.

**•4–113.** Replace the two forces acting on the post by a resultant force and couple moment at point $O$. Express the results in Cartesian vector form.

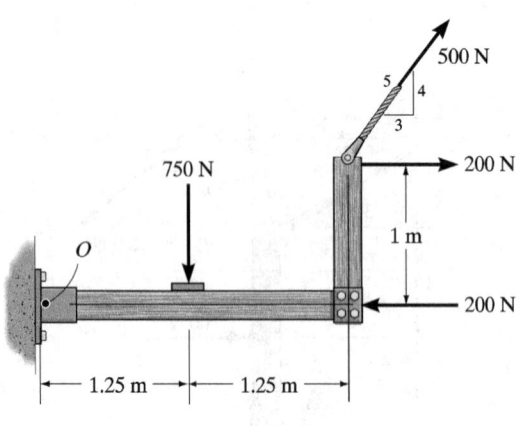

Prob. 4–111

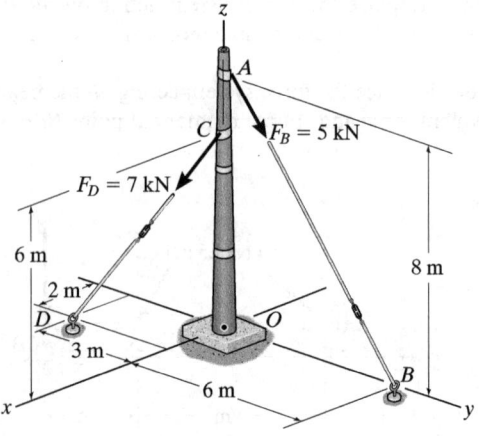

Prob. 4–113

**4–114.** The three forces act on the pipe assembly. If $F_1 = 50$ N and $F_2 = 80$ N, replace this force system by an equivalent resultant force and couple moment acting at $O$. Express the results in Cartesian vector form.

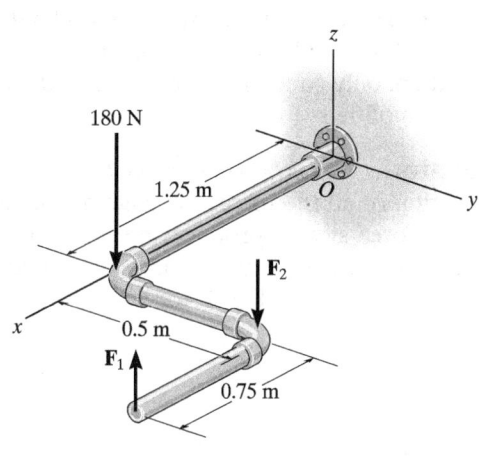

Prob. 4–114

**\*4–116.** Replace the force system acting on the pipe assembly by a resultant force and couple moment at point $O$. Express the results in Cartesian vector form.

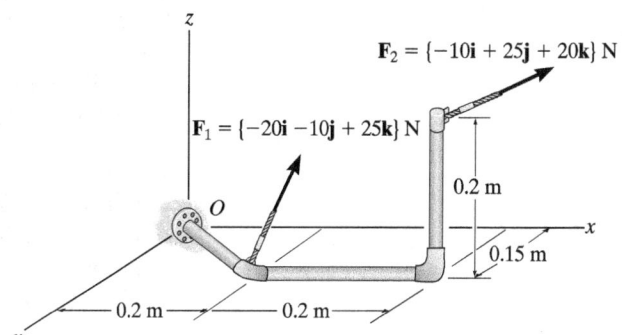

Prob. 4–116

**4–115.** Handle forces $\mathbf{F}_1$ and $\mathbf{F}_2$ are applied to the electric drill. Replace this force system by an equivalent resultant force and couple moment acting at point $O$. Express the results in Cartesian vector form.

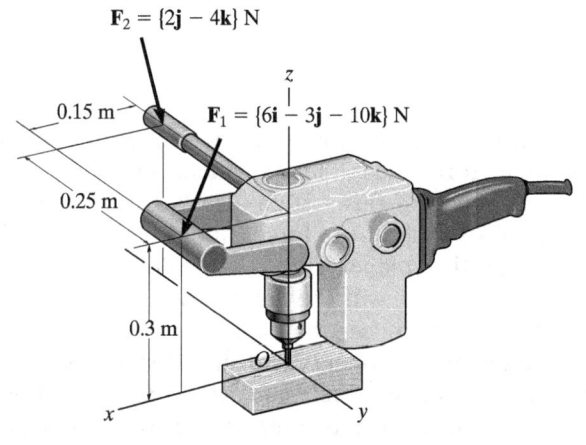

Prob. 4–115

**•4–117.** The slab is to be hoisted using the three slings shown. Replace the system of forces acting on the slings by an equivalent force and couple moment at point $O$. The force $\mathbf{F}_1$ is vertical.

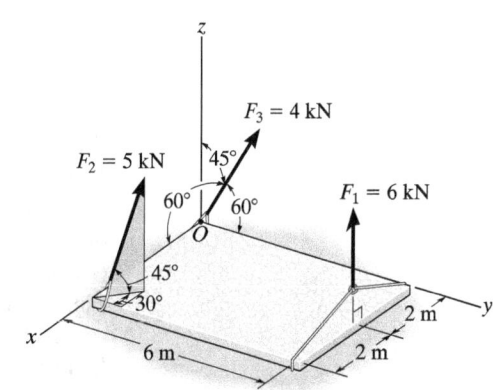

Prob. 4–117

## 4.8 Further Simplification of a Force and Couple System

In the preceding section, we developed a way to reduce a force and couple moment system acting on a rigid body into an equivalent resultant force $\mathbf{F}_R$ acting at a specific point $O$ and a resultant couple moment $(\mathbf{M}_R)_O$. The force system can be further reduced to an equivalent single resultant force provided the lines of action of $\mathbf{F}_R$ and $(\mathbf{M}_R)_O$ are *perpendicular* to each other. Because of this condition, only concurrent, coplanar, and parallel force systems can be further simplified.

**Concurrent Force System.** Since a *concurrent force system* is one in which the lines of action of all the forces intersect at a common point $O$, Fig. 4–40a, then the force system produces no moment about this point. As a result, the equivalent system can be represented by a single resultant force $\mathbf{F}_R = \Sigma \mathbf{F}$ acting at $O$, Fig. 4–40b.

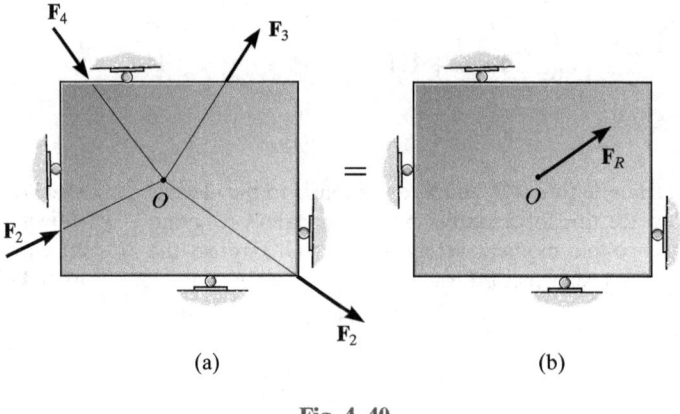

Fig. 4–40

**Coplanar Force System.** In the case of a *coplanar force system*, the lines of action of all the forces lie in the same plane, Fig. 4–41a, and so the resultant force $\mathbf{F}_R = \Sigma \mathbf{F}$ of this system also lies in this plane. Furthermore, the moment of each of the forces about any point $O$ is directed perpendicular to this plane. Thus, the resultant moment $(\mathbf{M}_R)_O$ and resultant force $\mathbf{F}_R$ will be *mutually perpendicular*, Fig. 4–41b. The resultant moment can be replaced by moving the resultant force $\mathbf{F}_R$ a perpendicular or moment arm distance $d$ away from point $O$ such that $\mathbf{F}_R$ produces the *same moment* $(\mathbf{M}_R)_O$ about point $O$, Fig. 4–41c. This distance $d$ can be determined from the scalar equation $(M_R)_O = F_R d = \Sigma M_O$ or $d = (M_R)_O / F_R$.

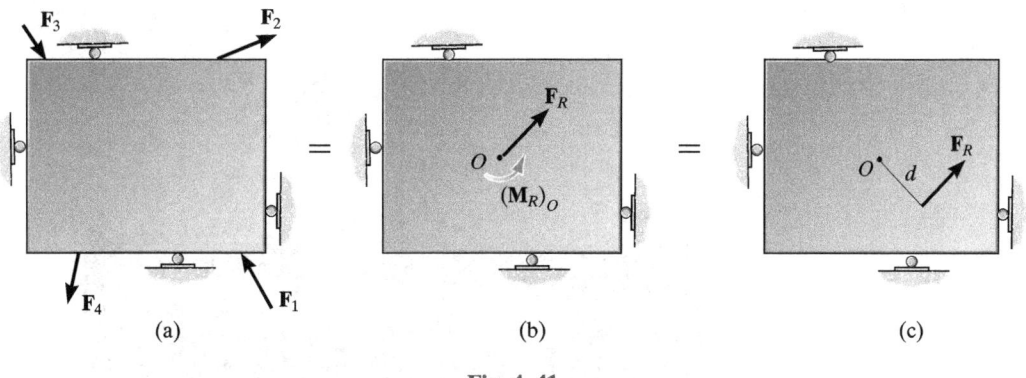

Fig. 4–41

**Parallel Force System.** The *parallel force system* shown in Fig. 4–42a consists of forces that are all parallel to the $z$ axis. Thus, the resultant force $\mathbf{F}_R = \Sigma \mathbf{F}$ at point $O$ must also be parallel to this axis, Fig. 4–42b. The moment produced by each force lies in the plane of the plate, and so the resultant couple moment, $(\mathbf{M}_R)_O$, will also lie in this plane, along the moment axis $a$ since $\mathbf{F}_R$ and $(\mathbf{M}_R)_O$ are mutually perpendicular. As a result, the force system can be further reduced to an equivalent single resultant force $\mathbf{F}_R$, acting through point $P$ located on the perpendicular $b$ axis, Fig. 4–42c. The distance $d$ along this axis from point $O$ requires $(M_R)_O = F_R d = \Sigma M_O$ or $d = \Sigma M_O / F_R$.

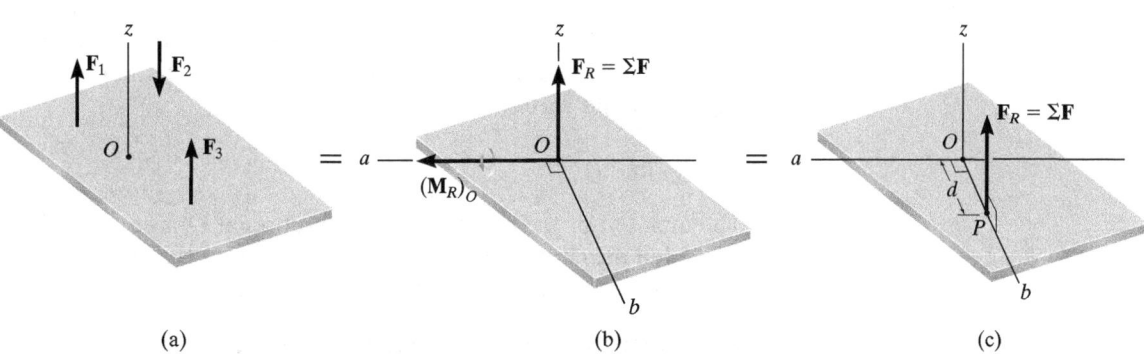

Fig. 4–42

The four cable forces are all concurrent at point $O$ on this bridge tower. Consequently they produce no resultant moment there, only a resultant force $\mathbf{F}_R$. Note that the designers have positioned the cables so that $\mathbf{F}_R$ is directed *along* the bridge tower directly to the support, so that it does not cause any bending of the tower.

### Procedure for Analysis

The technique used to reduce a coplanar or parallel force system to a single resultant force follows a similar procedure outlined in the previous section.

- Establish the $x$, $y$, $z$ axes and locate the resultant force $\mathbf{F}_R$ an arbitrary distance away from the origin of the coordinates.

Force Summation.

- The resultant force is equal to the sum of all the forces in the system.

- For a coplanar force system, resolve each force into its $x$ and $y$ components. Positive components are directed along the positive $x$ and $y$ axes, and negative components are directed along the negative $x$ and $y$ axes.

Moment Summation.

- The moment of the resultant force about point $O$ is equal to the sum of all the couple moments in the system plus the moments of all the forces in the system about $O$.

- This moment condition is used to find the location of the resultant force from point $O$.

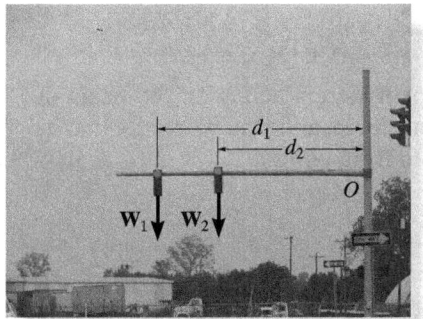 

Here the weights of the traffic lights are replaced by their resultant force $W_R = W_1 + W_2$ which acts at a distance $d = (W_1 d_1 + W_2 d_2)/W_R$ from $O$. Both systems are equivalent.

**Reduction to a Wrench** In general, a three-dimensional force and couple moment system will have an equivalent resultant force $\mathbf{F}_R$ acting at point $O$ and a resultant couple moment $(\mathbf{M}_R)_O$ that are *not perpendicular* to one another, as shown in Fig. 4–43a. Although a force system such as this cannot be further reduced to an equivalent single resultant force, the resultant couple moment $(\mathbf{M}_R)_O$ can be resolved into components parallel and perpendicular to the line of action of $\mathbf{F}_R$, Fig. 4–43a. The perpendicular component $\mathbf{M}_\perp$ can be replaced if we move $\mathbf{F}_R$ to point $P$, a distance $d$ from point $O$ along the $b$ axis, Fig. 4–43b. As seen, this axis is perpendicular to both the $a$ axis and the line of action of $\mathbf{F}_R$. The location of $P$ can be determined from $d = M_\perp / F_R$. Finally, because $\mathbf{M}_\parallel$ is a free vector, it can be moved to point $P$, Fig. 4–43c. This combination of a resultant force $\mathbf{F}_R$ and collinear couple moment $\mathbf{M}_\parallel$ will tend to translate and rotate the body about its axis and is referred to as a *wrench* or *screw*. A wrench is the simplest system that can represent any general force and couple moment system acting on a body.

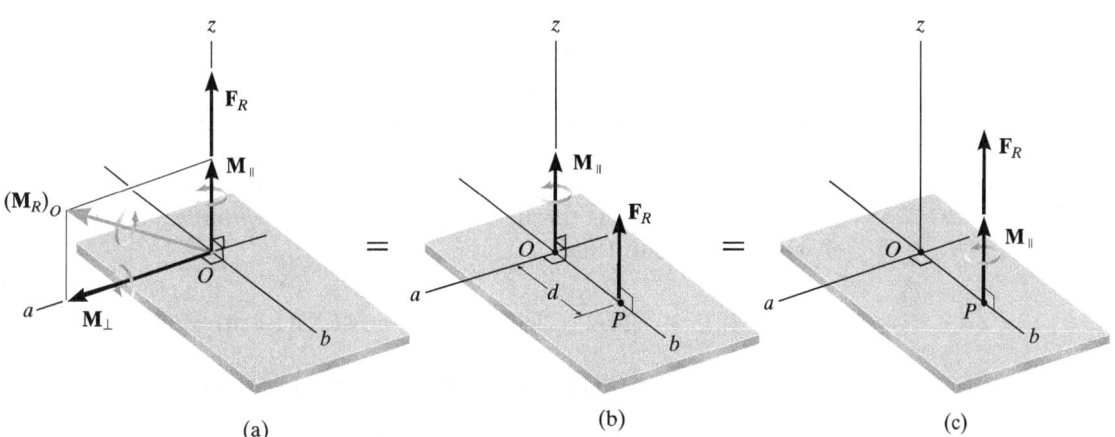

Fig. 4–43

# EXAMPLE 4.17

Replace the force and couple moment system acting on the beam in Fig. 4–44a by an equivalent resultant force, and find where its line of action intersects the beam, measured from point $O$.

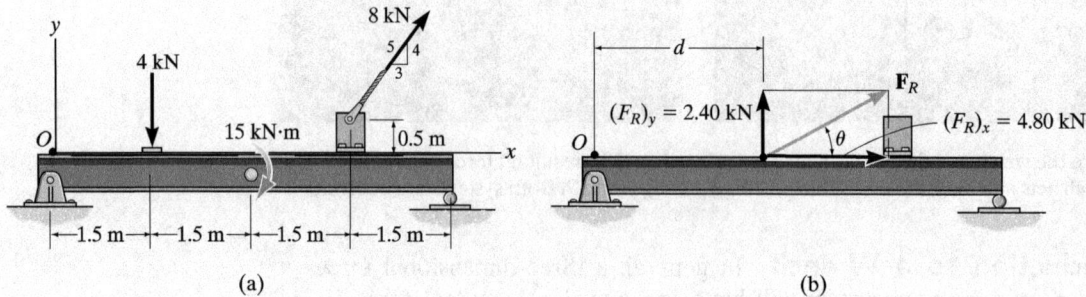

Fig. 4–44

## SOLUTION

**Force Summation.** Summing the force components,

$$\xrightarrow{+} (F_R)_x = \Sigma F_x; \quad (F_R)_x = 8 \text{ kN}\left(\tfrac{3}{5}\right) = 4.80 \text{ kN} \rightarrow$$

$$+\uparrow (F_R)_y = \Sigma F_y; \quad (F_R)_y = -4 \text{ kN} + 8 \text{ kN}\left(\tfrac{4}{5}\right) = 2.40 \text{ kN}\uparrow$$

From Fig. 4–44b, the magnitude of $\mathbf{F}_R$ is

$$F_R = \sqrt{(4.80 \text{ kN})^2 + (2.40 \text{ kN})^2} = 5.37 \text{ kN} \qquad Ans.$$

The angle $\theta$ is

$$\theta = \tan^{-1}\left(\frac{2.40 \text{ kN}}{4.80 \text{ kN}}\right) = 26.6° \qquad Ans.$$

**Moment Summation.** We must equate the moment of $\mathbf{F}_R$ about point $O$ in Fig. 4–44b to the sum of the moments of the force and couple moment system about point $O$ in Fig. 4–44a. Since the line of action of $(\mathbf{F}_R)_x$ acts through point $O$, only $(\mathbf{F}_R)_y$ produces a moment about this point. Thus,

$$\zeta + (M_R)_O = \Sigma M_O; \quad 2.40 \text{ kN}(d) = -(4 \text{ kN})(1.5 \text{ m}) - 15 \text{ kN·m}$$

$$-[8 \text{ kN}\left(\tfrac{3}{5}\right)](0.5 \text{ m}) + [8 \text{ kN}\left(\tfrac{4}{5}\right)](4.5 \text{ m})$$

$$d = 2.25 \text{ m} \qquad Ans.$$

## EXAMPLE 4.18

The jib crane shown in Fig. 4–45a is subjected to three coplanar forces. Replace this loading by an equivalent resultant force and specify where the resultant's line of action intersects the column $AB$ and boom $BC$.

### SOLUTION

**Force Summation.** Resolving the 2.50-kN force into $x$ and $y$ components and summing the force components yield

$\xrightarrow{+} F_{R_x} = \Sigma F_x;\quad F_{R_x} = -2.50\text{ kN}(\tfrac{3}{5}) - 1.75\text{ kN} = -3.25\text{ kN} = 3.25\text{ kN} \leftarrow$

$+\uparrow F_{R_y} = \Sigma F_y;\quad F_{R_y} = -2.50\text{ kN}(\tfrac{4}{5}) - 0.60\text{ kN} = -2.60\text{ kN} = 2.60\text{ kN} \downarrow$

As shown by the vector addition in Fig. 4–45b,

$$F_R = \sqrt{(3.25\text{ kN})^2 + (2.60\text{ kN})^2} = 4.16\text{ kN} \quad Ans.$$

$$\theta = \tan^{-1}\left(\frac{2.60\text{ kN}}{3.25\text{ kN}}\right) = 38.7° \; \theta \searrow \quad Ans.$$

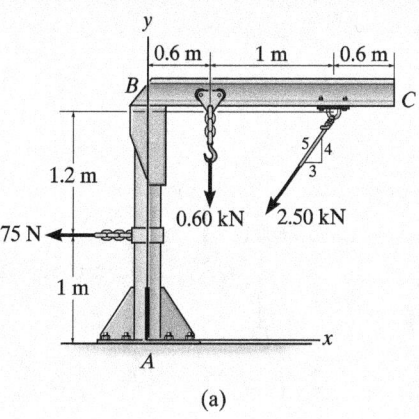

**Moment Summation.** Moments will be summed about point $A$. Assuming the line of action of $\mathbf{F}_R$ intersects $AB$ at a distance $y$ from $A$, Fig. 4–45b, we have

$\zeta + M_{R_A} = \Sigma M_A;\quad 3.25\text{ kN}(y) + 2.60\text{ kN}(0)$

$= 1.75\text{ kN}(1\text{ m}) - 0.60\text{ kN}(0.6\text{ m}) + 2.50\text{ kN}(\tfrac{3}{5})(2.2\text{ m}) - 2.50\text{ kN}(\tfrac{4}{5})(1.6\text{ m})$

$$y = 0.458\text{ m} \quad Ans.$$

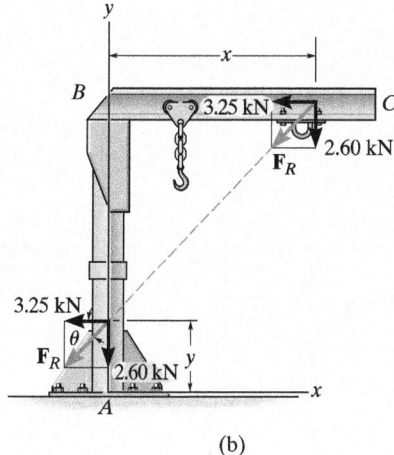

Fig. 4–45

By the principle of transmissibility, $\mathbf{F}_R$ can be placed at a distance $x$ where it intersects $BC$, Fig. 4–45b. In this case we have

$\zeta + M_{R_A} = \Sigma M_A;\quad 3.25\text{ kN}(2.2\text{ m}) - 2.60\text{ kN}(x)$

$= 1.75\text{ kN}(1\text{ m}) - 0.60\text{ kN}(0.6\text{ m}) + 2.50\text{ kN}(\tfrac{3}{5})(2.2\text{ m}) - 2.50\text{ kN}(\tfrac{4}{5})(1.6\text{ m})$

$$x = 2.177\text{ m} \quad Ans.$$

## EXAMPLE 4.19

The slab in Fig. 4–46a is subjected to four parallel forces. Determine the magnitude and direction of a resultant force equivalent to the given force system and locate its point of application on the slab.

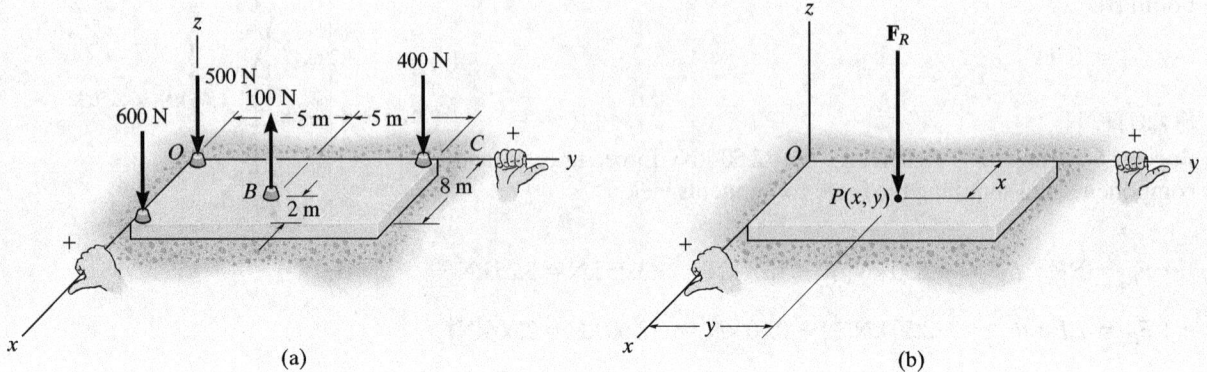

Fig. 4–46

### SOLUTION (SCALAR ANALYSIS)

**Force Summation.** From Fig. 4–46a, the resultant force is

$+\uparrow F_R = \Sigma F;$   $-F_R = -600 \text{ N} + 100 \text{ N} - 400 \text{ N} - 500 \text{ N}$

$= -1400 \text{ N} = 1400 \text{ N} \downarrow$   *Ans.*

**Moment Summation.** We require the moment about the $x$ axis of the resultant force, Fig. 4–46b, to be equal to the sum of the moments about the $x$ axis of all the forces in the system, Fig. 4–46a. The moment arms are determined from the $y$ coordinates since these coordinates represent the *perpendicular distances* from the $x$ axis to the lines of action of the forces. Using the right-hand rule, we have

$(M_R)_x = \Sigma M_x;$

$-(1400 \text{ N})y = 600 \text{ N}(0) + 100 \text{ N}(5 \text{ m}) - 400 \text{ N}(10 \text{ m}) + 500 \text{ N}(0)$

$-1400y = -3500$   $y = 2.50 \text{ m}$   *Ans.*

In a similar manner, a moment equation can be written about the $y$ axis using moment arms defined by the $x$ coordinates of each force.

$(M_R)_y = \Sigma M_y;$

$(1400 \text{ N})x = 600 \text{ N}(8 \text{ m}) - 100 \text{ N}(6 \text{ m}) + 400 \text{ N}(0) + 500 \text{ N}(0)$

$1400x = 4200$

$x = 3 \text{ m}$   *Ans.*

**NOTE:** A force of $F_R = 1400$ N placed at point $P(3.00 \text{ m}, 2.50 \text{ m})$ on the slab, Fig. 4–46b, is therefore equivalent to the parallel force system acting on the slab in Fig. 4–46a.

## EXAMPLE 4.20

Replace the force system in Fig. 4–47a by an equivalent resultant force and specify its point of application on the pedestal.

### SOLUTION

**Force Summation.** Here we will demonstrate a vector analysis. Summing forces,

$$\mathbf{F}_R = \Sigma \mathbf{F}; \quad \mathbf{F}_R = \mathbf{F}_A + \mathbf{F}_B + \mathbf{F}_C$$
$$= \{-300\mathbf{k}\} \text{ kN} + \{-500\mathbf{k}\} \text{ kN} + \{100\mathbf{k}\} \text{ kN}$$
$$= \{-700\mathbf{k}\} \text{ kN} \quad \textit{Ans.}$$

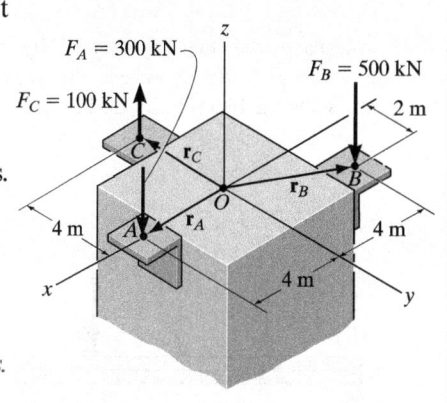

**Location.** Moments will be summed about point $O$. The resultant force $\mathbf{F}_R$ is assumed to act through point $P(x, y, 0)$, Fig. 4–47b. Thus

$$(\mathbf{M}_R)_O = \Sigma \mathbf{M}_O;$$
$$\mathbf{r}_P \times \mathbf{F}_R = (\mathbf{r}_A \times \mathbf{F}_A) + (\mathbf{r}_B \times \mathbf{F}_B) + (\mathbf{r}_C \times \mathbf{F}_C)$$
$$(x\mathbf{i} + y\mathbf{j}) \times (-700\mathbf{k}) = [(4\mathbf{i}) \times (-300\mathbf{k})]$$
$$+ [(-4\mathbf{i} + 2\mathbf{j}) \times (-500\mathbf{k})] + [(-4\mathbf{j}) \times (100\mathbf{k})]$$
$$-700x(\mathbf{i} \times \mathbf{k}) - 700y(\mathbf{j} \times \mathbf{k}) = -1200(\mathbf{i} \times \mathbf{k}) + 2000(\mathbf{i} \times \mathbf{k})$$
$$- 1000(\mathbf{j} \times \mathbf{k}) - 400(\mathbf{j} \times \mathbf{k})$$
$$700x\mathbf{j} - 700y\mathbf{i} = 1200\mathbf{j} - 2000\mathbf{j} - 1000\mathbf{i} - 400\mathbf{i}$$

Equating the **i** and **j** components,

$$-700y = -1400 \quad (1)$$
$$y = 2 \text{ m} \quad \textit{Ans.}$$
$$700x = -800 \quad (2)$$
$$x = -1.14 \text{ m} \quad \textit{Ans.}$$

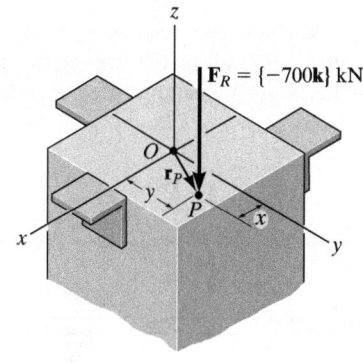

**Fig. 4–47**

The negative sign indicates that the $x$ coordinate of point $P$ is negative.

**NOTE:** It is also possible to establish Eq. 1 and 2 directly by summing moments about the $x$ and $y$ axes. Using the right-hand rule, we have

$$(M_R)_x = \Sigma M_x; \quad -700y = -100 \text{ kN}(4 \text{ m}) - 500 \text{ kN}(2 \text{ m})$$

$$(M_R)_y = \Sigma M_y; \quad 700x = 300 \text{ kN}(4 \text{ m}) - 500 \text{ kN}(4 \text{ m})$$

# FUNDAMENTAL PROBLEMS

**F4–31.** Replace the loading system by an equivalent resultant force and specify where the resultant's line of action intersects the beam measured from $O$.

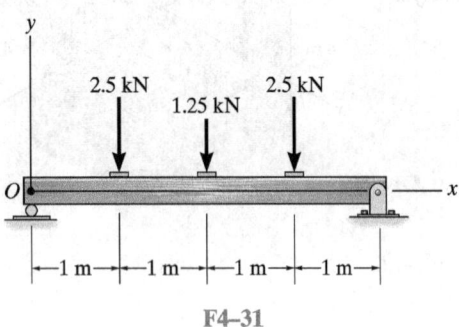

F4–31

**F4–32.** Replace the loading system by an equivalent resultant force and specify where the resultant's line of action intersects the member measured from $A$.

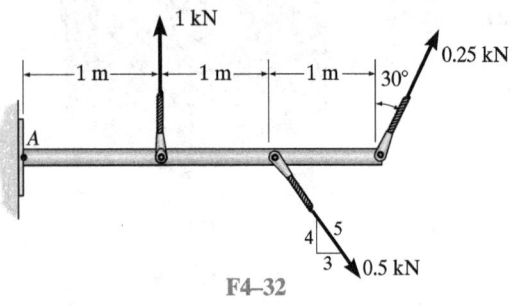

F4–32

**F4–33.** Replace the loading system by an equivalent resultant force and specify where the resultant's line of action intersects the member measured from $A$.

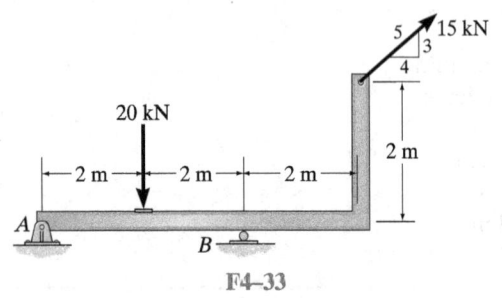

F4–33

**F4–34.** Replace the loading system by an equivalent resultant force and specify where the resultant's line of action intersects the member $AB$ measured from $A$.

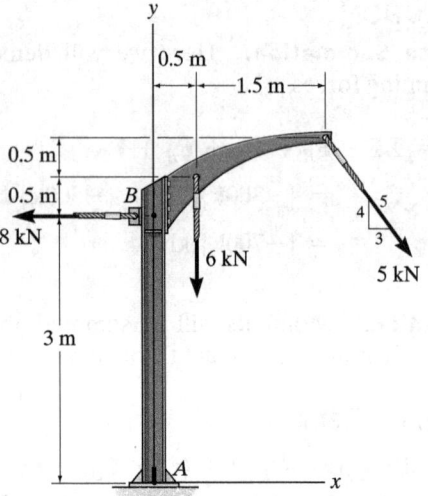

F4–34

**F4–35.** Replace the loading shown by an equivalent single resultant force and specify the $x$ and $y$ coordinates of its line of action.

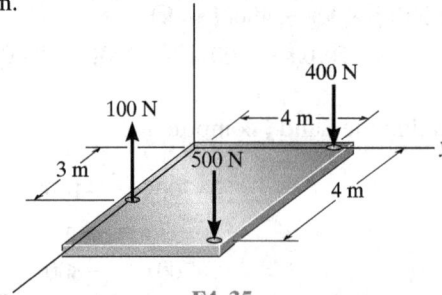

F4–35

**F4–36.** Replace the loading shown by an equivalent single resultant force and specify the $x$ and $y$ coordinates of its line of action.

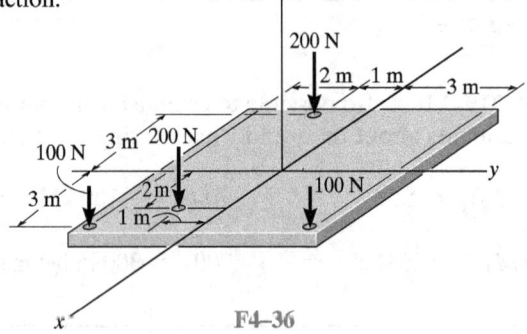

F4–36

## PROBLEMS

**4–118.** The weights of the various components of the truck are shown. Replace this system of forces by an equivalent resultant force and specify its location measured from $B$.

**4–119.** The weights of the various components of the truck are shown. Replace this system of forces by an equivalent resultant force and specify its location measured from point $A$.

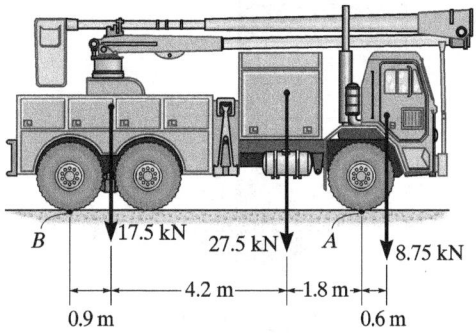

Probs. 4–118/119

**•4–121.** The system of four forces acts on the roof truss. Determine the equivalent resultant force and specify its location along $AB$, measured from point $A$.

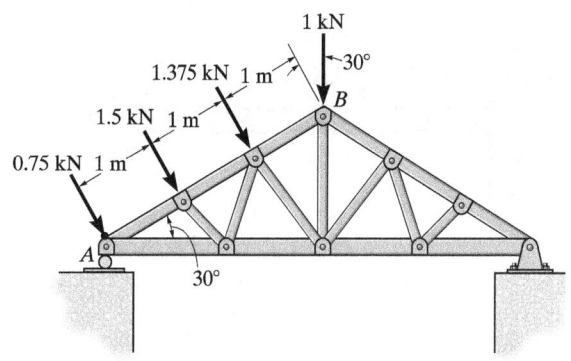

Prob. 4–121

**4–122.** Replace the force and couple system acting on the frame by an equivalent resultant force and specify where the resultant's line of action intersects member $AB$, measured from $A$.

**4–123.** Replace the force and couple system acting on the frame by an equivalent resultant force and specify where the resultant's line of action intersects member $BC$, measured from $B$.

***4–120.** The system of parallel forces acts on the top of the *Warren truss*. Determine the equivalent resultant force of the system and specify its location measured from point $A$.

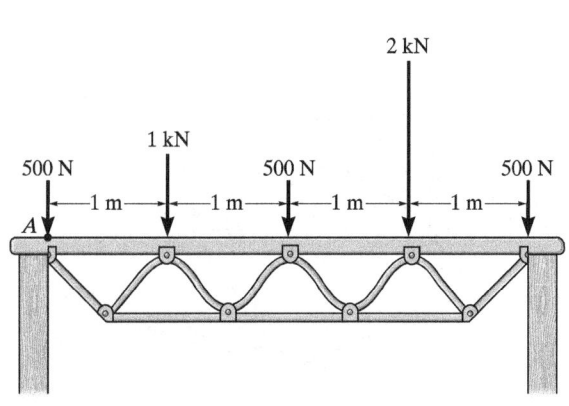

Prob. 4–120

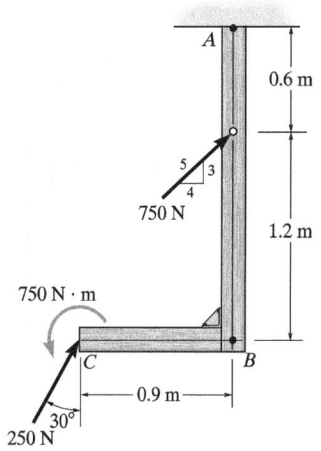

Probs. 4–122/123

**\*4–124.** Replace the force and couple moment system acting on the overhang beam by a resultant force, and specify its location along $AB$ measured from point $A$.

**4–127.** Replace the force system acting on the post by a resultant force, and specify where its line of action intersects the post $AB$ measured from point $A$.

**\*4–128.** Replace the force system acting on the post by a resultant force, and specify where its line of action intersects the post $AB$ measured from point $B$.

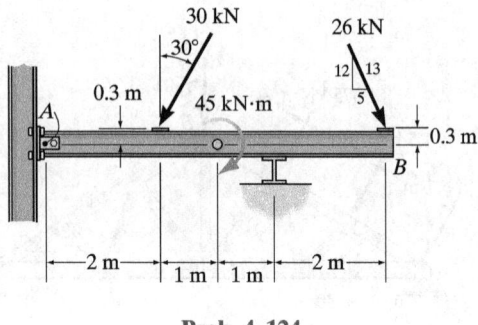

Prob. 4–124

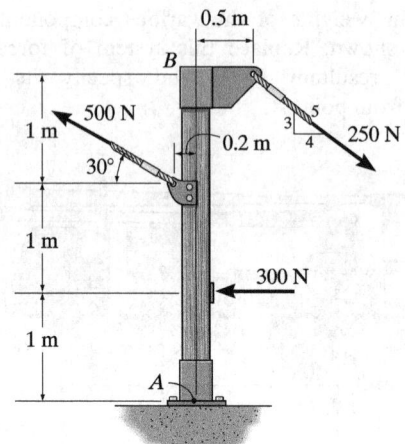

Probs. 4–127/128

**•4–125.** Replace the force system acting on the frame by an equivalent resultant force and specify where the resultant's line of action intersects member $AB$, measured from point $A$.

**4–126.** Replace the force system acting on the frame by an equivalent resultant force and specify where the resultant's line of action intersects member $BC$, measured from point $B$.

**•4–129.** The building slab is subjected to four parallel column loadings. Determine the equivalent resultant force and specify its location $(x, y)$ on the slab. Take $F_1 = 30$ kN, $F_2 = 40$ kN.

**4–130.** The building slab is subjected to four parallel column loadings. Determine the equivalent resultant force and specify its location $(x, y)$ on the slab. Take $F_1 = 20$ kN, $F_2 = 50$ kN.

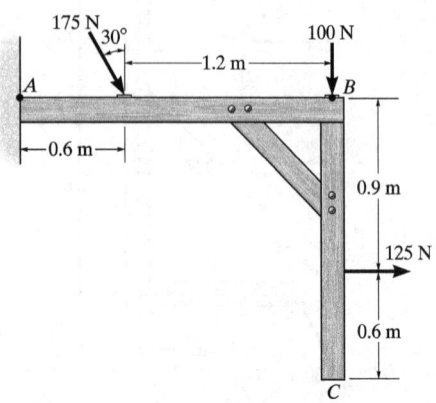

Probs. 4–125/126

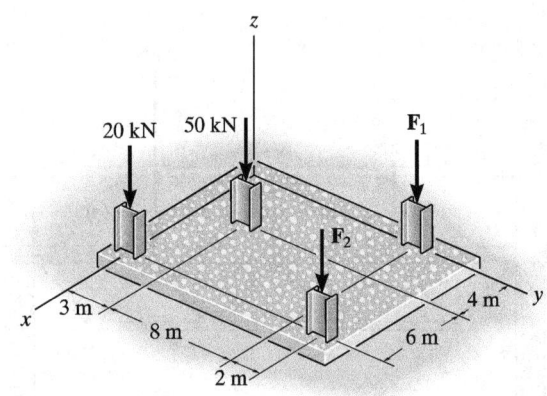

Probs. 4–129/130

**4–131.** The tube supports the four parallel forces. Determine the magnitudes of forces $\mathbf{F}_C$ and $\mathbf{F}_D$ acting at $C$ and $D$ so that the equivalent resultant force of the force system acts through the midpoint $O$ of the tube.

**4–134.** If $F_A = 40$ kN and $F_B = 35$ kN, determine the magnitude of the resultant force and specify the location of its point of application $(x, y)$ on the slab.

**4–135.** If the resultant force is required to act at the center of the slab, determine the magnitude of the column loadings $\mathbf{F}_A$ and $\mathbf{F}_B$ and the magnitude of the resultant force.

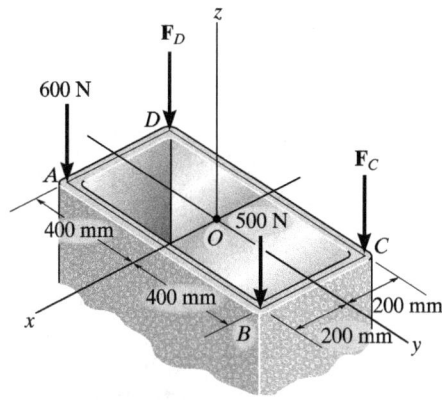

Prob. 4–131

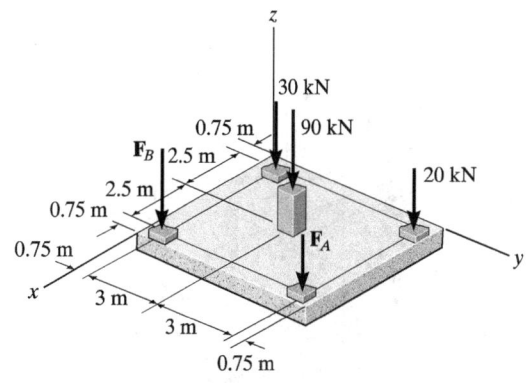

Probs. 4–134/135

*__4–132.__ Three parallel bolting forces act on the circular plate. Determine the resultant force, and specify its location $(x, z)$ on the plate. $F_A = 1000$ N, $F_B = 500$ N, and $F_C = 2000$ N.

•**4–133.** The three parallel bolting forces act on the circular plate. If the force at $A$ has a magnitude of $F_A = 1000$ N, determine the magnitudes of $\mathbf{F}_B$ and $\mathbf{F}_C$ so that the resultant force $\mathbf{F}_R$ of the system has a line of action that coincides with the $y$ axis. *Hint:* This requires $\Sigma M_x = 0$ and $\Sigma M_z = 0$.

*__4–136.__ Replace the parallel force system acting on the plate by a resultant force and specify its location on the $x$–$z$ plane.

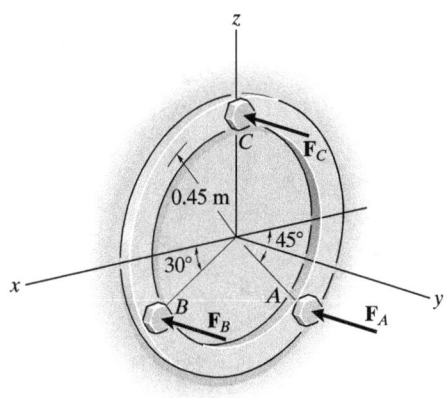

Probs. 4–132/133

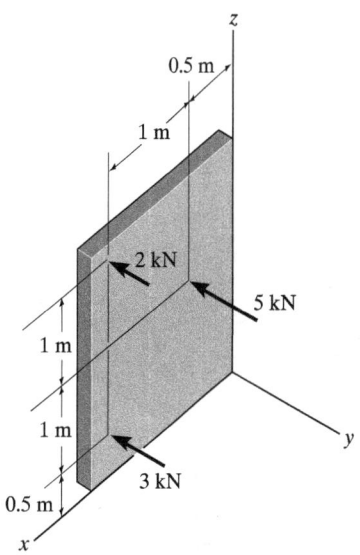

Prob. 4–136

•**4–137.** If $F_A = 7$ kN and $F_B = 5$ kN, represent the force system acting on the corbels by a resultant force, and specify its location on the $x$–$y$ plane.

**4–138.** Determine the magnitudes of $\mathbf{F}_A$ and $\mathbf{F}_B$ so that the resultant force passes through point $O$ of the column.

\***4–140.** Replace the three forces acting on the plate by a wrench. Specify the magnitude of the force and couple moment for the wrench and the point $P(y, z)$ where its line of action intersects the plate.

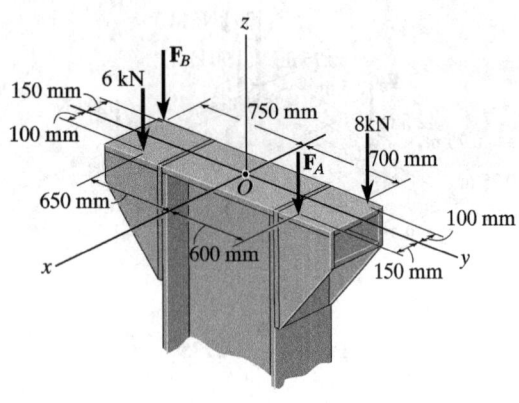

Probs. 4–137/138

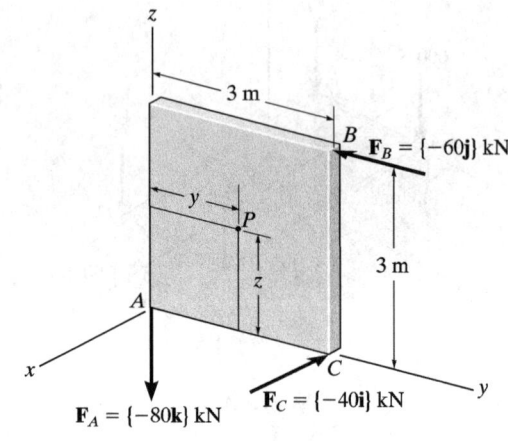

Prob. 4–140

**4–139.** Replace the force and couple moment system acting on the rectangular block by a wrench. Specify the magnitude of the force and couple moment of the wrench and where its line of action intersects the $x$–$y$ plane.

•**4–141.** Replace the three forces acting on the plate by a wrench. Specify the magnitude of the force and couple moment for the wrench and the point $P(x, y)$ where its line of action intersects the plate.

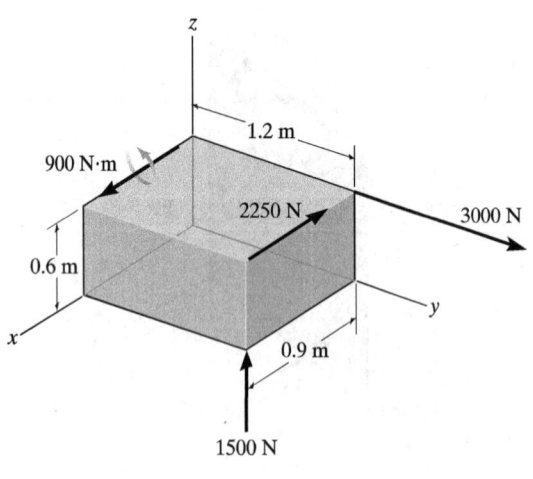

Prob. 4–139

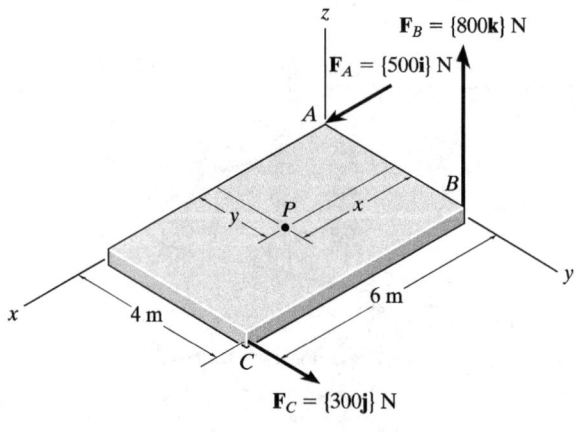

Prob. 4–141

## 4.9 Reduction of a Simple Distributed Loading

Sometimes, a body may be subjected to a loading that is distributed over its surface. For example, the pressure of the wind on the face of a sign, the pressure of water within a tank, or the weight of sand on the floor of a storage container, are all *distributed loadings*. The pressure exerted at each point on the surface indicates the intensity of the loading. It is measured using pascals Pa (or N/m²) in SI units or lb/ft² in the U.S. Customary system.

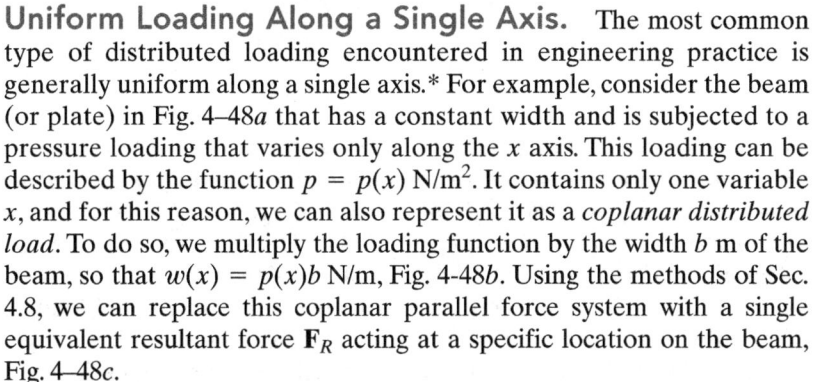

**Uniform Loading Along a Single Axis.** The most common type of distributed loading encountered in engineering practice is generally uniform along a single axis.* For example, consider the beam (or plate) in Fig. 4–48a that has a constant width and is subjected to a pressure loading that varies only along the $x$ axis. This loading can be described by the function $p = p(x)$ N/m². It contains only one variable $x$, and for this reason, we can also represent it as a *coplanar distributed load*. To do so, we multiply the loading function by the width $b$ m of the beam, so that $w(x) = p(x)b$ N/m, Fig. 4-48b. Using the methods of Sec. 4.8, we can replace this coplanar parallel force system with a single equivalent resultant force $\mathbf{F}_R$ acting at a specific location on the beam, Fig. 4–48c.

**Magnitude of Resultant Force.** From Eq. 4–17 ($F_R = \Sigma F$), the magnitude of $\mathbf{F}_R$ is equivalent to the sum of all the forces in the system. In this case, integration must be used since there is an infinite number of parallel forces $d\mathbf{F}$ acting on the beam, Fig. 4–48b. Since $d\mathbf{F}$ is acting on an element of length $dx$, and $w(x)$ is a force per unit length, then $dF = w(x)\, dx = dA$. In other words, the magnitude of $d\mathbf{F}$ is determined from the colored differential *area* $dA$ under the loading curve. For the entire length $L$,

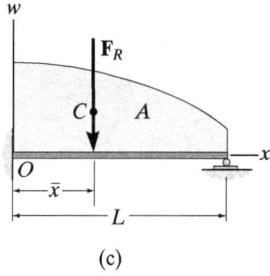

Fig. 4–48

$$+\downarrow F_R = \Sigma F; \qquad F_R = \int_L w(x)\, dx = \int_A dA = A \qquad (4\text{--}19)$$

Therefore, the magnitude of the resultant force is equal to the total area $A$ under the loading diagram, Fig. 4–48c.

---

*The more general case of a nonuniform surface loading acting on a body is considered in Sec. 9.5.

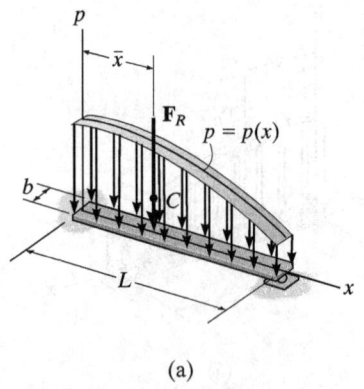

(a)

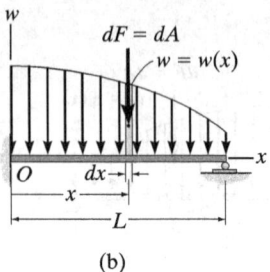

(b)

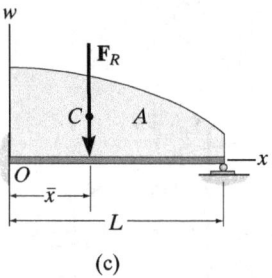

(c)

## Location of Resultant Force.

Applying Eq. 4–17 ($M_{R_O} = \Sigma M_O$), the location $\bar{x}$ of the line of action of $\mathbf{F}_R$ can be determined by equating the moments of the force resultant and the parallel force distribution about point $O$ (the $y$ axis). Since $d\mathbf{F}$ produces a moment of $x\, dF = xw(x)\, dx$ about $O$, Fig. 4–48b, then for the entire length, Fig. 4–48c,

$$\zeta + (M_R)_O = \Sigma M_O; \quad -\bar{x}F_R = -\int_L xw(x)\, dx$$

Solving for $\bar{x}$, using Eq. 4–19, we have

$$\bar{x} = \frac{\int_L xw(x)\, dx}{\int_L w(x)\, dx} = \frac{\int_A x\, dA}{\int_A dA} \quad (4\text{–}20)$$

This coordinate $\bar{x}$, locates the geometric center or *centroid* of the *area* under the distributed loading. *In other words, the resultant force has a line of action which passes through the centroid C (geometric center) of the area under the loading diagram*, Fig. 4–48c. Detailed treatment of the integration techniques for finding the location of the centroid for areas is given in Chapter 9. In many cases, however, the distributed-loading diagram is in the shape of a rectangle, triangle, or some other simple geometric form. The centroid location for such common shapes does not have to be determined from the above equation but can be obtained directly from the tabulation given on the inside back cover.

Once $\bar{x}$ is determined, $\mathbf{F}_R$ by symmetry passes through point $(\bar{x}, 0)$ on the surface of the beam, Fig. 4–48a. Therefore, in this case, the resultant force has a magnitude equal to the volume under the loading curve $p = p(x)$ and a line of action which passes through the centroid (geometric center) of this volume.

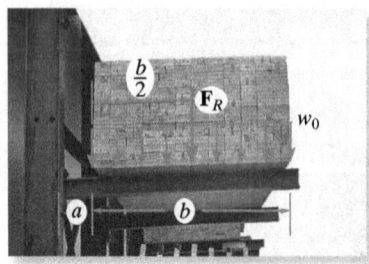

The beam supporting this stack of lumber is subjected to a uniform loading of $w_0$. The resultant force is therefore equal to the area under the loading diagram $F_R = w_0 b$. It acts through the centroid or geometric center of this area, $b/2$ from the support.

### Important Points

- Coplanar distributed loadings are defined by using a loading function $w = w(x)$ that indicates the intensity of the loading along the length of a member. This intensity is measured in N/m or lb/ft.

- The external effects caused by a coplanar distributed load acting on a body can be represented by a single resultant force.

- This resultant force is equivalent to the *area* under the loading diagram, and has a line of action that passes through the *centroid* or geometric center of this area.

## EXAMPLE 4.21

Determine the magnitude and location of the equivalent resultant force acting on the shaft in Fig. 4–49a.

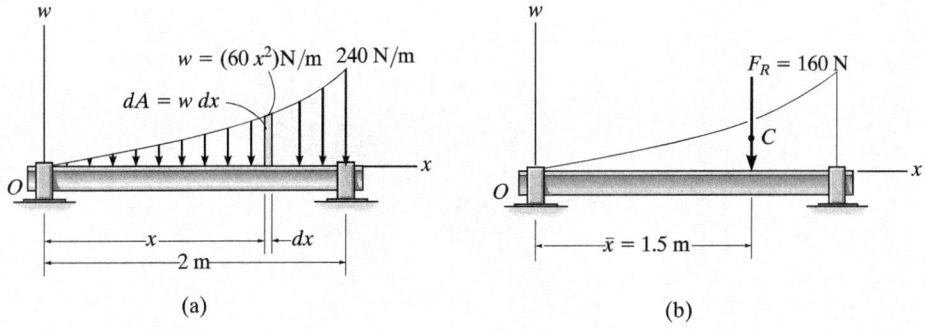

(a)

(b)

Fig. 4–49

### SOLUTION

Since $w = w(x)$ is given, this problem will be solved by integration.

The differential element has an area $dA = w\,dx = 60x^2\,dx$. Applying Eq. 4–19,

$+\downarrow F_R = \Sigma F;$

$$F_R = \int_A dA = \int_0^{2\,m} 60x^2\,dx = 60\left(\frac{x^3}{3}\right)\bigg|_0^{2\,m} = 60\left(\frac{2^3}{3} - \frac{0^3}{3}\right)$$

$$= 160\text{ N} \qquad\qquad\qquad Ans.$$

The location $\bar{x}$ of $\mathbf{F}_R$ measured from $O$, Fig. 4–49b, is determined from Eq. 4–20.

$$\bar{x} = \frac{\int_A x\,dA}{\int_A dA} = \frac{\int_0^{2\,m} x(60x^2)\,dx}{160\text{ N}} = \frac{60\left(\frac{x^4}{4}\right)\bigg|_0^{2\,m}}{160\text{ N}} = \frac{60\left(\frac{2^4}{4} - \frac{0^4}{4}\right)}{160\text{ N}}$$

$$= 1.5\text{ m} \qquad\qquad\qquad Ans.$$

**NOTE:** These results can be checked by using the table on the inside back cover, where it is shown that for an exparabolic area of length $a$, height $b$, and shape shown in Fig. 4–49a, we have

$$A = \frac{ab}{3} = \frac{2\text{ m}(240\text{ N/m})}{3} = 160\text{ N and } \bar{x} = \frac{3}{4}a = \frac{3}{4}(2\text{ m}) = 1.5\text{ m}$$

## EXAMPLE 4.22

A distributed loading of $p = (800x)$ Pa acts over the top surface of the beam shown in Fig. 4–50a. Determine the magnitude and location of the equivalent resultant force.

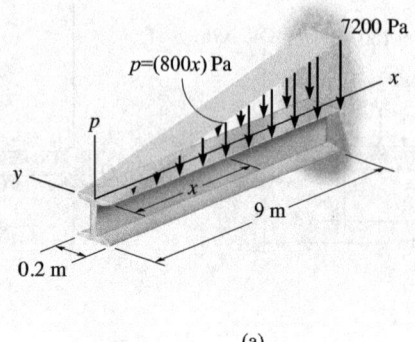

(a)

### SOLUTION
Since the loading intensity is uniform along the width of the beam (the $y$ axis), the loading can be viewed in two dimensions as shown in Fig. 4–50b. Here

$$w = (800x \text{ N/m}^2)(0.2 \text{ m})$$
$$= (160x) \text{ N/m}$$

At $x = 9$ m, note that $w = 1440$ N/m. Although we may again apply Eqs. 4–19 and 4–20 as in the previous example, it is simpler to use the table on the inside back cover.

The magnitude of the resultant force is equivalent to the area of the triangle.

$$F_R = \tfrac{1}{2}(9 \text{ m})(1440 \text{ N/m}) = 6480 \text{ N} = 6.48 \text{ kN} \qquad Ans.$$

The line of action of $\mathbf{F}_R$ passes through the *centroid C* of this triangle. Hence,

$$\bar{x} = 9 \text{ m} - \tfrac{1}{3}(9 \text{ m}) = 6 \text{ m} \qquad Ans.$$

The results are shown in Fig. 4–50c.

**NOTE:** We may also view the resultant $\mathbf{F}_R$ as *acting* through the *centroid* of the *volume* of the loading diagram $p = p(x)$ in Fig. 4–50a. Hence $\mathbf{F}_R$ intersects the $x$–$y$ plane at the point $(6 \text{ m}, 0)$. Furthermore, the magnitude of $\mathbf{F}_R$ is equal to the volume under the loading diagram; i.e.,

$$F_R = V = \tfrac{1}{2}(7200 \text{ N/m}^2)(9 \text{ m})(0.2 \text{ m}) = 6.48 \text{ kN} \qquad Ans.$$

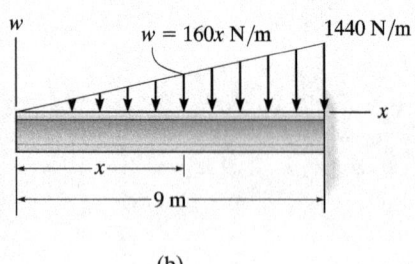

(b)

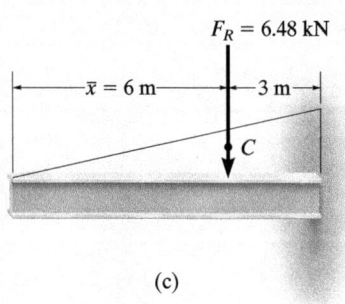

(c)

Fig. 4–50

## EXAMPLE 4.23

The granular material exerts the distributed loading on the beam as shown in Fig. 4–51a. Determine the magnitude and location of the equivalent resultant of this load.

### SOLUTION
The area of the loading diagram is a *trapezoid*, and therefore the solution can be obtained directly from the area and centroid formulas for a trapezoid listed on the inside back cover. Since these formulas are not easily remembered, instead we will solve this problem by using "composite" areas. Here we will divide the trapezoidal loading into a rectangular and triangular loading as shown in Fig. 4–51b. The magnitude of the force represented by each of these loadings is equal to its associated *area*,

$$F_1 = \tfrac{1}{2}(9 \text{ m})(50 \text{ kN/m}) = 225 \text{ kN}$$

$$F_2 = (9 \text{ m})(50 \text{ kN/m}) = 450 \text{ kN}$$

The lines of action of these parallel forces act through the *centroid* of their associated areas and therefore intersect the beam at

$$\bar{x}_1 = \tfrac{1}{3}(9 \text{ m}) = 3 \text{ m}$$

$$\bar{x}_2 = \tfrac{1}{2}(9 \text{ m}) = 4.5 \text{ m}$$

The two parallel forces $\mathbf{F}_1$ and $\mathbf{F}_2$ can be reduced to a single resultant $\mathbf{F}_R$. The magnitude of $\mathbf{F}_R$ is

$$+\downarrow F_R = \Sigma F; \qquad F_R = 225 + 450 = 675 \text{ kN} \qquad \textit{Ans.}$$

We can find the location of $\mathbf{F}_R$ with reference to point $A$, Fig. 4–51b and 4–51c. We require

$$\zeta + M_{R_A} = \Sigma M_A; \qquad \bar{x}(675) = 3(225) + 4.5(450)$$

$$\bar{x} = 4 \text{ m} \qquad \textit{Ans.}$$

**NOTE:** The trapezoidal area in Fig. 4–51a can also be divided into two triangular areas as shown in Fig. 4–51d. In this case

$$F_3 = \tfrac{1}{2}(9 \text{ m})(100 \text{ kN/m}) = 450 \text{ kN}$$

$$F_4 = \tfrac{1}{2}(9 \text{ m})(50 \text{ kN/m}) = 225 \text{ kN}$$

and

$$\bar{x}_3 = \tfrac{1}{3}(9 \text{ m}) = 3 \text{ m}$$

$$\bar{x}_4 = 9 \text{ m} - \tfrac{1}{3}(9 \text{ m}) = 6 \text{ m}$$

**NOTE:** Using these results, show again that $F_R = 675$ kN and $\bar{x} = 4$ m.

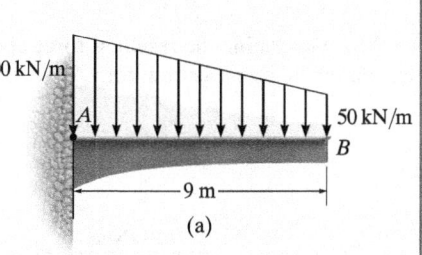
(a)

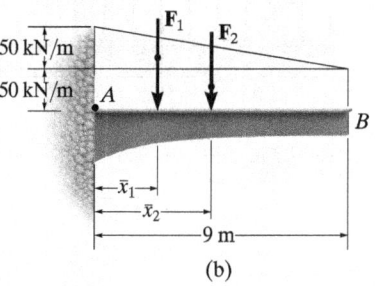
(b)

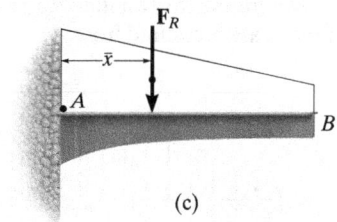
(c)

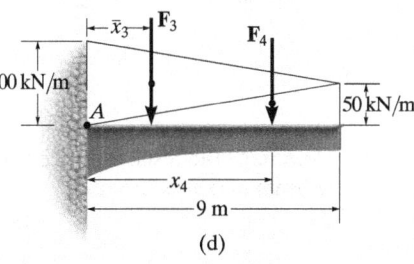
(d)

Fig. 4–51

## FUNDAMENTAL PROBLEMS

**F4–37.** Determine the resultant force and specify where it acts on the beam measured from $A$.

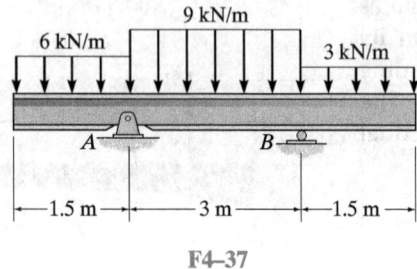

F4–37

**F4–38.** Determine the resultant force and specify where it acts on the beam measured from $A$.

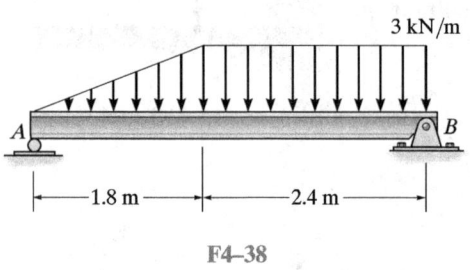

F4–38

**F4–39.** Determine the resultant force and specify where it acts on the beam measured from $A$.

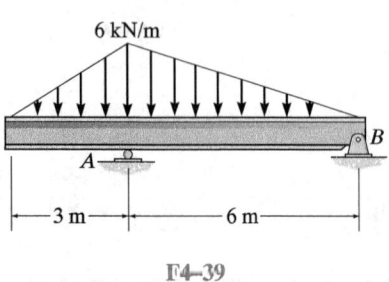

F4–39

**F4–40.** Determine the resultant force and specify where it acts on the beam measured from $A$.

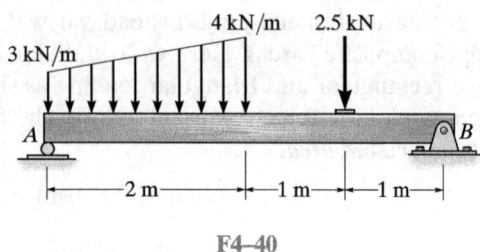

F4–40

**F4–41.** Determine the resultant force and specify where it acts on the beam measured from $A$.

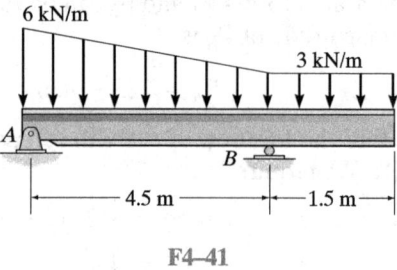

F4–41

**F4–42.** Determine the resultant force and specify where it acts on the beam measured from $A$.

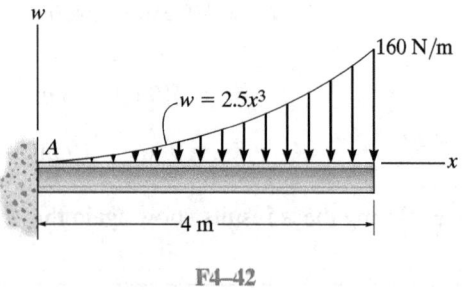

F4–42

## PROBLEMS

**4–142.** Replace the distributed loading with an equivalent resultant force, and specify its location on the beam measured from point $A$.

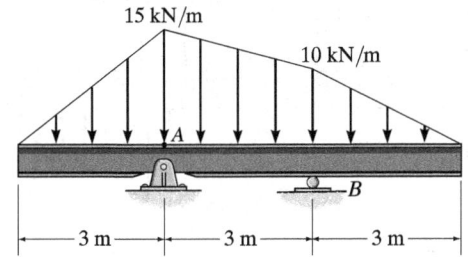

Prob. 4–142

**4–143.** Replace the distributed loading with an equivalent resultant force, and specify its location on the beam measured from point $A$.

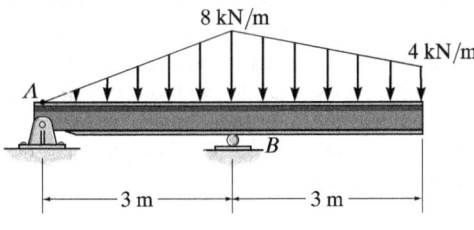

Prob. 4–143

**\*4–144.** Replace the distributed loading by an equivalent resultant force and specify its location, measured from point $A$.

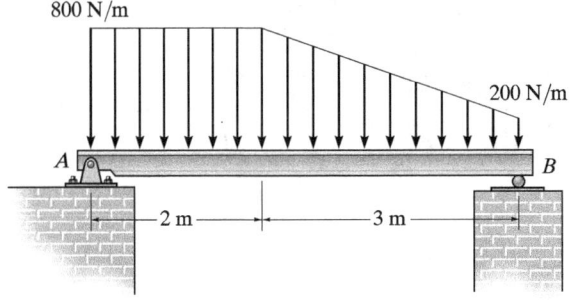

Prob. 4–144

**•4–145.** Replace the distributed loading with an equivalent resultant force, and specify its location on the beam measured from point $A$.

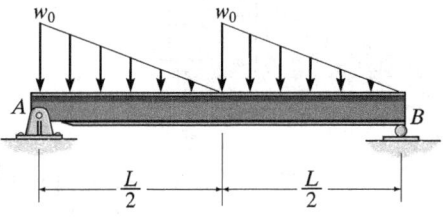

Prob. 4–145

**4–146.** The distribution of soil loading on the bottom of a building slab is shown. Replace this loading by an equivalent resultant force and specify its location, measured from point $O$.

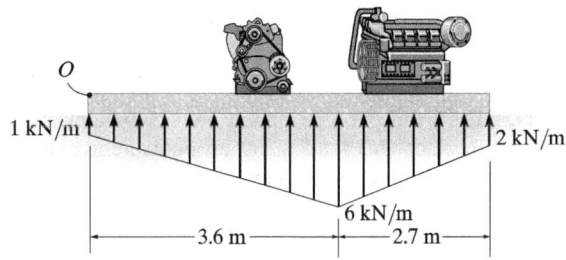

Prob. 4–146

**4–147.** Determine the intensities $w_1$ and $w_2$ of the distributed loading acting on the bottom of the slab so that this loading has an equivalent resultant force that is equal but opposite to the resultant of the distributed loading acting on the top of the plate.

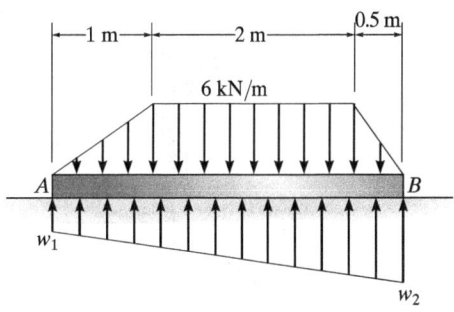

Prob. 4–147

**\*4–148.** The bricks on top of the beam and the supports at the bottom create the distributed loading shown in the second figure. Determine the required intensity $w$ and dimension $d$ of the right support so that the resultant force and couple moment about point $A$ of the system are both zero.

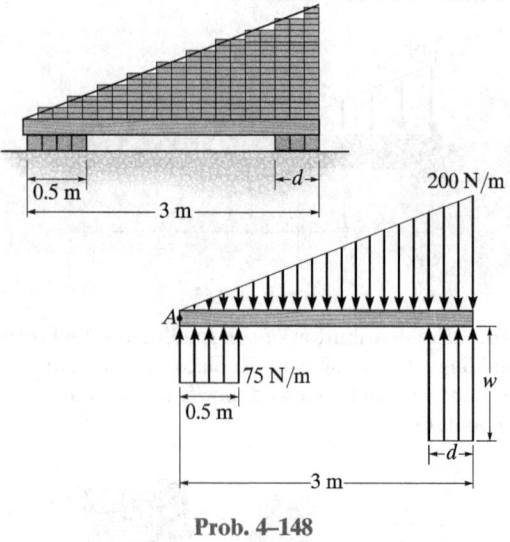

Prob. 4–148

**•4–149.** The wind pressure acting on a triangular sign is uniform. Replace this loading by an equivalent resultant force and couple moment at point $O$.

**4–150.** The beam is subjected to the distributed loading. Determine the length $b$ of the uniform load and its position $a$ on the beam such that the resultant force and couple moment acting on the beam are zero.

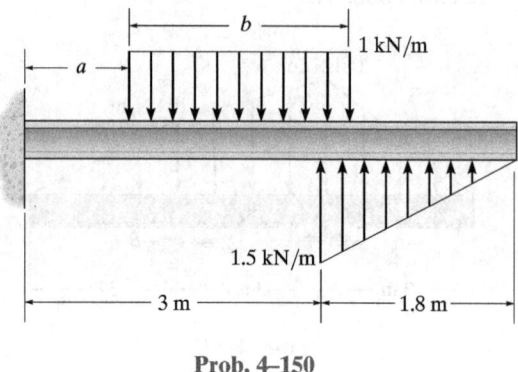

Prob. 4–150

**4–151.** Currently eighty-five percent of all neck injuries are caused by rear-end car collisions. To alleviate this problem, an automobile seat restraint has been developed that provides additional pressure contact with the cranium. During dynamic tests the distribution of load on the cranium has been plotted and shown to be parabolic. Determine the equivalent resultant force and its location, measured from point $A$.

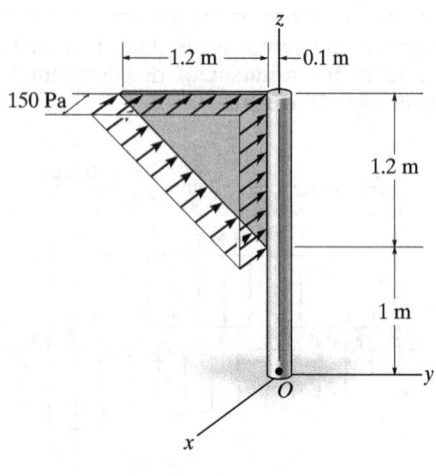

Prob. 4–149

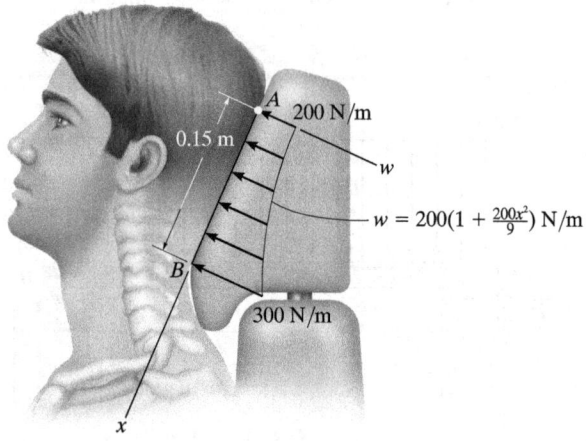

$w = 200(1 + \frac{200x^2}{9})$ N/m

Prob. 4–151

***4–152.** Wind has blown sand over a platform such that the intensity of the load can be approximated by the function $w = (0.5x^3)$ N/m. Simplify this distributed loading to an equivalent resultant force and specify its magnitude and location measured from $A$.

**4–154.** Replace the distributed loading with an equivalent resultant force, and specify its location on the beam measured from point $A$.

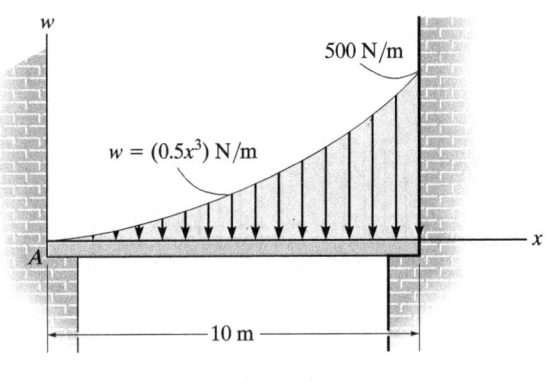

Prob. 4–152

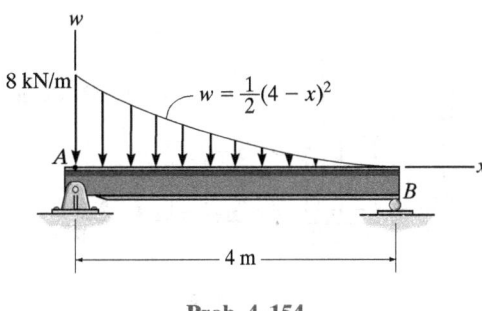

Prob. 4–154

**•4–153.** Wet concrete exerts a pressure distribution along the wall of the form. Determine the resultant force of this distribution and specify the height $h$ where the bracing strut should be placed so that it lies through the line of action of the resultant force. The wall has a width of 5 m.

**4–155.** Replace the loading by an equivalent resultant force and couple moment at point $A$.

**\*4–156.** Replace the loading by an equivalent resultant force and couple moment acting at point $B$.

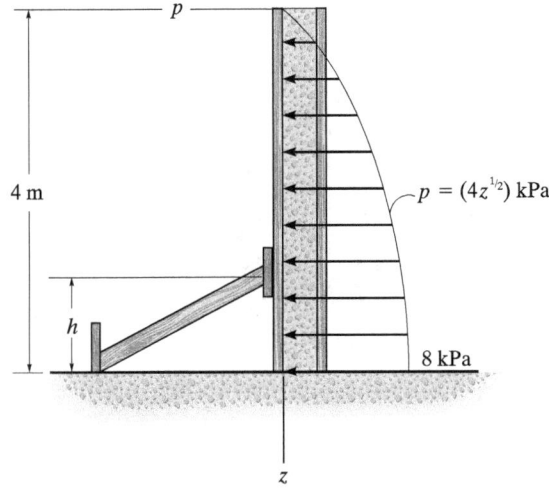

Prob. 4–153

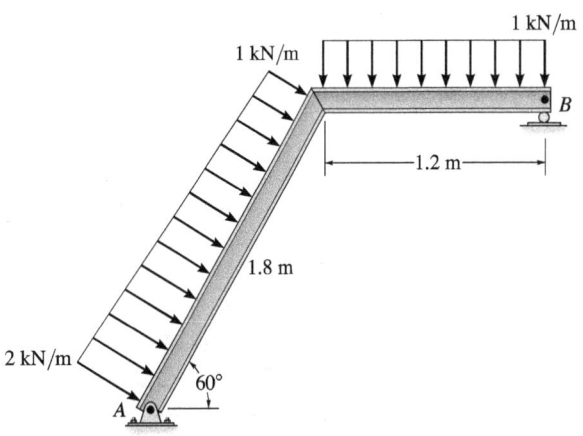

Probs. 4–155/156

•4–157. The lifting force along the wing of a jet aircraft consists of a uniform distribution along $AB$, and a semiparabolic distribution along $BC$ with origin at $B$. Replace this loading by a single resultant force and specify its location measured from point $A$.

*4–160. The distributed load acts on the beam as shown. Determine the magnitude of the equivalent resultant force and specify its location, measured from point $A$.

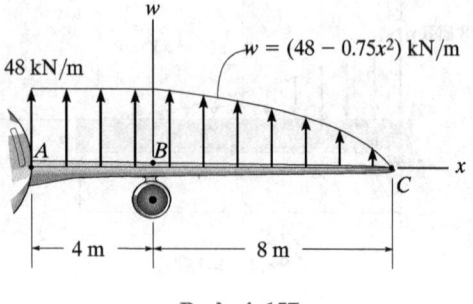

Prob. 4–157

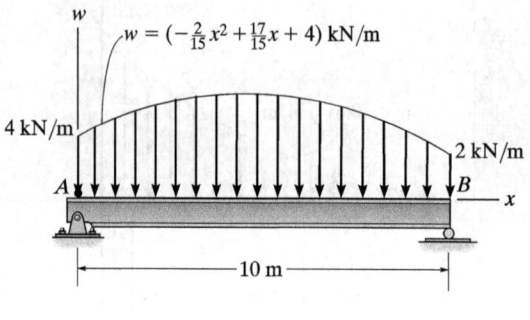

Prob. 4–160

4–158. The distributed load acts on the beam as shown. Determine the magnitude of the equivalent resultant force and specify where it acts, measured from point $A$.

4–159. The distributed load acts on the beam as shown. Determine the maximum intensity $w_{max}$. What is the magnitude of the equivalent resultant force? Specify where it acts, measured from point $B$.

•4–161. If the distribution of the ground reaction on the pipe per meter of length can be approximated as shown, determine the magnitude of the resultant force due to this loading.

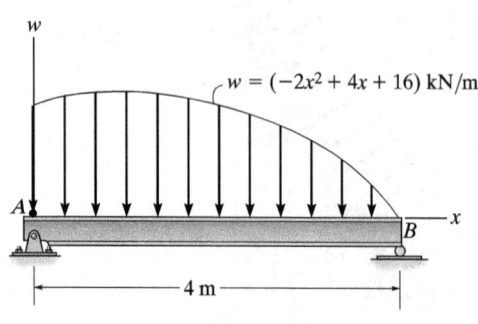

Probs. 4–158/159

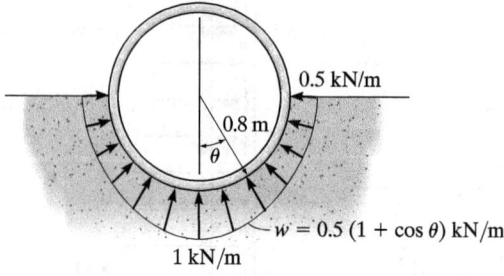

Prob. 4–161

# CHAPTER REVIEW

**Moment of Force—Scalar Definition**

A force produces a turning effect or moment about a point $O$ that does not lie on its line of action. In scalar form, the moment *magnitude* is the product of the force and the moment arm or perpendicular distance from point $O$ to the line of action of the force.

$$M_O = Fd$$

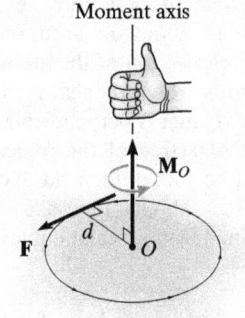

The *direction* of the moment is defined using the right-hand rule. $\mathbf{M}_O$ always acts along an axis perpendicular to the plane containing $\mathbf{F}$ and $d$, and passes through the point $O$.

Rather than finding $d$, it is normally easier to resolve the force into its $x$ and $y$ components, determine the moment of each component about the point, and then sum the results. This is called the principle of moments.

$$M_O = Fd = F_x y - F_y x$$

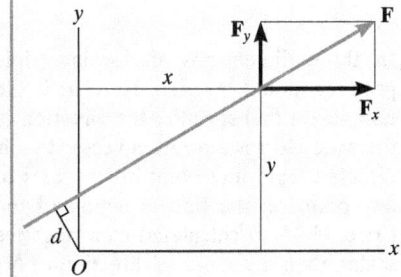

**Moment of a Force—Vector Definition**

Since three-dimensional geometry is generally more difficult to visualize, the vector cross product should be used to determine the moment. Here $\mathbf{M}_O = \mathbf{r} \times \mathbf{F}$, where $\mathbf{r}$ is a position vector that extends from point $O$ to any point $A$, $B$, or $C$ on the line of action of $\mathbf{F}$.

$$\mathbf{M}_O = \mathbf{r}_A \times \mathbf{F} = \mathbf{r}_B \times \mathbf{F} = \mathbf{r}_C \times \mathbf{F}$$

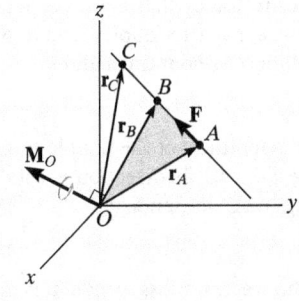

If the position vector $\mathbf{r}$ and force $\mathbf{F}$ are expressed as Cartesian vectors, then the cross product results from the expansion of a determinant.

$$\mathbf{M}_O = \mathbf{r} \times \mathbf{F} = \begin{vmatrix} \mathbf{i} & \mathbf{j} & \mathbf{k} \\ r_x & r_y & r_z \\ F_x & F_y & F_z \end{vmatrix}$$

## Moment about an Axis

If the moment of a force **F** is to be determined about an arbitrary axis $a$, then the projection of the moment onto the axis must be obtained. Provided the distance $d_a$ that is perpendicular to *both* the line of action of the force and the axis can be found, then the moment of the force about the axis can be determined from a scalar equation.

$$M_a = F d_a$$

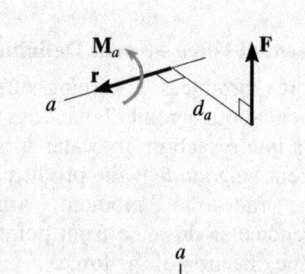

Note that when the line of action of **F** intersects the axis then the moment of **F** about the axis is zero. Also, when the line of action of **F** is parallel to the axis, the moment of **F** about the axis is zero.

In three dimensions, the scalar triple product should be used. Here $\mathbf{u}_a$ is the unit vector that specifies the direction of the axis, and **r** is a position vector that is directed from any point on the axis to any point on the line of action of the force. If $M_a$ is calculated as a negative scalar, then the sense of direction of $\mathbf{M}_a$ is opposite to $\mathbf{u}_a$.

$$M_a = \mathbf{u}_a \cdot (\mathbf{r} \times \mathbf{F}) = \begin{vmatrix} u_{a_x} & u_{a_y} & u_z \\ r_x & r_y & r_z \\ F_x & F_y & F_z \end{vmatrix}$$

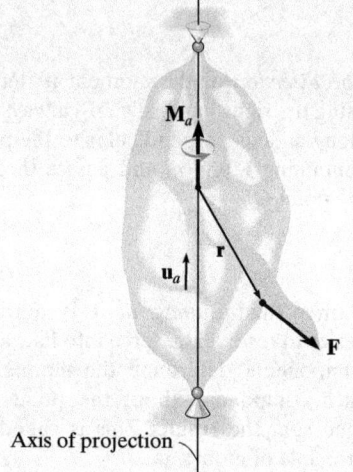

Axis of projection

## Couple Moment

A couple consists of two equal but opposite forces that act a perpendicular distance $d$ apart. Couples tend to produce a rotation without translation.

$$M = Fd$$

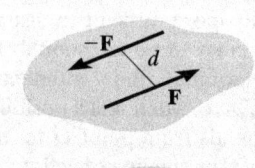

The magnitude of the couple moment is $M = Fd$, and its direction is established using the right-hand rule.

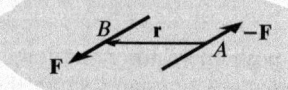

If the vector cross product is used to determine the moment of a couple, then **r** extends from any point on the line of action of one of the forces to any point on the line of action of the other force **F** that is used in the cross product.

$$\mathbf{M} = \mathbf{r} \times \mathbf{F}$$

# Chapter Review

**Simplification of a Force and Couple System**

Any system of forces and couples can be reduced to a single resultant force and resultant couple moment acting at a point. The resultant force is the sum of all the forces in the system, $\mathbf{F}_R = \Sigma \mathbf{F}$, and the resultant couple moment is equal to the sum of all the moments of the forces about the point and couple moments. $\mathbf{M}_{R_O} = \Sigma \mathbf{M}_O + \Sigma \mathbf{M}$.

Further simplification to a single resultant force is possible provided the force system is concurrent, coplanar, or parallel. To find the location of the resultant force from a point, it is necessary to equate the moment of the resultant force about the point to the moment of the forces and couples in the system about the same point.

If the resultant force and couple moment at a point are not perpendicular to one another, then this system can be reduced to a wrench, which consists of the resultant force and collinear couple moment.

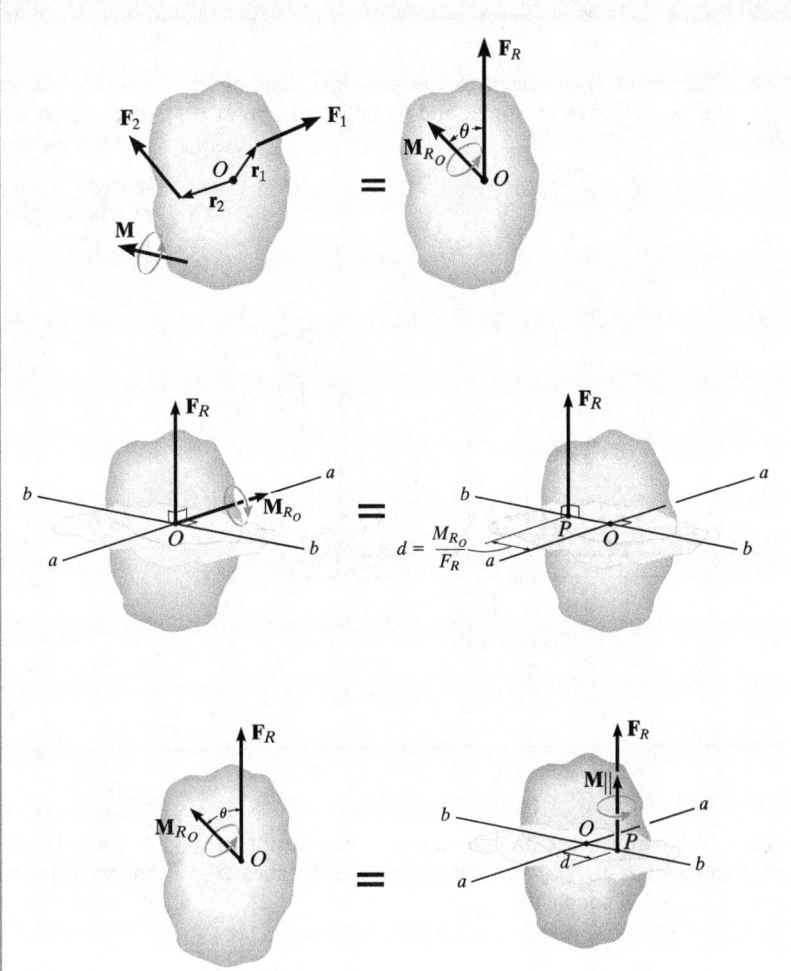

**Coplanar Distributed Loading**

A simple distributed loading can be represented by its resultant force, which is equivalent to the *area* under the loading curve. This resultant has a line of action that passes through the *centroid* or geometric center of the area or volume under the loading diagram.

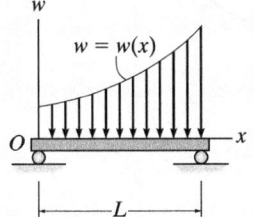 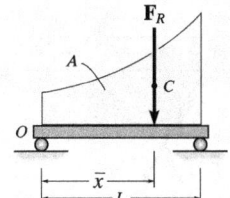

# REVIEW PROBLEMS

**4–162.** The beam is subjected to the parabolic loading. Determine an equivalent force and couple system at point $A$.

***4–164.** Determine the coordinate direction angles $\alpha, \beta, \gamma$ of $\mathbf{F}$, which is applied to the end of the pipe assembly, so that the moment of $\mathbf{F}$ about $O$ is zero.

**•4–165.** Determine the moment of the force $\mathbf{F}$ about point $O$. The force has coordinate direction angles of $\alpha = 60°$, $\beta = 120°, \gamma = 45°$. Express the result as a Cartesian vector.

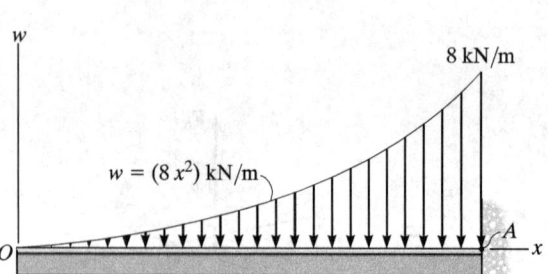

Prob. 4–162

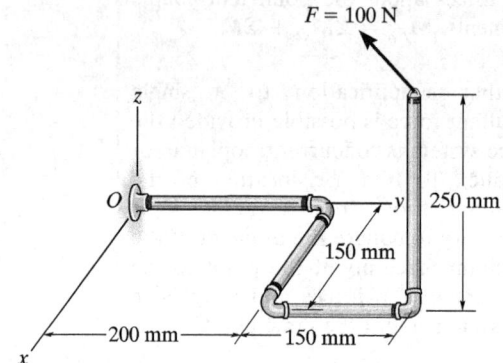

Probs. 4–164/165

**4–163.** Two couples act on the frame. If the resultant couple moment is to be zero, determine the distance $d$ between the 500-N couple forces.

**4–166.** The snorkel boom lift is extended into the position shown. If the worker weighs 800 N ($\approx$ 80 kg), determine the moment of this force about the connection at $A$.

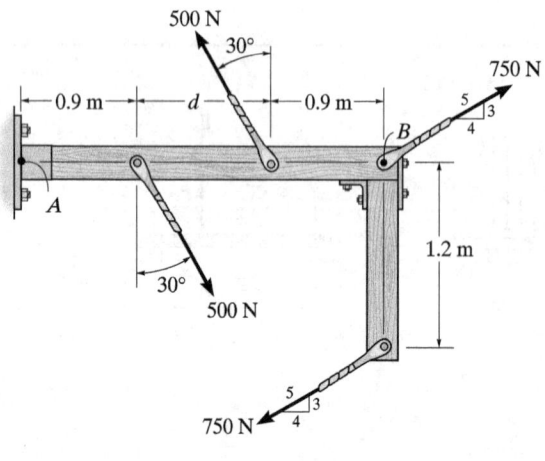

Prob. 4–163

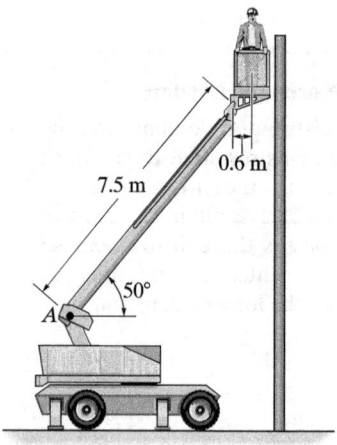

Prob. 4–166

REVIEW PROBLEMS 197

**4–167.** Determine the moment of the force $\mathbf{F}_C$ about the door hinge at $A$. Express the result as a Cartesian vector.

**\*4–168.** Determine the magnitude of the moment of the force $\mathbf{F}_C$ about the hinged axis $aa$ of the door.

**4–171.** Replace the force at $A$ by an equivalent resultant force and couple moment at point $P$. Express the results in Cartesian vector form.

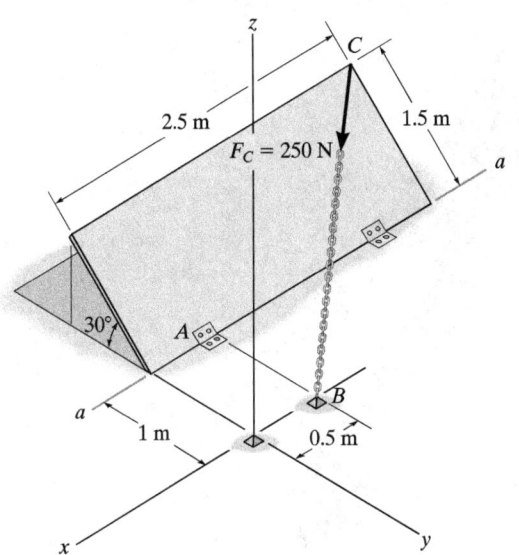

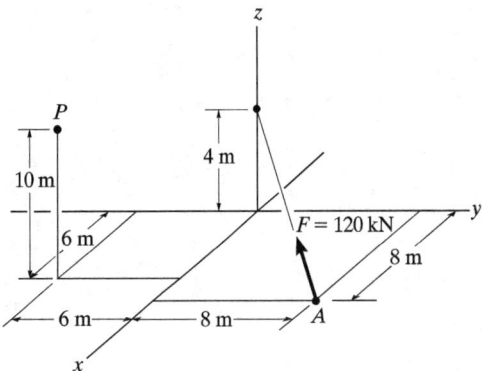

Prob. 4–171

Probs. 4–167/168

**•4–169.** Express the moment of the couple acting on the pipe assembly in Cartesian vector form. Solve the problem (a) using Eq. 4–13 and (b) summing the moment of each force about point $O$. Take $\mathbf{F} = \{25\mathbf{k}\}$ N.

**4–170.** If the couple moment acting on the pipe has a magnitude of $400\ \mathrm{N \cdot m}$, determine the magnitude $F$ of the vertical force applied to each wrench.

**\*4–172.** The horizontal 30-N force acts on the handle of the wrench. Determine the moment of this force about point $O$. Specify the coordinate direction angles $\alpha, \beta, \gamma$ of the moment axis.

**•4–173.** The horizontal 30-N force acts on the handle of the wrench. What is the magnitude of the moment of this force about the $z$ axis?

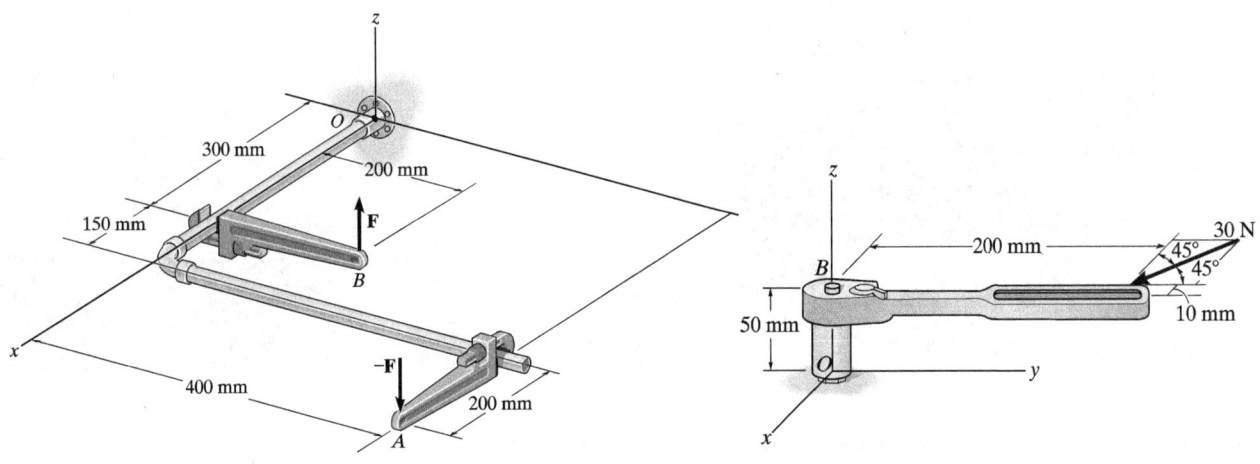

Probs. 4–169/170    Probs. 4–172/173

The crane is subjected to its weight and the load it supports. In order to calculate the support reactions on the crane, it is necessary to apply the principles of equilibrium.

# Equilibrium of a Rigid Body

## 5

### CHAPTER OBJECTIVES

- To develop the equations of equilibrium for a rigid body.
- To introduce the concept of the free-body diagram for a rigid body.
- To show how to solve rigid-body equilibrium problems using the equations of equilibrium.

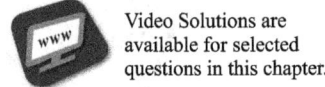

Video Solutions are available for selected questions in this chapter.

## 5.1 Conditions for Rigid-Body Equilibrium

In this section, we will develop both the necessary and sufficient conditions for the equilibrium of the rigid body in Fig. 5–1a. As shown, this body is subjected to an external force and couple moment system that is the result of the effects of gravitational, electrical, magnetic, or contact forces caused by adjacent bodies. The internal forces caused by interactions between particles within the body are not shown in this figure because these forces occur in equal but opposite collinear pairs and hence will cancel out, a consequence of Newton's third law.

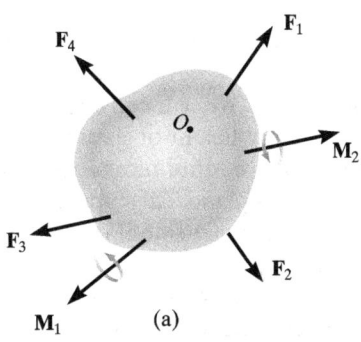

(a)

Fig. 5–1

(b)

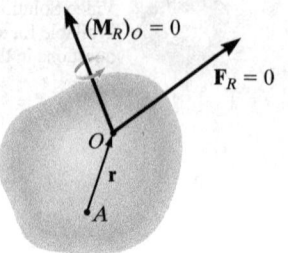

(c)

Fig. 5–1

Using the methods of the previous chapter, the force and couple moment system acting on a body can be reduced to an equivalent resultant force and resultant couple moment at any arbitrary point $O$ on or off the body, Fig. 5–1b. If this resultant force and couple moment are both equal to zero, then the body is said to be in *equilibrium*. Mathematically, the equilibrium of a body is expressed as

$$\mathbf{F}_R = \Sigma \mathbf{F} = \mathbf{0}$$
$$(\mathbf{M}_R)_O = \Sigma \mathbf{M}_O = \mathbf{0}$$
(5–1)

The first of these equations states that the sum of the forces acting on the body is equal to *zero*. The second equation states that the sum of the moments of all the forces in the system about point $O$, added to all the couple moments, is equal to *zero*. These two equations are not only necessary for equilibrium, they are also sufficient. To show this, consider summing moments about some other point, such as point $A$ in Fig. 5–1c. We require

$$\Sigma \mathbf{M}_A = \mathbf{r} \times \mathbf{F}_R + (\mathbf{M}_R)_O = \mathbf{0}$$

Since $\mathbf{r} \neq \mathbf{0}$, this equation is satisfied only if Eqs. 5–1 are satisfied, namely $\mathbf{F}_R = \mathbf{0}$ and $(\mathbf{M}_R)_O = \mathbf{0}$.

When applying the equations of equilibrium, we will assume that the body remains rigid. In reality, however, all bodies deform when subjected to loads. Although this is the case, most engineering materials such as steel and concrete are very rigid and so their deformation is usually very small. Therefore, when applying the equations of equilibrium, we can generally assume that the body will remain *rigid* and *not deform* under the applied load without introducing any significant error. This way the direction of the applied forces and their moment arms with respect to a fixed reference remain unchanged before and after the body is loaded.

## EQUILIBRIUM IN TWO DIMENSIONS

In the first part of the chapter, we will consider the case where the force system acting on a rigid body lies in or may be projected onto a *single* plane and, furthermore, any couple moments acting on the body are directed perpendicular to this plane. This type of force and couple system is often referred to as a two-dimensional or *coplanar* force system. For example, the airplane in Fig. 5–2 has a plane of symmetry through its center axis, and so the loads acting on the airplane are symmetrical with respect to this plane. Thus, each of the two wing tires will support the same load **T**, which is represented on the side (two-dimensional) view of the plane as 2**T**.

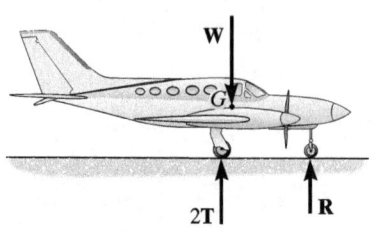

Fig. 5–2

## 5.2 Free-Body Diagrams

Successful application of the equations of equilibrium requires a complete specification of *all* the known and unknown external forces that act *on* the body. The best way to account for these forces is to draw a free-body diagram. This diagram is a sketch of the outlined shape of the body, which represents it as being *isolated* or "free" from its surroundings, i.e., a "free body." On this sketch it is necessary to show *all* the forces and couple moments that the surroundings exert *on the body* so that these effects can be accounted for when the equations of equilibrium are applied. *A thorough understanding of how to draw a free-body diagram is of primary importance for solving problems in mechanics.*

**Support Reactions.** Before presenting a formal procedure as to how to draw a free-body diagram, we will first consider the various types of reactions that occur at supports and points of contact between bodies subjected to coplanar force systems. As a general rule,

- If a support prevents the translation of a body in a given direction, then a force is developed on the body in that direction.

- If rotation is prevented, a couple moment is exerted on the body.

For example, let us consider three ways in which a horizontal member, such as a beam, is supported at its end. One method consists of a *roller* or cylinder, Fig. 5–3a. Since this support only prevents the beam from *translating* in the vertical direction, the roller will only exert a *force* on the beam in this direction, Fig. 5–3b.

The beam can be supported in a more restrictive manner by using a *pin*, Fig. 5–3c. The pin passes through a hole in the beam and two leaves which are fixed to the ground. Here the pin can prevent *translation* of the beam in *any direction* $\phi$, Fig. 5–3d, and so the pin must exert a *force* **F** on the beam in this direction. For purposes of analysis, it is generally easier to represent this resultant force **F** by its two rectangular components $\mathbf{F}_x$ and $\mathbf{F}_y$, Fig. 5–3e. If $F_x$ and $F_y$ are known, then $F$ and $\phi$ can be calculated.

The most restrictive way to support the beam would be to use a *fixed support* as shown in Fig. 5–3f. This support will prevent both *translation and rotation* of the beam. To do this a *force and couple moment* must be developed on the beam at its point of connection, Fig. 5–3g. As in the case of the pin, the force is usually represented by its rectangular components $\mathbf{F}_x$ and $\mathbf{F}_y$.

Table 5–1 lists other common types of supports for bodies subjected to coplanar force systems. (In all cases, the angle $\theta$ is assumed to be known.) Carefully study each of the symbols used to represent these supports and the types of reactions they exert on their contacting members.

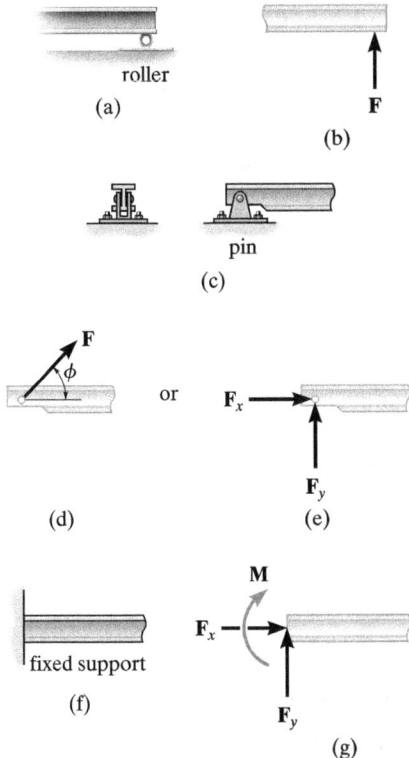

Fig. 5–3

## TABLE 5–1 Supports for Rigid Bodies Subjected to Two-Dimensional Force Systems

| Types of Connection | Reaction | Number of Unknowns |
|---|---|---|
| (1) cable | $F$ | One unknown. The reaction is a tension force which acts away from the member in the direction of the cable. |
| (2) weightless link | $F$ or $F$ | One unknown. The reaction is a force which acts along the axis of the link. |
| (3) roller | $F$ | One unknown. The reaction is a force which acts perpendicular to the surface at the point of contact. |
| (4) roller or pin in confined smooth slot | $F$ or $F$ | One unknown. The reaction is a force which acts perpendicular to the slot. |
| (5) rocker | $F$ | One unknown. The reaction is a force which acts perpendicular to the surface at the point of contact. |
| (6) smooth contacting surface | $F$ | One unknown. The reaction is a force which acts perpendicular to the surface at the point of contact. |
| (7) member pin connected to collar on smooth rod | $F$ or $F$ | One unknown. The reaction is a force which acts perpendicular to the rod. |

*continued*

## TABLE 5-1  Continued

| Types of Connection | Reaction | Number of Unknowns |
|---|---|---|
| (8) smooth pin or hinge | $F_y$, $F_x$ or $F$, $\phi$ | Two unknowns. The reactions are two components of force, or the magnitude and direction $\phi$ of the resultant force. Note that $\phi$ and $\theta$ are not necessarily equal [usually not, unless the rod shown is a link as in (2)]. |
| (9) member fixed connected to collar on smooth rod | $F$, $M$ | Two unknowns. The reactions are the couple moment and the force which acts perpendicular to the rod. |
| (10) fixed support | $F_y$, $F_x$, $M$ or $F$, $\phi$, $M$ | Three unknowns. The reactions are the couple moment and the two force components, or the couple moment and the magnitude and direction $\phi$ of the resultant force. |

Typical examples of actual supports are shown in the following sequence of photos. The numbers refer to the connection types in Table 5–1.

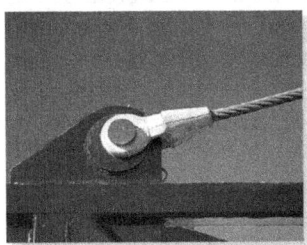

The cable exerts a force on the bracket in the direction of the cable. (1)

The rocker support for this bridge girder allows horizontal movement so the bridge is free to expand and contract due to a change in temperature. (5)

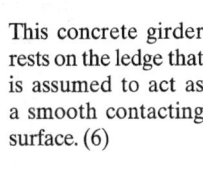

This concrete girder rests on the ledge that is assumed to act as a smooth contacting surface. (6)

This utility building is pin supported at the top of the column. (8)

The floor beams of this building are welded together and thus form fixed connections. (10)

**Internal Forces.** As stated in Sec. 5.1, the internal forces that act between adjacent particles in a body always occur in collinear pairs such that they have the same magnitude and act in opposite directions (Newton's third law). Since these forces cancel each other, they will not create an *external effect* on the body. It is for this reason that the internal forces should not be included on the free-body diagram if the entire body is to be considered. For example, the engine shown in Fig. 5–4a has a free-body diagram shown in Fig. 5–4b. The internal forces between all its connected parts, such as the screws and bolts, will cancel out because they form equal and opposite collinear pairs. Only the external forces $T_1$ and $T_2$, exerted by the chains and the engine weight $W$, are shown on the free-body diagram.

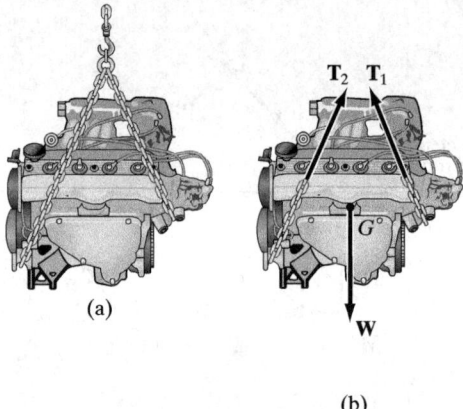

(a)

(b)

Fig. 5–4

**Weight and the Center of Gravity.** When a body is within a gravitational field, then each of its particles has a specified weight. It was shown in Sec. 4.8 that such a system of forces can be reduced to a single resultant force acting through a specified point. We refer to this force resultant as the *weight* $W$ of the body and to the location of its point of application as the *center of gravity*. The methods used for its determination will be developed in Chapter 9.

In the examples and problems that follow, if the weight of the body is important for the analysis, this force will be reported in the problem statement. Also, when the body is *uniform* or made from the same material, the center of gravity will be located at the body's *geometric center or centroid*; however, if the body consists of a nonuniform distribution of material, or has an unusual shape, then the location of its center of gravity $G$ will be given.

**Idealized Models.** When an engineer performs a force analysis of any object, he or she considers a corresponding analytical or idealized model that gives results that approximate as closely as possible the actual situation. To do this, careful choices have to be made so that selection of the type of supports, the material behavior, and the object's dimensions can be justified. This way one can feel confident that any design or analysis will yield results which can be trusted. In complex cases, this process may require developing several different models of the object that must be analyzed. In any case, this selection process requires both skill and experience.

The following two cases illustrate what is required to develop a proper model. In Fig. 5–5a, the steel beam is to be used to support the three roof joists of a building. For a force analysis, it is reasonable to assume the material (steel) is rigid since only very small deflections will occur when the beam is loaded. A bolted connection at A will allow for any slight rotation that occurs here when the load is applied, and so a *pin* can be considered for this support. At B, a *roller* can be considered since this support offers no resistance to horizontal movement. Building code is used to specify the roof loading A so that the joist loads **F** can be calculated. These forces will be larger than any actual loading on the beam since they account for extreme loading cases and for dynamic or vibrational effects. Finally, the weight of the beam is generally neglected when it is small compared to the load the beam supports. The idealized model of the beam is therefore shown with average dimensions $a$, $b$, $c$, and $d$ in Fig. 5–5b.

As a second case, consider the lift boom in Fig. 5–6a. By inspection, it is supported by a pin at A and by the hydraulic cylinder BC, which can be approximated as a weightless link. The material can be assumed rigid, and with its density known, the weight of the boom and the location of its center of gravity G are determined. When a design loading **P** is specified, the idealized model shown in Fig. 5–6b can be used for a force analysis. Average dimensions (not shown) are used to specify the location of the loads and the supports.

Idealized models of specific objects will be given in some of the examples throughout the text. It should be realized, however, that each case represents the reduction of a practical situation using simplifying assumptions like the ones illustrated here.

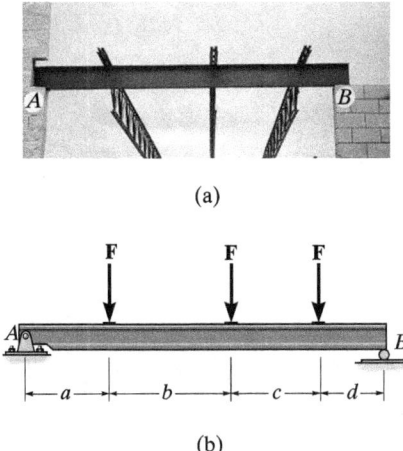

**Fig. 5–5**

(a)

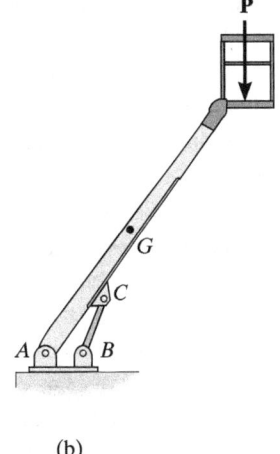

(b)

**Fig. 5–6**

## Procedure for Analysis

To construct a free-body diagram for a rigid body or any group of bodies considered as a single system, the following steps should be performed:

**Draw Outlined Shape.**

Imagine the body to be *isolated* or cut "free" from its constraints and connections and draw (sketch) its outlined shape.

**Show All Forces and Couple Moments.**

Identify all the known and unknown *external forces* and couple moments that *act on the body*. Those generally encountered are due to (1) applied loadings, (2) reactions occurring at the supports or at points of contact with other bodies (see Table 5–1), and (3) the weight of the body. To account for all these effects, it may help to trace over the boundary, carefully noting each force or couple moment acting on it.

**Identify Each Loading and Give Dimensions.**

The forces and couple moments that are known should be labeled with their proper magnitudes and directions. Letters are used to represent the magnitudes and direction angles of forces and couple moments that are unknown. Establish an $x, y$ coordinate system so that these unknowns, $A_x$, $A_y$, etc., can be identified. Finally, indicate the dimensions of the body necessary for calculating the moments of forces.

## Important Points

- No equilibrium problem should be solved without *first drawing the free-body diagram*, so as to account for all the forces and couple moments that act on the body.
- If a support *prevents translation* of a body in a particular direction, then the support exerts a *force* on the body in that direction.
- If *rotation is prevented*, then the support exerts a *couple moment* on the body.
- Study Table 5–1.
- Internal forces are never shown on the free-body diagram since they occur in equal but opposite collinear pairs and therefore cancel out.
- The weight of a body is an external force, and its effect is represented by a single resultant force acting through the body's center of gravity $G$.
- *Couple moments* can be placed anywhere on the free-body diagram since they are *free vectors*. *Forces* can act at any point along their lines of action since they are *sliding vectors*.

# EXAMPLE 5.1

Draw the free-body diagram of the uniform beam shown in Fig. 5–7a. The beam has a mass of 100 kg.

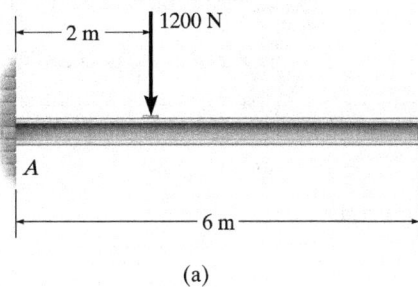

(a)

## SOLUTION

The free-body diagram of the beam is shown in Fig. 5–7b. Since the support at $A$ is fixed, the wall exerts three reactions *on the beam*, denoted as $\mathbf{A}_x$, $\mathbf{A}_y$, and $\mathbf{M}_A$. The magnitudes of these reactions are *unknown*, and their sense has been *assumed*. The weight of the beam, $W = 100(9.81)\text{ N} = 981\text{ N}$, acts through the beam's center of gravity $G$, which is 3 m from $A$ since the beam is uniform.

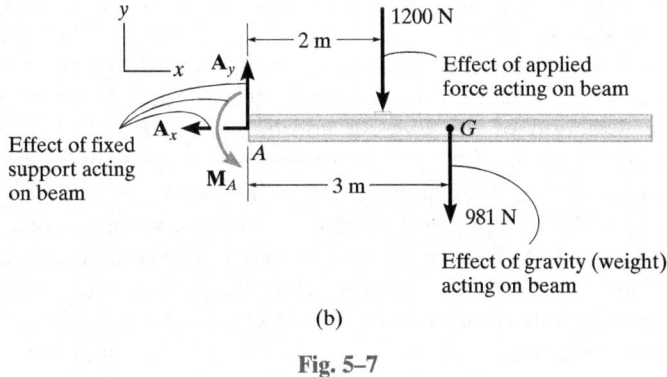

(b)

Fig. 5–7

## EXAMPLE 5.2

Draw the free-body diagram of the foot lever shown in Fig. 5–8a. The operator applies a vertical force to the pedal so that the spring is stretched 40 mm and the force in the short link at B is 100 N.

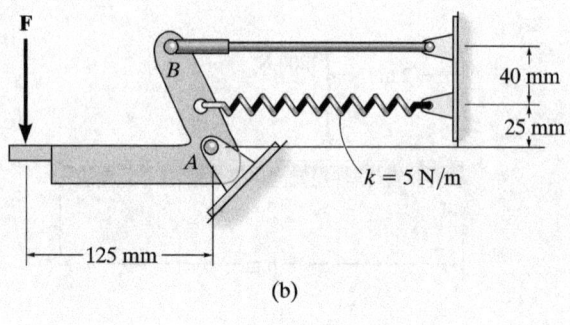

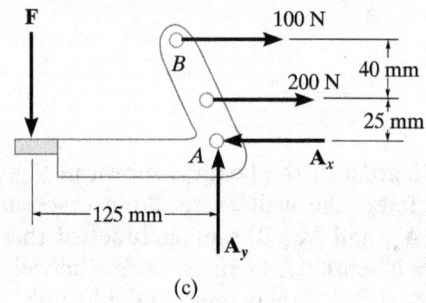

(a)

Fig. 5–8

### SOLUTION

By inspection of the photo the lever is loosely bolted to the frame at $A$. The rod at $B$ is pinned at its ends and acts as a "short link." After making the proper measurements, the idealized model of the lever is shown in Fig. 5–8b. From this, the free-body diagram is shown in Fig. 5–8c. The pin support at $A$ exerts force components $\mathbf{A}_x$ and $\mathbf{A}_y$ on the lever. The link at $B$ exerts a force of 100 N, acting in the direction of the link. In addition the spring also exerts a horizontal force on the lever. If the stiffness is measured and found to be $k = 5$ N/m, then since the stretch $s = 40$ mm, using Eq. 3–2, $F_s = ks = 5$ N/m (40 mm) = 200 N. Finally, the operator's shoe applies a vertical force of $\mathbf{F}$ on the pedal. The dimensions of the lever are also shown on the free-body diagram, since this information will be useful when computing the moments of the forces. As usual, the senses of the unknown forces at $A$ have been assumed. The correct senses will become apparent after solving the equilibrium equations.

## EXAMPLE 5.3

Two smooth pipes, each having a mass of 300 kg, are supported by the forked tines of the tractor in Fig. 5–9a. Draw the free-body diagrams for each pipe and both pipes together.

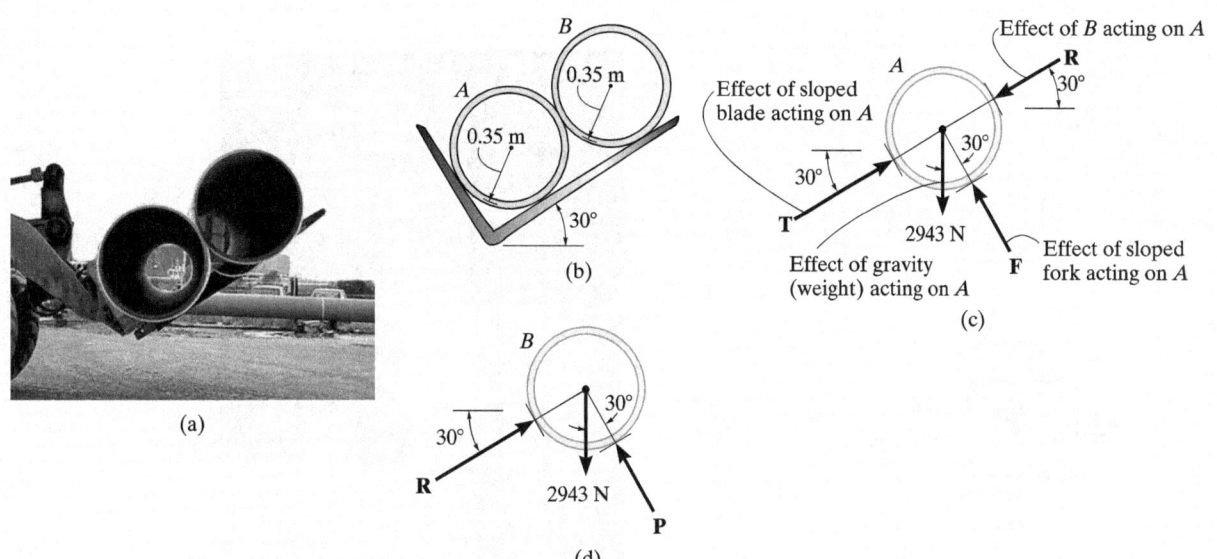

### SOLUTION
The idealized model from which we must draw the free-body diagrams is shown in Fig. 5–9b. Here the pipes are identified, the dimensions have been added, and the physical situation reduced to its simplest form.

The free-body diagram for pipe $A$ is shown in Fig. 5–9c. Its weight is $W = 300(9.81)$ N $= 2943$ N. Assuming all contacting surfaces are *smooth*, the reactive forces **T**, **F**, **R** act in a direction *normal* to the tangent at their surfaces of contact.

The free-body diagram of pipe $B$ is shown in Fig. 5–9d. Can you identify each of the three forces acting *on this pipe*? In particular, note that **R**, representing the force of $A$ on $B$, Fig. 5–9d, is equal and opposite to **R** representing the force of $B$ on $A$, Fig. 5–9c. This is a consequence of Newton's third law of motion.

The free-body diagram of both pipes combined ("system") is shown in Fig. 5–9e. Here the contact force **R**, which acts between $A$ and $B$, is considered as an *internal* force and hence is not shown on the free-body diagram. That is, it represents a pair of equal but opposite collinear forces which cancel each other.

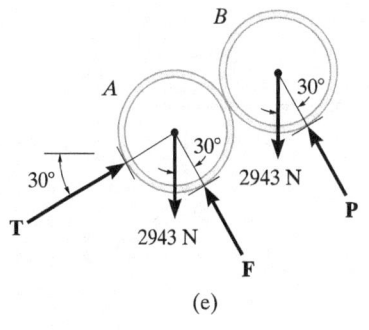

Fig. 5–9

## EXAMPLE 5.4

Draw the free-body diagram of the unloaded platform that is suspended off the edge of the oil rig shown in Fig. 5–10a. The platform has a mass of 200 kg.

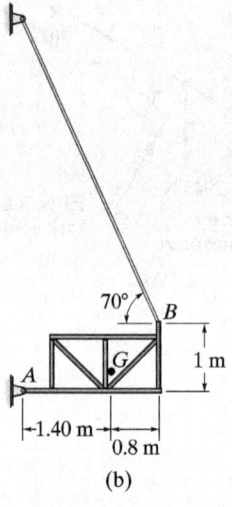

(b)

(a)

Fig. 5–10

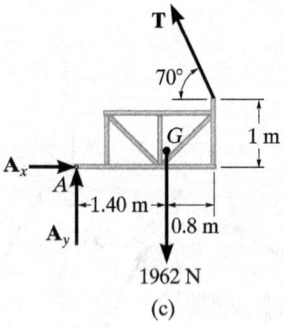

(c)

### SOLUTION

The idealized model of the platform will be considered in two dimensions because by observation the loading and the dimensions are all symmetrical about a vertical plane passing through its center, Fig. 5–10b. The connection at $A$ is considered to be a pin, and the cable supports the platform at $B$. The direction of the cable and average dimensions of the platform are listed, and the center of gravity $G$ has been determined. It is from this model that we have drawn the free-body diagram shown in Fig. 5–10c. The platform's weight is $200(9.81) = 1962$ N. The force components $\mathbf{A}_x$ and $\mathbf{A}_y$ along with the cable force $\mathbf{T}$ represent the reactions that *both* pins and *both* cables exert on the platform, Fig. 5–10a. Consequently, after the solution for these reactions, half their magnitude is developed at $A$ and half is developed at $B$.

## PROBLEMS

**•5–1.** Draw the free-body diagram of the 50-kg paper roll which has a center of mass at G and rests on the smooth blade of the paper hauler. Explain the significance of each force acting on the diagram. (See Fig. 5–7b.)

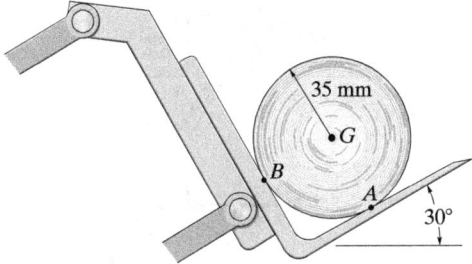

Prob. 5–1

**5–2.** Draw the free-body diagram of member $AB$, which is supported by a roller at $A$ and a pin at $B$. Explain the significance of each force on the diagram. (See Fig. 5–7b.)

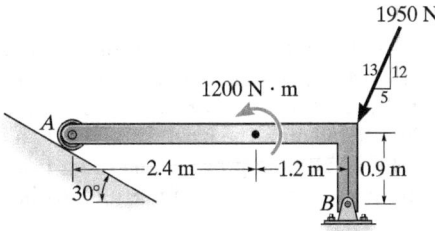

Prob. 5–2

**5–3.** Draw the free-body diagram of the dumpster $D$ of the truck, which has a weight of 25 kN and a center of gravity at $G$. It is supported by a pin at $A$ and a pin-connected hydraulic cylinder $BC$ (short link). Explain the significance of each force on the diagram. (See Fig. 5–7b.)

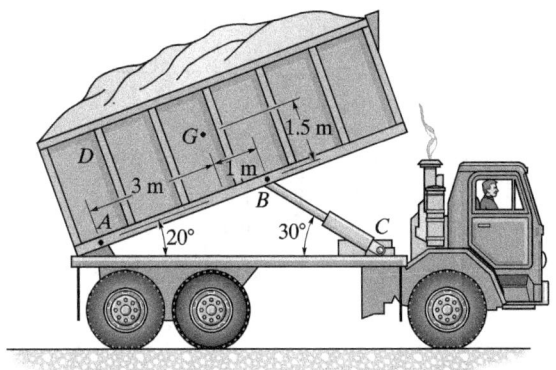

Prob. 5–3

**\*5–4.** Draw the free-body diagram of the beam which supports the 80-kg load and is supported by the pin at $A$ and a cable which wraps around the pulley at $D$. Explain the significance of each force on the diagram. (See Fig. 5–7b.)

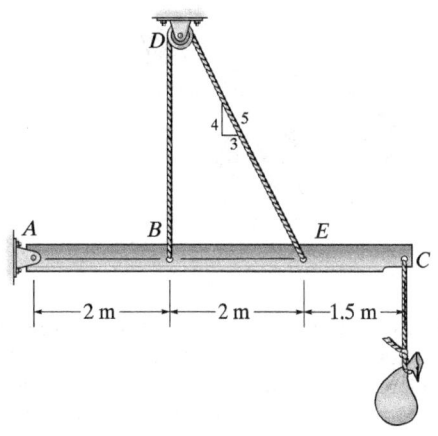

Prob. 5–4

**•5–5.** Draw the free-body diagram of the truss that is supported by the cable $AB$ and pin $C$. Explain the significance of each force acting on the diagram. (See Fig. 5–7b.)

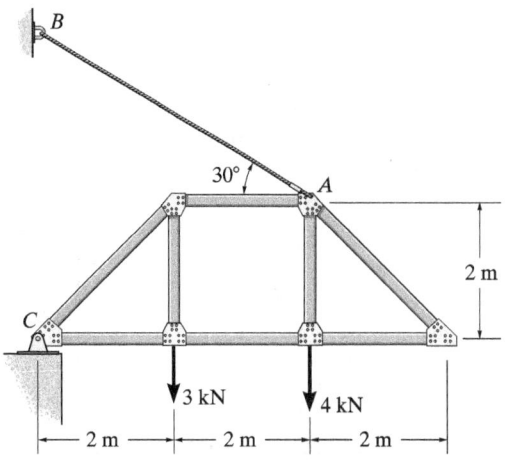

Prob. 5–5

**5–6.** Draw the free-body diagram of the crane boom $AB$ which has a weight of 3.25 kN and center of gravity at $G$. The boom is supported by a pin at $A$ and cable $BC$. The load of 6.25 kN is suspended from a cable attached at $B$. Explain the significance of each force acting on the diagram. (See Fig. 5–7b.)

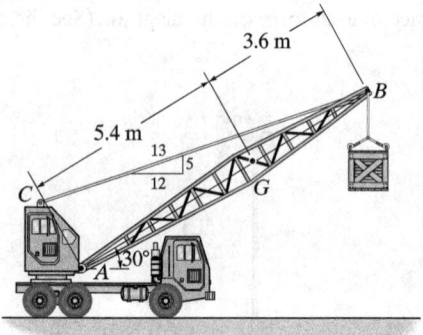

Prob. 5–6

**5–7.** Draw the free-body diagram of the "spanner wrench" subjected to the 100-N force. The support at $A$ can be considered a pin, and the surface of contact at $B$ is smooth. Explain the significance of each force on the diagram. (See Fig. 5–7b.)

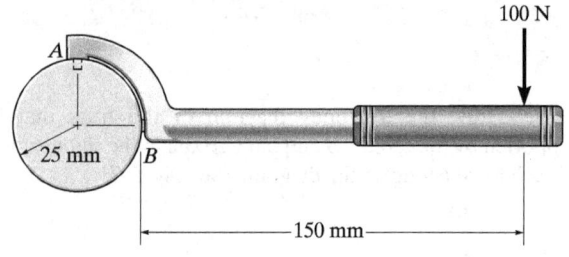

Prob. 5–7

**\*5–8.** Draw the free-body diagram of member $ABC$ which is supported by a smooth collar at $A$, roller at $B$, and short link $CD$. Explain the significance of each force acting on the diagram. (See Fig. 5–7b.)

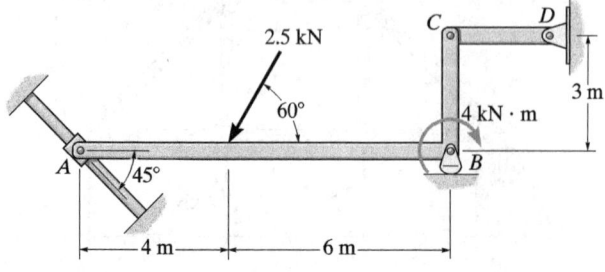

Prob. 5–8

**•5–9.** Draw the free-body diagram of the bar, which has a negligible thickness and smooth points of contact at $A$, $B$, and $C$. Explain the significance of each force on the diagram. (See Fig. 5–7b.)

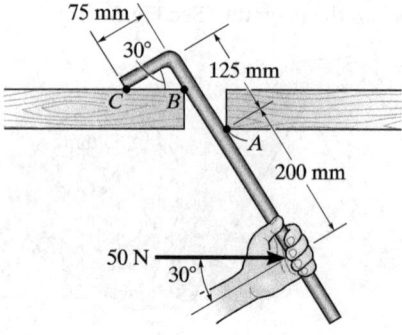

Prob. 5–9

**5–10.** Draw the free-body diagram of the winch, which consists of a drum of radius 100 mm. It is pin-connected at its center $C$, and at its outer rim is a ratchet gear having a mean radius of 150 mm. The pawl $AB$ serves as a two-force member (short link) and prevents the drum from rotating. Explain the significance of each force on the diagram. (See Fig. 5–7b.)

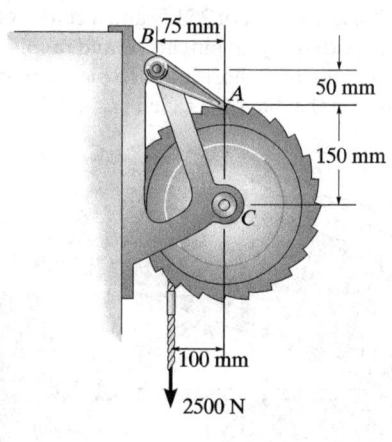

Prob. 5–10

## CONCEPTUAL PROBLEMS

**P5–1.** Draw the free-body diagram of the uniform trash bucket which has a significant weight. It is pinned at *A* and rests against the smooth horizontal member at *B*. Show your result in side view. Label any necessary dimensions.

P5–1

**P5–2.** Draw the free-body diagram of the outrigger *ABC* used to support a backhoe. The top pin *B* is connected to the hydraulic cylinder, which can be considered to be a short link (two-force member), the bearing shoe at *A* is smooth, and the outrigger is pinned to the frame at *C*.

P5–2

**P5–3.** Draw the free-body diagram of the wing on the passenger plane. The weights of the engine and wing are significant. The tires at *B* are smooth.

P5–3

**\*P5–4.** Draw the free-body diagram of the wheel and member *ABC* used as part of the landing gear on a jet plane. The hydraulic cylinder *AD* acts as a two-force member, and there is a pin connection at *B*.

P5–4

## 5.3 Equations of Equilibrium

In Sec. 5.1, we developed the two equations which are both necessary and sufficient for the equilibrium of a rigid body, namely, $\Sigma \mathbf{F} = \mathbf{0}$ and $\Sigma \mathbf{M}_O = \mathbf{0}$. When the body is subjected to a system of forces, which all lie in the $x$–$y$ plane, then the forces can be resolved into their $x$ and $y$ components. Consequently, the conditions for equilibrium in two dimensions are

$$\begin{aligned} \Sigma F_x &= 0 \\ \Sigma F_y &= 0 \\ \Sigma M_O &= 0 \end{aligned} \qquad (5\text{-}2)$$

Here $\Sigma F_x$ and $\Sigma F_y$ represent, respectively, the algebraic sums of the $x$ and $y$ components of all the forces acting on the body, and $\Sigma M_O$ represents the algebraic sum of the couple moments and the moments of all the force components about the $z$ axis, which is perpendicular to the $x$–$y$ plane and passes through the arbitrary point $O$.

Please refer to the Companion CD for the animation: *Equilibrium of a Free Body*

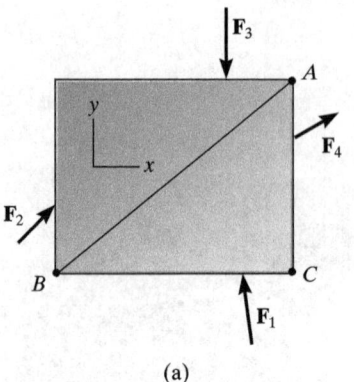

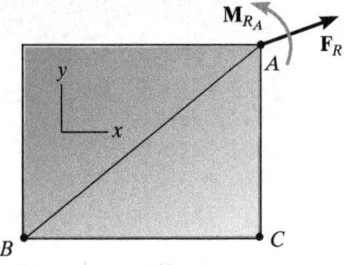

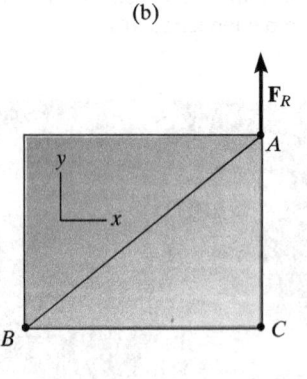

Fig. 5–11

### Alternative Sets of Equilibrium Equations.

Although Eqs. 5–2 are *most often* used for solving coplanar equilibrium problems, two *alternative* sets of three independent equilibrium equations may also be used. One such set is

$$\begin{aligned} \Sigma F_x &= 0 \\ \Sigma M_A &= 0 \\ \Sigma M_B &= 0 \end{aligned} \qquad (5\text{-}3)$$

When using these equations, it is required that a line passing through points $A$ and $B$ is *not parallel* to the $y$ axis. To prove that Eqs. 5–3 provide the *conditions* for equilibrium, consider the free-body diagram of the plate shown in Fig. 5–11$a$. Using the methods of Sec. 4.8, all the forces on the free-body diagram may be replaced by an equivalent resultant force $\mathbf{F}_R = \Sigma \mathbf{F}$, acting at point $A$, and a resultant couple moment $\mathbf{M}_{R_A} = \Sigma \mathbf{M}_A$, Fig. 5–11$b$. If $\Sigma M_A = 0$ is satisfied, it is necessary that $\mathbf{M}_{R_A} = \mathbf{0}$. Furthermore, in order that $\mathbf{F}_R$ satisfy $\Sigma F_x = 0$, it must have no component along the x axis, and therefore $\mathbf{F}_R$ must be parallel to the y axis, Fig. 5–11$c$. Finally, if it is required that $\Sigma M_B = 0$, where B does not lie on the line of action of $\mathbf{F}_R$, then $\mathbf{F}_R = \mathbf{0}$. Since Eqs. 5–3 show that both of these resultants are zero, indeed the body in Fig. 5–11$a$ must be in equilibrium.

A second alternative set of equilibrium equations is

$$\Sigma M_A = 0$$
$$\Sigma M_B = 0 \qquad (5\text{–}4)$$
$$\Sigma M_C = 0$$

Here it is necessary that points A, B, and C do not lie on the same line. To prove that these equations, when satisfied, ensure equilibrium, consider again the free-body diagram in Fig. 5–11b. If $\Sigma M_A = 0$ is to be satisfied, then $\mathbf{M}_{R_A} = \mathbf{0}$. $\Sigma M_C = 0$ is satisfied if the line of action of $\mathbf{F}_R$ passes through point C as shown in Fig. 5–11c. Finally, if we require $\Sigma M_B = 0$, it is necessary that $\mathbf{F}_R = \mathbf{0}$, and so the plate in Fig. 5–11a must then be in equilibrium.

## Procedure for Analysis

Coplanar force equilibrium problems for a rigid body can be solved using the following procedure.

### Free-Body Diagram.

- Establish the $x, y$ coordinate axes in any suitable orientation.
- Draw an outlined shape of the body.
- Show all the forces and couple moments acting on the body.
- Label all the loadings and specify their directions relative to the $x$ or $y$ axis. The sense of a force or couple moment having an *unknown* magnitude but known line of action can be *assumed*.
- Indicate the dimensions of the body necessary for computing the moments of forces.

### Equations of Equilibrium.

- Apply the moment equation of equilibrium, $\Sigma M_O = 0$, about a point ($O$) that lies at the intersection of the lines of action of two unknown forces. In this way, the moments of these unknowns are zero about $O$, and a *direct solution* for the third unknown can be determined.
- When applying the force equilibrium equations, $\Sigma F_x = 0$ and $\Sigma F_y = 0$, orient the $x$ and $y$ axes along lines that will provide the simplest resolution of the forces into their $x$ and $y$ components.
- If the solution of the equilibrium equations yields a negative scalar for a force or couple moment magnitude, this indicates that the sense is opposite to that which was assumed on the free-body diagram.

## EXAMPLE 5.5

Determine the horizontal and vertical components of reaction on the beam caused by the pin at B and the rocker at A as shown in Fig. 5–12a. Neglect the weight of the beam.

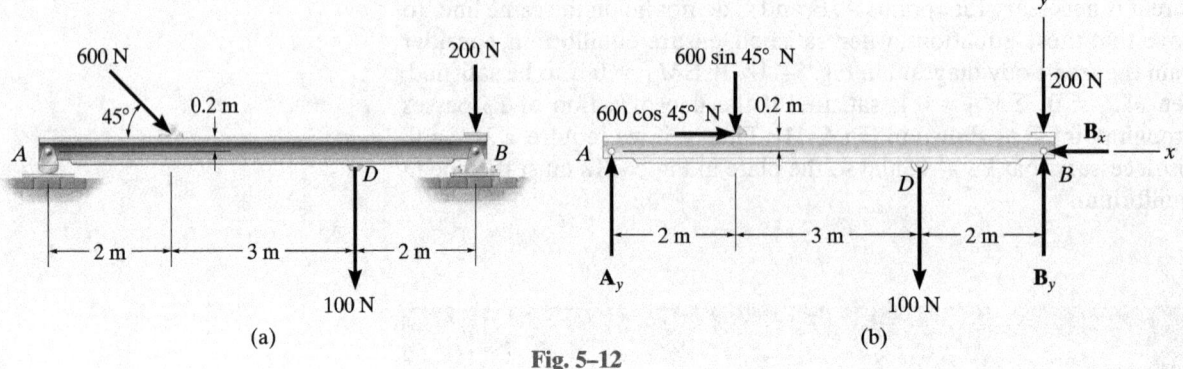

Fig. 5–12

### SOLUTION

**Free-Body Diagram.** Identify each of the forces shown on the free-body diagram of the beam, Fig. 5–12b. (See Example 5.1.) For simplicity, the 600-N force is represented by its $x$ and $y$ components as shown in Fig. 5–12b.

**Equations of Equilibrium.** Summing forces in the $x$ direction yields

$$\xrightarrow{+} \Sigma F_x = 0; \quad 600 \cos 45° \text{ N} - B_x = 0$$
$$B_x = 424 \text{ N} \quad \text{Ans.}$$

A direct solution for $\mathbf{A}_y$ can be obtained by applying the moment equation $\Sigma M_B = 0$ about point $B$.

$$\zeta + \Sigma M_B = 0; \quad 100 \text{ N}(2 \text{ m}) + (600 \sin 45° \text{ N})(5 \text{ m})$$
$$- (600 \cos 45° \text{ N})(0.2 \text{ m}) - A_y(7 \text{ m}) = 0$$
$$A_y = 319 \text{ N} \quad \text{Ans.}$$

Summing forces in the $y$ direction, using this result, gives

$$+\uparrow \Sigma F_y = 0; \quad 319 \text{ N} - 600 \sin 45° \text{ N} - 100 \text{ N} - 200 \text{ N} + B_y = 0$$
$$B_y = 405 \text{ N} \quad \text{Ans.}$$

**NOTE:** We can check this result by summing moments about point $A$.

$$\zeta + \Sigma M_A = 0; \quad -(600 \sin 45° \text{ N})(2 \text{ m}) - (600 \cos 45° \text{ N})(0.2 \text{ m})$$
$$-(100 \text{ N})(5 \text{ m}) - (200 \text{ N})(7 \text{ m}) + B_y(7 \text{ m}) = 0$$
$$B_y = 405 \text{ N} \quad \text{Ans.}$$

# EXAMPLE 5.6

The cord shown in Fig. 5–13a supports a force of 500 N and wraps over the frictionless pulley. Determine the tension in the cord at $C$ and the horizontal and vertical components of reaction at pin $A$.

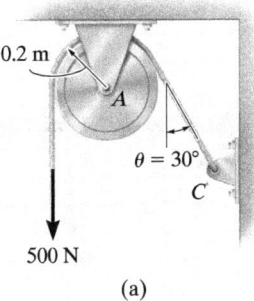

(a)

## SOLUTION

**Free-Body Diagrams.** The free-body diagrams of the cord and pulley are shown in Fig. 5–13b. Note that the principle of action, equal but opposite reaction must be carefully observed when drawing each of these diagrams: the cord exerts an unknown load distribution $p$ on the pulley at the contact surface, whereas the pulley exerts an equal but opposite effect on the cord. For the solution, however, it is simpler to *combine* the free-body diagrams of the pulley and this portion of the cord, so that the distributed load becomes *internal* to this "system" and is therefore eliminated from the analysis, Fig. 5–13c.

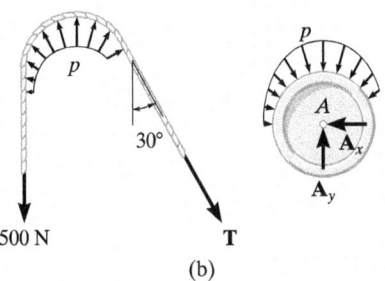

(b)

**Equations of Equilibrium.** Summing moments about point $A$ to eliminate $A_x$ and $A_y$, Fig. 5–13c, we have

$\zeta + \Sigma M_A = 0;$     500 N (0.2 m) $- T(0.2$ m$) = 0$

$T = 500$ N      Ans.

Using the result,

$\xrightarrow{+} \Sigma F_x = 0;$     $-A_x + 500 \sin 30° $ N $ = 0$

$A_x = 250$ N      Ans.

$+\uparrow \Sigma F_y = 0;$     $A_y - 500$ N $- 500 \cos 30°$ N $= 0$

$A_y = 933$ N      Ans.

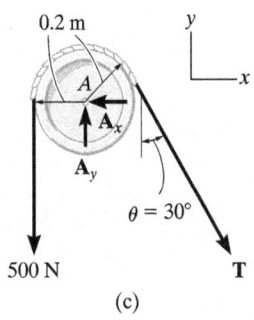

(c)

Fig. 5–13

**NOTE:** It is seen that the tension remains *constant* as the cord passes over the pulley. (This of course is true for *any angle* $\theta$ at which the cord is directed and for *any radius* $r$ of the pulley.)

## EXAMPLE 5.7

The member shown in Fig. 5–14a is pin-connected at $A$ and rests against a smooth support at $B$. Determine the horizontal and vertical components of reaction at the pin $A$.

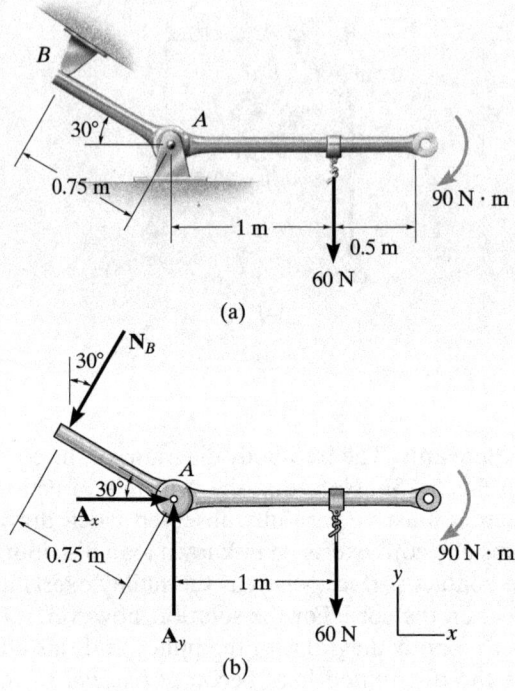

Fig. 5–14

### SOLUTION

**Free-Body Diagram.** As shown in Fig. 5–14b, the reaction $\mathbf{N}_B$ is perpendicular to the member at $B$. Also, horizontal and vertical components of reaction are represented at $A$.

**Equations of Equilibrium.** Summing moments about $A$, we obtain a direct solution for $N_B$,

$\zeta + \Sigma M_A = 0;$  $\quad -90 \text{ N} \cdot \text{m} - 60 \text{ N}(1 \text{ m}) + N_B(0.75 \text{ m}) = 0$

$$N_B = 200 \text{ N}$$

Using this result,

$\xrightarrow{+} \Sigma F_x = 0;$  $\quad A_x - 200 \sin 30° \text{ N} = 0$

$\quad A_x = 100 \text{ N}$  *Ans.*

$+\uparrow \Sigma F_y = 0;$  $\quad A_y - 200 \cos 30° \text{ N} - 60 \text{ N} = 0$

$\quad A_y = 233 \text{ N}$  *Ans.*

# EXAMPLE 5.8

The box wrench in Fig. 5–15a is used to tighten the bolt at $A$. If the wrench does not turn when the load is applied to the handle, determine the torque or moment applied to the bolt and the force of the wrench on the bolt.

## SOLUTION

**Free-Body Diagram.** The free-body diagram for the wrench is shown in Fig. 5–15b. Since the bolt acts as a "fixed support," it exerts force components $A_x$ and $A_y$ and a moment $M_A$ on the wrench at $A$.

**Equations of Equilibrium.**

$\stackrel{+}{\rightarrow} \Sigma F_x = 0;$   $A_x - 52\left(\frac{5}{13}\right) \text{N} + 30 \cos 60° \text{ N} = 0$

$\qquad A_x = 5.00 \text{ N} \qquad \qquad Ans.$

$+\uparrow \Sigma F_y = 0;$   $A_y - 52\left(\frac{12}{13}\right) \text{N} - 30 \sin 60° \text{ N} = 0$

$\qquad A_y = 74.0 \text{ N} \qquad \qquad Ans.$

$\zeta + \Sigma M_A = 0;$ $M_A - \left[52\left(\frac{12}{13}\right) \text{N}\right](0.3 \text{ m}) - (30 \sin 60° \text{ N})(0.7 \text{ m}) = 0$

$\qquad M_A = 32.6 \text{ N} \cdot \text{m} \qquad \qquad Ans.$

Note that $M_A$ must be *included* in this moment summation. This couple moment is a free vector and represents the twisting resistance of the bolt on the wrench. By Newton's third law, the wrench exerts an equal but opposite moment or torque on the bolt. Furthermore, the resultant force on the wrench is

$$F_A = \sqrt{(5.00)^2 + (74.0)^2} = 74.1 \text{ N} \qquad Ans.$$

**NOTE:** Although only *three* independent equilibrium equations can be written for a rigid body, it is a good practice to *check* the calculations using a fourth equilibrium equation. For example, the above computations may be verified in part by summing moments about point $C$:

$\zeta + \Sigma M_C = 0;$ $\left[52\left(\frac{12}{13}\right) \text{N}\right](0.4 \text{ m}) + 32.6 \text{ N} \cdot \text{m} - 74.0 \text{ N}(0.7 \text{ m}) = 0$

$\qquad 19.2 \text{ N} \cdot \text{m} + 32.6 \text{ N} \cdot \text{m} - 51.8 \text{ N} \cdot \text{m} = 0$

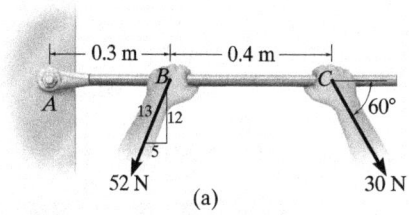

(a)

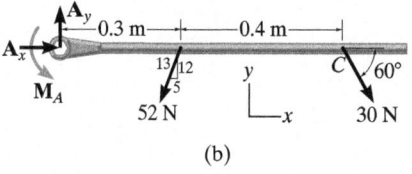

(b)

Fig. 5–15

## EXAMPLE 5.9

Please refer to the Companion CD for the animation: *Free-body Diagram for a Beam on Slanting Support*

Determine the horizontal and vertical components of reaction on the member at the pin $A$, and the normal reaction at the roller $B$ in Fig. 5–16$a$.

### SOLUTION

**Free-Body Diagram.** The free-body diagram is shown in Fig. 5–16$b$. The pin at $A$ exerts two components of reaction on the member, $\mathbf{A}_x$ and $\mathbf{A}_y$.

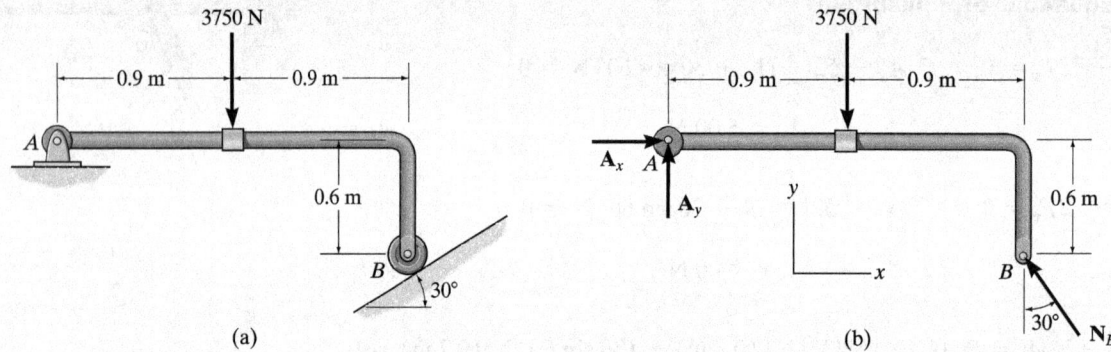

Fig. 5–16

**Equations of Equilibrium.** The reaction $N_B$ can be obtained *directly* by summing moments about point $A$ since $\mathbf{A}_x$ and $\mathbf{A}_y$ produce no moment about $A$.

$\zeta + \Sigma M_A = 0;$

$$[N_B \cos 30°](1.8 \text{ m}) - [N_B \sin 30°](0.6 \text{ m}) - 3750 \text{ N}(0.9 \text{ m}) = 0$$

$$N_B = 2681 \text{ N} \qquad \qquad Ans.$$

Using this result,

$\xrightarrow{+} \Sigma F_x = 0; \qquad A_x - (2681 \text{ N}) \sin 30° = 0$

$$A_x = 1340.5 \text{ N} \qquad \qquad Ans.$$

$+ \uparrow \Sigma F_y = 0; \qquad A_y + (2681 \text{ N}) \cos 30° - 3750 \text{ N} = 0$

$$A_y = 1428.2 \text{ N} \qquad \qquad Ans.$$

## EXAMPLE 5.10

The uniform smooth rod shown in Fig. 5–17a is subjected to a force and couple moment. If the rod is supported at A by a smooth wall and at B and C either at the top or bottom by rollers, determine the reactions at these supports. Neglect the weight of the rod.

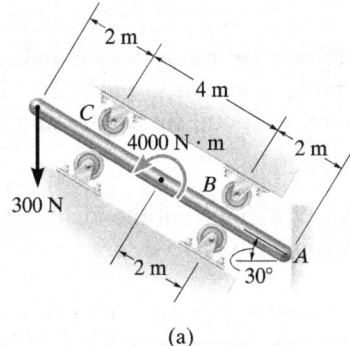

### SOLUTION

**Free-Body Diagram.** As shown in Fig. 5–17b, all the support reactions act normal to the surfaces of contact since these surfaces are smooth. The reactions at B and C are shown acting in the positive $y'$ direction. This assumes that only the rollers located on the bottom of the rod are used for support.

**Equations of Equilibrium.** Using the $x$, $y$ coordinate system in Fig. 5–17b, we have

$$\xrightarrow{+} \Sigma F_x = 0; \quad C_{y'} \sin 30° + B_{y'} \sin 30° - A_x = 0 \quad (1)$$

$$+\uparrow \Sigma F_y = 0; \quad -300 \text{ N} + C_{y'} \cos 30° + B_{y'} \cos 30° = 0 \quad (2)$$

$$\zeta + \Sigma M_A = 0; \quad -B_{y'}(2 \text{ m}) + 4000 \text{ N} \cdot \text{m} - C_{y'}(6 \text{ m})$$
$$+ (300 \cos 30° \text{ N})(8 \text{ m}) = 0 \quad (3)$$

When writing the moment equation, it should be noted that the line of action of the force component 300 sin 30° N passes through point $A$, and therefore this force is not included in the moment equation.

Solving Eqs. 2 and 3 simultaneously, we obtain

$$B_{y'} = -1000.0 \text{ N} = -1 \text{ kN} \qquad \textit{Ans.}$$
$$C_{y'} = 1346.4 \text{ N} = 1.35 \text{ kN} \qquad \textit{Ans.}$$

Since $B_{y'}$ is a negative scalar, the sense of $\mathbf{B}_{y'}$ is opposite to that shown on the free-body diagram in Fig. 5–17b. Therefore, the top roller at B serves as the support rather than the bottom one. Retaining the negative sign for $B_{y'}$ (Why?) and substituting the results into Eq. 1, we obtain

$$1346.4 \sin 30° \text{ N} + (-1000.0 \sin 30° \text{ N}) - A_x = 0$$
$$A_x = 173 \text{ N} \qquad \textit{Ans.}$$

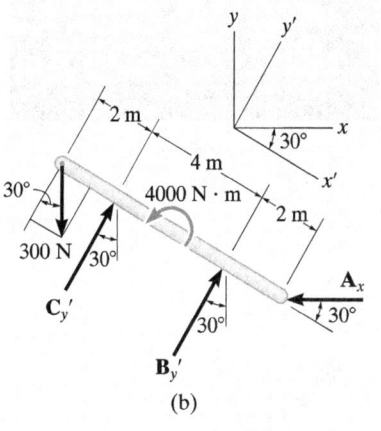

Fig. 5–17

## EXAMPLE 5.11

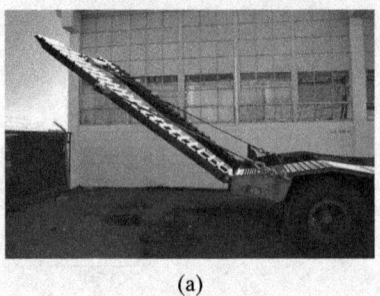

(a)

The uniform truck ramp shown in Fig. 5–18a has a weight of 1600 N and is pinned to the body of the truck at each side and held in the position shown by the two side cables. Determine the tension in the cables.

### SOLUTION
The idealized model of the ramp, which indicates all necessary dimensions and supports, is shown in Fig. 5–18b. Here the center of gravity is located at the midpoint since the ramp is considered to be uniform.

**Free-Body Diagram.** Working from the idealized model, the ramp's free-body diagram is shown in Fig. 5–18c.

**Equations of Equilibrium.** Summing moments about point $A$ will yield a direct solution for the cable tension. Using the principle of moments, there are several ways of determining the moment of **T** about $A$. If we use $x$ and $y$ components, with **T** applied at $B$, we have

$$\zeta + \Sigma M_A = 0; \quad -T \cos 20°(2 \sin 30° \text{ m}) + T \sin 20°(2 \cos 30° \text{ m})$$
$$+ 1600 \text{ N } (1.5 \cos 30° \text{ m}) = 0$$
$$T = 5985 \text{ N}$$

The simplest way to determine the moment of **T** about $A$ is to resolve it into components along and perpendicular to the ramp at $B$. Then the moment of the component along the ramp will be zero about $A$, so that

$$\zeta + \Sigma M_A = 0; \quad -T \sin 10°(2 \text{ m}) + 1600 \text{ N } (1.5 \cos 30° \text{ m}) = 0$$
$$T = 5985 \text{ N}$$

Since there are two cables supporting the ramp,

$$T' = \frac{T}{2} = 2992.5 \text{ N} \qquad Ans.$$

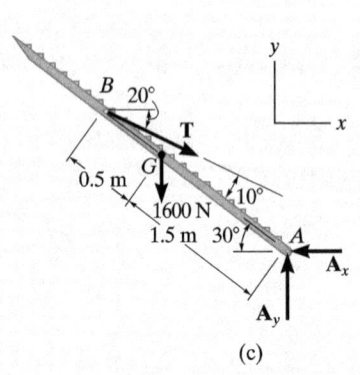

(c)

Fig. 5–18

**NOTE:** As an exercise, show that $A_x = 5624$ N and $A_y = 3647$ N.

# EXAMPLE 5.12

Determine the support reactions on the member in Fig. 5–19a. The collar at A is fixed to the member and can slide vertically along the vertical shaft.

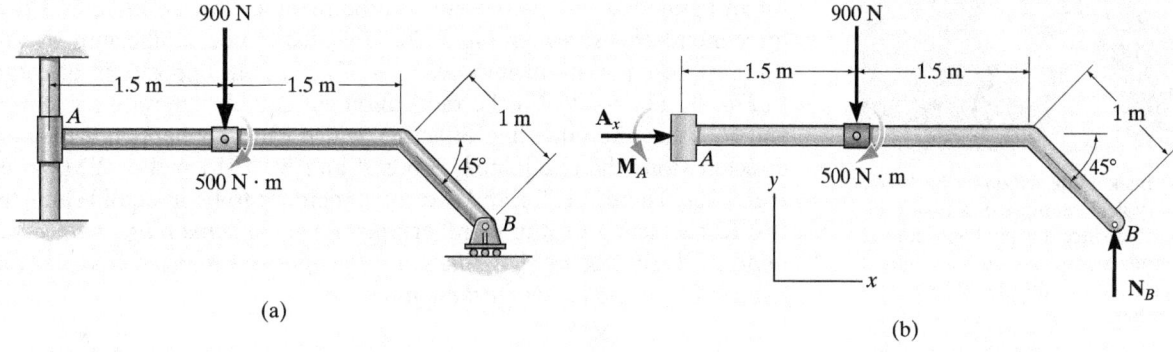

Fig. 5–19

## SOLUTION

**Free-Body Diagram.** The free-body diagram of the member is shown in Fig. 5–19b. The collar exerts a horizontal force $\mathbf{A}_x$ and moment $\mathbf{M}_A$ on the member. The reaction $\mathbf{N}_B$ of the roller on the member is vertical.

**Equations of Equilibrium.** The forces $A_x$ and $N_B$ can be determined directly from the force equations of equilibrium.

$\xrightarrow{+} \Sigma F_x = 0;$   $A_x = 0$   *Ans.*

$+\uparrow \Sigma F_y = 0;$   $N_B - 900 \text{ N} = 0$

$N_B - 900 \text{ N}$   *Ans.*

The moment $M_A$ can be determined by summing moments either about point A or point B.

$\zeta + \Sigma M_A = 0;$

$M_A - 900 \text{ N}(1.5 \text{ m}) - 500 \text{ N} \cdot \text{m} + 900 \text{ N} [3 \text{ m} + (1 \text{ m}) \cos 45°] = 0$

$M_A = -1486 \text{ N} \cdot \text{m} = 1.49 \text{ kN} \cdot \text{m} \downarrow$   *Ans.*

or

$\zeta + \Sigma M_B = 0;$   $M_A + 900 \text{ N} [1.5 \text{ m} + (1 \text{ m}) \cos 45°] - 500 \text{ N} \cdot \text{m} = 0$

$M_A = -1486 \text{ N} \cdot \text{m} = 1.49 \text{ kN} \cdot \text{m} \downarrow$   *Ans.*

The negative sign indicates that $\mathbf{M}_A$ has the opposite sense of rotation to that shown on the free-body diagram.

## 5.4 Two- and Three-Force Members

The solutions to some equilibrium problems can be simplified by recognizing members that are subjected to only two or three forces.

**Two-Force Members** As the name implies, a *two-force member* has forces applied at only two points on the member. An example of a two-force member is shown in Fig. 5–20a. To satisfy force equilibrium, $\mathbf{F}_A$ and $\mathbf{F}_B$ must be equal in magnitude, $F_A = F_B = F$, but opposite in direction ($\Sigma \mathbf{F} = \mathbf{0}$), Fig. 5–20b. Furthermore, moment equilibrium requires that $\mathbf{F}_A$ and $\mathbf{F}_B$ share the same line of action, which can only happen if they are directed along the line joining points $A$ and $B$ ($\Sigma \mathbf{M}_A = \mathbf{0}$ or $\Sigma \mathbf{M}_B = \mathbf{0}$), Fig. 5–20c. Therefore, for any two-force member to be in equilibrium, the two forces acting on the member *must have the same magnitude, act in opposite directions, and have the same line of action, directed along the line joining the two points where these forces act.*

The bucket link $AB$ on the back-hoe is a typical example of a two-force member since it is pin connected at its ends and, provided its weight is neglected, only the pin forces on this member.

The link used for this railroad car brake is a three-force member. Since the force $\mathbf{F}_B$ in the tie rod at $B$ and $\mathbf{F}_C$ from the link at $C$ are parallel, then for equilibrium the resultant force $\mathbf{F}_A$ at pin $A$ must also be parallel with these two forces.

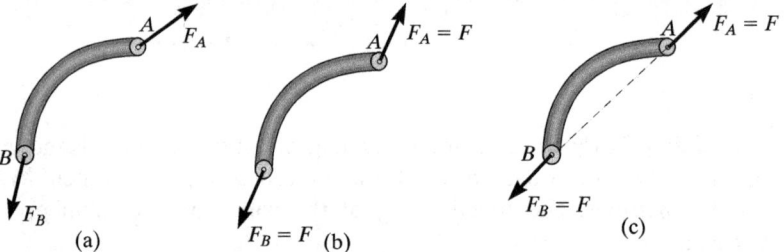

Two-force member

Fig. 5–20

**Three-Force Members** If a member is subjected to only *three forces*, it is called a *three-force member*. Moment equilibrium can be satisfied only if the three forces form a *concurrent* or *parallel* force system. To illustrate, consider the member subjected to the three forces $\mathbf{F}_1$, $\mathbf{F}_2$, and $\mathbf{F}_3$, shown in Fig. 5–21a. If the lines of action of $\mathbf{F}_1$ and $\mathbf{F}_2$ intersect at point $O$, then the line of action of $\mathbf{F}_3$ must *also* pass through point $O$ so that the forces satisfy $\Sigma \mathbf{M}_O = \mathbf{0}$. As a special case, if the three forces are all parallel, Fig. 5–21b, the location of the point of intersection, $O$, will approach infinity.

The boom and bucket on this lift is a three-force member, provided its weight is neglected. Here the lines of action of the weight of the worker, $\mathbf{W}$, and the force of the two-force member (hydraulic cylinder) at $B$, $\mathbf{F}_B$, intersect at $O$. For moment equilibrium, the resultant force at pin $A$, $\mathbf{F}_A$, must also be directed towards $O$.

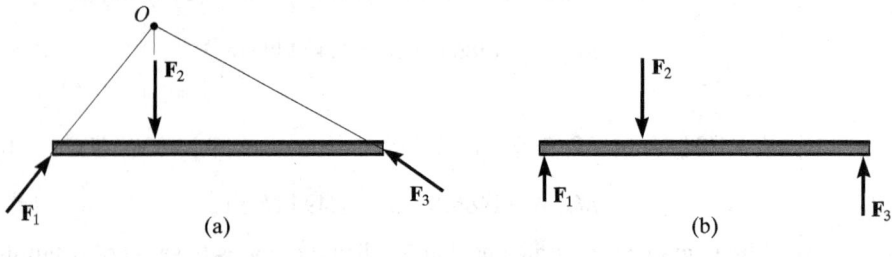

Three-force member

Fig. 5–21

## EXAMPLE 5.13

The lever $ABC$ is pin supported at $A$ and connected to a short link $BD$ as shown in Fig. 5–22a. If the weight of the members is negligible, determine the force of the pin on the lever at $A$.

### SOLUTION

**Free-Body Diagrams.** As shown in Fig. 5–22b, the short link $BD$ is a *two-force member*, so the resultant forces at pins $D$ and $B$ must be equal, opposite, and collinear. Although the magnitude of the force is unknown, the line of action is known since it passes through $B$ and $D$.

Lever $ABC$ is a *three-force member*, and therefore, in order to satisfy moment equilibrium, the three nonparallel forces acting on it must be concurrent at $O$, Fig. 5–22c. In particular, note that the force **F** on the lever at $B$ is equal but opposite to the force **F** acting at $B$ on the link. Why? The distance $CO$ must be 0.5 m since the lines of action of **F** and the 400-N force are known.

**Equations of Equilibrium.** By requiring the force system to be concurrent at $O$, since $\Sigma M_O = 0$, the angle $\theta$ which defines the line of action of $\mathbf{F}_A$ can be determined from trigonometry,

$$\theta = \tan^{-1}\left(\frac{0.7}{0.4}\right) = 60.3°$$

Using the $x$, $y$ axes and applying the force equilibrium equations,

$\xrightarrow{+} \Sigma F_x = 0;$    $F_A \cos 60.3° - F \cos 45° + 400 \text{ N} = 0$

$+\uparrow \Sigma F_y = 0;$    $F_A \sin 60.3° - F \sin 45° = 0$

Solving, we get

$$F_A = 1.07 \text{ kN} \quad \quad Ans.$$

$$F = 1.32 \text{ kN}$$

**NOTE:** We can also solve this problem by representing the force at $A$ by its two components $\mathbf{A}_x$ and $\mathbf{A}_y$ and applying $\Sigma M_A = 0$, $\Sigma F_x = 0$, $\Sigma F_y = 0$ to the lever. Once $A_x$ and $A_y$ are determined, we can get $F_A$ and $\theta$.

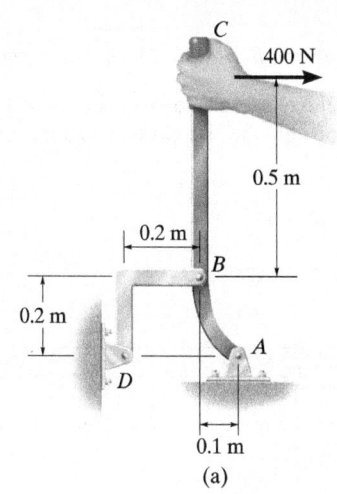

(a)

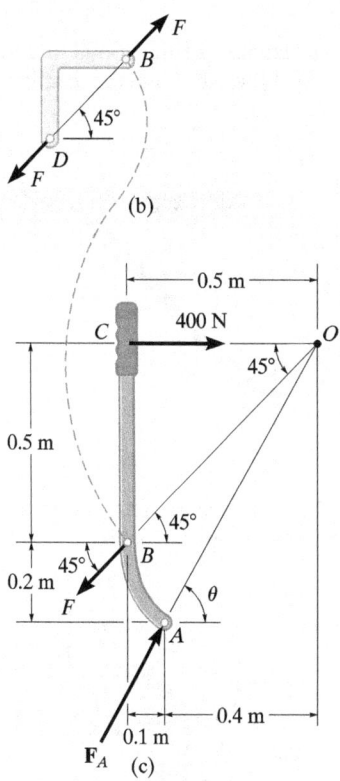

(b)

(c)

Fig. 5–22

# FUNDAMENTAL PROBLEMS

*All problem solutions must include an FBD.*

**F5–1.** Determine the horizontal and vertical components of reaction at the supports. Neglect the thickness of the beam.

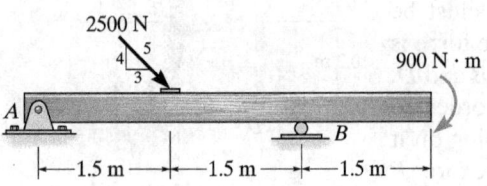

F5–1

**F5–2.** Determine the horizontal and vertical components of reaction at the pin $A$ and the reaction on the beam at $C$.

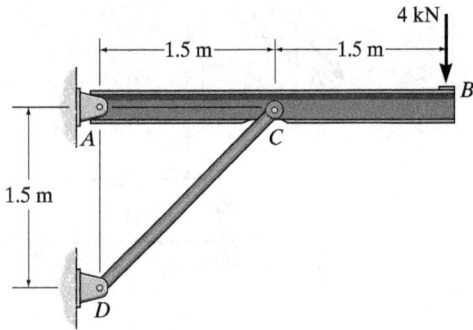

F5–2

**F5–3.** The truss is supported by a pin at $A$ and a roller at $B$. Determine the support reactions.

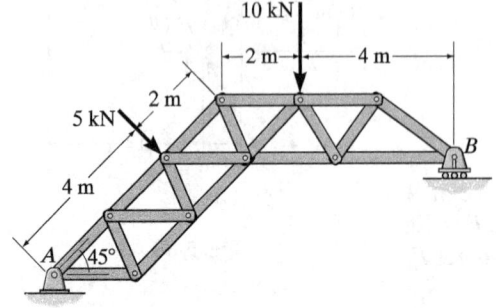

F5–3

**F5–4.** Determine the components of reaction at the fixed support $A$. Neglect the thickness of the beam.

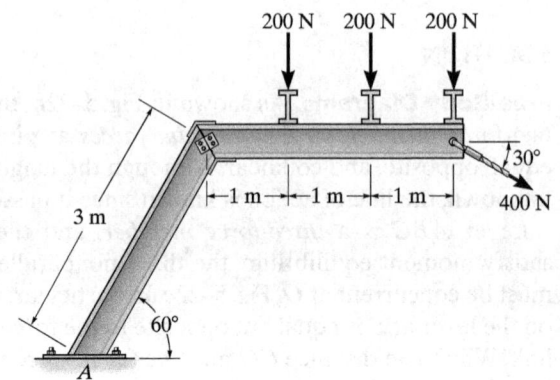

F5–4

**F5–5.** The 25-kg bar has a center of mass at $G$. If it is supported by a smooth peg at $C$, a roller at $A$, and cord $AB$, determine the reactions at these supports.

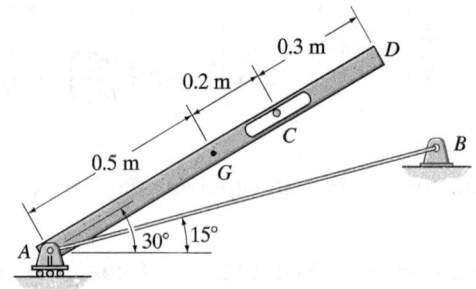

F5–5

**F5–6.** Determine the reactions at the smooth contact points $A$, $B$, and $C$ on the bar.

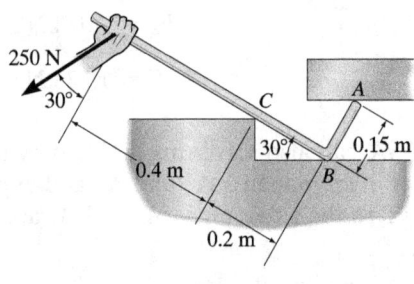

F5–6

## PROBLEMS

*All problem solutions must include an FBD.*

**5–11.** Determine the normal reactions at $A$ and $B$ in Prob. 5–1.

*****5–12.** Determine the tension in the cord and the horizontal and vertical components of reaction at support $A$ of the beam in Prob. 5–4.

**•5–13.** Determine the horizontal and vertical components of reaction at $C$ and the tension in the cable $AB$ for the truss in Prob. 5–5.

**5–14.** Determine the horizontal and vertical components of reaction at $A$ and the tension in cable $BC$ on the boom in Prob. 5–6.

**5–15.** Determine the horizontal and vertical components of reaction at $A$ and the normal reaction at $B$ on the spanner wrench in Prob. 5–7.

*****5–16.** Determine the normal reactions at $A$ and $B$ and the force in link $CD$ acting on the member in Prob. 5–8.

**•5–17.** Determine the normal reactions at the points of contact at $A$, $B$, and $C$ of the bar in Prob. 5–9.

**5–18.** Determine the horizontal and vertical components of reaction at pin $C$ and the force in the pawl of the winch in Prob. 5–10.

**5–19.** Compare the force exerted on the toe and heel of a 600-N woman when she is wearing regular shoes and stiletto heels. Assume all her weight is placed on one foot and the reactions occur at points $A$ and $B$ as shown.

*****5–20.** The train car has a weight of 120 kN and a center of gravity at $G$. It is suspended from its front and rear on the track by six tires located at $A$, $B$, and $C$. Determine the normal reactions on these tires if the track is assumed to be a smooth surface and an equal portion of the load is supported at both the front and rear tires.

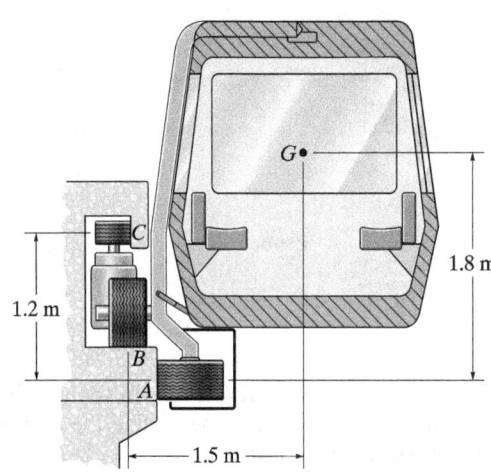

Prob. 5–20

**•5–21.** Determine the horizontal and vertical components of reaction at the pin $A$ and the tension developed in cable $BC$ used to support the steel frame.

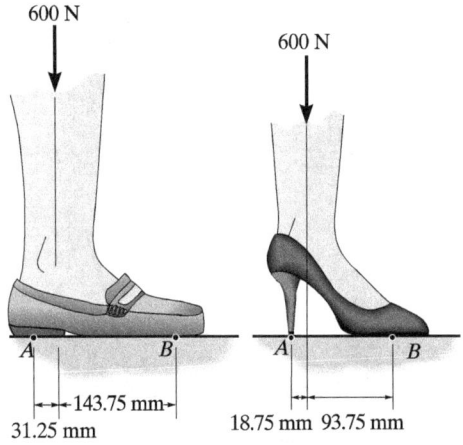

Prob. 5–19

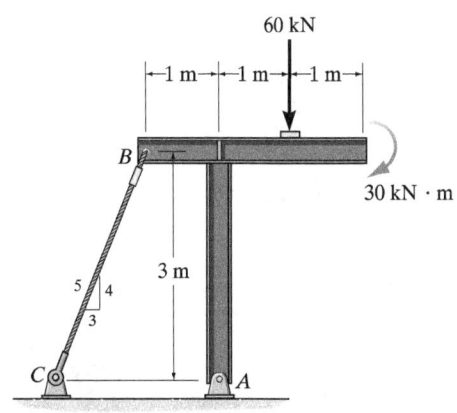

Prob. 5–21

**5–22.** The articulated crane boom has a weight of 625 N and center of gravity at *G*. If it supports a load of 3000 N, determine the force acting at the pin *A* and the force in the hydraulic cylinder *BC* when the boom is in the position shown.

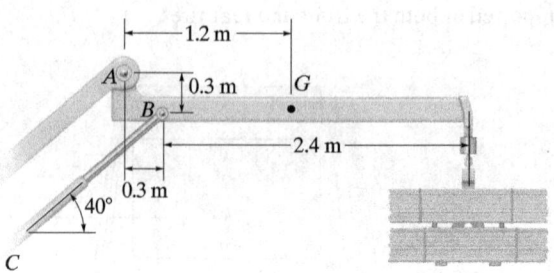

Prob. 5–22

**5–23.** The airstroke actuator at *D* is used to apply a force of *F* = 200 N on the member at *B*. Determine the horizontal and vertical components of reaction at the pin *A* and the force of the smooth shaft at *C* on the member.

**\*5–24.** The airstroke actuator at *D* is used to apply a force of **F** on the member at *B*. The normal reaction of the smooth shaft at *C* on the member is 300 N. Determine the magnitude of **F** and the horizontal and vertical components of reaction at pin *A*.

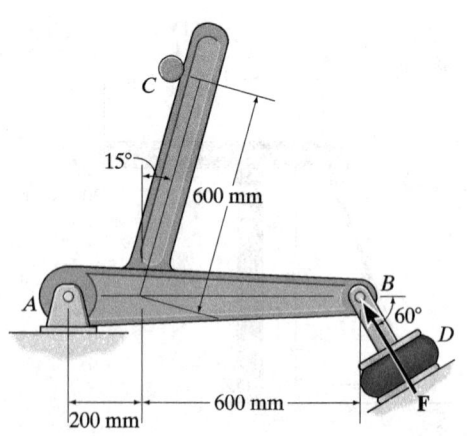

Probs. 5–23/24

**•5–25.** The 1500-N electrical transformer with center of gravity at *G* is supported by a pin at *A* and a smooth pad at *B*. Determine the horizontal and vertical components of reaction at the pin *A* and the reaction of the pad *B* on the transformer.

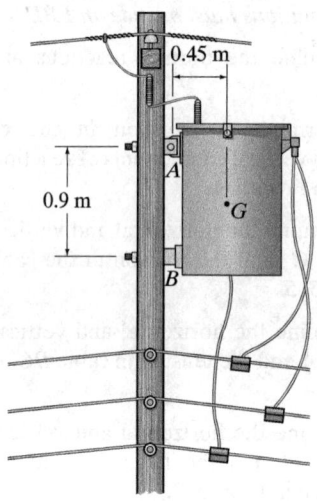

Prob. 5–25

**5–26.** A skeletal diagram of a hand holding a load is shown in the upper figure. If the load and the forearm have masses of 2 kg and 1.2 kg, respectively, and their centers of mass are located at $G_1$ and $G_2$, determine the force developed in the biceps *CD* and the horizontal and vertical components of reaction at the elbow joint *B*. The forearm supporting system can be modeled as the structural system shown in the lower figure.

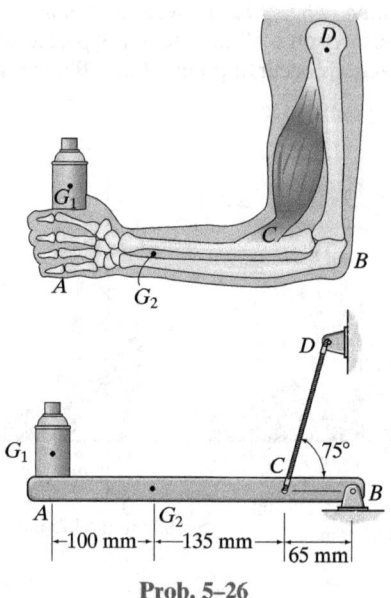

Prob. 5–26

**5–27.** As an airplane's brakes are applied, the nose wheel exerts two forces on the end of the landing gear as shown. Determine the horizontal and vertical components of reaction at the pin $C$ and the force in strut $AB$.

**•5–29.** The mass of 700 kg is suspended from a trolley which moves along the crane rail from $d = 1.7$ m to $d = 3.5$ m. Determine the force along the pin-connected knee strut $BC$ (short link) and the magnitude of force at pin $A$ as a function of position $d$. Plot these results of $F_{BC}$ and $F_A$ (vertical axis) versus $d$ (horizontal axis).

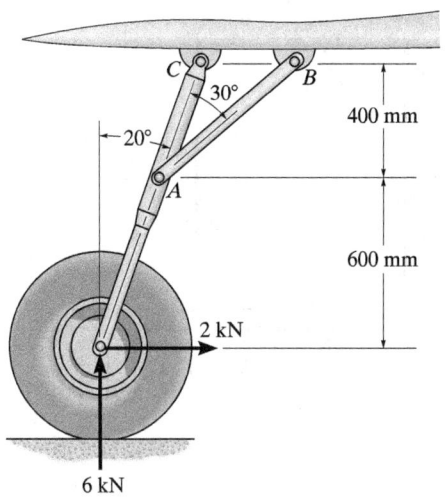

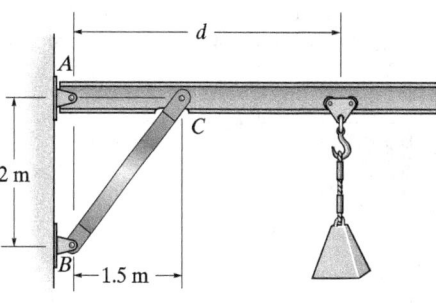

Prob. 5–29

**5–30.** If the force of $F = 500$ N is applied to the handle of the bar bender, determine the horizontal and vertical components of reaction at pin $A$ and the reaction of the roller $B$ on the smooth bar.

**5–31.** If the force of the smooth roller at $B$ on the bar bender is required to be 7.5 kN, determine the horizontal and vertical components of reaction at pin $A$ and the required magnitude of force $\mathbf{F}$ applied to the handle.

Prob. 5–27

**\*5–28.** The 1.4-Mg drainpipe is held in the tines of the fork lift. Determine the normal forces at $A$ and $B$ as functions of the blade angle $\theta$ and plot the results of force (vertical axis) versus $\theta$ (horizontal axis) for $0 \leq \theta \leq 90°$.

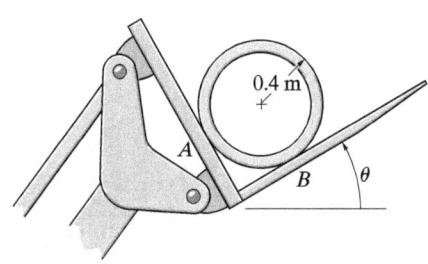

Prob. 5–28

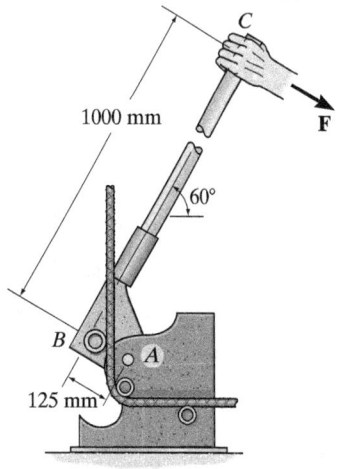

Probs. 5–30/31

**\*5–32.** The jib crane is supported by a pin at $C$ and rod $AB$. If the load has a mass of 2 Mg with its center of mass located at $G$, determine the horizontal and vertical components of reaction at the pin $C$ and the force developed in rod $AB$ on the crane when $x = 5$ m.

**•5–33.** The jib crane is supported by a pin at $C$ and rod $AB$. The rod can withstand a maximum tension of 40 kN. If the load has a mass of 2 Mg, with its center of mass located at $G$, determine its maximum allowable distance $x$ and the corresponding horizontal and vertical components of reaction at $C$.

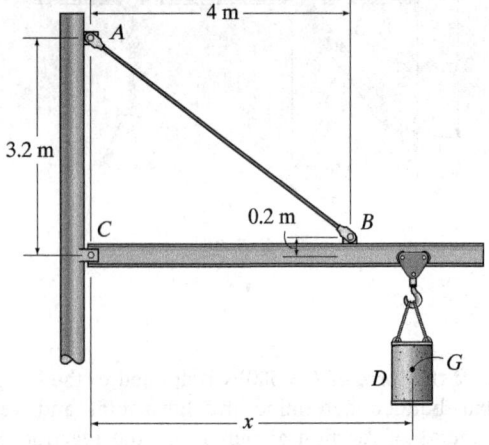

Probs. 5–32/33

**5–34.** Determine the horizontal and vertical components of reaction at the pin $A$ and the normal force at the smooth peg $B$ on the member.

**5–35.** The framework is supported by the member $AB$ which rests on the smooth floor. When loaded, the pressure distribution on $AB$ is linear as shown. Determine the length $d$ of member $AB$ and the intensity $w$ for this case.

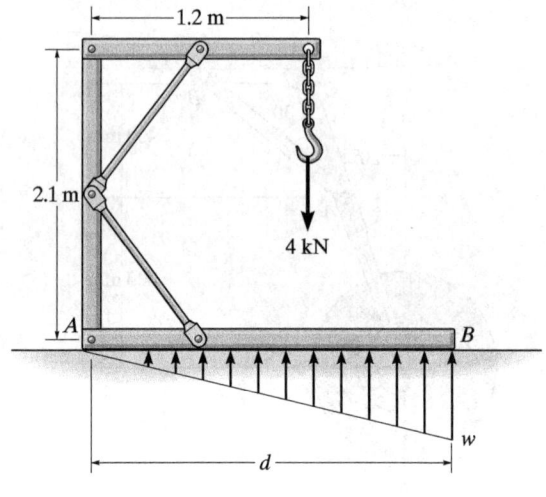

Prob. 5–35

**\*5–36.** Outriggers $A$ and $B$ are used to stabilize the crane from overturning when lifting large loads. If the load to be lifted is 3 Mg, determine the *maximum* boom angle $\theta$ so that the crane does not overturn. The crane has a mass of 5 Mg and center of mass at $G_C$, whereas the boom has a mass of 0.6 Mg and center of mass at $G_B$.

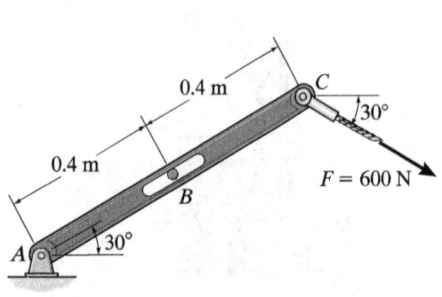

Prob. 5–34

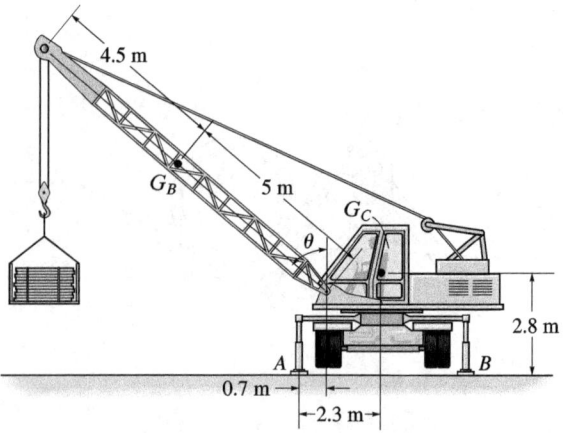

Prob. 5–36

•5–37. The wooden plank resting between the buildings deflects slightly when it supports the 50-kg boy. This deflection causes a triangular distribution of load at its ends, having maximum intensities of $w_A$ and $w_B$. Determine $w_A$ and $w_B$, each measured in N/m, when the boy is standing 3 m from one end as shown. Neglect the mass of the plank.

*5–40. The platform assembly has a weight of 1.25 kN and center of gravity at $G_1$. If it is intended to support a maximum load of 2 kN placed at point $G_2$, determine the smallest counterweight $W$ that should be placed at $B$ in order to prevent the platform from tipping over.

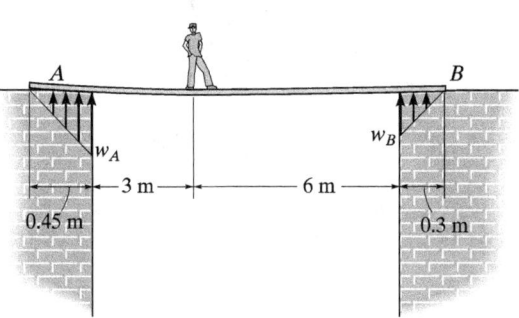

Prob. 5–37

5–38. Spring $CD$ remains in the horizontal position at all times due to the roller at $D$. If the spring is unstretched when $\theta = 0°$ and the bracket achieves its equilibrium position when $\theta = 30°$, determine the stiffness $k$ of the spring and the horizontal and vertical components of reaction at pin $A$.

5–39. Spring $CD$ remains in the horizontal position at all times due to the roller at $D$. If the spring is unstretched when $\theta = 0°$ and the stiffness is $k = 1.5$ kN/m, determine the smallest angle $\theta$ for equilibrium and the horizontal and vertical components of reaction at pin $A$.

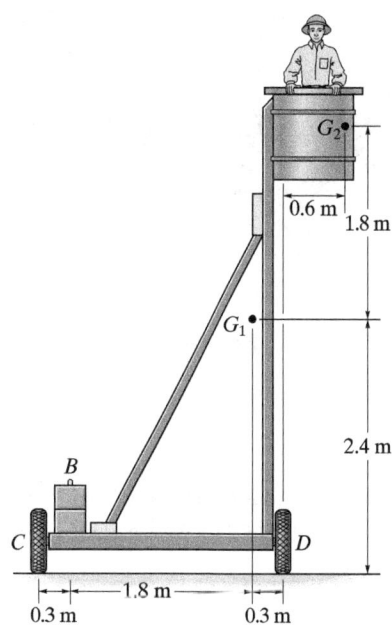

Prob. 5–40

•5–41. Determine the horizontal and vertical components of reaction at the pin $A$ and the reaction of the smooth collar $B$ on the rod.

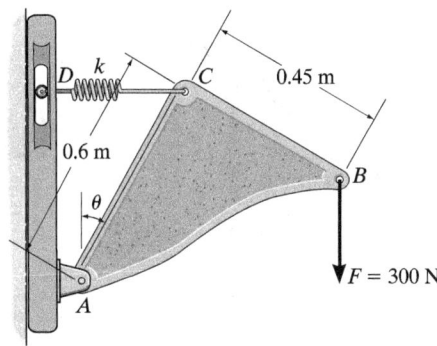

Probs. 5–38/39

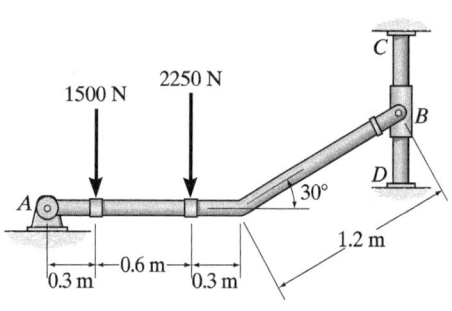

Prob. 5–41

**5–42.** Determine the support reactions of roller $A$ and the smooth collar $B$ on the rod. The collar is fixed to the rod $AB$, but is allowed to slide along rod $CD$.

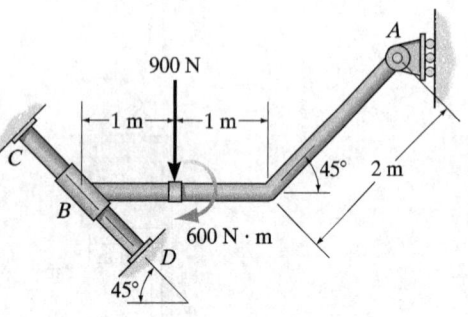

Prob. 5–42

**\*5–44.** Determine the horizontal and vertical components of force at the pin $A$ and the reaction at the rocker $B$ of the curved beam.

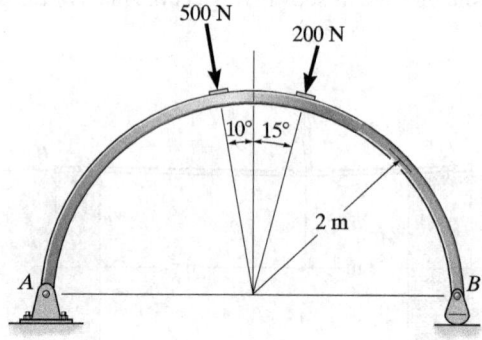

Prob. 5–44

**5–43.** The uniform rod $AB$ has a weight of 75 N. Determine the force in the cable when the rod is in the position shown.

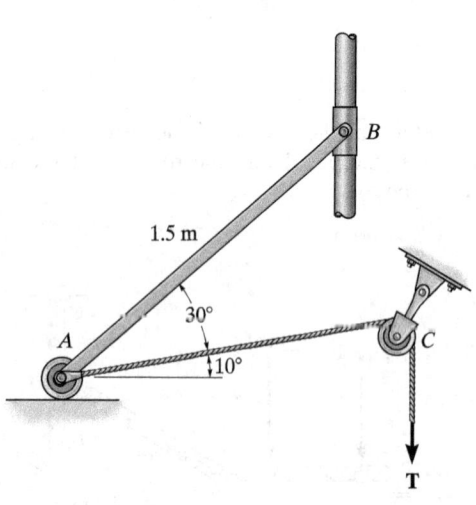

Prob. 5–43

**•5–45.** The floor crane and the driver have a total weight of 12.5 kN with a center of gravity at $G$. If the crane is required to lift the 2.5-kN drum, determine the normal reaction on *both* the wheels at $A$ and *both* the wheels at $B$ when the boom is in the position shown.

**5–46.** The floor crane and the driver have a total weight of 12.5 kN with a center of gravity at $G$. Determine the largest weight of the drum that can be lifted without causing the crane to overturn when its boom is in the position shown.

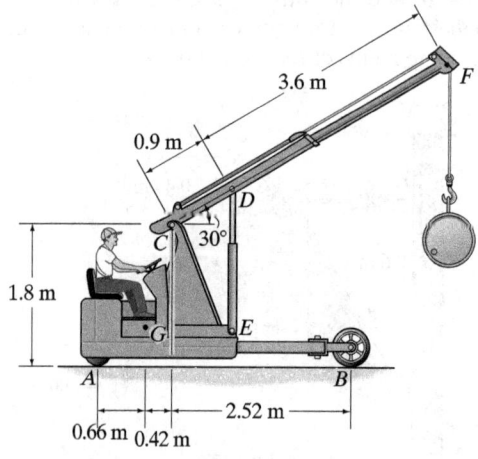

Probs. 5–45/46

**5–47.** The motor has a weight of 4.25 kN. Determine the force that each of the chains exerts on the supporting hooks at A, B, and C. Neglect the size of the hooks and the thickness of the beam.

**5–50.** The winch cable on a tow truck is subjected to a force of $T = 6$ kN when the cable is directed at $\theta = 60°$. Determine the magnitudes of the total brake frictional force **F** for the rear set of wheels B and the total normal forces at *both* front wheels A and both rear wheels B for equilibrium. The truck has a total mass of 4 Mg and mass center at G.

**5–51.** Determine the minimum cable force $T$ and critical angle $\theta$ which will cause the tow truck to start tipping, i.e., for the normal reaction at A to be zero. Assume that the truck is braked and will not slip at B. The truck has a total mass of 4 Mg and mass center at G.

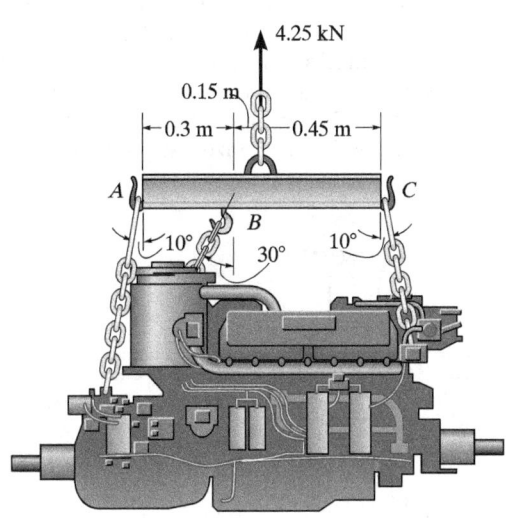

Prob. 5–47

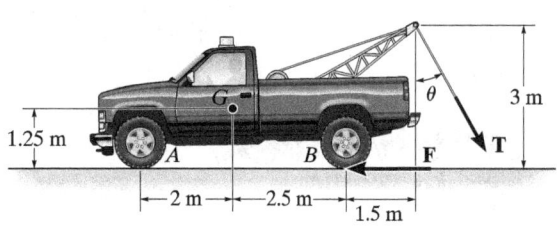

Probs. 5–50/51

***5–48.** Determine the force $P$ needed to pull the 50-kg roller over the smooth step. Take $\theta = 60°$.

**•5–49.** Determine the magnitude and direction $\theta$ of the minimum force $P$ needed to pull the 50-kg roller over the smooth step.

***5–52.** Three uniform books, each having a weight $W$ and length $a$, are stacked as shown. Determine the maximum distance $d$ that the top book can extend out from the bottom one so the stack does not topple over.

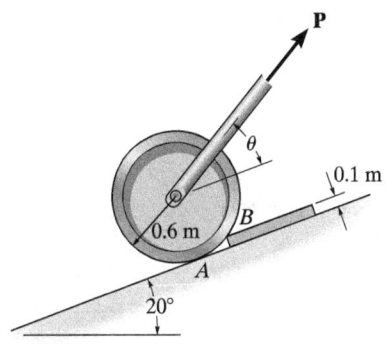

Probs. 5–48/49

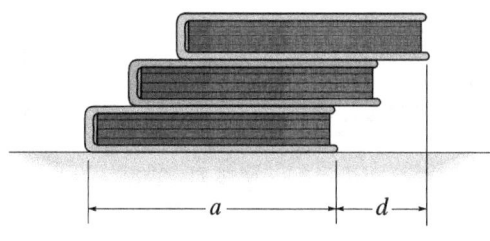

Prob. 5–52

**•5–53.** Determine the angle $\theta$ at which the link $ABC$ is held in equilibrium if member $BD$ moves 50 mm to the right. The springs are originally unstretched when $\theta = 0°$. Each spring has the stiffness shown. The springs remain horizontal since they are attached to roller guides.

**5–55.** The horizontal beam is supported by springs at its ends. Each spring has a stiffness of $k = 5$ kN/m and is originally unstretched so that the beam is in the horizontal position. Determine the angle of tilt of the beam if a load of 800 N is applied at point $C$ as shown.

**\*5–56.** The horizontal beam is supported by springs at its ends. If the stiffness of the spring at $A$ is $k_A = 5$ kN/m, determine the required stiffness of the spring at $B$ so that if the beam is loaded with the 800 N it remains in the horizontal position. The springs are originally constructed so that the beam is in the horizontal position when it is unloaded.

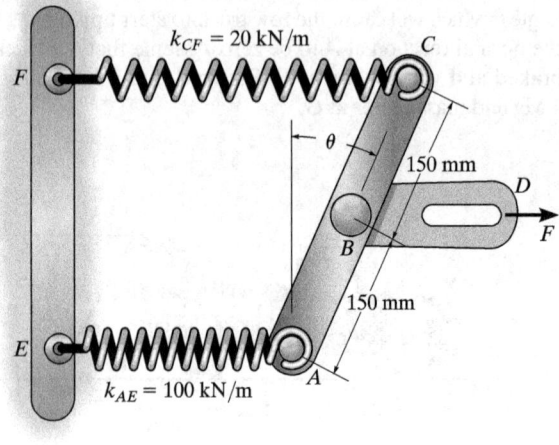

Prob. 5–53

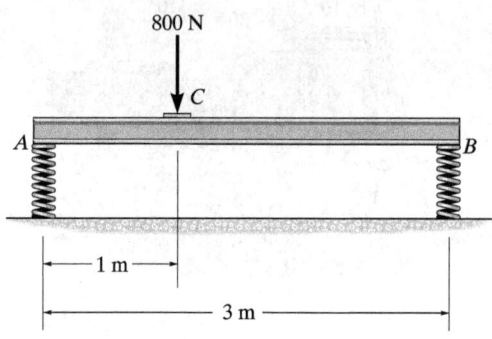

Probs. 5–55/56

**5–54.** The uniform rod $AB$ has a weight of 75 N and the spring is unstretched when $\theta = 0°$. If $\theta = 30°$, determine the stiffness $k$ of the spring.

**•5–57.** The smooth disks $D$ and $E$ have a weight of 1 kN and 0.5 kN, respectively. If a horizontal force of $P = 1$ kN is applied to the center of disk $E$, determine the normal reactions at the points of contact with the ground at $A$, $B$, and $C$.

**5–58.** The smooth disks $D$ and $E$ have a weight of 1 kN and 0.5 kN, respectively. Determine the largest horizontal force $P$ that can be applied to the center of disk $E$ without causing the disk $D$ to move up the incline.

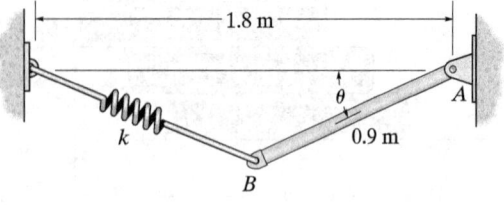

Prob. 5–54

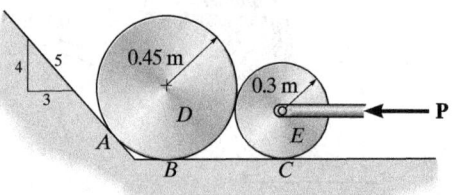

Probs. 5–57/58

**5–59.** A man stands out at the end of the diving board, which is supported by two springs $A$ and $B$, each having a stiffness of $k = 15$ kN/m. In the position shown the board is horizontal. If the man has a mass of 40 kg, determine the angle of tilt which the board makes with the horizontal after he jumps off. Neglect the weight of the board and assume it is rigid.

**•5–61.** If spring $BC$ is unstretched with $\theta = 0°$ and the bell crank achieves its equilibrium position when $\theta = 15°$, determine the force **F** applied perpendicular to segment $AD$ and the horizontal and vertical components of reaction at pin $A$. Spring $BC$ remains in the horizontal postion at all times due to the roller at $C$.

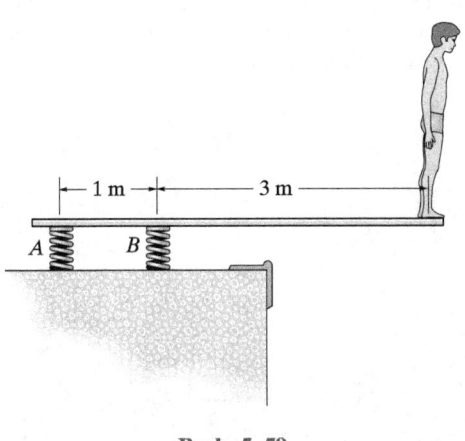

Prob. 5–59

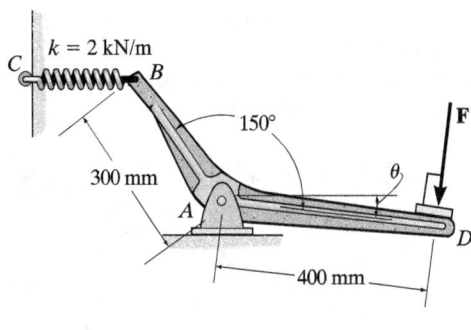

Prob. 5–61

**\*5–60.** The uniform rod has a length $l$ and weight $W$. It is supported at one end $A$ by a smooth wall and the other end by a cord of length $s$ which is attached to the wall as shown. Show that for equilibrium it is required that $h = [(s^2 - l^2)/3]^{1/2}$.

**5–62.** The thin rod of length $l$ is supported by the smooth tube. Determine the distance $a$ needed for equilibrium if the applied load is **P**.

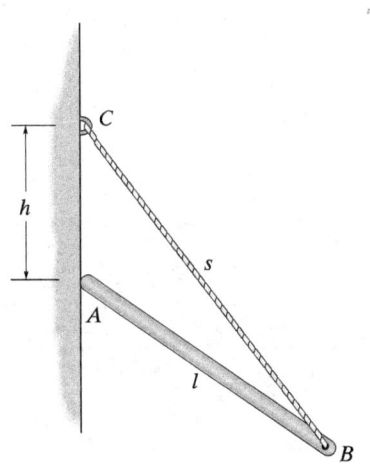

Prob. 5–60

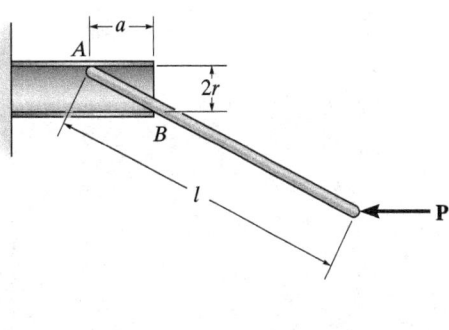

Prob. 5–62

# CONCEPTUAL PROBLEMS

**P5–5.** The tie rod is used to support this overhang at the entrance of a building. If it is pin connected to the building wall at $A$ and to the center of the overhang $B$, determine if the force in the rod will increase, decrease, or remain the same if (a) the support at $A$ is moved to a lower position $D$, and (b) the support at $B$ is moved to the outer position $C$. Explain your answer with an equilibrium analysis, using dimensions and loads. Assume the overhang is pin supported from the building wall.

P5–5

**P5–6.** The man attempts to pull the four wheeler up the incline and onto the truck bed. From the position shown, is it more effective to keep the rope attached at $A$, or would it be better to attach it to the axle of the front wheels at $B$? Draw a free-body diagram and do an equilibrium analysis to explain your answer.

P5–6

**P5–7.** Like all aircraft, this jet plane rests on three wheels. Why not use an additional wheel at the tail for better support? (Can you think of any other reason for not including this wheel?) If there was a fourth tail wheel, draw a free-body diagram of the plane from a side (2 D) view, and show why one would not be able to determine all the wheel reactions using the equations of equilibrium.

P5–7

**\*P5–8.** Where is the best place to arrange most of the logs in the wheelbarrow so that it minimizes the amount of force on the backbone of the person transporting the load? Do an equilibrium analysis to explain your answer.

P5–8

# EQUILIBRIUM IN THREE DIMENSIONS

## 5.5 Free-Body Diagrams

The first step in solving three-dimensional equilibrium problems, as in the case of two dimensions, is to draw a free-body diagram. Before we can do this, however, it is first necessary to discuss the types of reactions that can occur at the supports.

**Support Reactions.** The reactive forces and couple moments acting at various types of supports and connections, when the members are viewed in three dimensions, are listed in Table 5–2. It is important to recognize the symbols used to represent each of these supports and to understand clearly how the forces and couple moments are developed. As in the two-dimensional case:

- A force is developed by a support that restricts the translation of its attached member.
- A couple moment is developed when rotation of the attached member is prevented.

For example, in Table 5–2, item (4), the ball-and-socket joint prevents any translation of the connecting member; therefore, a force must act on the member at the point of connection. This force has three components having unknown magnitudes, Fx, Fy, Fz. Provided these components are known, one can obtain the magnitude of force, $F = \sqrt{F_x^2 + F_y^2 + F_z^2}$, and the force's orientation defined by its coordinate direction angles $\alpha$, $\beta$, $\gamma$, Eqs. 2–7.* Since the connecting member is allowed to rotate freely about any axis, no couple moment is resisted by a ball-and-socket joint.

It should be noted that the *single* bearing supports in items (5) and (7), the *single* pin (8), and the *single* hinge (9) are shown to resist both force and couple-moment components. If, however, these supports are used in conjunction with *other* bearings, pins, or hinges to hold a rigid body in equilibrium and the supports are *properly aligned* when connected to the body, then the *force reactions* at these supports *alone* are adequate for supporting the body. In other words, the couple moments become redundant and are not shown on the free-body diagram. The reason for this should become clear after studying the examples which follow.

Typical examples of actual supports that are referenced to Table 5–2 are shown in the following sequence of photos.

---

*The three unknowns may also be represented as an unknown force magnitude $F$ and two unknown coordinate direction angles. The third direction angle is obtained using the identity $\cos^2 \alpha + \cos^2 \beta + \cos^2 \gamma = 1$, Eq. 2–8.

## TABLE 5-2 Supports for Rigid Bodies Subjected to Three-Dimensional Force Systems

| Types of Connection | Reaction | Number of Unknowns |
|---|---|---|
| (1) cable | $F$ | One unknown. The reaction is a force which acts away from the member in the known direction of the cable. |
| (2) smooth surface support | $F$ | One unknown. The reaction is a force which acts perpendicular to the surface at the point of contact. |
| (3) roller | $F$ | One unknown. The reaction is a force which acts perpendicular to the surface at the point of contact. |
| (4) ball and socket | $F_x$, $F_y$, $F_z$ | Three unknowns. The reactions are three rectangular force components. |
| (5) single journal bearing | $M_x$, $M_z$, $F_x$, $F_z$ | Four unknowns. The reactions are two force and two couple-moment components which act perpendicular to the shaft. Note: The couple moments are generally not applied if the body is supported elsewhere. See the examples. |

*continued*

5.5 FREE-BODY DIAGRAMS     239

**TABLE 5–2 Continued**

| Types of Connection | Reaction | Number of Unknowns |
|---|---|---|
| (6) single journal bearing with square shaft | $M_z$, $F_z$, $M_y$, $M_x$, $F_x$ | Five unknowns. The reactions are two force and three couple-moment components. *Note*: The couple moments are generally not applied if the body is supported elsewhere. See the examples. |
| (7) single thrust bearing | $M_z$, $F_y$, $F_z$, $M_x$, $F_x$ | Five unknowns. The reactions are three force and two couple-moment components. *Note*: The couple moments are generally not applied if the body is supported elsewhere. See the examples. |
| (8) single smooth pin | $M_z$, $F_z$, $F_y$, $M_y$, $F_x$ | Five unknowns. The reactions are three force and two couple-moment components. *Note*: The couple moments are generally not applied if the body is supported elsewhere. See the examples. |
| (9) single hinge | $M_z$, $F_z$, $F_y$, $F_x$, $M_x$ | Five unknowns. The reactions are three force and two couple-moment components. *Note*: The couple moments are generally not applied if the body is supported elsewhere. See the examples. |
| (10) fixed support | $M_z$, $F_z$, $F_x$, $F_y$, $M_y$, $M_x$ | Six unknowns. The reactions are three force and three couple-moment components. |

**Free-Body Diagrams.** The general procedure for establishing the free-body diagram of a rigid body has been outlined in Sec. 5.2.

This ball-and-socket joint provides a connection for the housing of an earth grader to its frame. (4)

This journal bearing supports the end of the shaft. (5)

This thrust bearing is used to support the drive shaft on a machine. (7)

This pin is used to support the end of the strut used on a tractor. (8)

Essentially it requires first "isolating" the body by drawing its outlined shape. This is followed by a careful *labeling* of *all* the forces and couple moments with reference to an established $x, y, z$ coordinate system. It is suggested to show the unknown components of reaction as acting on the free-body diagram in the *positive sense*. In this way, if any negative values are obtained, they will indicate that the components act in the negative coordinate directions.

## EXAMPLE 5.14

Consider the two rods and plate, along with their associated free-body diagrams shown in Fig. 5–23. The *x, y, z* axes are established on the diagram and the unknown reaction components are indicated in the *positive sense*. The weight is neglected.

### SOLUTION

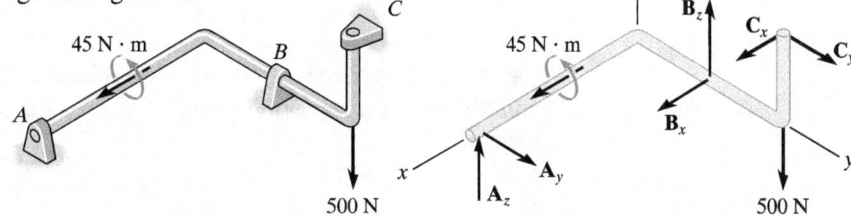

Properly aligned journal bearings at *A, B, C*.

The force reactions developed by the bearings are sufficient for equilibrium since they prevent the shaft from rotating about each of the coordinate axes.

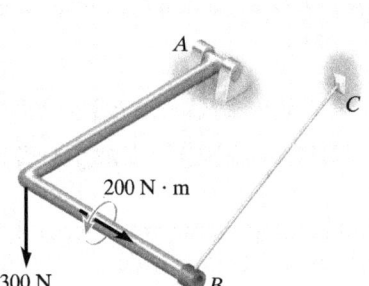

Pin at *A* and cable *BC*.

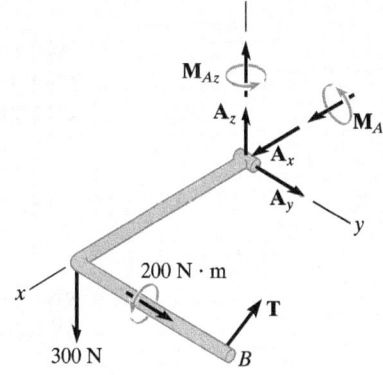

Moment components are developed by the pin on the rod to prevent rotation about the *x* and *z* axes.

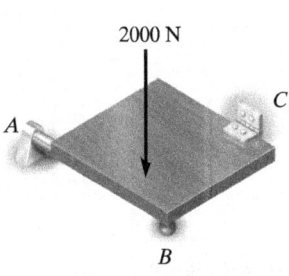

Properly aligned journal bearing at *A* and hinge at *C*. Roller at *B*.

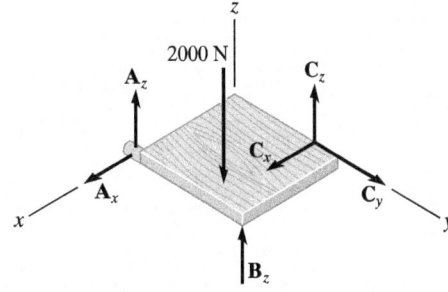

Only force reactions are developed by the bearing and hinge on the plate to prevent rotation about each coordinate axis. No moments at the hinge are developed.

Fig. 5–23

## 5.6 Equations of Equilibrium

As stated in Sec. 5.1, the conditions for equilibrium of a rigid body subjected to a three-dimensional force system require that both the *resultant* force and *resultant* couple moment acting on the body be equal to *zero*.

**Vector Equations of Equilibrium.** The two conditions for equilibrium of a rigid body may be expressed mathematically in vector form as

$$\Sigma \mathbf{F} = 0$$
$$\Sigma \mathbf{M}_O = 0 \qquad (5\text{--}5)$$

where $\Sigma \mathbf{F}$ is the vector sum of all the external forces acting on the body and $\Sigma \mathbf{M}_O$ is the sum of the couple moments and the moments of all the forces about any point $O$ located either on or off the body.

**Scalar Equations of Equilibrium.** If all the external forces and couple moments are expressed in Cartesian vector form and substituted into Eqs. 5–5, we have

$$\Sigma \mathbf{F} = \Sigma F_x \mathbf{i} + \Sigma F_y \mathbf{j} + \Sigma F_z \mathbf{k} = \mathbf{0}$$
$$\Sigma \mathbf{M}_O = \Sigma M_x \mathbf{i} + \Sigma M_y \mathbf{j} + \Sigma M_z \mathbf{k} = \mathbf{0}$$

Since the **i**, **j**, and **k** components are independent from one another, the above equations are satisfied provided

$$\Sigma F_x = 0$$
$$\Sigma F_y = 0 \qquad (5\text{--}6a)$$
$$\Sigma F_z = 0$$

and

$$\Sigma M_x = 0$$
$$\Sigma M_y = 0 \qquad (5\text{--}6b)$$
$$\Sigma M_z = 0$$

These *six scalar equilibrium equations* may be used to solve for at most six unknowns shown on the free-body diagram. Equations 5–6a require the sum of the external force components acting in the $x$, $y$, and $z$ directions to be zero, and Eqs. 5–6b require the sum of the moment components about the $x$, $y$, and $z$ axes to be zero.

## 5.7 Constraints and Statical Determinacy

To ensure the equilibrium of a rigid body, it is not only necessary to satisfy the equations of equilibrium, but the body must also be properly held or constrained by its supports. Some bodies may have more supports than are necessary for equilibrium, whereas others may not have enough or the supports may be arranged in a particular manner that could cause the body to move. Each of these cases will now be discussed.

**Redundant Constraints.** When a body has redundant supports, that is, more supports than are necessary to hold it in equilibrium, it becomes statically indeterminate. *Statically indeterminate* means that there will be more unknown loadings on the body than equations of equilibrium available for their solution. For example, the beam in Fig. 5–24a and the pipe assembly in Fig. 5–24b, shown together with their free-body diagrams, are both statically indeterminate because of additional (or redundant) support reactions. For the beam there are five unknowns, $M_A$, $A_x$, $A_y$, $B_y$, and $C_y$, for which only three equilibrium equations can be written ($\Sigma F_x = 0$, $\Sigma F_y = 0$, and $\Sigma M_O = 0$, Eqs. 5–2). The pipe assembly has eight unknowns, for which only six equilibrium equations can be written, Eqs. 5–6.

The additional equations needed to solve statically indeterminate problems of the type shown in Fig. 5–24 are generally obtained from the deformation conditions at the points of support. These equations involve the physical properties of the body which are studied in subjects dealing with the mechanics of deformation, such as "mechanics of materials."*

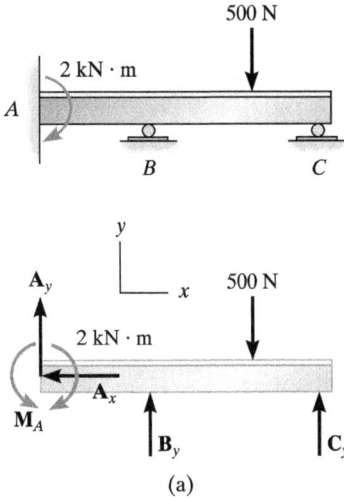

(a)

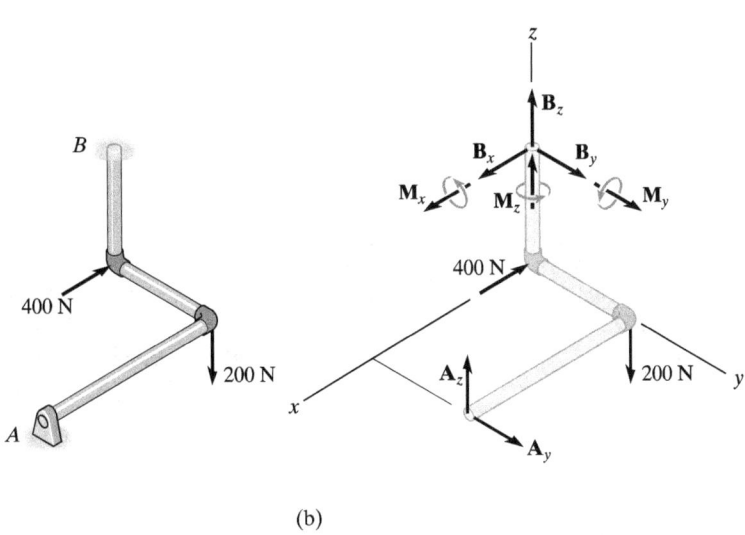

(b)

*See R. C. Hibbeler, *Mechanics of Materials*, 7th edition, Pearson Education/Prentice Hall, Inc.

**Fig. 5–24**

## Improper Constraints.

Having the same number of unknown reactive forces as available equations of equilibrium does not always guarantee that a body will be stable when subjected to a particular loading. For example, the pin support at $A$ and the roller support at $B$ for the beam in Fig. 5–25a are placed in such a way that the lines of action of the reactive forces are *concurrent* at point $A$. Consequently, the applied loading $\mathbf{P}$ will cause the beam to rotate slightly about $A$, and so the beam is improperly constrained, $\Sigma M_A \neq 0$.

In three dimensions, a body will be improperly constrained if the lines of action of all the reactive forces intersect a common axis. For example, the reactive forces at the ball-and-socket supports at $A$ and $B$ in Fig. 5–25b all intersect the axis passing through $A$ and $B$. Since the moments of these forces about $A$ and $B$ are all zero, then the loading $\mathbf{P}$ will rotate the member about the $AB$ axis, $\Sigma M_{AB} \neq 0$.

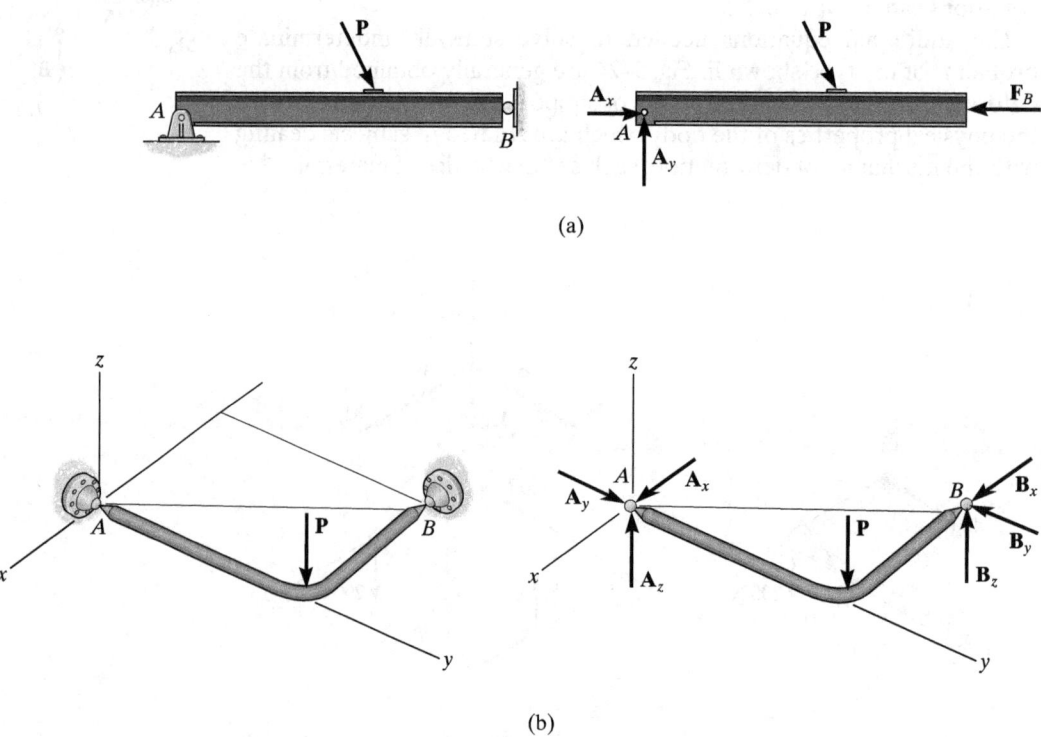

(a)

(b)

Fig. 5–25

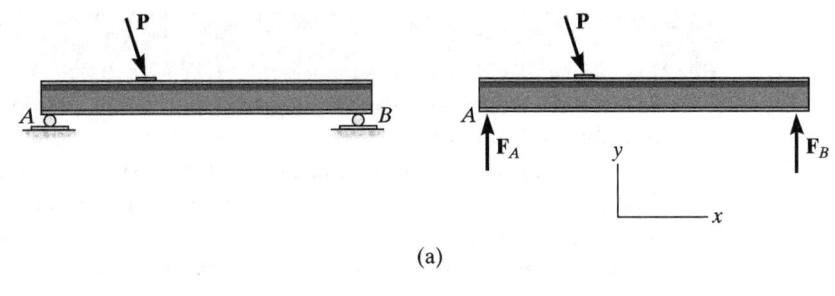

(a)

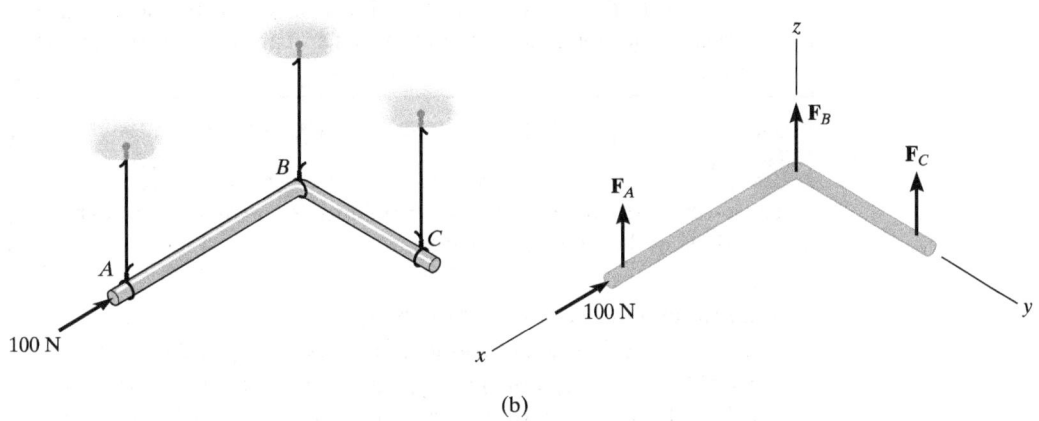

(b)

Fig. 5–26

Another way in which improper constraining leads to instability occurs when the *reactive forces* are all *parallel*. Two- and three-dimensional examples of this are shown in Fig. 5–26. In both cases, the summation of forces along the $x$ axis will not equal zero.

In some cases, a body may have *fewer* reactive forces than equations of equilibrium that must be satisfied. The body then becomes only *partially constrained*. For example, consider member $AB$ in Fig. 5–27a with its corresponding free-body diagram in Fig. 5–27b. Here $\Sigma F_y = 0$ will not be satisfied for the loading conditions and therefore equilibrium will not be maintained.

To summarize these points, a body is considered *improperly constrained* if all the reactive forces intersect at a common point or pass through a common axis, or if all the reactive forces are parallel. In engineering practice, these situations should be avoided at all times since they will cause an unstable condition.

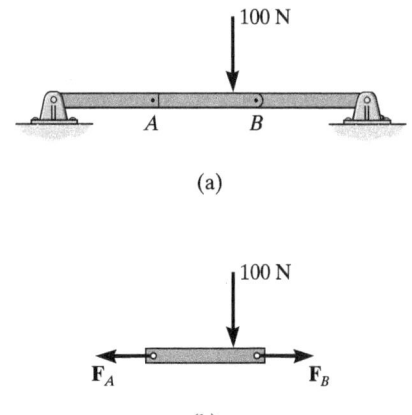

Fig. 5–27

## Important Points

- Always draw the free-body diagram first when solving any equilibrium problem.
- If a support *prevents translation* of a body in a specific direction, then the support exerts a *force* on the body in that direction.
- If a support *prevents rotation about an axis*, then the support exerts a *couple moment* on the body about the axis.
- If a body is subjected to more unknown reactions than available equations of equilibrium, then the problem is *statically indeterminate*.
- A stable body requires that the lines of action of the reactive forces do not intersect a common axis and are not parallel to one another.

## Procedure for Analysis

Three-dimensional equilibrium problems for a rigid body can be solved using the following procedure.

### Free-Body Diagram.

- Draw an outlined shape of the body.
- Show all the forces and couple moments acting on the body.
- Establish the origin of the $x, y, z$ axes at a convenient point and orient the axes so that they are parallel to as many of the external forces and moments as possible.
- Label all the loadings and specify their directions. In general, show all the *unknown* components having a *positive sense* along the $x, y, z$ axes.
- Indicate the dimensions of the body necessary for computing the moments of forces.

### Equations of Equilibrium.

- If the $x, y, z$ force and moment components seem easy to determine, then apply the six scalar equations of equilibrium; otherwise use the vector equations.
- It is not necessary that the set of axes chosen for force summation coincide with the set of axes chosen for moment summation. Actually, an axis in any arbitrary direction may be chosen for summing forces and moments.
- Choose the direction of an axis for moment summation such that it intersects the lines of action of as many unknown forces as possible. Realize that the moments of forces passing through points on this axis and the moments of forces which are parallel to the axis will then be zero.
- If the solution of the equilibrium equations yields a negative scalar for a force or couple moment magnitude, it indicates that the sense is opposite to that assumed on the free-body diagram.

## EXAMPLE 5.15

The homogeneous plate shown in Fig. 5–28a has a mass of 100 kg and is subjected to a force and couple moment along its edges. If it is supported in the horizontal plane by a roller at $A$, a ball-and-socket joint at $B$, and a cord at $C$, determine the components of reaction at these supports.

### SOLUTION (SCALAR ANALYSIS)

**Free-Body Diagram.** There are five unknown reactions acting on the plate, as shown in Fig. 5–28b. Each of these reactions is assumed to act in a positive coordinate direction.

**Equations of Equilibrium.** Since the three-dimensional geometry is rather simple, a *scalar analysis* provides a *direct solution* to this problem. A force summation along each axis yields

$\Sigma F_x = 0;$    $B_x = 0$    *Ans.*

$\Sigma F_y = 0;$    $B_y = 0$    *Ans.*

$\Sigma F_z = 0;$    $A_z + B_z + T_C - 300 \text{ N} - 981 \text{ N} = 0$    (1)

Recall that the moment of a force about an axis is equal to the product of the force magnitude and the perpendicular distance (moment arm) from the line of action of the force to the axis. Also, forces that are parallel to an axis or pass through it create no moment about the axis. Hence, summing moments about the positive $x$ and $y$ axes, we have

$\Sigma M_x = 0;$    $T_C(2 \text{ m}) - 981 \text{ N}(1 \text{ m}) + B_z(2 \text{ m}) = 0$    (2)

$\Sigma M_y = 0;$

$300 \text{ N}(1.5 \text{ m}) + 981 \text{ N}(1.5 \text{ m}) - B_z(3 \text{ m}) - A_z(3 \text{ m}) - 200 \text{ N} \cdot \text{m} = 0$    (3)

The components of the force at $B$ can be eliminated if moments are summed about the $x'$ and $y'$ axes. We obtain

$\Sigma M_{x'} = 0;$    $981 \text{ N}(1 \text{ m}) + 300 \text{ N}(2 \text{ m}) - A_z(2 \text{ m}) = 0$    (4)

$\Sigma M_{y'} = 0;$

$-300 \text{ N}(1.5 \text{ m}) - 981 \text{ N}(1.5 \text{ m}) - 200 \text{ N} \cdot \text{m} + T_C(3 \text{ m}) = 0$    (5)

Solving Eqs. 1 through 3 or the more convenient Eqs. 1, 4, and 5 yields

$A_z = 790 \text{ N}$    $B_z = -217 \text{ N}$    $T_C = 707 \text{ N}$    *Ans.*

The negative sign indicates that $\mathbf{B}_z$ acts downward.

**NOTE:** The solution of this problem does not require a summation of moments about the $z$ axis. The plate is partially constrained since the supports cannot prevent it from turning about the $z$ axis if a force is applied to it in the $x$–$y$ plane.

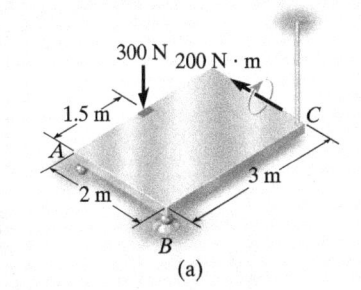

(a)

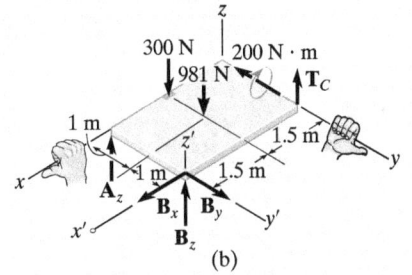

(b)

Fig. 5–28

## EXAMPLE 5.16

Determine the components of reaction that the ball-and-socket joint at $A$, the smooth journal bearing at $B$, and the roller support at $C$ exert on the rod assembly in Fig. 5–29a.

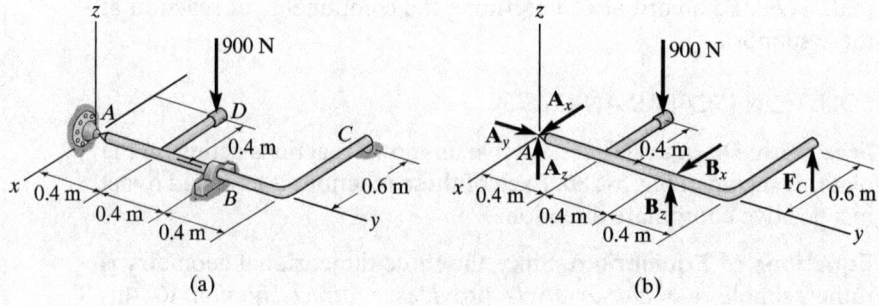

Fig. 5–29

### SOLUTION

**Free-Body Diagram.** As shown on the free-body diagram, Fig. 5–29b, the reactive forces of the supports will prevent the assembly from rotating about each coordinate axis, and so the journal bearing at $B$ only exerts reactive forces on the member.

**Equations of Equilibrium.** A direct solution for $A_y$ can be obtained by summing forces along the $y$ axis.

$\Sigma F_y = 0;$   $A_y = 0$   *Ans.*

The force $F_C$ can be determined directly by summing moments about the $y$ axis.

$\Sigma M_y = 0;$   $F_C(0.6 \text{ m}) - 900 \text{ N}(0.4 \text{ m}) = 0$
$F_C = 600 \text{ N}$   *Ans.*

Using this result, $B_z$ can be determined by summing moments about the $x$ axis.

$\Sigma M_x = 0;$   $B_z(0.8 \text{ m}) + 600 \text{ N}(1.2 \text{ m}) - 900 \text{ N}(0.4 \text{ m}) = 0$
$B_z = -450 \text{ N}$   *Ans.*

The negative sign indicates that $\mathbf{B}_z$ acts downward. The force $B_x$ can be found by summing moments about the $z$ axis.

$\Sigma M_z = 0;$   $-B_x(0.8 \text{ m}) = 0$   $B_x = 0$   *Ans.*

Thus,

$\Sigma F_x = 0;$   $A_x + 0 = 0$   $A_x = 0$   *Ans.*

Finally, using the results of $B_z$ and $F_C$.

$\Sigma F_z = 0;$   $A_z + (-450 \text{ N}) + 600 \text{ N} - 900 \text{ N} = 0$
$A_z = 750 \text{ N}$   *Ans.*

## EXAMPLE 5.17

The boom is used to support the 375-N ($\approx$ 37.5 kg) flowerpot in Fig. 5–30a. Determine the tension developed in wires $AB$ and $AC$.

### SOLUTION

**Free-Body Diagram.** The free-body diagram of the boom is shown in Fig. 5–30b.

**Equations of Equilibrium.** We will use a vector analysis.

$$\mathbf{F}_{AB} = F_{AB}\left(\frac{\mathbf{r}_{AB}}{r_{AB}}\right) = F_{AB}\left(\frac{\{0.2\mathbf{i} - 0.6\mathbf{j} + 0.3\mathbf{k}\} \text{ m}}{\sqrt{(0.2 \text{ m})^2 + (-0.6 \text{ m})^2 + (0.3 \text{ m})^2}}\right)$$

$$= \tfrac{2}{7} F_{AB}\mathbf{i} - \tfrac{6}{7} F_{AB}\mathbf{j} + \tfrac{3}{7} F_{AB}\mathbf{k}$$

$$\mathbf{F}_{AC} = F_{AC}\left(\frac{\mathbf{r}_{AC}}{r_{AC}}\right) = F_{AC}\left(\frac{\{-0.2\mathbf{i} - 0.6\mathbf{j} + 0.3\mathbf{k}\} \text{ m}}{\sqrt{(-0.2 \text{ m})^2 + (-0.6 \text{ m})^2 + (0.3 \text{ m})^2}}\right)$$

$$= -\tfrac{2}{7} F_{AC}\mathbf{i} - \tfrac{6}{7} F_{AC}\mathbf{j} + \tfrac{3}{7} F_{AC}\mathbf{k}$$

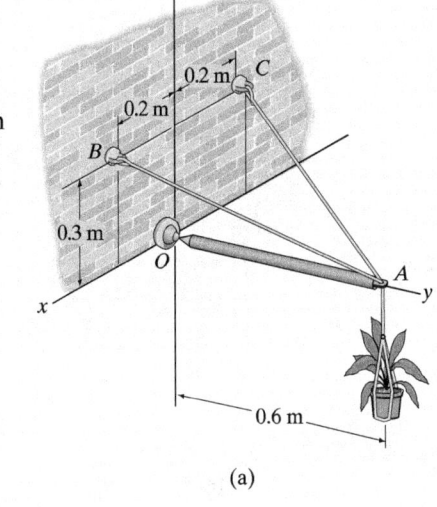

(a)

We can eliminate the force reaction at $O$ by writing the moment equation of equilibrium about point $O$.

$$\Sigma \mathbf{M}_O = \mathbf{0}; \qquad \mathbf{r}_A \times (\mathbf{F}_{AB} + \mathbf{F}_{AC} + \mathbf{W}) = \mathbf{0}$$

$$(0.6\mathbf{j}) \times \left[\left(\tfrac{2}{7} F_{AB}\mathbf{i} - \tfrac{6}{7} F_{AB}\mathbf{j} + \tfrac{3}{7} F_{AB}\mathbf{k}\right) + \left(-\tfrac{2}{7} F_{AC}\mathbf{i} - \tfrac{6}{7} F_{AC}\mathbf{j} + \tfrac{3}{7} F_{AC}\mathbf{k}\right) + (-375\mathbf{k})\right] = \mathbf{0}$$

$$\left(\tfrac{18}{7} F_{AB} + \tfrac{18}{7} F_{AC} - 2250\right)\mathbf{i} + \left(-\tfrac{12}{7} F_{AB} + \tfrac{12}{7} F_{AC}\right)\mathbf{k} = \mathbf{0}$$

$$\Sigma M_x = 0; \qquad \tfrac{18}{7} F_{AB} + \tfrac{18}{7} F_{AC} - 2250 = 0 \qquad (1)$$

$$\Sigma M_y = 0; \qquad 0 = 0$$

$$\Sigma M_z = 0; \qquad -\tfrac{12}{7} F_{AB} + \tfrac{12}{7} F_{AC} = 0 \qquad (2)$$

Solving Eqs. (1) and (2) simultaneously,

$$F_{AB} = F_{AC} = 437.5 \text{ N} \qquad \textit{Ans.}$$

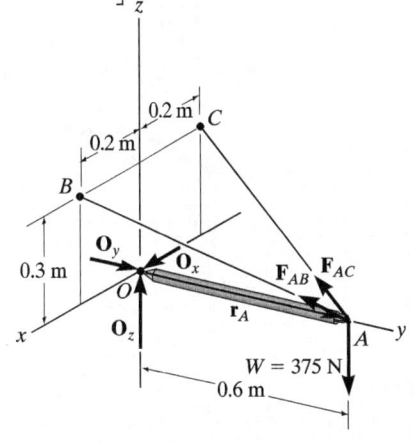

(b)

Fig. 5–30

# EXAMPLE 5.18

Rod $AB$ shown in Fig. 5–31$a$ is subjected to the 200-N force. Determine the reactions at the ball-and-socket joint $A$ and the tension in the cables $BD$ and $BE$.

### SOLUTION (*VECTOR ANALYSIS*)

**Free-Body Diagram.** Fig. 5–31$b$.

**Equations of Equilibrium.** Representing each force on the free-body diagram in Cartesian vector form, we have

$$\mathbf{F}_A = A_x\mathbf{i} + A_y\mathbf{j} + A_z\mathbf{k}$$
$$\mathbf{T}_E = T_E\mathbf{i}$$
$$\mathbf{T}_D = T_D\mathbf{j}$$
$$\mathbf{F} = \{-200\mathbf{k}\}\text{ N}$$

Applying the force equation of equilibrium.

$\Sigma \mathbf{F} = \mathbf{0}$;  $\qquad \mathbf{F}_A + \mathbf{T}_E + \mathbf{T}_D + \mathbf{F} = \mathbf{0}$
$(A_x + T_E)\mathbf{i} + (A_y + T_D)\mathbf{j} + (A_z - 200)\mathbf{k} = \mathbf{0}$

$\Sigma F_x = 0$;  $\qquad A_x + T_E = 0$ \hfill (1)
$\Sigma F_y = 0$;  $\qquad A_y + T_D = 0$ \hfill (2)
$\Sigma F_z = 0$;  $\qquad A_z - 200 = 0$ \hfill (3)

Summing moments about point $A$ yields

$\Sigma \mathbf{M}_A = \mathbf{0}$;  $\qquad \mathbf{r}_C \times \mathbf{F} + \mathbf{r}_B \times (\mathbf{T}_E + \mathbf{T}_D) = \mathbf{0}$

Since $\mathbf{r}_C = \frac{1}{2}\mathbf{r}_B$, then

$(0.5\mathbf{i} + 1\mathbf{j} - 1\mathbf{k}) \times (-200\mathbf{k}) + (1\mathbf{i} + 2\mathbf{j} - 2\mathbf{k}) \times (T_E\mathbf{i} + T_D\mathbf{j}) = \mathbf{0}$

Expanding and rearranging terms gives

$(2T_D - 200)\mathbf{i} + (-2T_E + 100)\mathbf{j} + (T_D - 2T_E)\mathbf{k} = \mathbf{0}$

$\Sigma M_x = 0$;  $\qquad 2T_D - 200 = 0$ \hfill (4)
$\Sigma M_y = 0$;  $\qquad -2T_E + 100 = 0$ \hfill (5)
$\Sigma M_z = 0$;  $\qquad T_D - 2T_E = 0$ \hfill (6)

Solving Eqs. 1 through 5, we get

$T_D = 100$ N  \hfill *Ans.*
$T_E = 50$ N   \hfill *Ans.*
$A_x = -50$ N  \hfill *Ans.*
$A_y = -100$ N \hfill *Ans.*
$A_z = 200$ N  \hfill *Ans.*

**NOTE:** The negative sign indicates that $\mathbf{A}_x$ and $\mathbf{A}_y$ have a sense which is opposite to that shown on the free-body diagram, Fig. 5–31$b$.

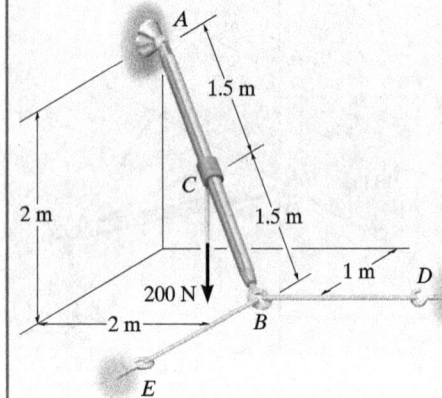

(a)

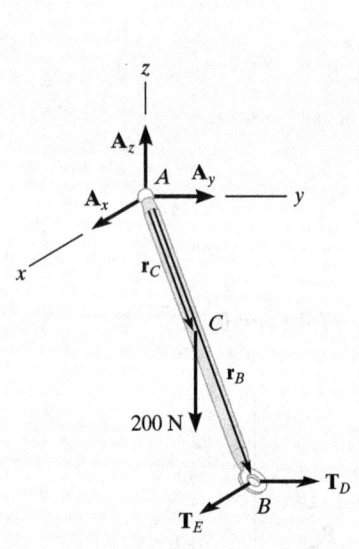

(b)

Fig. 5–31

## EXAMPLE 5.19

The bent rod in Fig. 5–32a is supported at $A$ by a journal bearing, at $D$ by a ball-and-socket joint, and at $B$ by means of cable $BC$. Using only *one equilibrium equation*, obtain a direct solution for the tension in cable $BC$. The bearing at $A$ is capable of exerting force components only in the $z$ and $y$ directions since it is properly aligned on the shaft.

### SOLUTION (*VECTOR ANALYSIS*)

**Free-Body Diagram.** As shown in Fig. 5–32b, there are six unknowns.

**Equations of Equilibrium.** The cable tension $\mathbf{T}_B$ may be obtained *directly* by summing moments about an axis that passes through points $D$ and $A$. Why? The direction of this axis is defined by the unit vector $\mathbf{u}$, where

$$\mathbf{u} = \frac{\mathbf{r}_{DA}}{r_{DA}} = -\frac{1}{\sqrt{2}}\mathbf{i} - \frac{1}{\sqrt{2}}\mathbf{j}$$
$$= -0.7071\mathbf{i} - 0.7071\mathbf{j}$$

Hence, the sum of the moments about this axis is zero provided

$$\Sigma M_{DA} = \mathbf{u} \cdot \Sigma(\mathbf{r} \times \mathbf{F}) = 0$$

Here $\mathbf{r}$ represents a position vector drawn from *any point* on the axis $DA$ to any point on the line of action of force $\mathbf{F}$ (see Eq. 4–11). With reference to Fig. 5–32b, we can therefore write

$$\mathbf{u} \cdot (\mathbf{r}_B \times \mathbf{T}_B + \mathbf{r}_E \times \mathbf{W}) = \mathbf{0}$$

$$(-0.7071\mathbf{i} - 0.7071\mathbf{j}) \cdot \left[(-1\mathbf{j}) \times (T_B\mathbf{k}) + (-0.5\mathbf{j}) \times (-981\mathbf{k})\right] = \mathbf{0}$$

$$(-0.7071\mathbf{i} - 0.7071\mathbf{j}) \cdot [(-T_B + 490.5)\mathbf{i}] = \mathbf{0}$$

$$-0.7071(-T_B + 490.5) + 0 + 0 = 0$$

$$T_B = 490.5 \text{ N} \qquad \textit{Ans.}$$

Since the moment arms from the axis to $\mathbf{T}_B$ and $\mathbf{W}$ are easy to obtain, we can also determine this result using a scalar analysis. As shown in Fig. 5–32b,

$$\Sigma M_{DA} = 0; \quad T_B(1 \text{ m} \sin 45°) - 981 \text{ N}(0.5 \text{ m} \sin 45°) = 0$$

$$T_B = 490.5 \text{ N} \qquad \textit{Ans.}$$

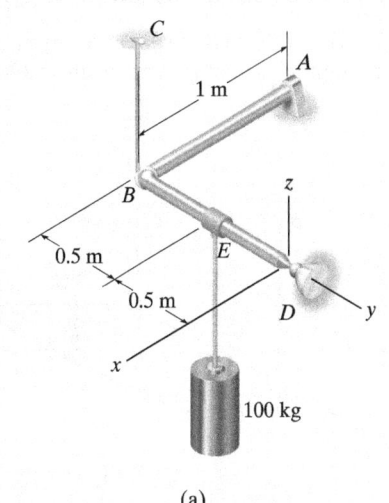

(a)

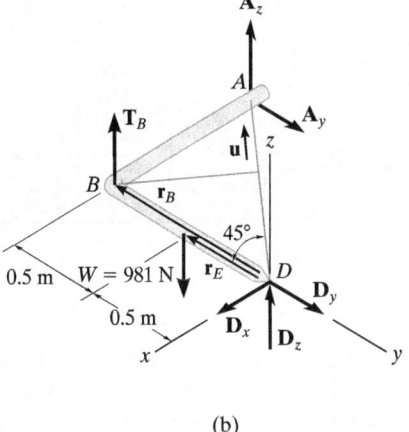

(b)

Fig. 5–32

# FUNDAMENTAL PROBLEMS

*All problem solutions must include an FBD.*

**F5–7.** The uniform plate has a weight of 2.5 kN. Determine the tension in each of the supporting cables.

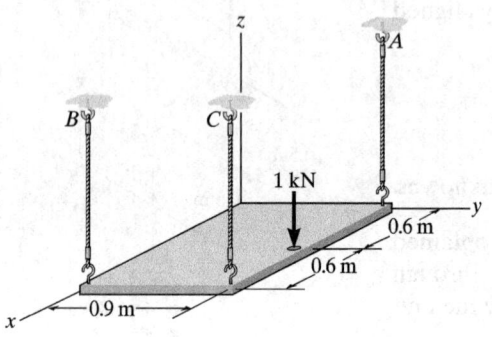

F5–7

**F5–8.** Determine the reactions at the roller support $A$, the ball-and-socket joint $D$, and the tension in cable $BC$ for the plate.

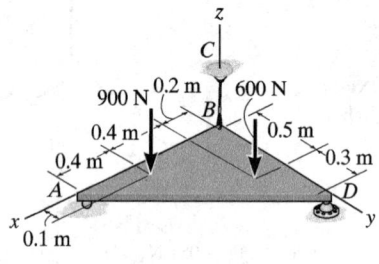

F5–8

**F5–9.** The rod is supported by smooth journal bearings at $A$, $B$ and $C$ and is subjected to the two forces. Determine the reactions at these supports.

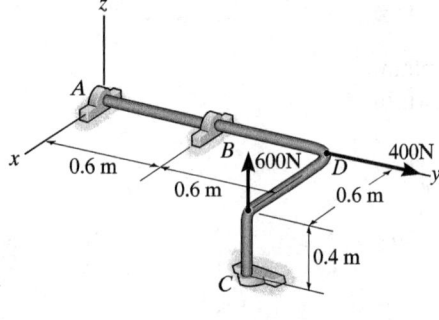

F5–9

**F5–10.** Determine the support reactions at the smooth journal bearings $A$, $B$, and $C$ of the pipe assembly.

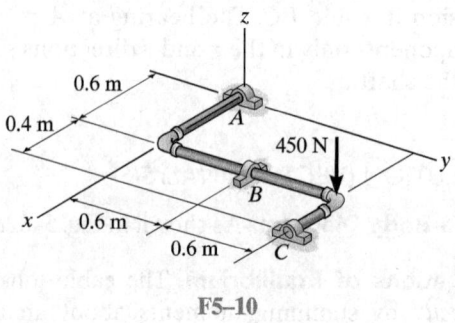

F5–10

**F5–11.** Determine the force developed in cords $BD$, $CE$, and $CF$ and the reactions of the ball-and-socket joint $A$ on the block.

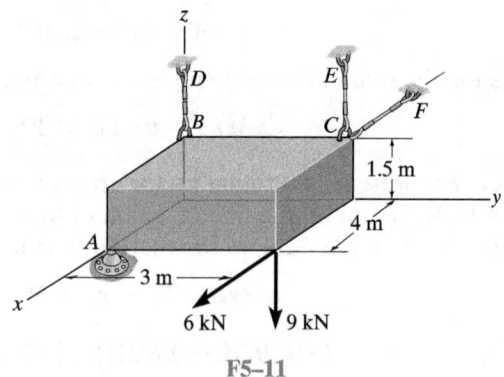

F5–11

**F5–12.** Determine the components of reaction that the thrust bearing $A$ and cable $BC$ exert on the bar.

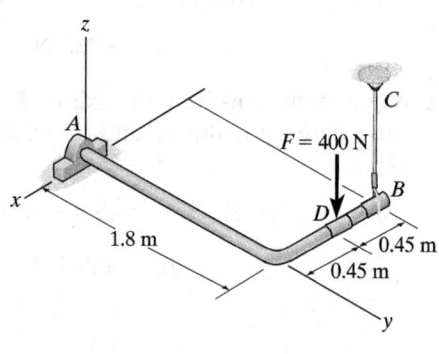

F5–12

## PROBLEMS

*All problem solutions must include an FBD.*

**5–63.** The cart supports the uniform crate having a mass of 85 kg. Determine the vertical reactions on the three casters at A, B, and C. The caster at B is not shown. Neglect the mass of the cart.

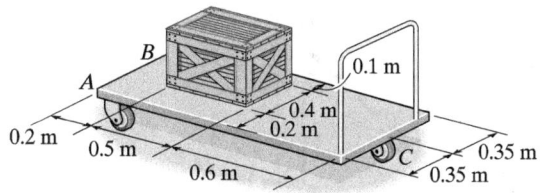

Prob. 5–63

**\*5–64.** The pole for a power line is subjected to the two cable forces of 300 N, each force lying in a plane parallel to the x–y plane. If the tension in the guy wire AB is 400 N, determine the x, y, z components of reaction at the fixed base of the pole, O.

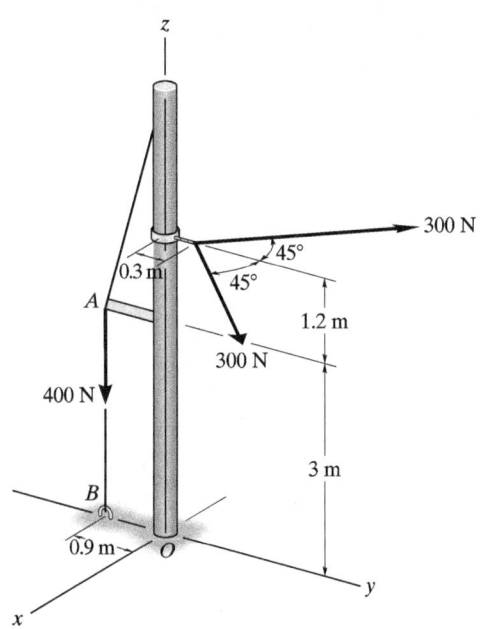

Prob. 5–64

**•5–65.** If $P = 6$ kN, $x = 0.75$ m and $y = 1$ m, determine the tension developed in cables AB, CD, and EF. Neglect the weight of the plate.

**5–66.** Determine the location x and y of the point of application of force **P** so that the tension developed in cables AB, CD, and EF is the same. Neglect the weight of the plate.

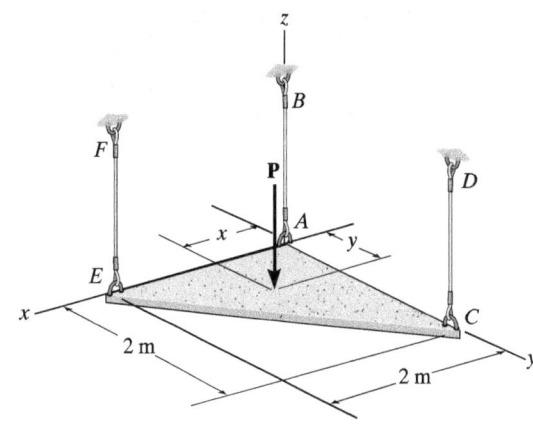

Probs. 5–65/66

**5–67.** Due to an unequal distribution of fuel in the wing tanks, the centers of gravity for the airplane fuselage A and wings B and C are located as shown. If these components have weights $W_A = 225$ kN, $W_B = 40$ kN, and $W_C = 30$ kN, determine the normal reactions of the wheels D, E, and F on the ground.

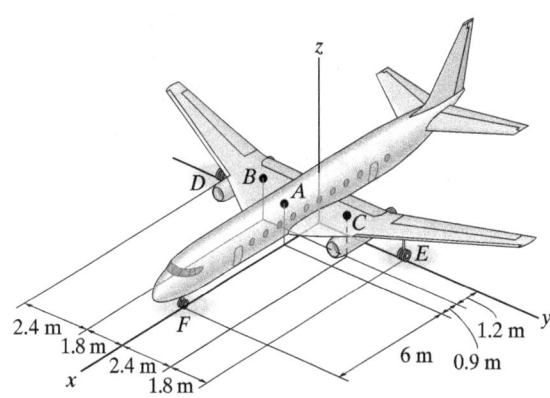

Prob. 5–67

*5–68. Determine the magnitude of force **F** that must be exerted on the handle at $C$ to hold the 75-kg crate in the position shown. Also, determine the components of reaction at the thrust bearing $A$ and smooth journal bearing $B$.

5–70. Determine the tension in cables $BD$ and $CD$ and the $x$, $y$, $z$ components of reaction at the ball-and-socket joint at $A$.

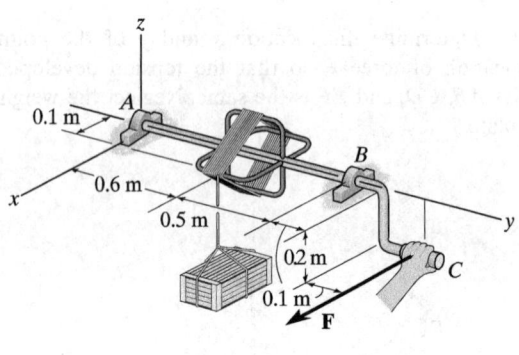

Prob. 5–68

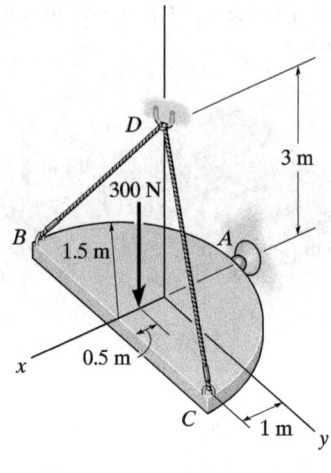

Prob. 5–70

•5–69. The shaft is supported by three smooth journal bearings at $A$, $B$, and $C$. Determine the components of reaction at these bearings.

5–71. The rod assembly is used to support the 1.25-kN ($\approx$ 125 kg) cylinder. Determine the components of reaction at the ball-and-socket joint $A$, the smooth journal bearing $E$, and the force developed along rod $CD$. The connections at $C$ and $D$ are ball-and-socket joints.

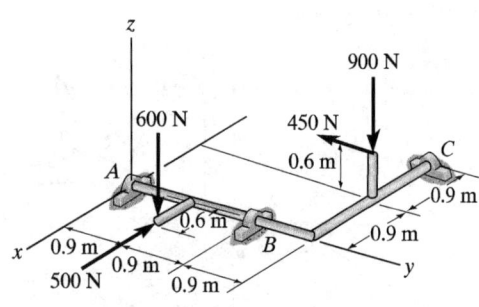

Prob. 5–69

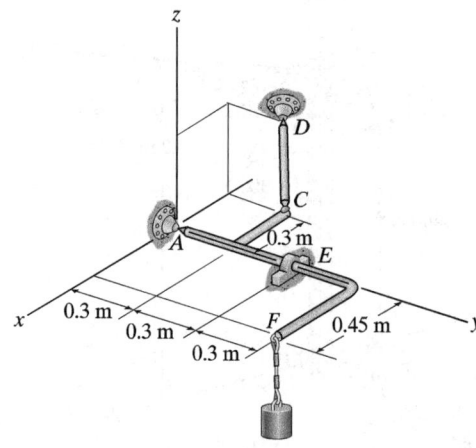

Prob. 5–71

**\*5–72.** Determine the components of reaction acting at the smooth journal bearings $A$, $B$, and $C$.

**5–74.** If the load has a weight of 200 kN, determine the $x$, $y$, $z$ components of reaction at the ball-and-socket joint $A$ and the tension in each of the wires.

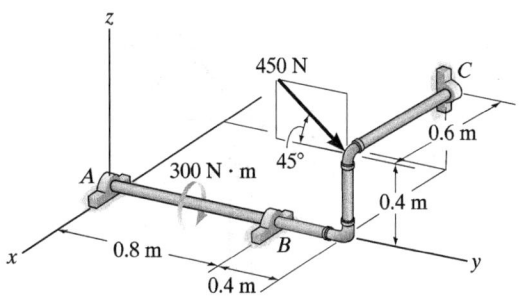

Prob. 5–72

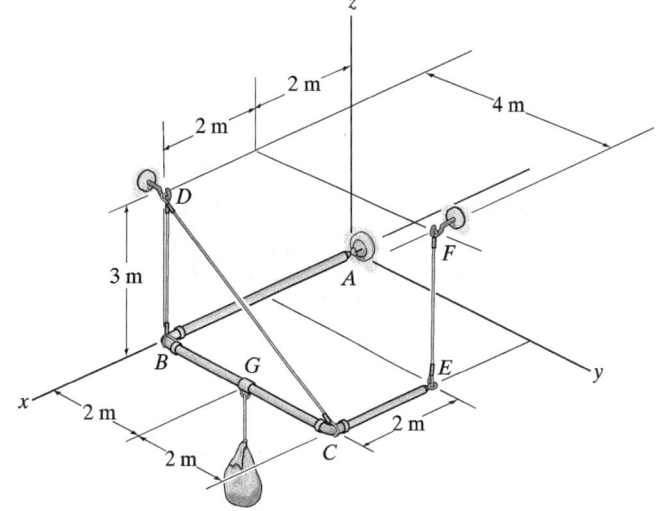

Prob. 5–74

**•5–73.** Determine the force components acting on the ball-and-socket at $A$, the reaction at the roller $B$ and the tension on the cord $CD$ needed for equilibrium of the quarter circular plate.

**5–75.** If the cable can be subjected to a maximum tension of 1.5 kN, determine the maximum force $F$ which may be applied to the plate. Compute the $x$, $y$, $z$ components of reaction at the hinge $A$ for this loading.

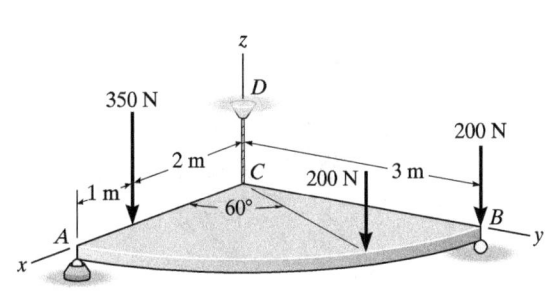

Prob. 5–73

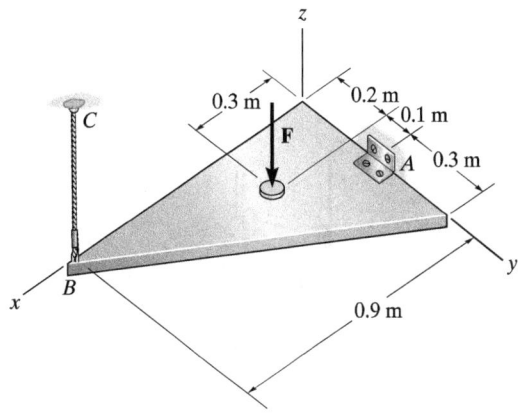

Prob. 5–75

**\*5–76.** The member is supported by a pin at $A$ and a cable $BC$. If the load at $D$ is 1.5 kN, determine the $x$, $y$, $z$ components of reaction at the pin $A$ and the tension in cable $BC$.

**5–79.** The boom is supported by a ball-and-socket joint at $A$ and a guy wire at $B$. If the 5-kN loads lie in a plane which is parallel to the $x$–$y$ plane, determine the $x, y, z$ components of reaction at $A$ and the tension in the cable at $B$.

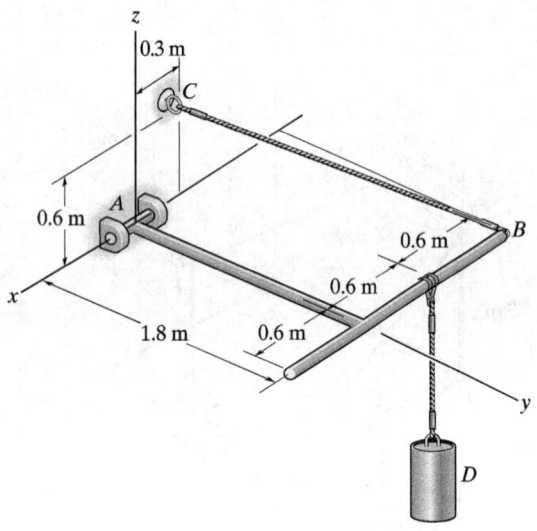

Prob. 5–76

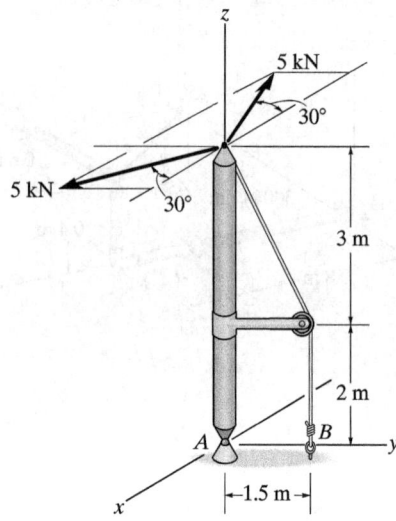

Prob. 5–79

**•5–77.** The plate has a weight of $W$ with a center of gravity at $G$. Determine the distance $d$ along line $GH$ where the vertical force $P = 0.75W$ will cause the tension in wire $CD$ to become zero.

**\*5–80.** The circular door has a weight of 275 N and a center of gravity at $G$. Determine the $x$, $y$, $z$ components of reaction at the hinge $A$ and the force acting along strut $CB$ needed to hold the door in equilibrium. Set $\theta = 45°$.

**5–78.** The plate has a weight of $W$ with a center of gravity at $G$. Determine the tension developed in wires $AB$, $CD$, and $EF$ if the force $P = 0.75W$ is applied at $d = L/2$.

**•5–81.** The circular door has a weight of 275 N and a center of gravity at $G$. Determine the $x$, $y$, $z$ components of reaction at the hinge $A$ and the force acting along strut $CB$ needed to hold the door in equilibrium. Set $\theta = 90°$.

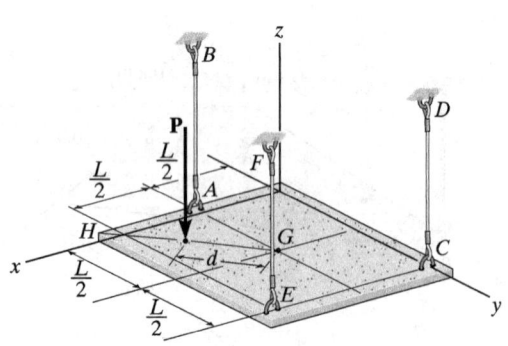

Probs. 5–77/78

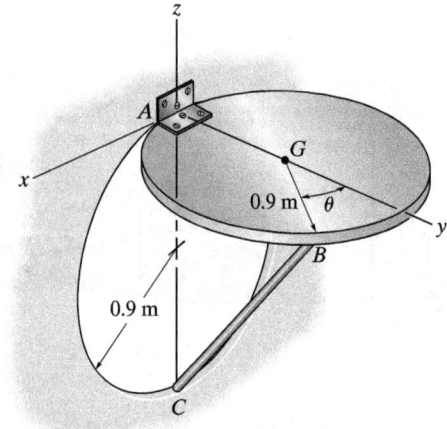

Probs. 5–80/81

**5–82.** Member $AB$ is supported at $B$ by a cable and at $A$ by a smooth fixed *square* rod which fits loosely through the square hole of the collar. If $\mathbf{F} = \{2\mathbf{i} - 4\mathbf{j} - 7.5\mathbf{k}\}$ kN, determine the $x$, $y$, $z$ components of reaction at $A$ and the tension in the cable.

**5–83.** Member $AB$ is supported at $B$ by a cable and at $A$ by a smooth fixed *square* rod which fits loosely through the square hole of the collar. Determine the tension in cable $BC$ if the force $\mathbf{F} = \{-4.5\mathbf{k}\}$ kN.

**•5–85.** The circular plate has a weight $W$ and center of gravity at its center. If it is supported by three vertical cords tied to its edge, determine the largest distance $d$ from the center to where any vertical force $\mathbf{P}$ can be applied so as not to cause the force in any one of the cables to become zero.

**5–86.** Solve Prob. 5–85 if the plate's weight $W$ is neglected.

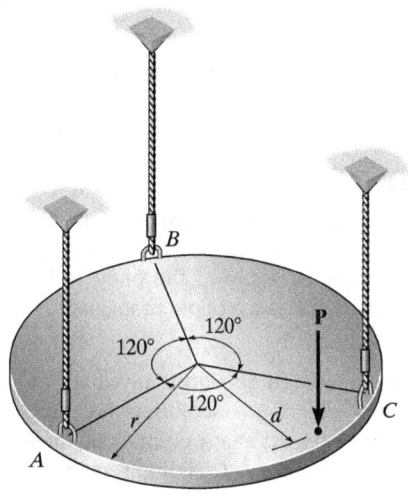

Probs. 5–85/86

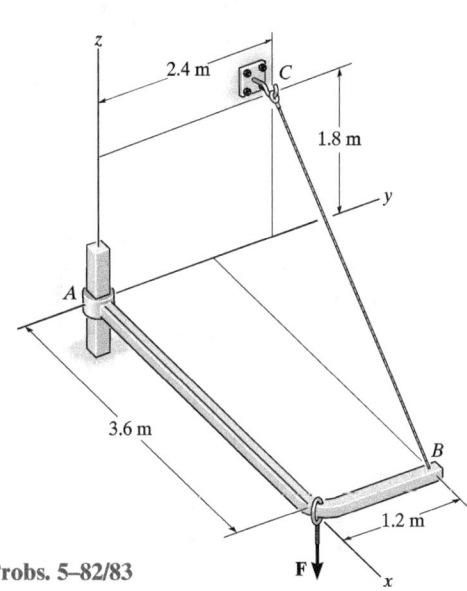

Probs. 5–82/83

**\*5–84.** Determine the largest weight of the oil drum that the floor crane can support without overturning. Also, what are the vertical reactions at the smooth wheels $A$, $B$, and $C$ for this case. The floor crane has a weight of 1.5 kN, with its center of gravity located at $G$.

**5–87.** A uniform square table having a weight $W$ and sides $a$ is supported by three vertical legs. Determine the smallest vertical force $\mathbf{P}$ that can be applied to its top that will cause it to tip over.

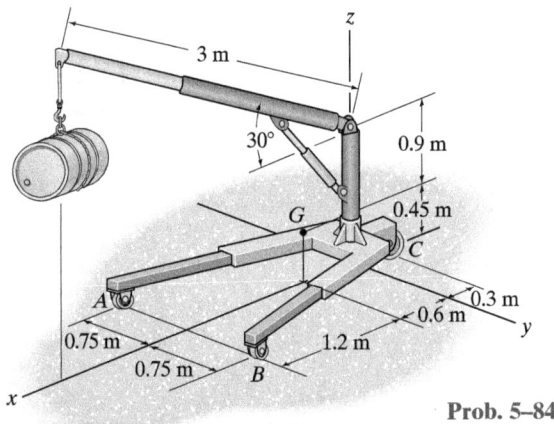

Prob. 5–84

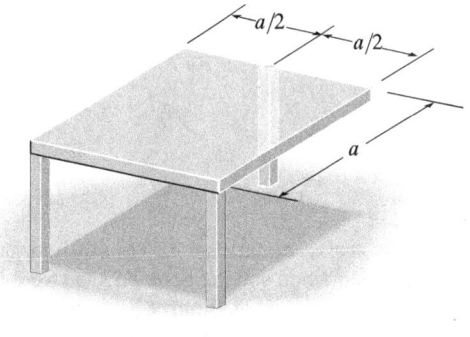

Prob. 5–87

# CHAPTER REVIEW

**Equilibrium**

A body in equilibrium does not rotate but can translate with constant velocity, or it does not move at all.

$$\Sigma \mathbf{F} = 0$$
$$\Sigma \mathbf{M} = 0$$

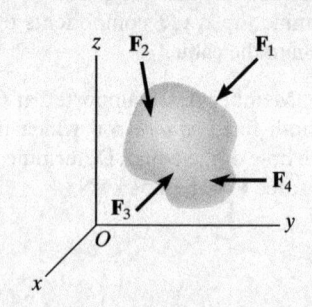

**Two Dimensions**

Before analyzing the equilibrium of a body, it is first necessary to draw its free-body diagram. This is an outlined shape of the body, which shows all the forces and couple moments that act on it.

Couple moments can be placed anywhere on a free-body diagram since they are free vectors. Forces can act at any point along their line of action since they are sliding vectors.

Angles used to resolve forces, and dimensions used to take moments of the forces, should also be shown on the free-body diagram.

Some common types of supports and their reactions are shown below in two dimensions.

Remember that a support will exert a force on the body in a particular direction if it prevents translation of the body in that direction, and it will exert a couple moment on the body if it prevents rotation.

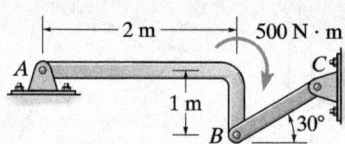

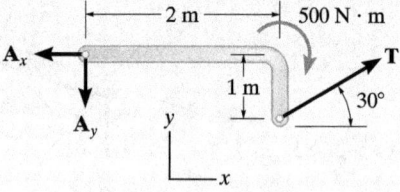

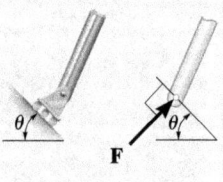
roller

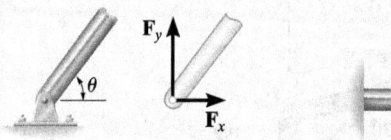
smooth pin or hinge

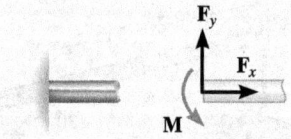
fixed support

The three scalar equations of equilibrium can be applied when solving problems in two dimensions, since the geometry is easy to visualize.

$$\Sigma F_x = 0$$
$$\Sigma F_y = 0$$
$$\Sigma M_O = 0$$

For the most direct solution, try to sum forces along an axis that will eliminate as many unknown forces as possible. Sum moments about a point $A$ that passes through the line of action of as many unknown forces as possible.

$\Sigma F_x = 0;$
$A_x - P_2 = 0 \qquad A_x = P_2$

$\Sigma M_A = 0;$
$P_2 d_2 + B_y d_B - P_1 d_1 = 0$

$B_y = \dfrac{P_1 d_1 - P_2 d_2}{d_B}$

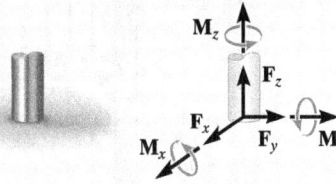

## Three Dimensions

Some common types of supports and their reactions are shown here in three dimensions.

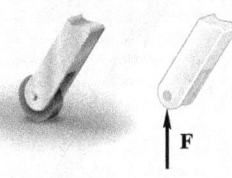

roller

ball and socket

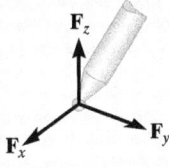

fixed support

In three dimensions, it is often advantageous to use a Cartesian vector analysis when applying the equations of equilibrium. To do this, first express each known and unknown force and couple moment shown on the free-body diagram as a Cartesian vector. Then set the force summation equal to zero. Take moments about a point $O$ that lies on the line of action of as many unknown force components as possible. From point $O$ direct position vectors to each force, and then use the cross product to determine the moment of each force.

$\Sigma \mathbf{F} = \mathbf{0}$
$\Sigma \mathbf{M}_O = \mathbf{0}$

$\Sigma F_x = 0$
$\Sigma F_y = 0$
$\Sigma F_z = 0$

$\Sigma M_x = 0$
$\Sigma M_y = 0$
$\Sigma M_z = 0$

The six scalar equations of equilibrium are established by setting the respective $\mathbf{i}, \mathbf{j}$, and $\mathbf{k}$ components of these force and moment summations equal to zero.

## Determinacy and Stability

If a body is supported by a minimum number of constraints to ensure equilibrium, then it is statically determinate. If it has more constraints than required, then it is statically indeterminate.

To properly constrain the body, the reactions must not all be parallel to one another or concurrent.

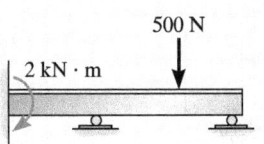

Statically indeterminate, five reactions, three equilibrium equations

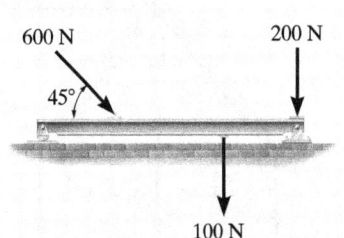

Proper constraint, statically determinate

## REVIEW PROBLEMS

**\*5–88.** Determine the horizontal and vertical components of reaction at pin $A$ and force in the cable $BC$. Neglect the thickness of the members.

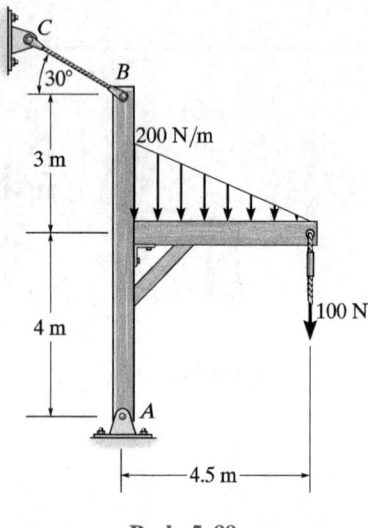

Prob. 5–88

**5–91.** Determine the normal reaction at the roller $A$ and horizontal and vertical components at pin $B$ for equilibrium of the member.

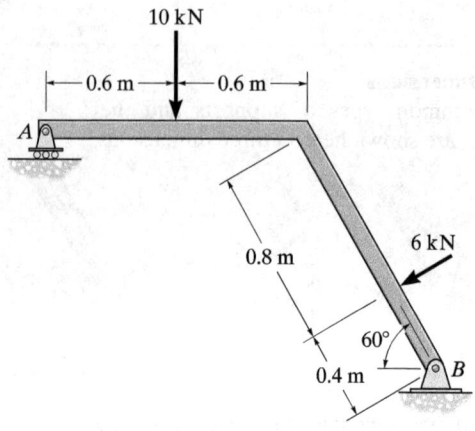

Prob. 5–91

**•5–89.** Determine the horizontal and vertical components of reaction at pin $A$ and the reaction at roller $B$ required to support the truss. Set $F = 600$ N.

**5–90.** If the roller at $B$ can sustain a maximum load of 3 kN, determine the largest magnitude of each of the three forces $\mathbf{F}$ that can be supported by the truss.

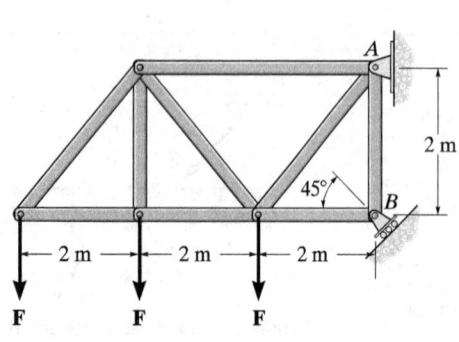

Probs. 5–89/90

**\*5–92.** The shaft assembly is supported by two smooth journal bearings $A$ and $B$ and a short link $DC$. If a couple moment is applied to the shaft as shown, determine the components of force reaction at the journal bearings and the force in the link. The link lies in a plane parallel to the $y$–$z$ plane and the bearings are properly aligned on the shaft.

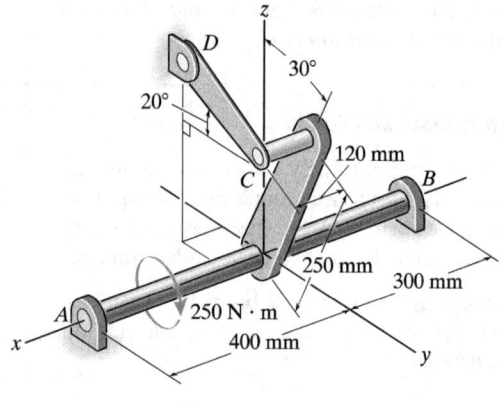

Prob. 5–92

REVIEW PROBLEMS    261

•5–93. Determine the reactions at the supports $A$ and $B$ of the frame.

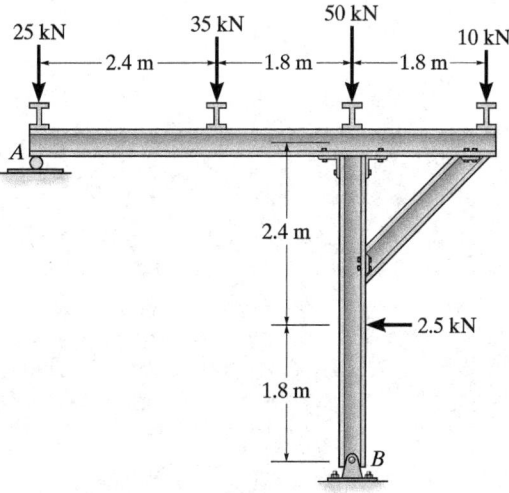

Prob. 5–93

5–94. A skeletal diagram of the lower leg is shown in the lower figure. Here it can be noted that this portion of the leg is lifted by the quadriceps muscle attached to the hip at $A$ and to the patella bone at $B$. This bone slides freely over cartilage at the knee joint. The quadriceps is further extended and attached to the tibia at $C$. Using the mechanical system shown in the upper figure to model the lower leg, determine the tension in the quadriceps at $C$ and the magnitude of the resultant force at the femur (pin), $D$, in order to hold the lower leg in the position shown. The lower leg has a mass of 3.2 kg and a mass center at $G_1$; the foot has a mass of 1.6 kg and a mass center at $G_2$.

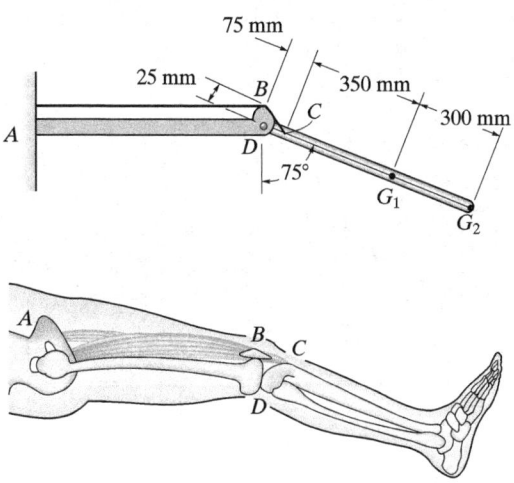

Prob. 5–94

5–95. A vertical force of 400 N acts on the crankshaft. Determine the horizontal equilibrium force $P$ that must be applied to the handle and the $x, y, z$ components of force at the smooth journal bearing $A$ and the thrust bearing $B$. The bearings are properly aligned and exert only force reactions on the shaft.

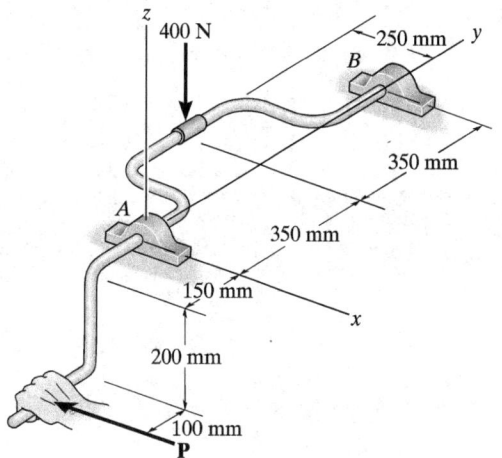

Prob. 5–95

*5–96. The symmetrical shelf is subjected to a uniform load of 4 kPa. Support is provided by a bolt (or pin) located at each end $A$ and $A'$ and by the symmetrical brace arms, which bear against the smooth wall on both sides at $B$ and $B'$. Determine the force resisted by each bolt at the wall and the normal force at $B$ for equilibrium.

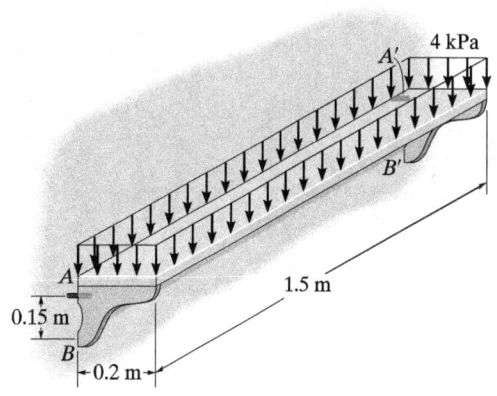

Prob. 5–96

The forces within the members of each truss bridge must be determined if the members are to be properly designed.

# Structural Analysis

## CHAPTER OBJECTIVES

- To show how to determine the forces in the members of a truss using the method of joints and the method of sections.
- To analyze the forces acting on the members of frames and machines composed of pin-connected members.

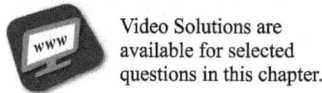

Video Solutions are available for selected questions in this chapter.

## 6.1 Simple Trusses

A *truss* is a structure composed of slender members joined together at their end points. The members commonly used in construction consist of wooden struts or metal bars. In particular, *planar* trusses lie in a single plane and are often used to support roofs and bridges. The truss shown in Fig. 6–1a is an example of a typical roof-supporting truss. In this figure, the roof load is transmitted to the truss *at the joints* by means of a series of *purlins*. Since this loading acts in the same plane as the truss, Fig. 6–1b, the analysis of the forces developed in the truss members will be two-dimensional.

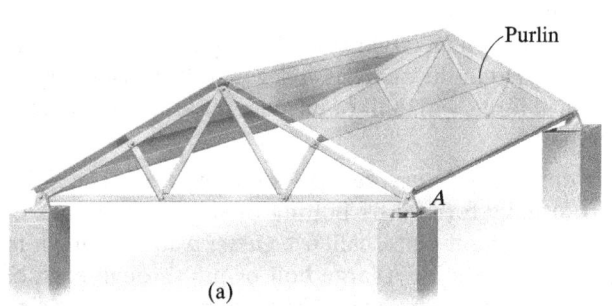

(a)

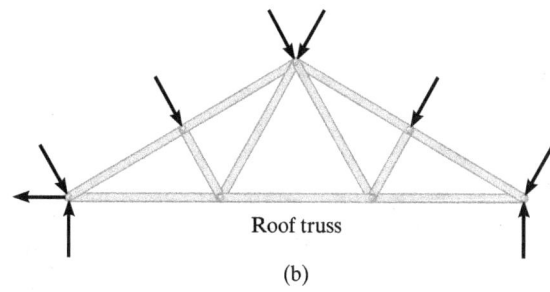

Roof truss

(b)

Fig. 6–1

**264** CHAPTER 6 STRUCTURAL ANALYSIS

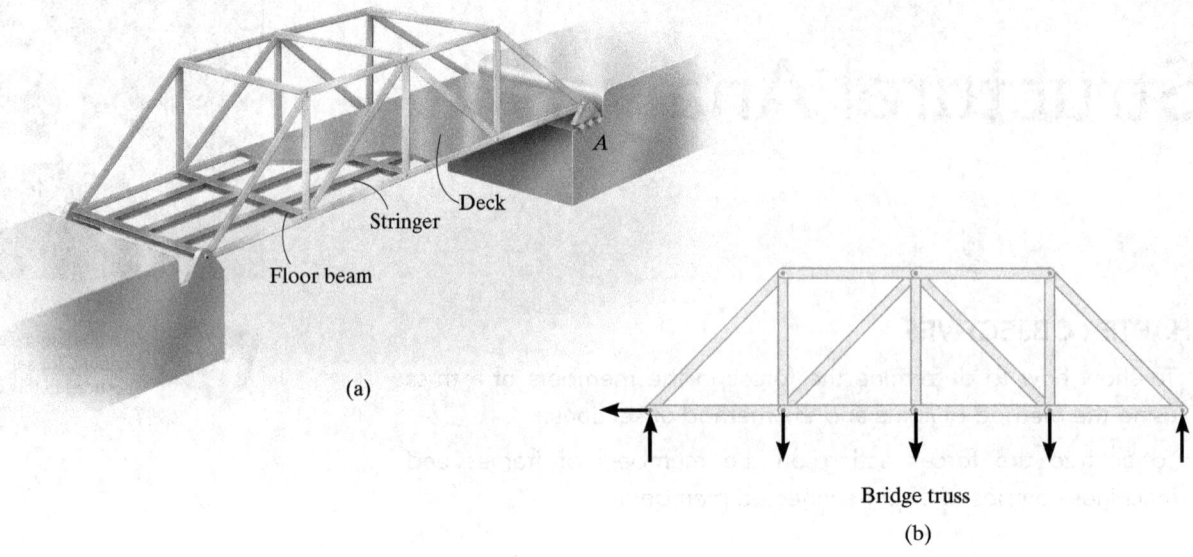

Bridge truss
(b)

Fig. 6–2

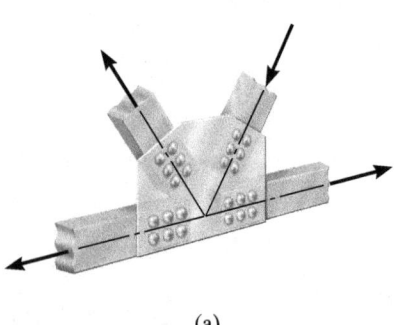

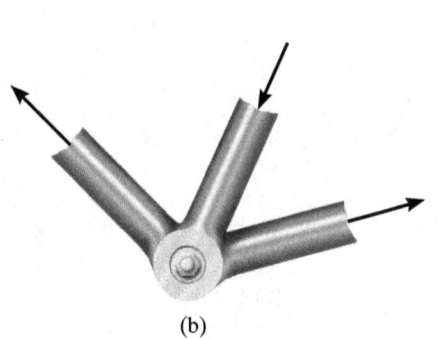

Fig. 6–3

In the case of a bridge, such as shown in Fig. 6–2a, the load on the deck is first transmitted to *stringers*, then to *floor beams*, and finally to the *joints* of the two supporting side trusses. Like the roof truss, the bridge truss loading is also coplanar, Fig. 6–2b.

When bridge or roof trusses extend over large distances, a rocker or roller is commonly used for supporting one end, for example, joint $A$ in Figs. 6–1a and 6–2a. This type of support allows freedom for expansion or contraction of the members due to a change in temperature or application of loads.

**Assumptions for Design.** To design both the members and the connections of a truss, it is necessary first to determine the *force* developed in each member when the truss is subjected to a given loading. To do this we will make two important assumptions:

- *All loadings are applied at the joints.* In most situations, such as for bridge and roof trusses, this assumption is true. Frequently the weight of the members is neglected because the force supported by each member is usually much larger than its weight. However, if the weight is to be included in the analysis, it is generally satisfactory to apply it as a vertical force, with half of its magnitude applied at each end of the member.

- *The members are joined together by smooth pins.* The joint connections are usually formed by bolting or welding the ends of the members to a common plate, called a *gusset plate*, as shown in Fig. 6–3a, or by simply passing a large bolt or pin through each of the members, Fig. 6–3b. We can assume these connections act as pins provided the center lines of the joining members are *concurrent*, as in Fig. 6–3.

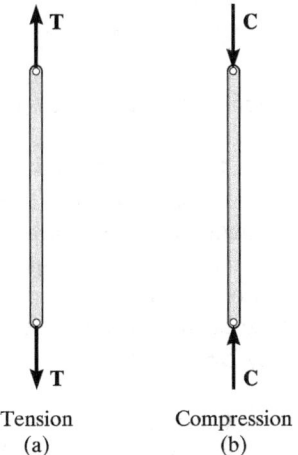

Tension
(a)

Compression
(b)

Fig. 6–4

Because of these two assumptions, *each truss member will act as a two-force member*, and therefore the force acting at each end of the member will be directed along the axis of the member. If the force tends to *elongate* the member, it is a *tensile force* (T), Fig. 6–4a; whereas if it tends to *shorten* the member, it is a *compressive force* (C), Fig. 6–4b. In the actual design of a truss it is important to state whether the nature of the force is tensile or compressive. Often, compression members must be made *thicker* than tension members because of the buckling or column effect that occurs when a member is in compression.

**Simple Truss.** If three members are pin connected at their ends they form a *triangular truss* that will be *rigid*, Fig. 6–5. Attaching two more members and connecting these members to a new joint $D$ forms a larger truss, Fig. 6–6. This procedure can be repeated as many times as desired to form an even larger truss. If a truss can be constructed by expanding the basic triangular truss in this way, it is called a *simple truss*.

The use of metal gusset plates in the construction of these Warren trusses is clearly evident.

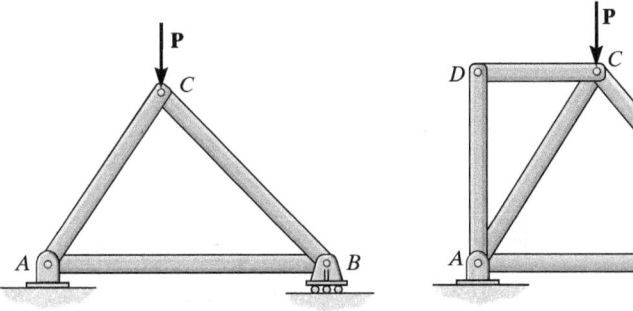

Fig. 6–5

Fig. 6–6

## 6.2 The Method of Joints

Please refer to the Companion CD for the animation: *Free-body Diagram For an A-frame*

In order to analyze or design a truss, it is necessary to determine the force in each of its members. One way to do this is to use the method of joints. This method is based on the fact that if the entire truss is in equilibrium, then each of its joints is also in equilibrium. Therefore, if the free-body diagram of each joint is drawn, the force equilibrium equations can then be used to obtain the member forces acting on each joint. Since the members of a *plane truss* are straight two-force members lying in a single plane, each joint is subjected to a force system that is *coplanar and concurrent*. As a result, only $\Sigma F_x = 0$ and $\Sigma F_y = 0$ need to be satisfied for equilibrium.

For example, consider the pin at joint $B$ of the truss in Fig. 6–7a. Three forces act on the pin, namely, the 500-N force and the forces exerted by members $BA$ and $BC$. The free-body diagram of the pin is shown in Fig. 6–7b. Here, $\mathbf{F}_{BA}$ is "pulling" on the pin, which means that member $BA$ is in *tension;* whereas $\mathbf{F}_{BC}$ is "pushing" on the pin, and consequently member $BC$ is in *compression*. These effects are clearly demonstrated by isolating the joint with small segments of the member connected to the pin, Fig. 6–7c. The pushing or pulling on these small segments indicates the effect of the member being either in compression or tension.

When using the method of joints, always start at a joint having at least one known force and at most two unknown forces, as in Fig. 6–7b. In this way, application of $\Sigma F_x = 0$ and $\Sigma F_y = 0$ yields two algebraic equations which can be solved for the two unknowns. When applying these equations, the correct sense of an unknown member force can be determined using one of two possible methods.

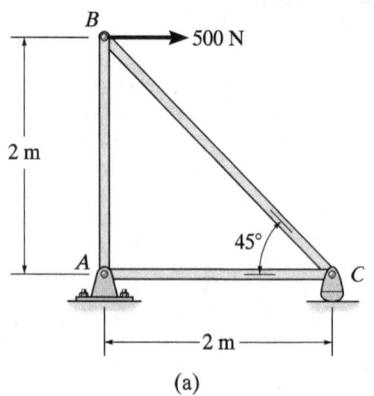
(a)

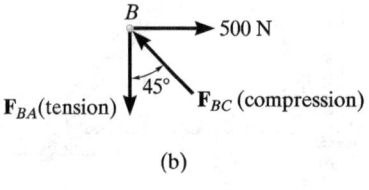
(b)

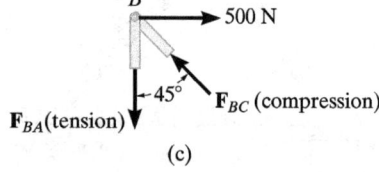
(c)

Fig. 6–7

- The *correct* sense of direction of an unknown member force can, in many cases, be determined "by inspection." For example, $\mathbf{F}_{BC}$ in Fig. 6–7b must push on the pin (compression) since its horizontal component, $F_{BC} \sin 45°$, must balance the 500-N force ($\Sigma F_x = 0$). Likewise, $\mathbf{F}_{BA}$ is a tensile force since it balances the vertical component, $F_{BC} \cos 45°$ ($\Sigma F_y = 0$). In more complicated cases, the sense of an unknown member force can be *assumed;* then, after applying the equilibrium equations, the assumed sense can be verified from the numerical results. A *positive* answer indicates that the sense is *correct*, whereas a *negative* answer indicates that the sense shown on the free-body diagram must be *reversed*.

- *Always assume* the *unknown member forces* acting on the joint's free-body diagram to be in *tension;* i.e., the forces "pull" on the pin. If this is done, then numerical solution of the equilibrium equations will yield *positive scalars for members in tension and negative scalars for members in compression*. Once an unknown member force is found, use its *correct* magnitude and sense (T or C) on subsequent joint free-body diagrams.

The forces in the members of this simple roof truss can be determined using the method of joints.

### Procedure for Analysis

The following procedure provides a means for analyzing a truss using the method of joints.

- Draw the free-body diagram of a joint having at least one known force and at most two unknown forces. (If this joint is at one of the supports, then it may be necessary first to calculate the external reactions at the support.)

- Use one of the two methods described above for establishing the sense of an unknown force.

- Orient the $x$ and $y$ axes such that the forces on the free-body diagram can be easily resolved into their $x$ and $y$ components and then apply the two force equilibrium equations $\Sigma F_x = 0$ and $\Sigma F_y = 0$. Solve for the two unknown member forces and verify their correct sense.

- Using the calculated results, continue to analyze each of the other joints. Remember that a member in *compression* "pushes" on the joint and a member in *tension* "pulls" on the joint. Also, be sure to choose a joint having at most two unknowns and at least one known force.

## EXAMPLE 6.1

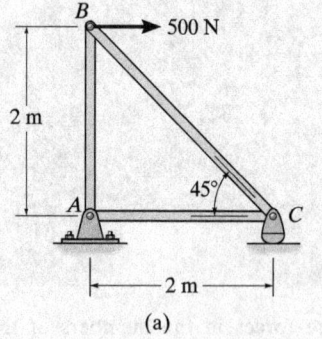

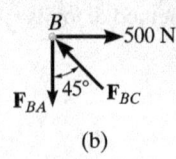

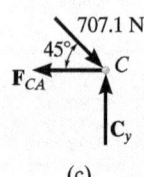

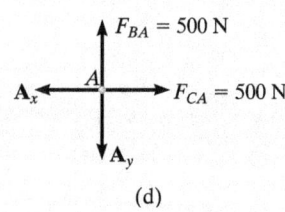

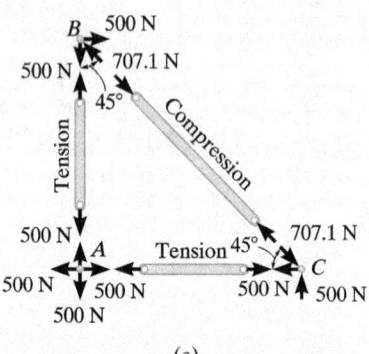

Fig. 6–8

Determine the force in each member of the truss shown in Fig. 6–8a and indicate whether the members are in tension or compression.

### SOLUTION

Since we should have no more than two unknown forces at the joint and at least one known force acting there, we will begin our analysis at joint B.

**Joint B.** The free-body diagram of the joint at B is shown in Fig. 6–8b. Applying the equations of equilibrium, we have

$\xrightarrow{+} \Sigma F_x = 0;$  $500 \text{ N} - F_{BC} \sin 45° = 0$  $F_{BC} = 707.1 \text{ N (C)}$  *Ans.*

$+\uparrow \Sigma F_y = 0;$  $F_{BC} \cos 45° - F_{BA} = 0$  $F_{BA} = 500 \text{ N (T)}$  *Ans.*

Since the force in member BC has been calculated, we can proceed to analyze joint C to determine the force in member CA and the support reaction at the rocker.

**Joint C.** From the free-body diagram of joint C, Fig. 6–8c, we have

$\xrightarrow{+} \Sigma F_x = 0;$  $-F_{CA} + 707.1 \cos 45° \text{ N} = 0$  $F_{CA} = 500 \text{ N (T)}$  *Ans.*

$+\uparrow \Sigma F_y = 0;$  $C_y - 707.1 \sin 45° \text{ N} = 0$  $C_y = 500 \text{ N}$  *Ans.*

**Joint A.** Although it is not necessary, we can determine the components of the support reactions at joint A using the results of $F_{CA}$ and $F_{BA}$. From the free-body diagram, Fig. 6–8d, we have

$\xrightarrow{+} \Sigma F_x = 0;$  $500 \text{ N} - A_x = 0$  $A_x = 500 \text{ N}$

$+\uparrow \Sigma F_y = 0;$  $500 \text{ N} - A_y = 0$  $A_y = 500 \text{ N}$

**NOTE:** The results of the analysis are summarized in Fig. 6–8e. Note that the free-body diagram of each joint (or pin) shows the effects of all the connected members and external forces applied to the joint, whereas the free-body diagram of each member shows only the effects of the end joints on the member.

# EXAMPLE 6.2

Determine the force in each member of the truss in Fig. 6–9a and indicate if the members are in tension or compression.

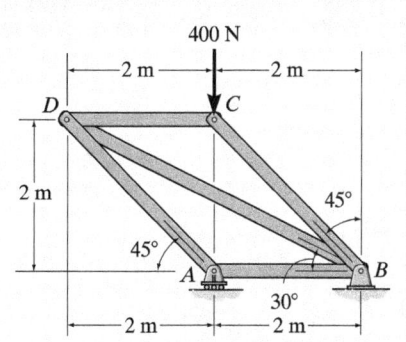

(a)

## SOLUTION

Since joint C has one known and only two unknown forces acting on it, it is possible to start at this joint, then analyze joint D, and finally joint A. This way the support reactions will not have to be determined prior to starting the analysis.

**Joint C.** By inspection of the force equilibrium, Fig. 6–9b, it can be seen that both members BC and CD must be in compression.

$+\uparrow \Sigma F_y = 0;$ $\qquad F_{BC} \sin 45° - 400 \text{ N} = 0$

$\qquad\qquad\qquad\qquad F_{BC} = 565.69 \text{ N} = 566 \text{ N (C)} \qquad$ Ans.

$\xrightarrow{+} \Sigma F_x = 0;$ $\qquad F_{CD} - (565.69 \text{ N}) \cos 45° = 0$

$\qquad\qquad\qquad\qquad F_{CD} = 400 \text{ N (C)} \qquad$ Ans.

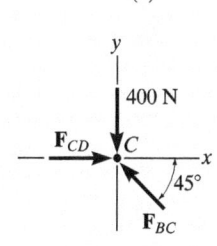

(b)

**Joint D.** Using the result $F_{CD} = 400$ N (C), the force in members BD and AD can be found by analyzing the equilibrium of joint D. We will assume $\mathbf{F}_{AD}$ and $\mathbf{F}_{BD}$ are both tensile forces, Fig. 6–9c. The $x'$, $y'$ coordinate system will be established so that the $x'$ axis is directed along $\mathbf{F}_{BD}$. This way, we will eliminate the need to solve two equations simultaneously. Now $\mathbf{F}_{AD}$ can be obtained *directly* by applying $\Sigma F_{y'} = 0$.

$+\nearrow \Sigma F_{y'} = 0;$ $\qquad - F_{AD} \sin 15° - 400 \sin 30° = 0$

$\qquad\qquad\qquad\qquad F_{AD} = -772.74 \text{ N} = 773 \text{ N (C)} \qquad$ Ans.

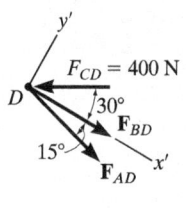

(c)

The negative sign indicates that $\mathbf{F}_{AD}$ is a compressive force. Using this result,

$+\searrow \Sigma F_{x'} = 0;$ $\quad F_{BD} + (-772.74 \cos 15°) - 400 \cos 30° = 0$

$\qquad\qquad\qquad\qquad F_{BD} = 1092.82 \text{ N} = 1.09 \text{ kN (T)} \qquad$ Ans.

**Joint A.** The force in member AB can be found by analyzing the equilibrium of joint A, Fig. 6–9d. We have

$\xrightarrow{+} \Sigma F_x = 0;$ $\qquad (772.74 \text{ N}) \cos 45° - F_{AB} = 0$

$\qquad\qquad\qquad\qquad F_{AB} = 546.41 \text{ N (C)} = 546 \text{ N (C)} \qquad$ Ans.

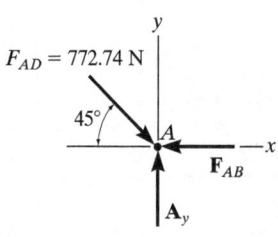

(d)

Fig. 6–9

# EXAMPLE 6.3

Determine the force in each member of the truss shown in Fig. 6–10a. Indicate whether the members are in tension or compression.

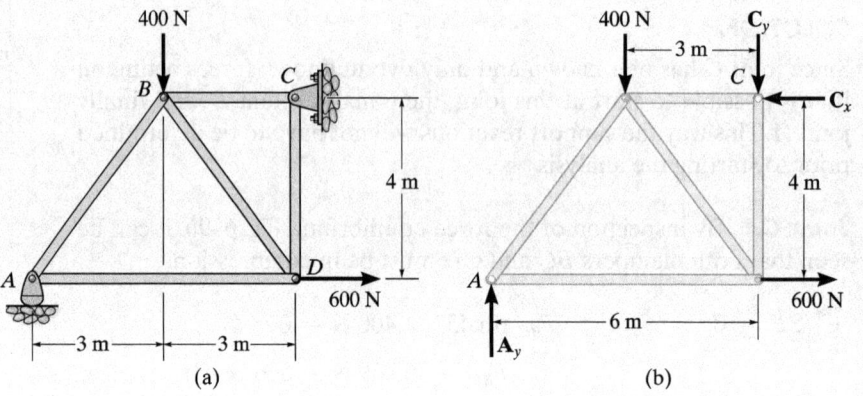

Fig. 6–10

## SOLUTION

**Support Reactions.** No joint can be analyzed until the support reactions are determined, because each joint has more than three unknown forces acting on it. A free-body diagram of the entire truss is given in Fig. 6–10b. Applying the equations of equilibrium, we have

$\xrightarrow{+} \Sigma F_x = 0;$      $600 \text{ N} - C_x = 0$      $C_x = 600 \text{ N}$

$\zeta + \Sigma M_C = 0;$      $-A_y(6 \text{ m}) + 400 \text{ N}(3 \text{ m}) + 600 \text{ N}(4 \text{ m}) = 0$

$A_y = 600 \text{ N}$

$+\uparrow \Sigma F_y = 0;$      $600 \text{ N} - 400 \text{ N} - C_y = 0$      $C_y = 200 \text{ N}$

The analysis can now start at either joint A or C. The choice is arbitrary since there are one known and two unknown member forces acting on the pin at each of these joints.

**Joint A.** (Fig. 6–10c). As shown on the free-body diagram, $\mathbf{F}_{AB}$ is assumed to be compressive and $\mathbf{F}_{AD}$ is tensile. Applying the equations of equilibrium, we have

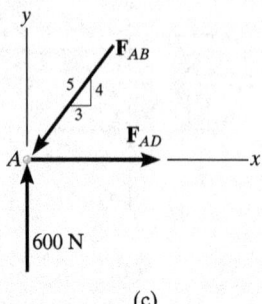

$+\uparrow \Sigma F_y = 0;$      $600 \text{ N} - \frac{4}{5}F_{AB} = 0$      $F_{AB} = 750 \text{ N}$   (C)    *Ans.*

$\xrightarrow{+} \Sigma F_x = 0;$      $F_{AD} - \frac{3}{5}(750 \text{ N}) = 0$      $F_{AD} = 450 \text{ N}$   (T)    *Ans.*

**Joint D.** (Fig. 6–10d). Using the result for $F_{AD}$ and summing forces in the horizontal direction, Fig. 6–10d, we have

$\xrightarrow{+} \Sigma F_x = 0;$   $-450 \text{ N} + \frac{3}{5} F_{DB} + 600 \text{ N} = 0$   $F_{DB} = -250 \text{ N}$

The negative sign indicates that $\mathbf{F}_{DB}$ acts in the *opposite sense* to that shown in Fig. 6–10d.* Hence,

$$F_{DB} = 250 \text{ N (T)} \qquad Ans.$$

To determine $\mathbf{F}_{DC}$, we can either correct the sense of $\mathbf{F}_{DB}$ on the free-body diagram, and then apply $\Sigma F_y = 0$, or apply this equation and retain the negative sign for $F_{DB}$, i.e.,

$+\uparrow \Sigma F_y = 0;$   $-F_{DC} - \frac{4}{5}(-250 \text{ N}) = 0$   $F_{DC} = 200 \text{ N}$ (C)   *Ans.*

**Joint C.** (Fig. 6–10e).

$\xrightarrow{+} \Sigma F_x = 0;$   $F_{CB} - 600 \text{ N} = 0$   $F_{CB} = 600 \text{ N}$ (C)   *Ans.*
$+\uparrow \Sigma F_y = 0;$   $200 \text{ N} - 200 \text{ N} = 0$   (check)

**NOTE:** The analysis is summarized in Fig. 6–10f, which shows the free-body diagram for each joint and member.

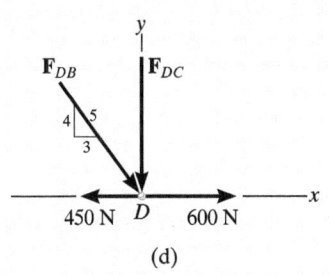

(d)

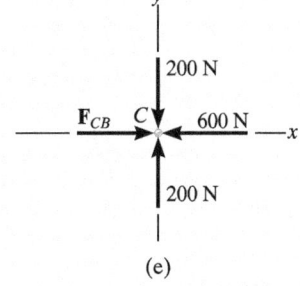

(e)

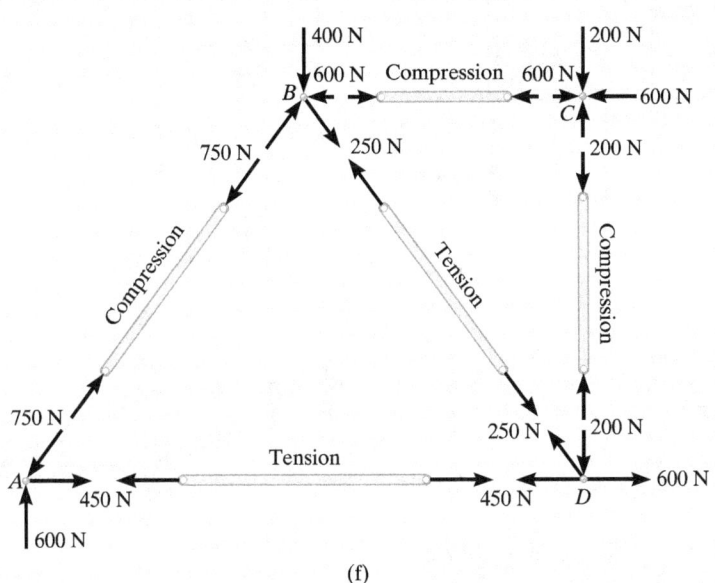

(f)

*The proper sense could have been determined by inspection, prior to applying $\Sigma F_x = 0$.

## 6.3 Zero-Force Members

Truss analysis using the method of joints is greatly simplified if we can first identify those members which support *no loading*. These *zero-force members* are used to increase the stability of the truss during construction and to provide added support if the loading is changed.

The zero-force members of a truss can generally be found *by inspection* of each of the joints. For example, consider the truss shown in Fig. 6–11a. If a free-body diagram of the pin at joint $A$ is drawn, Fig. 6–11b, it is seen that members $AB$ and $AF$ are zero-force members. (We could not have come to this conclusion if we had considered the free-body diagrams of joints $F$ or $B$ simply because there are five unknowns at each of these joints.) In a similar manner, consider the free-body diagram of joint $D$, Fig. 6–11c. Here again it is seen that $DC$ and $DE$ are zero-force members. From these observations, we can conclude that *if only two members form a truss joint and no external load or support reaction is applied to the joint, the two members must be zero-force members*. The load on the truss in Fig. 6–11a is therefore supported by only five members as shown in Fig. 6–11d.

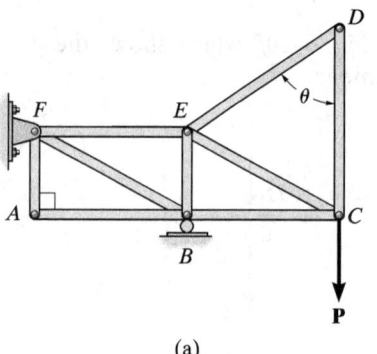

(a)

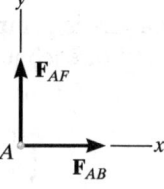

$\xrightarrow{+} \Sigma F_x = 0; \quad F_{AB} = 0$
$+\uparrow \Sigma F_y = 0; \quad F_{AF} = 0$

(b)

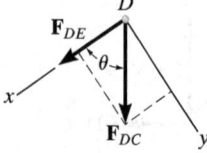

$+\searrow \Sigma F_y = 0; \quad F_{DC} \sin \theta = 0; \quad F_{DC} = 0 \text{ since } \sin \theta \neq 0$
$+\swarrow \Sigma F_x = 0; \quad F_{DE} + 0 = 0; \quad F_{DE} = 0$

(c)

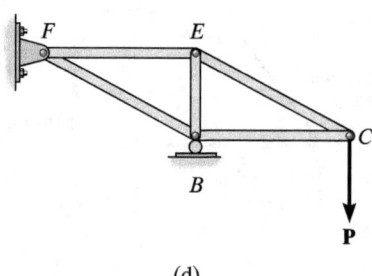

(d)

Fig. 6–11

## 6.3 ZERO-FORCE MEMBERS

Now consider the truss shown in Fig. 6–12a. The free-body diagram of the pin at joint $D$ is shown in Fig. 6–12b. By orienting the $y$ axis along members $DC$ and $DE$ and the $x$ axis along member $DA$, it is seen that $DA$ is a zero-force member. Note that this is also the case for member $CA$, Fig. 6–12c. In general then, *if three members form a truss joint for which two of the members are collinear, the third member is a zero-force member provided no external force or support reaction is applied to the joint.* The truss shown in Fig. 6–12d is therefore suitable for supporting the load **P**.

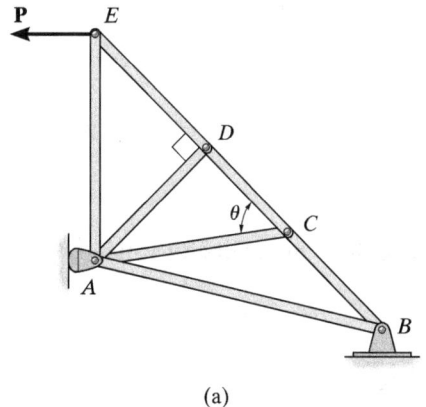

(a)

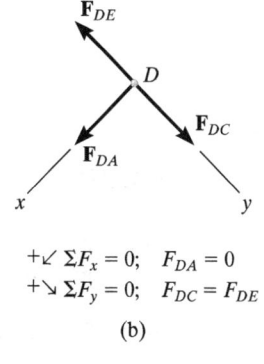

$+\swarrow \Sigma F_x = 0; \quad F_{DA} = 0$
$+\searrow \Sigma F_y = 0; \quad F_{DC} = F_{DE}$

(b)

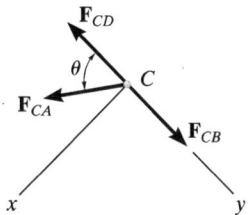

$+\swarrow \Sigma F_x = 0; \quad F_{CA} \sin\theta = 0; \quad F_{CA} = 0 \text{ since } \sin\theta \neq 0;$
$+\searrow \Sigma F_y = 0; \quad F_{CB} = F_{CD}$

(c)

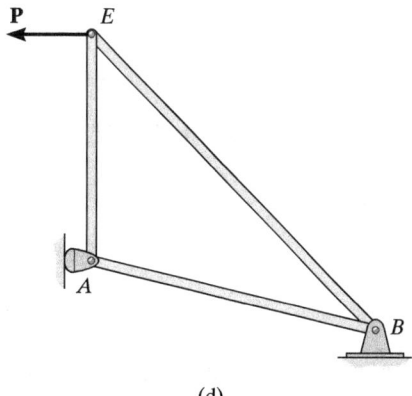

(d)

Fig. 6–12

## EXAMPLE 6.4

Using the method of joints, determine all the zero-force members of the *Fink roof truss* shown in Fig. 6–13a. Assume all joints are pin connected.

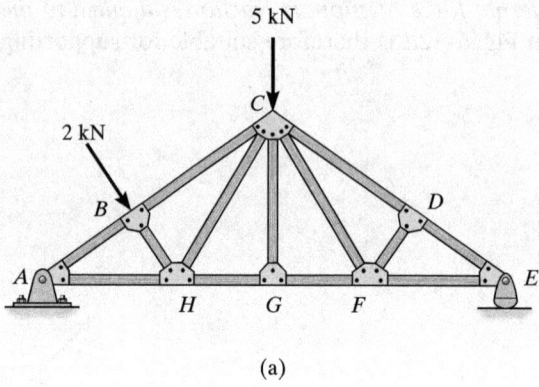

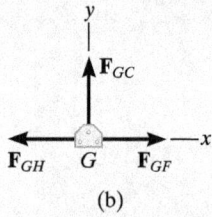

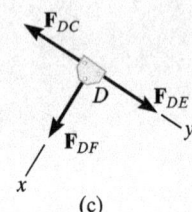

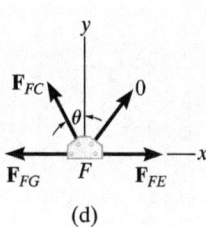

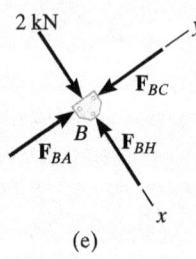

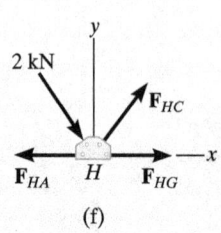

Fig. 6–13

### SOLUTION

Look for joint geometries that have three members for which two are collinear. We have

**Joint G.** (Fig. 6–13b).

$+\uparrow \Sigma F_y = 0;$    $F_{GC} = 0$    Ans.

Realize that we could not conclude that $GC$ is a zero-force member by considering joint $C$, where there are five unknowns. The fact that $GC$ is a zero-force member means that the 5-kN load at $C$ must be supported by members $CB$, $CH$, $CF$, and $CD$.

**Joint D.** (Fig. 6–13c).

$+\swarrow \Sigma F_x = 0;$    $F_{DF} = 0$    Ans.

**Joint F.** (Fig. 6–13d).

$+\uparrow \Sigma F_y = 0;$    $F_{FC} \cos \theta = 0$    Since $\theta \neq 90°$, $F_{FC} = 0$    Ans.

**NOTE:** If joint $B$ is analyzed, Fig. 6–13e,

$+\searrow \Sigma F_x = 0;$    $2 \text{ kN} - F_{BH} = 0$    $F_{BH} = 2 \text{ kN}$    (C)

Also, $F_{HC}$ must satisfy $\Sigma F_y = 0$, Fig. 6–13f, and therefore $HC$ is *not* a zero-force member.

## FUNDAMENTAL PROBLEMS

**F6–1.** Determine the force in each member of the truss. State if the members are in tension or compression.

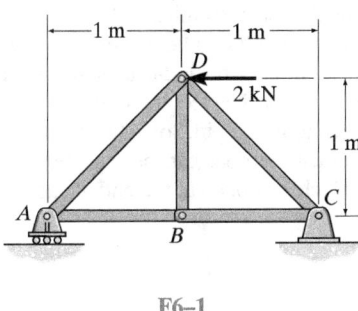

F6–1

**F6–2.** Determine the force in each member of the truss. State if the members are in tension or compression.

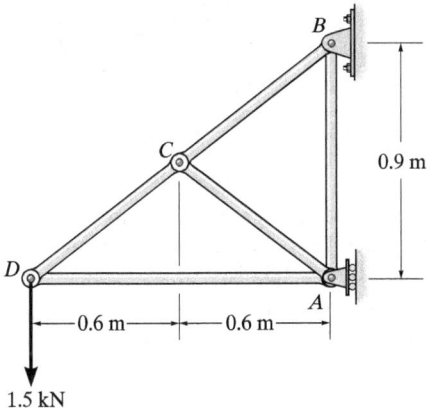

F6–2

**F6–3.** Determine the force in members $AE$ and $DC$. State if the members are in tension or compression.

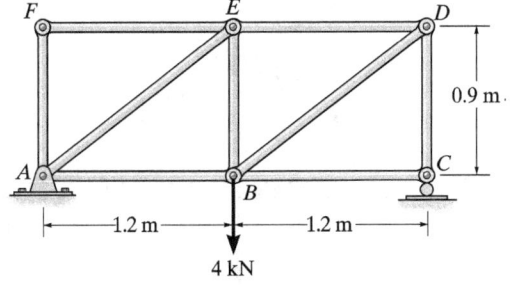

F6–3

**F6–4.** Determine the greatest load $P$ that can be applied to the truss so that none of the members are subjected to a force exceeding either 2 kN in tension or 1.5 kN in compression.

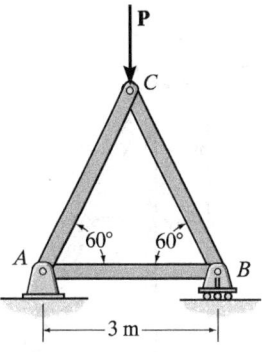

F6–4

**F6–5.** Identify the zero-force members in the truss.

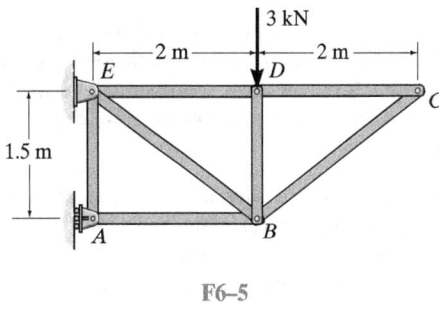

F6–5

**F6–6.** Determine the force in each member of the truss. State if the members are in tension or compression.

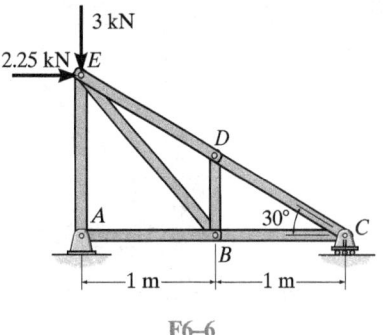

F6–6

## PROBLEMS

**•6–1.** Determine the force in each member of the truss, and state if the members are in tension or compression.

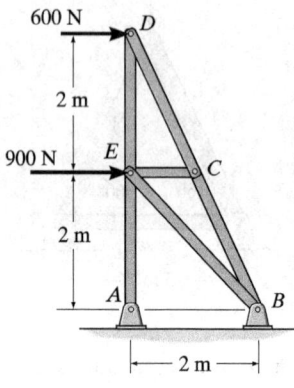

Prob. 6–1

**6–2.** The truss, used to support a balcony, is subjected to the loading shown. Approximate each joint as a pin and determine the force in each member. State whether the members are in tension or compression. Set $P_1 = 3$ kN, $P_2 = 2$ kN.

**6–3.** The truss, used to support a balcony, is subjected to the loading shown. Approximate each joint as a pin and determine the force in each member. State whether the members are in tension or compression. Set $P_1 = 4$ kN, $P_2 = 0$.

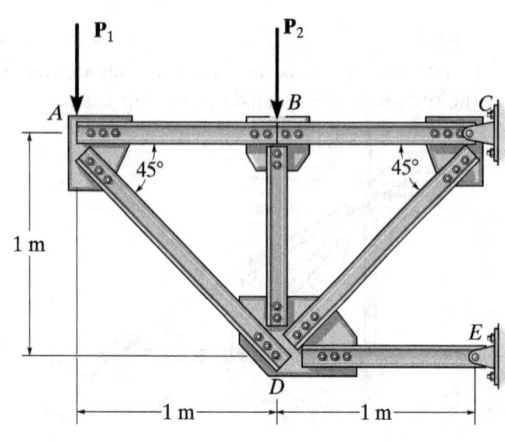

Probs. 6–2/3

**\*6–4.** Determine the force in each member of the truss and state if the members are in tension or compression. Assume each joint as a pin. Set $P = 4$ kN.

**•6–5.** Assume that each member of the truss is made of steel having a mass per length of 4 kg/m. Set $P = 0$, determine the force in each member, and indicate if the members are in tension or compression. Neglect the weight of the gusset plates and assume each joint is a pin. Solve the problem by assuming the weight of each member can be represented as a vertical force, half of which is applied at the end of each member.

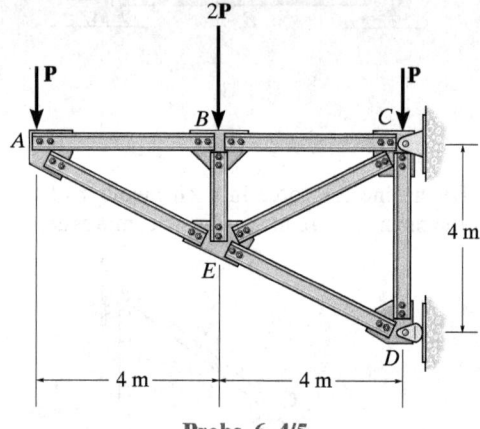

Probs. 6–4/5

**6–6.** Determine the force in each member of the truss and state if the members are in tension or compression. Set $P_1 = 2$ kN and $P_2 = 1.5$ kN.

**6–7.** Determine the force in each member of the truss and state if the members are in tension or compression. Set $P_1 = P_2 = 4$ kN.

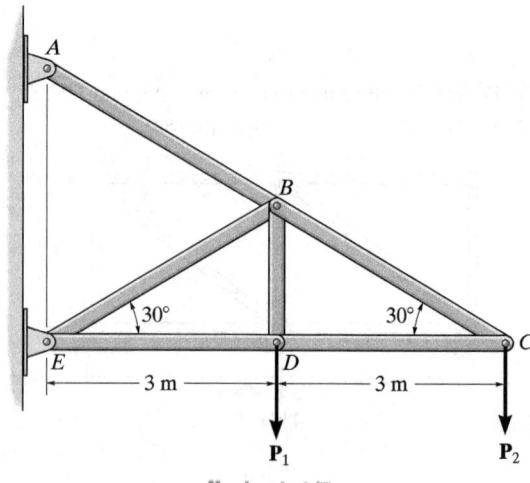

Probs. 6–6/7

***6–8.** Determine the force in each member of the truss, and state if the members are in tension or compression. Set $P = 4$ kN.

**•6–9.** Remove the 2.5-kN force and then determine the greatest force $P$ that can be applied to the truss so that none of the members are subjected to a force exceeding either 4 kN in tension or 3 kN in compression.

***6–12.** Determine the force in each member of the truss and state if the members are in tension or compression. Set $P_1 = 1200$ N, $P_2 = 500$ N.

**•6–13.** Determine the largest load $P_2$ that can be applied to the truss so that the force in any member does not exceed 2.5 kN (T) or 1.75 kN (C). Take $P_1 = 0$.

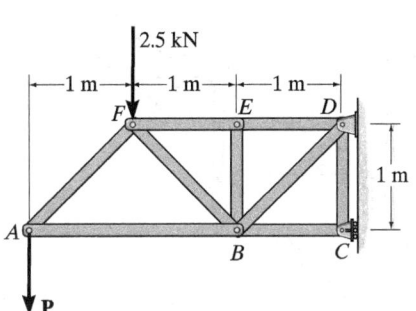

Probs. 6–8/9

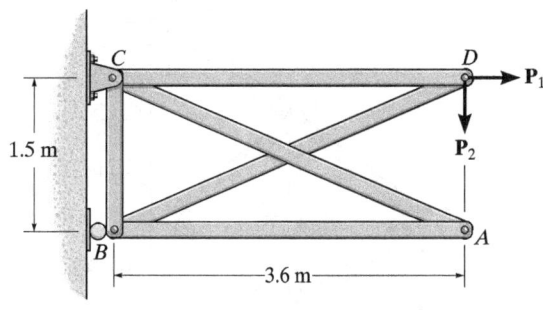

Probs. 6–12/13

**6–10.** Determine the force in each member of the truss and state if the members are in tension or compression. Set $P_1 = 4$ kN, $P_2 = 0$.

**6–11.** Determine the force in each member of the truss and state if the members are in tension or compression. Set $P_1 = 3$ kN, $P_2 = 2$ kN.

**6–14.** Determine the force in each member of the truss, and state if the members are in tension or compression. Set $P = 12.5$ kN.

**6–15.** Remove the 6-kN forces and determine the greatest force $P$ that can be applied to the truss so that none of the members are subjected to a force exceeding either 10 kN in tension or 7.5 kN in compression.

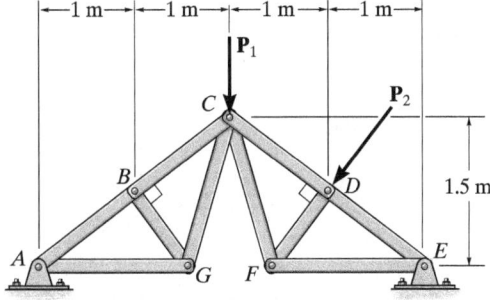

Probs. 6–10/11

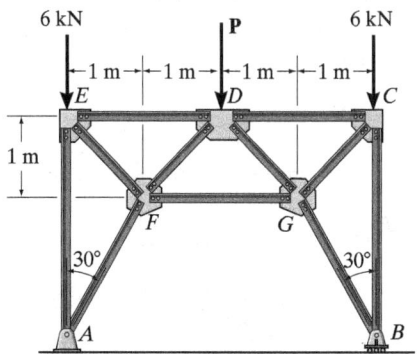

Probs. 6–14/15

**\*6–16.** Determine the force in each member of the truss, and state if the members are in tension or compression. Set $P = 5$ kN.

**•6–17.** Determine the greatest force $P$ that can be applied to the truss so that none of the members are subjected to a force exceeding either 2.5 kN in tension or 2 kN in compression.

**\*6–20.** Determine the force in each member of the truss and state if the members are in tension or compression. The load has a mass of 40 kg.

**•6–21.** Determine the largest mass $m$ of the suspended block so that the force in any member does not exceed 30 kN (T) or 25 kN (C).

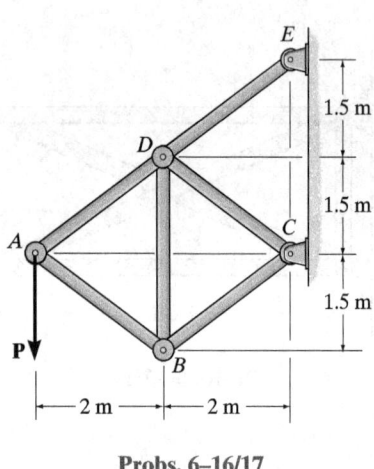

Probs. 6–16/17

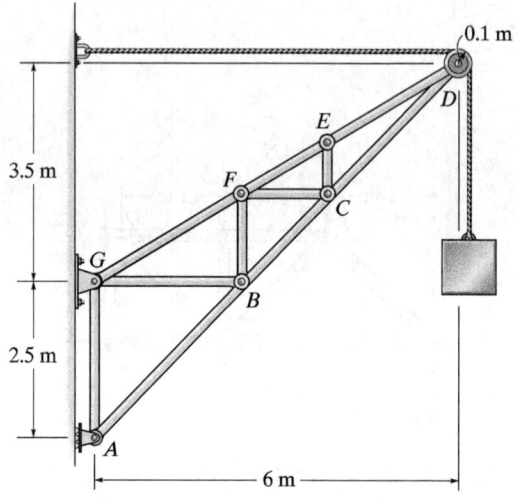

Probs. 6–20/21

**6–18.** Determine the force in each member of the truss, and state if the members are in tension or compression.

**6–19.** The truss is fabricated using members having a weight of 0.2 kN/m. Remove the external forces from the truss, and determine the force in each member due to the weight of the members. State whether the members are in tension or compression. Assume that the total force acting on a joint is the sum of half of the weight of every member connected to the joint.

**6–22.** Determine the force in each member of the truss, and state if the members are in tension or compression.

**6–23.** The truss is fabricated using uniform members having a mass of 5 kg/m. Remove the external forces from the truss, and determine the force in each member due to the weight of the truss. State whether the members are in tension or compression. Assume that the total force acting on a joint is the sum of half of the weight of every member connected to the joint.

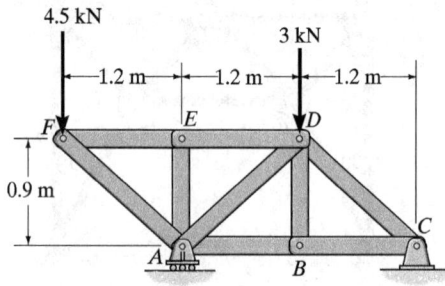

Probs. 6 18/19

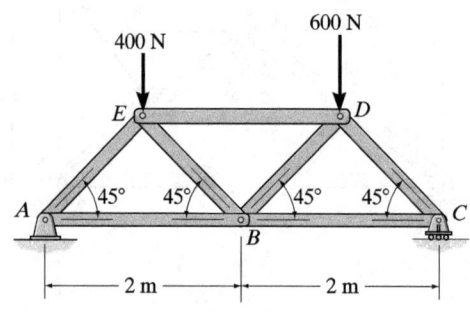

Probs. 6–22/23

**\*6–24.** Determine the force in each member of the truss, and state if the members are in tension or compression. Set $P = 4$ kN.

**•6–25.** Determine the greatest force $P$ that can be applied to the truss so that none of the members are subjected to a force exceeding either 1.5 kN in tension or 1 kN in compression.

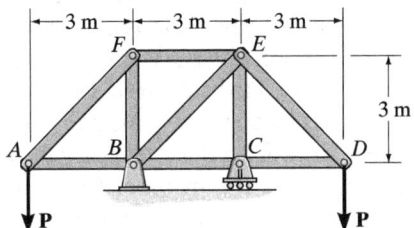

Probs. 6–24/25

**6–26.** A sign is subjected to a wind loading that exerts horizontal forces of 1.5 kN on joints $B$ and $C$ of one of the side supporting trusses. Determine the force in each member of the truss and state if the members are in tension or compression.

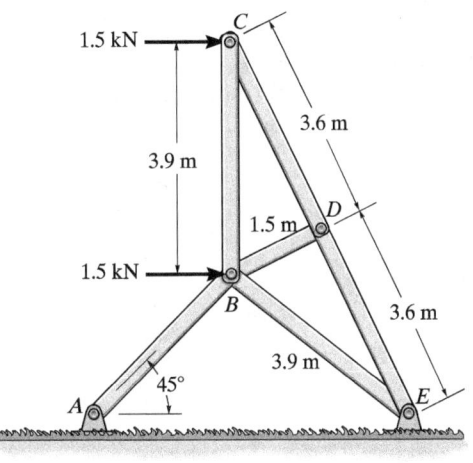

Prob. 6–26

**6–27.** Determine the force in each member of the double scissors truss in terms of the load $P$ and state if the members are in tension or compression.

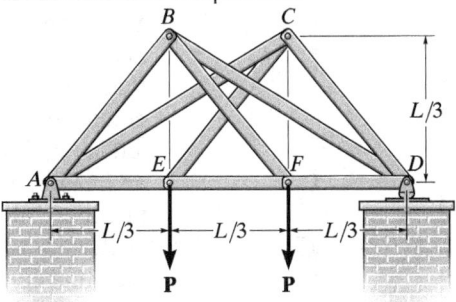

Prob. 6–27

**\*6–28.** Determine the force in each member of the truss in terms of the load $P$, and indicate whether the members are in tension or compression.

**•6–29.** If the maximum force that any member can support is 4 kN in tension and 3 kN in compression, determine the maximum force $P$ that can be applied at joint $B$. Take $d = 1$ m.

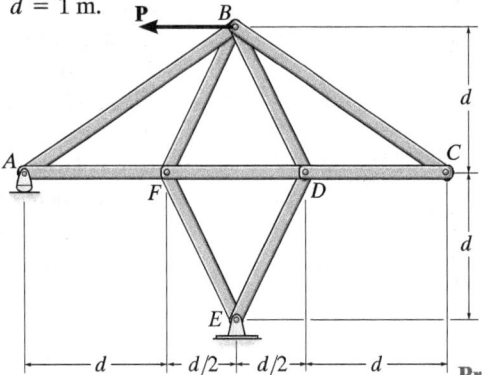

Probs. 6–28/29

**6–30.** The two-member truss is subjected to the force of 1.5 kN. Determine the range of $\theta$ for application of the load so that the force in either member does not exceed 2 kN (T) or 1 kN (C).

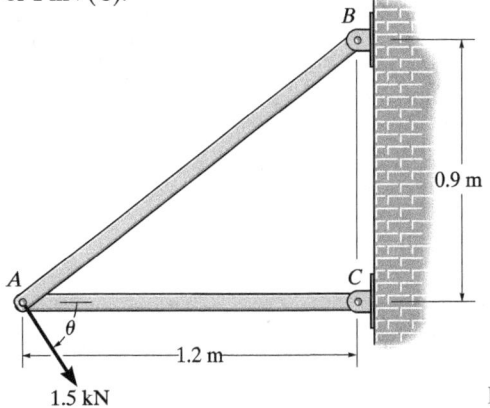

Prob. 6–30

## 6.4 The Method of Sections

When we need to find the force in only a few members of a truss, we can analyze the truss using the *method of sections*. It is based on the principle that if the truss is in equilibrium then any segment of the truss is also in equilibrium. For example, consider the two truss members shown on the left in Fig. 6–14. If the forces within the members are to be determined, then an imaginary section, indicated by the blue line, can be used to cut each member into two parts and thereby "expose" each internal force as "external" to the free-body diagrams shown on the right. Clearly, it can be seen that equilibrium requires that the member in tension (T) be subjected to a "pull," whereas the member in compression (C) is subjected to a "push."

The method of sections can also be used to "cut" or section the members of an entire truss. If the section passes through the truss and the free-body diagram of either of its two parts is drawn, we can then apply the equations of equilibrium to that part to determine the member forces at the "cut section." Since only *three* independent equilibrium equations ($\Sigma F_x = 0$, $\Sigma F_y = 0$, $\Sigma M_O = 0$) can be applied to the free-body diagram of any segment, then we should try to select a section that, in general, passes through not more than *three* members in which the forces are unknown. For example, consider the truss in Fig. 6–15a. If the forces in members $BC$, $GC$, and $GF$ are to be determined, then section $aa$ would be appropriate. The free-body diagrams of the two segments are shown in Figs. 6–15b and 6–15c. Note that the line of action of each member force is specified from the *geometry* of the truss, since the force in a member is along its axis. Also, the member forces acting on one part of the truss are equal but opposite to those acting on the other part—Newton's third law. Members $BC$ and $GC$ are assumed to be in *tension* since they are subjected to a "pull," whereas $GF$ in *compression* since it is subjected to a "push."

The three unknown member forces $\mathbf{F}_{BC}$, $\mathbf{F}_{GC}$, and $\mathbf{F}_{GF}$ can be obtained by applying the three equilibrium equations to the free-body diagram in Fig. 6–15b. If, however, the free-body diagram in Fig. 6–15c is considered, the three support reactions $\mathbf{D}_x$, $\mathbf{D}_y$ and $\mathbf{E}_x$ will have to be known, because only three equations of equilibrium are available. (This, of course, is done in the usual manner by considering a free-body diagram of the *entire truss*.)

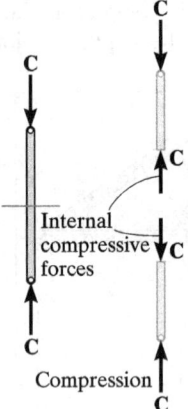

**Fig. 6–14**

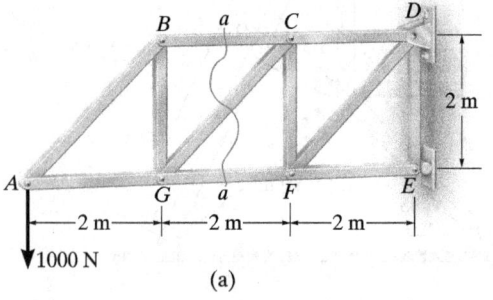

(a)

**Fig. 6–15**

When applying the equilibrium equations, we should carefully consider ways of writing the equations so as to yield a *direct solution* for each of the unknowns, rather than having to solve simultaneous equations. For example, using the truss segment in Fig. 6–15b and summing moments about C would yield a direct solution for $\mathbf{F}_{GF}$ since $\mathbf{F}_{BC}$ and $\mathbf{F}_{GC}$ create zero moment about C. Likewise, $\mathbf{F}_{BC}$ can be directly obtained by summing moments about G. Finally, $\mathbf{F}_{GC}$ can be found directly from a force summation in the vertical direction since $\mathbf{F}_{GF}$ and $\mathbf{F}_{BC}$ have no vertical components. This ability to *determine directly* the force in a particular truss member is one of the main advantages of using the method of sections.*

As in the method of joints, there are two ways in which we can determine the correct sense of an unknown member force:

The forces in selected members of this Pratt truss can readily be determined using the method of sections.

- The correct sense of an unknown member force can in many cases be determined "by inspection." For example, $\mathbf{F}_{BC}$ is a tensile force as represented in Fig. 6–15b since moment equilibrium about G requires that $\mathbf{F}_{BC}$ creates a moment opposite to that of the 1000-N force. Also, $\mathbf{F}_{GC}$ is tensile since its vertical component must balance the 1000-N force which acts downward. In more complicated cases, the sense of an unknown member force may be *assumed*. If the solution yields a *negative* scalar, it indicates that the force's sense is *opposite* to that shown on the free-body diagram.

- *Always assume* that the unknown member forces at the cut section are *tensile* forces, i.e., "pulling" on the member. By doing this, the numerical solution of the equilibrium equations will yield *positive scalars for members in tension and negative scalars for members in compression.*

---

*Notice that if the method of joints were used to determine, say, the force in member GC, it would be necessary to analyze joints A, B, and G in sequence.

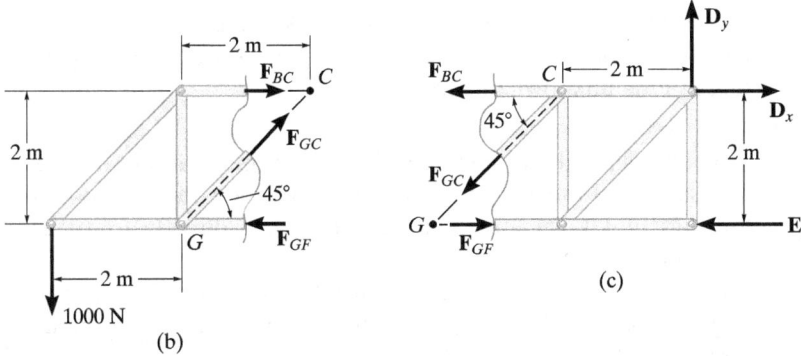

Fig. 6–15

Simple trusses are often used in the construction of large cranes in order to reduce the weight of the boom and tower.

## Procedure for Analysis

The forces in the members of a truss may be determined by the method of sections using the following procedure.

### Free-Body Diagram.

- Make a decision on how to "cut" or section the truss through the members where forces are to be determined.
- Before isolating the appropriate section, it may first be necessary to determine the truss's support reactions. If this is done then the three equilibrium equations will be available to solve for member forces at the section.
- Draw the free-body diagram of that segment of the sectioned truss which has the least number of forces acting on it.
- Use one of the two methods described above for establishing the sense of the unknown member forces.

### Equations of Equilibrium.

- Moments should be summed about a point that lies at the intersection of the lines of action of two unknown forces, so that the third unknown force can be determined directly from the moment equation.
- If two of the unknown forces are *parallel*, forces may be summed *perpendicular* to the direction of these unknowns to determine *directly* the third unknown force.

## EXAMPLE 6.5

Determine the force in members $GE$, $GC$, and $BC$ of the truss shown in Fig. 6–16a. Indicate whether the members are in tension or compression.

Please refer to the Companion CD for the animation: *Free-body Diagram for a Truss*

### SOLUTION
Section $aa$ in Fig. 6–16a has been chosen since it cuts through the *three* members whose forces are to be determined. In order to use the method of sections, however, it is *first* necessary to determine the external reactions at $A$ or $D$. Why? A free-body diagram of the entire truss is shown in Fig. 6–16b. Applying the equations of equilibrium, we have

$\xrightarrow{+} \Sigma F_x = 0;$  $\qquad 400 \text{ N} - A_x = 0 \qquad A_x = 400 \text{ N}$

$\zeta + \Sigma M_A = 0;$  $\quad -1200 \text{ N}(8 \text{ m}) - 400 \text{ N}(3 \text{ m}) + D_y(12 \text{ m}) = 0$

$\qquad\qquad\qquad D_y = 900 \text{ N}$

$+\uparrow \Sigma F_y = 0;$  $\quad A_y - 1200 \text{ N} + 900 \text{ N} = 0 \qquad A_y = 300 \text{ N}$

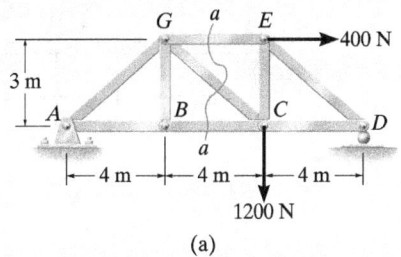

(a)

**Free-Body Diagram.** For the analysis the free-body diagram of the left portion of the sectioned truss will be used, since it involves the least number of forces, Fig. 6–16c.

**Equations of Equilibrium.** Summing moments about point $G$ eliminates $\mathbf{F}_{GE}$ and $\mathbf{F}_{GC}$ and yields a direct solution for $F_{BC}$.

$\zeta + \Sigma M_G = 0;$  $\quad -300 \text{ N}(4 \text{ m}) - 400 \text{ N}(3 \text{ m}) + F_{BC}(3 \text{ m}) = 0$

$\qquad\qquad F_{BC} = 800 \text{ N} \quad \text{(T)} \qquad\qquad\qquad Ans.$

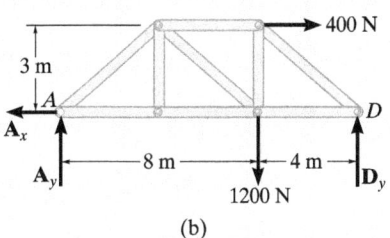

(b)

In the same manner, by summing moments about point $C$ we obtain a direct solution for $F_{GE}$.

$\zeta + \Sigma M_C = 0;$  $\quad -300 \text{ N}(8 \text{ m}) + F_{GE}(3 \text{ m}) = 0$

$\qquad\qquad F_{GE} = 800 \text{ N} \quad \text{(C)} \qquad\qquad\qquad Ans.$

Since $\mathbf{F}_{BC}$ and $\mathbf{F}_{GE}$ have no vertical components, summing forces in the $y$ direction directly yields $F_{GC}$, i.e.,

$+\uparrow \Sigma F_y = 0;$  $\quad 300 \text{ N} - \frac{3}{5} F_{GC} = 0$

$\qquad\qquad F_{GC} = 500 \text{ N} \quad \text{(T)} \qquad\qquad\qquad Ans.$

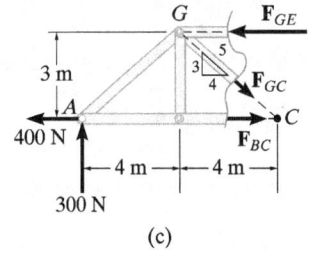

(c)

Fig. 6–16

**NOTE:** Here it is possible to tell, by inspection, the proper direction for each unknown member force. For example, $\Sigma M_C = 0$ requires $\mathbf{F}_{GE}$ to be *compressive* because it must balance the moment of the 300-N force about $C$.

## EXAMPLE 6.6

Determine the force in member $CF$ of the truss shown in Fig. 6–17a. Indicate whether the member is in tension or compression. Assume each member is pin connected.

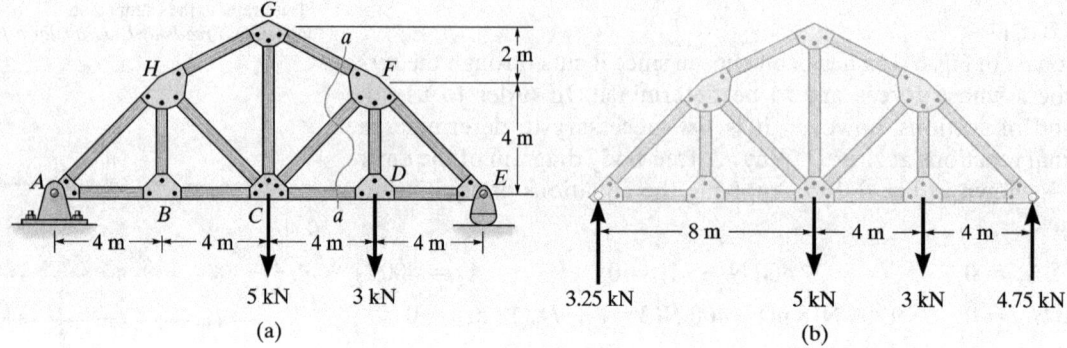

Fig. 6–17

### SOLUTION

**Free-Body Diagram.** Section $aa$ in Fig. 6–17a will be used since this section will "expose" the internal force in member $CF$ as "external" on the free-body diagram of either the right or left portion of the truss. It is first necessary, however, to determine the support reactions on either the left or right side. Verify the results shown on the free-body diagram in Fig. 6–17b.

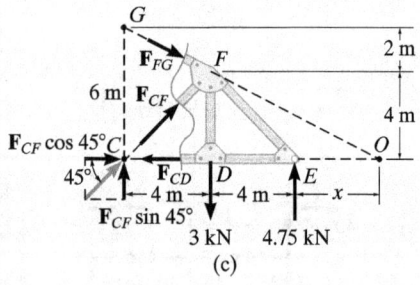

The free-body diagram of the right portion of the truss, which is the easiest to analyze, is shown in Fig. 6–17c. There are three unknowns, $F_{FG}$, $F_{CF}$, and $F_{CD}$.

**Equations of Equilibrium.** We will apply the moment equation about point $O$ in order to eliminate the two unknowns $F_{FG}$ and $F_{CD}$. The location of point $O$ measured from $E$ can be determined from proportional triangles, i.e., $4/(4 + x) = 6/(8 + x)$, $x = 4$ m. Or, stated in another manner, the slope of member $GF$ has a drop of 2 m to a horizontal distance of 4 m. Since $FD$ is 4 m, Fig. 6–17c, then from $D$ to $O$ the distance must be 8 m.

An easy way to determine the moment of $\mathbf{F}_{CF}$ about point $O$ is to use the principle of transmissibility and slide $\mathbf{F}_{CF}$ to point $C$, and then resolve $\mathbf{F}_{CF}$ into its two rectangular components. We have

$\zeta + \Sigma M_O = 0;$
$-F_{CF} \sin 45°(12 \text{ m}) + (3 \text{ kN})(8 \text{ m}) - (4.75 \text{ kN})(4 \text{ m}) = 0$
$F_{CF} = 0.589 \text{ kN}$ (C) *Ans.*

## EXAMPLE 6.7

Determine the force in member *EB* of the roof truss shown in Fig. 6–18a. Indicate whether the member is in tension or compression.

### SOLUTION

**Free-Body Diagrams.** By the method of sections, any imaginary section that cuts through *EB*, Fig. 6–18a, will also have to cut through three other members for which the forces are unknown. For example, section *aa* cuts through *ED*, *EB*, *FB*, and *AB*. If a free-body diagram of the left side of this section is considered, Fig. 6–18b, it is possible to obtain $\mathbf{F}_{ED}$ by summing moments about *B* to eliminate the other three unknowns; however, $\mathbf{F}_{EB}$ cannot be determined from the remaining two equilibrium equations. One possible way of obtaining $\mathbf{F}_{EB}$ is first to determine $\mathbf{F}_{ED}$ from section *aa*, then use this result on section *bb*, Fig. 6–18a, which is shown in Fig. 6–18c. Here the force system is concurrent and our sectioned free-body diagram is the same as the free-body diagram for the joint at *E*.

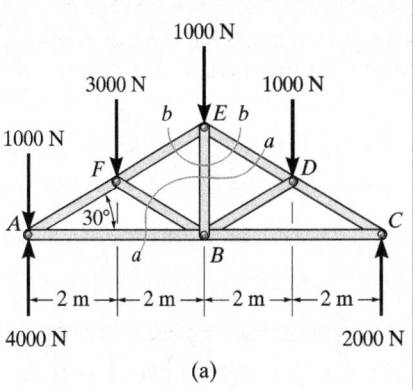

(a)

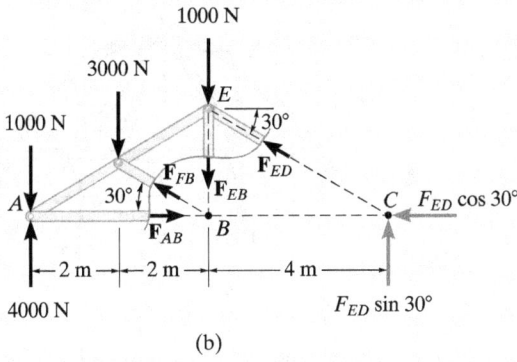

(b)

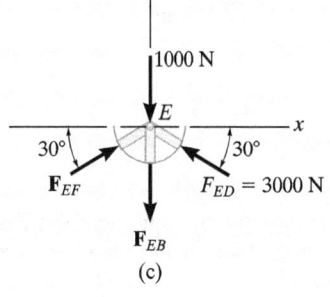

(c)

**Fig. 6–18**

**Equations of Equilibrium.** In order to determine the moment of $\mathbf{F}_{ED}$ about point *B*, Fig. 6–18b, we will use the principle of transmissibility and slide the force to point *C* and then resolve it into its rectangular components as shown. Therefore,

$\zeta + \Sigma M_B = 0;$  $1000 \text{ N}(4 \text{ m}) + 3000 \text{ N}(2 \text{ m}) - 4000 \text{ N}(4 \text{ m})$
$\qquad + F_{ED} \sin 30°(4 \text{ m}) = 0$
$\qquad\qquad F_{ED} = 3000 \text{ N} \quad (C)$

Considering now the free-body diagram of section *bb*, Fig. 6–18c, we have

$\xrightarrow{+} \Sigma F_x = 0;$  $F_{EF} \cos 30° - 3000 \cos 30° \text{ N} = 0$
$\qquad\qquad F_{EF} = 3000 \text{ N} \quad (C)$
$+\uparrow \Sigma F_y = 0;$  $2(3000 \sin 30° \text{ N}) - 1000 \text{ N} - F_{EB} = 0$
$\qquad\qquad F_{EB} = 2000 \text{ N} \quad (T) \qquad\qquad Ans.$

# FUNDAMENTAL PROBLEMS

**F6–7.** Determine the force in members $BC$, $CF$, and $FE$. State if the members are in tension or compression.

**F6–10.** Determine the force in members $EF$, $CF$, and $BC$ of the truss. State if the members are in tension or compression.

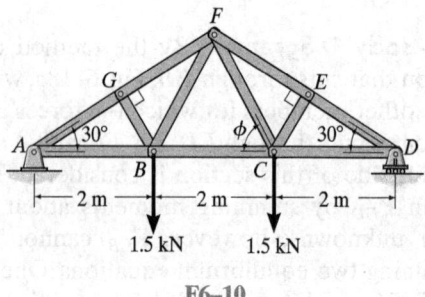

F6–10

**F6–11.** Determine the force in members $GF$, $GD$, and $CD$ of the truss. State if the members are in tension or compression.

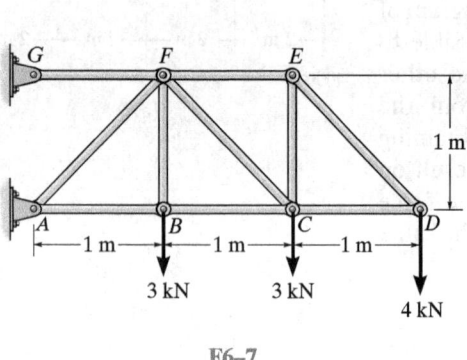

F6–7

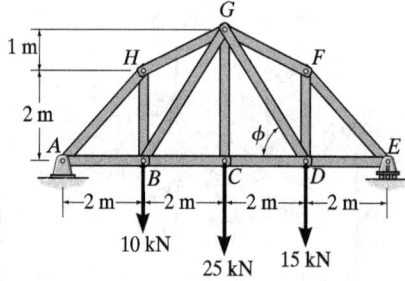

F6–11

**F6–8.** Determine the force in members $LK$, $KC$, and $CD$ of the Pratt truss. State if the members are in tension or compression.

**F6–9.** Determine the force in members $KJ$, $KD$, and $CD$ of the Pratt truss. State if the members are in tension or compression.

**F6–12.** Determine the force in members $DC$, $HI$, and $JI$ of the truss. State if the members are in tension or compression.

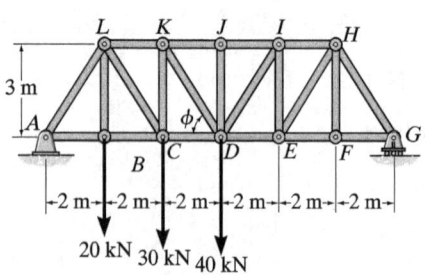

F6–8/9

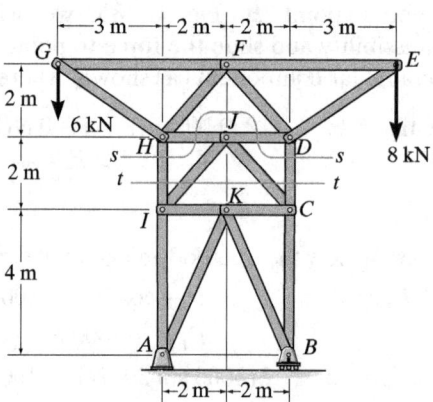

F6–12

## PROBLEMS

**6–31.** The internal drag truss for the wing of a light airplane is subjected to the forces shown. Determine the force in members *BC*, *BH*, and *HC*, and state if the members are in tension or compression.

**6–34.** Determine the force in members *JK*, *CJ*, and *CD* of the truss, and state if the members are in tension or compression.

**6–35.** Determine the force in members *HI*, *FI*, and *EF* of the truss, and state if the members are in tension or compression.

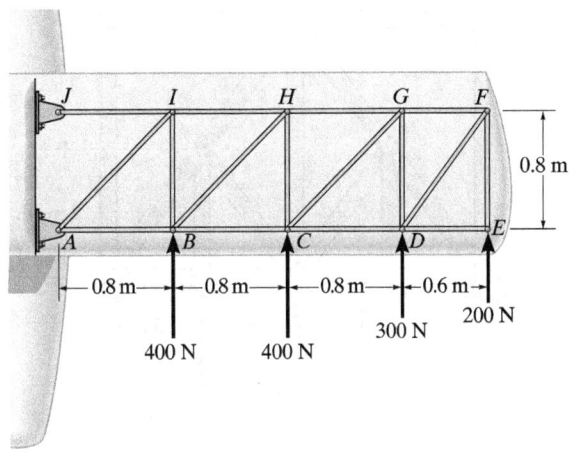

Prob. 6–31

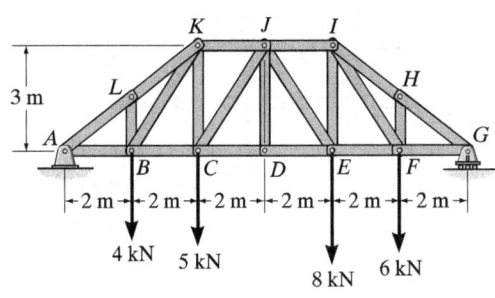

Probs. 6–34/35

*****6–32.** The *Howe bridge truss* is subjected to the loading shown. Determine the force in members *HD*, *CD*, and *GD*, and state if the members are in tension or compression.

•**6–33.** The *Howe bridge truss* is subjected to the loading shown. Determine the force in members *HI*, *HB*, and *BC*, and state if the members are in tension or compression.

*****6–36.** Determine the force in members *BC*, *CG*, and *GF* of the *Warren* truss. Indicate if the members are in tension or compression.

•**6–37.** Determine the force in members *CD*, *CF*, and *FG* of the *Warren* truss. Indicate if the members are in tension or compression.

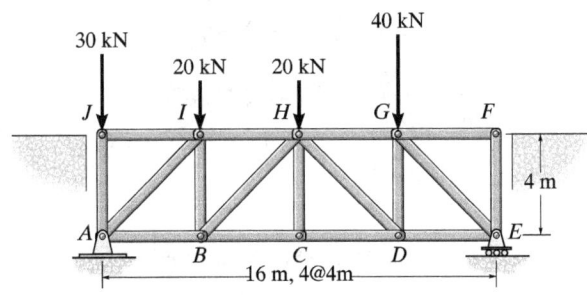

Probs. 6–32/33

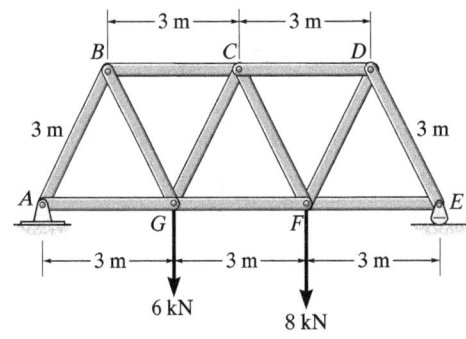

Probs. 6–36/37

**6–38.** Determine the force in members $DC$, $HC$, and $HI$ of the truss, and state if the members are in tension or compression.

**6–39.** Determine the force in members $ED$, $EH$, and $GH$ of the truss, and state if the members are in tension or compression.

**6–42.** Determine the force in members $IC$ and $CG$ of the truss and state if these members are in tension or compression. Also, indicate all zero-force members.

**6–43.** Determine the force in members $JE$ and $GF$ of the truss and state if these members are in tension or compression. Also, indicate all zero-force members.

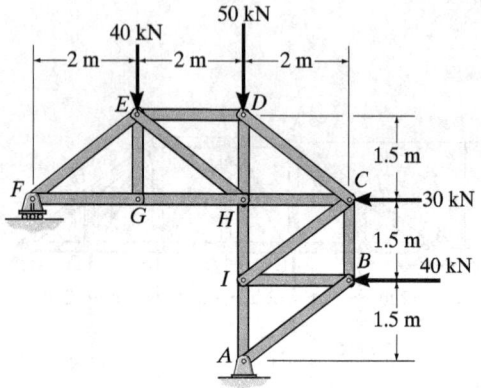

Probs. 6–38/39

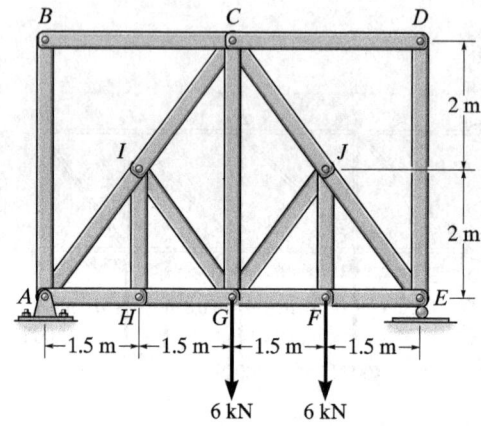

Probs. 6–42/43

**\*6–40.** Determine the force in members $GF$, $GD$, and $CD$ of the truss and state if the members are in tension or compression.

**•6–41.** Determine the force in members $BG$, $BC$, and $HG$ of the truss and state if the members are in tension or compression.

**\*6–44.** Determine the force in members $JI$, $EF$, $EI$, and $JE$ of the truss, and state if the members are in tension or compression.

**•6–45.** Determine the force in members $CD$, $LD$, and $KL$ of the truss, and state if the members are in tension or compression.

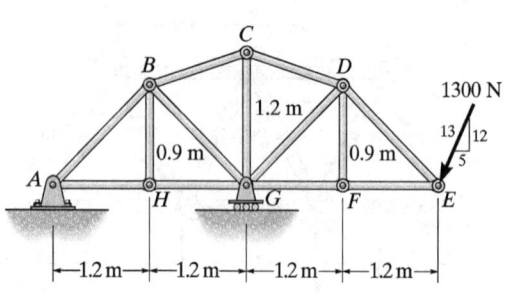

Probs. 6–40/41

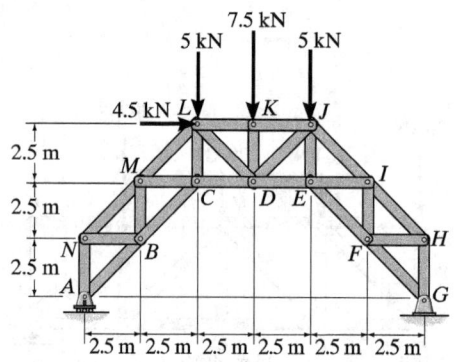

Probs. 6–44/45

**6–46.** Determine the force developed in members *BC* and *CH* of the roof truss and state if the members are in tension or compression.

**6–47.** Determine the force in members *CD* and *GF* of the truss and state if the members are in tension or compression. Also indicate all zero-force members.

**6–50.** Determine the force in each member of the truss and state if the members are in tension or compression. Set $P_1 = 20$ kN, $P_2 = 10$ kN.

**6–51.** Determine the force in each member of the truss and state if the members are in tension or compression. Set $P_1 = 40$ kN, $P_2 = 20$ kN.

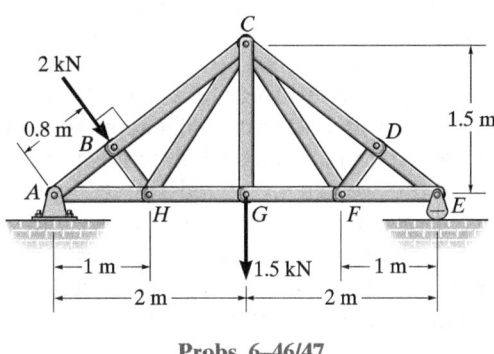

Probs. 6–46/47

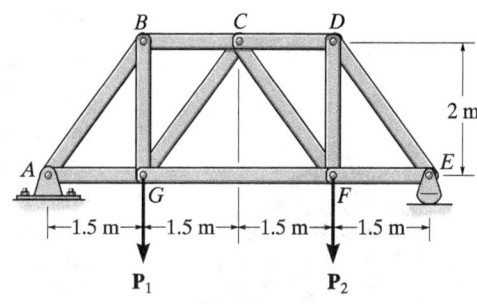

Probs. 6–50/51

**\*6–48.** Determine the force in members *IJ*, *EJ*, and *CD* of the *Howe* truss, and state if the members are in tension or compression.

**•6–49.** Determine the force in members *KJ*, *KC*, and *BC* of the *Howe* truss, and state if the members are in tension or compression.

**\*6–52.** Determine the force in members *KJ*, *NJ*, *ND*, and *CD* of the *K truss*. Indicate if the members are in tension or compression. *Hint:* Use sections *aa* and *bb*.

**•6–53.** Determine the force in members *JI* and *DE* of the *K truss*. Indicate if the members are in tension or compression.

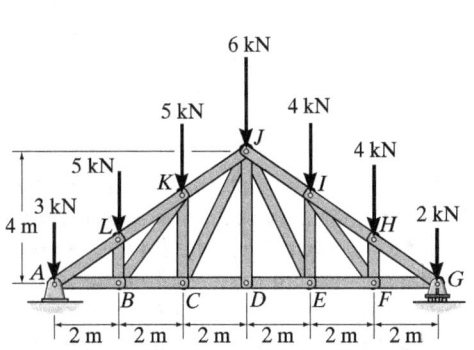

Probs. 6–48/49

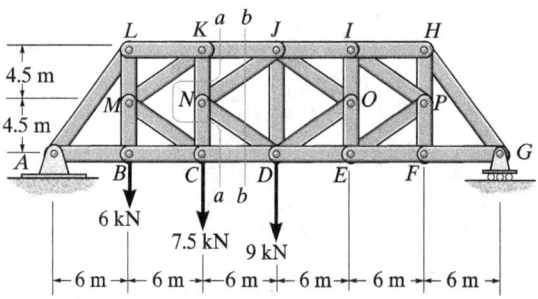

Probs. 6–52/53

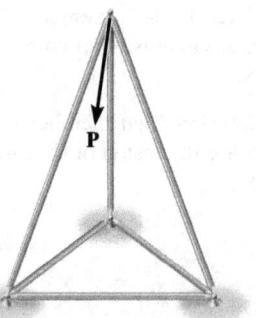

Fig. 6–19

## *6.5 Space Trusses

A *space truss* consists of members joined together at their ends to form a stable three-dimensional structure. The simplest form of a space truss is a *tetrahedron*, formed by connecting six members together, as shown in Fig. 6–19. Any additional members added to this basic element would be redundant in supporting the force **P**. A *simple space truss* can be built from this basic tetrahedral element by adding three additional members and a joint, and continuing in this manner to form a system of multiconnected tetrahedrons.

### Assumptions for Design

The members of a space truss may be treated as two-force members provided the external loading is applied at the joints and the joints consist of ball-and-socket connections. These assumptions are justified if the welded or bolted connections of the joined members intersect at a common point and the weight of the members can be neglected. In cases where the weight of a member is to be included in the analysis, it is generally satisfactory to apply it as a vertical force, half of its magnitude applied at each end of the member.

Typical roof-supporting space truss. Notice the use of ball-and-socket joints for the connections.

### Procedure for Analysis

Either the method of joints or the method of sections can be used to determine the forces developed in the members of a simple space truss.

#### Method of Joints.

If the forces in *all* the members of the truss are to be determined, then the method of joints is most suitable for the analysis. Here it is necessary to apply the three equilibrium equations $\Sigma F_x = 0$, $\Sigma F_y = 0$, $\Sigma F_z = 0$ to the forces acting at each joint. Remember that the solution of many simultaneous equations can be avoided if the force analysis begins at a joint having at least one known force and at most three unknown forces. Also, if the three-dimensional geometry of the force system at the joint is hard to visualize, it is recommended that a Cartesian vector analysis be used for the solution.

#### Method of Sections.

If only a *few* member forces are to be determined, the method of sections can be used. When an imaginary section is passed through a truss and the truss is separated into two parts, the force system acting on one of the segments must satisfy the *six* equilibrium equations: $\Sigma F_x = 0$, $\Sigma F_y = 0$, $\Sigma F_z = 0$, $\Sigma M_x = 0$, $\Sigma M_y = 0$, $\Sigma M_z = 0$ (Eqs. 5–6). By proper choice of the section and axes for summing forces and moments, many of the unknown member forces in a space truss can be computed *directly*, using a single equilibrium equation.

For economic reasons, large electrical transmission towers are often constructed using space trusses.

## EXAMPLE 6.8

Determine the forces acting in the members of the space truss shown in Fig. 6–20a. Indicate whether the members are in tension or compression.

### SOLUTION

Since there are one known force and three unknown forces acting at joint $A$, the force analysis of the truss will begin at this joint.

**Joint A.** (Fig. 6–20b). Expressing each force acting on the free-body diagram of joint $A$ as a Cartesian vector, we have

$$\mathbf{P} = \{-4\mathbf{j}\} \text{ kN}, \qquad \mathbf{F}_{AB} = F_{AB}\mathbf{j}, \quad \mathbf{F}_{AC} = -F_{AC}\mathbf{k},$$

$$\mathbf{F}_{AE} = F_{AE}\left(\frac{\mathbf{r}_{AE}}{r_{AE}}\right) = F_{AE}(0.577\mathbf{i} + 0.577\mathbf{j} - 0.577\mathbf{k})$$

For equilibrium,

$$\Sigma \mathbf{F} = \mathbf{0}; \qquad \mathbf{P} + \mathbf{F}_{AB} + \mathbf{F}_{AC} + \mathbf{F}_{AE} = \mathbf{0}$$

$$-4\mathbf{j} + F_{AB}\mathbf{j} - F_{AC}\mathbf{k} + 0.577F_{AE}\mathbf{i} + 0.577F_{AE}\mathbf{j} - 0.577F_{AE}\mathbf{k} = \mathbf{0}$$

$$\Sigma F_x = 0; \qquad\qquad 0.577 F_{AE} = 0$$
$$\Sigma F_y = 0; \qquad -4 + F_{AB} + 0.577 F_{AE} = 0$$
$$\Sigma F_z = 0; \qquad\qquad -F_{AC} - 0.577 F_{AE} = 0$$
$$F_{AC} = F_{AE} = 0 \qquad \text{Ans.}$$
$$F_{AB} = 4 \text{ kN} \quad (T) \qquad \text{Ans.}$$

Since $F_{AB}$ is known, joint $B$ can be analyzed next.

**Joint B.** (Fig. 6–20c).

$$\Sigma F_x = 0; \qquad -R_B \cos 45° + 0.707 F_{BE} = 0$$
$$\Sigma F_y = 0; \qquad -4 + R_B \sin 45° = 0$$
$$\Sigma F_z = 0; \qquad 2 + F_{BD} - 0.707 F_{BE} = 0$$
$$R_B = F_{BE} = 5.66 \text{ kN} \quad (T), \qquad F_{BD} = 2 \text{ kN} \quad (C) \qquad \text{Ans.}$$

The *scalar* equations of equilibrium may also be applied directly to the forces acting on the free-body diagrams of joints $D$ and $C$ since the force components are easily determined. Show that

$$F_{DE} = F_{DC} = F_{CE} = 0 \qquad \text{Ans.}$$

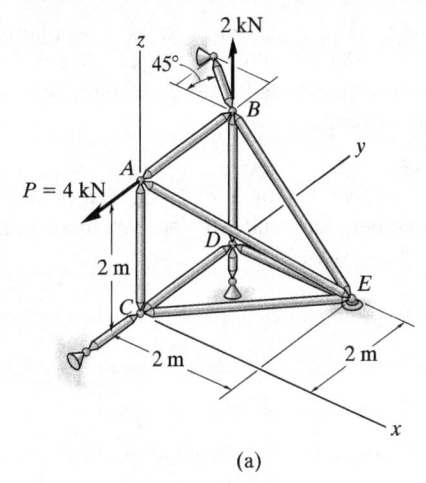

(a)

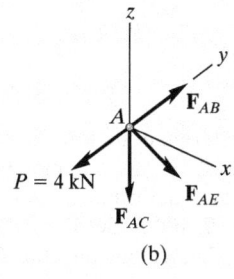

(b)

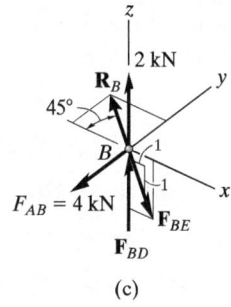

(c)

Fig. 6–20

# PROBLEMS

**6–54.** The space truss supports a force $\mathbf{F} = \{-500\mathbf{i} + 600\mathbf{j} + 400\mathbf{k}\}$ N. Determine the force in each member, and state if the members are in tension or compression.

**6–55.** The space truss supports a force $\mathbf{F} = \{600\mathbf{i} + 450\mathbf{j} - 750\mathbf{k}\}$ N. Determine the force in each member, and state if the members are in tension or compression.

**6–58.** Determine the force in members $BE$, $DF$, and $BC$ of the space truss and state if the members are in tension or compression.

**6–59.** Determine the force in members $AB$, $CD$, $ED$, and $CF$ of the space truss and state if the members are in tension or compression.

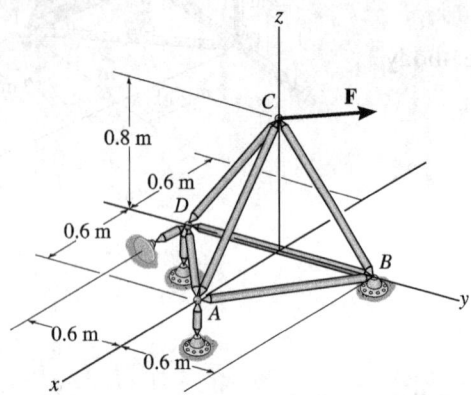

Probs. 6–54/55

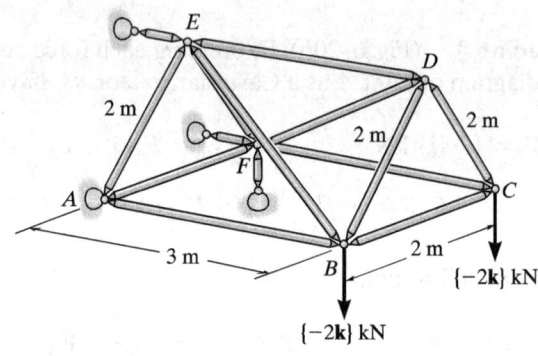

Probs. 6–58/59

**\*6–56.** Determine the force in each member of the space truss and state if the members are in tension or compression. The truss is supported by ball-and-socket joints at $A$, $B$, and $E$. Set $\mathbf{F} = \{800\mathbf{j}\}$ N. *Hint:* The support reaction at $E$ acts along member $EC$. Why?

**•6–57.** Determine the force in each member of the space truss and state if the members are in tension or compression. The truss is supported by ball-and-socket joints at $A$, $B$, and $E$. Set $\mathbf{F} = \{-200\mathbf{i} + 400\mathbf{j}\}$ N. *Hint:* The support reaction at $E$ acts along member $EC$. Why?

**\*6–60.** Determine the force in the members $AB$, $AE$, $BC$, $BF$, $BD$, and $BE$ of the space truss, and state if the members are in tension or compression.

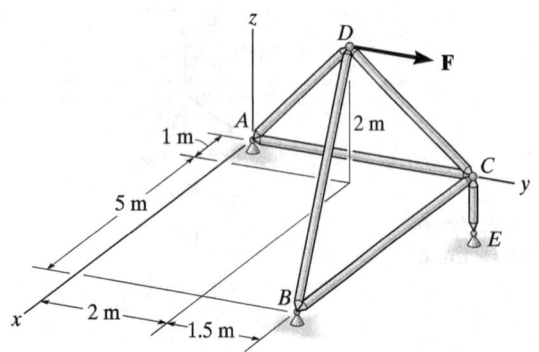

Probs. 6–56/57

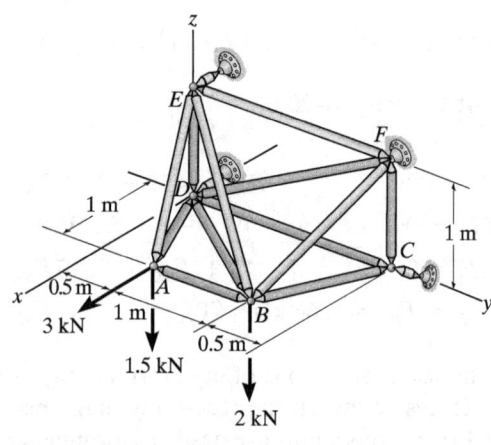

Prob. 6–60

•**6–61.** Determine the force in the members *EF*, *DF*, *CF*, and *CD* of the space truss, and state if the members are in tension or compression.

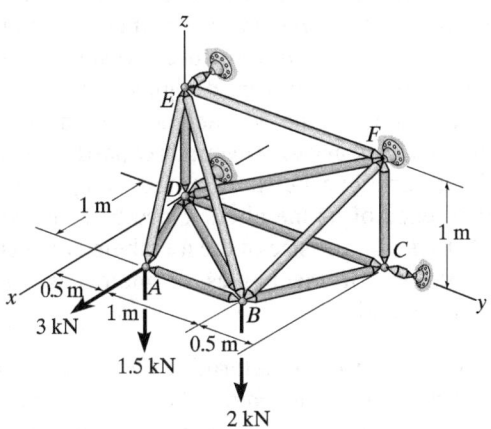

Prob. 6–61

**6–62.** If the truss supports a force of $F = 200$ N, determine the force in each member and state if the members are in tension or compression.

**6–63.** If each member of the space truss can support a maximum force of 600 N in compression and 800 N in tension, determine the greatest force $F$ the truss can support.

\***6–64.** Determine the force developed in each member of the space truss and state if the members are in tension or compression. The crate has a weight of 750 N.

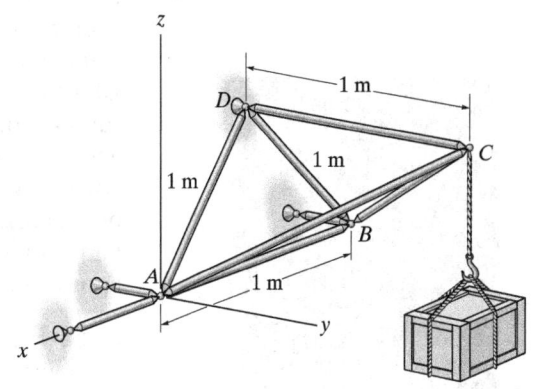

Prob. 6–64

•**6–65.** Determine the force in members *FE* and *ED* of the space truss and state if the members are in tension or compression. The truss is supported by a ball-and-socket joint at *C* and short links at *A* and *B*.

**6–66.** Determine the force in members *GD*, *GE*, and *FD* of the space truss and state if the members are in tension or compression.

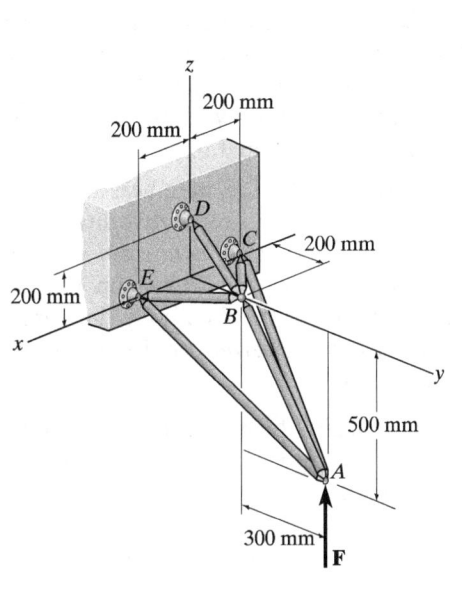

Probs. 6–62/63

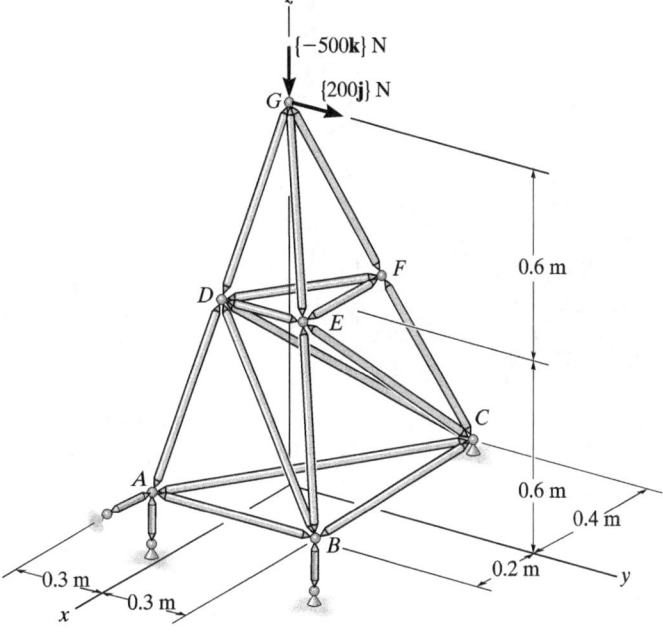

Probs. 6–65/66

## 6.6 Frames and Machines

Frames and machines are two types of structures which are often composed of pin-connected *multiforce members*, i.e., members that are subjected to more than two forces. *Frames* are used to support loads, whereas *machines* contain moving parts and are designed to transmit and alter the effect of forces. Provided a frame or machine contains no more supports or members than are necessary to prevent its collapse, the forces acting at the joints and supports can be determined by applying the equations of equilibrium to each of its members. Once these forces are obtained, it is then possible to *design* the size of the members, connections, and supports using the theory of mechanics of materials and an appropriate engineering design code.

**Free-Body Diagrams.** In order to determine the forces acting at the joints and supports of a frame or machine, the structure must be disassembled and the free-body diagrams of its parts must be drawn. The following important points *must* be observed:

- Isolate each part by drawing its *outlined shape*. Then show all the forces and/or couple moments that act on the part. Make sure to *label* or *identify* each known and unknown force and couple moment with reference to an established $x, y$ coordinate system. Also, indicate any dimensions used for taking moments. Most often the equations of equilibrium are easier to apply if the forces are represented by their rectangular components. As usual, the sense of an unknown force or couple moment can be assumed.

- Identify all the two-force members in the structure and represent their free-body diagrams as having two equal but opposite collinear forces acting at their points of application. (See Sec. 5.4.) By recognizing the two-force members, we can avoid solving an unnecessary number of equilibrium equations.

- Forces common to any two *contacting* members act with equal magnitudes but opposite sense on the respective members. If the two members are treated as a *"system" of connected members*, then these forces are *"internal"* and are *not shown* on the *free-body diagram of the system*; however, if the free-body diagram of *each member* is drawn, the forces are *"external"* and *must* be shown on each of the free-body diagrams.

The following examples graphically illustrate how to draw the free-body diagrams of a dismembered frame or machine. In all cases, the weight of the members is neglected.

This large crane is a typical example of a framework.

Common tools such as these pliers act as simple machines. Here the applied force on the handles creates a much larger force at the jaws.

## EXAMPLE 6.9

For the frame shown in Fig. 6–21a, draw the free-body diagram of (a) each member, (b) the pin at B, and (c) the two members connected together.

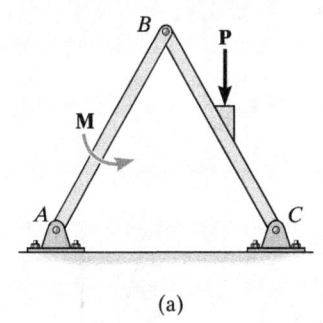

(a)

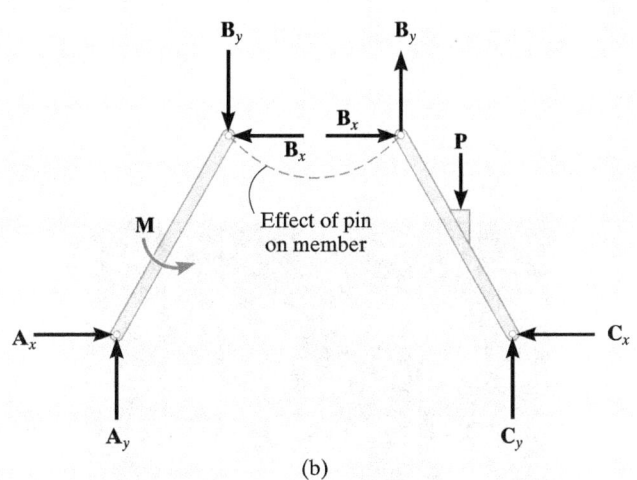

(b)

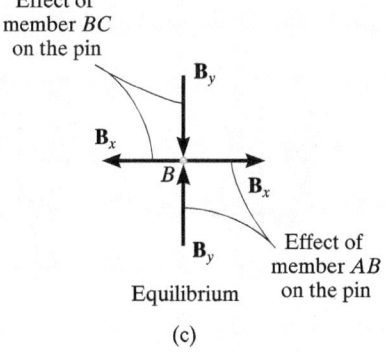

(c)

### SOLUTION

**Part (a).** By inspection, members $BA$ and $BC$ are *not* two-force members. Instead, as shown on the free-body diagrams, Fig. 6–21b, $BC$ is subjected to a force from the pins at $B$ and $C$ and the external force **P**. Likewise, $AB$ is subjected to a force from the pins at $A$ and $B$ and the external couple moment **M**. The pin forces are represented by their $x$ and $y$ components.

**Part (b).** The pin at $B$ is subjected to only *two forces*, i.e., the force of member $BC$ and the force of member $AB$. For *equilibrium*, these forces or their respective components must be equal but opposite, Fig. 6–21c. Realize that Newton's third law is applied between the pin and its connected members, i.e., the effect of the pin on the two members, Fig. 6–21b, and the equal but opposite effect of the two members on the pin, Fig. 6–21c.

**Part (c).** The free-body diagram of both members connected together, yet removed from the supporting pins at $A$ and $C$, is shown in Fig. 6–21d. The force components $\mathbf{B}_x$ and $\mathbf{B}_y$ are *not shown* on this diagram since they are *internal* forces (Fig. 6–21b) and therefore cancel out. Also, to be consistent when later applying the equilibrium equations, the unknown force components at $A$ and $C$ must act in the *same sense* as those shown in Fig. 6–21b.

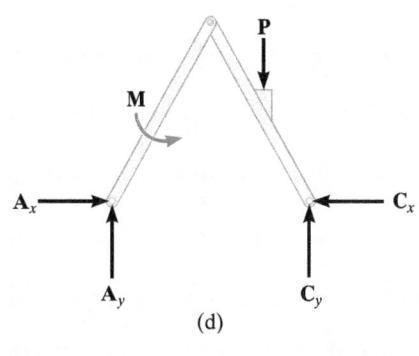

Fig. 6–21

## EXAMPLE 6.10

A constant tension in the conveyor belt is maintained by using the device shown in Fig. 6–22a. Draw the free-body diagrams of the frame and the cylinder that the belt surrounds. The suspended block has a weight of $W$.

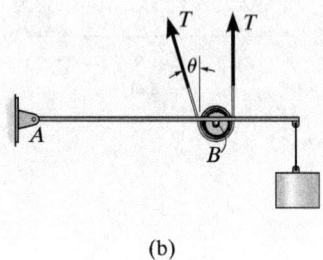

(b)

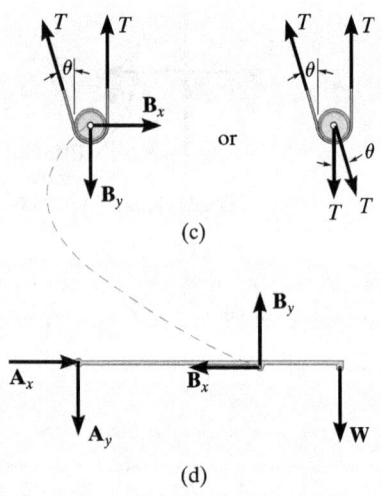

(c)

(d)

(a)

Fig. 6–22

### SOLUTION

The idealized model of the device is shown in Fig. 6–22b. Here the angle $\theta$ is assumed to be known. From this model, the free-body diagrams of the cylinder and frame are shown in Figs. 6–22c and 6–22d, respectively. Note that the force that the pin at $B$ exerts on the cylinder can be represented by either its horizontal and vertical components $\mathbf{B}_x$ and $\mathbf{B}_y$, which can be determined by using the force equations of equilibrium applied to the cylinder, or by the two components $T$, which provide equal but opposite moments on the cylinder and thus keep it from turning. Also, realize that once the pin reactions at $A$ have been determined, half of their values act on each side of the frame since pin connections occur on each side, Fig. 6–22a.

## EXAMPLE 6.11

For the frame shown in Fig. 6–23a, draw the free-body diagrams of (a) the entire frame including the pulleys and cords, (b) the frame without the pulleys and cords, and (c) each of the pulleys.

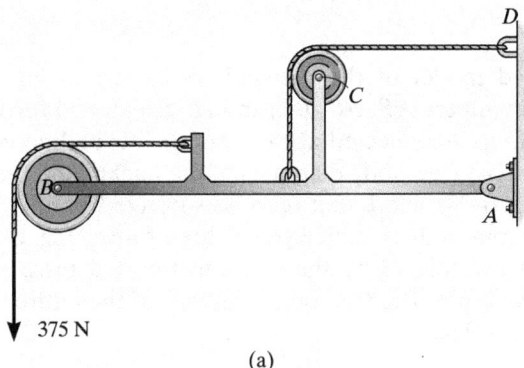

(a)

### SOLUTION

**Part (a).** When the entire frame including the pulleys and cords is considered, the interactions at the points where the pulleys and cords are connected to the frame become pairs of *internal* forces which cancel each other and therefore are not shown on the free-body diagram, Fig. 6–23b.

**Part (b).** When the cords and pulleys are removed, their effect *on the frame* must be shown, Fig. 6–23c.

**Part (c).** The force components $B_x$, $B_y$, $C_x$, $C_y$ of the pins on the pulleys, Fig. 6–23d, are equal but opposite to the force components exerted by the pins on the frame, Fig. 6–23c. Why?

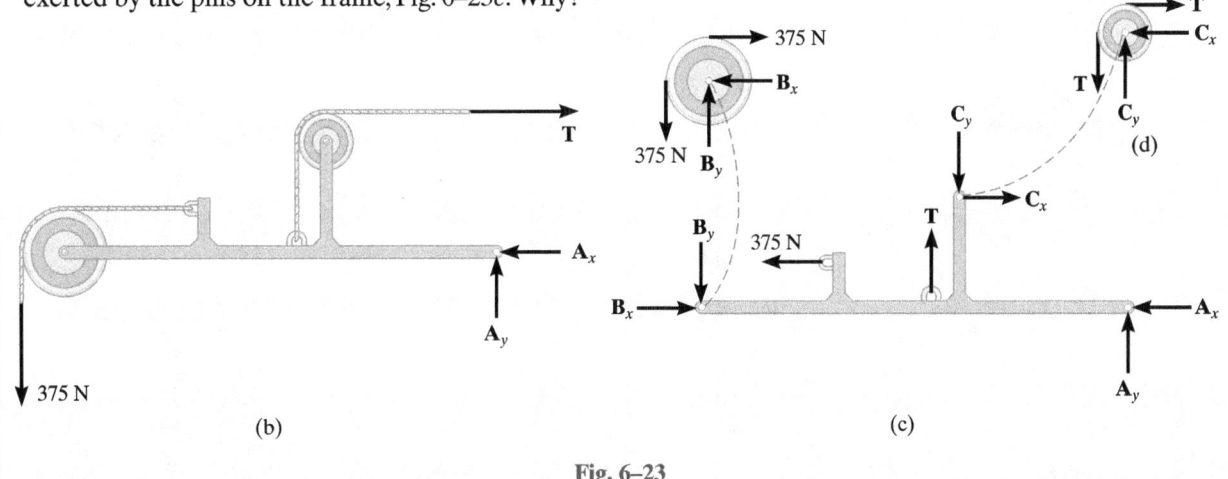

Fig. 6–23

## EXAMPLE 6.12

(a)

Draw the free-body diagrams of the bucket and the vertical boom of the backhoe shown in the photo, Fig. 6–24a. The bucket and its contents have a weight $W$. Neglect the weight of the members.

### SOLUTION

The idealized model of the assembly is shown in Fig. 6–24b. By inspection, members $AB$, $BC$, $BE$, and $HI$ are all two-force members since they are pin connected at their end points and no other forces act on them. The free-body diagrams of the bucket and the boom are shown in Fig. 6–24c. Note that pin $C$ is subjected to only two forces, whereas the pin at $B$ is subjected to three forces, Fig. 6–24d. These three forces are related by the two equations of force equilibrium applied to each pin. The free-body diagram of the entire assembly is shown in Fig. 6–24e.

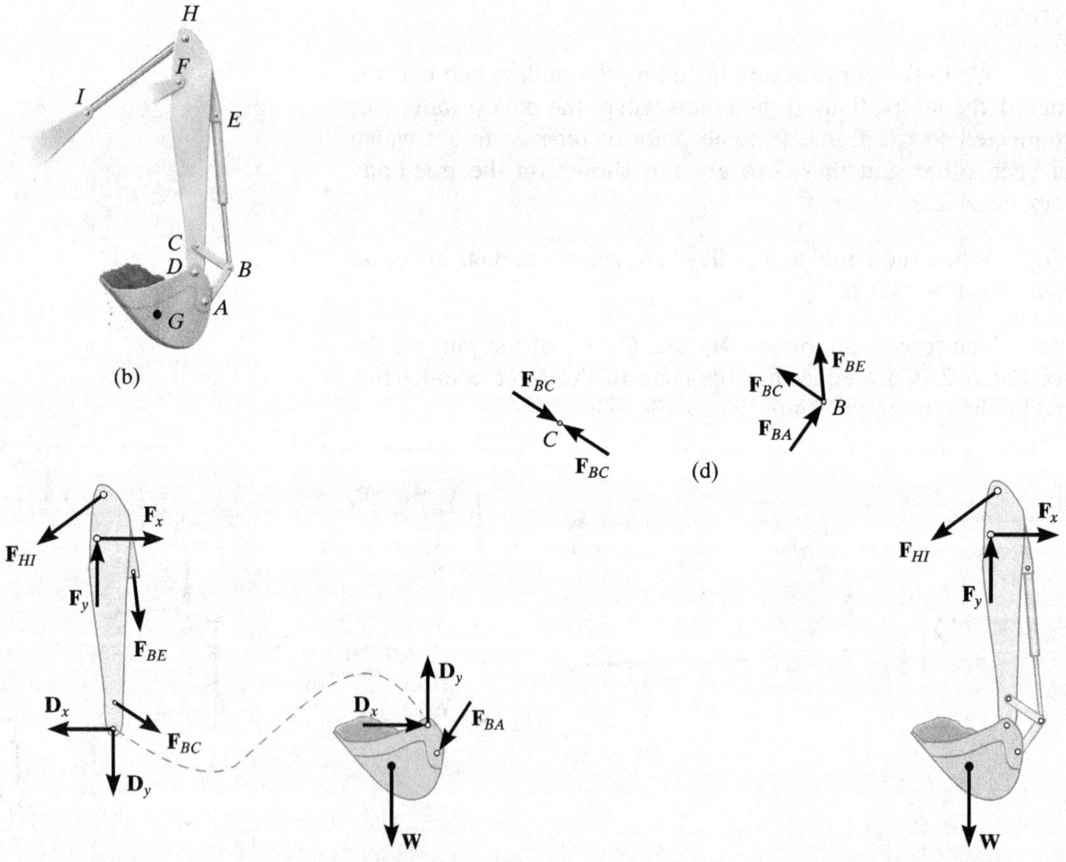

Fig. 6–24

## EXAMPLE 6.13

Draw the free-body diagram of each part of the smooth piston and link mechanism used to crush recycled cans, which is shown in Fig. 6–25a.

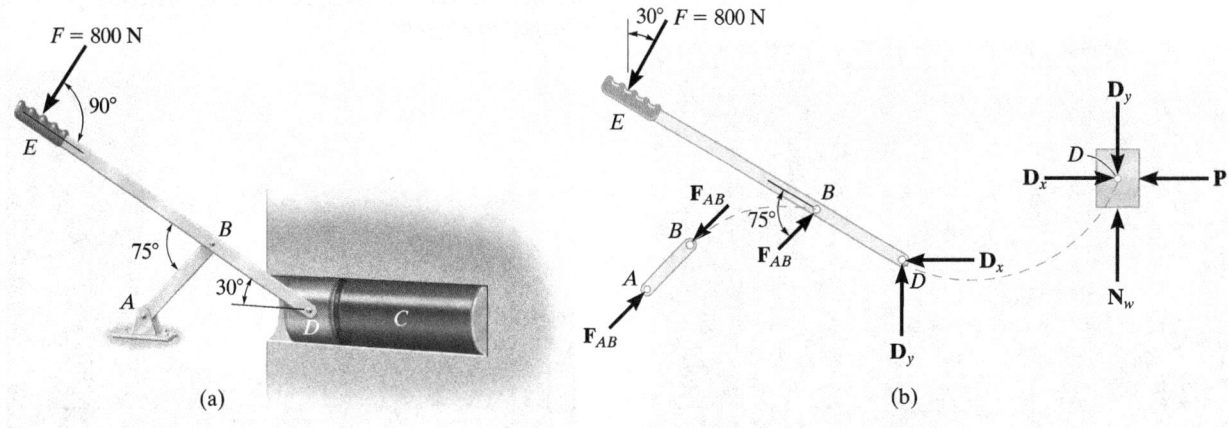

**Fig. 6–25**

### SOLUTION
By inspection, member $AB$ is a two-force member. The free-body diagrams of the parts are shown in Fig. 6–25b. Since the pins at $B$ and $D$ connect only two parts together, the forces there are shown as equal but opposite on the separate free-body diagrams of their connected members. In particular, four components of force act on the piston: $\mathbf{D}_x$ and $\mathbf{D}_y$ represent the effect of the pin (or lever $EBD$), $\mathbf{N}_w$ is the *resultant force* of the support, and $\mathbf{P}$ is the resultant compressive force caused by the can $C$.

**NOTE:** A free-body diagram of the entire assembly is shown in Fig. 6–25c. Here the forces between the components are internal and are not shown on the free-body diagram.

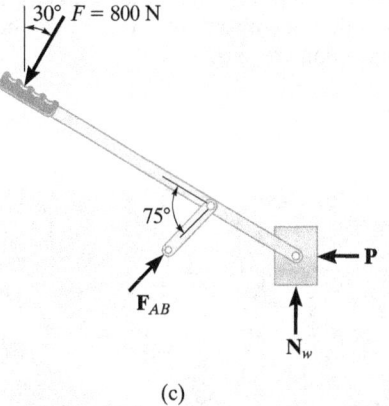

*Before proceeding, it is highly recommended that you cover the solutions to the previous examples and attempt to draw the requested free-body diagrams. When doing so, make sure the work is neat and that all the forces and couple moments are properly labeled. When finished, challenge yourself and solve the following four problems.*

## CONCEPTUAL PROBLEMS

**P6–1.** Draw the free-body diagrams of each of the crane boom segments *AB*, *BC*, and *BD*. Only the weights of *AB* and *BC* are significant. Assume *A* and *B* are pins.

P6–1

**P6–2.** Draw the free-body diagrams of the boom *ABCD* and the stick *EDFGH* of the backhoe. The weights of these two members are significant. Neglect the weights of all the other members, and assume all indicated points of connection are pins.

P6–2

**P6–3.** Draw the free-body diagrams of the boom *ABCDF* and the stick *FGH* of the bucket lift. Neglect the weights of the member. The bucket weighs *W*. The two force members are *BI*, *CE*, *DE* and *GE*. Assume all indicated points of connection are pins.

P6–3

**P6–4.** To operate the can crusher, one pushes down on the lever arm *ABC* which rotates about the fixed pin at *B*. This moves the side links *CD* downward, which causes the guide plate *E* to also move downward and thereby crush the can. Draw the free-body diagrams of the lever, side link, and guide plate. Make up some reasonable numbers and do an equilibrium analysis to show how much an applied vertical force at the handle is magnified when it is transmitted to the can. Assume all points of connection are pins and the guides for the plate are smooth.

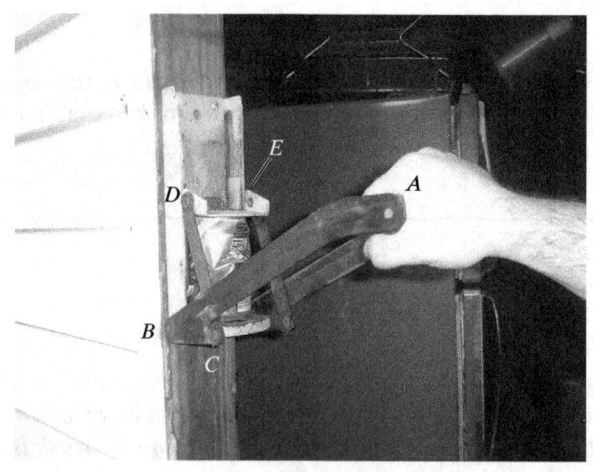

P6–4

## Procedure for Analysis

The joint reactions on frames or machines (structures) composed of multiforce members can be determined using the following procedure.

Free-Body Diagram.

- Draw the free-body diagram of the entire frame or machine, a portion of it, or each of its members. The choice should be made so that it leads to the most direct solution of the problem.

- When the free-body diagram of a group of members of a frame or machine is drawn, the forces between the connected parts of this group are internal forces and are not shown on the free-body diagram of the group.

- Forces common to two members, which are in contact, act with equal magnitude but opposite sense on the respective free-body diagrams of the members.

- Two-force members, regardless of their shape, have equal but opposite collinear forces acting at the ends of the member.

- In many cases it is possible to tell by inspection the proper sense of the unknown forces acting on a member; however, if this seems difficult, the sense can be assumed.

- Remember that a couple moment is a free vector and can act at any point on the free-body diagram. Also, a force is a sliding vector and can act at any point along its line of action.

Equations of Equilibrium.

- Count the number of unknowns and compare it to the total number of equilibrium equations that are available. In two dimensions, there are three equilibrium equations that can be written for each member.

- Sum moments about a point that lies at the intersection of the lines of action of as many of the unknown forces as possible.

- If the solution of a force or couple moment magnitude is found to be negative, it means the sense of the force is the reverse of that shown on the free-body diagram.

## EXAMPLE 6.14

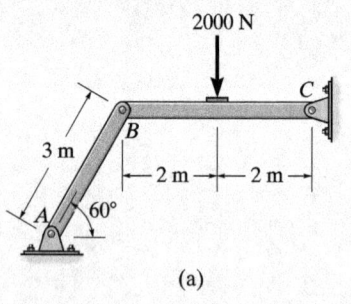

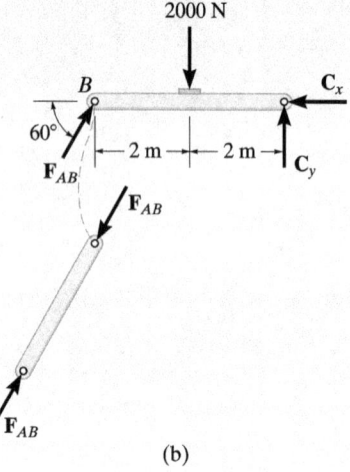

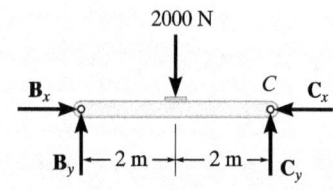

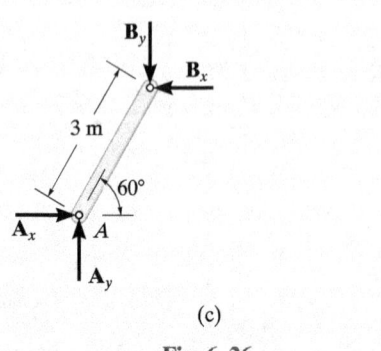

Fig. 6–26

Determine the horizontal and vertical components of force which the pin at $C$ exerts on member $BC$ of the frame in Fig. 6–26a.

### SOLUTION I

**Free-Body Diagrams.** By inspection, it can be seen that $AB$ is a two-force member. The free-body diagrams are shown in Fig. 6–26b.

**Equations of Equilibrium.** The *three unknowns* can be determined by applying the three equations of equilibrium to member $CB$.

$\zeta + \Sigma M_C = 0;$  $2000 \text{ N}(2 \text{ m}) - (F_{AB} \sin 60°)(4 \text{ m}) = 0$  $F_{AB} = 1154.7 \text{ N}$

$\xrightarrow{+} \Sigma F_x = 0;$  $1154.7 \cos 60° \text{ N} - C_x = 0$  $C_x = 577 \text{ N}$  *Ans.*

$+\uparrow \Sigma F_y = 0;$  $1154.7 \sin 60° \text{ N} - 2000 \text{ N} + C_y = 0$  $C_y = 1000 \text{ N}$ *Ans.*

### SOLUTION II

**Free-Body Diagrams.** If one does not recognize that $AB$ is a two-force member, then more work is involved in solving this problem. The free-body diagrams are shown in Fig. 6–26c.

**Equations of Equilibrium.** The *six unknowns* are determined by applying the three equations of equilibrium to each member.

Member $AB$

$\zeta + \Sigma M_A = 0;$  $B_x(3 \sin 60° \text{ m}) - B_y(3 \cos 60° \text{ m}) = 0$ (1)

$\xrightarrow{+} \Sigma F_x = 0;$  $A_x - B_x = 0$ (2)

$+\uparrow \Sigma F_y = 0;$  $A_y - B_y = 0$ (3)

Member $BC$

$\zeta + \Sigma M_C = 0;$  $2000 \text{ N}(2 \text{ m}) - B_y(4 \text{ m}) = 0$ (4)

$\xrightarrow{+} \Sigma F_x = 0;$  $B_x - C_x = 0$ (5)

$+\uparrow \Sigma F_y = 0;$  $B_y - 2000 \text{ N} + C_y = 0$ (6)

The results for $C_x$ and $C_y$ can be determined by solving these equations in the following sequence: 4, 1, 5, then 6. The results are

$B_y = 1000 \text{ N}$

$B_x = 577 \text{ N}$

$C_x = 577 \text{ N}$  *Ans.*

$C_y = 1000 \text{ N}$  *Ans.*

By comparison, Solution I is simpler since the requirement that $F_{AB}$ in Fig. 6–26b be equal, opposite, and collinear at the ends of member $AB$ automatically satisfies Eqs. 1, 2, and 3 above and therefore eliminates the need to write these equations. As a result, *save yourself some time and effort by always identifying the two-force members before starting the analysis!*

## EXAMPLE 6.15

The compound beam shown in Fig. 6–27a is pin connected at B. Determine the components of reaction at its supports. Neglect its weight and thickness.

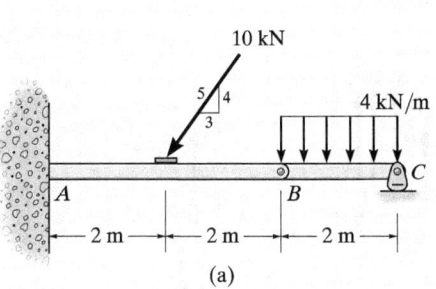
(a)

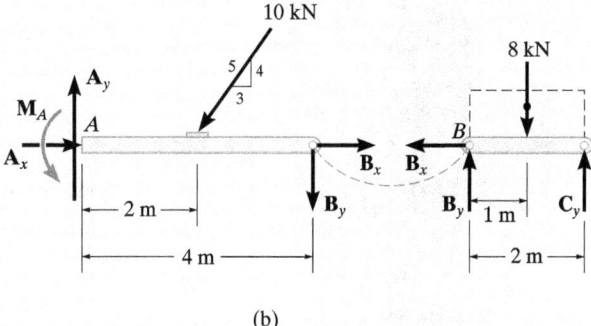
(b)

Fig. 6–27

### SOLUTION

**Free-Body Diagrams.** By inspection, if we consider a free-body diagram of the *entire beam ABC*, there will be three unknown reactions at A and one at C. These four unknowns cannot all be obtained from the three available equations of equilibrium, and so for the solution it will become necessary to dismember the beam into its two segments, as shown in Fig. 6–27b.

**Equations of Equilibrium.** The six unknowns are determined as follows:

*Segment BC*

$\xrightarrow{+} \Sigma F_x = 0;$ $\qquad B_x = 0$

$\zeta + \Sigma M_B = 0;$ $\qquad -8 \text{ kN}(1 \text{ m}) + C_y(2 \text{ m}) = 0$

$+ \uparrow \Sigma F_y = 0;$ $\qquad B_y - 8 \text{ kN} + C_y = 0$

*Segment AB*

$\xrightarrow{+} \Sigma F_x = 0;$ $\qquad A_x - (10 \text{ kN})(\tfrac{3}{5}) + B_x = 0$

$\zeta + \Sigma M_A = 0;$ $\qquad M_A - (10 \text{ kN})(\tfrac{4}{5})(2 \text{ m}) - B_y(4 \text{ m}) = 0$

$+ \uparrow \Sigma F_y = 0;$ $\qquad A_y - (10 \text{ kN})(\tfrac{4}{5}) - B_y = 0$

Solving each of these equations successively, using previously calculated results, we obtain

$A_x = 6 \text{ kN}$ $\qquad A_y = 12 \text{ kN}$ $\qquad M_A = 32 \text{ kN} \cdot \text{m}$ *Ans.*

$B_x = 0$ $\qquad B_y = 4 \text{ kN}$

$C_y = 4 \text{ kN}$ $\qquad$ *Ans.*

## EXAMPLE 6.16

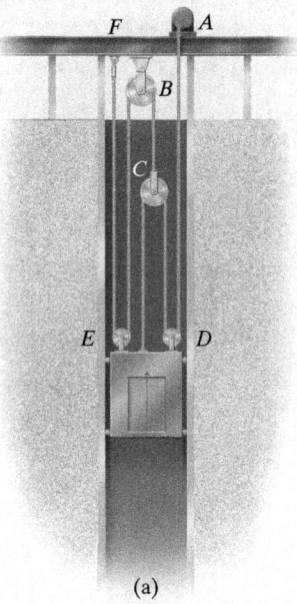

A 500-kg elevator car in Fig. 6–28a is being hoisted by motor $A$ using the pulley system shown. If the car is traveling with a constant speed, determine the force developed in the two cables. Neglect the mass of the cable and pulleys.

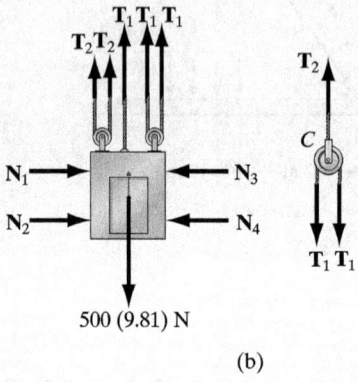

(a)

(b)

Fig. 6–28

### SOLUTION

**Free-Body Diagram.** We can solve this problem using the free-body diagrams of the elevator car and pulley $C$, Fig. 6–28b. The tensile forces developed in the cables are denoted as $T_1$ and $T_2$.

**Equations of Equilibrium.** For pulley $C$,

$$+\uparrow \Sigma F_y = 0; \qquad T_2 - 2T_1 = 0 \quad \text{or} \quad T_2 = 2T_1 \qquad (1)$$

For the elevator car,

$$+\uparrow \Sigma F_y = 0; \qquad 3T_1 + 2T_2 - 500(9.81) \text{ N} = 0 \qquad (2)$$

Substituting Eq. (1) into Eq. (2) yields

$$3T_1 + 2(2T_1) - 500(9.81) \text{ N} = 0$$

$$T_1 = 700.71 \text{ N} = 701 \text{ N} \qquad Ans.$$

Substituting this result into Eq. (1),

$$T_2 = 2(700.71) \text{ N} = 1401 \text{ N} = 1.40 \text{ kN} \qquad Ans.$$

## EXAMPLE 6.17

The smooth disk shown in Fig. 6–29a is pinned at $D$ and has a weight of 200 N. Neglecting the weights of the other members, determine the horizontal and vertical components of reaction at pins $B$ and $D$.

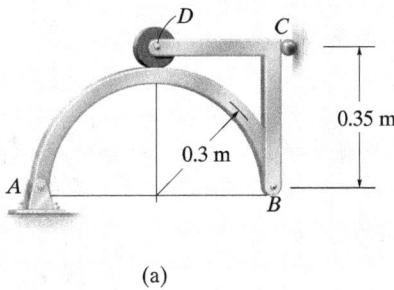

(a)

### SOLUTION

**Free-Body Diagrams.** The free-body diagrams of the entire frame and each of its members are shown in Fig. 6–29b.

**Equations of Equilibrium.** The eight unknowns can of course be obtained by applying the eight equilibrium equations to each member—three to member $AB$, three to member $BCD$, and two to the disk. (Moment equilibrium is automatically satisfied for the disk.) If this is done, however, all the results can be obtained only from a simultaneous solution of some of the equations. (Try it and find out.) To avoid this situation, it is best first to determine the three support reactions on the *entire* frame; then, using these results, the remaining five equilibrium equations can be applied to two other parts in order to solve successively for the other unknowns.

*Entire Frame*

$\zeta + \Sigma M_A = 0;$  $-200\text{ N }(0.3\text{ m}) + C_x(0.35\text{ m}) = 0$  $C_x = 171\text{ N}$

$\xrightarrow{+} \Sigma F_x = 0;$  $A_x - 171\text{ N} = 0$  $A_x = 171\text{ N}$

$+\uparrow \Sigma F_y = 0;$  $A_y - 200\text{ N} = 0$  $A_y = 200\text{ N}$

*Member AB*

$\xrightarrow{+} \Sigma F_x = 0;$  $171\text{ N} - B_x = 0$  $B_x = 171\text{ N}$  *Ans.*

$\zeta + \Sigma M_B = 0;$  $-200\text{ N }(0.6\text{ m}) + N_D(0.3\text{ m}) = 0$  $N_D = 400\text{ N}$

$+\uparrow \Sigma F_y = 0;$  $200\text{ N} - 400\text{ N} + B_y = 0$  $B_y = 200\text{ N}$  *Ans.*

*Disk*

$\xrightarrow{+} \Sigma F_x = 0;$  $D_x = 0$  *Ans.*

$+\uparrow \Sigma F_y = 0;$  $400\text{ N} - 200\text{ N} - D_y = 0$  $D_y = 200\text{ N}$  *Ans.*

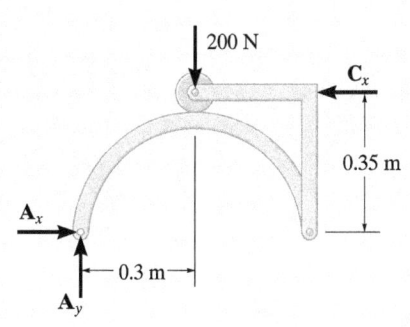

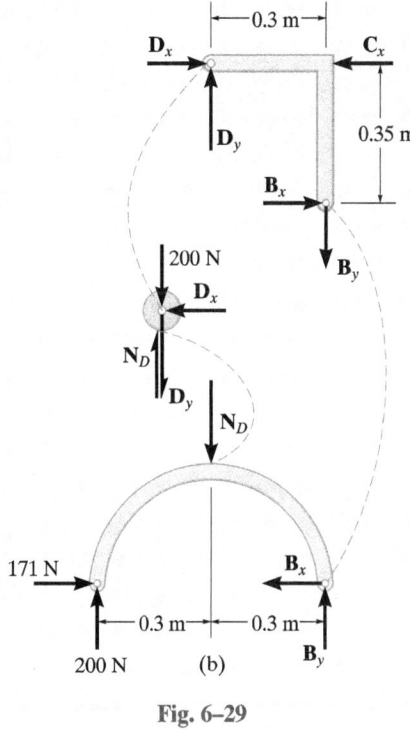

(b)

Fig. 6–29

## EXAMPLE 6.18

Determine the tension in the cables and also the force **P** required to support the 600-N force using the frictionless pulley system shown in Fig. 6–30a.

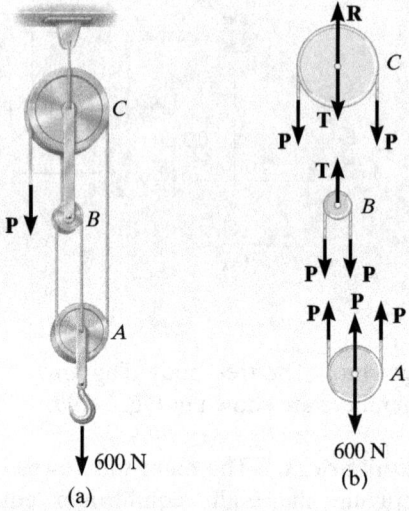

Fig. 6–30

### SOLUTION

**Free-Body Diagram.** A free-body diagram of each pulley *including* its pin and a portion of the contacting cable is shown in Fig. 6–30b. Since the cable is *continuous*, it has a *constant tension P* acting throughout its length. The link connection between pulleys B and C is a two-force member, and therefore it has an unknown tension T acting on it. Notice that the *principle of action, equal but opposite reaction* must be carefully observed for forces **P** and **T** when the *separate* free-body diagrams are drawn.

**Equations of Equilibrium.** The three unknowns are obtained as follows:

Pulley A

$+\uparrow \Sigma F_y = 0;$   $3P - 600 \text{ N} = 0$   $P = 200 \text{ N}$   *Ans.*

Pulley B

$+\uparrow \Sigma F_y = 0;$   $T - 2P = 0$   $T = 400 \text{ N}$   *Ans.*

Pulley C

$+\uparrow \Sigma F_y = 0;$   $R - 2P - T = 0$   $R = 800 \text{ N}$   *Ans.*

## EXAMPLE 6.19

The two planks in Fig. 6–31a are connected together by cable $BC$ and a smooth spacer $DE$. Determine the reactions at the smooth supports $A$ and $F$, and also find the force developed in the cable and spacer.

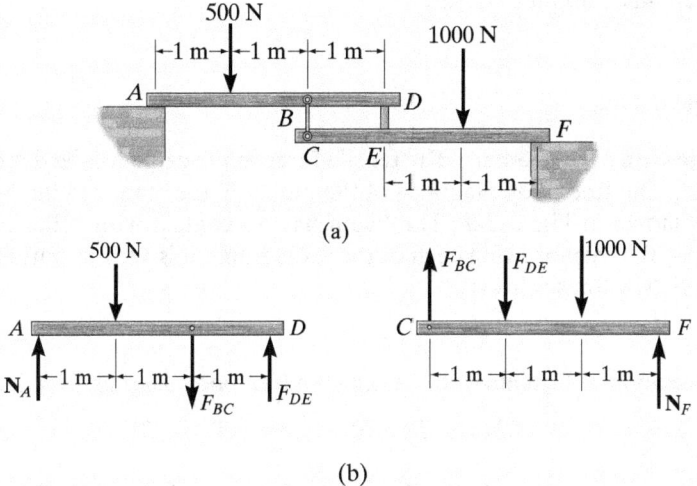

(a)

(b)

Fig. 6–31

### SOLUTION

**Free-Body Diagrams.** The free-body diagram of each plank is shown in Fig. 6–31b. It is important to apply Newton's third law to the interaction forces as shown.

**Equations of Equilibrium.** For plank $AD$,

$\zeta + \Sigma M_A = 0;$   $F_{DE}(3 \text{ m}) - F_{BC}(2 \text{ m}) - 500 \text{ N} (1 \text{ m}) = 0$

For plank $CF$,

$\zeta + \Sigma M_F = 0;$   $F_{DE}(2 \text{ m}) - F_{BC}(3 \text{ m}) + 1000 \text{ N} (1 \text{ m}) = 0$

Solving simultaneously,

$$F_{DE} = 700 \text{ N} \quad F_{BC} = 800 \text{ N} \qquad Ans.$$

Using these results, for plank $AD$,

$+\uparrow \Sigma F_y = 0;$   $N_A + 700 \text{ N} - 800 \text{ N} - 500 \text{ N} = 0$

$\qquad N_A = 600 \text{ N} \qquad Ans.$

And for plank $CF$,

$+\uparrow \Sigma F_y = 0;$   $N_F + 800 \text{ N} - 700 \text{ N} - 1000 \text{ N} = 0$

$\qquad N_F = 900 \text{ N} \qquad Ans.$

## EXAMPLE 6.20

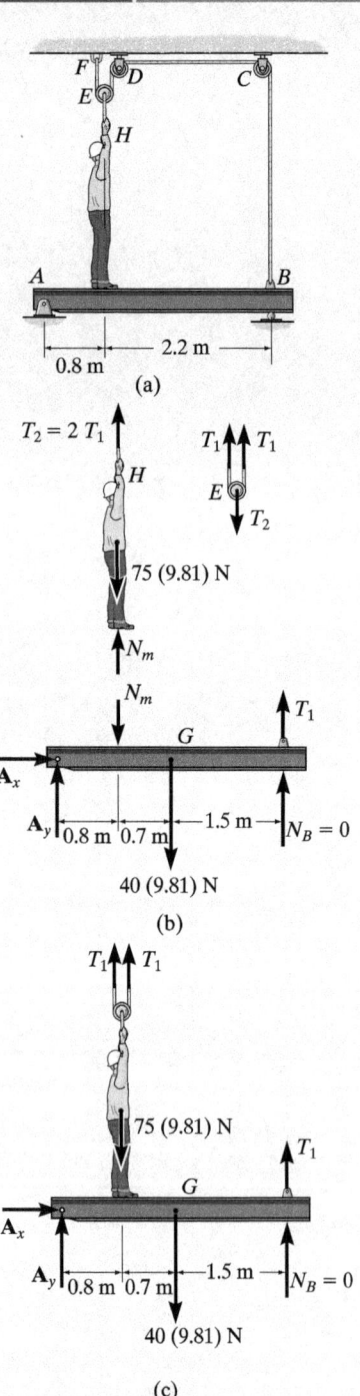

Fig. 6–32

The 75-kg man in Fig. 6–32a attempts to lift the 40-kg uniform beam off the roller support at B. Determine the tension developed in the cable attached to B and the normal reaction of the man on the beam when this is about to occur.

### SOLUTION

**Free-Body Diagrams.** The tensile force in the cable will be denoted as $T_1$. The free-body diagrams of the pulley $E$, the man, and the beam are shown in Fig. 6–32b. The beam has no contact with roller $B$, so $N_B = 0$. When drawing each of these diagrams, it is very important to apply Newton's third law.

**Equations of Equilibrium.** Using the free-body diagram of pulley $E$,

$$+\uparrow \Sigma F_y = 0; \qquad 2T_1 - T_2 = 0 \quad \text{or} \quad T_2 = 2T_1 \qquad (1)$$

Referring to the free-body diagram of the man using this result,

$$+\uparrow \Sigma F_y = 0; \qquad N_m + 2T_1 - 75(9.81) \text{ N} = 0 \qquad (2)$$

Summing moments about point $A$ on the beam,

$$\zeta+\Sigma M_A = 0; \quad T_1(3 \text{ m}) - N_m(0.8 \text{ m}) - [40(9.81) \text{ N}](1.5 \text{ m}) = 0 \quad (3)$$

Solving Eqs. 2 and 3 simultaneously for $T_1$ and $N_m$, then using Eq. (1) for $T_2$, we obtain

$$T_1 = 256 \text{ N} \qquad N_m = 224 \text{ N} \qquad T_2 = 512 \text{ N} \qquad Ans.$$

### SOLUTION II

A direct solution for $T_1$ can be obtained by considering the beam, the man, and pulley $E$ as a *single system*. The free-body diagram is shown in Fig. 6–32c. Thus,

$$\zeta+\Sigma M_A = 0; \quad 2T_1(0.8 \text{ m}) - [75(9.81) \text{ N}](0.8 \text{ m})$$
$$-[40(9.81) \text{ N}](1.5 \text{ m}) + T_1(3 \text{ m}) = 0$$
$$T_1 = 256 \text{ N} \qquad Ans.$$

With this result, Eqs. 1 and 2 can then be used to find $N_m$ and $T_2$.

# EXAMPLE 6.21

The frame in Fig. 6–33a supports the 50-kg cylinder. Determine the horizontal and vertical components of reaction at $A$ and the force at $C$.

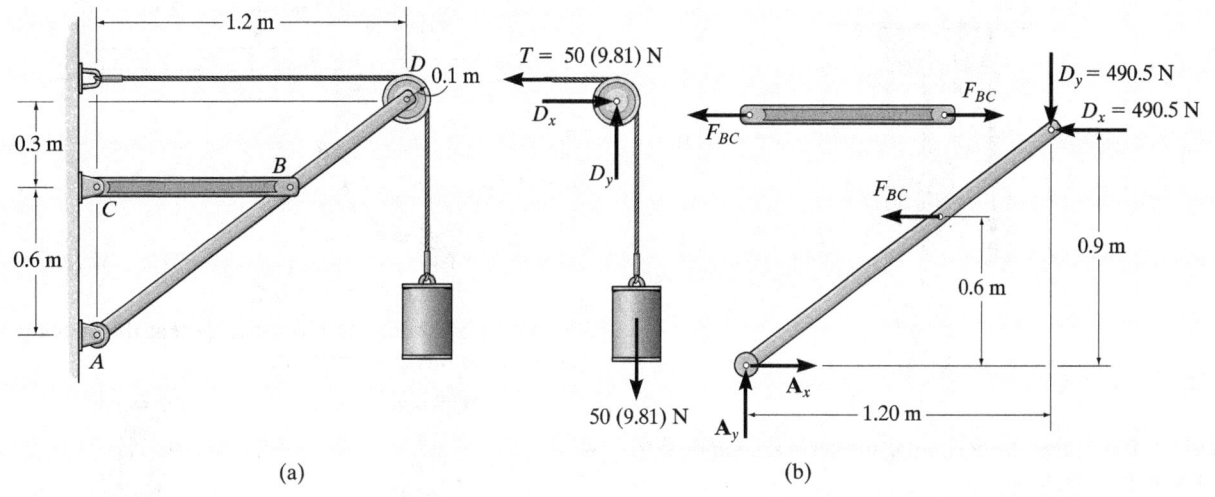

Fig. 6–33

## SOLUTION

**Free-Body Diagrams.** The free-body diagram of pulley $D$, along with the cylinder and a portion of the cord (a system), is shown in Fig. 6–33b. Member $BC$ is a two-force member as indicated by its free-body diagram. The free-body diagram of member $ABD$ is also shown.

**Equations of Equilibrium.** We will begin by analyzing the equilibrium of the pulley. The moment equation of equilibrium is automatically satisfied with $T = 50(9.81)$ N, and so

$\xrightarrow{+} \Sigma F_x = 0; \quad D_x - 50(9.81) \text{ N} = 0 \quad D_x = 490.5 \text{ N}$

$+\uparrow \Sigma F_y = 0; \quad D_y - 50(9.81) \text{ N} = 0 \quad D_y = 490.5 \text{ N}$ *Ans.*

Using these results, $F_{BC}$ can be determined by summing moments about point $A$ on member $ABD$.

$\zeta + \Sigma M_A = 0; \; F_{BC} (0.6 \text{ m}) + 490.5 \text{ N}(0.9 \text{ m}) - 490.5 \text{ N}(1.20 \text{ m}) = 0$

$$F_{BC} = 245.25 \text{ N} \qquad Ans.$$

Now $A_x$ and $A_y$ can be determined by summing forces.

$\xrightarrow{+} \Sigma F_x = 0; \quad A_x - 245.25 \text{ N} - 490.5 \text{ N} = 0 \quad A_x = 736 \text{ N}$ *Ans.*

$+\uparrow \Sigma F_y = 0; \qquad A_y - 490.5 \text{ N} = 0 \qquad A_y = 490.5 \text{ N}$ *Ans.*

## FUNDAMENTAL PROBLEMS

**F6–13.** Determine the force $P$ needed to hold the 300-N weight in equilibrium.

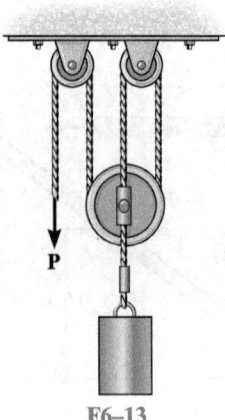

F6–13

**F6–14.** Determine the horizontal and vertical components of reaction at pin $C$.

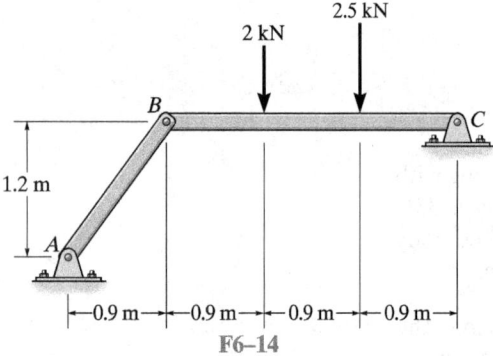

F6–14

**F6–15.** If a 100-N force is applied to the handles of the pliers, determine the clamping force exerted on the smooth pipe $B$ and the magnitude of the resultant force at pin $A$.

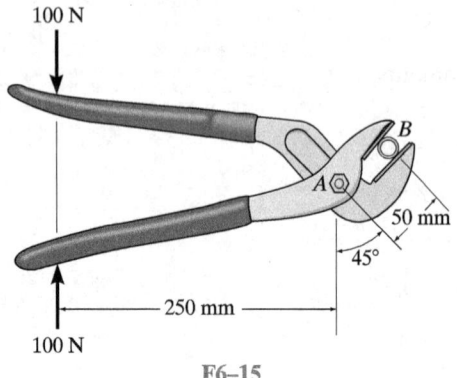

F6–15

**F6–16.** Determine the horizontal and vertical components of reaction at pin $C$.

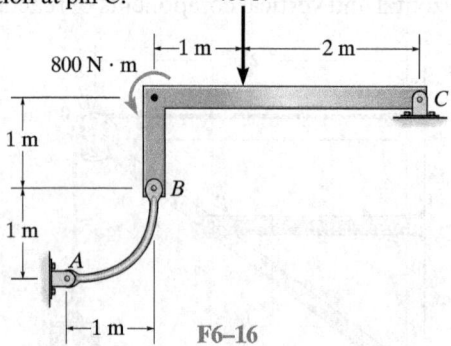

F6–16

**F6–17.** Determine the normal force that the 500-N plate $A$ exerts on the 150-N plate $B$.

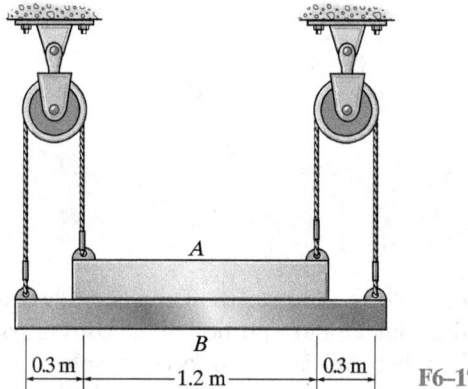

F6–17

**F6–18.** Determine the force $P$ needed to lift the load. Also, determine the proper placement $x$ of the hook for equilibrium. Neglect the weight of the beam.

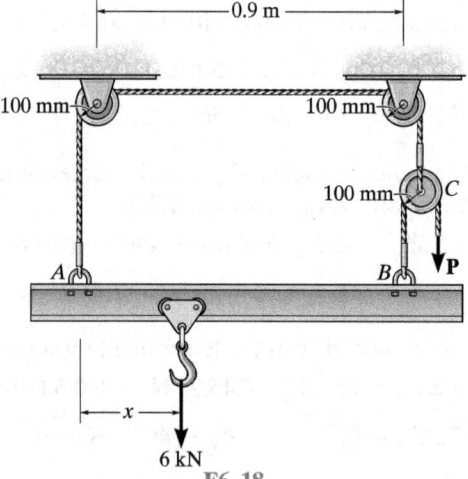

F6–18

**6–67.** Determine the force **P** required to hold the 100-kg weight in equilibrium.

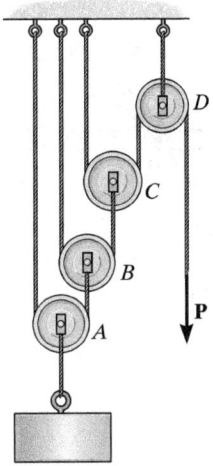

Prob. 6–67

**•6–69.** Determine the force **P** required to hold the 50-kg mass in equilibrium.

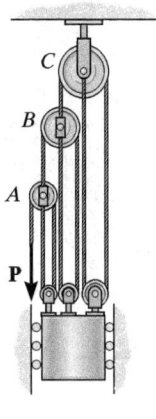

Prob. 6–69

**\*6–68.** Determine the force **P** required to hold the 150-kg crate in equilibrium.

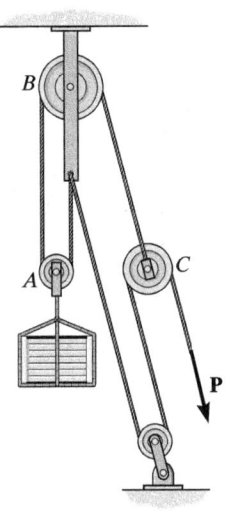

Prob. 6–68

**6–70.** Determine the force **P** needed to hold the 10-kg block in equilibrium.

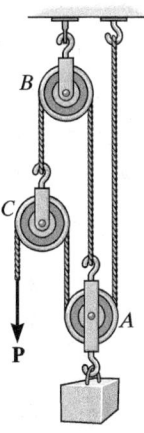

Prob. 6–70

**6–71.** Determine the force **P** needed to support the 50-kg weight. Each pulley has a weight of 50 N. Also, what are the cord reactions at $A$ and $B$?

**•6–73.** If the peg at $B$ is smooth, determine the components of reaction at pin $A$ and fixed support $C$.

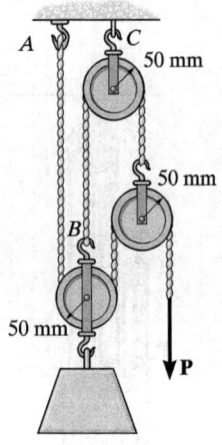

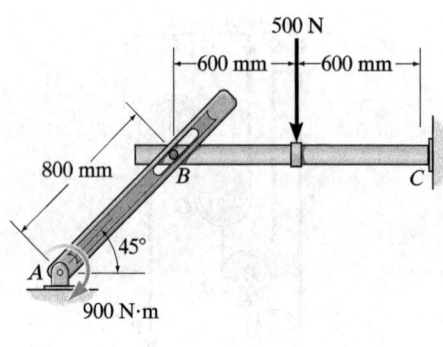

Prob. 6–71

Prob. 6–73

**\*6–72.** The cable and pulleys are used to lift the 300-kg stone. Determine the force that must be exerted on the cable at $A$ and the corresponding magnitude of the resultant force the pulley at $C$ exerts on pin $B$ when the cables are in the position shown.

**6–74.** Determine the horizontal and vertical components of reaction at pins $A$ and $C$.

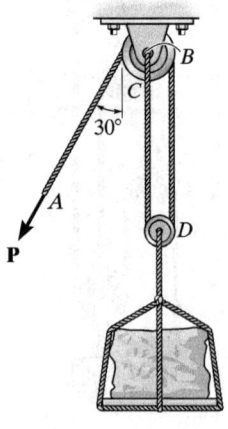

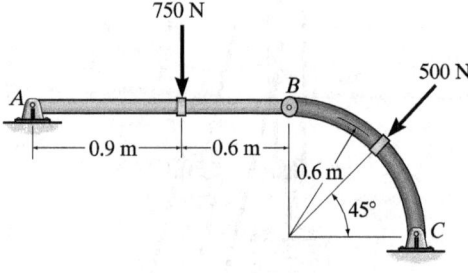

Prob. 6–72

Prob. 6–74

**6–75.** The compound beam is fixed at $A$ and supported by rockers at $B$ and $C$. There are hinges (pins) at $D$ and $E$. Determine the components of reaction at the supports.

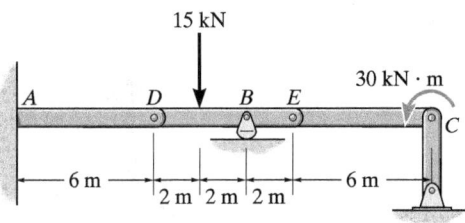

Prob. 6–75

**\*6–76.** The compound beam is pin-supported at $C$ and supported by rollers at $A$ and $B$. There is a hinge (pin) at $D$. Determine the components of reaction at the supports. Neglect the thickness of the beam.

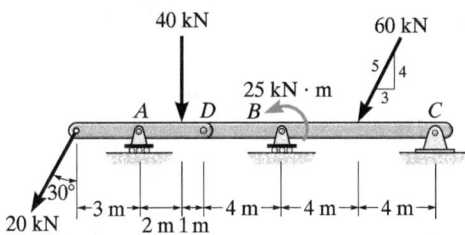

Prob. 6–76

**•6–77.** The compound beam is supported by a rocker at $B$ and is fixed to the wall at $A$. If it is hinged (pinned) together at $C$, determine the components of reaction at the supports. Neglect the thickness of the beam.

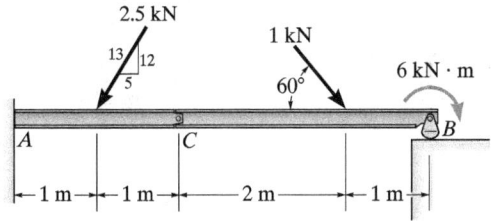

Prob. 6–77

**6–78.** Determine the horizontal and vertical components of reaction at pins $A$ and $C$ of the two-member frame.

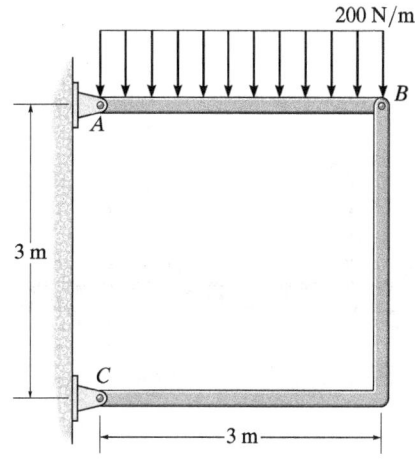

Prob. 6–78

**6–79.** If a force of $F = 50$ N acts on the rope, determine the cutting force on the smooth tree limb at $D$ and the horizontal and vertical components of force acting on pin $A$. The rope passes through a small pulley at $C$ and a smooth ring at $E$.

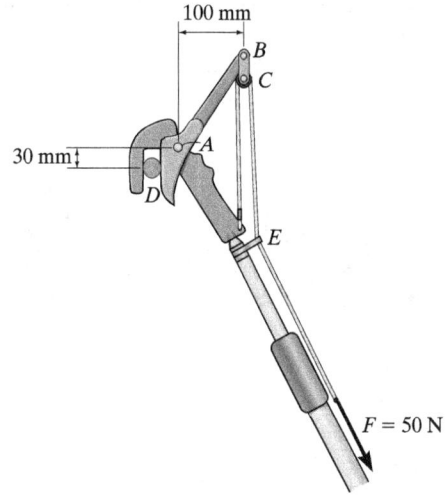

Prob. 6–79

**\*6–80.** Two beams are connected together by the short link BC. Determine the components of reaction at the fixed support A and at pin D.

**6–82.** If the 300-kg drum has a center of mass at point G, determine the horizontal and vertical components of force acting at pin A and the reactions on the smooth pads C and D. The grip at B on member DAB resists both horizontal and vertical components of force at the rim of the drum.

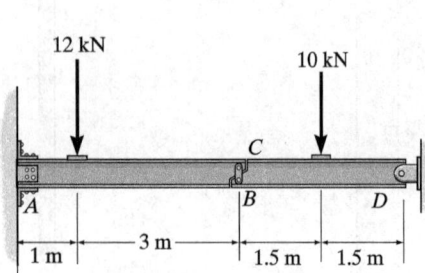

Prob. 6–80

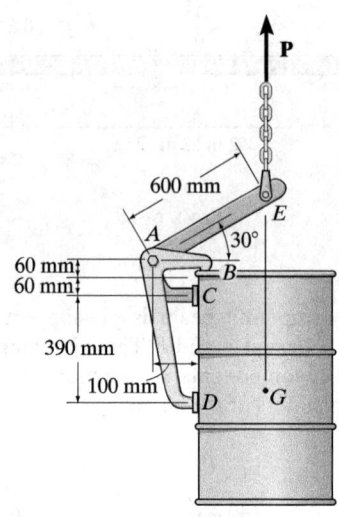

Prob. 6–82

**•6–81.** The bridge frame consists of three segments which can be considered pinned at A, D, and E, rocker supported at C and F, and roller supported at B. Determine the horizontal and vertical components of reaction at all these supports due to the loading shown.

**6–83.** Determine the horizontal and vertical components of reaction that pins A and C exert on the two-member arch.

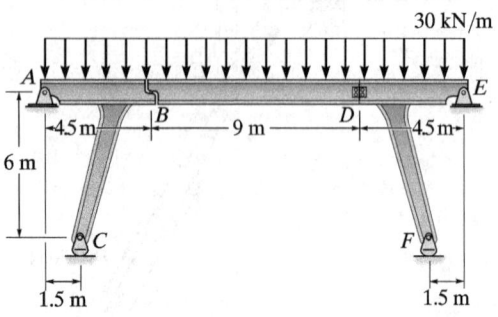

Prob. 6–81

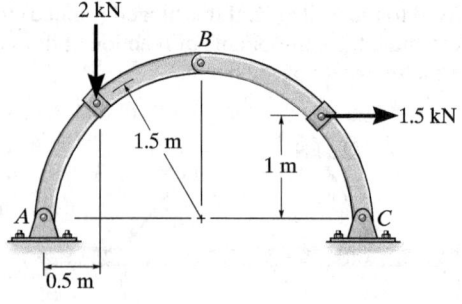

Prob. 6–83

**\*6–84.** The truck and the tanker have weights of 40 kN and 100 kN respectively. Their respective centers of gravity are located at points $G_1$ and $G_2$. If the truck is at rest, determine the reactions on both wheels at $A$, at $B$, and at $C$. The tanker is connected to the truck at the turntable $D$ which acts as a pin.

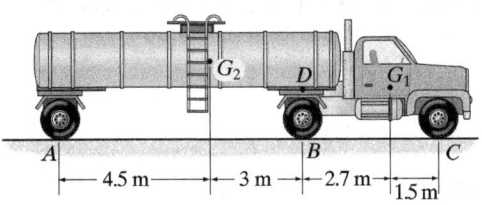

Prob. 6–84

**•6–85.** The platform scale consists of a combination of third and first class levers so that the load on one lever becomes the effort that moves the next lever. Through this arrangement, a small weight can balance a massive object. If $x = 450$ mm, determine the required mass of the counterweight $S$ required to balance a 90-kg load, $L$.

**6–86.** The platform scale consists of a combination of third and first class levers so that the load on one lever becomes the effort that moves the next lever. Through this arrangement, a small weight can balance a massive object. If $x = 450$ mm and, the mass of the counterweight $S$ is 2 kg, determine the mass of the load $L$ required to maintain the balance.

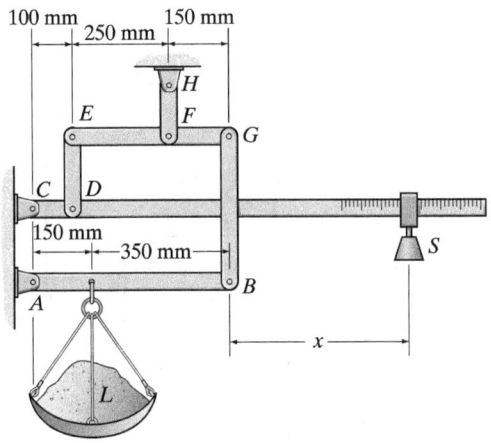

Probs. 6–85/86

**6–87.** The hoist supports the 125-kg engine. Determine the force the load creates in member $DB$ and in member $FB$, which contains the hydraulic cylinder $H$.

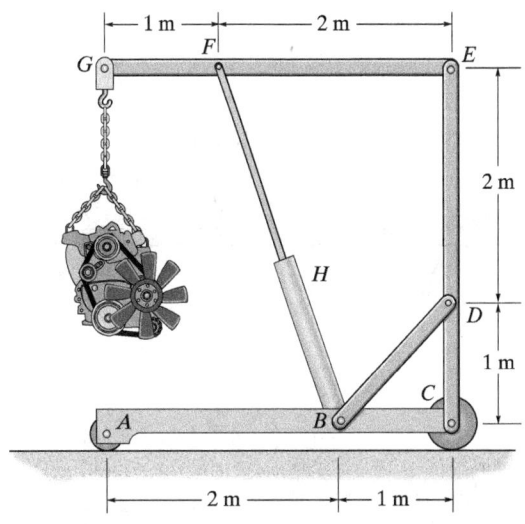

Prob. 6–87

**\*6–88.** The frame is used to support the 100-kg cylinder $E$. Determine the horizontal and vertical components of reaction at $A$ and $D$.

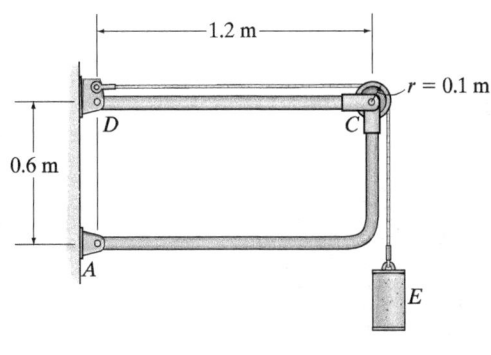

Prob. 6–88

•6–89. Determine the horizontal and vertical components of reaction which the pins exert on member AB of the frame.

6–90. Determine the horizontal and vertical components of reaction which the pins exert on member EDC of the frame.

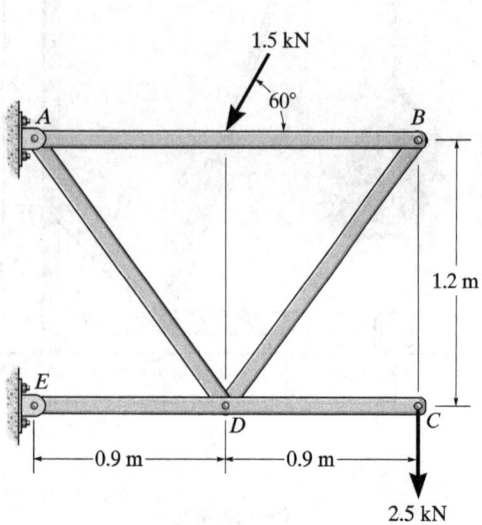

Probs. 6–89/90

6–91. The clamping hooks are used to lift the uniform smooth 500-kg plate. Determine the resultant compressive force that the hook exerts on the plate at A and B, and the pin reaction at C.

*6–92. The wall crane supports a load of 3.5 kN. Determine the horizontal and vertical components of reaction at pins A and D. Also, what is the force in the cable at the winch W?

•6–93. The wall crane supports a load of 3.5 kN. Determine the horizontal and vertical components of reaction at pins A and D. Also, what is the force in the cable at the winch W? The jib ABC has a weight of 500 N and member BD has a weight of 200 N. Each member is uniform and has a center of gravity at its center.

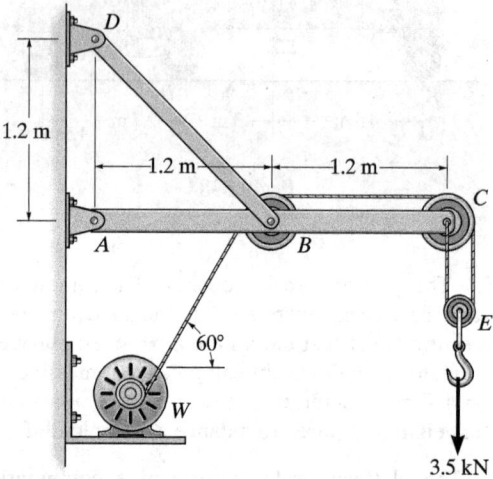

Probs. 6–92/93

6–94. The lever-actuated scale consists of a series of compound levers. If a load of weight $W = 750$ N is placed on the platform, determine the required weight of the counterweight S to balance the load. Is it necessary to place the load symmetrically on the platform? Explain.

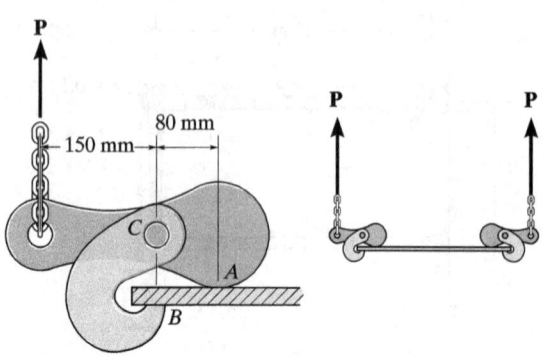

Prob. 6–91

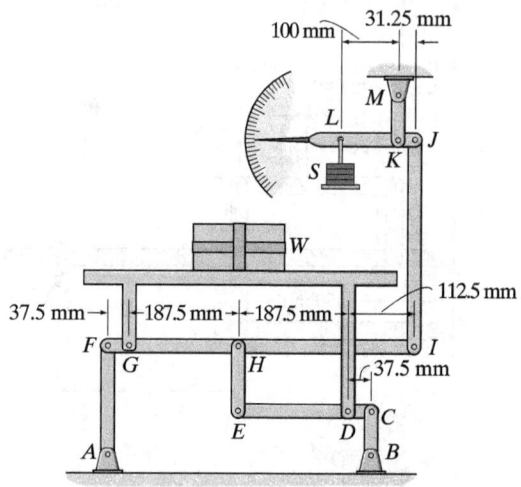

Prob. 6–94

**6–95.** If $P = 75$ N, determine the force $F$ that the toggle clamp exerts on the wooden block.

**\*6–96.** If the wooden block exerts a force of $F = 600$ N on the toggle clamp, determine the force $P$ applied to the handle.

**6–98.** A 300-kg counterweight, with center of mass at $G$, is mounted on the pitman crank $AB$ of the oil-pumping unit. If a force of $F = 5$ kN is to be developed in the fixed cable attached to the end of the walking beam $DEF$, determine the torque $M$ that must be supplied by the motor.

**6–99.** A 300-kg counterweight, with center of mass at $G$, is mounted on the pitman crank $AB$ of the oil-pumping unit. If the motor supplies a torque of $M = 2500$ N · m, determine the force **F** developed in the fixed cable attached to the end of the walking beam $DEF$.

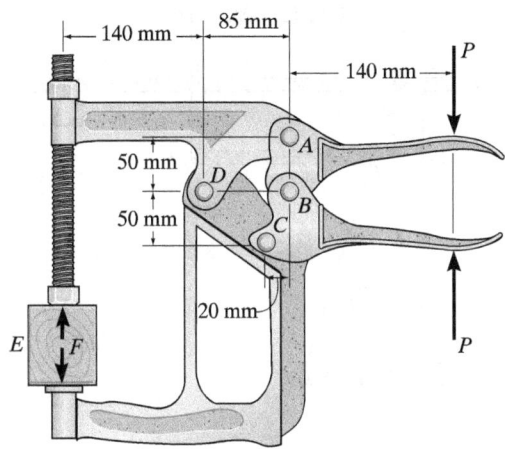

Probs. 6–95/96

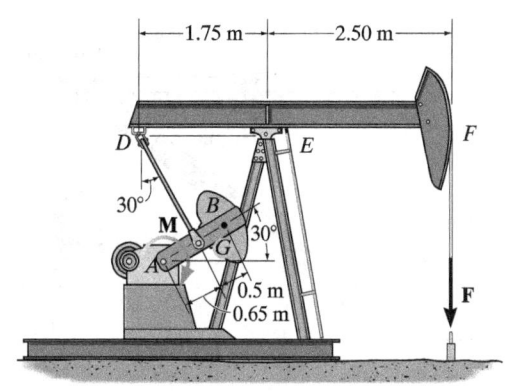

Probs. 6–98/99

**•6–97.** The pipe cutter is clamped around the pipe $P$. If the wheel at $A$ exerts a normal force of $F_A = 80$ N on the pipe, determine the normal forces of wheels $B$ and $C$ on the pipe. The three wheels each have a radius of 7 mm and the pipe has an outer radius of 10 mm.

**\*6–100.** The two-member structure is connected at $C$ by a pin, which is fixed to $BDE$ and passes through the smooth slot in member $AC$. Determine the horizontal and vertical components of reaction at the supports.

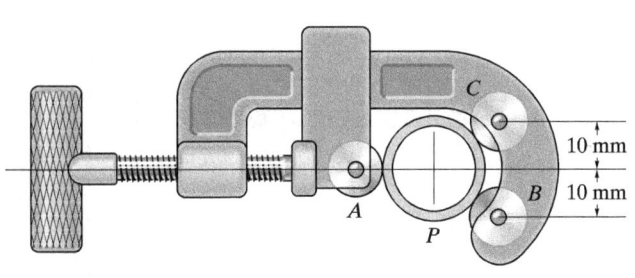

Prob. 6–97

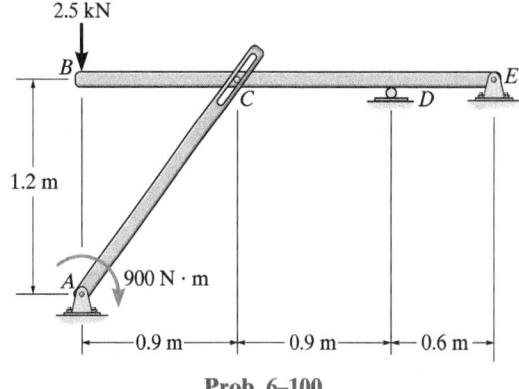

Prob. 6–100

**•6–101.** The frame is used to support the 50-kg cylinder. Determine the horizontal and vertical components of reaction at $A$ and $D$.

**6–102.** The frame is used to support the 50-kg cylinder. Determine the force of the pin at $C$ on member $ABC$ and on member $CD$.

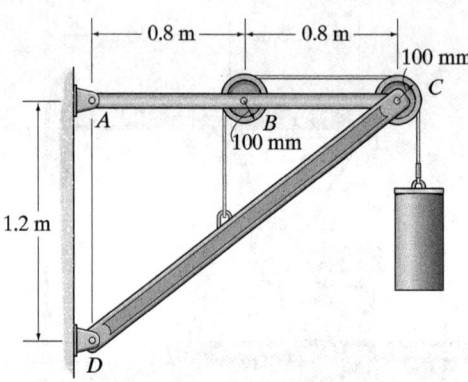

Probs. 6–101/102

**\*6–104.** The compound arrangement of the pan scale is shown. If the mass on the pan is 4 kg, determine the horizontal and vertical components at pins $A$, $B$, and $C$ and the distance $x$ of the 25-g mass to keep the scale in balance.

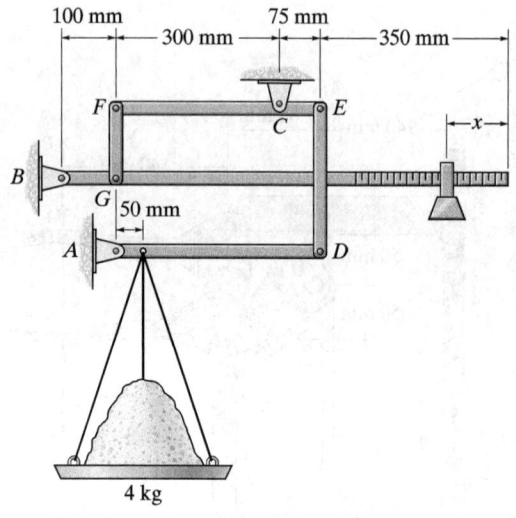

Prob. 6–104

**6–103.** Determine the reactions at the fixed support $E$ and the smooth support $A$. The pin, attached to member $BD$, passes through a smooth slot at $D$.

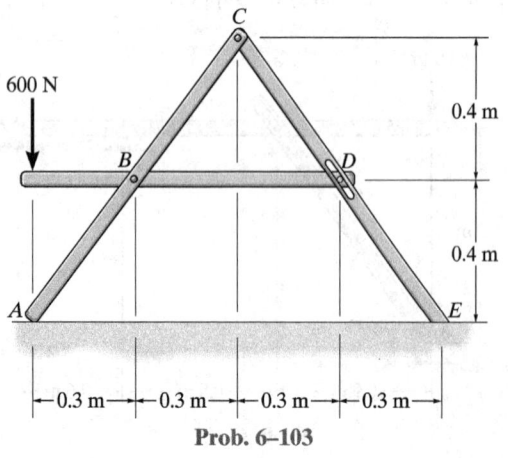

Prob. 6–103

**•6–105.** Determine the horizontal and vertical components of reaction that the pins at $A$, $B$, and $C$ exert on the frame. The cylinder has a mass of 80 kg.

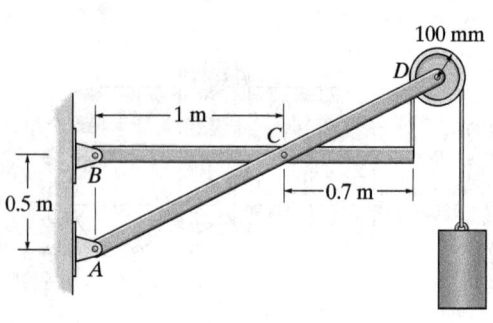

Prob. 6–105

**6–106.** The bucket of the backhoe and its contents have a weight of 6 kN and a center of gravity at G. Determine the forces of the hydraulic cylinder AB and in links AC and AD in order to hold the load in the position shown. The bucket is pinned at E.

**•6–109.** If a clamping force of 300 N is required at A, determine the amount of force **F** that must be applied to the handle of the toggle clamp.

**6–110.** If a force of $F = 350$ N is applied to the handle of the toggle clamp, determine the resulting clamping force at A.

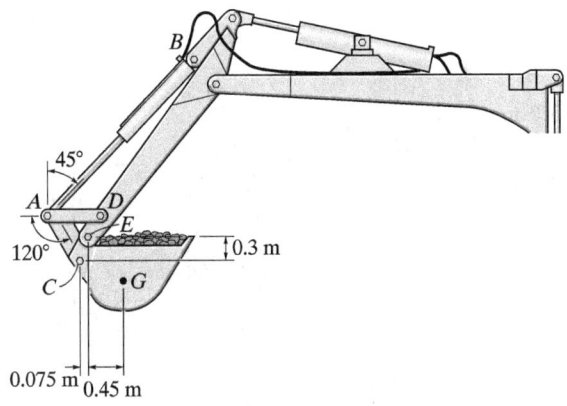

Prob. 6–106

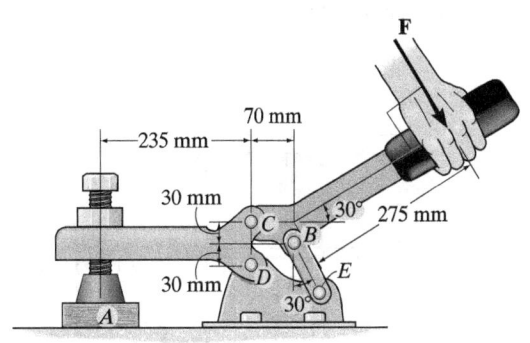

Probs. 6–109/110

**6–107.** A man having a weight of 875 N ($\approx$ 87.5 kg) attempts to hold himself using one of the two methods shown. Determine the total force he must exert on bar AB in each case and the normal reaction he exerts on the platform at C. Neglect the weight of the platform.

**\*6–108.** A man having a weight of 875 N ($\approx$ 87.5 kg) attempts to hold himself using one of the two methods shown. Determine the total force he must exert on bar AB in each case and the normal reaction he exerts on the platform at C. The platform has a weight of 150 N ($\approx$ 15 kg).

**6–111.** Two smooth tubes A and B, each having the same weight, W, are suspended from a common point O by means of equal-length cords. A third tube, C, is placed between A and B. Determine the greatest weight of C without upsetting equilibrium.

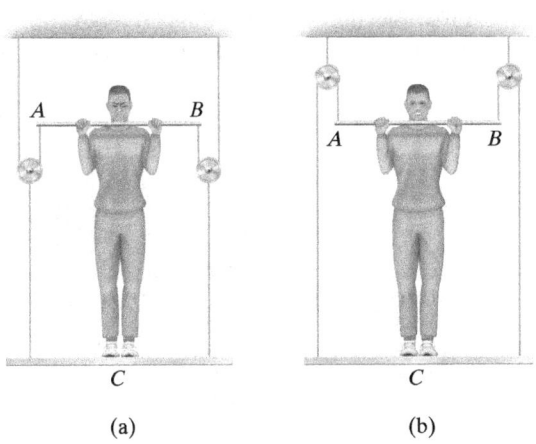

Probs. 6–107/108

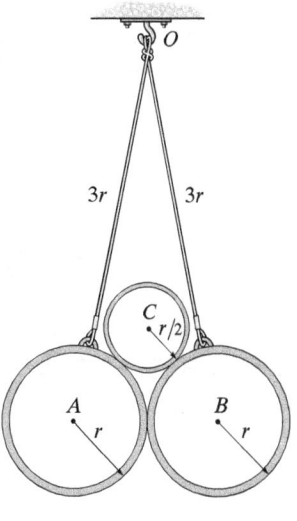

Prob. 6–111

**\*6–112.** The handle of the sector press is fixed to gear $G$, which in turn is in mesh with the sector gear $C$. Note that $AB$ is pinned at its ends to gear $C$ and the underside of the table $EF$, which is allowed to move vertically due to the smooth guides at $E$ and $F$. If the gears only exert tangential forces between them, determine the compressive force developed on the cylinder $S$ when a vertical force of 40 N is applied to the handle of the press.

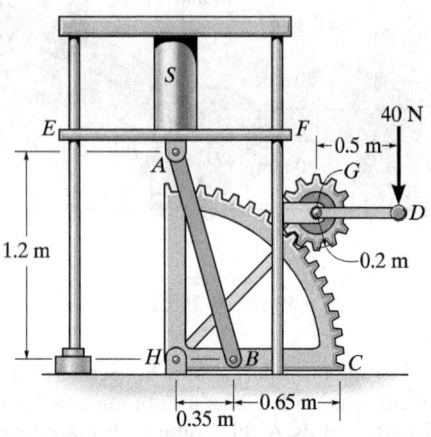

Prob. 6–112

**•6–113.** Show that the weight $W_1$ of the counterweight at $H$ required for equilibrium is $W_1 = (b/a)W$, and so it is independent of the placement of the load $W$ on the platform.

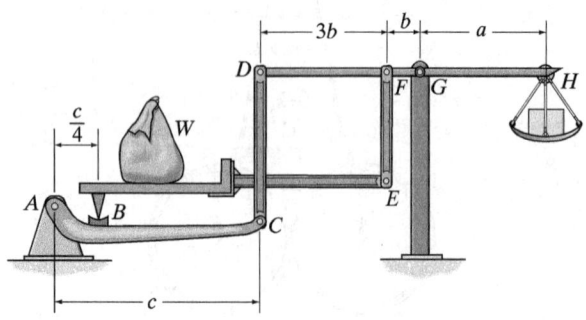

Prob. 6–113

**6–114.** The tractor shovel carries a 500-kg load of soil, having a center of mass at $G$. Compute the forces developed in the hydraulic cylinders $IJ$ and $BC$ due to this loading.

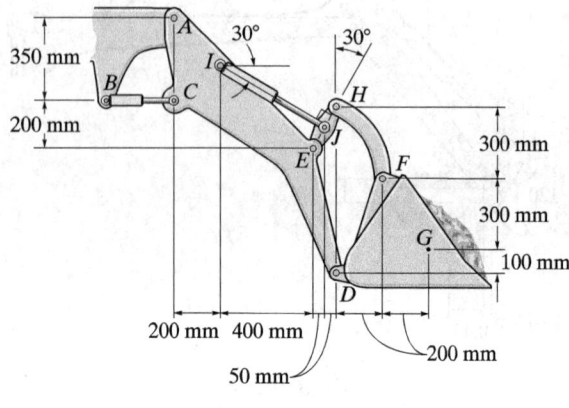

Prob. 6–114

**6–115.** If a force of $P = 100$ N is applied to the handle of the toggle clamp, determine the horizontal clamping force $N_E$ that the clamp exerts on the smooth wooden block at $E$.

**\*6–116.** If the horizontal clamping force that the toggle clamp exerts on the smooth wooden block at $E$ is $N_E = 200$ N, determine the force **P** applied to the handle of the clamp.

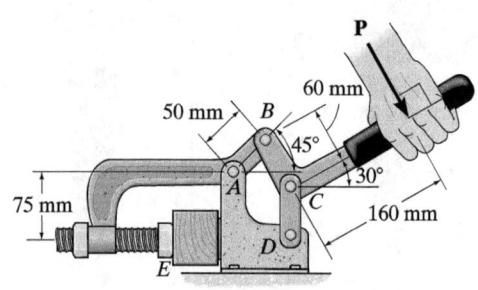

Probs. 6–115/116

**•6–117.** The engine hoist is used to support the 200-kg engine. Determine the force acting in the hydraulic cylinder $AB$, the horizontal and vertical components of force at the pin $C$, and the reactions at the fixed support $D$.

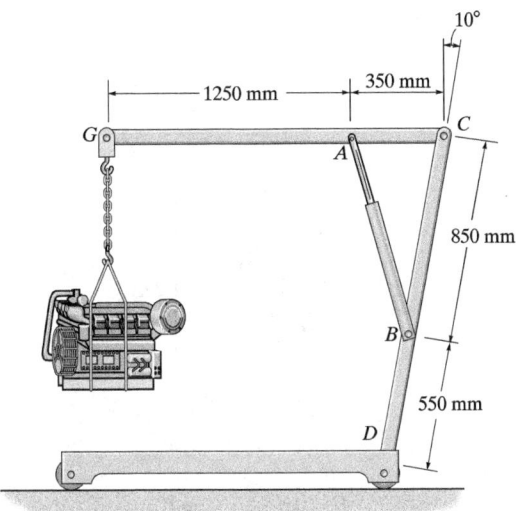

Prob. 6–117

**6–118.** Determine the force that the smooth roller $C$ exerts on member $AB$. Also, what are the horizontal and vertical components of reaction at pin $A$? Neglect the weight of the frame and roller.

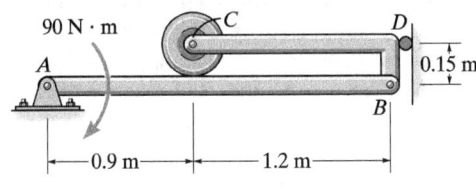

Prob. 6–118

**6–119.** Determine the horizontal and vertical components of reaction which the pins exert on member $ABC$.

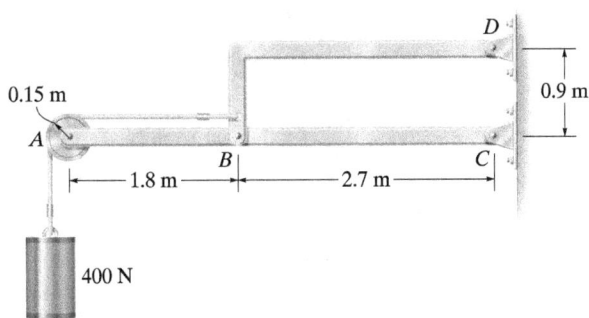

Prob. 6–119

**\*6–120.** Determine the couple moment **M** that must be applied to member $DC$ for equilibrium of the quick-return mechanism. Express the result in terms of the angles $\phi$ and $\theta$, dimension $L$, and the applied *vertical force* **P**. The block at $C$ is confined to slide within the slot of member $AB$.

**•6–121.** Determine the couple moment **M** that must be applied to member $DC$ for equilibrium of the quick-return mechanism. Express the result in terms of the angles $\phi$ and $\theta$, dimension $L$, and the applied force **P**, which should be changed in the figure and instead be directed horizontally to the right. The block at $C$ is confined to slide within the slot of member $AB$.

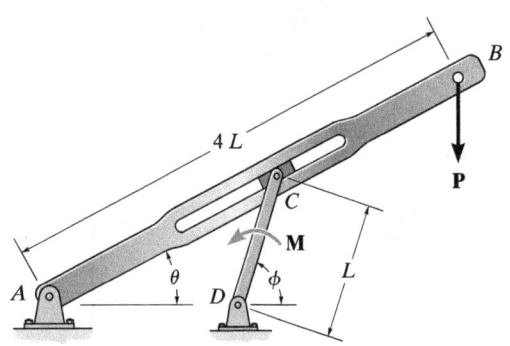

Probs. 6–120/121

**6–122.** The kinetic sculpture requires that each of the three pinned beams be in perfect balance at all times during its slow motion. If each member has a uniform weight of 50 N/m and length of 0.9 m, determine the necessary counterweights $W_1$, $W_2$, and $W_3$ which must be added to the ends of each member to keep the system in balance for any position. Neglect the size of the counterweights.

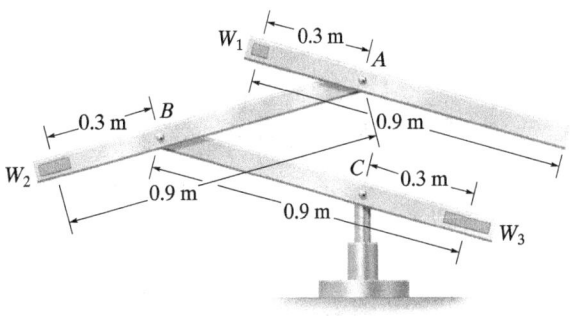

Prob. 6–122

**6–123.** The four-member "A" frame is supported at $A$ and $E$ by smooth collars and at $G$ by a pin. All the other joints are ball-and-sockets. If the pin at $G$ will fail when the resultant force there is 800 N, determine the largest vertical force $P$ that can be supported by the frame. Also, what are the $x, y, z$ force components which member $BD$ exerts on members $EDC$ and $ABC$? The collars at $A$ and $E$ and the pin at $G$ only exert force components on the frame.

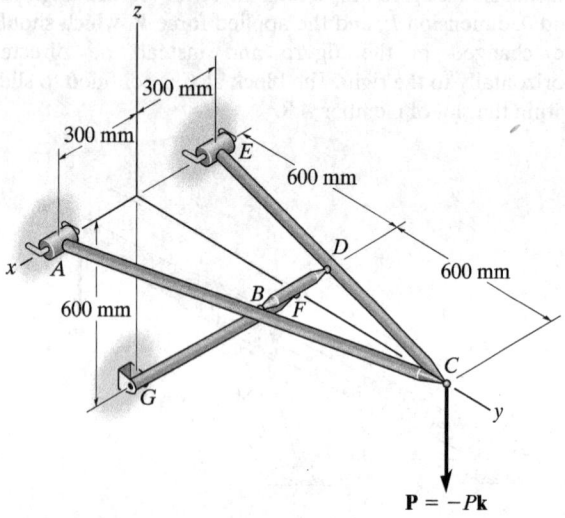

Prob. 6–123

**\*6–124.** The structure is subjected to the loading shown. Member $AD$ is supported by a cable $AB$ and roller at $C$ and fits through a smooth circular hole at $D$. Member $ED$ is supported by a roller at $D$ and a pole that fits in a smooth snug circular hole at $E$. Determine the $x, y, z$ components of reaction at $E$ and the tension in cable $AB$.

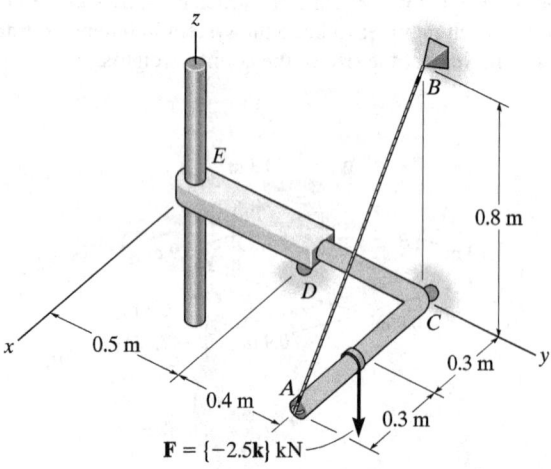

Prob. 6–124

**•6–125.** The three-member frame is connected at its ends using ball-and-socket joints. Determine the $x, y, z$ components of reaction at $B$ and the tension in member $ED$. The force acting at $D$ is $\mathbf{F} = \{135\mathbf{i} + 200\mathbf{j} - 180\mathbf{k}\}$ kN.

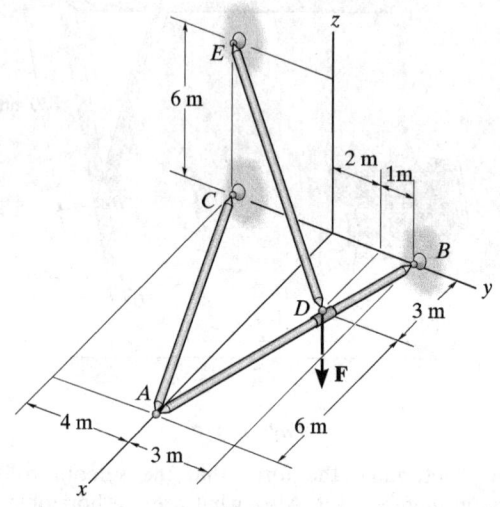

Prob. 6–125

**6–126.** The structure is subjected to the loadings shown. Member $AB$ is supported by a ball-and-socket at $A$ and smooth collar at $B$. Member $CD$ is supported by a pin at $C$. Determine the $x, y, z$ components of reaction at $A$ and $C$.

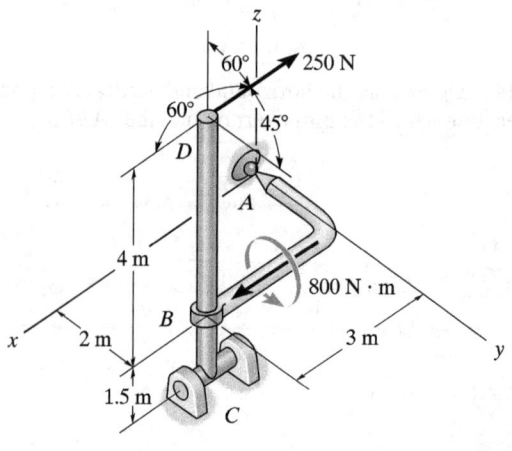

Prob. 6–126

# CHAPTER REVIEW

**Simple Truss**

A simple truss consists of triangular elements connected together by pinned joints. The forces within its members can be determined by assuming the members are all two-force members, connected concurrently at each joint. The members are either in tension or compression, or carry no force.

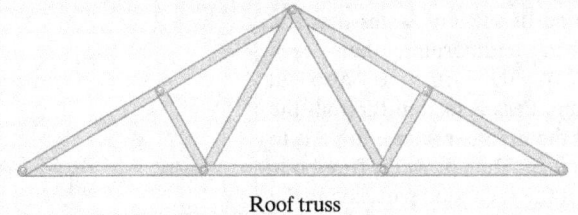

Roof truss

**Method of Joints**

The method of joints states that if a truss is in equilibrium, then each of its joints is also in equilibrium. For a plane truss, the concurrent force system at each joint must satisfy force equilibrium.

To obtain a numerical solution for the forces in the members, select a joint that has a free-body diagram with at most two unknown forces and one known force. (This may require first finding the reactions at the supports.)

Once a member force is determined, use its value and apply it to an adjacent joint.

Remember that forces that are found to *pull* on the joint are *tensile forces*, and those that *push* on the joint are *compressive forces*.

To avoid a simultaneous solution of two equations, set one of the coordinate axes along the line of action of one of the unknown forces and sum forces perpendicular to this axis. This will allow a direct solution for the other unknown.

The analysis can also be simplified by first identifying all the zero-force members.

$\Sigma F_x = 0$

$\Sigma F_y = 0$

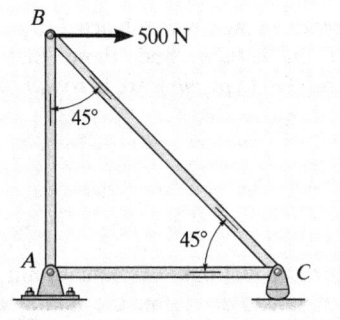

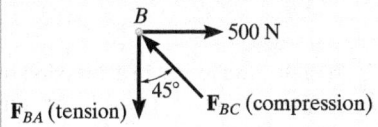

## Method of Sections

The method of sections states that if a truss is in equilibrium, then each segment of the truss is also in equilibrium. Pass a section through the truss and the member whose force is to be determined. Then draw the free-body diagram of the sectioned part having the least number of forces on it.

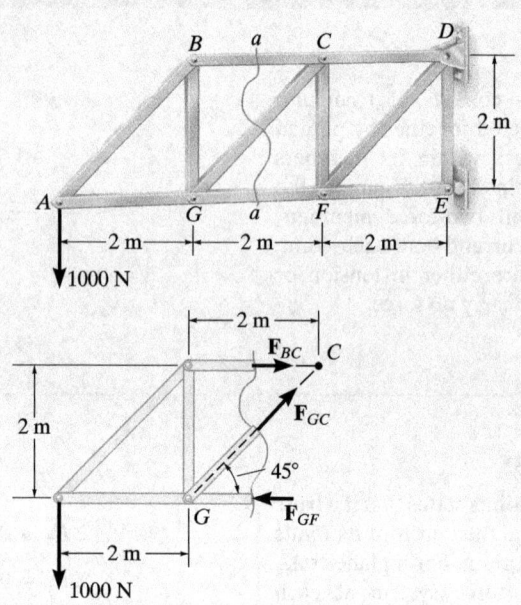

Sectioned members subjected to *pulling* are in *tension*, and those that are subjected to *pushing* are in *compression*.

Three equations of equilibrium are available to determine the unknowns.

$$\Sigma F_x = 0$$
$$\Sigma F_y = 0$$
$$\Sigma M_O = 0$$

If possible, sum forces in a direction that is perpendicular to two of the three unknown forces. This will yield a direct solution for the third force.

$$+\uparrow \Sigma F_y = 0$$
$$-1000 \text{ N} + F_{GC} \sin 45° = 0$$
$$F_{GC} = 1.41 \text{ kN (T)}$$

Sum moments about the point where the lines of action of two of the three unknown forces intersect, so that the third unknown force can be determined directly.

$$\zeta + \Sigma M_C = 0$$
$$1000 \text{ N}(4 \text{ m}) - F_{GF}(2 \text{ m}) = 0$$
$$F_{GF} = 2 \text{ kN (C)}$$

## Space Truss

A space truss is a three-dimensional truss built from tetrahedral elements, and is analyzed using the same methods as for plane trusses. The joints are assumed to be ball-and-socket connections.

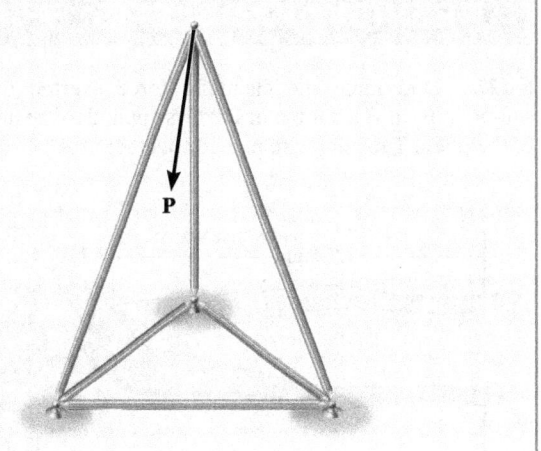

## Frames and Machines

Frames and machines are structures that contain one or more multiforce members, that is, members with three or more forces or couples acting on them. Frames are designed to support loads, and machines transmit and alter the effect of forces.

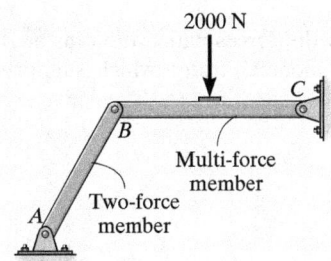

The forces acting at the joints of a frame or machine can be determined by drawing the free-body diagrams of each of its members or parts. The principle of action–reaction should be carefully observed when indicating these forces on the free-body diagram of each adjacent member or pin. For a coplanar force system, there are three equilibrium equations available for each member.

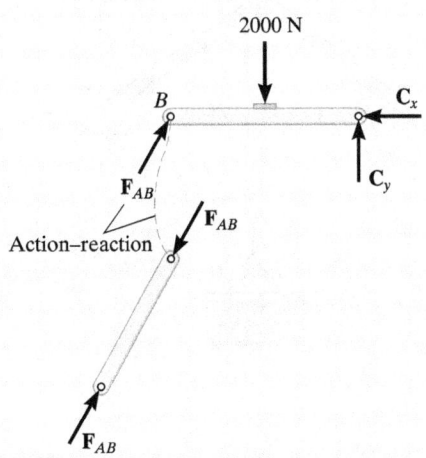

To simplify the analysis, be sure to recognize all two-force members. They have equal but opposite collinear forces at their ends.

## REVIEW PROBLEMS

**6–127.** Determine the clamping force exerted on the smooth pipe at $B$ if a force of 100 N is applied to the handles of the pliers. The pliers are pinned together at $A$.

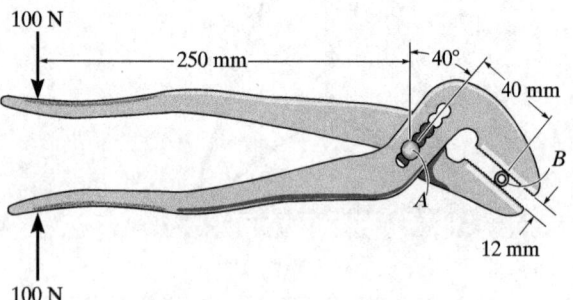

Prob. 6–127

**\*6–128.** Determine the forces which the pins at $A$ and $B$ exert on the two-member frame which supports the 100-kg crate.

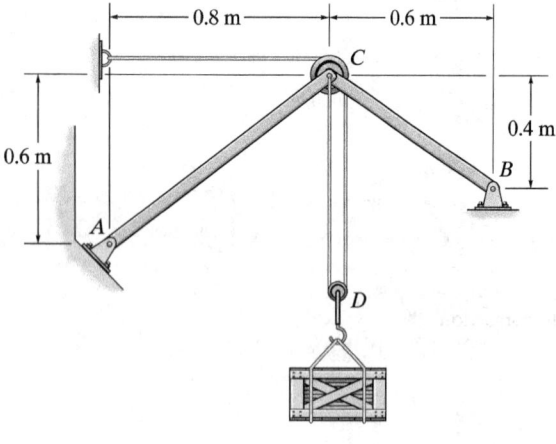

Prob. 6–128

**•6–129.** Determine the force in each member of the truss and state if the members are in tension or compression.

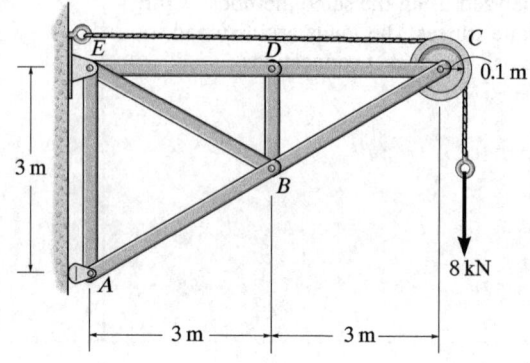

Prob. 6–129

**6–130.** The space truss is supported by a ball-and-socket joint at $D$ and short links at $C$ and $E$. Determine the force in each member and state if the members are in tension or compression. Take $\mathbf{F}_1 = \{-500\mathbf{k}\}$ kN and $\mathbf{F}_2 = \{400\mathbf{j}\}$ kN.

**6–131.** The space truss is supported by a ball-and-socket joint at $D$ and short links at $C$ and $E$. Determine the force in each member and state if the members are in tension or compression. Take $\mathbf{F}_1 = \{200\mathbf{i} + 300\mathbf{j} - 500\mathbf{k}\}$ kN and $\mathbf{F}_2 = \{400\mathbf{j}\}$ kN.

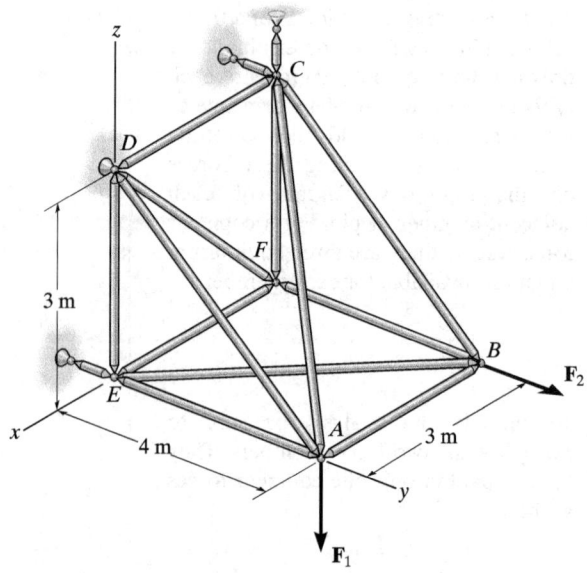

Probs. 6–130/131

**\*6–132.** Determine the horizontal and vertical components of reaction that the pins $A$ and $B$ exert on the two-member frame. Set $F = 0$.

**•6–133.** Determine the horizontal and vertical components of reaction that pins $A$ and $B$ exert on the two-member frame. Set $F = 500$ N.

**6–135.** Determine the horizontal and vertical components of reaction at the pin supports $A$ and $E$ of the compound beam assembly.

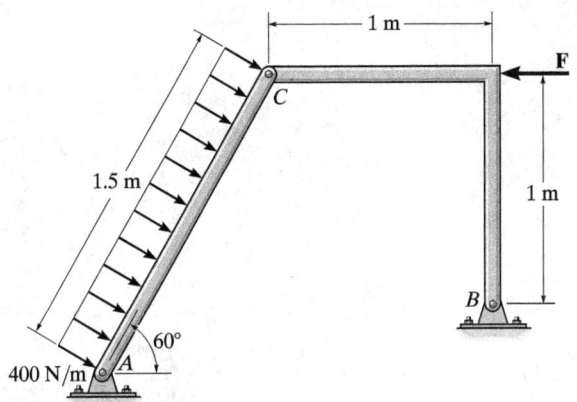

Probs. 6–132/133

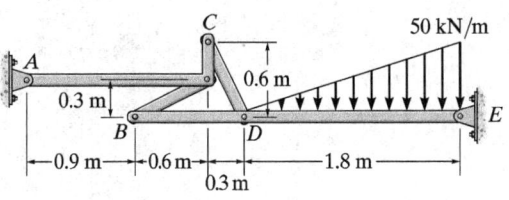

Prob. 6–135

**6–134.** The two-bar mechanism consists of a lever arm $AB$ and smooth link $CD$, which has a fixed smooth collar at its end $C$ and a roller at the other end $D$. Determine the force **P** needed to hold the lever in the position $\theta$. The spring has a stiffness $k$ and unstretched length $2L$. The roller contacts either the top or bottom portion of the horizontal guide.

**\*6–136.** Determine the force in members $AB$, $AD$, and $AC$ of the space truss and state if the members are in tension or compression.

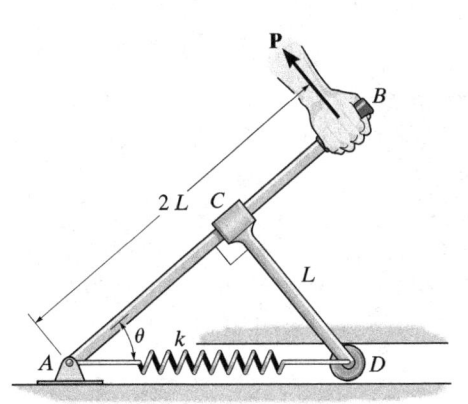

Prob. 6–134

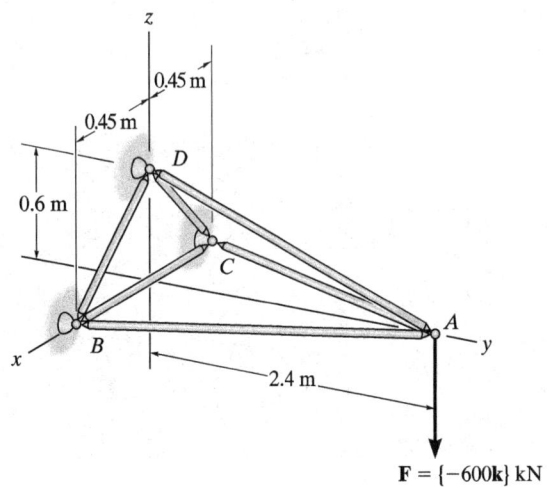

Prob. 6–136

The effective design of a brake system, such as the one for this bicycle, requires an efficient capacity for the mechanism to resist frictional forces. In this chapter, we will study the nature of friction and show how these forces are considered in engineering analysis and design.

# Friction

# 7

## CHAPTER OBJECTIVES

- To introduce the concept of dry friction and show how to analyze the equilibrium of rigid bodies subjected to this force.
- To present specific applications of frictional force analysis on wedges, screws, belts, and bearings.
- To investigate the concept of rolling resistance.

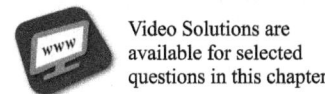

Video Solutions are available for selected questions in this chapter.

## 7.1 Characteristics of Dry Friction

*Friction* is a force that resists the movement of two contacting surfaces that slide relative to one another. This force always acts *tangent* to the surface at the points of contact and is directed so as to oppose the possible or existing motion between the surfaces.

In this chapter, we will study the effects of *dry friction*, which is sometimes called *Coulomb friction* since its characteristics were studied extensively by C. A. Coulomb in 1781. Dry friction occurs between the contacting surfaces of bodies when there is no lubricating fluid.*

The heat generated by the abrasive action of friction can be noticed when using this grinder to sharpen a metal blade.

*Another type of friction, called fluid friction, is studied in fluid mechanics.

# 330    CHAPTER 7    FRICTION

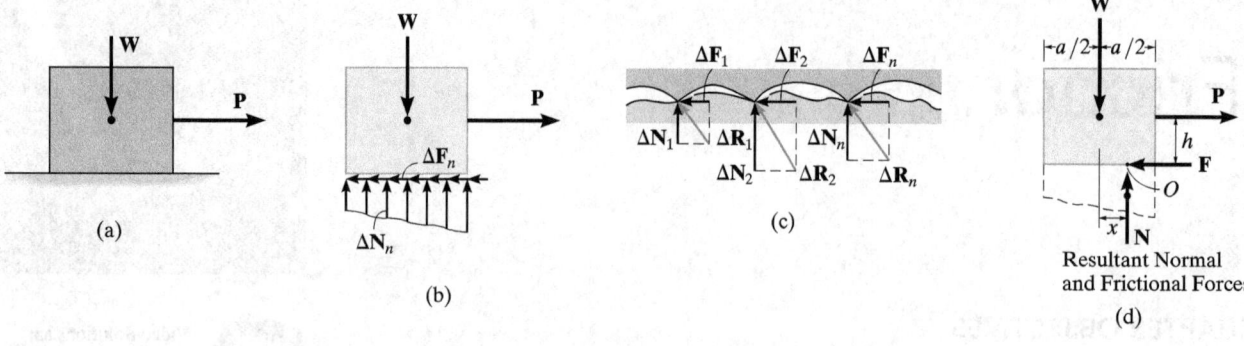

**Fig. 7–1**

Regardless of the weight of the rake or shovel that is suspended, the device has been designed so that the small roller holds the handle in equilibrium due to frictional forces that develop at the points of contact, $A, B, C$.

**Theory of Dry Friction.** The theory of dry friction can be explained by considering the effects caused by pulling horizontally on a block of uniform weight **W** which is resting on a rough horizontal surface that is *nonrigid or deformable*, Fig. 7–1a. The upper portion of the block, however, can be considered rigid. As shown on the free-body diagram of the block, Fig. 7–1b, the floor exerts an uneven *distribution* of both *normal force* $\Delta \mathbf{N}_n$ and *frictional force* $\Delta \mathbf{F}_n$ along the contacting surface. For equilibrium, the normal forces must act *upward* to balance the block's weight **W**, and the frictional forces act to the left to prevent the applied force **P** from moving the block to the right. Close examination of the contacting surfaces between the floor and block reveals how these frictional and normal forces develop, Fig. 7–1c. It can be seen that many microscopic irregularities exist between the two surfaces and, as a result, reactive forces $\Delta \mathbf{R}_n$ are developed at each point of contact.* As shown, each reactive force contributes both a frictional component $\Delta \mathbf{F}_n$ and a normal component $\Delta \mathbf{N}_n$.

**Equilibrium.** The effect of the *distributed* normal and frictional loadings is indicated by their *resultants* **N** and **F** on the free-body diagram, Fig. 7–1d. Notice that **N** acts a distance $x$ to the right of the line of action of **W**, Fig. 7–1d. This location, which coincides with the centroid or geometric center of the normal force distribution in Fig. 7–1b, is necessary in order to balance the "tipping effect" caused by **P**. For example, if **P** is applied at a height $h$ from the surface, Fig. 7–1d, then moment equilibrium about point $O$ is satisfied if $Wx = Ph$ or $x = Ph/W$.

---

*Besides mechanical interactions as explained here, which is referred to as a classical approach, a detailed treatment of the nature of frictional forces must also include the effects of temperature, density, cleanliness, and atomic or molecular attraction between the contacting surfaces. See J. Krim, *Scientific American*, October, 1996.

## 7.1 CHARACTERISTICS OF DRY FRICTION

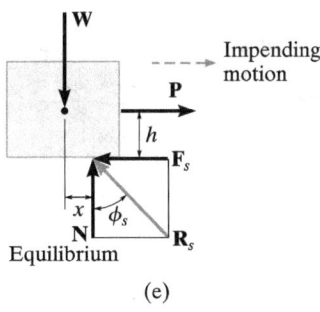

Equilibrium

(e)

**Fig. 7–1**

**Impending Motion.** In cases where the surfaces of contact are rather "slippery," the frictional force **F** may *not* be great enough to balance **P**, and consequently the block will tend to slip. In other words, as *P* is slowly increased, *F* correspondingly increases until it attains a certain *maximum value* $F_s$, called the *limiting static frictional force*, Fig. 7–1e. When this value is reached, the block is in *unstable equilibrium* since any further increase in *P* will cause the block to move. Experimentally, it has been determined that this limiting static frictional force $F_s$ is *directly proportional* to the resultant normal force *N*. Expressed mathematically,

$$F_s = \mu_s N \qquad (7\text{--}1)$$

where the constant of proportionality, $\mu_s$ (mu "sub" *s*), is called the *coefficient of static friction*.

Thus, when the block is on the *verge of sliding*, the normal force **N** and frictional force $\mathbf{F}_s$ combine to create a resultant $\mathbf{R}_s$, Fig. 7–1e. The angle $\phi_s$ (phi "sub" *s*) that $\mathbf{R}_s$ makes with **N** is called the *angle of static friction*. From the figure,

$$\phi_s = \tan^{-1}\left(\frac{F_s}{N}\right) = \tan^{-1}\left(\frac{\mu_s N}{N}\right) = \tan^{-1}\mu_s$$

Typical values for $\mu_s$ are given in Table 7–1. Note that these values can vary since experimental testing was done under variable conditions of roughness and cleanliness of the contacting surfaces. For applications, therefore, it is important that both caution and judgment be exercised when selecting a coefficient of friction for a given set of conditions. When a more accurate calculation of $F_s$ is required, the coefficient of friction should be determined directly by an experiment that involves the two materials to be used.

**Table 7–1**
**Typical Values for $\mu_s$**

| Contact Materials | Coefficient of Static Friction ($\mu_s$) |
|---|---|
| Metal on ice | 0.03–0.05 |
| Wood on wood | 0.30–0.70 |
| Leather on wood | 0.20–0.50 |
| Leather on metal | 0.30–0.60 |
| Aluminum on aluminium | 1.10–1.70 |

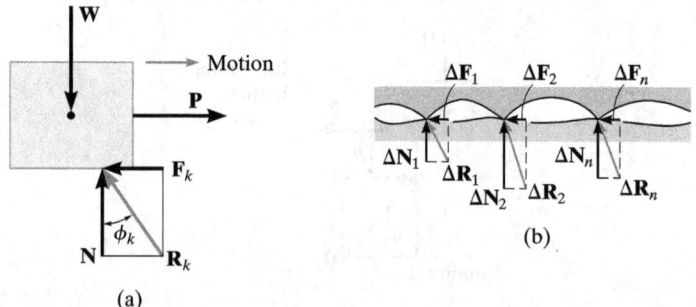

Fig. 7–2

**Motion.** If the magnitude of **P** acting on the block is increased so that it becomes slightly greater than $F_s$, the frictional force at the contacting surface will drop to a smaller value $F_k$, called the *kinetic frictional force*. The block will begin to slide with increasing speed, Fig. 7–2a. As this occurs, the block will "ride" on top of these peaks at the points of contact, as shown in Fig. 7–2b. The continued breakdown of the surface is the dominant mechanism creating kinetic friction.

Experiments with sliding blocks indicate that the magnitude of the kinetic friction force is directly proportional to the magnitude of the resultant normal force, expressed mathematically as

$$F_k = \mu_k N \tag{7-2}$$

Here the constant of proportionality, $\mu_k$, is called the *coefficient of kinetic friction*. Typical values for $\mu_k$ are approximately 25 percent *smaller* than those listed in Table 7–1 for $\mu_s$.

As shown in Fig. 7–2a, in this case, the resultant force at the surface of contact, $\mathbf{R}_k$, has a line of action defined by $\phi_k$. This angle is referred to as the *angle of kinetic friction*, where

$$\phi_k = \tan^{-1}\left(\frac{F_k}{N}\right) = \tan^{-1}\left(\frac{\mu_k N}{N}\right) = \tan^{-1}\mu_k$$

By comparison, $\phi_s \geq \phi_k$.

The above effects regarding friction can be summarized by referring to the graph in Fig. 7–3, which shows the variation of the frictional force $F$ versus the applied load $P$. Here, the frictional force is categorized in three different ways:

- $F$ is a *static frictional force* if equilibrium is maintained.

- $F$ is a *limiting static frictional force* $F_s$ when it reaches a maximum value needed to maintain equilibrium.

- $F$ is termed a *kinetic frictional force* $F_k$ when sliding occurs at the contacting surface.

Notice also from the graph that for very large values of $P$ or for high speeds, aerodynamic effects will cause $F_k$ and likewise $\mu_k$ to begin to decrease.

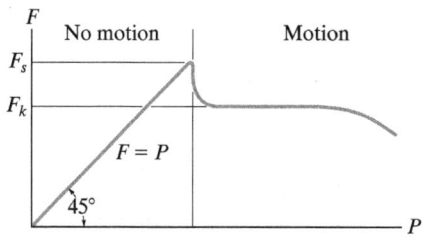

Fig. 7–3

## Characteristics of Dry Friction.
As a result of *experiments* that pertain to the foregoing discussion, we can state the following rules which apply to bodies subjected to dry friction.

- The frictional force acts *tangent* to the contacting surfaces in a direction *opposed* to the *motion* or tendency for motion of one surface relative to another.

- The maximum static frictional force $F_s$ that can be developed is independent of the area of contact, provided the normal pressure is not very low nor great enough to severely deform or crush the contacting surfaces of the bodies.

- The maximum static frictional force is generally greater than the kinetic frictional force for any two surfaces of contact. However, if one of the bodies is moving with a *very low velocity* over the surface of another, $F_k$ becomes approximately equal to $F_s$, i.e., $\mu_s \approx \mu_k$.

- When *slipping* at the surface of contact is *about to occur*, the maximum static frictional force is proportional to the normal force, such that $F_s = \mu_s N$.

- When *slipping* at the surface of contact is *occurring*, the kinetic frictional force is proportional to the normal force, such that $F_k = \mu_k N$.

# 7.2 Problems Involving Dry Friction

If a rigid body is in equilibrium when it is subjected to a system of forces that includes the effect of friction, the force system must satisfy not only the equations of equilibrium but *also* the laws that govern the frictional forces.

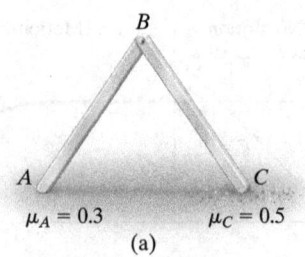

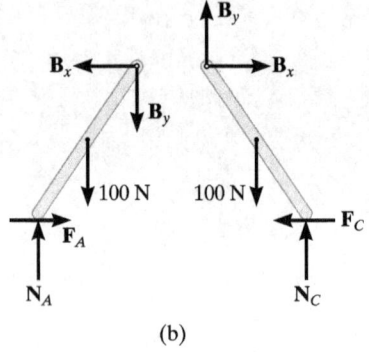

Fig. 7–4

**Types of Friction Problems.** In general, there are three types of mechanics problems involving dry friction. They can easily be classified once free-body diagrams are drawn and the total number of unknowns are identified and compared with the total number of available equilibrium equations.

**No Apparent Impending Motion.** Problems in this category are strictly equilibrium problems, which require the number of unknowns to be *equal* to the number of available equilibrium equations. Once the frictional forces are determined from the solution, however, their numerical values must be checked to be sure they satisfy the inequality $F \leq \mu_s N$; otherwise, slipping will occur and the body will not remain in equilibrium. A problem of this type is shown in Fig. 7–4a. Here, we must determine the frictional forces at $A$ and $C$ to check if the equilibrium position of the two-member frame can be maintained. If the bars are uniform and have known weights of 100 N each, then the free-body diagrams are as shown in Fig. 7–4b. There are six unknown force components which can be determined *strictly* from the six equilibrium equations (three for each member). Once $F_A$, $N_A$, $F_C$, and $N_C$ are determined, then the bars will remain in equilibrium provided $F_A \leq 0.3 N_A$ and $F_C \leq 0.5 N_C$ are satisfied.

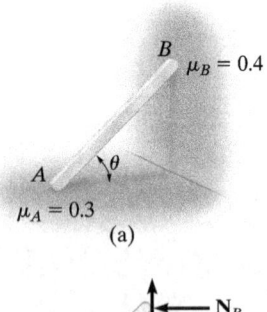

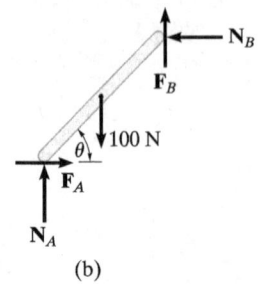

Fig. 7–5

**Impending Motion at All Points of Contact.** In this case, the total number of unknowns will *equal* the total number of available equilibrium equations *plus* the total number of available frictional equations, $F = \mu N$. When *motion is impending* at the points of contact, then $F_s = \mu_s N$; whereas if the body is *slipping*, then $F_k = \mu_k N$. For example, consider the problem of finding the smallest angle $\theta$ at which the 100-N bar in Fig. 7–5a can be placed against the wall without slipping. The free-body diagram is shown in Fig. 7–5b. Here, the *five* unknowns are determined from the *three* equilibrium equations and *two* static frictional equations which apply at *both* points of contact, so that $F_A = 0.3 N_A$ and $F_B = 0.4 N_B$.

**Impending Motion at Some Points of Contact.** Here, the number of unknowns will be *less* than the number of available equilibrium equations plus the number of available frictional equations or conditional equations for tipping. As a result, several possibilities for motion or impending motion will exist and the problem will involve a determination of the kind of motion which actually occurs. For example, consider the two-member frame in Fig. 7–6a. In this problem, we wish to determine the horizontal force $P$ needed to cause movement. If each member has a weight of 100 N, then the free-body diagrams are as shown in Fig. 7–6b. There are *seven* unknowns. For a unique solution, we must satisfy the *six* equilibrium equations (three for each member) and only *one* of two possible static frictional equations. This means that as $P$ increases it will either cause slipping at $A$ and no slipping at $C$, so that $F_A = 0.3N_A$ and $F_C \le 0.5N_C$; or slipping occurs at $C$ and no slipping at $A$, in which case $F_C = 0.5N_C$ and $F_A \le 0.3N_A$. The actual situation can be determined by calculating $P$ for each case and then choosing the case for which $P$ is *smaller*. If in both cases the *same value* for $P$ is calculated, which in practice would be highly improbable, then slipping at both points occurs simultaneously; i.e., the *seven unknowns* would satisfy *eight equations*.

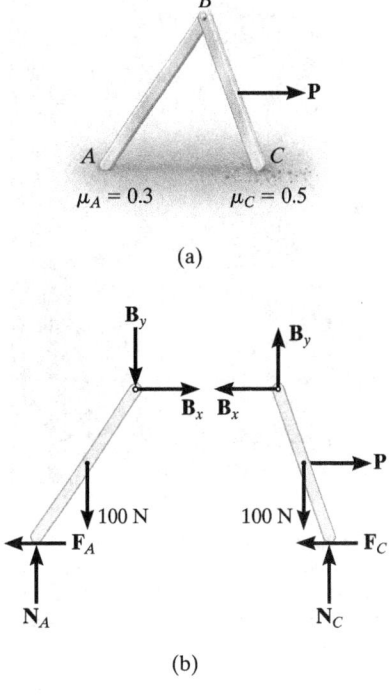

Fig. 7–6

 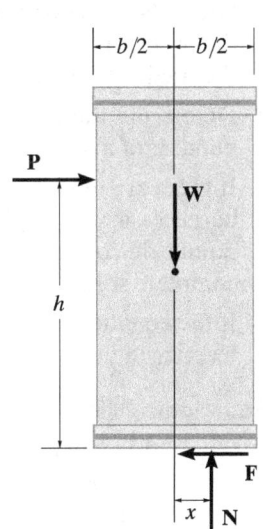 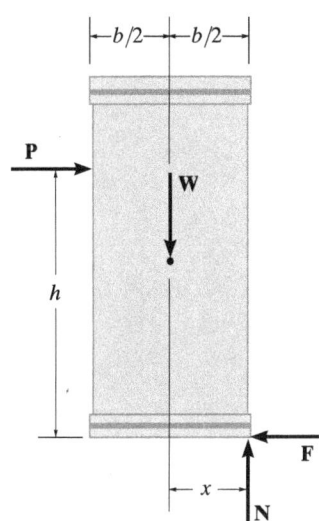

Consider pushing on the uniform crate that has a weight $W$ and sits on the rough surface. As shown on the first free-body diagram, if the magnitude of **P** is small, the crate will remain in equilibrium. As $P$ increases the crate will either be on the verge of slipping on the surface ($F = \mu_s N$), or if the surface is very rough (large $\mu_s$) then the resultant normal force will shift to the corner, $x = b/2$, as shown on the second free-body diagram. At this point, the crate will begin to tip over. The crate also has a greater chance of tipping if **P** is applied at a greater height $h$ above the surface, or if its width $b$ is smaller.

**Equilibrium Versus Frictional Equations.** Whenever we solve problems where the friction force $F$ is to be an "equilibrium force" and satisfies the inequality $F < \mu_s N$, then we can assume the sense of direction of $F$ on the free-body diagram. The correct sense is made known *after* solving the equations of equilibrium for $F$. If $F$ is a negative scalar, the sense of **F** is the reverse of that which was assumed. This convenience of *assuming* the sense of **F** is possible because the equilibrium equations equate to zero with the *components of vectors* acting in the *same direction*. However, in cases where the frictional equation $F = \mu N$ is used in the solution of a problem, the convenience of *assuming* the sense of **F** is *lost*, since the frictional equation relates only the *magnitudes* of two *perpendicular* vectors. Consequently, **F** *must always* be shown acting with its *correct sense* on the free-body diagram, *whenever* the frictional equation is used for the solution of a problem.

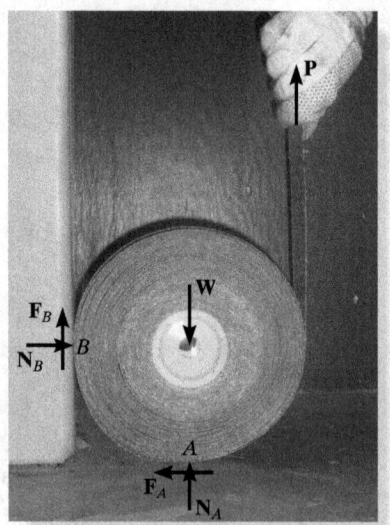

The applied vertical force **P** on this roll must be large enough to overcome the resistance of friction at the contacting surfaces $A$ and $B$ in order to cause rotation.

## Procedure for Analysis

Equilibrium problems involving dry friction can be solved using the following procedure.

### Free-Body Diagrams.

- Draw the necessary free-body diagrams, and unless it is stated in the problem that impending motion or slipping occurs, *always* show the frictional forces as unknowns (i.e., *do not assume* $F = \mu N$).

- Determine the number of unknowns and compare this with the number of available equilibrium equations.

- If there are more unknowns than equations of equilibrium, it will be necessary to apply the frictional equation at some, if not all, points of contact to obtain the extra equations needed for a complete solution.

- If the equation $F = \mu N$ is to be used, it will be necessary to show **F** acting in the correct sense of direction on the free-body diagram.

### Equations of Equilibrium and Friction.

- Apply the equations of equilibrium and the necessary frictional equations (or conditional equations if tipping is possible) and solve for the unknowns.

- If the problem involves a three-dimensional force system such that it becomes difficult to obtain the force components or the necessary moment arms, apply the equations of equilibrium using Cartesian vectors.

## EXAMPLE 7.1

The uniform crate shown in Fig. 7–7a has a mass of 20 kg. If a force $P = 80$ N is applied to the crate, determine if it remains in equilibrium. The coefficient of static friction is $\mu_s = 0.3$.

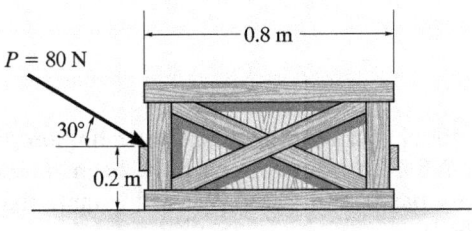

(a)

Fig. 7–7

### SOLUTION

**Free-Body Diagram.** As shown in Fig. 7–7b, the *resultant* normal force $\mathbf{N}_C$ must act a distance $x$ from the crate's center line in order to counteract the tipping effect caused by $\mathbf{P}$. There are *three unknowns*, $F$, $N_C$, and $x$, which can be determined strictly from the *three* equations of equilibrium.

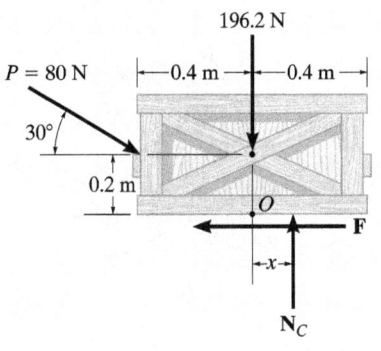

(b)

### Equations of Equilibrium.

$\stackrel{+}{\rightarrow} \Sigma F_x = 0;$  $\qquad 80 \cos 30° \text{ N} - F = 0$

$+\uparrow \Sigma F_y = 0;$  $\quad -80 \sin 30° \text{ N} + N_C - 196.2 \text{ N} = 0$

$\zeta + \Sigma M_O = 0;$  $\quad 80 \sin 30° \text{ N}(0.4 \text{ m}) - 80 \cos 30° \text{ N}(0.2 \text{ m}) + N_C(x) = 0$

Solving,

$$F = 69.3 \text{ N}$$
$$N_C = 236 \text{ N}$$
$$x = -0.00908 \text{ m} = -9.08 \text{ mm}$$

Since $x$ is negative it indicates the *resultant* normal force acts (slightly) to the *left* of the crate's center line. No tipping will occur since $x < 0.4$ m. Also, the *maximum* frictional force which can be developed at the surface of contact is $F_{max} = \mu_s N_C = 0.3(236 \text{ N}) = 70.8$ N. Since $F = 69.3$ N $< 70.8$ N, the crate will *not slip*, although it is very close to doing so.

## EXAMPLE 7.2

(a)

It is observed that when the bed of the dump truck is raised to an angle of $\theta = 25°$ the vending machines will begin to slide off the bed, Fig. 7–8a. Determine the static coefficient of friction between a vending machine and the surface of the truckbed.

### SOLUTION
An idealized model of a vending machine resting on the truckbed is shown in Fig. 7–8b. The dimensions have been measured and the center of gravity has been located. We will assume that the vending machine weighs $W$.

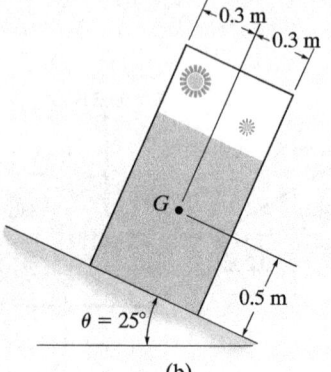

(b)

**Free-Body Diagram.** As shown in Fig. 7–8c, the dimension $x$ is used to locate the position of the resultant normal force **N**. There are four unknowns, $N, F, \mu_s$, and $x$.

**Equations of Equilibrium.**

$$+\searrow \Sigma F_x = 0; \qquad W \sin 25° - F = 0 \qquad (1)$$
$$+\nearrow \Sigma F_y = 0; \qquad N - W \cos 25° = 0 \qquad (2)$$
$$\zeta + \Sigma M_O = 0; \quad -W \sin 25°(0.5 \text{ m}) + W \cos 25°(x) = 0 \qquad (3)$$

Since slipping impends at $\theta = 25°$, using Eqs. 1 and 2, we have

$$F_s = \mu_s N; \qquad W \sin 25° = \mu_s(W \cos 25°)$$
$$\mu_s = \tan 25° = 0.466 \qquad \textit{Ans.}$$

The angle of $\theta = 25°$ is referred to as the *angle of repose*, and by comparison, it is equal to the angle of static friction, $\theta = \phi_s$. Notice from the calculation that $\theta$ is independent of the weight of the vending machine, and so knowing $\theta$ provides a convenient method for determining the coefficient of static friction.

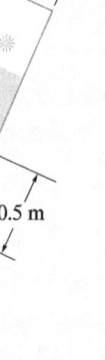

(c)

**Fig. 7–8**

**NOTE:** From Eq. 3, we find $x = 0.233$ m. Since $0.233 \text{ m} < 0.5 \text{ m}$, indeed the vending machine will slip before it can tip as observed in Fig. 7–8a.

## EXAMPLE 7.3

The uniform 10-kg ladder in Fig. 7–9a rests against the smooth wall at B, and the end A rests on the rough horizontal plane for which the coefficient of static friction is $\mu_s = 0.3$. Determine the angle of inclination $\theta$ of the ladder and the normal reaction at B if the ladder is on the verge of slipping.

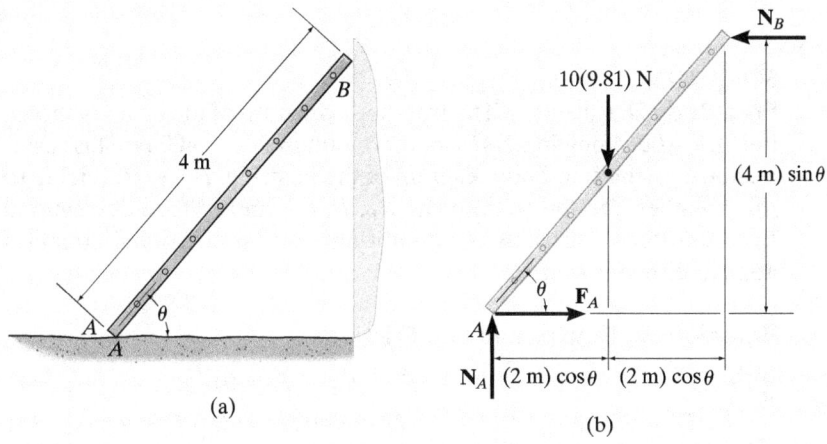

Fig. 7–9

### SOLUTION
**Free-Body Diagram.** As shown on the free-body diagram, Fig. 7–9b, the frictional force $\mathbf{F}_A$ must act to the right since impending motion at A is to the left.

**Equations of Equilibrium and Friction.** Since the ladder is on the verge of slipping, then $F_A = \mu_s N_A = 0.3 N_A$. By inspection, $N_A$ can be obtained directly.

$+\uparrow \Sigma F_y = 0;$ $\qquad N_A - 10(9.81)\text{ N} = 0 \qquad N_A = 98.1 \text{ N}$

Using this result, $F_A = 0.3(98.1 \text{ N}) = 29.43 \text{ N}$. Now $N_B$ can be found.

$\xrightarrow{+} \Sigma F_x = 0;$ $\qquad 29.43 \text{ N} - N_B = 0$

$\qquad\qquad N_B = 29.43 \text{ N} = 29.4 \text{ N} \qquad\qquad Ans.$

Finally, the angle $\theta$ can be determined by summing moments about point A.

$\zeta + \Sigma M_A = 0;$ $\qquad (29.43 \text{ N})(4 \text{ m}) \sin \theta - [10(9.81) \text{ N}](2 \text{ m}) \cos \theta = 0$

$\qquad\qquad \dfrac{\sin \theta}{\cos \theta} = \tan \theta = 1.6667$

$\qquad\qquad \theta = 59.04° = 59.0° \qquad\qquad Ans.$

## EXAMPLE 7.4

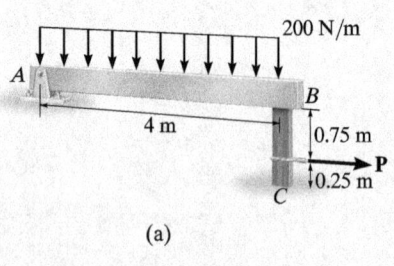

(a)

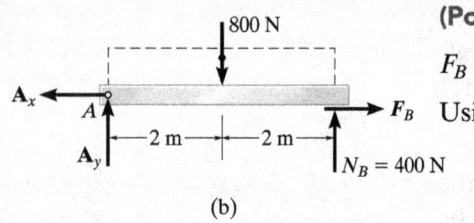

(b)

Fig. 7-10

Beam $AB$ is subjected to a uniform load of 200 N/m and is supported at $B$ by post $BC$, Fig. 7–10a. If the coefficients of static friction at $B$ and $C$ are $\mu_B = 0.2$ and $\mu_C = 0.5$, determine the force **P** needed to pull the post out from under the beam. Neglect the weight of the members and the thickness of the beam.

### SOLUTION

**Free-Body Diagrams.** The free-body diagram of the beam is shown in Fig. 7–10b. Applying $\Sigma M_A = 0$, we obtain $N_B = 400$ N. This result is shown on the free-body diagram of the post, Fig. 7–10c. Referring to this member, the *four* unknowns $F_B$, $P$, $F_C$, and $N_C$ are determined from the *three* equations of equilibrium and *one* frictional equation applied either at $B$ or $C$.

### Equations of Equilibrium and Friction.

$$\xrightarrow{+} \Sigma F_x = 0; \qquad P - F_B - F_C = 0 \qquad (1)$$

$$+\uparrow \Sigma F_y = 0; \qquad N_C - 400 \text{ N} = 0 \qquad (2)$$

$$\zeta + \Sigma M_C = 0; \qquad -P(0.25 \text{ m}) + F_B(1 \text{ m}) = 0 \qquad (3)$$

**(Post Slips at B and Rotates about C.)** This requires $F_C \leq \mu_C N_C$ and

$$F_B = \mu_B N_B; \qquad F_B = 0.2(400 \text{ N}) = 80 \text{ N}$$

Using this result and solving Eqs. 1 through 3, we obtain

$$P = 320 \text{ N}$$
$$F_C = 240 \text{ N}$$
$$N_C = 400 \text{ N}$$

Since $F_C = 240$ N $> \mu_C N_C = 0.5(400$ N$) = 200$ N, slipping at $C$ occurs. Thus, the other case of movement must be investigated.

**(Post Slips at C and Rotates about B.)** Here $F_B \leq \mu_B N_B$ and

$$F_C = \mu_C N_C; \qquad F_C = 0.5 N_C \qquad (4)$$

Solving Eqs. 1 through 4 yields

$$P = 267 \text{ N} \qquad \qquad Ans.$$
$$N_C = 400 \text{ N}$$
$$F_C = 200 \text{ N}$$
$$F_B = 66.7 \text{ N}$$

Obviously, this case occurs first since it requires a *smaller* value for $P$.

# EXAMPLE 7.5

Blocks $A$ and $B$ have a mass of 3 kg and 9 kg, respectively, and are connected to the weightless links shown in Fig. 7–11a. Determine the largest vertical force **P** that can be applied at the pin $C$ without causing any movement. The coefficient of static friction between the blocks and the contacting surfaces is $\mu_s = 0.3$.

## SOLUTION

**Free-Body Diagram.** The links are two-force members and so the free-body diagrams of pin $C$ and blocks $A$ and $B$ are shown in Fig. 7–11b. Since the horizontal component of $\mathbf{F}_{AC}$ tends to move block $A$ to the left, $\mathbf{F}_A$ must act to the right. Similarly, $\mathbf{F}_B$ must act to the left to oppose the tendency of motion of block $B$ to the right, caused by $\mathbf{F}_{BC}$. There are seven unknowns and six available force equilibrium equations, two for the pin and two for each block, so that *only one* frictional equation is needed.

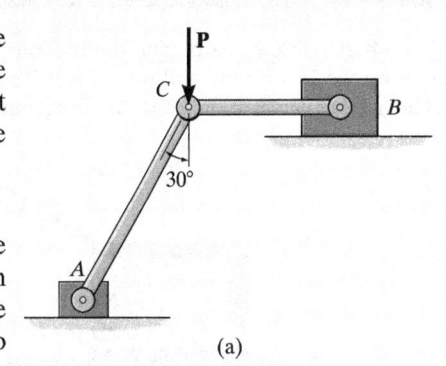

(a)

**Equations of Equilibrium and Friction.** The force in links $AC$ and $BC$ can be related to $P$ by considering the equilibrium of pin $C$.

$+\uparrow \Sigma F_y = 0;$ $\quad F_{AC} \cos 30° - P = 0;$ $\quad F_{AC} = 1.155P$

$\xrightarrow{+} \Sigma F_x = 0;$ $\quad 1.155P \sin 30° - F_{BC} = 0;$ $\quad F_{BC} = 0.5774P$

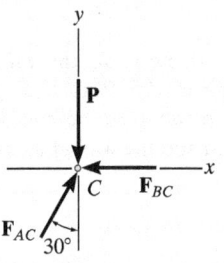

Using the result for $F_{AC}$, for block $A$,

$\xrightarrow{+} \Sigma F_x = 0;$ $\quad F_A - 1.155P \sin 30° = 0;$ $\quad F_A = 0.5774P \quad (1)$

$+\uparrow \Sigma F_y = 0;$ $\quad N_A - 1.155P \cos 30° - 3(9.81 \text{ N}) = 0;$

$\quad N_A = P + 29.43 \text{ N} \quad (2)$

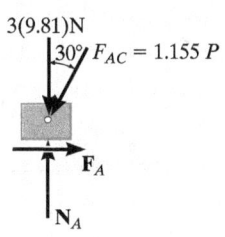

Using the result for $F_{BC}$, for block $B$,

$\xrightarrow{+} \Sigma F_x = 0;$ $\quad (0.5774P) - F_B = 0;$ $\quad F_B = 0.5774P \quad (3)$

$+\uparrow \Sigma F_y = 0;$ $\quad N_B - 9(9.81) \text{ N} = 0;$ $\quad N_B = 88.29 \text{ N}$

Movement of the system may be caused by the initial slipping of *either* block $A$ or block $B$. If we assume that block $A$ slips first, then

$$F_A = \mu_s N_A = 0.3 N_A \quad (4)$$

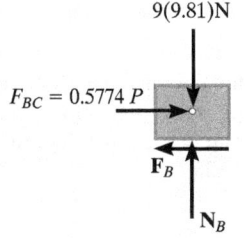

Substituting Eqs. 1 and 2 into Eq. 4,

$$0.5774P = 0.3(P + 29.43)$$
$$P = 31.8 \text{ N} \quad \textit{Ans.}$$

Substituting this result into Eq. 3, we obtain $F_B = 18.4$ N. Since the maximum static frictional force at $B$ is $(F_B)_{max} = \mu_s N_B = 0.3(88.29 \text{ N}) = 26.5 \text{ N} > F_B$, block $B$ will not slip. Thus, the above assumption is correct. Notice that if the inequalities were not satisfied, we would have to assume slipping of block $B$ and then solve for $P$.

(b)

Fig. 7–11

## FUNDAMENTAL PROBLEMS

**F7–1.** If $P = 200$ N, determine the friction that developed between the 50-kg crate and the ground. The coefficient of static friction between the crate and the ground is $\mu_s = 0.3$.

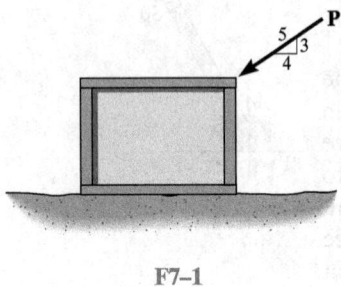

F7–1

**F7–2.** Determine the minimum force $P$ to prevent the 30-kg rod $AB$ from sliding. The contact surface at $B$ is smooth, whereas the coefficient of static friction between the rod and the wall at $A$ is $\mu_s = 0.2$.

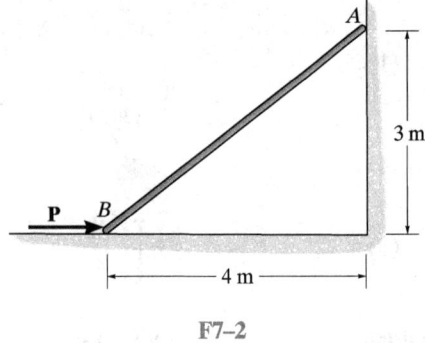

F7–2

**F7–3.** Determine the maximum force $P$ that can be applied without causing the two 50-kg crates to move. The coefficient of static friction between each crate and the ground is $\mu_s = 0.25$.

F7–3

**F7–4.** If the coefficient of static friction at contact points $A$ and $B$ is $\mu_s = 0.3$, determine the maximum force $P$ that can be applied without causing the 100-kg spool to move.

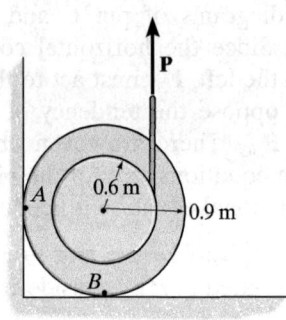

F7–4

**F7–5.** Determine the minimum force $P$ that can be applied without causing movement of the 125-kg crate which has a center of gravity at $G$. The coefficient of static friction at the floor is $\mu_s = 0.4$.

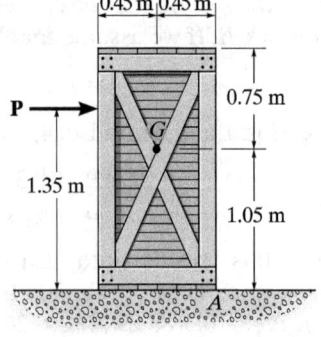

F7–5

## PROBLEMS

**•7–1.** Determine the minimum horizontal force $P$ required to hold the crate from sliding down the plane. The crate has a mass of 50 kg and the coefficient of static friction between the crate and the plane is $\mu_s = 0.25$.

**7–2.** Determine the minimum force $P$ required to push the crate up the plane. The crate has a mass of 50 kg and the coefficient of static friction between the crate and the plane is $\mu_s = 0.25$.

**7–3.** A horizontal force of $P = 100$ N is just sufficient to hold the crate from sliding down the plane, and a horizontal force of $P = 350$ N is required to just push the crate up the plane. Determine the coefficient of static friction between the plane and the crate, and find the mass of the crate.

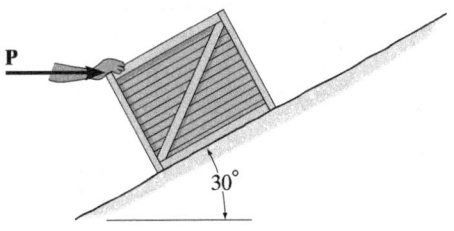

Probs. 7–1/2/3

**\*7–4.** If the coefficient of static friction at $A$ is $\mu_s = 0.4$ and the collar at $B$ is smooth so it only exerts a horizontal force on the pipe, determine the minimum distance $x$ so that the bracket can support the cylinder of any mass without slipping. Neglect the mass of the bracket.

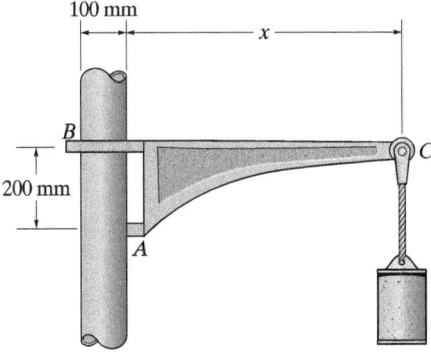

Prob. 7–4

**•7–5.** The 90-kg man climbs up the ladder and stops at the position shown after he senses that the ladder is on the verge of slipping. Determine the inclination $\theta$ of the ladder if the coefficient of static friction between the friction pad $A$ and the ground is $\mu_s = 0.4$. Assume the wall at $B$ is smooth. The center of gravity for the man is at $G$. Neglect the weight of the ladder.

**7–6.** The 90-kg man climbs up the ladder and stops at the position shown after he senses that the ladder is on the verge of slipping. Determine the coefficient of static friction between the friction pad at $A$ and ground if the inclination of the ladder is $\theta = 60°$ and the wall at $B$ is smooth. The center of gravity for the man is at $G$. Neglect the weight of the ladder.

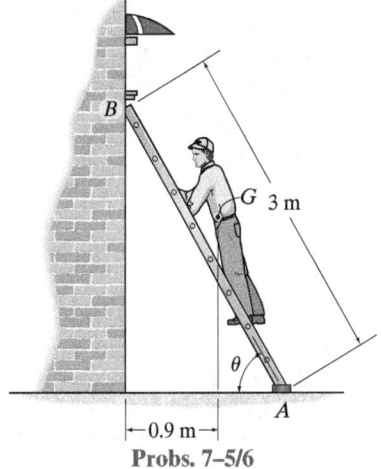

Probs. 7–5/6

**7–7.** The uniform thin pole has a weight of 150 N and a length of 7.8 m. If it is placed against the smooth wall and on the rough floor in the position $d = 3$ m, will it remain in this position when it is released? The coefficient of static friction is $\mu_s = 0.3$.

**\*7–8.** The uniform pole has a weight of 150 N and a length of 7.8 m. Determine the maximum distance $d$ it can be placed from the smooth wall and not slip. The coefficient of static friction between the floor and the pole is $\mu_s = 0.3$.

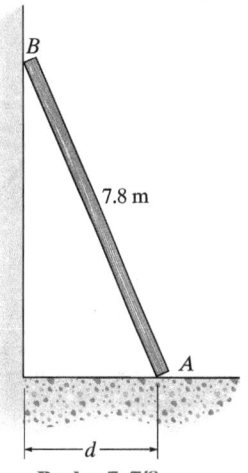

Probs. 7–7/8

**•7–9.** If the coefficient of static friction at all contacting surfaces is $\mu_s$, determine the inclination $\theta$ at which the identical blocks, each of weight $W$, begin to slide.

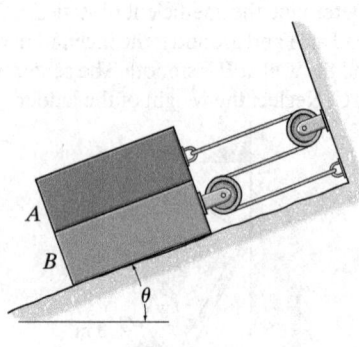

Prob. 7–9

**7–10.** The uniform 10-kg ladder rests on the rough floor for which the coefficient of static friction is $\mu_s = 0.8$ and against the smooth wall at $B$. Determine the horizontal force $P$ the man must exert on the ladder in order to cause it to move.

**7–11.** The uniform 10-kg ladder rests on the rough floor for which the coefficient of static friction is $\mu_s = 0.4$ and against the smooth wall at $B$. Determine the horizontal force $P$ the man must exert on the ladder in order to cause it to move.

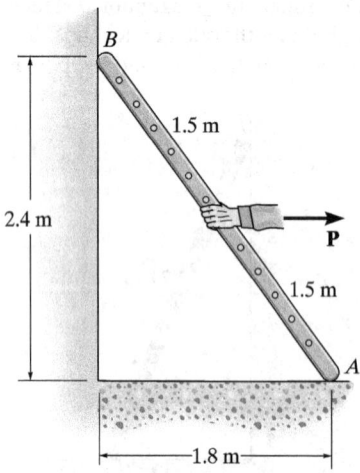

Probs. 7 10/11

**\*7–12.** The coefficients of static and kinetic friction between the drum and brake bar are $\mu_s = 0.4$ and $\mu_k = 0.3$, respectively. If $M = 50$ N·m and $P = 85$ N determine the horizontal and vertical components of reaction at the pin $O$. Neglect the weight and thickness of the brake. The drum has a mass of 25 kg.

**•7–13.** The coefficient of static friction between the drum and brake bar is $\mu_s = 0.4$. If the moment $M = 35$ N·m, determine the smallest force $P$ that needs to be applied to the brake bar in order to prevent the drum from rotating. Also, determine the corresponding horizontal and vertical components of reaction at pin $O$. Neglect the weight and thickness of the brake bar. The drum has a mass of 25 kg.

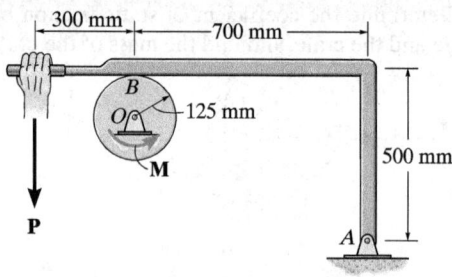

Probs. 7–12/13

**7–14.** Determine the minimum coefficient of static friction between the uniform 50-kg spool and the wall so that the spool does not slip.

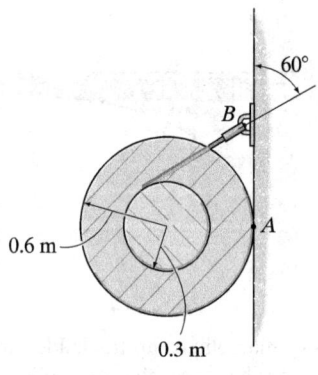

Prob. 7–14

**7–15.** The spool has a mass of 200 kg and rests against the wall and on the floor. If the coefficient of static friction at $B$ is $(\mu_s)_B = 0.3$, the coefficient of kinetic friction is $(\mu_k)_B = 0.2$, and the wall is smooth, determine the friction force developed at $B$ when the vertical force applied to the cable is $P = 800$ N.

**7–18.** The tongs are used to lift the 150-kg crate, whose center of mass is at $G$. Determine the least coefficient of static friction at the pivot blocks so that the crate can be lifted.

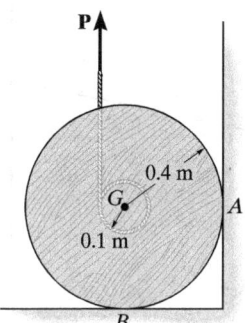

Prob. 7–15

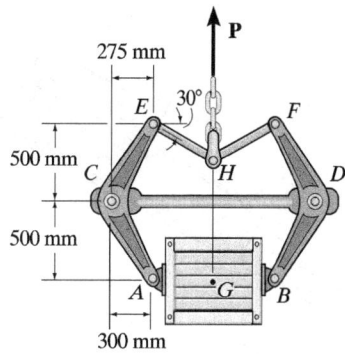

Prob. 7–18

*****7–16.** The 40-kg boy stands on the beam and pulls on the cord with a force large enough to just cause him to slip. If the coefficient of static friction between his shoes and the beam is $(\mu_s)_D = 0.4$, determine the reactions at $A$ and $B$. The beam is uniform and has a weight of 500 N. Neglect the size of the pulleys and the thickness of the beam.

**•7–17.** The 40-kg boy stands on the beam and pulls with a force of 200 N. If $(\mu_s)_D = 0.4$, determine the frictional force between his shoes and the beam and the reactions at $A$ and $B$. The beam is uniform and has a weight of 500 N. Neglect the size of the pulleys and the thickness of the beam.

**7–19.** Two blocks $A$ and $B$ have a weight of 50 N and 30 N, respectively. They are resting on the incline for which the coefficients of static friction are $\mu_A = 0.15$ and $\mu_B = 0.25$. Determine the incline angle $\theta$ for which both blocks begin to slide. Also, find the required stretch or compression in the connecting spring for this to occur. The spring has a stiffness of $k = 40$ N/m.

*****7–20.** Two blocks $A$ and $B$ have a weight of 50 N and 30 N, respectively. They are resting on the incline for which the coefficients of static friction are $\mu_A = 0.15$ and $\mu_B = 0.25$. Determine the angle $\theta$ which will cause motion of one of the blocks. What is the friction force under each of the blocks when this occurs? The spring has a stiffness of $k = 40$ N/m and is originally unstretched.

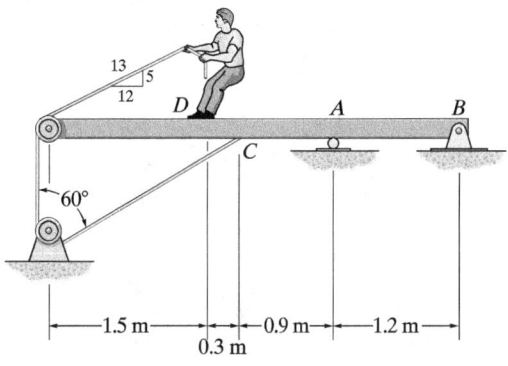

Probs. 7–16/17

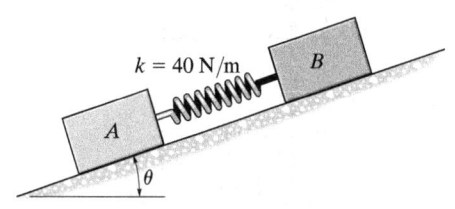

Probs. 7–19/20

**•7–21.** Crates $A$ and $B$ weigh 1000 N and 750 N, respectively. They are connected together with a cable and placed on the inclined plane. If the angle $\theta$ is gradually increased, determine $\theta$ when the crates begin to slide. The coefficients of static friction between the crates and the plane are $\mu_A = 0.25$ and $\mu_B = 0.35$.

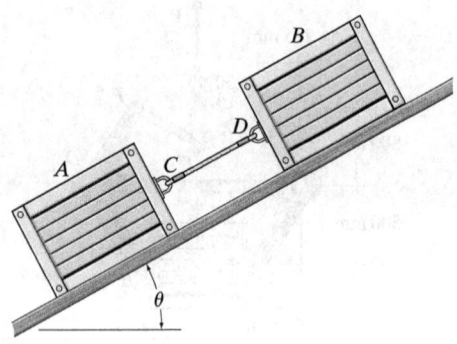

Prob. 7–21

**7–22.** A man attempts to support a stack of books horizontally by applying a compressive force of $F = 120$ N to the ends of the stack with his hands. If each book has a mass of 0.95 kg, determine the greatest number of books that can be supported in the stack. The coefficient of static friction between the man's hands and a book is $(\mu_s)_h = 0.6$ and between any two books $(\mu_s)_b = 0.4$.

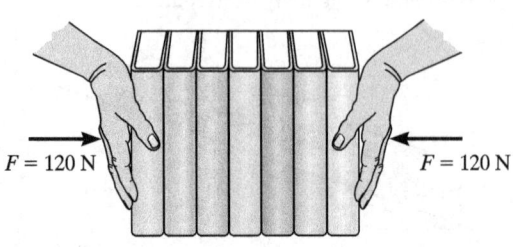

Prob. 7–22

**7–23.** The paper towel dispenser carries two rolls of paper. The one in use is called the stub roll $A$ and the other is the fresh roll $B$. They weigh 10 N and 25 N, respectively. If the coefficients of static friction at the points of contact $C$ and $D$ are $(\mu_s)_C = 0.2$ and $(\mu_s)_D = 0.5$, determine the initial vertical force $P$ that must be applied to the paper on the stub roll in order to pull down a sheet. The stub roll is pinned in the center, whereas the fresh roll is not. Neglect friction at the pin.

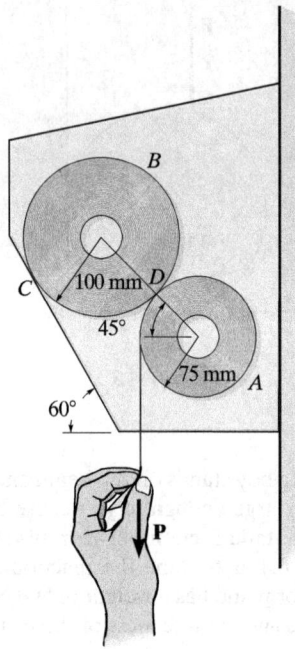

Prob. 7–23

**\*7–24.** The drum has a weight of 500 N and rests on the floor for which the coefficient of static friction is $\mu_s = 0.6$. If $a = 0.6$ m and $b = 0.9$ m, determine the smallest magnitude of the force $P$ that will cause impending motion of the drum.

**•7–25.** The drum has a weight of 500 N and rests on the floor for which the coefficient of static friction is $\mu_s = 0.5$. If $a = 0.9$ m and $b = 1.2$ m, determine the smallest magnitude of the force $P$ that will cause impending motion of the drum.

Probs. 7–24/25

**7–26.** The refrigerator has a weight of 900 N ($\approx$ 90 kg) and rests on a tile floor for which $\mu_s = 0.25$. If the man pushes horizontally on the refrigerator in the direction shown, determine the smallest magnitude of horizontal force needed to move it. Also, if the man has a weight of 750 N ($\approx$ 75 kg), determine the smallest coefficient of friction between his shoes and the floor so that he does not slip.

**7–27.** The refrigerator has a weight of 900 N ($\approx$ 90 kg) and rests on a tile floor for which $\mu_s = 0.25$. Also, the man has a weight of 750 N ($\approx$ 75 kg) and the coefficient of static friction between the floor and his shoes is $\mu_s = 0.6$. If he pushes horizontally on the refrigerator, determine if he can move it. If so, does the refrigerator slip or tip?

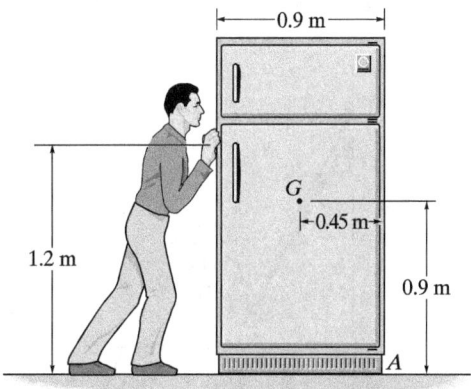

Probs. 7–26/27

**\*7–28.** Determine the minimum force $P$ needed to push the two 75-kg cylinders up the incline. The force acts parallel to the plane and the coefficients of static friction of the contacting surfaces are $\mu_A = 0.3$, $\mu_B = 0.25$, and $\mu_C = 0.4$. Each cylinder has a radius of 150 mm.

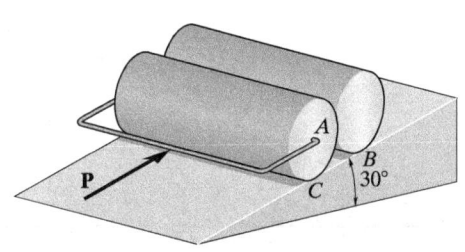

Prob. 7–28

**•7–29.** If the center of gravity of the stacked tables is at $G$, and the stack weighs 500 N ($\approx$ 50 kg), determine the smallest force $P$ the boy must push on the stack in order to cause movement. The coefficient of static friction at $A$ and $B$ is $\mu_s = 0.3$. The tables are locked together.

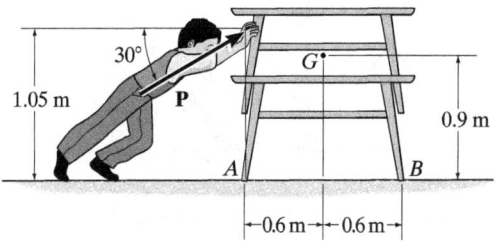

Prob. 7–29

**7–30.** The tractor has a weight of 40 kN with center of gravity at $G$. Determine if it can push the 250-kg log up the incline. The coefficient of static friction between the log and the ground is $\mu_s = 0.5$, and between the rear wheels of the tractor and the ground $\mu_s' = 0.8$. The front wheels are free to roll. Assume the engine can develop enough torque to cause the rear wheels to slip.

**7–31.** The tractor has a weight of 40 kN with center of gravity at $G$. Determine the greatest weight of the log that can be pushed up the incline. The coefficient of static friction between the log and the ground is $\mu_s = 0.5$, and between the rear wheels of the tractor and the ground $\mu_s' = 0.7$. The front wheels are free to roll. Assume the engine can develop enough torque to cause the rear wheels to slip.

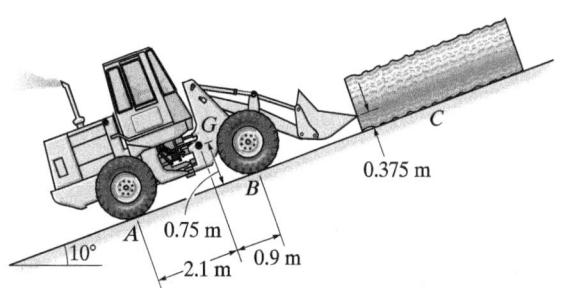

Probs. 7–30/31

***7–32.** The 50-kg uniform pole is on the verge of slipping at $A$ when $\theta = 45°$. Determine the coefficient of static friction at $A$.

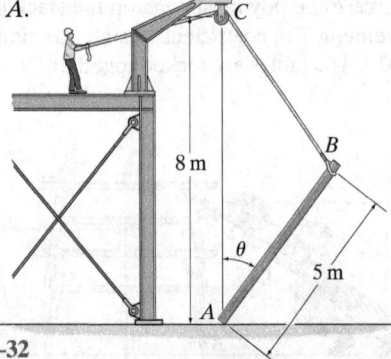

Prob. 7–32

**•7–33.** A force of $P = 100$ N is applied perpendicular to the handle of the gooseneck wrecking bar as shown. If the coefficient of static friction between the bar and the wood is $\mu_s = 0.5$, determine the normal force of the tines at $A$ on the upper board. Assume the surface at $C$ is smooth.

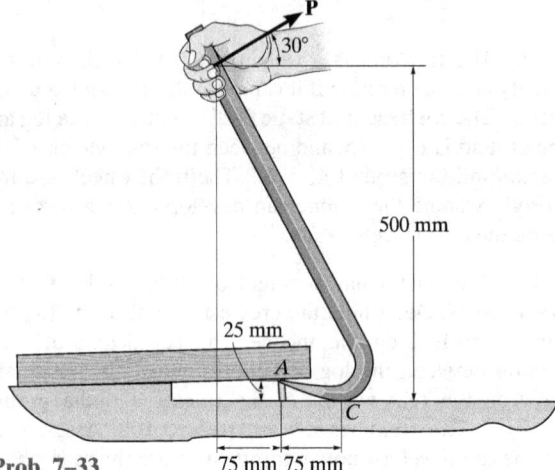

Prob. 7–33

**7–34.** The thin rod has a weight $W$ and rests against the floor and wall for which the coefficients of static friction are $\mu_A$ and $\mu_B$, respectively. Determine the smallest value of $\theta$ for which the rod will not move.

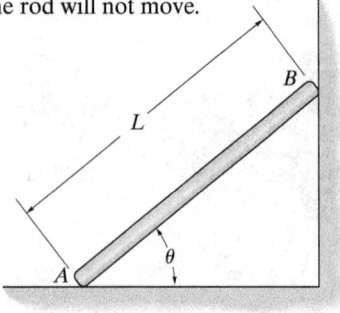

Prob. 7–34

**7–35.** A roll of paper has a uniform weight of 3.75 N ($\approx$ 0.375 kg) and is suspended from the wire hanger so that it rests against the wall. If the hanger has a negligible weight and the bearing at $O$ can be considered frictionless, determine the force $P$ needed to start turning the roll if $\theta = 30°$. The coefficient of static friction between the wall and the paper is $\mu_s = 0.25$.

***7–36.** A roll of paper has a uniform weight of 3.75 N ($\approx$ 0.375 kg) and is suspended from the wire hanger so that it rests against the wall. If the hanger has a negligible weight and the bearing at $O$ can be considered frictionless, determine the minimum force $P$ and the associated angle $\theta$ needed to start turning the roll. The coefficient of static friction between the wall and the paper is $\mu_s = 0.25$.

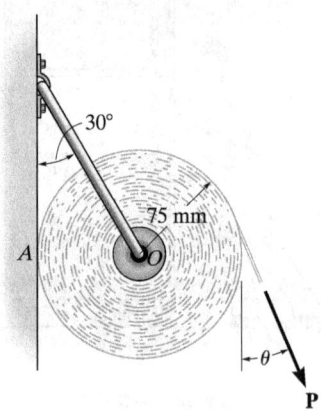

Probs. 7–35/36

**•7–37.** If the coefficient of static friction between the chain and the inclined plane is $\mu_s = \tan \theta$, determine the overhang length $b$ so that the chain is on the verge of slipping up the plane. The chain weighs $w$ per unit length.

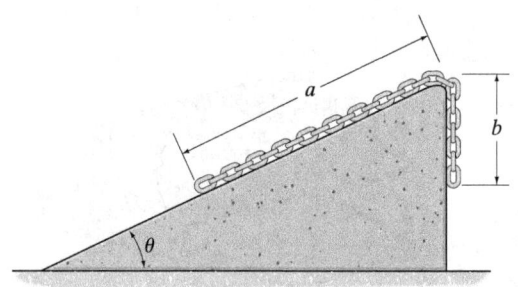

Prob. 7–37

**7–38.** Determine the maximum height $h$ in meters to which the girl can walk up the slide without supporting herself by the rails or by her left leg. The coefficient of static friction between the girl's shoes and the slide is $\mu_s = 0.8$.

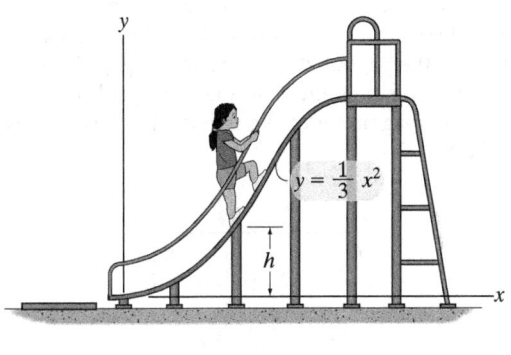

Prob. 7–38

**•7–41.** The clamp is used to tighten the connection between two concrete drain pipes. Determine the least coefficient of static friction at $A$ or $B$ so that the clamp does not slip regardless of the force in the shaft $CD$.

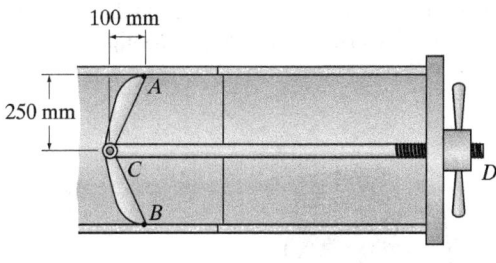

Prob. 7–41

**7–39.** If the coefficient of static friction at $B$ is $\mu_s = 0.3$, determine the largest angle $\theta$ and the minimum coefficient of static friction at $A$ so that the roller remains self-locking, regardless of the magnitude of force **P** applied to the belt. Neglect the weight of the roller and neglect friction between the belt and the vertical surface.

**\*7–40.** If $\theta = 30°$, determine the minimum coefficient of static friction at $A$ and $B$ so that the roller remains self-locking, regardless of the magnitude of force **P** applied to the belt. Neglect the weight of the roller and neglect friction between the belt and the vertical surface.

**7–42.** The coefficient of static friction between the 150-kg crate and the ground is $\mu_s = 0.3$, while the coefficient of static friction between the 80-kg man's shoes and the ground is $\mu_s' = 0.4$. Determine if the man can move the crate.

**7–43.** If the coefficient of static friction between the crate and the ground is $\mu_s = 0.3$, determine the minimum coefficient of static friction between the man's shoes and the ground so that the man can move the crate.

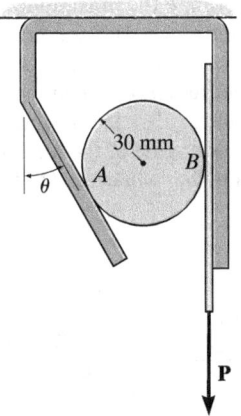

Probs. 7–39/40

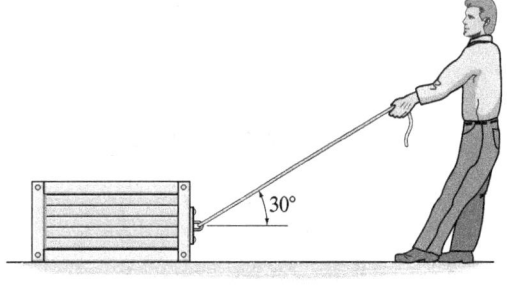

Probs. 7–42/43

**\*7-44.** The 3-Mg rear-wheel-drive skid loader has a center of mass at $G$. Determine the largest number of crates that can be pushed by the loader if each crate has a mass of 500 kg. The coefficient of static friction between a crate and the ground is $\mu_s = 0.3$, and the coefficient of static friction between the rear wheels of the loader and the ground is $\mu'_s = 0.5$. The front wheels are free to roll. Assume that the engine of the loader is powerful enough to generate a torque that will cause the rear wheels to slip.

**7-47.** Block $C$ has a mass of 50 kg and is confined between two walls by smooth rollers. If the block rests on top of the 40-kg spool, determine the minimum cable force $P$ needed to move the spool. The cable is wrapped around the spool's inner core. The coefficients of static friction at $A$ and $B$ are $\mu_A = 0.3$ and $\mu_B = 0.6$.

**\*7-48.** Block $C$ has a mass of 50 kg and is confined between two walls by smooth rollers. If the block rests on top of the 40-kg spool, determine the required coefficients of static friction at $A$ and $B$ so that the spool slips at $A$ and $B$ when the magnitude of the applied force is increased to $P = 300$ N.

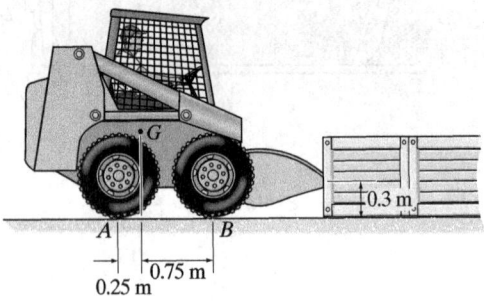

Prob. 7-44

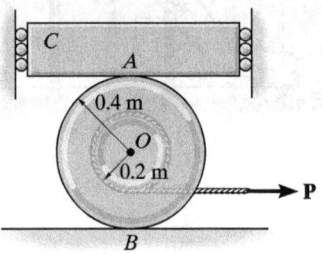

Probs. 7-47/48

**•7-45.** The 45-kg disk rests on the surface for which the coefficient of static friction is $\mu_A = 0.2$. Determine the largest couple moment $M$ that can be applied to the bar without causing motion.

**7-46.** The 45-kg disk rests on the surface for which the coefficient of static friction is $\mu_A = 0.15$. If $M = 50$ N·m, determine the friction force at $A$.

**•7-49.** The 3-Mg four-wheel-drive truck (SUV) has a center of mass at $G$. Determine the maximum mass of the log that can be towed by the truck. The coefficient of static friction between the log and the ground is $\mu_s = 0.8$, and the coefficient of static friction between the wheels of the truck and the ground is $\mu'_s = 0.4$. Assume that the engine of the truck is powerful enough to generate a torque that will cause all the wheels to slip.

**7-50.** A 3-Mg front-wheel-drive truck (SUV) has a center of mass at $G$. Determine the maximum mass of the log that can be towed by the truck. The coefficient of static friction between the log and the ground is $\mu_s = 0.8$, and the coefficient of static friction between the front wheels of the truck and the ground is $\mu'_s = 0.4$. The rear wheels are free to roll. Assume that the engine of the truck is powerful enough to generate a torque that will cause the front wheels to slip.

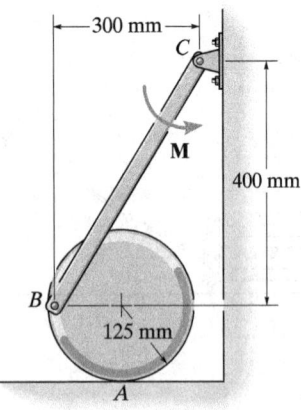

Probs. 7-45/46

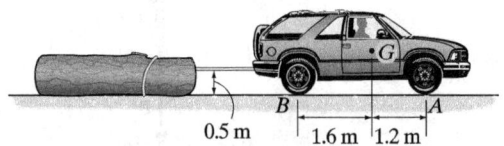

Probs. 7-49/50

**7–51.** If the coefficients of static friction at contact points $A$ and $B$ are $\mu_s = 0.3$ and $\mu_s' = 0.4$ respectively, determine the smallest force $P$ that will cause the 150-kg spool to have impending motion.

**\*7–52.** If the coefficients of static friction at contact points $A$ and $B$ are $\mu_s = 0.4$ and $\mu_s' = 0.2$ respectively, determine the smallest force $P$ that will cause the 150-kg spool to have impending motion.

**7–55.** If the 35-kg girl is at position $d = 1.2$ m, determine the minimum coefficient of static friction $\mu_s$ at contact points $A$ and $B$ so that the plank does not slip. Neglect the weight of the plank.

**\*7–56.** If the coefficient of static friction at the contact points $A$ and $B$ is $\mu_s = 0.4$, determine the minimum distance $d$ where a 35-kg girl can stand on the plank without causing it to slip. Neglect the weight of the plank.

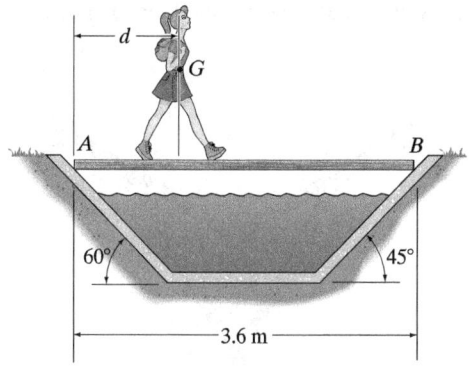

Probs. 7–55/56

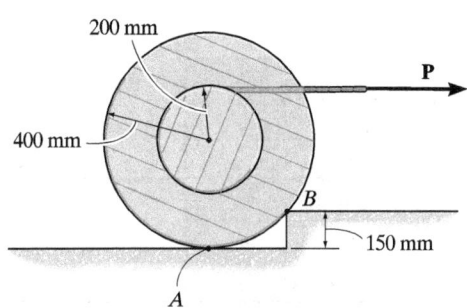

Probs. 7–51/52

**•7–53.** The carpenter slowly pushes the uniform board horizontally over the top of the saw horse. The board has a uniform weight of 50 N/m, and the saw horse has a weight of 75 N and a center of gravity at $G$. Determine if the saw horse will stay in position, slip, or tip if the board is pushed forward when $d = 3$ m. The coefficients of static friction are shown in the figure.

**7–54.** The carpenter slowly pushes the uniform board horizontally over the top of the saw horse. The board has a uniform weight of 50 N/m, and the saw horse has a weight of 75 N and a center of gravity at $G$. Determine if the saw horse will stay in position, slip, or tip if the board is pushed forward when $d = 4.2$ m. The coefficients of static friction are shown in the figure.

**•7–57.** If each box weighs 75 kg, determine the least horizontal force $P$ that the man must exert on the top box in order to cause motion. The coefficient of static friction between the boxes is $\mu_s = 0.5$, and the coefficient of static friction between the box and the floor is $\mu_s' = 0.2$.

**7–58.** If each box weighs 75 kg, determine the least horizontal force $P$ that the man must exert on the top box in order to cause motion. The coefficient of static friction between the boxes is $\mu_s = 0.65$, and the coefficient of static friction between the box and the floor is $\mu_s' = 0.35$.

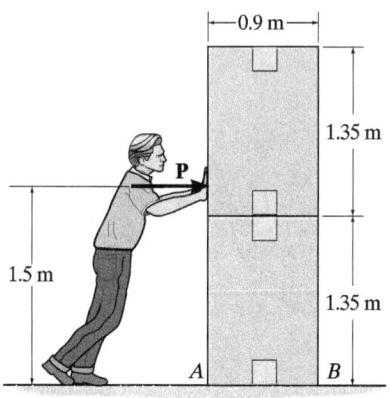

Probs. 7–57/58

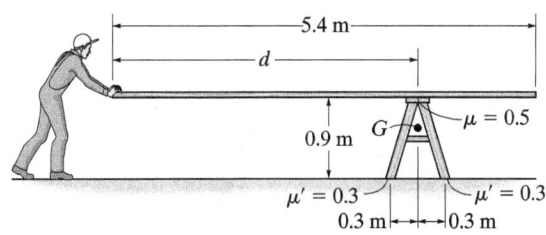

Probs. 7–53/54

**7–59.** If the coefficient of static friction between the collars $A$ and $B$ and the rod is $\mu_s = 0.6$, determine the maximum angle $\theta$ for the system to remain in equilibrium, regardless of the weight of cylinder $D$. Links $AC$ and $BC$ have negligible weight and are connected together at $C$ by a pin.

**\*7–60.** If $\theta = 15°$, determine the minimum coefficient of static friction between the collars $A$ and $B$ and the rod required for the system to remain in equilibrium, regardless of the weight of cylinder $D$. Links $AC$ and $BC$ have negligible weight and are connected together at $C$ by a pin.

**7–62.** Blocks $A$, $B$, and $C$ have weights of 250 N, 125 N, and 75 N, respectively. Determine the smallest horizontal force $P$ that will cause impending motion. The coefficient of static friction between $A$ and $B$ is $\mu_s = 0.3$, between $B$ and $C$, $\mu'_s = 0.4$, and between block $C$ and the ground, $\mu''_s = 0.35$.

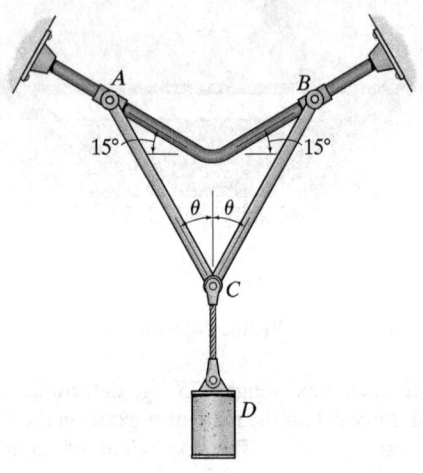

Probs. 7–59/60

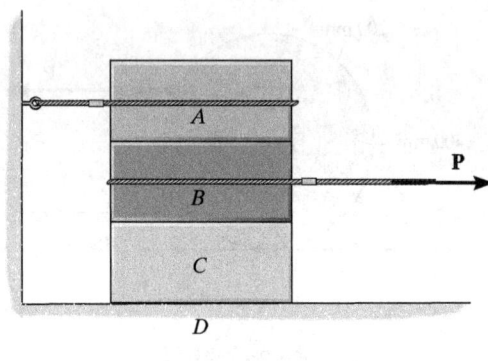

Prob. 7–62

**•7–61.** Each of the cylinders has a mass of 50 kg. If the coefficients of static friction at the points of contact are $\mu_A = 0.5$, $\mu_B = 0.5$, $\mu_C = 0.5$, and $\mu_D = 0.6$, determine the smallest couple moment $M$ needed to rotate cylinder $E$.

**7–63.** Determine the smallest force $P$ that will cause impending motion. The crate and wheel have a mass of 50 kg and 25 kg, respectively. The coefficient of static friction between the crate and the ground is $\mu_s = 0.2$, and between the wheel and the ground $\mu'_s = 0.5$.

**\*7–64.** Determine the smallest force $P$ that will cause impending motion. The crate and wheel have a mass of 50 kg and 25 kg, respectively. The coefficient of static friction between the crate and the ground is $\mu_s = 0.5$, and between the wheel and the ground $\mu'_s = 0.3$.

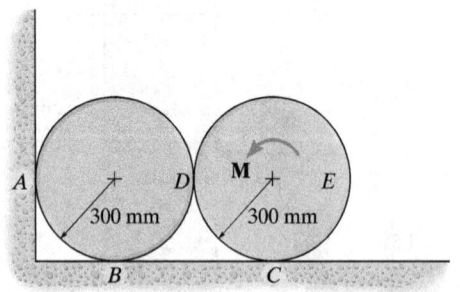

Prob. 7–61

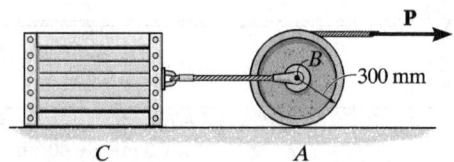

Probs. 7–63/64

## CONCEPTUAL PROBLEMS

**P7–1.** Is it more effective to move the load forward at constant velocity with the boom fully extended as shown, or should the boom be fully retracted? Power is supplied to the rear wheels. The front wheels are free to roll. Do an equilibrium analysis to explain your answer.

P7–1

**P7–2.** The lug nut on the free-turning wheel is to be removed using the wrench. Which is the most effective way to apply force to the wrench? Also, why is it best to keep the car tire on the ground rather than first jacking it up? Explain your answers with an equilibrium analysis.

P7–2

**P7–3.** The rope is used to tow the refrigerator. Is it best to pull slightly up on the rope as shown, pull horizontally, or pull somewhat downwards? Also, is it best to attach the rope at a high position as shown, or at a lower position? Do an equilibrium analysis to explain your answer.

**P7–4.** The rope is used to tow the refrigerator. In order to prevent yourself from slipping while towing, is it best to pull up as shown, pull horizontally, or pull downwards on the rope? Do an equilibrium analysis to explain your answer.

P7–3/4

**P7–5.** Is it easier to tow the load by applying a force along the tow bar when it is in an almost horizontal position as shown, or is it better to pull on the bar when it has a steeper slope? Do an equilibrium analysis to explain your answer.

P7–5

## 7.3 Wedges

A *wedge* is a simple machine that is often used to transform an applied force into much larger forces, directed at approximately right angles to the applied force. Wedges also can be used to slightly move or adjust heavy loads.

Consider, for example, the wedge shown in Fig. 7–12a, which is used to *lift* the block by applying a force to the wedge. Free-body diagrams of the block and wedge are shown in Fig. 7–12b. Here, we have excluded the weight of the wedge since it is usually *small* compared to the weight **W** of the block. Also, note that the frictional forces **F**$_1$ and **F**$_2$ must oppose the motion of the wedge. Likewise, the frictional force **F**$_3$ of the wall on the block must act downward so as to oppose the block's upward motion. The locations of the resultant normal forces are not important in the force analysis since neither the block nor wedge will "tip." Hence the moment equilibrium equations will not be considered. There are seven unknowns, consisting of the applied force **P** needed to cause motion of the wedge, and six normal and frictional forces. The seven available equations consist of four force equilibrium equations, $\Sigma F_x = 0$, $\Sigma F_y = 0$ applied to the wedge and block, and three frictional equations, $F = \mu N$, applied at the surface of contact.

If the block is to be *lowered*, then the frictional forces will all act in a sense opposite to that shown in Fig. 7–12b. Provided the coefficient of friction is very *small* or the wedge angle $\theta$ is *large*, then the applied force **P** must act to the right to hold the block. Otherwise, **P** may have a reverse sense of direction in order to *pull* on the wedge to remove it. If **P** is *not applied* and friction forces hold the block in place, then the wedge is referred to as *self-locking*.

Please refer to the Companion CD for the animation: *Free-body Diagram For Wedges*

Wedges are often used to adjust the elevation of structural or mechanical parts. Also, they provide stability for objects such as this pipe.

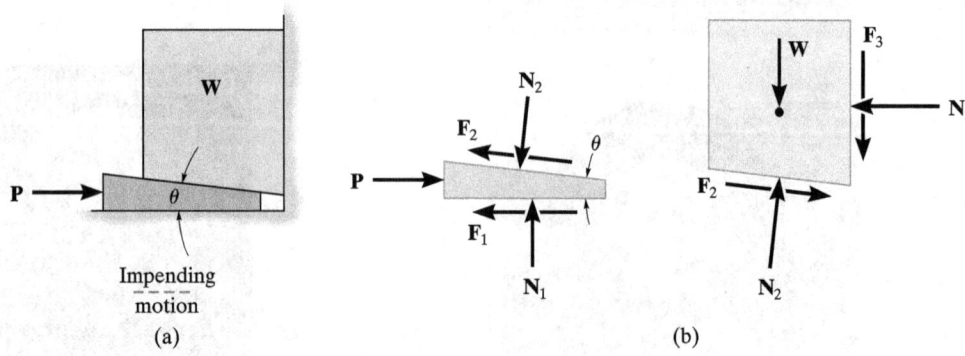

Fig. 7–12

## EXAMPLE 7.6

The uniform stone in Fig. 7–13a has a mass of 500 kg and is held in the horizontal position using a wedge at B. If the coefficient of static friction is $\mu_s = 0.3$ at the surfaces of contact, determine the minimum force P needed to remove the wedge. Assume that the stone does not slip at A.

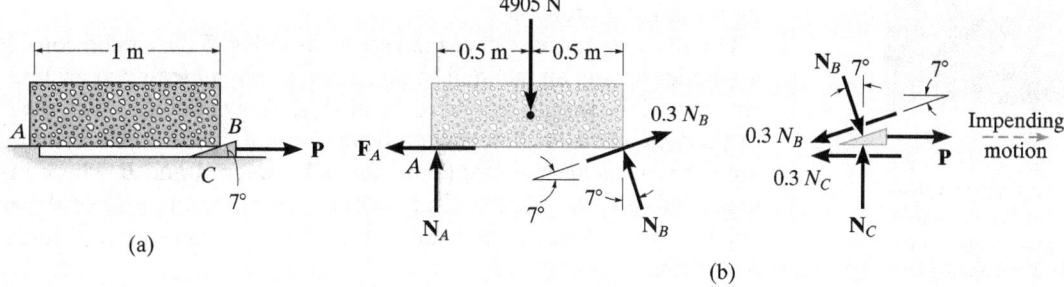

(a)

(b)

Fig. 7–13

### SOLUTION

The minimum force P requires $F = \mu_s N$ at the surfaces of contact with the wedge. The free-body diagrams of the stone and wedge are shown in Fig. 7–13b. On the wedge, the friction force opposes the impending motion, and on the stone at A, $F_A \leq \mu_s N_A$, since slipping does not occur there. There are five unknowns. Three equilibrium equations for the stone and two for the wedge are available for solution. From the free-body diagram of the stone,

$\zeta + \Sigma M_A = 0;$  $-4905 \text{ N}(0.5 \text{ m}) + (N_B \cos 7° \text{ N})(1 \text{ m})$
$+ (0.3 N_B \sin 7° \text{ N})(1 \text{ m}) = 0$
$N_B = 2383.1 \text{ N}$

Using this result for the wedge, we have

$+\uparrow \Sigma F_y = 0;$  $N_C - 2383.1 \cos 7° \text{ N} - 0.3(2383.1 \sin 7° \text{ N}) = 0$
$N_C = 2452.5 \text{ N}$

$\xrightarrow{+} \Sigma F_x = 0;$  $2383.1 \sin 7° \text{ N} - 0.3(2383.1 \cos 7° \text{ N}) +$
$P - 0.3(2452.5 \text{ N}) = 0$
$P = 1154.9 \text{ N} = 1.15 \text{ kN}$  *Ans.*

**NOTE:** Since P is positive, indeed the wedge must be pulled out. If P was zero, the wedge would remain in place (self-locking) and the frictional forces developed at B and C would satisfy $F_B < \mu_s N_B$ and $F_C < \mu_s N_C$.

## 7.4 Frictional Forces on Screws

In most cases screws are used as fasteners; however, in many types of machines they are incorporated to transmit power or motion from one part of the machine to another. A *square-threaded screw* is commonly used for the latter purpose, especially when large forces are applied along its axis. In this section, we will analyze the forces acting on square-threaded screws. The analysis of other types of screws, such as the V-thread, is based on these same principles.

For analysis, a square-threaded screw, as in Fig. 7–14, can be considered a cylinder having an inclined square ridge or *thread* wrapped around it. If we unwind the thread by one revolution, as shown in Fig. 7–14b, the slope or the *lead angle* $\theta$ is determined from $\theta = \tan^{-1}(l/2\pi r)$. Here, $l$ and $2\pi r$ are the vertical and horizontal distances between $A$ and $B$, where $r$ is the mean radius of the thread. The distance $l$ is called the *lead* of the screw and it is equivalent to the distance the screw advances when it turns one revolution.

Square-threaded screws find applications on valves, jacks, and vises, where particularly large forces must be developed along the axis of the screw.

**Upward Impending Motion.** Let us now consider the case of a square-threaded screw that is subjected to upward impending motion caused by the applied torsional moment **M**, Fig. 7–15.* A free-body diagram of the *entire unraveled thread* can be represented as a block as shown in Fig. 7–14a. The force **W** is the vertical force acting on the thread or the axial force applied to the shaft, Fig. 7–15, and $M/r$ is the resultant horizontal force produced by the couple moment $M$ about the axis of the shaft. The reaction **R** of the groove on the thread, has both frictional and normal components, where $F = \mu_s N$. The angle of static friction is $\phi_s = \tan^{-1}(F/N) = \tan^{-1}\mu_s$. Applying the force equations of equilibrium along the horizontal and vertical axes, we have

$$\xrightarrow{+} \Sigma F_x = 0; \quad M/r - R \sin(\phi_s + \theta) = 0$$

$$+\uparrow \Sigma F_y = 0; \quad R \cos(\phi_s + \theta) - W = 0$$

Eliminating $R$ from these equations, we obtain

$$\boxed{M = rW \tan(\phi_s + \theta)} \quad (7\text{–}3)$$

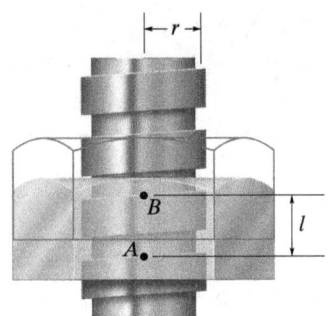

(a)

**Fig. 7–14**

*For applications, **M** is developed by applying a horizontal force **P** at a right angle to the end of a lever that would be fixed to the screw.

## 7.4 Frictional Forces on Screws

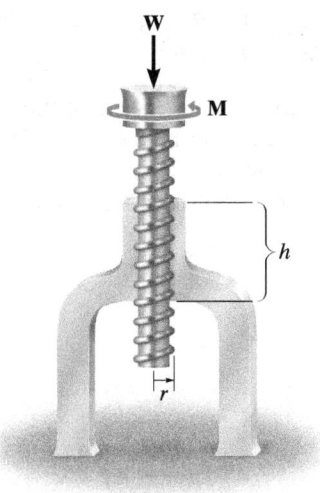

Fig. 7–15

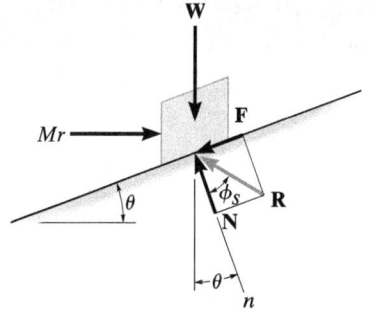

Upward screw motion
(a)

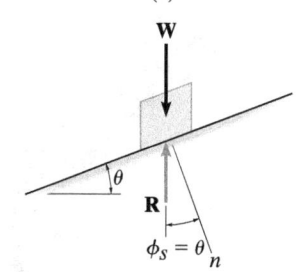

Self-locking screw ($\theta = \phi_S$)
(on the verge of rotating downward)
(b)

**Self-Locking Screw.** A screw is said to be *self-locking* if it remains in place under any axial load **W** when the moment **M** is removed. For this to occur, the direction of the frictional force must be reversed so that **R** acts on the other side of **N**. Here, the angle of static friction $\phi_s$ becomes greater than or equal to $\theta$, Fig. 7–16d. If $\phi_s = \theta$, Fig. 7–16b, then **R** will act vertically to balance **W**, and the screw will be on the verge of winding downward.

**Downward Impending Motion.** ($\phi_s > \theta$). If a screw is self-locking, a couple moment **M'** must be applied to the screw in the opposite direction to wind the screw downward ($\phi_s > \theta$). This causes a reverse horizontal force $M'/r$ that pushes the thread down as indicated in Fig. 7–16c. Using the same procedure as before, we obtain

$$M' = rW \tan(\theta - \phi_s) \qquad (7\text{--}4)$$

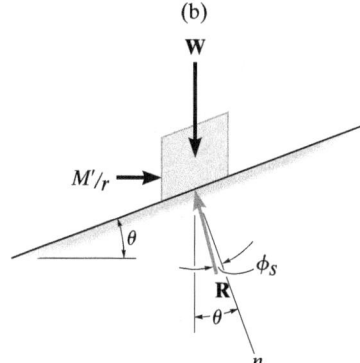

Downward screw motion ($\theta > \phi_S$)
(c)

**Downward Impending Motion.** ($\phi_s < \theta$). If the screw is not self-locking, it is necessary to apply a moment **M''** to prevent the screw from winding downward ($\phi_s < \theta$). Here, a horizontal force $M''/r$ is required to push against the thread to prevent it from sliding down the plane, Fig. 7–16d. Thus, the magnitude of the moment **M''** required to prevent this unwinding is

$$M'' = Wr \tan(\phi_s - \theta) \qquad (7\text{--}5)$$

If *motion of the screw* occurs, Eqs. 7–3, 7–4, and 7–5 can be applied by simply replacing $\phi_s$ with $\phi_k$.

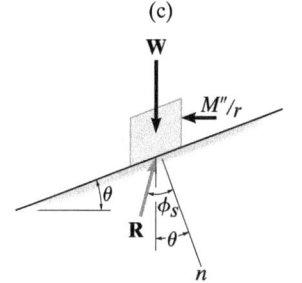

Downward screw motion ($\theta < \phi_S$)
(d)

Fig. 7–16

## EXAMPLE 7.7

The turnbuckle shown in Fig. 7–17 has a square thread with a mean radius of 5 mm and a lead of 2 mm. If the coefficient of static friction between the screw and the turnbuckle is $\mu_s = 0.25$, determine the moment **M** that must be applied to draw the end screws closer together.

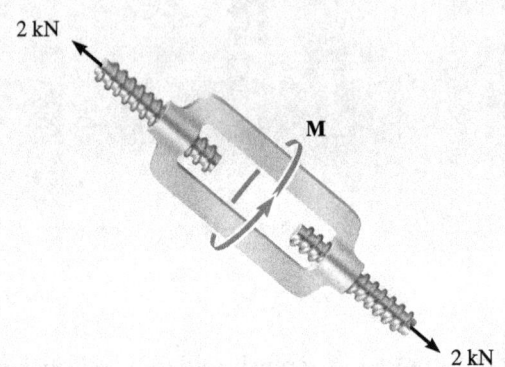

Fig. 7–17

### SOLUTION
The moment can be obtained by applying Eq. 7–3. Since friction at *two screws* must be overcome, this requires

$$M = 2[Wr \tan(\theta + \phi)] \tag{1}$$

Here $W = 2000$ N, $r = 5$ mm, $\phi_s = \tan^{-1} \mu_s = \tan^{-1}(0.25) = 14.04°$, and $\theta = \tan^{-1}(l/2\pi r) = \tan^{-1}(2 \text{ mm}/[2\pi(5 \text{ mm})]) = 3.64°$. Substituting these values into Eq. 1 and solving gives

$$M = 2[(2000 \text{ N})(5 \text{ mm}) \tan(14.04° + 3.64°)]$$

$$= 6374.7 \text{ N} \cdot \text{mm} = 6.37 \text{ N} \cdot \text{m} \qquad Ans.$$

**NOTE:** When the moment is *removed*, the turnbuckle will be self-locking; i.e., it will not unscrew since $\phi_s > \theta$.

## PROBLEMS

**•7–65.** Determine the smallest horizontal force $P$ required to pull out wedge $A$. The crate has a weight of 1500 N ($\approx$ 150 kg) and the coefficient of static friction at all contacting surfaces is $\mu_s = 0.3$. Neglect the weight of the wedge.

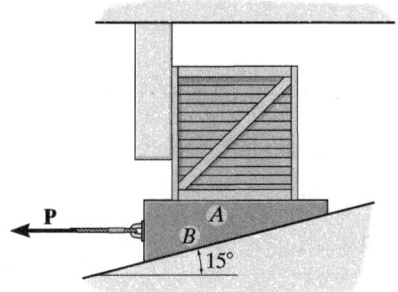

Prob. 7–65

**7–66.** Determine the smallest horizontal force $P$ required to lift the 200-kg crate. The coefficient of static friction at all contacting surfaces is $\mu_s = 0.3$. Neglect the mass of the wedge.

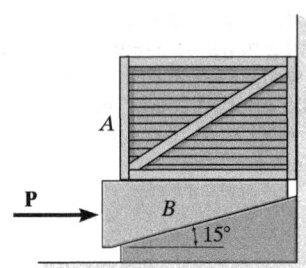

Prob. 7–66

**7–67.** Determine the smallest horizontal force $P$ required to lift the 100-kg cylinder. The coefficients of static friction at the contact points $A$ and $B$ are $(\mu_s)_A = 0.6$ and $(\mu_s)_B = 0.2$, respectively; and the coefficient of static friction between the wedge and the ground is $\mu_s = 0.3$.

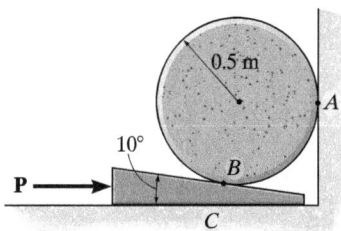

Prob. 7–67

**\*7–68.** The wedge has a negligible weight and a coefficient of static friction $\mu_s = 0.35$ with all contacting surfaces. Determine the largest angle $\theta$ so that it is "self-locking." This requires no slipping for any magnitude of the force **P** applied to the joint.

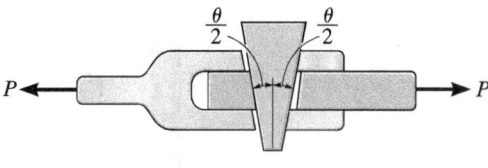

Prob. 7–68

**•7–69.** Determine the smallest horizontal force $P$ required to just move block $A$ to the right if the spring force is 600 N and the coefficient of static friction at all contacting surfaces on $A$ is $\mu_s = 0.3$. The sleeve at $C$ is smooth. Neglect the mass of $A$ and $B$.

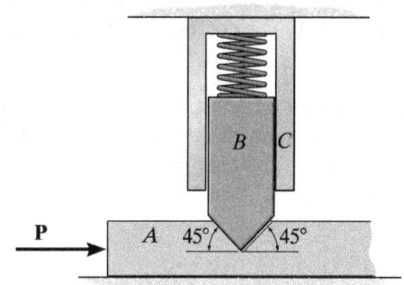

Prob. 7–69

**7–70.** The three stone blocks have weights of $W_A = 3000$ N, $W_B = 750$ N, and $W_C = 2500$ N. Determine the smallest horizontal force $P$ that must be applied to block $C$ in order to move this block. The coefficient of static friction between the blocks is $\mu_s = 0.3$, and between the floor and each block $\mu_s' = 0.5$.

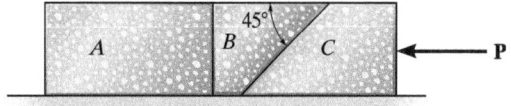

Prob. 7–70

**7–71.** Determine the smallest horizontal force $P$ required to move the wedge to the right. The coefficient of static friction at all contacting surfaces is $\mu_s = 0.3$. Set $\theta = 15°$ and $F = 400$ N. Neglect the weight of the wedge.

**\*7–72.** If the horizontal force $\mathbf{P}$ is removed, determine the largest angle $\theta$ that will cause the wedge to be self-locking regardless of the magnitude of force $\mathbf{F}$ applied to the handle. The coefficient of static friction at all contacting surfaces is $\mu_s = 0.3$.

**7–75.** If the uniform concrete block has a mass of 500 kg, determine the smallest horizontal force $P$ needed to move the wedge to the left. The coefficient of static friction between the wedge and the concrete and the wedge and the floor is $\mu_s = 0.3$. The coefficient of static friction between the concrete and floor is $\mu_s' = 0.5$.

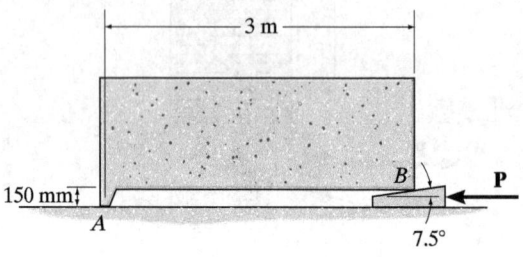

Prob. 7–75

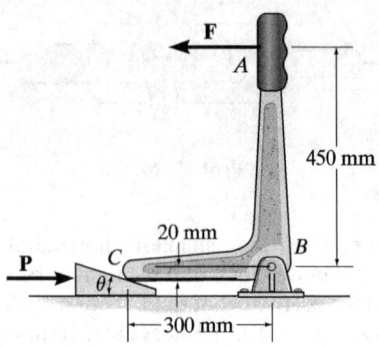

Probs. 7–71/72

**•7–73.** Determine the smallest vertical force $P$ required to hold the wedge between the two identical cylinders, each having a weight of $W$. The coefficient of static friction at all contacting surfaces is $\mu_s = 0.1$.

**7–74.** Determine the smallest vertical force $P$ required to push the wedge between the two identical cylinders, each having a weight of $W$. The coefficient of static friction at all contacting surfaces is $\mu_s = 0.3$.

**\*7–76.** The wedge blocks are used to hold the specimen in a tension testing machine. Determine the largest design angle $\theta$ of the wedges so that the specimen will not slip regardless of the applied load. The coefficients of static friction are $\mu_A = 0.1$ at $A$ and $\mu_B = 0.6$ at $B$. Neglect the weight of the blocks.

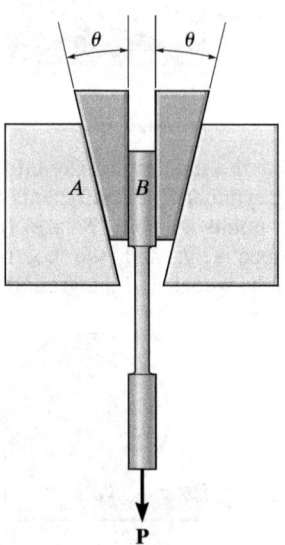

Prob. 7–76

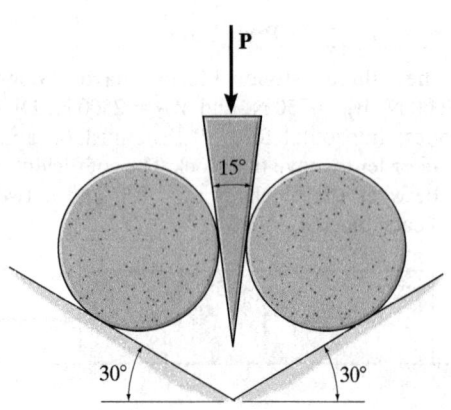

Probs. 7–73/74

•**7–77.** The square threaded screw of the clamp has a mean diameter of 14 mm and a lead of 6 mm. If $\mu_s = 0.2$ for the threads, and the torque applied to the handle is 1.5 N·m, determine the compressive force $F$ on the block.

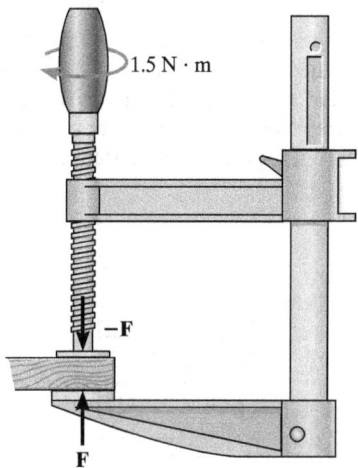

Prob. 7–77

**7–78.** The device is used to pull the battery cable terminal $C$ from the post of a battery. If the required pulling force is 425 N, determine the torque $M$ that must be applied to the handle on the screw to tighten it. The screw has square threads, a mean diameter of 5 mm, a lead of 2 mm, and the coefficient of static friction is $\mu_s = 0.5$.

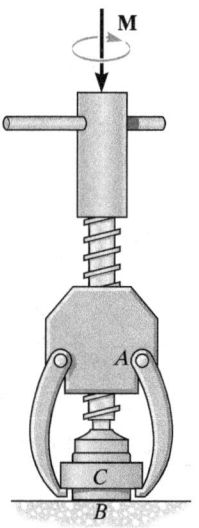

Prob. 7–78

**7–79.** The jacking mechanism consists of a link that has a square-threaded screw with a mean diameter of 12 mm and a lead of 5 mm, and the coefficient of static friction is $\mu_s = 0.4$. Determine the torque $M$ that should be applied to the screw to start lifting the 30-kN load acting at the end of member $ABC$.

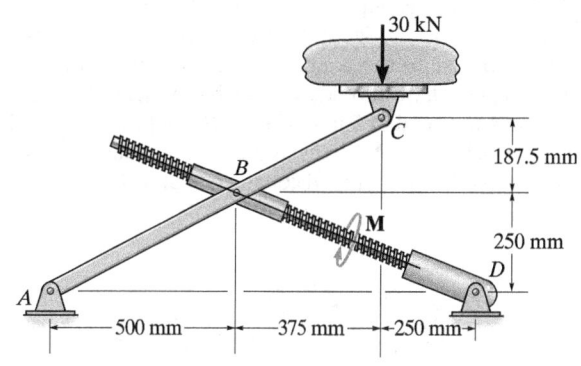

Prob. 7–79

***7–80.** Determine the magnitude of the horizontal force **P** that must be applied to the handle of the bench vise in order to produce a clamping force of 600 N on the block. The single square-threaded screw has a mean diameter of 25 mm and a lead of 7.5 mm. The coefficient of static friction is $\mu_s = 0.25$.

•**7–81.** Determine the clamping force exerted on the block if a force of $P = 30$ N is applied to the lever of the bench vise. The single square-threaded screw has a mean diameter of 25 mm and a lead of 7.5 mm. The coefficient of static friction is $\mu_s = 0.25$.

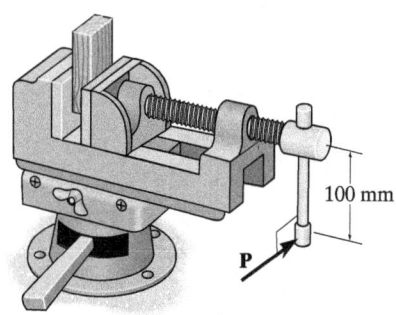

Probs. 7–80/81

**7–82.** Determine the required horizontal force that must be applied perpendicular to the handle in order to develop a 900-N clamping force on the pipe. The single square-threaded screw has a mean diameter of 25 mm and a lead of 5 mm. The coefficient of static friction is $\mu_s = 0.4$. *Note:* The screw is a two-force member since it is contained within pinned collars at $A$ and $B$.

**7–83.** If the clamping force on the pipe is 900 N, determine the horizontal force that must be applied perpendicular to the handle in order to loosen the screw. The single square-threaded screw has a mean diameter of 25 mm and a lead of 5 mm. The coefficient of static friction is $\mu_s = 0.4$. *Note:* The screw is a two-force member since it is contained within pinned collars at $A$ and $B$.

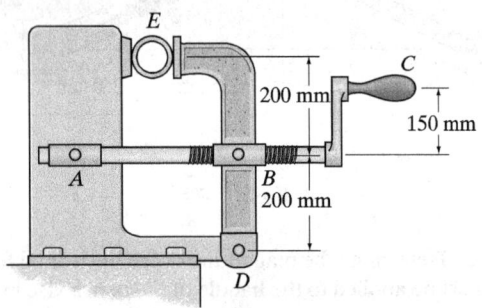

Probs. 7–82/83

**\*7–84.** The clamp provides pressure from several directions on the edges of the board. If the square-threaded screw has a lead of 3 mm, mean radius of 10 mm, and the coefficient of static friction is $\mu_s = 0.4$, determine the horizontal force developed on the board at $A$ and the vertical forces developed at $B$ and $C$ if a torque of $M = 1.5$ N·m is applied to the handle to tighten it further. The blocks at $B$ and $C$ are pin connected to the board.

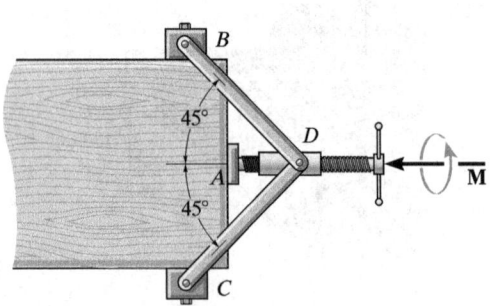

Prob. 7–84

**•7–85.** If the jack supports the 200-kg crate, determine the horizontal force that must be applied perpendicular to the handle at $E$ to lower the crate. Each single square-threaded screw has a mean diameter of 25 mm and a lead of 7.5 mm. The coefficient of static friction is $\mu_s = 0.25$.

**7–86.** If the jack is required to lift the 200-kg crate, determine the horizontal force that must be applied perpendicular to the handle at $E$. Each single square-threaded screw has a mean diameter of 25 mm and a lead of 7.5 mm. The coefficient of static friction is $\mu_s = 0.25$.

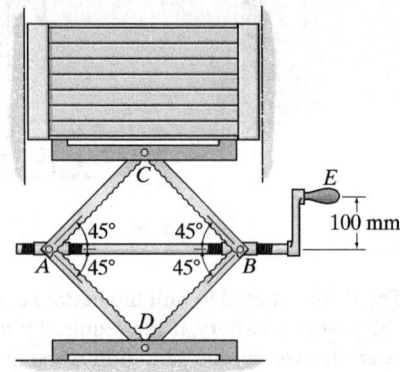

Probs. 7–85/86

**7–87.** The machine part is held in place using the double-end clamp. The bolt at $B$ has square threads with a mean radius of 4 mm and a lead of 2 mm, and the coefficient of static friction with the nut is $\mu_s = 0.5$. If a torque of $M = 0.4$ N·m is applied to the nut to tighten it, determine the normal force of the clamp at the smooth contacts $A$ and $C$.

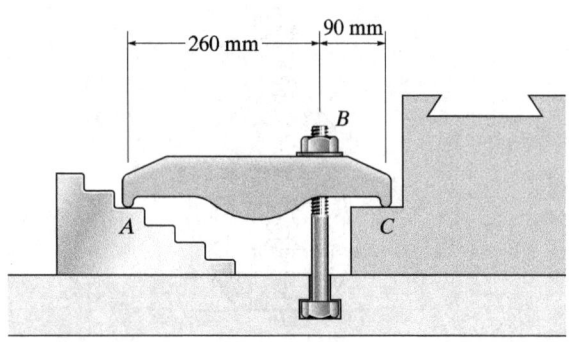

Prob. 7–87

## 7.5 Frictional Forces on Flat Belts

Whenever belt drives or band brakes are designed, it is necessary to determine the frictional forces developed between the belt and its contacting surface. In this section we will analyze the frictional forces acting on a flat belt, although the analysis of other types of belts, such as the V-belt, is based on similar principles.

Consider the flat belt shown in Fig. 7–18a, which passes over a fixed curved surface. The total angle of belt to surface contact in radians is $\beta$, and the coefficient of friction between the two surfaces is $\mu$. We wish to determine the tension $T_2$ in the belt, which is needed to pull the belt counterclockwise over the surface, and thereby overcome both the frictional forces at the surface of contact and the tension $T_1$ in the other end of the belt. Obviously, $T_2 > T_1$.

**Frictional Analysis.** A free-body diagram of the belt segment in contact with the surface is shown in Fig. 7–18b. As shown, the normal and frictional forces, acting at different points along the belt, will vary both in magnitude and direction. Due to this *unknown* distribution, the analysis of the problem will first require a study of the forces acting on a differential element of the belt.

A free-body diagram of an element having a length $ds$ is shown in Fig. 7–18c. Assuming either impending motion or motion of the belt, the magnitude of the frictional force $dF = \mu\, dN$. This force opposes the sliding motion of the belt, and so it will increase the magnitude of the tensile force acting in the belt by $dT$. Applying the two force equations of equilibrium, we have

$$\searrow + \Sigma F_x = 0; \qquad T\cos\left(\frac{d\theta}{2}\right) + \mu\, dN - (T + dT)\cos\left(\frac{d\theta}{2}\right) = 0$$

$$+\nearrow \Sigma F_y = 0; \qquad dN - (T + dT)\sin\left(\frac{d\theta}{2}\right) - T\sin\left(\frac{d\theta}{2}\right) = 0$$

Since $d\theta$ is of *infinitesimal size*, $\sin(d\theta/2) = d\theta/2$ and $\cos(d\theta/2) = 1$. Also, the *product* of the two infinitesimals $dT$ and $d\theta/2$ may be neglected when compared to infinitesimals of the first order. As a result, these two equations become

$$\mu\, dN = dT$$

and

$$dN = T\, d\theta$$

Eliminating $dN$ yields

$$\frac{dT}{T} = \mu\, d\theta$$

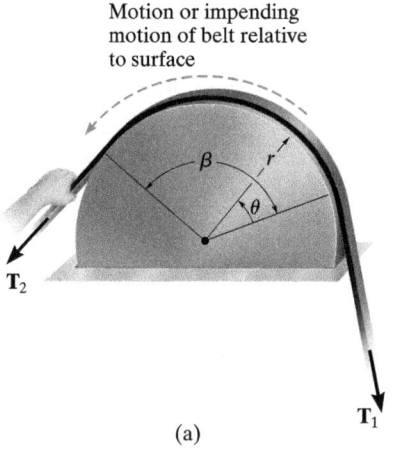

(a)

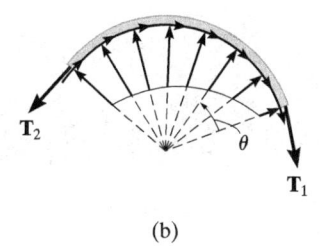

(b)

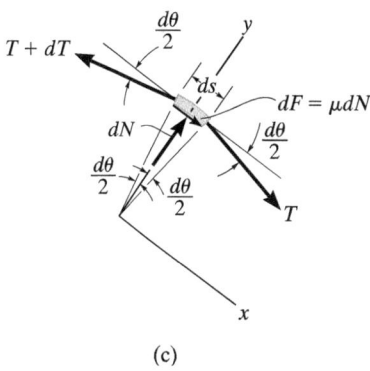

(c)

Fig. 7–18

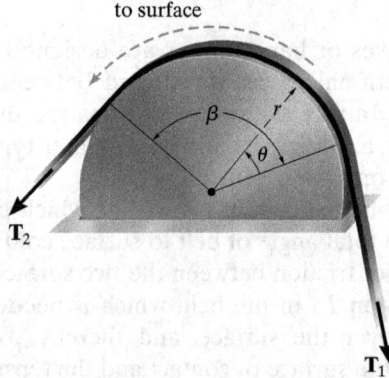

Integrating this equation between all the points of contact that the belt makes with the drum, and noting that $T = T_1$ at $\theta = 0$ and $T = T_2$ at $\theta = \beta$, yields

$$\int_{T_1}^{T_2} \frac{dT}{T} = \mu \int_0^\beta d\theta$$

$$\ln \frac{T_2}{T_1} = \mu \beta$$

Solving for $T_2$, we obtain

$$T_2 = T_1 e^{\mu \beta} \tag{7-6}$$

where

Flat or V-belts are often used to transmit the torque developed by a motor to a wheel attached to a pump, fan or blower.

$T_2, T_1$ = belt tensions; $T_1$ opposes the direction of motion (or impending motion) of the belt measured relative to the surface, while $T_2$ acts in the direction of the relative belt motion (or impending motion); because of friction, $T_2 > T_1$

$\mu$ = coefficient of static or kinetic friction between the belt and the surface of contact

$\beta$ = angle of belt to surface contact, measured in radians

$e$ = 2.718..., base of the natural logarithm

Note that $T_2$ is *independent* of the *radius* of the drum, and instead it is a function of the angle of belt to surface contact, $\beta$. As a result, this equation is valid for flat belts passing over any curved contacting surface.

## EXAMPLE 7.8

The maximum tension that can be developed in the cord shown in Fig. 7–19a is 500 N. If the pulley at A is free to rotate and the coefficient of static friction at the fixed drums B and C is $\mu_s = 0.25$, determine the largest mass of the cylinder that can be lifted by the cord.

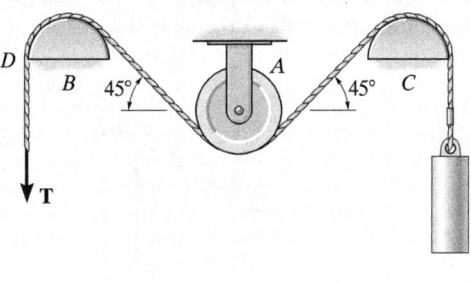

(a)

### SOLUTION

Lifting the cylinder, which has a weight $W = mg$, causes the cord to move counterclockwise over the drums at B and C; hence, the maximum tension $T_2$ in the cord occurs at D. Thus, $F = T_2 = 500$ N. A section of the cord passing over the drum at B is shown in Fig. 7–19b. Since $180° = \pi$ rad the angle of contact between the drum and the cord is $\beta = (135°/180°)\pi = 3\pi/4$ rad. Using Eq. 7–6, we have

$$T_2 = T_1 e^{\mu_s \beta}; \qquad 500 \text{ N} = T_1 e^{0.25[(3/4)\pi]}$$

Hence,

$$T_1 = \frac{500 \text{ N}}{e^{0.25[(3/4)\pi]}} = \frac{500 \text{ N}}{1.80} = 277.4 \text{ N}$$

Since the pulley at A is free to rotate, equilibrium requires that the tension in the cord remains the *same* on both sides of the pulley.

The section of the cord passing over the drum at C is shown in Fig. 7–19c. The weight $W < 277.4$ N. Why? Applying Eq. 7–6, we obtain

$$T_2 = T_1 e^{\mu_s \beta}; \qquad 277.4 \text{ N} = W e^{0.25[(3/4)\pi]}$$

$$W = 153.9 \text{ N}$$

so that

$$m = \frac{W}{g} = \frac{153.9 \text{ N}}{9.81 \text{ m/s}^2}$$

$$= 15.7 \text{ kg} \qquad \textit{Ans.}$$

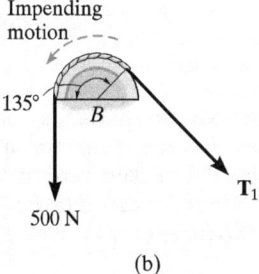

(b)

(c)

Fig. 7–19

# PROBLEMS

***7–88.** Blocks $A$ and $B$ weigh 250 N and 150 N, respectively. Using the coefficients of static friction indicated, determine the greatest weight of block $D$ without causing motion.

**•7–89.** Blocks $A$ and $B$ weigh 375 N each, and $D$ weighs 150 N. Using the coefficients of static friction indicated, determine the frictional force between blocks $A$ and $B$ and between block $A$ and the floor $C$.

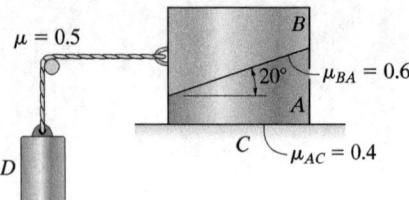

Probs. 7–88/89

**7–90.** A cylinder having a mass of 250 kg is to be supported by the cord which wraps over the pipe. Determine the smallest vertical force $F$ needed to support the load if the cord passes (a) once over the pipe, $\beta = 180°$, and (b) two times over the pipe, $\beta = 540°$. Take $\mu_s = 0.2$.

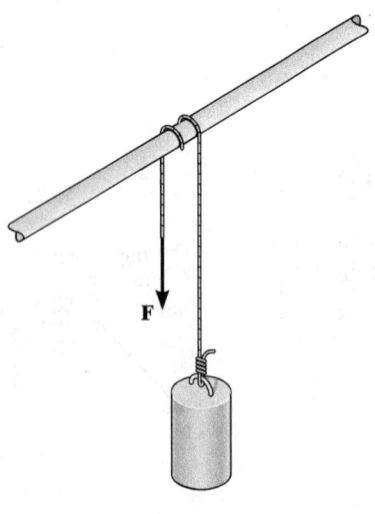

Prob. 7–90

**7–91.** A cylinder having a mass of 250 kg is to be supported by the cord which wraps over the pipe. Determine the largest vertical force $F$ that can be applied to the cord without moving the cylinder. The cord passes (a) once over the pipe, $\beta = 180°$, and (b) two times over the pipe, $\beta = 540°$. Take $\mu_s = 0.2$.

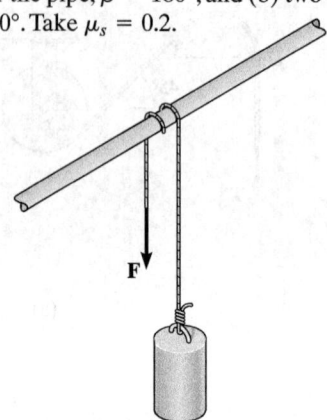

Prob. 7–91

**\*7–92.** The boat has a weight of 2500 N ($\approx$ 250 kg) and is held in position off the side of a ship by the spars at $A$ and $B$. A man having a weight of 650 N ($\approx$ 65 kg) gets in the boat, wraps a rope around an overhead boom at $C$, and ties it to the end of the boat as shown. If the boat is disconnected from the spars, determine the *minimum number of half turns* the rope must make around the boom so that the boat can be safely lowered into the water at constant velocity. Also, what is the normal force between the boat and the man? The coefficient of kinetic friction between the rope and the boom is $\mu_s = 0.15$. *Hint:* The problem requires that the normal force between the man's feet and the boat be as small as possible.

Prob. 7–92

•**7–93.** The 50-kg boy at $A$ is suspended from the cable that passes over the quarter circular cliff rock. Determine if it is possible for the 92.5-kg woman to hoist him up; and if this is possible, what smallest force must she exert on the horizontal cable? The coefficient of static friction between the cable and the rock is $\mu_s = 0.2$, and between the shoes of the woman and the ground is $\mu'_s = 0.8$.

**7–94.** The 50-kg boy at $A$ is suspended from the cable that passes over the quarter circular cliff rock. What horizontal force must the woman at $A$ exert on the cable in order to let the boy descend at constant velocity? The coefficients of static and kinetic friction between the cable and the rock are $\mu_s = 0.4$ and $\mu_k = 0.35$, respectively.

Probs. 7–93/94

**7–95.** A 10-kg cylinder $D$, which is attached to a small pulley $B$, is placed on the cord as shown. Determine the smallest angle $\theta$ so that the cord does not slip over the peg at $C$. The cylinder at $E$ has a mass of 10 kg, and the coefficient of static friction between the cord and the peg is $\mu_s = 0.1$.

*__7–96.__ A 10-kg cylinder $D$, which is attached to a small pulley $B$, is placed on the cord as shown. Determine the largest angle $\theta$ so that the cord does not slip over the peg at $C$. The cylinder at $E$ has a mass of 10 kg, and the coefficient of static friction between the cord and the peg is $\mu_s = 0.1$.

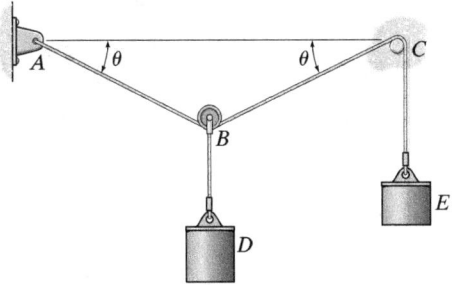

Probs. 7–95/96

•**7–97.** Determine the smallest lever force $P$ needed to prevent the wheel from rotating if it is subjected to a torque of $M = 250$ N·m. The coefficient of static friction between the belt and the wheel is $\mu_s = 0.3$. The wheel is pin connected at its center, $B$.

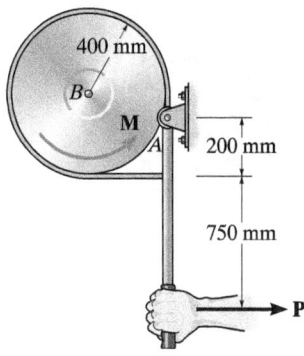

Prob. 7–97

**7–98.** If a force of $P = 200$ N is applied to the handle of the bell crank, determine the maximum torque $M$ that can be resisted so that the flywheel is not on the verge of rotating clockwise. The coefficient of static friction between the brake band and the rim of the wheel is $\mu_s = 0.3$.

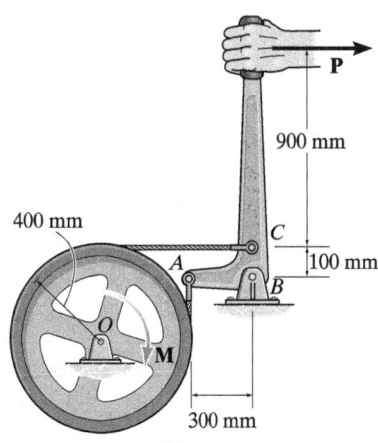

Prob. 7–98

**7–99.** Show that the frictional relationship between the belt tensions, the coefficient of friction $\mu$, and the angular contacts $\alpha$ and $\beta$ for the V-belt is $T_2 = T_1 e^{\mu\beta/\sin(\alpha/2)}$.

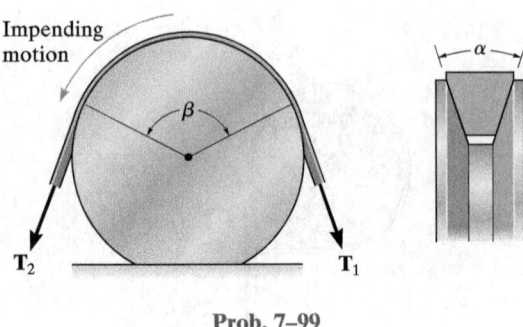

Prob. 7–99

**\*7–100.** Determine the force developed in spring $AB$ in order to hold the wheel from rotating when it is subjected to a couple moment of $M = 200\ \text{N} \cdot \text{m}$. The coefficient of static friction between the belt and the rim of the wheel is $\mu_s = 0.2$, and between the belt and peg $C$, $\mu_s' = 0.4$. The pulley at $B$ is free to rotate.

**•7–101.** If the tension in the spring is $F_{AB} = 2.5\ \text{kN}$, determine the largest couple moment that can be applied to the wheel without causing it to rotate. The coefficient of static friction between the belt and the wheel is $\mu_s = 0.2$, and between the belt the peg $\mu_s' = 0.4$. The pulley $B$ is free to rotate.

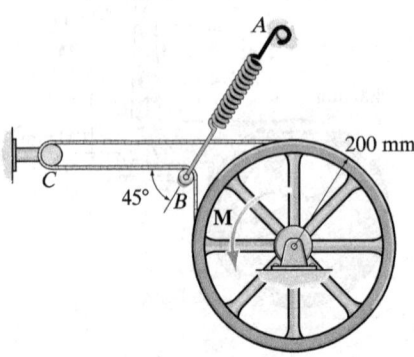

Probs. 7–100/101

**7–102.** The simple band brake is constructed so that the ends of the friction strap are connected to the pin at $A$ and the lever arm at $B$. If the wheel is subjected to a torque of $M = 120\ \text{N} \cdot \text{m}$, determine the smallest force $P$ applied to the lever that is required to hold the wheel stationary. The coefficient of static friction between the strap and wheel is $\mu_s = 0.5$.

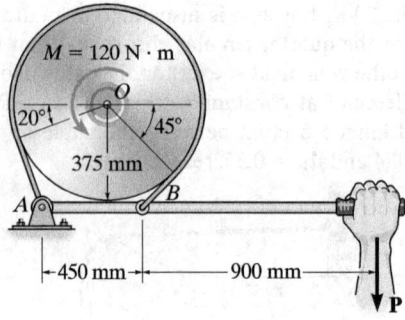

Prob. 7–102

**7–103.** A 90-kg farmer tries to restrain the cow from escaping by wrapping the rope two turns around the tree trunk as shown. If the cow exerts a force of 1250 N on the rope, determine if the farmer can successfully restrain the cow. The coefficient of static friction between the rope and the tree trunk is $\mu_s = 0.15$, and between the farmer's shoes and the ground is $\mu_s' = 0.3$.

Prob. 7–103

**\*7–104.** The uniform 25-kg beam is supported by the rope which is attached to the end of the beam, wraps over the rough peg, and is then connected to the 50-kg block. If the coefficient of static friction between the beam and the block, and between the rope and the peg, is $\mu_s = 0.4$, determine the maximum distance that the block can be placed from $A$ and still remain in equilibrium. Assume the block will not tip.

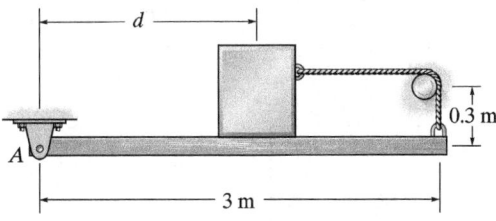

Prob. 7–104

**•7–105.** The 80-kg man tries to lower the 150-kg crate using a rope that passes over the rough peg. Determine the least number of full turns in addition to the basic wrap (165°) around the peg to do the job. The coefficients of static friction between the rope and the peg and between the man's shoes and the ground are $\mu_s = 0.1$ and $\mu'_s = 0.4$, respectively.

**7–106.** If the rope wraps three full turns plus the basic wrap (165°) around the peg, determine if the 80-kg man can keep the 300-kg crate from moving. The coefficients of static friction between the rope and the peg and between the man's shoes and the ground are $\mu_s = 0.1$ and $\mu'_s = 0.4$, respectively.

**7–107.** The drive pulley $B$ in a video tape recorder is on the verge of slipping when it is subjected to a torque of $M = 0.005$ N·m. If the coefficient of static friction between the tape and the drive wheel and between the tape and the fixed shafts $A$ and $C$ is $\mu_s = 0.1$, determine the tensions $T_1$ and $T_2$ developed in the tape for equilibrium.

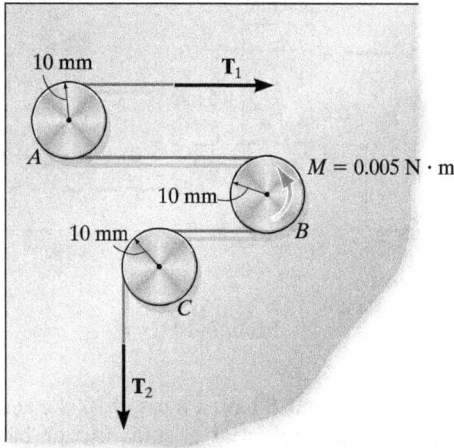

Prob. 7–107

**\*7–108.** Determine the maximum number of 25-kg packages that can be placed on the belt without causing the belt to slip at the drive wheel $A$ which is rotating with a constant angular velocity. Wheel $B$ is free to rotate. Also, find the corresponding torsional moment $\mathbf{M}$ that must be supplied to wheel $A$. The conveyor belt is pre-tensioned with the 1500-N horizontal force. The coefficient of kinetic friction between the belt and platform $P$ is $\mu_k = 0.2$, and the coefficient of static friction between the belt and the rim of each wheel is $\mu_s = 0.35$.

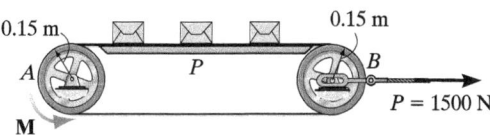

Prob. 7–108

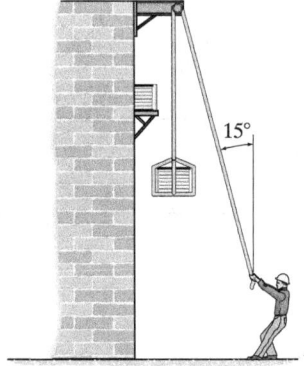

Probs. 7–105/106

**•7–109.** Blocks $A$ and $B$ have a mass of 7 kg and 10 kg, respectively. Using the coefficients of static friction indicated, determine the largest vertical force $P$ which can be applied to the cord without causing motion.

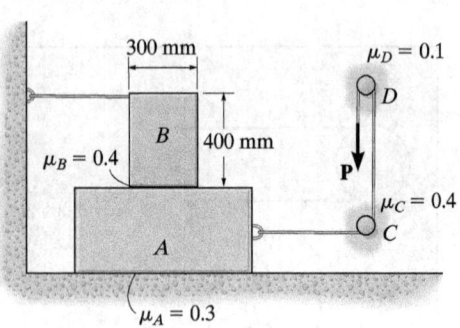

Prob. 7–109

**7–110.** Blocks $A$ and $B$ have a mass of 100 kg and 150 kg, respectively. If the coefficient of static friction between $A$ and $B$ and between $B$ and $C$ is $\mu_s = 0.25$, and between the ropes and the pegs $D$ and $E$ is $\mu'_s = 0.5$, determine the smallest force $F$ needed to cause motion of block $B$ if $P = 30$ N.

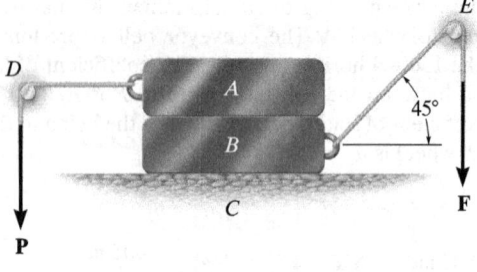

Prob. 7–110

**7–111.** Block $A$ has a weight of 500 N ($\approx$ 50 kg) and rests on a surface for which $\mu_s = 0.25$. If the coefficient of static friction between the cord and the fixed peg at $C$ is $\mu_s = 0.3$, determine the greatest weight of the suspended cylinder $B$ without causing motion.

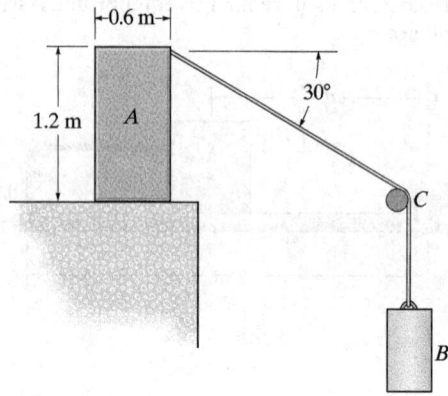

Prob. 7–111

**\*7–112.** Block $A$ has a mass of 50 kg and rests on surface $B$ for which $\mu_s = 0.25$. If the coefficient of static friction between the cord and the fixed peg at $C$ is $\mu'_s = 0.3$, determine the greatest mass of the suspended cylinder $D$ without causing motion.

**•7–113.** Block $A$ has a mass of 50 kg and rests on surface $B$ for which $\mu_s = 0.25$. If the mass of the suspended cylinder $D$ is 4 kg, determine the frictional force acting on $A$ and check if motion occurs. The coefficient of static friction between the cord and the fixed peg at $C$ is $\mu'_s = 0.3$.

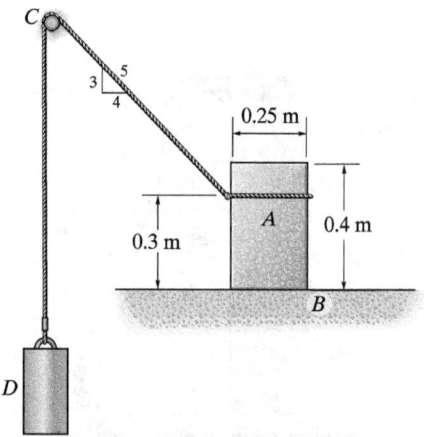

Prob. 7–112/113

## *7.6 Frictional Forces on Collar Bearings, Pivot Bearings, and Disks

*Pivot* and *collar bearings* are commonly used in machines to support an axial load on a rotating shaft. Typical examples are shown in Fig. 7–20. Provided these bearings are not lubricated, or are only partially lubricated, the laws of dry friction may be applied to determine the moment needed to turn the shaft when it supports an axial force.

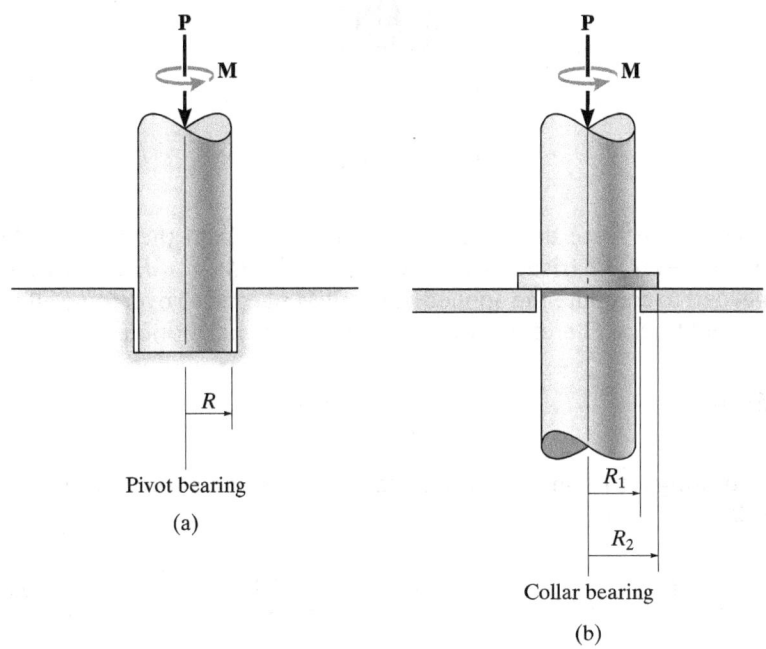

Fig. 7–20

**Frictional Analysis.** The collar bearing on the shaft shown in Fig. 7–21 is subjected to an axial force $\mathbf{P}$ and has a total bearing or contact area $\pi(R_2^2 - R_1^2)$. Provided the bearing is new and evenly supported, then the normal pressure $p$ on the bearing will be *uniformly distributed* over this area. Since $\Sigma F_z = 0$, then $p$, measured as a force per unit area, is $p = P/\pi(R_2^2 - R_1^2)$.

The moment needed to cause impending rotation of the shaft can be determined from moment equilibrium about the $z$ axis. A differential area element $dA = (r\,d\theta)(dr)$, shown in Fig. 7–21, is subjected to both a normal force $dN = p\,dA$ and an associated frictional force,

$$dF = \mu_s\,dN = \mu_s p\,dA = \frac{\mu_s P}{\pi(R_2^2 - R_1^2)}\,dA$$

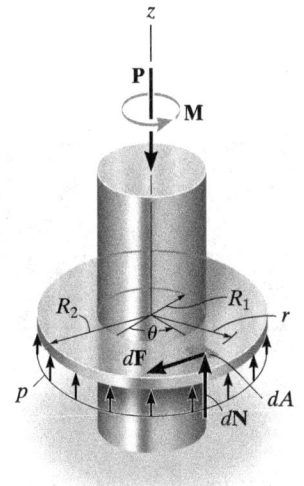

Fig. 7–21

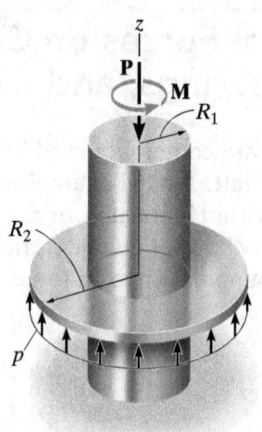

**Fig. 7–21 (Repeated)**

The normal force does not create a moment about the z axis of the shaft; however, the frictional force does; namely, $dM = r\, dF$. Integration is needed to compute the applied moment **M** needed to overcome all the frictional forces. Therefore, for impending rotational motion,

$$\Sigma M_z = 0; \qquad M - \int_A r\, dF = 0$$

Substituting for $dF$ and $dA$ and integrating over the entire bearing area yields

$$M = \int_{R_1}^{R_2}\int_0^{2\pi} r\left[\frac{\mu_s P}{\pi(R_2^2 - R_1^2)}\right](r\, d\theta\, dr) = \frac{\mu_s P}{\pi(R_2^2 - R_1^2)} \int_{R_1}^{R_2} r^2\, dr \int_0^{2\pi} d\theta$$

or

$$M = \frac{2}{3}\mu_s P\left(\frac{R_2^3 - R_1^3}{R_2^2 - R_1^2}\right) \qquad (7\text{–}7)$$

The moment developed at the end of the shaft, when it is *rotating* at constant speed, can be found by substituting $\mu_k$ for $\mu_s$ in Eq. 7–7.

In the case of a pivot bearing, Fig. 7–20a, then $R_2 = R$ and $R_1 = 0$, and Eq. 7–7 reduces to

$$M = \frac{2}{3}\mu_s PR \qquad (7\text{–}8)$$

Remember that Eqs. 7–7 and 7–8 apply only for bearing surfaces subjected to *constant pressure*. If the pressure is not uniform, a variation of the pressure as a function of the bearing area must be determined before integrating to obtain the moment. The following example illustrates this concept.

The motor that turns the disk of this sanding machine develops a torque that must overcome the frictional forces acting on the disk.

## EXAMPLE 7.9

The uniform bar shown in Fig. 7–22a has a weight of 20 N ($\approx$ 2 kg). If it is assumed that the normal pressure acting at the contacting surface varies linearly along the length of the bar as shown, determine the couple moment **M** required to rotate the bar. Assume that the bar's width is negligible in comparison to its length. The coefficient of static friction is equal to $\mu_s = 0.3$.

### SOLUTION
A free-body diagram of the bar is shown in Fig. 7–22b. The intensity $w_0$ of the distributed load at the center ($x = 0$) is determined from vertical force equilibrium, Fig. 7–22a.

$$+\uparrow \Sigma F_z = 0; \quad -20 \text{ N} + 2\left[\frac{1}{2}(0.5 \text{ m})w_0\right] = 0 \quad w_0 = 40 \text{ N/m}$$

Since $w = 0$ at $x = 0.5$ m, the distributed load expressed as a function of $x$ is

$$w = (40 \text{ N/m})\left(1 - \frac{x}{0.5 \text{ m}}\right) = 40 - 80x$$

The magnitude of the normal force acting on a differential segment of area having a length $dx$ is therefore

$$dN = w\,dx = (40 - 80x)dx$$

The magnitude of the frictional force acting on the same element of area is

$$dF = \mu_s\,dN = 0.3(40 - 80x)dx$$

Hence, the moment created by this force about the z axis is

$$dM = x\,dF = 0.3(40x - 80x^2)dx$$

The summation of moments about the z axis of the bar is determined by integration, which yields

$$\Sigma M_z = 0; \quad M - 2\int_0^{0.5}(0.3)(40x - 80x^2)\,dx = 0$$

$$M = 6\left(x^2 - \frac{4x^3}{3}\right)\Big|_0^{0.5}$$

$$M = 0.5 \text{ N} \cdot \text{m} \quad \textit{Ans.}$$

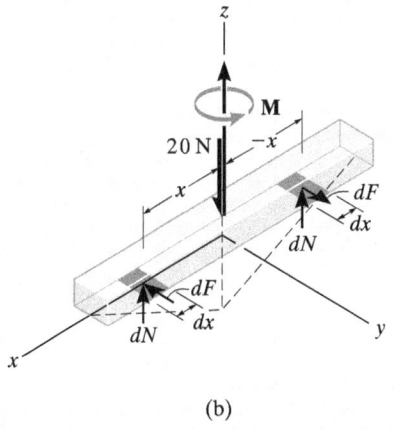

Fig. 7–22

## 7.7 Frictional Forces on Journal Bearings

When a shaft or axle is subjected to lateral loads, a *journal bearing* is commonly used for support. Provided the bearing is not lubricated, or is only partially lubricated, a reasonable analysis of the frictional resistance on the bearing can be based on the laws of dry friction.

**Frictional Analysis.** A typical journal-bearing support is shown in Fig. 7–23a. As the shaft rotates, the contact point moves up the wall of the bearing to some point $A$ where slipping occurs. If the vertical load acting at the end of the shaft is **P**, then the bearing reactive force **R** acting at $A$ will be equal and opposite to **P**, Fig. 7–23b. The moment needed to maintain constant rotation of the shaft can be found by summing moments about the $z$ axis of the shaft; i.e.,

$$\Sigma M_z = 0; \qquad M - (R \sin \phi_k)r = 0$$

or

$$M = Rr \sin \phi_k \qquad (7\text{–}9)$$

where $\phi_k$ is the angle of kinetic friction defined by $\tan \phi_k = F/N = \mu_k N/N = \mu_k$. In Fig. 7–23c, it is seen that $r \sin \phi_k = r_f$. The dashed circle with radius $r_f$ is called the *friction circle*, and as the shaft rotates, the reaction **R** will always be tangent to it. If the bearing is partially lubricated, $\mu_k$ is small, and therefore $\sin \phi_k \approx \tan \phi_k \approx \mu_k$. Under these conditions, a reasonable *approximation* to the moment needed to overcome the frictional resistance becomes

$$M \approx Rr\mu_k \qquad (7\text{–}10)$$

In practice, this type of journal bearing is not suitable for long service since friction between the shaft and bearing will wear down the surfaces. Instead, designers will incorporate "ball bearings" or "rollers" in journal bearings to minimize frictional losses.

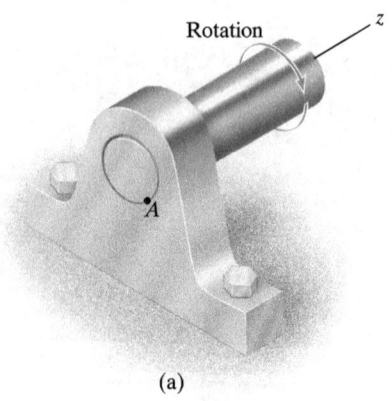

Unwinding the cable from this spool requires overcoming friction from the supporting shaft.

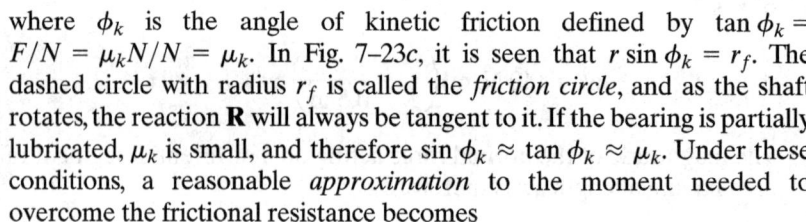

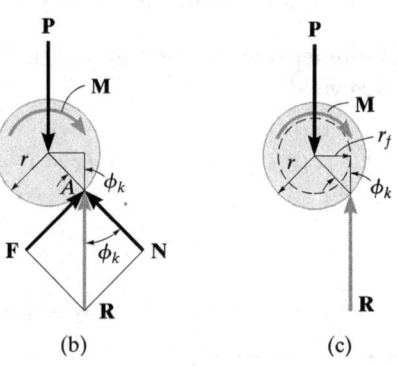

(a)

(b)

(c)

Fig. 7–23

## EXAMPLE 7.10

The 100-mm-diameter pulley shown in Fig. 7–24a fits loosely on a 10-mm-diameter shaft for which the coefficient of static friction is $\mu_s = 0.4$. Determine the minimum tension $T$ in the belt needed to (a) raise the 100-kg block and (b) lower the block. Assume that no slipping occurs between the belt and pulley and neglect the weight of the pulley.

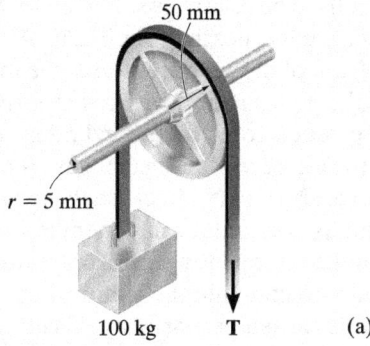

### SOLUTION

**Part (a).** A free-body diagram of the pulley is shown in Fig. 7–24b. When the pulley is subjected to belt tensions of 981 N each, it makes contact with the shaft at point $P_1$. As the tension $T$ is *increased*, the contact point will move around the shaft to point $P_2$ before motion impends. From the figure, the friction circle has a radius $r_f = r \sin \phi_s$. Using the simplification that $\sin \phi_s \approx \tan \phi_s \approx \mu_s$ then $r_f \approx r\mu_s = (5 \text{ mm})(0.4) = 2$ mm, so that summing moments about $P_2$ gives

$\zeta + \Sigma M_{P_2} = 0;$   $981 \text{ N}(52 \text{ mm}) - T(48 \text{ mm}) = 0$
$T = 1063 \text{ N} = 1.06 \text{ kN}$   *Ans.*

If a more exact analysis is used, then $\phi_s = \tan^{-1} 0.4 = 21.8°$. Thus, the radius of the friction circle would be $r_f = r \sin \phi_s = 5 \sin 21.8° = 1.86$ mm. Therefore,

$\zeta + \Sigma M_{P_2} = 0;$
$981 \text{ N}(50 \text{ mm} + 1.86 \text{ mm}) - T(50 \text{ mm} - 1.86 \text{ mm}) = 0$
$T = 1057 \text{ N} = 1.06 \text{ kN}$   *Ans.*

**Part (b).** When the block is lowered, the resultant force **R** acting on the shaft passes through point $P_3$ as shown in Fig. 7–24c. Summing moments about this point yields

$\zeta + \Sigma M_{P_3} = 0;$   $981 \text{ N}(48 \text{ mm}) - T(52 \text{ mm}) = 0$
$T = 906 \text{ N}$   *Ans.*

**NOTE:** The difference between raising and lowering the block is thus 157 N.

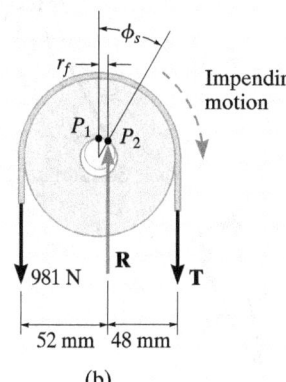

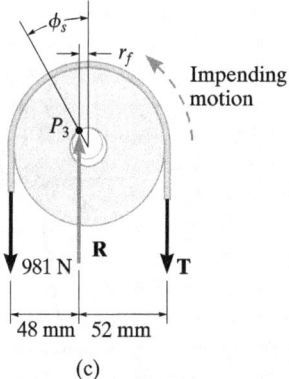

Fig. 7–24

## *7.8 Rolling Resistance

When a *rigid* cylinder rolls at constant velocity along a *rigid* surface, the normal force exerted by the surface on the cylinder acts perpendicular to the tangent at the point of contact, as shown in Fig. 7–25a. Actually, however, no materials are perfectly rigid, and therefore the reaction of the surface on the cylinder consists of a distribution of normal pressure. For example, consider the cylinder to be made of a very hard material, and the surface on which it rolls to be relatively soft. Due to its weight, the cylinder compresses the surface underneath it, Fig. 7–25b. As the cylinder rolls, the surface material in front of the cylinder *retards* the motion since it is being *deformed*, whereas the material in the rear is *restored* from the deformed state and therefore tends to *push* the cylinder forward. The normal pressures acting on the cylinder in this manner are represented in Fig. 7–25b by their resultant forces $\mathbf{N}_d$ and $\mathbf{N}_r$. Because the magnitude of the force of *deformation*, $\mathbf{N}_d$, and its horizontal component is *always greater* than that of *restoration*, $\mathbf{N}_r$, and consequently a horizontal driving force $\mathbf{P}$ must be applied to the cylinder to maintain the motion. Fig. 7–25b.*

Rolling resistance is caused primarily by this effect, although it is also, to a lesser degree, the result of surface adhesion and relative microsliding between the surfaces of contact. Because the actual force $\mathbf{P}$ needed to overcome these effects is difficult to determine, a simplified method will be developed here to explain one way engineers have analyzed this phenomenon. To do this, we will consider the resultant of the *entire* normal pressure, $\mathbf{N} = \mathbf{N}_d + \mathbf{N}_r$, acting on the cylinder, Fig. 7–25c. As shown in Fig. 7–25d, this force acts at an angle $\theta$ with the vertical. To keep the cylinder in equilibrium, i.e., rolling at a constant rate, it is necessary that $\mathbf{N}$ be *concurrent* with the driving force $\mathbf{P}$ and the weight $\mathbf{W}$. Summing moments about point $A$ gives $Wa = P(r \cos \theta)$. Since the deformations are generally very small in relation to the cylinder's radius, $\cos \theta \approx 1$; hence,

$$Wa \approx Pr$$

or

$$\boxed{P \approx \frac{Wa}{r}} \quad (7\text{–}11)$$

The distance $a$ is termed the *coefficient of rolling resistance*, which has the dimension of length. For instance, $a \approx 0.5$ mm for a wheel rolling on a rail, both of which are made of mild steel. For hardened

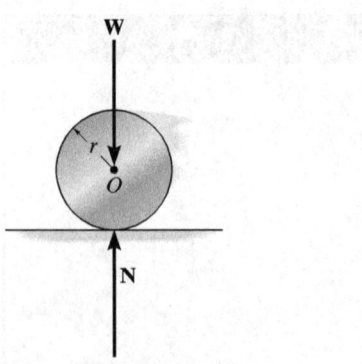

Rigid surface of contact
(a)

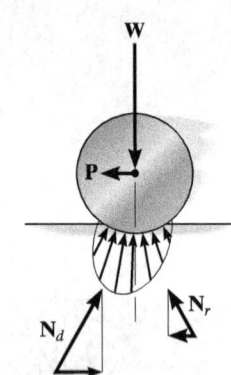

Soft surface of contact
(b)

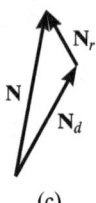

(c)

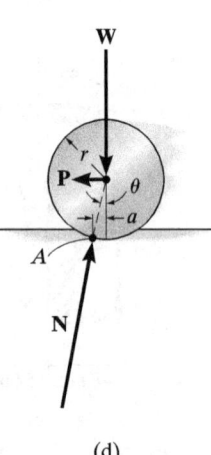

(d)

Fig. 7–25

---

*Actually, the deformation force $\mathbf{N}_d$ causes *energy* to be stored in the material as its magnitude is increased, whereas the restoration force $\mathbf{N}_r$, as its magnitude is decreased, allows some of this energy to be released. The remaining energy is *lost* since it is used to heat up the surface, and if the cylinder's weight is very large, it accounts for permanent deformation of the surface. Work must be done by the horizontal force $\mathbf{P}$ to make up for this loss.

steel ball bearings on steel, $a \approx 0.1$ mm. Experimentally, though, this factor is difficult to measure, since it depends on such parameters as the rate of rotation of the cylinder, the elastic properties of the contacting surfaces, and the surface finish. For this reason, little reliance is placed on the data for determining $a$. The analysis presented here does, however, indicate why a heavy load ($W$) offers greater resistance to motion ($P$) than a light load under the same conditions. Furthermore, since $Wa/r$ is generally very small compared to $\mu_k W$, the force needed to *roll* a cylinder over the surface will be much less than that needed to *slide* it across the surface. It is for this reason that a roller or ball bearings are often used to minimize the frictional resistance between moving parts.

Rolling resistance of railroad wheels on the rails is small since steel is very stiff. By comparison, the rolling resistance of the wheels of a tractor in a wet field is very large.

## EXAMPLE 7.11

A 10-kg steel wheel shown in Fig. 7–26a has a radius of 100 mm and rests on an inclined plane made of soft wood. If $\theta$ is increased so that the wheel begins to roll down the incline with constant velocity when $\theta = 1.2°$, determine the coefficient of rolling resistance.

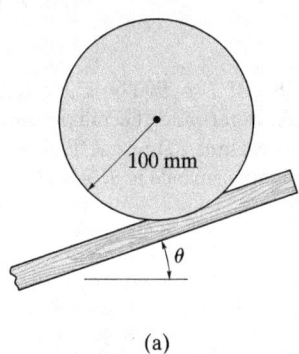

(a)

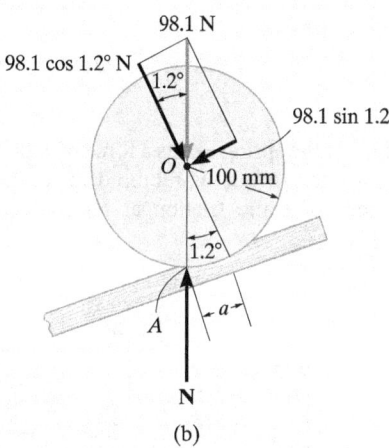

(b)

Fig. 7–26

### SOLUTION
As shown on the free-body diagram, Fig. 7–26b, when the wheel has impending motion, the normal reaction **N** acts at point $A$ defined by the dimension $a$. Resolving the weight into components parallel and perpendicular to the incline, and summing moments about point $A$, yields

$\zeta + \Sigma M_A = 0;$

$-(98.1 \cos 1.2° \text{ N})(a) + (98.1 \sin 1.2° \text{ N})(100 \cos 1.2° \text{ mm}) = 0$

Solving, we obtain

$a = 2.09$ mm          *Ans.*

## PROBLEMS

**7–114.** The collar bearing uniformly supports an axial force of $P = 4$ kN. If the coefficient of static friction is $\mu_s = 0.3$, determine the torque $M$ required to overcome friction.

**7–115.** The collar bearing uniformly supports an axial force of $P = 2.5$ kN. If a torque of $M = 5$ N·m is applied to the shaft and causes it to rotate at constant velocity, determine the coefficient of kinetic friction at the surface of contact.

**•7–117.** The *disk clutch* is used in standard transmissions of automobiles. If four springs are used to force the two plates $A$ and $B$ together, determine the force in each spring required to transmit a moment of $M = 1$ kN·m across the plates. The coefficient of static friction between $A$ and $B$ is $\mu_s = 0.3$.

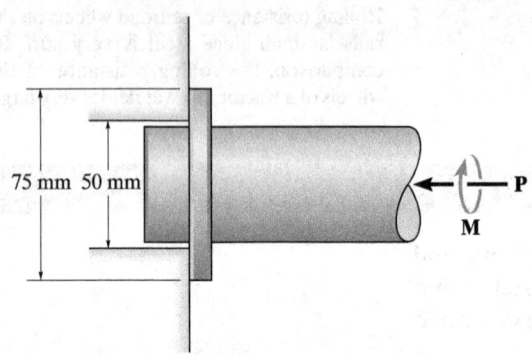

Probs. 7–114/115

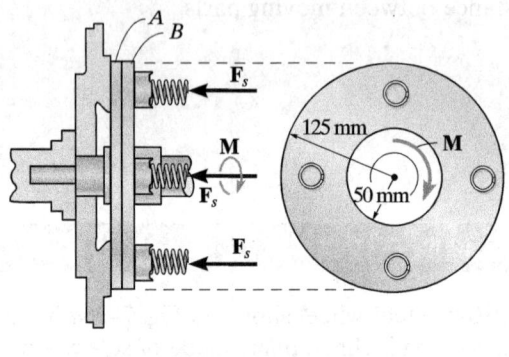

Prob. 7–117

**\*7–116.** If the spring exerts a force of 4500 N on the block, determine the torque $M$ required to rotate the shaft. The coefficient of static friction at all contacting surfaces is $\mu_s = 0.3$.

**7–118.** If $P = 900$ N is applied to the handle of the bell crank, determine the maximum torque $M$ the cone clutch can transmit. The coefficient of static friction at the contacting surface is $\mu_s = 0.3$.

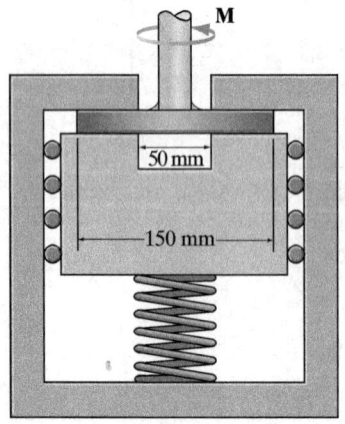

Prob. 7–116

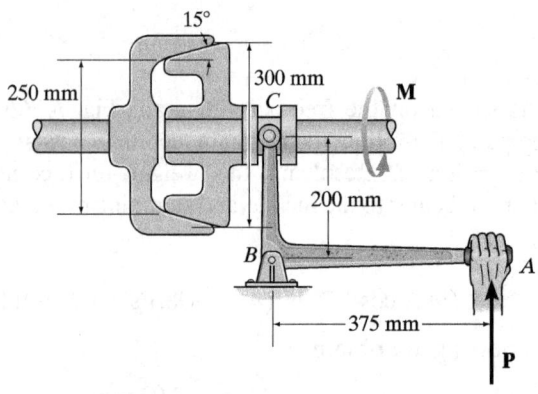

Prob. 7–118

**7–119.** Because of wearing at the edges, the pivot bearing is subjected to a conical pressure distribution at its surface of contact. Determine the torque $M$ required to overcome friction and turn the shaft, which supports an axial force **P**. The coefficient of static friction is $\mu_s$. For the solution, it is necessary to determine the peak pressure $p_0$ in terms of $P$ and the bearing radius $R$.

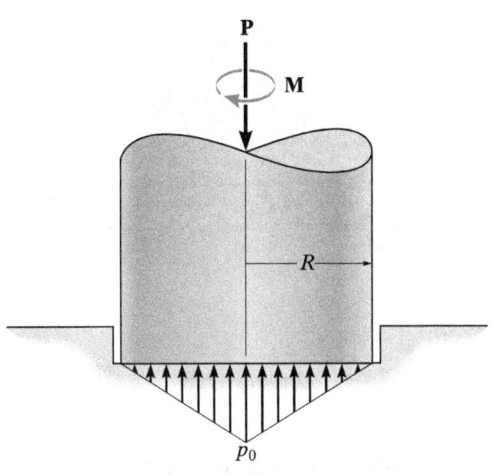

Prob. 7–119

**•7–121.** The shaft is subjected to an axial force **P**. If the reactive pressure on the conical bearing is uniform, determine the torque $M$ that is just sufficient to rotate the shaft. The coefficient of static friction at the contacting surface is $\mu_s$.

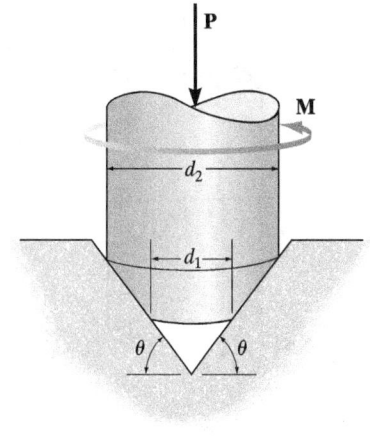

Prob. 7–121

**\*7–120.** The pivot bearing is subjected to a parabolic pressure distribution at its surface of contact. If the coefficient of static friction is $\mu_s$, determine the torque $M$ required to overcome friction and turn the shaft if it supports an axial force **P**.

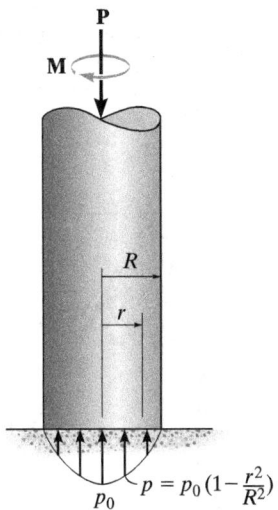

Prob. 7–120

**7–122.** The tractor is used to push the 7.5-kN ($\approx$ 750 kg) pipe. To do this it must overcome the frictional forces at the ground, caused by sand. Assuming that the sand exerts a pressure on the bottom of the pipe as shown, and the coefficient of static friction between the pipe and the sand is $\mu_s = 0.3$, determine the horizontal force required to push the pipe forward. Also, determine the peak pressure $p_0$.

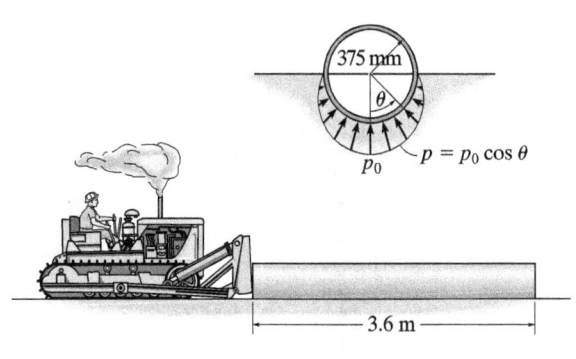

Prob. 7–122

**7–123.** The conical bearing is subjected to a constant pressure distribution at its surface of contact. If the coefficient of static friction is $\mu_s$, determine the torque $M$ required to overcome friction if the shaft supports an axial force **P**.

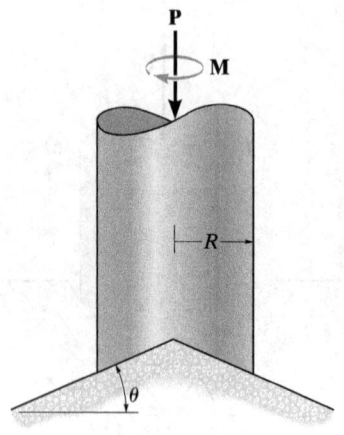

Prob. 7–123

***7–124.** Assuming that the variation of pressure at the bottom of the pivot bearing is defined as $p = p_0(R_2/r)$, determine the torque $M$ needed to overcome friction if the shaft is subjected to an axial force **P**. The coefficient of static friction is $\mu_s$. For the solution, it is necessary to determine $p_0$ in terms of $P$ and the bearing dimensions $R_1$ and $R_2$.

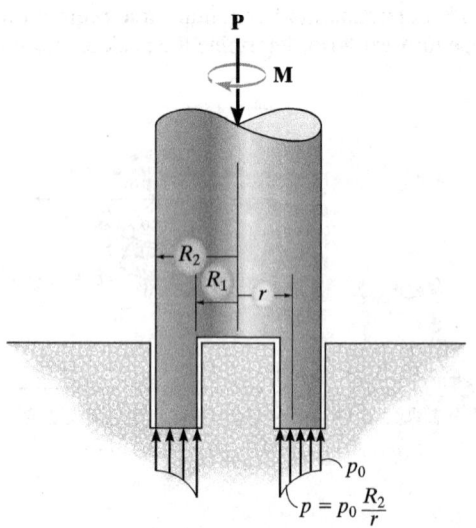

Prob. 7–124

**•7–125.** The shaft of radius $r$ fits loosely on the journal bearing. If the shaft transmits a vertical force **P** to the bearing and the coefficient of kinetic friction between the shaft and the bearing is $\mu_k$, determine the torque $M$ required to turn the shaft with constant velocity.

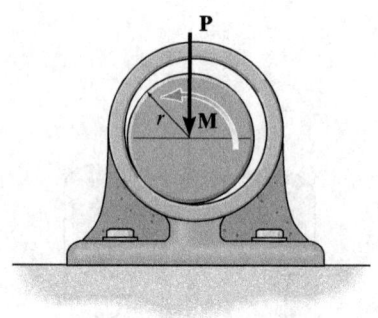

Prob. 7–125

**7–126.** The pulley is supported by a 25-mm-diameter pin. If the pulley fits loosely on the pin, determine the smallest force $P$ required to raise the bucket. The bucket has a mass of 20 kg and the coefficient of static friction between the pulley and the pin is $\mu_s = 0.3$. Neglect the mass of the pulley and assume that the cable does not slip on the pulley.

**7–127.** The pulley is supported by a 25-mm-diameter pin. If the pulley fits loosely on the pin, determine the largest force $P$ that can be applied to the rope and yet lower the bucket. The bucket has a mass of 20 kg and the coefficient of static friction between the pulley and the pin is $\mu_s = 0.3$. Neglect the mass of the pulley and assume that the cable does not slip on the pulley.

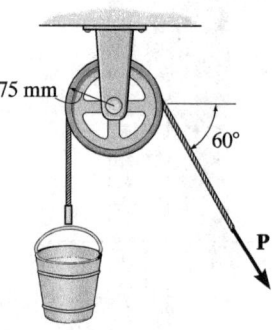

Probs. 7–126/127

**\*7–128.** The cylinders are suspended from the end of the bar which fits loosely into a 40-mm-diameter pin. If A has a mass of 10 kg, determine the required mass of B which is just sufficient to keep the bar from rotating clockwise. The coefficient of static friction between the bar and the pin is $\mu_s = 0.3$. Neglect the mass of the bar.

**•7–129.** The cylinders are suspended from the end of the bar which fits loosely into a 40-mm-diameter pin. If A has a mass of 10 kg, determine the required mass of B which is just sufficient to keep the bar from rotating counterclockwise. The coefficient of static friction between the bar and the pin is $\mu_s = 0.3$. Neglect the mass of the bar.

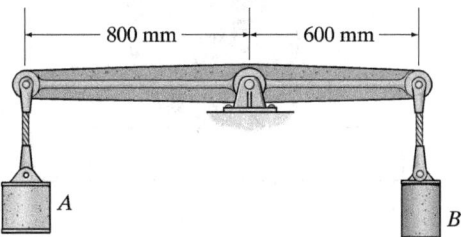

Probs. 7–128/129

**7–130.** The connecting rod is attached to the piston by a 20-mm-diameter pin at B and to the crank shaft by a 50-mm-diameter bearing A. If the piston is moving downwards, and the coefficient of static friction at the contact points is $\mu_s = 0.2$, determine the radius of the friction circle at each connection.

**7–131.** The connecting rod is attached to the piston by a 20-mm-diameter pin at B and to the crank shaft by a 50-mm-diameter bearing A. If the piston is moving upwards, and the coefficient of static friction at the contact points is $\mu_s = 0.3$, determine the radius of the friction circle at each connection.

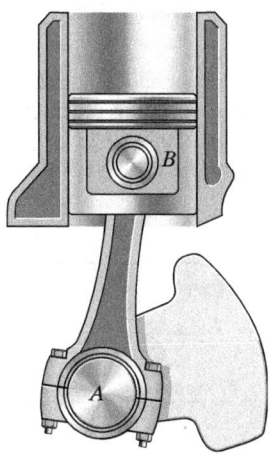

Probs. 7–130/131

**\*7–132.** The 5-kg pulley has a diameter of 240 mm and the axle has a diameter of 40 mm. If the coefficient of kinetic friction between the axle and the pulley is $\mu_k = 0.15$, determine the vertical force P on the rope required to lift the 80-kg block at constant velocity.

**•7–133.** Solve Prob. 7–132 if the force **P** is applied horizontally to the right.

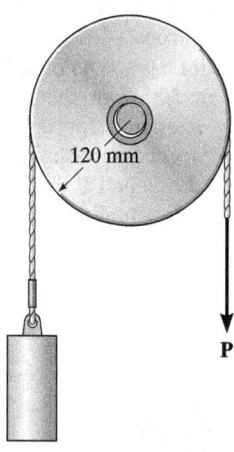

Probs. 7–132/133

**7–134.** The bell crank fits loosely into a 12-mm-diameter pin. Determine the required force P which is just sufficient to rotate the bell crank clockwise. The coefficient of static friction between the pin and the bell crank is $\mu_s = 0.3$.

**7–135.** The bell crank fits loosely into a 12-mm-diameter pin. If $P = 205$ N, the bell crank is then on the verge of rotating counterclockwise. Determine the coefficient of static friction between the pin and the bell crank.

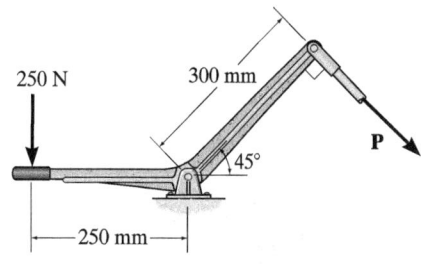

Probs. 7–134/135

**\*7–136.** The wagon together with the load weighs 750 N (≈ 75 kg). If the coefficient of rolling resistance is $a = 0.75$ mm, determine the force $P$ required to pull the wagon with constant velocity.

Prob. 7–136

**•7–137.** The lawn roller has a mass of 80 kg. If the arm $BA$ is held at an angle of 30° from the horizontal and the coefficient of rolling resistance for the roller is 25 mm, determine the force $P$ needed to push the roller at constant speed. Neglect friction developed at the axle, $A$, and assume that the resultant force **P** acting on the handle is applied along arm $BA$.

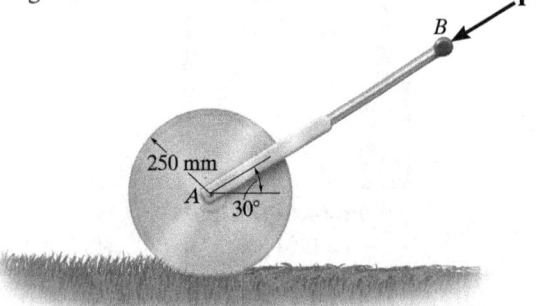

Prob. 7–137

**7–138.** Determine the force $P$ required to overcome rolling resistance and pull the 50-kg roller up the inclined plane with constant velocity. The coefficient of rolling resistance is $a = 15$ mm.

**7–139.** Determine the force $P$ required to overcome rolling resistance and support the 50-kg roller if it rolls down the inclined plane with constant velocity. The coefficient of rolling resistance is $a = 15$ mm.

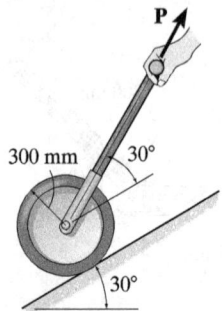

Probs. 7–138/139

**\*7–140.** The cylinder is subjected to a load that has a weight W. If the coefficients of rolling resistance for the cylinder's top and bottom surfaces are $a_A$ and $a_B$, respectively, show that a horizontal force having a magnitude of $P = [W(a_A + a_B)]/2r$ is required to move the load and thereby roll the cylinder forward. Neglect the weight of the cylinder.

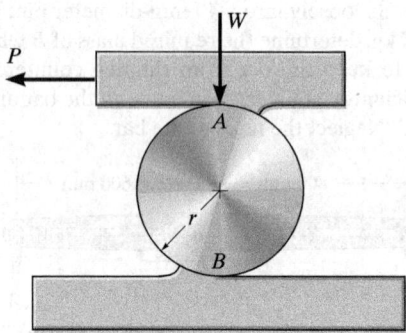

Prob. 7–140

**•7–141.** The 1.2-Mg steel beam is moved over a level surface using a series of 30-mm-diameter rollers for which the coefficient of rolling resistance is 0.4 mm at the ground and 0.2 mm at the bottom surface of the beam. Determine the horizontal force $P$ needed to push the beam forward at a constant speed. *Hint:* Use the result of Prob. 7–140.

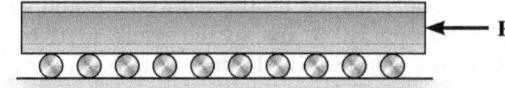

Prob. 7–141

**7–142.** Determine the smallest horizontal force $P$ that must be exerted on the 100-kg block to move it forward. The rollers each weigh 250 N (≈ 25 kg), and the coefficient of rolling resistance at the top and bottom surfaces is $a = 5$ mm.

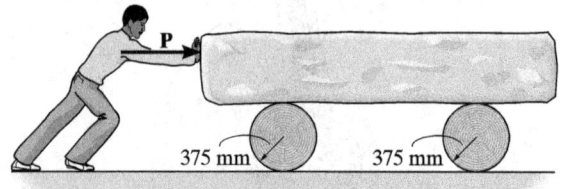

Prob. 7–142

# CHAPTER REVIEW

**Dry Friction**

Frictional forces exist between two rough surfaces of contact. These forces act on a body so as to oppose its motion or tendency of motion.

A static frictional force approaches a maximum value of $F_s = \mu_s N$, where $\mu_s$ is the *coefficient of static friction*. In this case, motion between the contacting surfaces is *impending*.

If slipping occurs, then the friction force remains essentially constant and equal to $F_k = \mu_k N$. Here $\mu_k$ is the *coefficient of kinetic friction*.

The solution of a problem involving friction requires first drawing the free-body diagram of the body. If the unknowns cannot be determined strictly from the equations of equilibrium, and the possibility of slipping occurs, then the friction equation should be applied at the appropriate points of contact in order to complete the solution.

It may also be possible for slender objects, like crates, to tip over, and this situation should also be investigated.

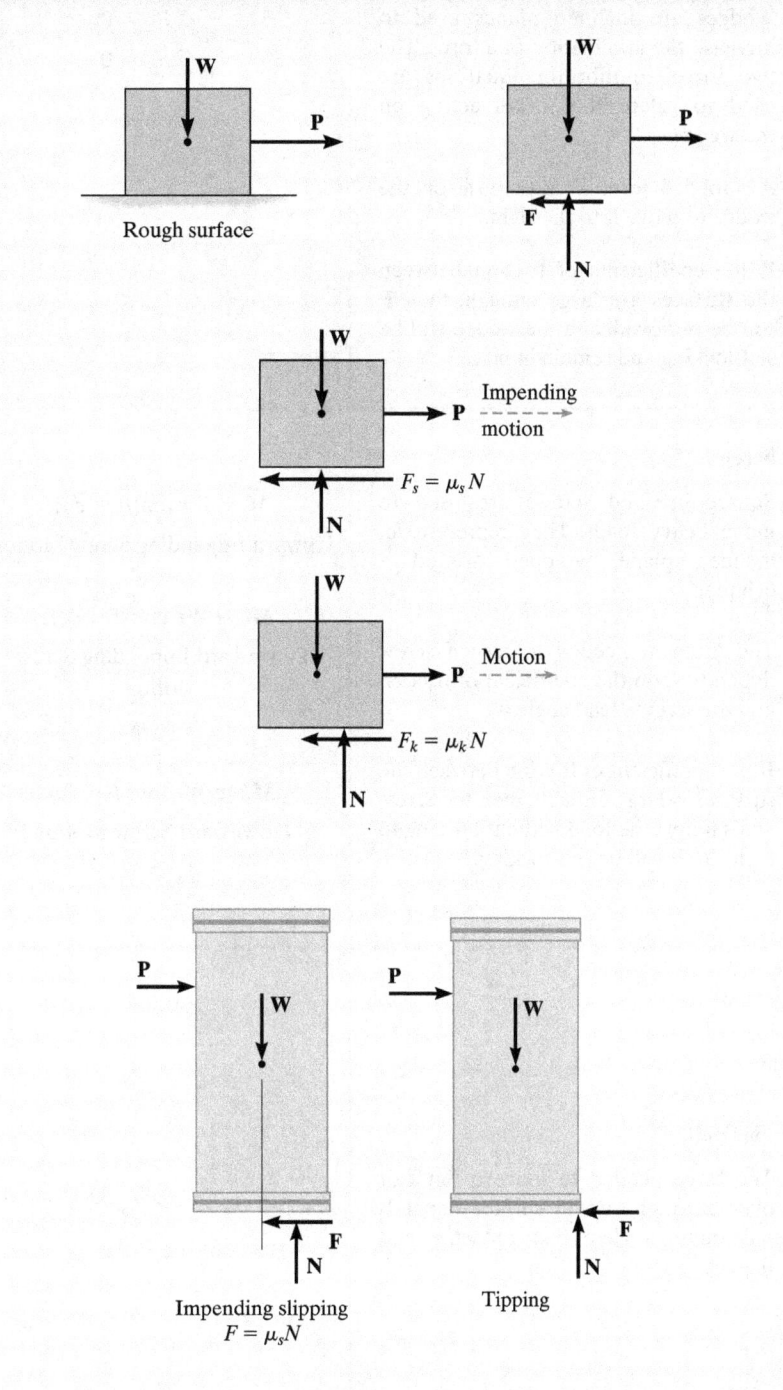

## Wedges

Wedges are inclined planes used to increase the application of a force. The two force equilibrium equations are used to relate the forces acting on the wedge.

An applied force **P** must push on the wedge to move it to the right.

If the coefficients of friction between the surfaces are large enough, then **P** can be removed, and the wedge will be self-locking and remain in place.

$\Sigma F_x = 0$

$\Sigma F_y = 0$

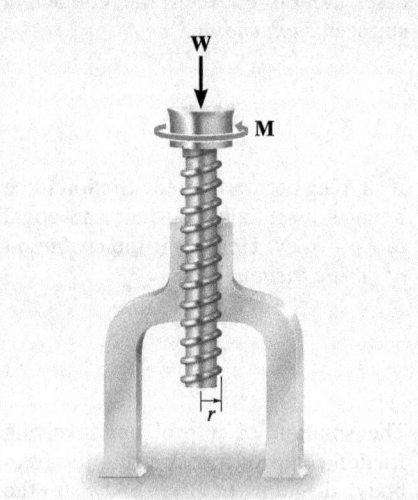

## Screws

Square-threaded screws are used to move heavy loads. They represent an inclined plane, wrapped around a cylinder.

The moment needed to turn a screw depends upon the coefficient of friction and the screw's lead angle $\theta$.

If the coefficient of friction between the surfaces is large enough, then the screw will support the load without tending to turn, i.e., it will be self-locking.

$M = Wr \tan(\theta + \phi_s)$
Upward Impending Screw Motion

$M' = Wr \tan(\theta - \phi_s)$
Downward Impending Screw Motion
$\theta > \phi$

$M'' = Wr \tan(\phi - \theta_s)$
Downward Screw Motion
$\phi_s > \theta$

## Flat Belts

The force needed to move a flat belt over a rough curved surface depends only on the angle of belt contact, $\beta$, and the coefficient of friction.

$T_2 = T_1 e^{\mu\beta}$

$T_2 > T_1$

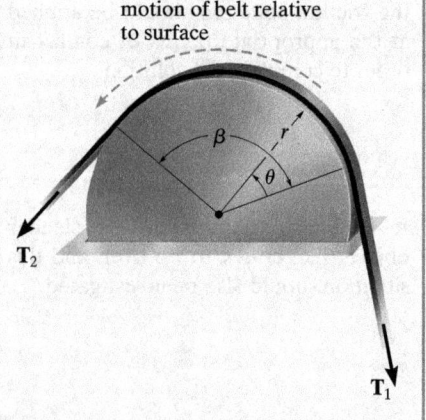

## Collar Bearings and Disks

The frictional analysis of a collar bearing or disk requires looking at a differential element of the contact area. The normal force acting on this element is determined from force equilibrium along the shaft, and the moment needed to turn the shaft at a constant rate is determined from moment equilibrium about the shaft's axis.

If the pressure on the surface of a collar bearing is uniform, then integration gives the result shown.

$$M = \frac{2}{3}\mu_s P\left(\frac{R_2^3 - R_1^3}{R_2^2 - R_1^2}\right)$$

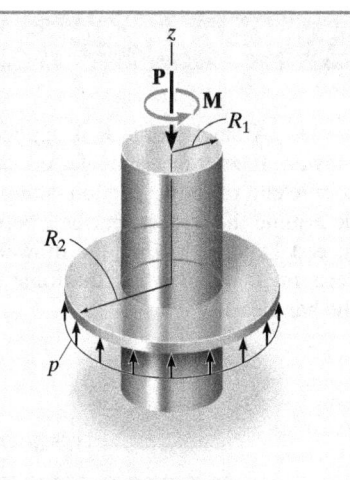

## Journal Bearings

When a moment is applied to a shaft in a nonlubricated or partially lubricated journal bearing, the shaft will tend to roll up the side of the bearing until slipping occurs. This defines the radius of a friction circle, and from it the moment needed to turn the shaft can be determined.

$$M = Rr \sin \phi_k$$

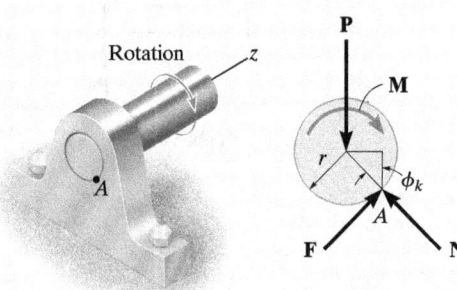

## Rolling Resistance

The resistance of a wheel to rolling over a surface is caused by localized *deformation* of the two materials in contact. This causes the resultant normal force acting on the rolling body to be inclined so that it provides a component that acts in the opposite direction of the applied force **P** causing the motion. This effect is characterized using the *coefficient of rolling resistance, a*, which is determined from experiment.

$$P \approx \frac{Wa}{r}$$

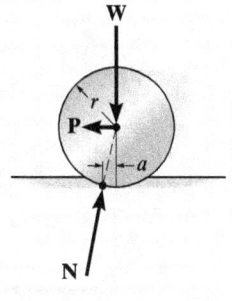

## REVIEW PROBLEMS

**7–143.** A single force **P** is applied to the handle of the drawer. If friction is neglected at the bottom and the coefficient of static friction along the sides is $\mu_s = 0.4$, determine the largest spacing $s$ between the symmetrically placed handles so that the drawer does not bind at the corners $A$ and $B$ when the force **P** is applied to one of the handles.

**•7–145.** The truck has a mass of 1.25 Mg and a center of mass at $G$. Determine the greatest load it can pull if (a) the truck has rear-wheel drive while the front wheels are free to roll, and (b) the truck has four-wheel drive. The coefficient of static friction between the wheels and the ground is $\mu_s = 0.5$, and between the crate and the ground is $\mu_s' = 0.4$.

**7–146.** Solve Prob. 7–145 if the truck and crate are traveling up a 10° incline.

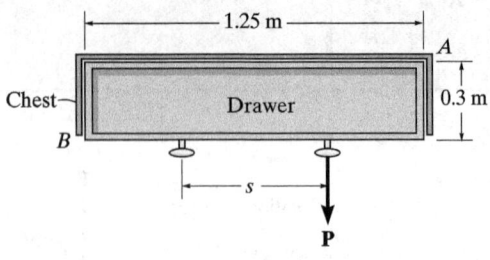

Prob. 7–143

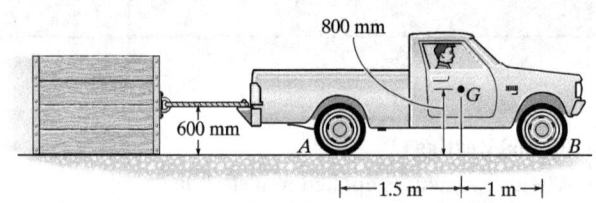

Probs. 7–145/146

**\*7–144.** The semicircular thin hoop of weight $W$ and center of gravity at $G$ is suspended by the small peg at $A$. A horizontal force **P** is slowly applied at $B$. If the hoop begins to slip at $A$ when $\theta = 30°$, determine the coefficient of static friction between the hoop and the peg.

**7–147.** If block $A$ has a mass of 1.5 kg, determine the largest mass of block $B$ without causing motion of the system. The coefficient of static friction between the blocks and inclined planes is $\mu_s = 0.2$.

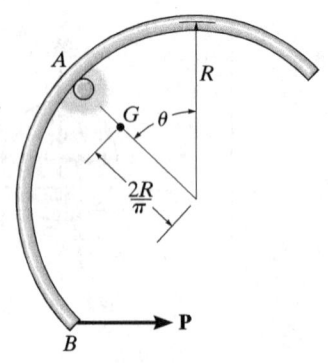

Prob. 7–144

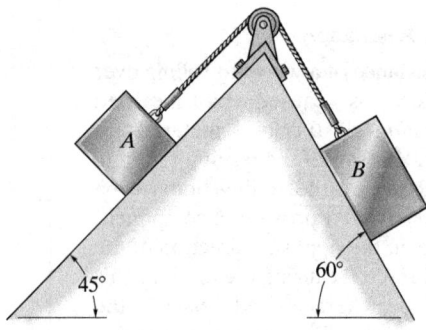

Prob. 7–147

REVIEW PROBLEMS 387

**\*7–148.** The cone has a weight $W$ and center of gravity at $G$. If a horizontal force **P** is gradually applied to the string attached to its vertex, determine the maximum coefficient of static friction for slipping to occur.

**7–151.** A roofer, having a mass of 70 kg, walks slowly in an upright position down along the surface of a dome that has a radius of curvature of $r = 20$ m. If the coefficient of static friction between his shoes and the dome is $\mu_s = 0.7$, determine the angle $\theta$ at which he first begins to slip.

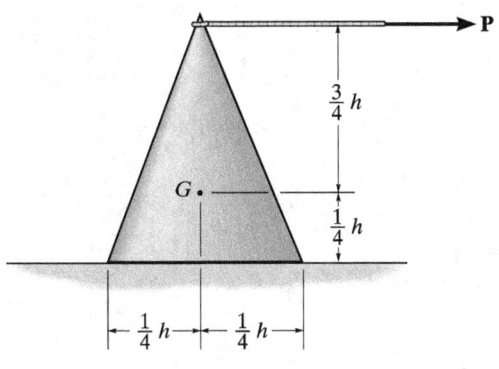

Prob. 7–148

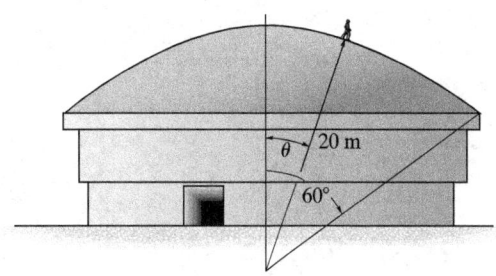

Prob. 7–151

**•7–149.** The tractor pulls on the fixed tree stump. Determine the torque that must be applied by the engine to the rear wheels to cause them to slip. The front wheels are free to roll. The tractor weighs 17.4 kN and has a center of gravity at $G$. The coefficient of static friction between the rear wheels and the ground is $\mu_s = 0.5$.

**7–150.** The tractor pulls on the fixed tree stump. If the coefficient of static friction between the rear wheels and the ground is $\mu_s = 0.6$, determine if the rear wheels slip or the front wheels lift off the ground as the engine provides torque to the rear wheels. What is the torque needed to cause this motion? The front wheels are free to roll. The tractor weighs 12.5 kN and has a center of gravity at $G$.

**\*7–152.** Column $D$ is subjected to a vertical load of 40 kN. It is supported on two identical wedges $A$ and $B$ for which the coefficient of static friction at the contacting surfaces between $A$ and $B$ and between $B$ and $C$ is $\mu_s = 0.4$. Determine the force $P$ needed to raise the column and the equilibrium force $P'$ needed to hold wedge $A$ stationary. The contacting surface between $A$ and $D$ is smooth.

**•7–153.** Column $D$ is subjected to a vertical load of 40 kN. It is supported on two identical wedges $A$ and $B$ for which the coefficient of static friction at the contacting surfaces between $A$ and $B$ and between $B$ and $C$ is $\mu_s = 0.4$. If the forces **P** and **P'** are removed, are the wedges self-locking? The contacting surface between $A$ and $D$ is smooth.

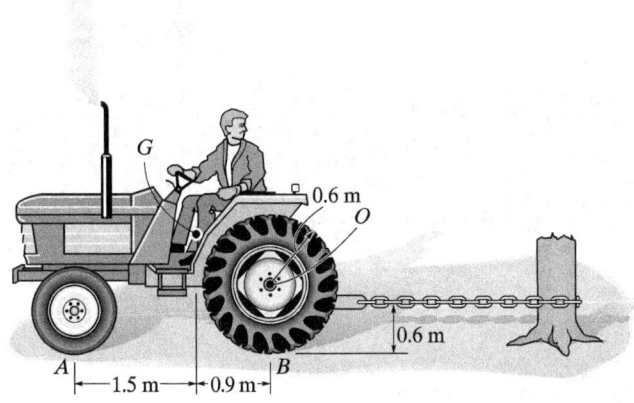

Probs. 7–149/150

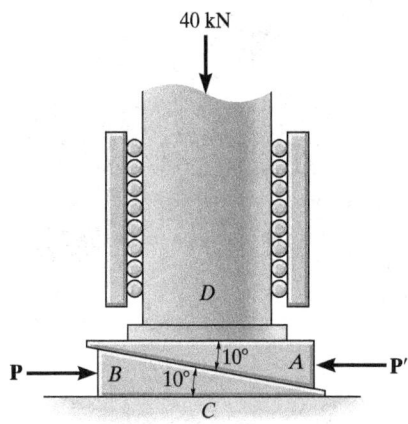

Probs. 7–152/153

Equilibrium and stability of this articulated crane boom as a function of the boom position can be analyzed using methods based on work and energy, which are explained in this chapter.

# Virtual Work

## 8

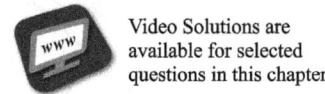

Video Solutions are available for selected questions in this chapter.

### CHAPTER OBJECTIVES

- To introduce the principle of virtual work and show how it applies to finding the equilibrium configuration of a system of pin-connected members.

- To establish the potential-energy function and use the potential-energy method to investigate the type of equilibrium or stability of a rigid body or system of pin-connected members.

## 8.1  Definition of Work

The *principle of virtual work* was proposed by the Swiss mathematician Jean Bernoulli in the eighteenth century. It provides an alternative method for solving problems involving the equilibrium of a particle, a rigid body, or a system of connected rigid bodies. Before we discuss this principle, however, we must first define the work produced by a force and by a couple moment.

# Chapter 8 Virtual Work

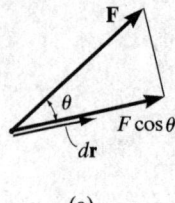

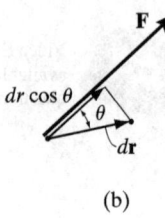

Fig. 8–1

**Work of a Force.** A force does work when it undergoes a displacement in the direction of its line of action. Consider, for example, the force **F** in Fig. 8–1a that undergoes a differential displacement $d\mathbf{r}$. If $\theta$ is the angle between the force and the displacement, then the component of **F** in the direction of the displacement is $F \cos \theta$. And so the work produced by **F** is

$$dU = F \, dr \cos \theta$$

Notice that this expression is also the product of the force $F$ and the component of displacement in the direction of the force, $dr \cos \theta$, Fig. 8–1b. If we use the definition of the dot product (Eq. 2–14), the work can also be written as

$$dU = \mathbf{F} \cdot d\mathbf{r}$$

As the above equations indicate, work is a *scalar*, and like other scalar quantities, it has a magnitude that can either be *positive* or *negative*.

In the SI system, the unit of work is a *joule* (J), which is the work produced by a 1-N force that displaces through a distance of 1 m in the direction of the force (1 J = 1 N·m).

The moment of a force has this same combination of units; however, the concepts of moment and work are in no way related. A moment is a vector quantity, whereas work is a scalar.

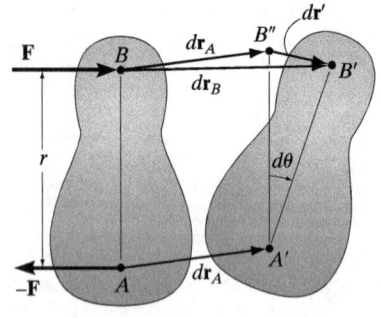

Fig. 8–2

**Work of a Couple Moment.** The rotation of a couple moment also produces work. Consider the rigid body in Fig. 8–2, which is acted upon by the couple forces **F** and –**F** that produce a couple moment **M** having a magnitude $M = Fr$. When the body undergoes the differential displacement shown, points $A$ and $B$ move $d\mathbf{r}_A$ and $d\mathbf{r}_B$ to their final positions $A'$ and $B'$, respectively. Since $d\mathbf{r}_B = d\mathbf{r}_A + d\mathbf{r}'$, this movement can be thought of as a *translation* $d\mathbf{r}_A$, where $A$ and $B$ move to $A'$ and $B''$, and a *rotation* about $A'$, where the body rotates through the angle $d\theta$ about $A$. The couple forces do no work during the translation $d\mathbf{r}_A$ because each force undergoes the same amount of displacement in opposite directions, thus canceling out the work. During rotation, however, **F** is displaced $dr'' = r \, d\theta$, and so it does work $dU = F \, dr'' = F r \, d\theta$. Since $M = Fr$, the work of the couple moment **M** is therefore

$$dU = M \, d\theta$$

If **M** and $d\theta$ have the same sense, the work is *positive*; however, if they have the opposite sense, the work will be *negative*.

**Virtual Work.** The definitions of the work of a force and a couple have been presented in terms of *actual movements* expressed by differential displacements having magnitudes of $dr$ and $d\theta$. Consider now an *imaginary* or *virtual movement* of a body in static equilibrium, which indicates a displacement or rotation that is *assumed* and *does not actually exist*. These movements are first-order differential quantities and will be denoted by the symbols $\delta r$ and $\delta\theta$ (delta $r$ and delta $\theta$), respectively. The *virtual work* done by a force having a virtual displacement $\delta r$ is

$$\delta U = F \cos\theta \, \delta r \qquad (8\text{--}1)$$

Similarly, when a couple undergoes a virtual rotation $\delta\theta$ in the plane of the couple forces, the *virtual work* is

$$\delta U = M \, \delta\theta \qquad (8\text{--}2)$$

## 8.2 Principle of Virtual Work

The principle of virtual work states that, if a body is in equilibrium, the algebraic sum of the virtual work done by all the forces and couple moments acting on the body is zero for any virtual displacement of the body. Thus,

$$\delta U = 0 \qquad (8\text{--}3)$$

For example, consider the free-body diagram of the particle (ball) that rests on the floor, Fig. 8–3. If we "imagine" the ball to be displaced downwards a virtual amount $\delta y$, then the weight does positive virtual work, $W\,\delta y$, and the normal force does negative virtual work, $-N\,\delta y$. For equilibrium, the total virtual work must be zero so that $\delta U = W\,\delta y - N\,\delta y = (W-N)\,\delta y = 0$. Since $\delta y \neq 0$, then $N = W$ as required by applying $\Sigma F_y = 0$.

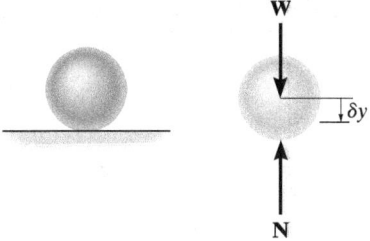

**Fig. 8–3**

In a similar manner, we can also apply the virtual-work equation $\delta U = 0$ to a rigid body subjected to a coplanar force system. Here, separate virtual translations in the $x$ and $y$ directions and a virtual rotation about an axis perpendicular to the $x$–$y$ plane that passes through an arbitrary point $O$, will correspond to the three equilibrium equations, $\Sigma F_x = 0$, $\Sigma F_y = 0$, and $\Sigma M_O = 0$. When writing these equations, it is *not necessary* to include the work done by the *internal forces* acting within the body since a rigid body *does not deform* when subjected to an external loading, and furthermore, when the body moves through a virtual displacement, the internal forces occur in equal but opposite collinear pairs, so that the corresponding work done by each pair of forces will cancel.

To demonstrate an application, consider the simply supported beam in Fig. 8–4a. When the beam is given a virtual rotation $\delta\theta$ about point $B$, Fig. 8–4b, the only forces that do work are $\mathbf{P}$ and $\mathbf{A}_y$. Since $\delta y = l\,\delta\theta$ and $\delta y' = (l/2)\,\delta\theta$, the virtual work equation for this case is $\delta U = A_y(l\,\delta\theta) - P(l/2)\,\delta\theta = (A_y l - Pl/2)\,\delta\theta = 0$. Since $\delta\theta \neq 0$, then $A_y = P/2$. Excluding $\delta\theta$, notice that the terms in parentheses actually represent the application of $\Sigma M_B = 0$.

As seen from the above two examples, no added advantage is gained by solving particle and rigid-body equilibrium problems using the principle of virtual work. This is because, for each application of the virtual-work equation, the virtual displacement, common to every term, factors out, leaving an equation that could have been obtained in a more *direct manner* by simply applying an equation of equilibrium.

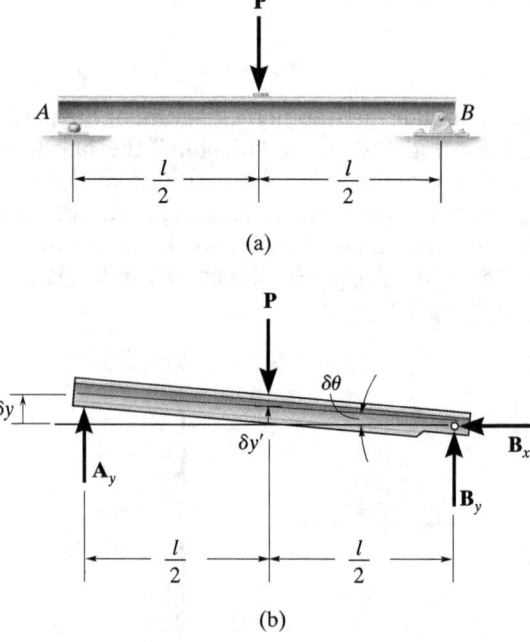

Fig. 8–4

# 8.3 Principle of Virtual Work for a System of Connected Rigid Bodies

The method of virtual work is particularly effective for solving equilibrium problems that involve a system of several *connected* rigid bodies, such as the ones shown in Fig. 8–5.

Each of these systems is said to have only one degree of freedom since the arrangement of the links can be completely specified using only one coordinate $\theta$. In other words, with this single coordinate and the length of the members, we can locate the position of the forces **F** and **P**.

In this text, we will only consider the application of the principle of virtual work to systems containing one degree of freedom*. Because they are less complicated, they will serve as a way to approach the solution of more complex problems involving systems with many degrees of freedom. The procedure for solving problems involving a system of frictionless connected rigid bodies follows.

Please refer to the Companion CD for the animation: *Principle of Virtual Work*

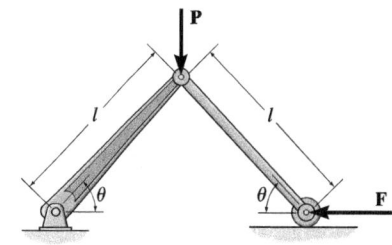

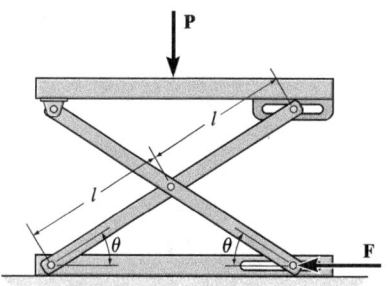

**Fig. 8–5**

## Important Points

- A force does work when it moves through a displacement in the direction of the force. A couple moment does work when it moves through a collinear rotation. Specifically, positive work is done when the force or couple moment and its displacement have the same sense of direction.
- The principle of virtual work is generally used to determine the equilibrium configuration for a system of multiply connected members.
- A virtual displacement is imaginary; i.e., it does not really happen. It is a differential displacement that is given in the positive direction of a position coordinate.
- Forces or couple moments that do not virtually displace do no virtual work.

This scissors lift has one degree of freedom. Without the need for dismembering the mechanism, the force in the hydraulic cylinder $AB$ required to provide the lift can be determined *directly* by using the principle of virtual work.

---

*This method of applying the principle of virtual work is sometimes called the *method of virtual displacements* because a virtual displacement is applied, resulting in the calculation of a real force. Although it is not used here, we can also apply the principle of virtual work as a *method of virtual forces*. This method is often used to apply a virtual force and then determine the displacements of points on deformable bodies. See R. C. Hibbeler, *Mechanics of Materials*, 7th edition, Pearson/Prentice Hall, 2007.

## Procedure for Analysis

**Free-Body Diagram.**

- Draw the free-body diagram of the entire system of connected bodies and define the *coordinate q*.

- Sketch the "deflected position" of the system on the free-body diagram when the system undergoes a *positive* virtual displacement $\delta q$.

**Virtual Displacements.**

- Indicate *position coordinates s*, each measured from a *fixed point* on the free-body diagram. These coordinates are directed to the forces that do work.

- Each of these coordinate axes should be *parallel* to the line of action of the force to which it is directed, so that the virtual work along the coordinate axis can be calculated.

- Relate each of the position coordinates $s$ to the coordinate $q$; then *differentiate* these expressions in order to express each virtual displacement $\delta s$ in terms of $\delta q$.

**Virtual-Work Equation.**

- Write the *virtual-work equation* for the system assuming that, whether possible or not, each position coordinate $s$ undergoes a *positive* virtual displacement $\delta s$. If a force or couple moment is in the same direction as the positive virtual displacement, the work is positive. Otherwise, it is negative.

- Express the work of *each* force and couple moment in the equation in terms of $\delta q$.

- Factor out this common displacement from all the terms, and solve for the unknown force, couple moment, or equilibrium position $q$.

## EXAMPLE 8.1

Determine the angle $\theta$ for equilibrium of the two-member linkage shown in Fig. 8–6a. Each member has a mass of 10 kg.

### SOLUTION

**Free-Body Diagram.** The system has only one degree of freedom since the location of both links can be specified by the single coordinate ($q =$) $\theta$. As shown on the free-body diagram in Fig. 8–6b, when $\theta$ has a *positive* (clockwise) virtual rotation $\delta\theta$, only the force **F** and the two 98.1-N weights do work. (The reactive forces $\mathbf{D}_x$ and $\mathbf{D}_y$ are fixed, and $\mathbf{B}_y$ does not displace along its line of action.)

**Virtual Displacements.** If the origin of coordinates is established at the *fixed* pin support $D$, then the position of **F** and **W** can be specified by the *position coordinates* $x_B$ and $y_w$. In order to determine the work, note that, as required, these coordinates are parallel to the lines of action of their associated forces. Expressing these position coordinates in terms of $\theta$ and taking the derivatives yields

$$x_B = 2(1 \cos \theta) \text{ m} \qquad \delta x_B = -2 \sin \theta \, \delta\theta \text{ m} \qquad (1)$$
$$y_w = \tfrac{1}{2}(1 \sin \theta) \text{ m} \qquad \delta y_w = 0.5 \cos \theta \, \delta\theta \text{ m} \qquad (2)$$

It is seen by the *signs* of these equations, and indicated in Fig. 8–6b, that an *increase* in $\theta$ (i.e., $\delta\theta$) causes a *decrease* in $x_B$ and an *increase* in $y_w$.

**Virtual-Work Equation.** If the virtual displacements $\delta x_B$ and $\delta y_w$ were *both positive*, then the forces **W** and **F** would do positive work since the forces and their corresponding displacements would have the same sense. Hence, the virtual-work equation for the displacement $\delta\theta$ is

$$\delta U = 0; \qquad W \, \delta y_w + W \, \delta y_w + F \, \delta x_B = 0 \qquad (3)$$

Substituting Eqs. 1 and 2 into Eq. 3 in order to relate the virtual displacements to the common virtual displacement $\delta\theta$ yields

$$98.1(0.5 \cos \theta \, \delta\theta) + 98.1(0.5 \cos \theta \, \delta\theta) + 25(-2 \sin \theta \, \delta\theta) = 0$$

Notice that the "negative work" done by **F** (force in the opposite sense to displacement) has actually been *accounted for* in the above equation by the "negative sign" of Eq. 1. Factoring out the *common displacement* $\delta\theta$ and solving for $\theta$, noting that $\delta\theta \neq 0$, yields

$$(98.1 \cos \theta - 50 \sin \theta) \, \delta\theta = 0$$

$$\theta = \tan^{-1} \frac{98.1}{50} = 63.0° \qquad \textit{Ans.}$$

**NOTE:** If this problem had been solved using the equations of equilibrium, it would be necessary to dismember the links and apply three scalar equations to *each* link. The principle of virtual work, by means of calculus, has eliminated this task so that the answer is obtained directly.

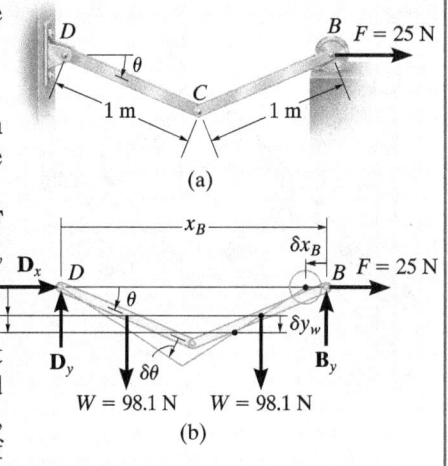

Fig. 8–6

## EXAMPLE 8.2

Determine the required force $P$ in Fig. 8–7a needed to maintain equilibrium of the scissors linkage when $\theta = 60°$. The spring is unstretched when $\theta = 30°$. Neglect the mass of the links.

### SOLUTION

**Free-Body Diagram.** Only $\mathbf{F}_s$ and $\mathbf{P}$ do work when $\theta$ undergoes a *positive* virtual displacement $\delta\theta$, Fig. 8–7b. For the arbitrary position $\theta$, the spring is stretched $(0.3 \text{ m}) \sin \theta - (0.3 \text{ m}) \sin 30°$, so that

$$F_s = ks = 5000 \text{ N/m} \left[(0.3 \text{ m}) \sin \theta - (0.3 \text{ m}) \sin 30°\right]$$
$$= (1500 \sin \theta - 750) \text{ N}$$

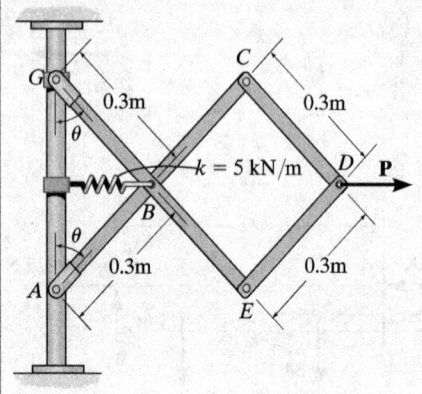

(a)

**Virtual Displacements.** The position coordinates, $x_B$ and $x_D$, measured from the *fixed point A*, are used to locate $\mathbf{F}_s$ and $\mathbf{P}$. These coordinates are parallel to the line of action of their corresponding forces. Expressing $x_B$ and $x_D$ in terms of the angle $\theta$ using trigonometry,

$$x_B = (0.3 \text{ m}) \sin \theta$$
$$x_D = 3[(0.3 \text{ m}) \sin \theta] = (0.9 \text{ m}) \sin \theta$$

Differentiating, we obtain the virtual displacements of points $B$ and $D$.

$$\delta x_B = 0.3 \cos \theta \, \delta\theta \quad (1)$$
$$\delta x_D = 0.9 \cos \theta \, \delta\theta \quad (2)$$

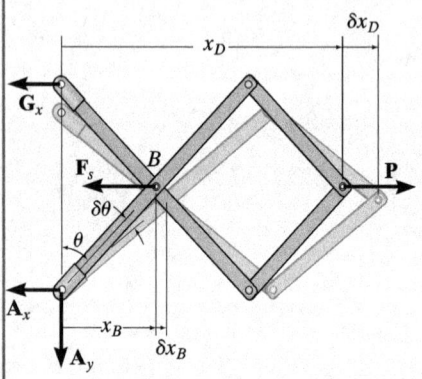

(b)

Fig. 8–7

**Virtual-Work Equation.** Force $\mathbf{P}$ does positive work since it acts in the positive sense of its virtual displacement. The spring force $\mathbf{F}_s$ does negative work since it acts opposite to its positive virtual displacement. Thus, the virtual-work equation becomes

$$\delta U = 0; \qquad -F_s \delta x_B + P \delta x_D = 0$$

$$-[1500 \sin \theta - 750](0.3 \cos \theta \, \delta\theta) + P(0.9 \cos \theta \, \delta\theta) = 0$$

$$[0.9P + 225 - 450 \sin \theta] \cos \theta \, \delta\theta = 0$$

Since $\cos \theta \, \delta\theta \neq 0$, then this equation requires

$$P = 500 \sin \theta - 250$$

When $\theta = 60°$,

$$P = 500 \sin 60° - 250 = 183 \text{ N} \qquad Ans.$$

## EXAMPLE 8.3

If the box in Fig. 8–8a has a mass of 10 kg, determine the couple moment $M$ needed to maintain equilibrium when $\theta = 60°$. Neglect the mass of the members.

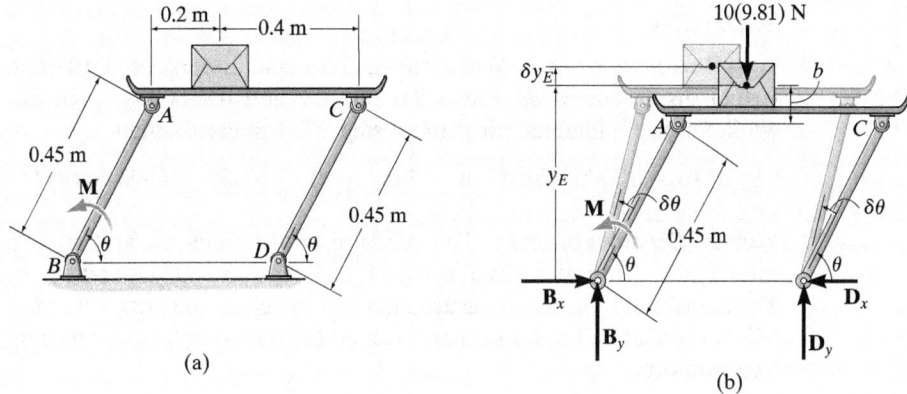

Fig. 8–8

### SOLUTION

**Free-Body Diagram.** When $\theta$ undergoes a positive virtual displacement $\delta\theta$, only the couple moment **M** and the weight of the box do work, Fig. 8–8b.

**Virtual Displacements.** The position coordinate $y_E$, measured from the *fixed point B*, locates the weight, 10(9.81) N. Here,

$$y_E = (0.45 \text{ m}) \sin \theta + b$$

where $b$ is a constant distance. Differentiating this equation, we obtain

$$\delta y_E = 0.45 \text{ m} \cos \theta \, \delta\theta \qquad (1)$$

**Virtual-Work Equation.** The virtual-work equation becomes

$$\delta U = 0; \qquad M\delta\theta - [10(9.81) \text{ N}]\delta y_E = 0$$

Substituting Eq. 1 into this equation

$$M\delta\theta - 10(9.81) \text{ N}(0.45 \text{ m} \cos \theta \, \delta\theta) = 0$$

$$\delta\theta(M - 44.145 \cos \theta) = 0$$

Since $\delta\theta \neq 0$, then

$$M - 44.145 \cos \theta = 0$$

Since it is required that $\theta = 60°$, then

$$M = 44.145 \cos 60° = 22.1 \text{ N} \cdot \text{m} \qquad \textit{Ans.}$$

## EXAMPLE 8.4

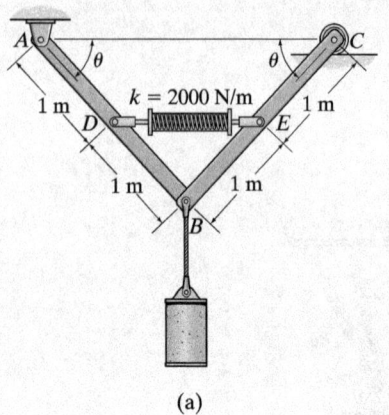

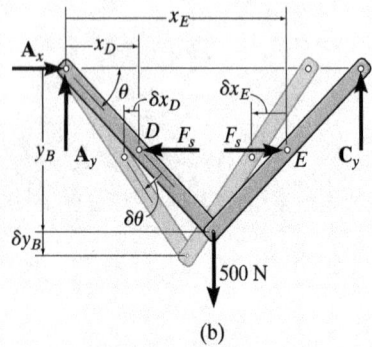

Fig. 8–9

The mechanism in Fig. 8–9a supports the 500-N ($\approx$ 50 kg) cylinder. Determine the angle $\theta$ for equilibrium if the spring has an unstretched length of 2 m when $\theta = 0°$. Neglect the mass of the members.

### SOLUTION

**Free-Body Diagram.** When the mechanism undergoes a positive virtual displacement $\delta\theta$, Fig. 8–9b, only $\mathbf{F}_s$ and the 500-N force do work. Since the final length of the spring is $2(1 \text{ m} \cos \theta)$, then

$$F_s = ks = (2000 \text{ N/m})(2 \text{ m} - 2 \text{ m} \cos \theta) = (4000 - 4000 \cos \theta) \text{ N}$$

**Virtual Displacements.** The position coordinates $x_D$ and $x_E$ are established from the *fixed point A* to locate $\mathbf{F}_s$ at $D$ and at $E$. The coordinate $y_B$, also measured from $A$, specifies the position of the 500-N force at $B$. The coordinates can be expressed in terms of $\theta$ using trigonometry.

$$x_D = (1 \text{ m}) \cos \theta$$
$$x_E = 3[(1 \text{ m}) \cos \theta] = (3 \text{ m}) \cos \theta$$
$$y_B = (2 \text{ m}) \sin \theta$$

Differentiating, we obtain the virtual displacements of points $D$, $E$, and $B$ as

$$\delta x_D = -1 \sin \theta \, \delta\theta \tag{1}$$
$$\delta x_E = -3 \sin \theta \, \delta\theta \tag{2}$$
$$\delta y_B = 2 \cos \theta \, \delta\theta \tag{3}$$

**Virtual-Work Equation.** The virtual-work equation is written as if all virtual displacements are positive, thus

$$\delta U = 0; \quad F_s \delta x_E + 500 \delta y_B - F_s \delta x_D = 0$$

$$(4000 - 4000 \cos \theta)(-3 \sin \theta \, \delta\theta) + 500(2 \cos \theta \, \delta\theta)$$
$$-(4000 - 4000 \cos \theta)(-1 \sin \theta \, \delta\theta) = 0$$

$$\delta\theta \, (8000 \sin \theta \cos \theta - 8000 \sin \theta + 1000 \cos \theta) = 0$$

Since $\delta\theta \neq 0$, then

$$8000 \sin \theta \cos \theta - 8000 \sin \theta + 1000 \cos \theta = 0$$

Solving by trial and error,

$$\theta = 34.9° \qquad \qquad Ans.$$

## FUNDAMENTAL PROBLEMS

**F8–1.** Determine the required magnitude of force **P** to maintain equilibrium of the linkage at $\theta = 60°$. Each link has a mass of 20 kg.

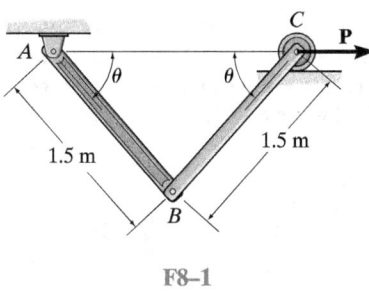

F8–1

**F8–2.** Determine the magnitude of force **P** required to hold the 50-kg smooth rod in equilibrium at $\theta = 60°$.

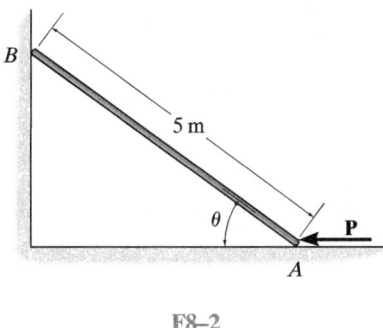

F8–2

**F8–3.** The linkage is subjected to a force of $P = 2$ kN. Determine the angle $\theta$ for equilibrium. The spring is unstretched when $\theta = 0°$. Neglect the mass of the links.

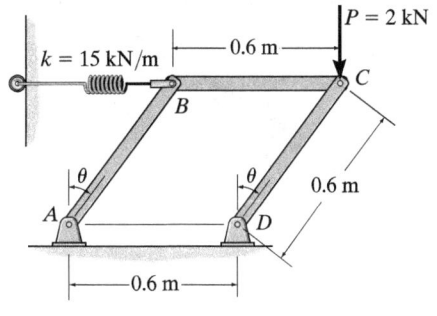

F8–3

**F8–4.** The linkage is subjected to a force of $P = 6$ kN. Determine the angle $\theta$ for equilibrium. The spring is unstretched at $\theta = 60°$. Neglect the mass of the links.

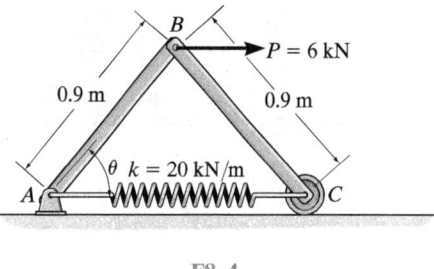

F8–4

**F8–5.** Determine the angle $\theta$ where the 50-kg bar is in equilibrium. The spring is unstretched at $\theta = 60°$.

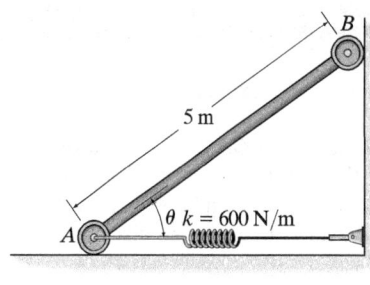

F8–5

**F8–6.** The scissors linkage is subjected to a force of $P = 150$ N. Determine the angle $\theta$ for equilibrium. The spring is unstretched at $\theta = 0°$. Neglect the mass of the links.

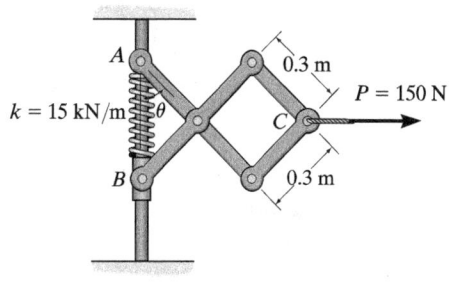

F8–6

## PROBLEMS

•**8–1.** The 200-kg crate is on the lift table at the position $\theta = 30°$. Determine the force in the hydraulic cylinder $AD$ for equilibrium. Neglect the mass of the lift table's components.

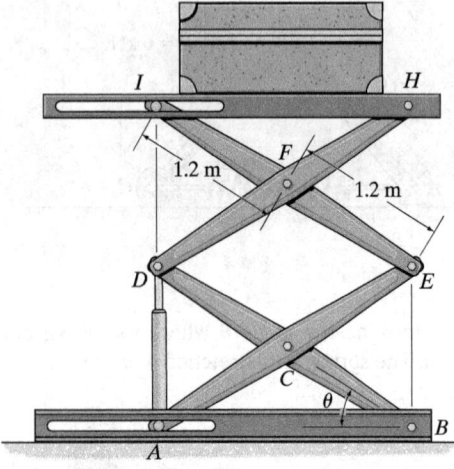

Prob. 8–1

**8–2.** The uniform rod $OA$ has a weight of 50 N. When the rod is in a vertical position, $\theta = 0°$, the spring is unstretched. Determine the angle $\theta$ for equilibrium if the end of the spring wraps around the periphery of the disk as the disk turns.

**8–3.** The "Nuremberg scissors" is subjected to a horizontal force of $P = 600$ N. Determine the angle $\theta$ for equilibrium. The spring has a stiffness of $k = 15$ kN/m and is unstretched when $\theta = 15°$.

*__8–4.__ The "Nuremberg scissors" is subjected to a horizontal force of $P = 600$ N. Determine the stiffness $k$ of the spring for equilibrium when $\theta = 60°$. The spring is unstretched when $\theta = 15°$.

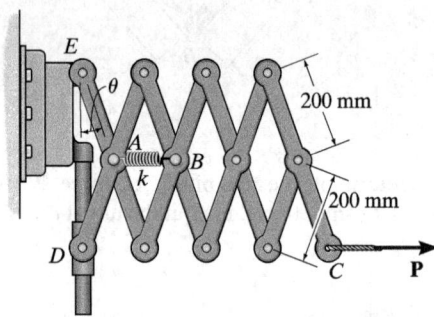

Probs. 8–3/4

•**8–5.** Determine the force developed in the spring required to keep the 50-N ($\approx$ 5-kg) uniform rod $AB$ in equilibrium when $\theta = 35°$.

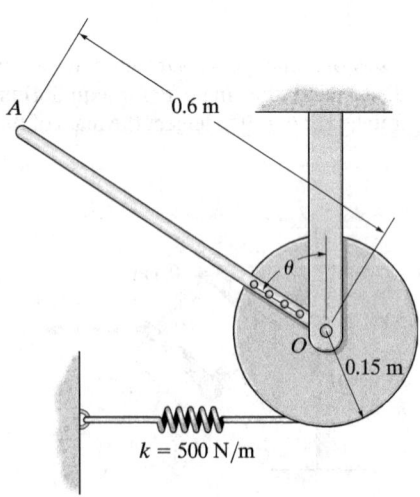

Prob. 8–2

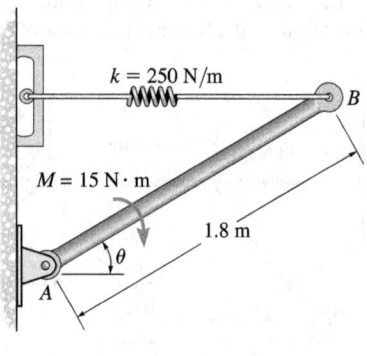

Prob. 8–5

**8–6.** If a force of $P = 25$ N is applied to the handle of the mechanism, determine the force the screw exerts on the cork of the bottle. The screw is attached to the pin at $A$ and passes through the collar that is attached to the bottle neck at $B$.

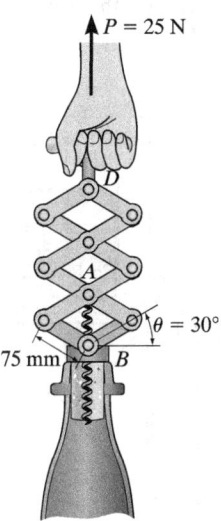

Prob. 8–6

**8–7.** The pin-connected mechanism is constrained at $A$ by a pin and at $B$ by a roller. If $P = 50$ N, determine the angle $\theta$ for equilibrium. The spring is unstretched when $\theta = 45°$. Neglect the weight of the members.

**\*8–8.** The pin-connected mechanism is constrained by a pin at $A$ and a roller at $B$. Determine the force $P$ that must be applied to the roller to hold the mechanism in equilibrium when $\theta = 30°$. The spring is unstretched when $\theta = 45°$. Neglect the weight of the members.

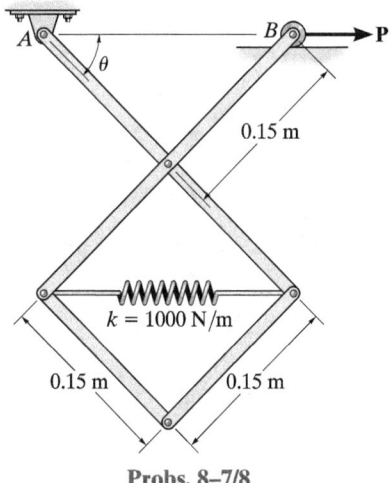

Probs. 8–7/8

**•8–9.** If a force $P = 100$ N is applied to the lever arm of the toggle press, determine the clamping force developed in the block when $\theta = 45°$. Neglect the weight of the block.

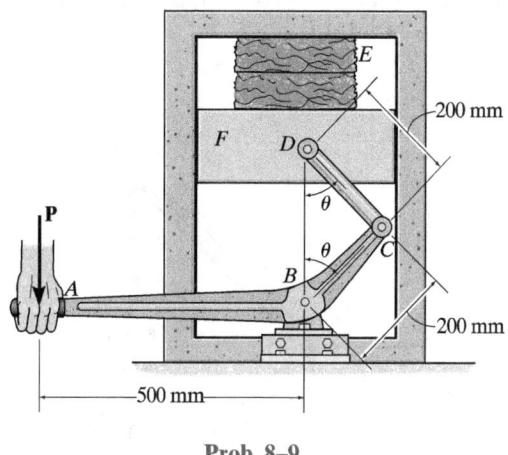

Prob. 8–9

**8–10.** When the forces are applied to the handles of the bottle opener, determine the pulling force developed on the cork.

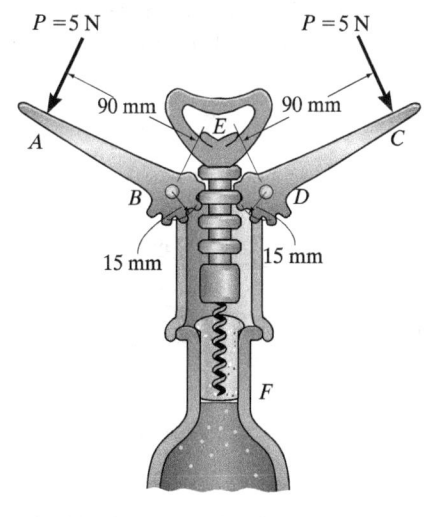

Prob. 8–10

**8–11.** If the spring has a stiffness $k$ and an unstretched length $l_0$, determine the force $P$ when the mechanism is in the position shown. Neglect the weight of the members.

**\*8–12.** Solve Prob. 8–11 if the force **P** is applied vertically downward at $B$.

**8–14.** The truck is weighed on the highway inspection scale. If a known mass $m$ is placed a distance $s$ from the fulcrum $B$ of the scale, determine the mass of the truck $m_t$ if its center of gravity is located at a distance $d$ from point $C$. When the scale is empty, the weight of the lever $ABC$ balances the scale $CDE$.

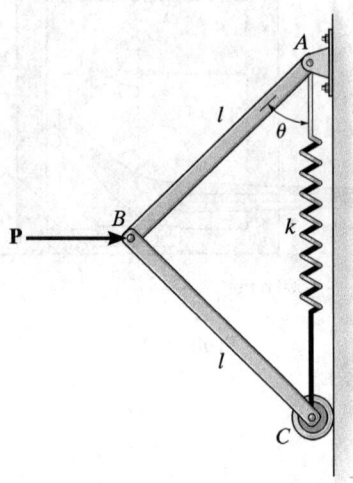

Probs. 8–11/12

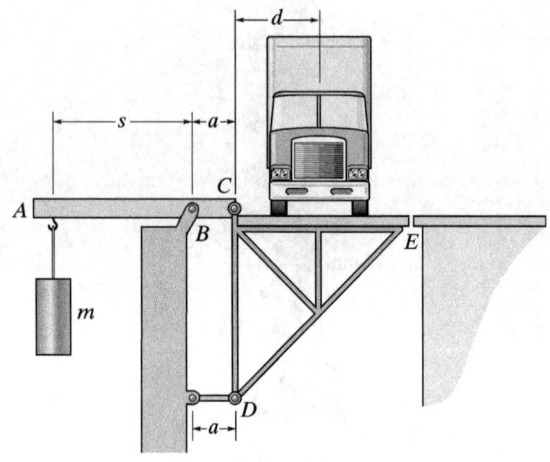

Prob. 8–14

**•8–13.** Determine the angles $\theta$ for equilibrium of the 20-N ($\approx$ 2-kg) disk using the principle of virtual work. Neglect the weight of the rod. The spring is unstretched when $\theta = 0°$ and always remains in the vertical position due to the roller guide.

**8–15.** The assembly is used for exercise. It consists of four pin-connected bars, each of length $L$, and a spring of stiffness $k$ and unstretched length $a$ ($< 2L$). If horizontal forces are applied to the handles so that $\theta$ is slowly decreased, determine the angle $\theta$ at which the magnitude of **P** becomes a maximum.

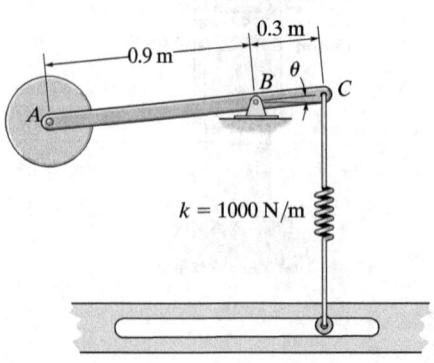

Prob. 8–13

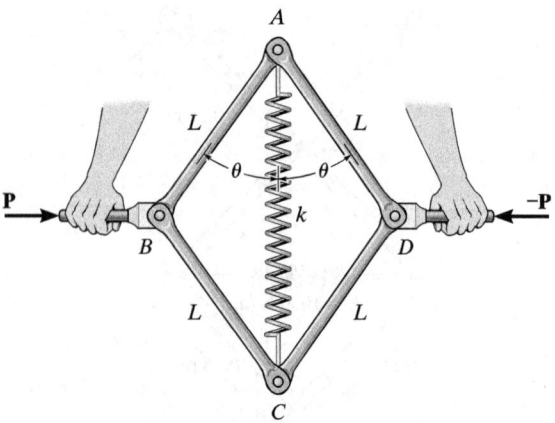

Prob. 8–15

**\*8–16.** A 5-kg uniform serving table is supported on each side by pairs of two identical links, $AB$ and $CD$, and springs $CE$. If the bowl has a mass of 1 kg, determine the angle $\theta$ where the table is in equilibrium. The springs each have a stiffness of $k = 200$ N/m and are unstretched when $\theta = 90°$. Neglect the mass of the links.

**•8–17.** A 5-kg uniform serving table is supported on each side by two pairs of identical links, $AB$ and $CD$, and springs $CE$. If the bowl has a mass of 1 kg and is in equilibrium when $\theta = 45°$, determine the stiffness $k$ of each spring. The springs are unstretched when $\theta = 90°$. Neglect the mass of the links.

**8–19.** The spring is unstretched when $\theta = 45°$ and has a stiffness of $k = 20$ kN/m. Determine the angle $\theta$ for equilibrium if each of the cylinders weighs 250 N. Neglect the weight of the members. The spring remains horizontal at all times due to the roller.

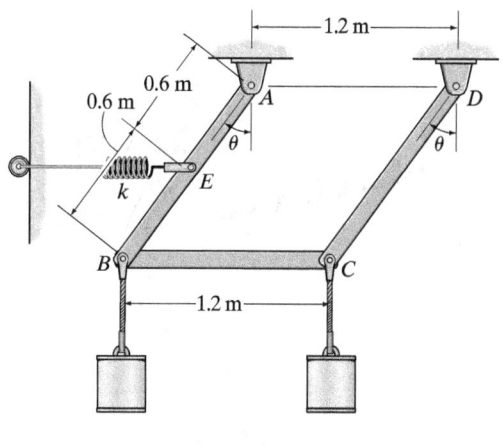

Prob. 8–19

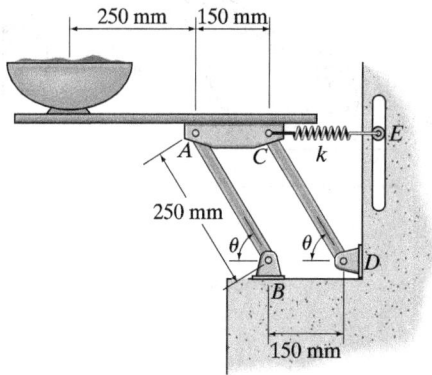

Probs. 8–16/17

**8–18.** If a vertical force of $P = 50$ N is applied to the handle of the toggle clamp, determine the clamping force exerted on the pipe.

**\*8–20.** The machine shown is used for forming metal plates. It consists of two toggles $ABC$ and $DEF$, which are operated by the hydraulic cylinder. The toggles push the moveable bar $G$ forward, pressing the plate into the cavity. If the force which the plate exerts on the head is $P = 8$ kN, determine the force $F$ in the hydraulic cylinder when $\theta = 30°$.

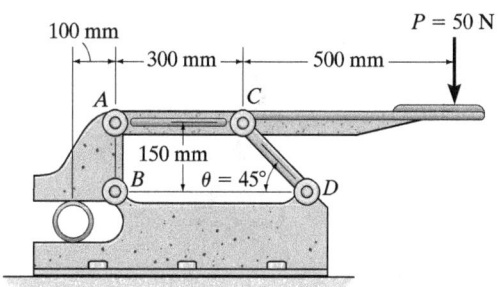

Prob. 8–18

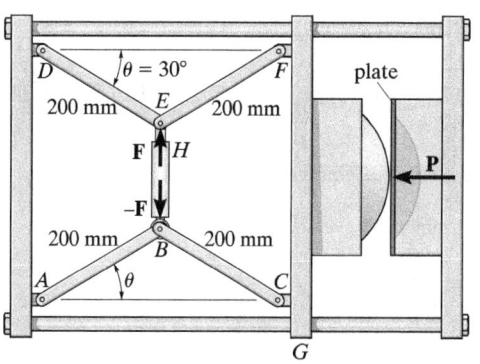

Prob. 8–20

•**8–21.** The vent plate is supported at B by a pin. If it weighs 75 N and has a center of gravity at G, determine the stiffness k of the spring so that the plate remains in equilibrium at θ = 30°. The spring is unstretched when θ = 0°.

***8–24.** Determine the magnitude of the couple moment M required to support the 20-kg cylinder in the configuration shown. The smooth peg at B can slide freely within the slot. Neglect the mass of the members.

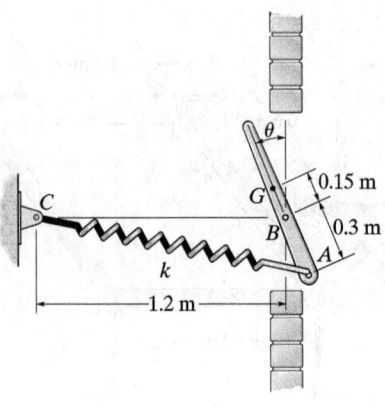

Prob. 8–21

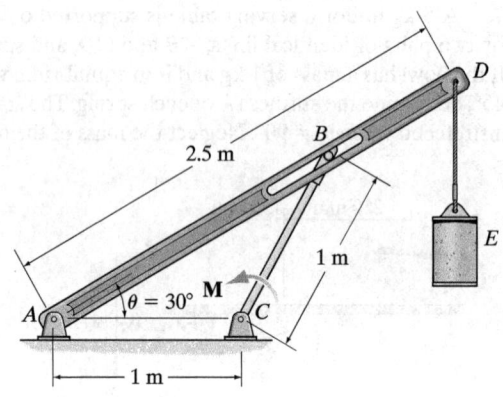

Prob. 8–24

**8–22.** Determine the weight of block G required to balance the differential lever when the 100-N (≈ 10-kg) load F is placed on the pan. The lever is in balance when the load and block are not on the lever. Take x = 300 mm.

**8–23.** If load F weighs 100 N and block G weighs 10 N, determine its position x for equilibrium of the differential lever. The lever is in balance when the load and block are not on the lever.

•**8–25.** The crankshaft is subjected to a torque of M = 75 N · m. Determine the vertical compressive force F applied to the piston for equilibrium when θ = 60°.

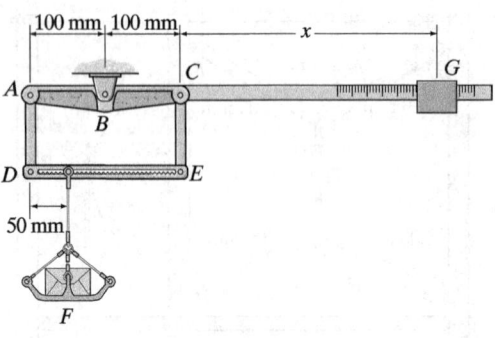

Probs. 8–22/23

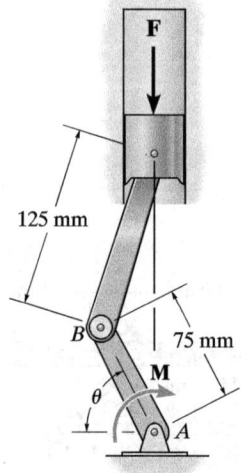

Prob. 8–25

## *8.4 Conservative Forces

If the work of a force only depends upon its initial and final positions, and is *independent* of the path it travels, then the force is referred to as a *conservative force*. The weight of a body and the force of a spring are two examples of conservative forces.

**Weight.** Consider a block of weight **W** that travels along the path in Fig. 8–10a. When it is displaced up the path by an amount $d\mathbf{r}$, then the work is $dU = \mathbf{W} \cdot d\mathbf{r}$ or $dU = -W(dr \cos\theta) = -W\,dy$, as shown in Fig. 8–10b. In this case, the work is *negative* since **W** acts in the opposite sense of $dy$. Thus, if the block moves from $A$ to $B$, through the vertical displacement $h$, the work is

$$U = -\int_0^h W\,dy = -Wh$$

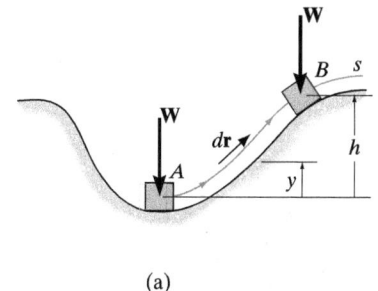

(a)

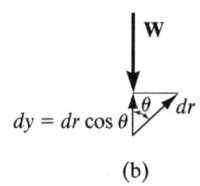

(b)

Fig. 8–10

The weight of a body is therefore a conservative force, since the work done by the weight depends only on the *vertical displacement* of the body, and is independent of the path along which the body travels.

**Spring Force.** Now consider the linearly elastic spring in Fig. 8–11, which undergoes a displacement $ds$. The work done by the spring force on the block is $dU = -F_s\,ds = -ks\,ds$. The work is *negative* because $\mathbf{F}_s$ acts in the opposite sense to that of $ds$. Thus, the work of $\mathbf{F}_s$ when the block is displaced from $s = s_1$ to $s = s_2$ is

$$U = -\int_{s_1}^{s_2} ks\,ds = -\left(\tfrac{1}{2}ks_2^2 - \tfrac{1}{2}ks_1^2\right)$$

Here, the work depends only on the spring's initial and final positions, $s_1$ and $s_2$, measured from the spring's unstretched position. Since this result is independent of the path taken by the block as it moves, then a spring force is also a *conservative force*.

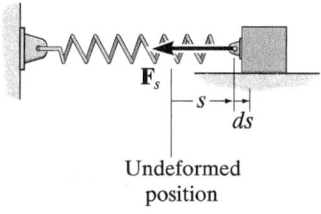

Undeformed position

Fig. 8–11

**Friction.** In contrast to a conservative force, consider the force of *friction* exerted on a sliding body by a fixed surface. The work done by the frictional force depends on the path; the longer the path, the greater the work. Consequently, frictional forces are *nonconservative*, and most of the work done by them is dissipated from the body in the form of heat.

## *8.5  Potential Energy

When a conservative force acts on a body, it gives the body the capacity to do work. This capacity, measured as *potential energy*, depends on the location of the body relative to a fixed reference position or datum.

**Gravitational Potential Energy.** If a body is located a distance *y above* a fixed horizontal reference or datum as in Fig. 8–12, the weight of the body has *positive* gravitational potential energy $V_g$ since **W** has the capacity of doing positive work when the body is moved back down to the datum. Likewise, if the body is located a distance *y below* the datum, $V_g$ is *negative* since the weight does negative work when the body is moved back up to the datum. At the datum, $V_g = 0$.

Measuring *y* as *positive upward*, the gravitational potential energy of the body's weight **W** is therefore

$$V_g = Wy \qquad (8\text{–}4)$$

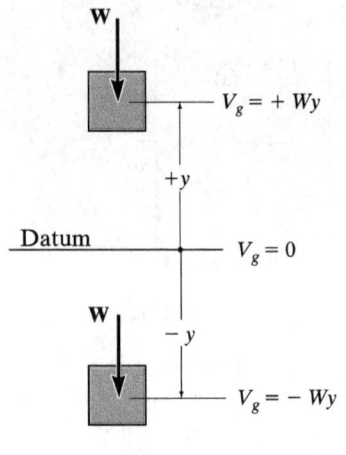

**Fig. 8–12**

**Elastic Potential Energy.** When a spring is either elongated or compressed by an amount *s* from its unstretched position (the datum), the energy stored in the spring is called *elastic potential energy*. It is determined from

$$V_e = \tfrac{1}{2} k s^2 \qquad (8\text{–}5)$$

This energy is always a positive quantity since the spring force acting on the attached body does *positive* work on the body as the force returns the body to the spring's unstretched position, Fig. 8–13.

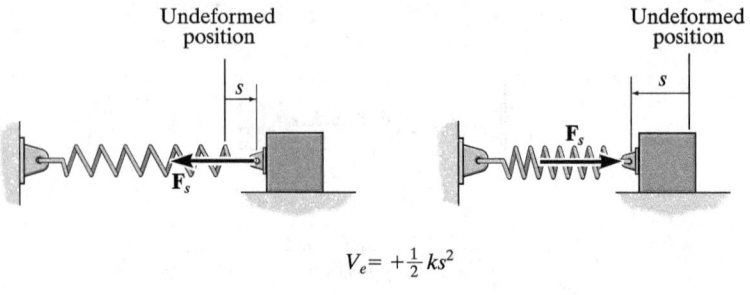

$$V_e = +\tfrac{1}{2} k s^2$$

**Fig. 8–13**

## Potential Function.

In the general case, if a body is subjected to *both* gravitational and elastic forces, the *potential energy or potential function V* of the body can be expressed as the algebraic sum

$$V = V_g + V_e \qquad (8\text{-}6)$$

where measurement of $V$ depends on the location of the body with respect to a selected datum in accordance with Eqs. 8–4 and 8–5.

In particular, if a *system* of frictionless connected rigid bodies has a *single degree of freedom*, such that its vertical position from the datum is defined by the coordinate $q$, then the potential function for the system can be expressed as $V = V(q)$. The work done by all the weight and spring forces acting on the system in moving it from $q_1$ to $q_2$, is measured by the *difference* in $V$; i.e.,

$$U_{1-2} = V(q_1) - V(q_2) \qquad (8\text{-}7)$$

For example, the potential function for a system consisting of a block of weight **W** supported by a spring, as in Fig. 8–14, can be expressed in terms of the coordinate $(q =) y$, measured from a fixed datum located at the unstretched length of the spring. Here

$$V = V_g + V_e$$
$$= -Wy + \tfrac{1}{2}ky^2 \qquad (8\text{-}8)$$

If the block moves from $y_1$ to $y_2$, then applying Eq. 8–7 the work of **W** and $\mathbf{F}_s$ is

$$U_{1-2} = V(y_1) - V(y_2) = -W(y_1 - y_2) + \tfrac{1}{2}ky_1^2 - \tfrac{1}{2}ky_2^2$$

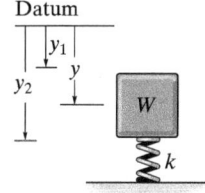

(a)

**Fig. 8–14**

## *8.6 Potential-Energy Criterion for Equilibrium

If a frictionless connected system has one degree of freedom, and its position is defined by the coordinate $q$, then if it displaces from $q$ to $q + dq$, Eq. 8–7 becomes

$$dU = V(q) - V(q + dq)$$

or

$$dU = -dV$$

If the system is in equilibrium and undergoes a *virtual displacement* $\delta q$, rather than an actual displacement $dq$, then the above equation becomes $\delta U = -\delta V$. However, the principle of virtual work requires that $\delta U = 0$, and therefore, $\delta V = 0$, and so we can write $\delta V = (dV/dq)\delta q = 0$. Since $\delta q \neq 0$, this expression becomes

$$\boxed{\frac{dV}{dq} = 0} \qquad (8\text{–}9)$$

Hence, *when a frictionless connected system of rigid bodies is in equilibrium, the first derivative of its potential function is zero.* For example, using Eq. 8–8 we can determine the equilibrium position for the spring and block in Fig. 8–14a. We have

$$\frac{dV}{dy} = -W + ky = 0$$

Hence, the equilibrium position $y = y_{eq}$ is

$$y_{eq} = \frac{W}{k}$$

Of course, this *same result* can be obtained by applying $\Sigma F_y = 0$ to the forces acting on the free-body diagram of the block, Fig. 8–14b.

(b)

Fig. 8–14

## *8.7 Stability of Equilibrium Configuration

The potential function $V$ of a system can also be used to investigate the stability of the equilibrium configuration, which is classified as *stable*, *neutral*, or *unstable*.

**Stable Equilibrium.** A system is said to be *stable* if a system has a tendency to return to its original position when a small displacement is given to the system. The potential energy of the system in this case is at its *minimum*. A simple example is shown in Fig. 8–15a. When the disk is given a small displacement, its center of gravity $G$ will always move (rotate) back to its equilibrium position, which is at the *lowest point* of its path. This is where the potential energy of the disk is at its *minimum*.

**Neutral Equilibrium.** A system is said to be in *neutral equilibrium* if the system still remains in equilibrium when the system is given a small displacement away from its original position. In this case, the potential energy of the system is *constant*. Neutral equilibrium is shown in Fig. 8–15b, where a disk is pinned at $G$. Each time the disk is rotated, a new equilibrium position is established and the potential energy remains unchanged.

**Unstable Equilibrium.** A system is said to be *unstable* if it has a tendency to be *displaced further away* from its original equilibrium position when it is given a small displacement. The potential energy of the system in this case is a *maximum*. An unstable equilibrium position of the disk is shown in Fig. 8–15c. Here, the disk will rotate away from its equilibrium position when its center of gravity is slightly displaced. At this *highest point*, its potential energy is at a *maximum*.

The counterweight at $A$ balances the weight of the deck $B$ of this simple lift bridge. By applying the method of potential energy, we can study the stability of the structure for various equilibrium positions of the deck.

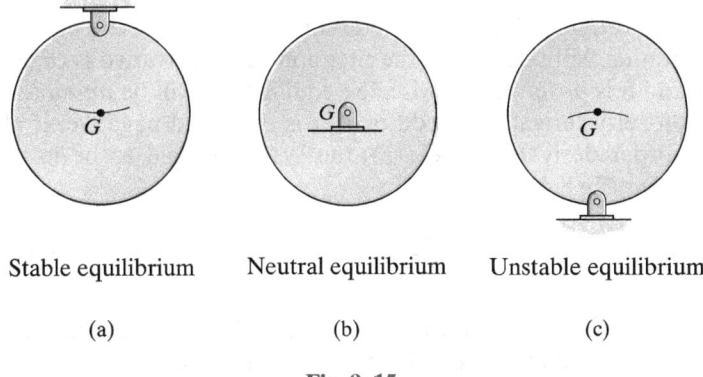

| Stable equilibrium | Neutral equilibrium | Unstable equilibrium |
| (a) | (b) | (c) |

**Fig. 8–15**

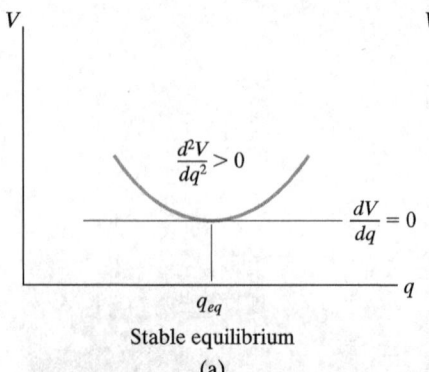
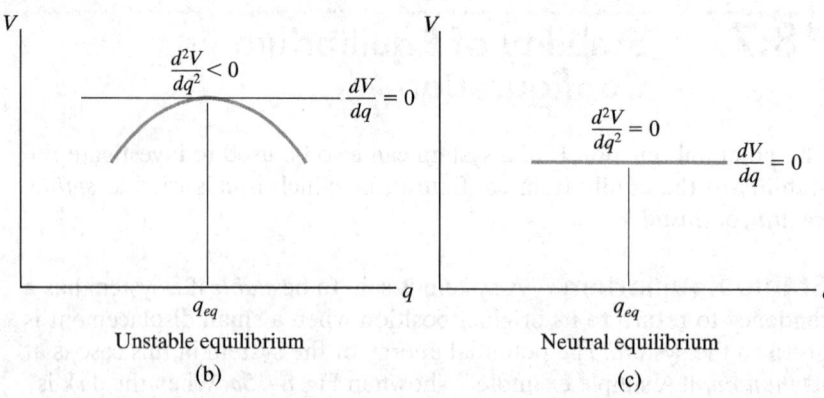

Stable equilibrium
(a)

Unstable equilibrium
(b)

Neutral equilibrium
(c)

Fig. 8–16

## One-Degree-of-Freedom System.

If a system has only one degree of freedom, and its position is defined by the coordinate $q$, then the potential function $V$ for the system in terms of $q$ can be plotted, Fig. 8–16. Provided the system is in *equilibrium*, then $dV/dq$, which represents the slope of this function, must be equal to zero. An investigation of stability at the equilibrium configuration therefore requires that the second derivative of the potential function be evaluated.

If $d^2V/dq^2$ is greater than zero, Fig. 8-16 $a$, the potential energy of the system will be a *minimum*. This indicates that the equilibrium configuration is *stable*. Thus,

$$\frac{dV}{dq} = 0, \qquad \frac{d^2V}{dq^2} > 0 \qquad \text{stable equilibrium} \qquad (8\text{–}10)$$

If $d^2V/dq^2$ is less than zero, Fig. 8-16 $b$, the potential energy of the system will be a *maximum*. This indicates an *unstable* equilibrium configuration. Thus,

$$\frac{dV}{dq} = 0, \qquad \frac{d^2V}{dq^2} < 0 \qquad \text{unstable equilibrium} \qquad (8\text{–}11)$$

Finally, if $d^2V/dq^2$ is equal to zero, it will be necessary to investigate the higher order derivatives to determine the stability. The equilibrium configuration will be *stable* if the first non-zero derivative is of an *even* order and it is *positive*. Likewise, the equilibrium will be *unstable* if this first non-zero derivative is odd or if it is even and negative. If all the higher order derivatives are *zero*, the system is said to be in *neutral equilibrium*, Fig 8-16c. Thus,

$$\frac{dV}{dq} = \frac{d^2V}{dq^2} = \frac{d^3V}{dq^3} = \cdots = 0 \qquad \text{neutral equilibrium} \qquad (8\text{–}12)$$

This condition occurs only if the potential-energy function for the system is constant at or around the neighborhood of $q_{eq}$.

During high winds and when going around a curve, these sugar-cane trucks can become unstable and tip over since their center of gravity is high off the road when they are fully loaded.

## Procedure for Analysis

Using potential-energy methods, the equilibrium positions and the stability of a body or a system of connected bodies having a single degree of freedom can be obtained by applying the following procedure.

### Potential Function.

- Sketch the system so that it is in the *arbitrary position* specified by the coordinate $q$.

- Establish a horizontal *datum* through a *fixed point*\* and express the gravitational potential energy $V_g$ in terms of the weight $W$ of each member and its vertical distance $y$ from the datum, $V_g = Wy$.

- Express the elastic potential energy $V_e$ of the system in terms of the stretch or compression, $s$, of any connecting spring, $V_e = \frac{1}{2}ks^2$.

- Formulate the potential function $V = V_g + V_e$ and express the *position coordinates* $y$ and $s$ in terms of the single coordinate $q$.

### Equilibrium Position.

- The equilibrium position of the system is determined by taking the first derivative of $V$ and setting it equal to zero, $dV/dq = 0$.

### Stability.

- Stability at the equilibrium position is determined by evaluating the second or higher-order derivatives of $V$.

- If the second derivative is greater than zero, the system is stable; if all derivatives are equal to zero, the system is in neutral equilibrium; and if the second derivative is less than zero, the system is unstable.

---

\*The location of the datum is *arbitrary*, since only the *changes* or differentials of $V$ are required for investigation of the equilibrium position and its stability.

## EXAMPLE 8.5

The uniform link shown in Fig. 8–17a has a mass of 10 kg. If the spring is unstretched when $\theta = 0°$, determine the angle $\theta$ for equilibrium and investigate the stability at the equilibrium position.

### SOLUTION

**Potential Function.** The datum is established at the bottom of the link, Fig. 8–17b. When the link is located in the arbitrary position $\theta$, the spring increases its potential energy by stretching and the weight decreases its potential energy. Hence,

$$V = V_e + V_g = \frac{1}{2}ks^2 + Wy$$

Since $l = s + l\cos\theta$ or $s = l(1 - \cos\theta)$, and $y = (l/2)\cos\theta$, then

$$V = \frac{1}{2}kl^2(1 - \cos\theta)^2 + W\left(\frac{l}{2}\cos\theta\right)$$

**Equilibrium Position.** The first derivative of $V$ is

$$\frac{dV}{d\theta} = kl^2(1 - \cos\theta)\sin\theta - \frac{Wl}{2}\sin\theta = 0$$

or

$$l\left[kl(1 - \cos\theta) - \frac{W}{2}\right]\sin\theta = 0$$

This equation is satisfied provided

$$\sin\theta = 0 \qquad \theta = 0° \qquad \text{Ans.}$$

$$\theta = \cos^{-1}\left(1 - \frac{W}{2kl}\right) = \cos^{-1}\left[1 - \frac{10(9.81)}{2(200)(0.6)}\right] = 53.8° \text{ Ans.}$$

**Stability.** The second derivative of $V$ is

$$\frac{d^2V}{d\theta^2} = kl^2(1 - \cos\theta)\cos\theta + kl^2\sin\theta\sin\theta - \frac{Wl}{2}\cos\theta$$

$$= kl^2(\cos\theta - \cos 2\theta) - \frac{Wl}{2}\cos\theta$$

Substituting values for the constants, with $\theta = 0°$ and $\theta = 53.8°$, yields

$$\left.\frac{d^2V}{d\theta^2}\right|_{\theta=0°} = 200(0.6)^2(\cos 0° - \cos 0°) - \frac{10(9.81)(0.6)}{2}\cos 0°$$

$$= -29.4 < 0 \qquad \text{(unstable equilibrium at } \theta = 0°\text{)} \qquad \text{Ans.}$$

$$\left.\frac{d^2V}{d\theta^2}\right|_{\theta=53.8°} = 200(0.6)^2(\cos 53.8° - \cos 107.6°) - \frac{10(9.81)(0.6)}{2}\cos 53.8°$$

$$= 46.9 > 0 \qquad \text{(stable equilibrium at } \theta = 53.8°\text{)} \qquad \text{Ans.}$$

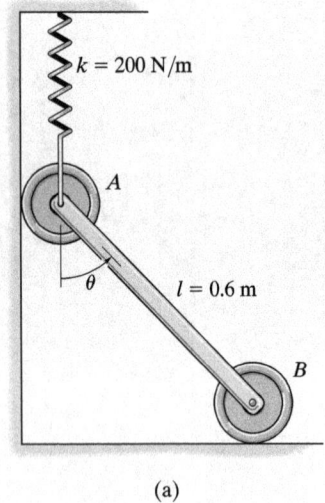

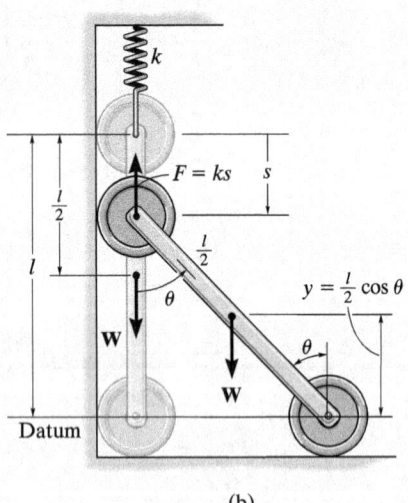

Fig 8–17

## EXAMPLE 8.6

If the spring AD in Fig. 8–18a has a stiffness of 18 kN/m and is unstretched when $\theta = 60°$, determine the angle $\theta$ for equilibrium. The load has a mass of 1.5 Mg. Investigate the stability at the equilibrium position.

### SOLUTION

**Potential Energy.** The gravitational potential energy for the load with respect to the fixed datum, shown in Fig. 8–18b, is

$$V_g = mgy = 1500(9.81) \text{ N}[(4 \text{ m}) \sin \theta + h] = 58\,860 \sin \theta + 14\,715h$$

where $h$ is a constant distance. From the geometry of the system, the elongation of the spring when the load is on the platform is
$s = (4 \text{ m}) \cos \theta - (4 \text{ m}) \cos 60° = (4 \text{ m}) \cos \theta - 2 \text{ m}$.

Thus, the elastic potential energy of the system is

$$V_e = \tfrac{1}{2}ks^2 = \tfrac{1}{2}(18\,000 \text{ N/m})(4 \text{ m} \cos \theta - 2 \text{ m})^2 = 9000(4 \cos \theta - 2)^2$$

The potential energy function for the system is therefore

$$V = V_g + V_e = 58\,860 \sin \theta + 14\,715h + 9000(4 \cos \theta - 2)^2 \quad (1)$$

**Equilibrium.** When the system is in equilibrium,

$$\frac{dV}{d\theta} = 58\,860 \cos \theta + 18\,000(4 \cos \theta - 2)(-4 \sin \theta) = 0$$

$$58\,860 \cos \theta - 288\,000 \sin \theta \cos \theta + 144\,000 \sin \theta = 0$$

Since $\sin 2\theta = 2 \sin \theta \cos \theta$,

$$58\,860 \cos \theta - 144\,000 \sin 2\theta + 144\,000 \sin \theta = 0$$

Solving by trial and error,

$$\theta = 28.18° \text{ and } \theta = 45.51° \quad \textit{Ans.}$$

**Stability.** Taking the second derivative of Eq. 1,

$$\frac{d^2V}{d\theta^2} = -58\,860 \sin \theta - 288\,000 \cos 2\theta + 144\,000 \cos \theta$$

Substituting $\theta = 28.18°$ yields

$$\frac{d^2V}{d\theta^2} = -60\,409 < 0 \qquad \text{Unstable} \qquad \textit{Ans.}$$

And for $\theta = 45.51°$,

$$\frac{d^2V}{d\theta^2} = 64\,073 > 0 \qquad \text{Stable} \qquad \textit{Ans.}$$

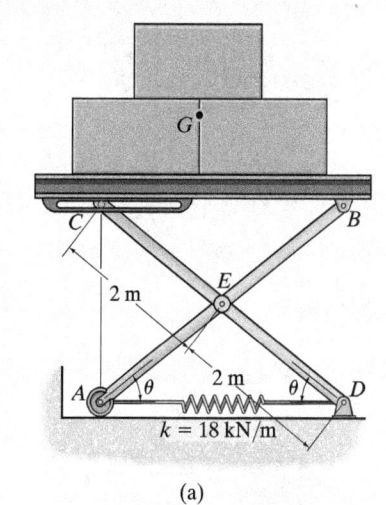

(a)

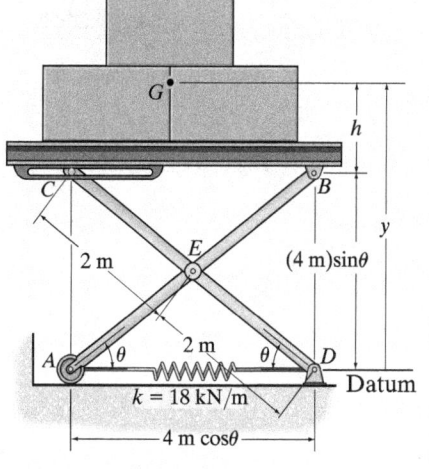

(b)

Fig 8–18

## EXAMPLE 8.7

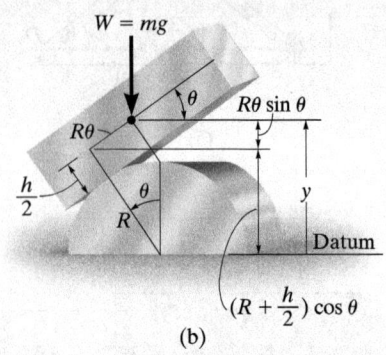

Fig 8-19

The uniform block having a mass $m$ rests on the top surface of the half cylinder, Fig. 8–19a. Show that this is a condition of unstable equilibrium if $h > 2R$.

### SOLUTION

**Potential Function.** The datum is established at the base of the cylinder, Fig. 8–19b. If the block is displaced by an amount $\theta$ from the equilibrium position, the potential function is

$$V = V_e + V_g$$
$$= 0 + mgy$$

From Fig. 8–19b,

$$y = \left(R + \frac{h}{2}\right)\cos\theta + R\theta\sin\theta$$

Thus,

$$V = mg\left[\left(R + \frac{h}{2}\right)\cos\theta + R\theta\sin\theta\right]$$

**Equilibrium Position.**

$$\frac{dV}{d\theta} = mg\left[-\left(R + \frac{h}{2}\right)\sin\theta + R\sin\theta + R\theta\cos\theta\right] = 0$$

$$= mg\left(-\frac{h}{2}\sin\theta + R\theta\cos\theta\right) = 0$$

Note that $\theta = 0°$ satisfies this equation.

**Stability.** Taking the second derivative of $V$ yields

$$\frac{d^2V}{d\theta^2} = mg\left(-\frac{h}{2}\cos\theta + R\cos\theta - R\theta\sin\theta\right)$$

At $\theta = 0°$,

$$\left.\frac{d^2V}{d\theta^2}\right|_{\theta=0°} = -mg\left(\frac{h}{2} - R\right)$$

Since all the constants are positive, the block is in unstable equilibrium provided $h > 2R$, because then $d^2V/d\theta^2 < 0$.

## PROBLEMS

**8–26.** If the potential energy for a conservative one-degree-of-freedom system is expressed by the relation $V = (4x^3 - x^2 - 3x + 10)$ J, where $x$ is given in meters, determine the equilibrium positions and investigate the stability at each position.

**8–27.** If the potential energy for a conservative one-degree-of-freedom system is expressed by the relation $V = (24 \sin \theta + 10 \cos 2\theta)$ J, $0° \leq \theta \leq 90°$, determine the equilibrium positions and investigate the stability at each position.

**\*8–28.** If the potential energy for a conservative one-degree-of-freedom system is expressed by the relation $V = (3y^3 + 2y^2 - 4y + 50)$ J, where $y$ is given in meters, determine the equilibrium positions and investigate the stability at each position.

**•8–29.** The 2-Mg bridge, with center of mass at point $G$, is lifted by two beams $CD$, located at each side of the bridge. If the 2-Mg counterweight $E$ is attached to the beams as shown, determine the angle $\theta$ for equilibrium. Neglect the weight of the beams and the tie rods.

**8–30.** The spring has a stiffness $k = 10$ kN/m and is unstretched when $\theta = 45°$. If the mechanism is in equilibrium when $\theta = 60°$, determine the weight of cylinder $D$. Neglect the weight of the members. Rod $AB$ remains horizontal at all times since the collar can slide freely along the vertical guide.

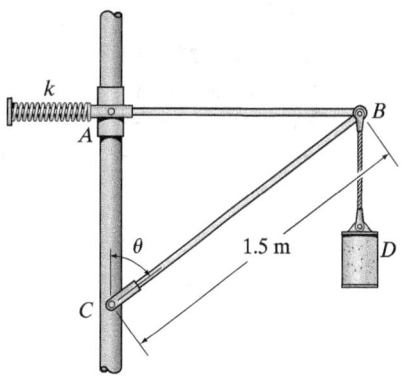

Prob. 8–30

**8–31.** If the springs at $A$ and $C$ have an unstretched length of 250 mm while the spring at $B$ has an unstretched length of 300 mm, determine the height $h$ of the platform when the system is in equilibrium. Investigate the stability of this equilibrium configuration. The package and the platform have a total weight of 750 N.

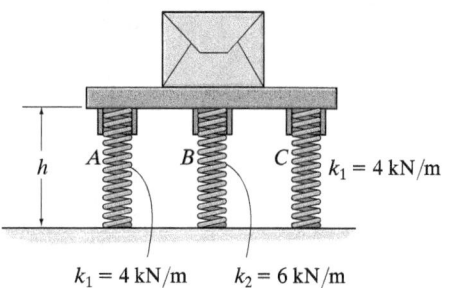

Prob. 8–31

Prob. 8–29

**\*8–32.** The spring is unstretched when $\theta = 45°$ and has a stiffness of $k = 20$ kN/m. Determine the angle $\theta$ for equilibrium if each of the cylinders weighs 250 N. Neglect the weight of the members.

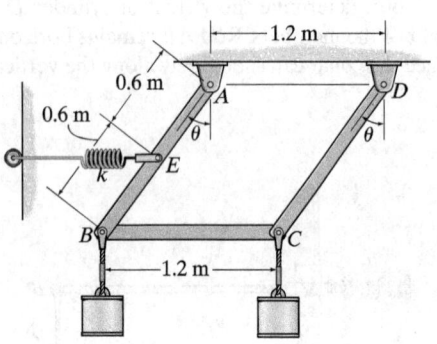

Prob. 8–32

**•8–33.** A 5-kg uniform serving table is supported on each side by pairs of two identical links, $AB$ and $CD$, and springs $CE$. If the bowl has a mass of 1 kg, determine the angle $\theta$ where the table is in equilibrium. The springs each have a stiffness of $k = 200$ N/m and are unstretched when $\theta = 90°$. Neglect the mass of the links.

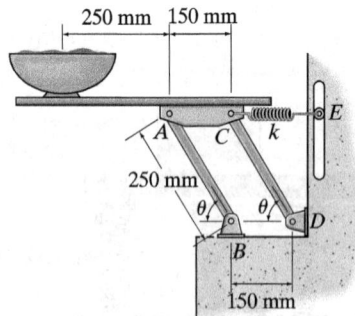

Prob. 8–33

**8–34.** If a 10-kg load $I$ is placed on the pan, determine the position $x$ of the 0.75-kg block $H$ for equilibrium. The scale is in balance when the weight and the load are not on the scale.

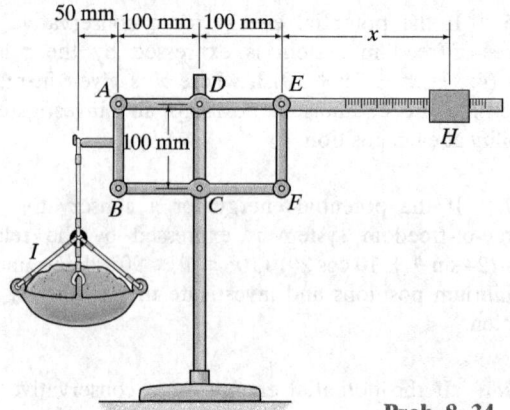

Prob. 8–34

**8–35.** Determine the angles $\theta$ for equilibrium of the 100-kg cylinder and investigate the stability of each position. The spring has a stiffness of $k = 6$ kN/m and an unstretched length of 225 mm.

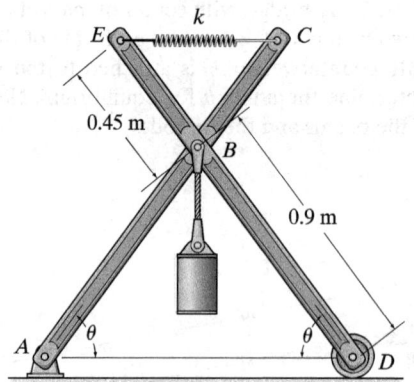

Prob. 8–35

**\*8–36.** Determine the angles $\theta$ for equilibrium of the 50-kg cylinder and investigate the stability of each position. The spring is uncompressed when $\theta = 60°$.

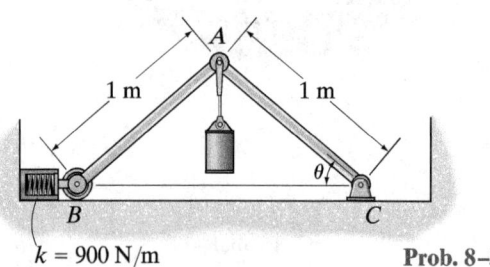

Prob. 8–36

•**8–37.** If the mechanism is in equilibrium when $\theta = 30°$, determine the mass of the bar *BC*. The spring has a stiffness of $k = 2$ kN/m and is uncompressed when $\theta = 0°$. Neglect the mass of the links.

**8–39.** The uniform link *AB* has a mass of 3 kg and is pin connected at both of its ends. The rod *BD*, having negligible weight, passes through a swivel block at *C*. If the spring has a stiffness of $k = 100$ N/m and is unstretched when $\theta = 0°$, determine the angle $\theta$ for equilibrium and investigate the stability at the equilibrium position. Neglect the size of the swivel block.

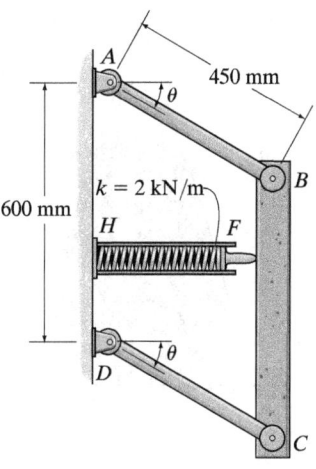

Prob. 8–37

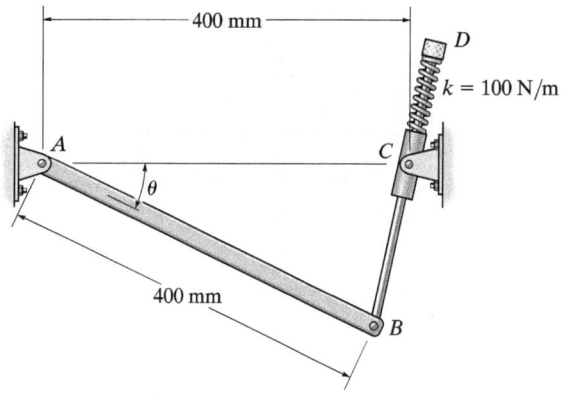

Prob. 8–39

**8–38.** The uniform rod *OA* weighs 100 N, and when the rod is in the vertical position, the spring is unstretched. Determine the position $\theta$ for equilibrium. Investigate the stability at the equilibrium position.

*__8–40.__ The truck has a mass of 20 Mg and a mass center at *G*. Determine the steepest grade $\theta$ along which it can park without overturning and investigate the stability in this position.

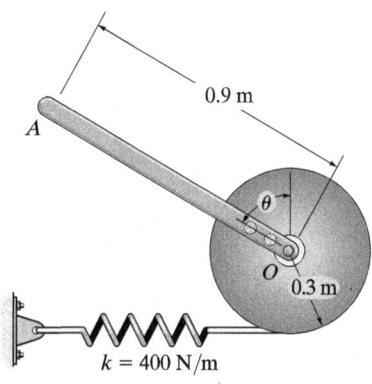

Prob. 8–38

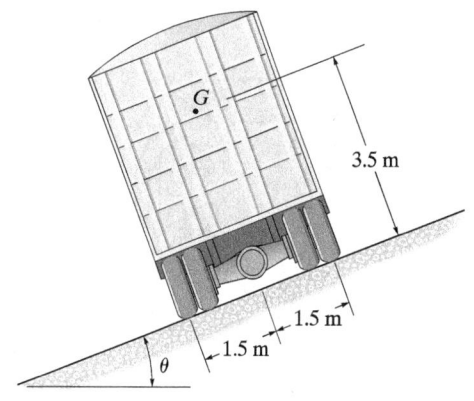

Prob. 8–40

•**8–41.** The cylinder is made of two materials such that it has a mass of $m$ and a center of gravity at point $G$. Show that when $G$ lies above the centroid $C$ of the cylinder, the equilibrium is unstable.

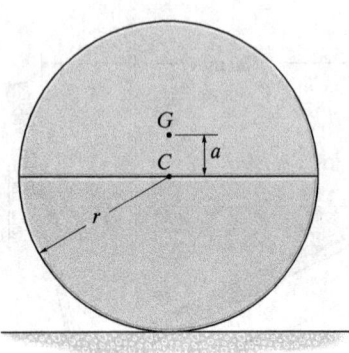

Prob. 8–41

**8–42.** The cap has a hemispherical bottom and a mass $m$. Determine the position $h$ of the center of mass $G$ so that the cup is in neutral equilibrium.

Prob. 8–42

**8–43.** Determine the height $h$ of the cone in terms of the radius $r$ of the hemisphere so that the assembly is in neutral equilibrium. Both the cone and the hemisphere are made from the same material.

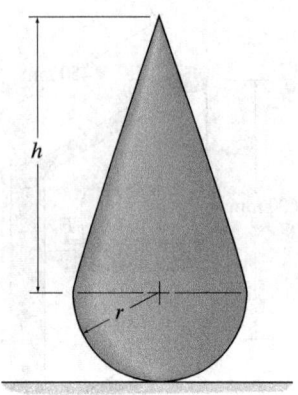

Prob. 8–43

**\*8–44.** A homogeneous block rests on top of the cylindrical surface. Derive the relationship between the radius of the cylinder, $r$, and the dimension of the block, $b$, for stable equilibrium. *Hint*: Establish the potential energy function for a small angle $\theta$, i.e., approximate $\sin \theta \approx 0$, and $\cos \theta \approx 1 - \theta^2/2$.

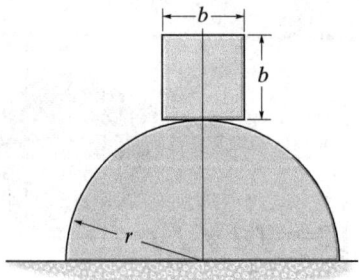

Prob. 8–44

**•8–45.** The homogeneous cone has a conical cavity cut into it as shown. Determine the depth $d$ of the cavity in terms of $h$ so that the cone balances on the pivot and remains in neutral equilibrium.

**\*8–48.** The assembly shown consists of a semicircular cylinder and a triangular prism. If the prism weighs 40 N and the cylinder weighs 10 N, investigate the stability when the assembly is resting in the equilibrium position.

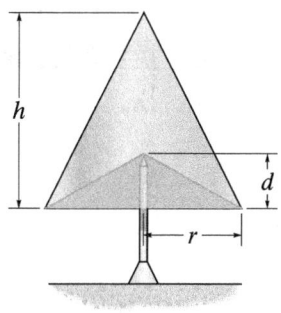

Prob. 8–45

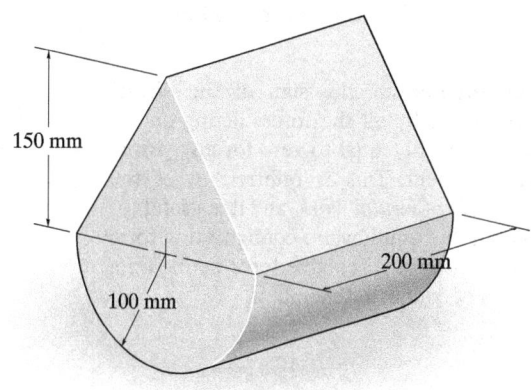

Prob. 8–48

**8–46.** The assembly shown consists of a semicylinder and a rectangular block. If the block weighs 40 N and the semicylinder weighs 10 N, investigate the stability when the assembly is resting in the equilibrium position. Set $h = 100$ mm.

**8–47.** The 10-N ($\approx$ 1-kg) semicylinder supports the block which has a specific weight of $\gamma = 13.5$ kN/m$^3$. Determine the height $h$ of the block which will produce neutral equilibrium in the position shown.

**•8–49.** A conical hole is drilled into the bottom of the cylinder, and it is then supported on the fulcrum at $A$. Determine the minimum distance $d$ in order for it to remain in stable equilibrium.

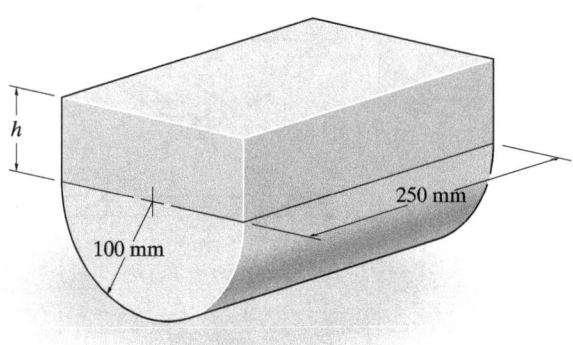

Probs. 8–46/47

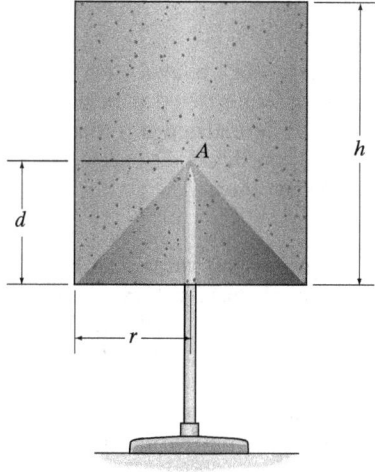

Prob. 8–49

# CHAPTER REVIEW

**Principle of Virtual Work**

The forces on a body will do *virtual work* when the body undergoes an *imaginary* differential displacement or rotation.

For equilibrium, the sum of the virtual work done by all the forces acting on the body must be equal to zero for any virtual displacement. This is referred to as the *principle of virtual work*, and it is useful for finding the equilibrium configuration for a mechanism or a reactive force acting on a series of connected members.

$\delta y, \delta y'$ – virtual displacements

$\delta \theta$ – virtual rotation

$\delta U = 0$

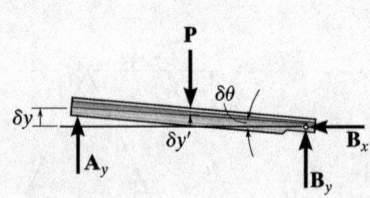

If the system of connected members has one degree of freedom, then its position can be specified by one independent coordinate such as $\theta$.

To apply the principle of virtual work, it is first necessary to use *position coordinates* to locate all the forces and moments on the mechanism that will do work when the mechanism undergoes a virtual movement $\delta \theta$.

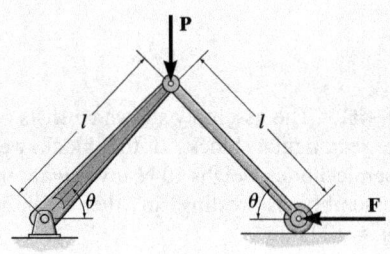

The coordinates are related to the independent coordinate $\theta$ and then these expressions are differentiated in order to relate the *virtual* coordinate displacements to the virtual displacement $\delta \theta$.

Finally, the equation of virtual work is written for the mechanism in terms of the common virtual displacement $\delta \theta$, and then it is set equal to zero. By factoring $\delta \theta$ out of the equation, it is then possible to determine either the unknown force or couple moment, or the equilibrium position $\theta$.

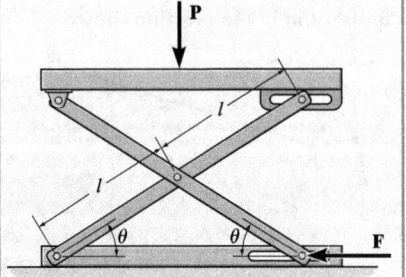

## Potential-Energy Criterion for Equilibrium

When a system is subjected only to conservative forces, such as weight and spring forces, then the equilibrium configuration can be determined using the *potential-energy function V* for the system.

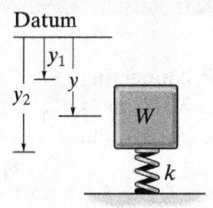

(a)

$$V = V_g + V_e = -W_y + \tfrac{1}{2}ky^2$$

The potential-energy function is established by expressing the weight and spring potential energy for the system in terms of the independent coordinate $q$.

Once the potential-energy function is formulated, its first derivative is set equal to zero. The solution yields the equilibrium position $q_{eq}$ for the system.

$$\frac{dV}{dq} = 0$$

The stability of the system can be investigated by taking the second derivative of $V$.

$$\frac{dV}{dq} = 0, \quad \frac{d^2V}{dq^2} > 0 \quad \text{stable equilibrium}$$

$$\frac{dV}{dq} = 0, \quad \frac{d^2V}{dq^2} < 0 \quad \text{unstable equilibrium}$$

$$\frac{dV}{dq} = \frac{d^2V}{dq^2} = \frac{d^3V}{dq^3} = \cdots = 0 \quad \text{neutral equilibrium}$$

# REVIEW PROBLEMS

**8–50.** The punch press consists of the ram $R$, connecting rod $AB$, and a flywheel. If a torque of $M = 50\,\text{N}\cdot\text{m}$ is applied to the flywheel, determine the force $F$ applied at the ram to hold the rod in the position $\theta = 60°$.

**\*8–52.** The uniform links $AB$ and $BC$ each weigh 10 N and the cylinder weighs 100 N. Determine the horizontal force $P$ required to hold the mechanism at $\theta = 45°$. The spring has an unstretched length of 150 mm.

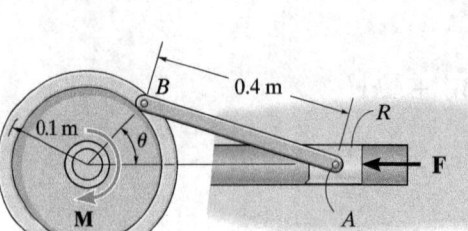

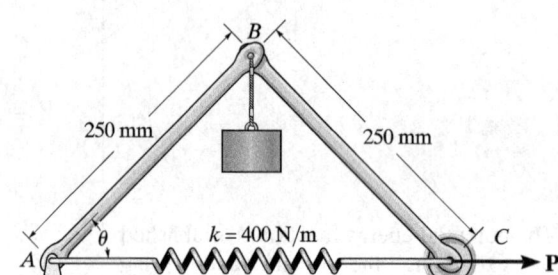

Prob. 8–50

Prob. 8–52

**8–51.** The uniform rod has a weight $W$. Determine the angle $\theta$ for equilibrium. The spring is uncompressed when $\theta = 90°$. Neglect the weight of the rollers.

**•8–53.** The spring attached to the mechanism has an unstretched length when $\theta = 90°$. Determine the position $\theta$ for equilibrium and investigate the stability of the mechanism at this position. Disk $A$ is pin connected to the frame at $B$ and has a weight of 100 N.

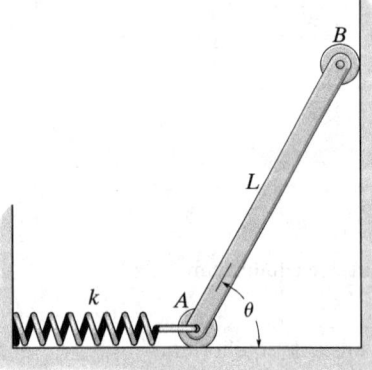

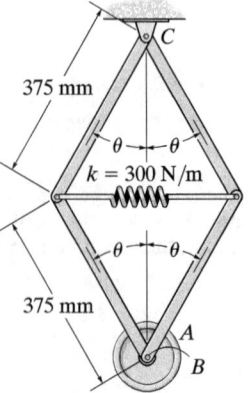

Prob. 8–51

Prob. 8–53

**8–54.** Determine the force $P$ that must be applied to the cord wrapped around the drum at $C$ which is necessary to lift the bucket having a mass $m$. Note that as the bucket is lifted, the pulley rolls on a cord that winds up on shaft $B$ and unwinds from shaft $A$.

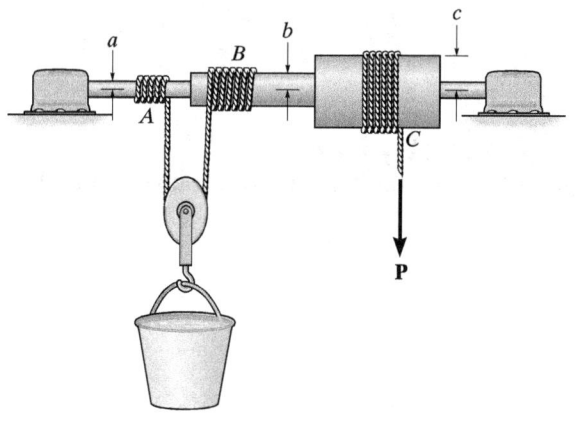

Prob. 8–54

**8–55.** The uniform bar $AB$ weighs 500 N. If both springs $DE$ and $BC$ are unstretched when $\theta = 90°$, determine the angle $\theta$ for equilibrium using the principle of potential energy. Investigate the stability at the equilibrium position. Both springs always remain in the horizontal position due to the roller guides at $C$ and $E$.

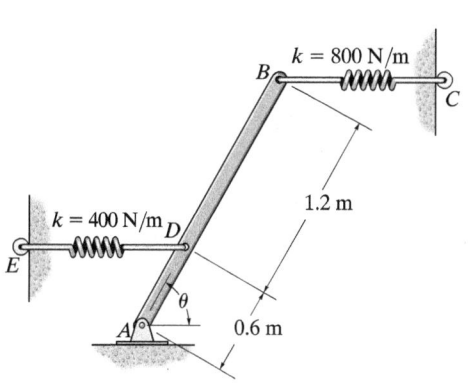

Prob. 8–55

**\*8–56.** The uniform rod $AB$ has a weight of 50 N. If the spring $DC$ is unstretched when $\theta = 0°$, determine the angle $\theta$ for equilibrium using the principle of virtual work. The spring always remains in the horizontal position due to the roller guide at $D$.

**•8–57.** Solve Prob. 8–56 using the principle of potential energy. Investigate the stability of the rod when it is in the equilibrium position.

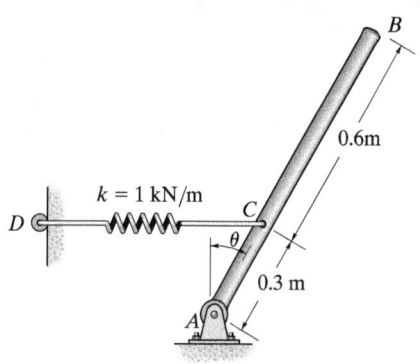

Probs. 8–56/57

**8–58.** Determine the height $h$ of block $B$ so that the rod is in neutral equilibrium. The springs are unstretched when the rod is in the vertical position. The block has a weight $W$.

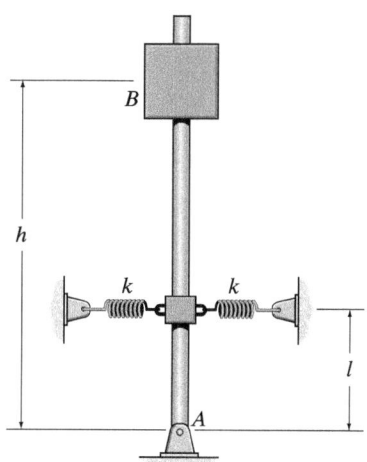

Prob. 8–58

# APPENDIX A

# Mathematical Review and Expressions

## Geometry and Trigonometry Review

The angles $\theta$ in Fig. A–1 are equal between the transverse and two parallel lines.

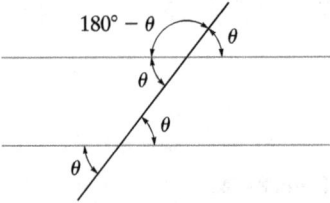

Fig. A–1

For a line and its normal, the angles $\theta$ in Fig. A–2 are equal.

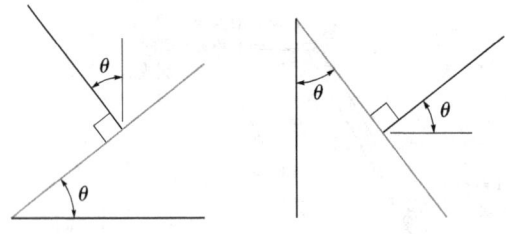

Fig. A–2

For the circle in Fig. A–3 $s = \theta r$, so that when $\theta = 360° = 2\pi$ rad then the circumference is $s = 2\pi r$. Also, since $180° = \pi$ rad, then $\theta \text{ (rad)} = (\pi/180°)\theta°$. The area of the circle is $A = \pi r^2$.

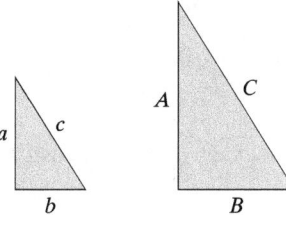

Fig. A–4

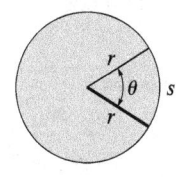

Fig. A–3

The sides of a similar triangle can be obtained by proportion as in Fig. A–4, where $\dfrac{a}{A} = \dfrac{b}{B} = \dfrac{c}{C}$.

For the right triangle in Fig. A–5, the Pythagorean theorem is

$$h = \sqrt{(o)^2 + (a)^2}$$

The trigonometric functions are

$$\sin \theta = \frac{o}{h}$$

$$\cos \theta = \frac{a}{h}$$

$$\tan \theta = \frac{o}{a}$$

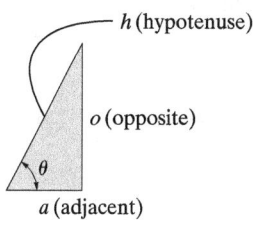

Fig. A–5

This is easily remembered as "soh, cah, toa", i.e., the sine is the opposite over the hypotenuse, etc. The other trigonometric functions follow from this.

$$\csc \theta = \frac{1}{\sin \theta} = \frac{h}{o}$$

$$\sec \theta = \frac{1}{\cos \theta} = \frac{h}{a}$$

$$\cot \theta = \frac{1}{\tan \theta} = \frac{a}{o}$$

## Trigonometric Identities

$\sin^2 \theta + \cos^2 \theta = 1$

$\sin(\theta \pm \phi) = \sin \theta \cos \phi \pm \cos \theta \sin \phi$

$\sin 2\theta = 2 \sin \theta \cos \theta$

$\cos(\theta \pm \phi) = \cos \theta \cos \phi \mp \sin \theta \sin \phi$

$\cos 2\theta = \cos^2 \theta - \sin^2 \theta$

$\cos \theta = \pm \sqrt{\dfrac{1 + \cos 2\theta}{2}}, \quad \sin \theta = \pm \sqrt{\dfrac{1 - \cos 2\theta}{2}}$

$\tan \theta = \dfrac{\sin \theta}{\cos \theta}$

$1 + \tan^2 \theta = \sec^2 \theta \qquad 1 + \cot^2 \theta = \csc^2 \theta$

## Quadratic Formula

If $ax^2 + bx + c = 0$, then $x = \dfrac{-b \pm \sqrt{b^2 - 4ac}}{2a}$

## Hyperbolic Functions

$\sinh x = \dfrac{e^x - e^{-x}}{2},$

$\cosh x = \dfrac{e^x + e^{-x}}{2},$

$\tanh x = \dfrac{\sinh x}{\cosh x}$

## Power-Series Expansions

$\sin x = x - \dfrac{x^3}{3!} + \cdots, \quad \cos x = 1 - \dfrac{x^2}{2!} + \cdots$

$\sinh x = x + \dfrac{x^3}{3!} + \cdots, \quad \cosh x = 1 + \dfrac{x^2}{2!} + \cdots$

## Derivatives

$\dfrac{d}{dx}(u^n) = nu^{n-1} \dfrac{du}{dx} \qquad \dfrac{d}{dx}(\sin u) = \cos u \dfrac{du}{dx}$

$\dfrac{d}{dx}(uv) = u \dfrac{dv}{dx} + v \dfrac{du}{dx} \qquad \dfrac{d}{dx}(\cos u) = -\sin u \dfrac{du}{dx}$

$\dfrac{d}{dx}\left(\dfrac{u}{v}\right) = \dfrac{v \dfrac{du}{dx} - u \dfrac{dv}{dx}}{v^2} \qquad \dfrac{d}{dx}(\tan u) = \sec^2 u \dfrac{du}{dx}$

$\dfrac{d}{dx}(\cot u) = -\csc^2 u \dfrac{du}{dx} \qquad \dfrac{d}{dx}(\sinh u) = \cosh u \dfrac{du}{dx}$

$\dfrac{d}{dx}(\sec u) = \tan u \sec u \dfrac{du}{dx} \qquad \dfrac{d}{dx}(\cosh u) = \sinh u \dfrac{du}{dx}$

$\dfrac{d}{dx}(\csc u) = -\csc u \cot u \dfrac{du}{dx}$

# Integrals

$$\int x^n \, dx = \frac{x^{n+1}}{n+1} + C, \, n \neq -1$$

$$\int \frac{dx}{a + bx} = \frac{1}{b}\ln(a + bx) + C$$

$$\int \frac{dx}{a + bx^2} = \frac{1}{2\sqrt{-ba}}\ln\left[\frac{a + x\sqrt{-ab}}{a - x\sqrt{-ab}}\right] + C, \quad ab < 0$$

$$\int \frac{x \, dx}{a + bx^2} = \frac{1}{2b}\ln(bx^2 + a) + C$$

$$\int \frac{x^2 \, dx}{a + bx^2} = \frac{x}{b} - \frac{a}{b\sqrt{ab}}\tan^{-1}\frac{x\sqrt{ab}}{a} + C, \, ab > 0$$

$$\int \sqrt{a + bx} \, dx = \frac{2}{3b}\sqrt{(a + bx)^3} + C$$

$$\int x\sqrt{a + bx} \, dx = \frac{-2(2a - 3bx)\sqrt{(a + bx)^3}}{15b^2} + C$$

$$\int x^2\sqrt{a + bx} \, dx =$$
$$\frac{2(8a^2 - 12abx + 15b^2 x^2)\sqrt{(a + bx)^3}}{105b^3} + C$$

$$\int \sqrt{a^2 - x^2} \, dx = \frac{1}{2}\left[x\sqrt{a^2 - x^2} + a^2 \sin^{-1}\frac{x}{a}\right] + C,$$
$$a > 0$$

$$\int x\sqrt{a^2 - x^2} \, dx = -\frac{1}{3}\sqrt{(a^2 - x^2)^3} + C$$

$$\int x^2\sqrt{a^2 - x^2} \, dx = -\frac{x}{4}\sqrt{(a^2 - x^2)^3}$$
$$+ \frac{a^2}{8}\left(x\sqrt{a^2 - x^2} + a^2 \sin^{-1}\frac{x}{a}\right) + C, \, a > 0$$

$$\int \sqrt{x^2 \pm a^2} \, dx =$$
$$\frac{1}{2}\left[x\sqrt{x^2 \pm a^2} \pm a^2 \ln\left(x + \sqrt{x^2 \pm a^2}\right)\right] + C$$

$$\int x\sqrt{x^2 \pm a^2} \, dx = \frac{1}{3}\sqrt{(x^2 \pm a^2)^3} + C$$

$$\int x^2\sqrt{x^2 \pm a^2} \, dx = \frac{x}{4}\sqrt{(x^2 \pm a^2)^3}$$
$$\mp \frac{a^2}{8}x\sqrt{x^2 \pm a^2} - \frac{a^4}{8}\ln\left(x + \sqrt{x^2 \pm a^2}\right) + C$$

$$\int \frac{dx}{\sqrt{a + bx}} = \frac{2\sqrt{a + bx}}{b} + C$$

$$\int \frac{x \, dx}{\sqrt{x^2 \pm a^2}} = \sqrt{x^2 \pm a^2} + C$$

$$\int \frac{dx}{\sqrt{a + bx + cx^2}} = \frac{1}{\sqrt{c}}\ln\left[\sqrt{a + bx + cx^2} \right.$$
$$\left. + x\sqrt{c} + \frac{b}{2\sqrt{c}}\right] + C, \, c > 0$$

$$= \frac{1}{\sqrt{-c}}\sin^{-1}\left(\frac{-2cx - b}{\sqrt{b^2 - 4ac}}\right) + C, \, c < 0$$

$$\int \sin x \, dx = -\cos x + C$$

$$\int \cos x \, dx = \sin x + C$$

$$\int x \cos(ax) \, dx = \frac{1}{a^2}\cos(ax) + \frac{x}{a}\sin(ax) + C$$

$$\int x^2 \cos(ax) \, dx = \frac{2x}{a^2}\cos(ax) + \frac{a^2 x^2 - 2}{a^3}\sin(ax) + C$$

$$\int e^{ax} \, dx = \frac{1}{a}e^{ax} + C$$

$$\int xe^{ax} \, dx = \frac{e^{ax}}{a^2}(ax - 1) + C$$

$$\int \sinh x \, dx = \cosh x + C$$

$$\int \cosh x \, dx = \sinh x + C$$

# Fundamental Problems
# Partial Solutions And Answers

## Chapter 2

**F2–1.**

$F_R = \sqrt{(2 \text{ kN})^2 + (6 \text{ kN})^2 - 2(2 \text{ kN})(6 \text{ kN}) \cos 105°}$
$= 6.798 \text{ kN} = 6.80 \text{ kN}$ *Ans.*

$\dfrac{\sin \phi}{6 \text{ kN}} = \dfrac{\sin 105°}{6.798 \text{ kN}}, \quad \phi = 58.49°$

$\theta = 45° + \phi = 45° + 58.49° = 103°$ *Ans.*

**F2–2.** $F_R = \sqrt{200^2 + 500^2 - 2(200)(500) \cos 140°}$
$= 666 \text{ N}$ *Ans.*

**F2–3.** $F_R = \sqrt{600^2 + 800^2 - 2(600)(800) \cos 60°}$
$= 721.11 \text{ N} = 721 \text{ N}$ *Ans.*

$\dfrac{\sin \alpha}{800} = \dfrac{\sin 60°}{721.11}; \quad \alpha = 73.90°$

$\phi = \alpha - 30° = 73.90° - 30° = 43.9°$ *Ans.*

**F2–4.** $\dfrac{F_u}{\sin 45°} = \dfrac{300}{\sin 105°}; \quad F_u = 219.6 \text{ N}$ *Ans.*

$\dfrac{F_v}{\sin 30°} = \dfrac{300}{\sin 105°}; \quad F_v = 155.3 \text{ N}$ *Ans.*

**F2–5.** $\dfrac{F_{AB}}{\sin 105°} = \dfrac{900}{\sin 30°}$

$F_{AB} = 1738.7 \text{ N}$ *Ans.*

$\dfrac{F_{AC}}{\sin 45°} = \dfrac{900}{\sin 30°}$

$F_{AC} = 1272.8 \text{ N}$ *Ans.*

**F2–6.** $\dfrac{F}{\sin 30°} = \dfrac{6}{\sin 105°} \quad F = 3.11 \text{ kN}$ *Ans.*

$\dfrac{F_v}{\sin 45°} = \dfrac{6}{\sin 105°} \quad F_v = 4.39 \text{ kN}$ *Ans.*

**F2–7.** $(F_1)_x = 0 \quad (F_1)_y = 300 \text{ N}$ *Ans.*

$(F_2)_x = -(450 \text{ N}) \cos 45° = -318 \text{ N}$ *Ans.*

$(F_2)_y = (450 \text{ N}) \sin 45° = 318 \text{ N}$ *Ans.*

$(F_3)_x = \left(\tfrac{3}{5}\right) 600 \text{ N} = 360 \text{ N}$ *Ans.*

$(F_3)_y = \left(\tfrac{4}{5}\right) 600 \text{ N} = 480 \text{ N}$ *Ans.*

**F2–8.** $F_{Rx} = 300 + 400 \cos 30° - 250\left(\tfrac{4}{5}\right) = 446.4 \text{ N}$

$F_{Ry} = 400 \sin 30° + 250\left(\tfrac{3}{5}\right) = 350 \text{ N}$

$F_R = \sqrt{(446.4)^2 + 350^2} = 567 \text{ N}$ *Ans.*

$\theta = \tan^{-1} \tfrac{350}{446.4} = 38.1° \measuredangle$ *Ans.*

**F2–9.**

$\xrightarrow{+}(F_R)_x = \Sigma F_x;$

$(F_R)_x = -(3.5 \text{ kN}) \cos 30° + 0 + \left(\tfrac{3}{5}\right)(3 \text{ kN})$
$= -1.231 \text{ kN}$

$+\uparrow (F_R)_y = \Sigma F_y;$

$(F_R)_y = -(3.5 \text{ kN}) \sin 30° - 2 \text{ kN} - \left(\tfrac{4}{5}\right)(3 \text{ kN})$
$= -6.15 \text{ kN}$

$F_R = \sqrt{(1.231 \text{ kN})^2 + (6.15 \text{ kN})^2} = 6.272 \text{ kN}$ *Ans.*

$\phi = \tan^{-1}\left(\tfrac{6.15}{1.231}\right) = 78.68°$

$\theta = 180° + \phi = 180° + 78.68° = 259°$ *Ans.*

**F2–10.** $\xrightarrow{+}(F_R)_x = \Sigma F_x;$

$750 \text{ N} = F \cos \theta + \left(\tfrac{5}{13}\right)(325 \text{ N}) + (600 \text{ N}) \cos 45°$

$+\uparrow (F_R)_y = \Sigma F_y;$

$0 = F \sin \theta + \left(\tfrac{12}{13}\right)(325 \text{ N}) - (600 \text{ N}) \sin 45°$

$\tan \theta = 0.6190 \quad \theta = 31.76° = 31.8° \measuredangle$ *Ans.*

$F = 236 \text{ N}$ *Ans.*

**F2–11.** $\xrightarrow{+}(F_R)_x = \Sigma F_x;$

$(400 \text{ N}) \cos 45° = F \cos \theta + 250 \text{ N} - \left(\tfrac{3}{5}\right) 450 \text{ N}$

$+\uparrow (F_R)_y = \Sigma F_y;$

$-(400 \text{ N}) \sin 45° = F \sin \theta - \left(\tfrac{4}{5}\right)(450 \text{ N})$

$\tan \theta = 0.2547 \quad \theta = 14.29° = 14.3° \measuredangle$ *Ans.*

$F = 312.5 \text{ N}$ *Ans.*

**F2–12.** $(F_R)_x = 15\left(\tfrac{4}{5}\right) + 0 + 15\left(\tfrac{4}{5}\right) = 24 \text{ kN} \rightarrow$

$(F_R)_y = 15\left(\tfrac{3}{5}\right) + 20 - 15\left(\tfrac{3}{5}\right) = 20 \text{ kN} \uparrow$

$F_R = 31.2 \text{ kN}$ *Ans.*

$\theta = 39.8°$ *Ans.*

**F2–13.** $F_x = 75 \cos 30° \sin 45° = 45.93 \text{ kN}$

$F_y = 75 \cos 30° \cos 45° = 45.93 \text{ kN}$

$F_z = -75 \sin 30° = -37.5 \text{ kN}$

$\alpha = \cos^{-1}\left(\tfrac{45.93}{75}\right) = 52.2°$ *Ans.*

$\beta = \cos^{-1}\left(\tfrac{45.93}{75}\right) = 52.2°$ *Ans.*

$\gamma = \cos^{-1}\left(\tfrac{-37.5}{75}\right) = 120°$ *Ans.*

**F2–14.** $\cos \beta = \sqrt{1 - \cos^2 120° - \cos^2 60°} = \pm 0.7071$
Require $\beta = 135°$.
$\mathbf{F} = F\mathbf{u}_F = (500 \text{ N})(-0.5\mathbf{i} - 0.7071\mathbf{j} + 0.5\mathbf{k})$
$= \{-250\mathbf{i} - 354\mathbf{j} + 250\mathbf{k}\} \text{ N}$  *Ans.*

**F2–15.** $\cos^2 \alpha + \cos^2 135° + \cos^2 120° = 1$
$\alpha = 60°$
$\mathbf{F} = F\mathbf{u}_F = (500 \text{ N})(0.5\mathbf{i} - 0.7071\mathbf{j} - 0.5\mathbf{k})$
$= \{250\mathbf{i} - 354\mathbf{j} - 250\mathbf{k}\} \text{ N}$  *Ans.*

**F2–16.** $F_z = (250 \text{ N}) \sin 45° = 176.78 \text{ N}$
$F' = (250 \text{ N}) \cos 45° = 176.78 \text{ N}$
$F_x = \left(\frac{3}{5}\right)(176.78 \text{ N}) = 106.1 \text{ N}$
$F_y = \left(\frac{4}{5}\right)(176.78 \text{ N}) = 141.4 \text{ N}$
$\mathbf{F} = \{-106.1\mathbf{i} + 141.4\mathbf{j} + 176.8\mathbf{k}\} \text{ N}$  *Ans.*

**F2–17.** $F_z = (750 \text{ N}) \sin 45° = 530.33 \text{ N}$
$F' = (750 \text{ N}) \cos 45° = 530.33 \text{ N}$
$F_x = (530.33 \text{ N}) \cos 60° = 265.1 \text{ N}$
$F_y = (530.33 \text{ N}) \sin 60° = 459.3 \text{ N}$
$\mathbf{F}_2 = \{265\mathbf{i} - 459\mathbf{j} + 530\mathbf{k}\} \text{ N}$  *Ans.*

**F2–18.** $\mathbf{F}_1 = \left(\frac{4}{5}\right)(2.5 \text{ kN})\mathbf{j} + \left(\frac{3}{5}\right)(2.5 \text{ kN})\mathbf{k}$
$= \{2\mathbf{j} + 1.5\mathbf{k}\} \text{ kN}$
$\mathbf{F}_2 = [(4 \text{ kN}) \cos 45°] \cos 30° \mathbf{i}$
$+ [(4 \text{ kN}) \cos 45°] \sin 30°\mathbf{j}$
$+ (4 \text{ kN}) \sin 45° (-\mathbf{k})$
$= \{2.45\mathbf{i} + 1.41\mathbf{j} - 2.83\mathbf{k}\} \text{ kN}$
$\mathbf{F}_R = \mathbf{F}_1 + \mathbf{F}_2 = \{2.45\mathbf{i} + 3.41\mathbf{j} - 1.33\mathbf{k}\} \text{ kN}$ *Ans.*

**F2–19.** $\mathbf{r}_{AB} = \{-6\mathbf{i} + 6\mathbf{j} + 3\mathbf{k}\} \text{ m}$  *Ans.*
$r_{AB} = \sqrt{(-6 \text{ m})^2 + (6 \text{ m})^2 + (3 \text{ m})^2} = 9 \text{ m}$  *Ans.*
$\alpha = 132°, \quad \beta = 48.2°, \quad \gamma = 70.5°$  *Ans.*

**F2–20.** $\mathbf{r}_{AB} = \{-2\mathbf{i} + 1\mathbf{j} + 2\mathbf{k}\} \text{ m}$  *Ans.*
$r_{AB} = \sqrt{(-2)^2 + (1)^2 + (2)^2} = 3 \text{ m}$  *Ans.*
$\alpha = \cos^{-1}\left(\frac{-2 \text{ m}}{3 \text{ m}}\right) = 131.8°$
$\theta = 180° - 131.8° = 48.2°$  *Ans.*

**F2–21.** $\mathbf{r}_B = \{2\mathbf{i} + 3\mathbf{j} - 6\mathbf{k}\} \text{ m}$
$\mathbf{F}_B = F_B \mathbf{u}_B$
$= (630 \text{ N})\left(\frac{2}{7}\mathbf{i} + \frac{3}{7}\mathbf{j} - \frac{6}{7}\mathbf{k}\right)$
$= \{180\mathbf{i} + 270\mathbf{j} - 540\mathbf{k}\} \text{ N}$  *Ans.*

**F2–22.** $\mathbf{F} = F\mathbf{u}_{AB} = 900\text{N}\left(-\frac{4}{9}\mathbf{i} + \frac{7}{9}\mathbf{j} - \frac{4}{9}\mathbf{k}\right)$
$= \{-400\mathbf{i} + 700\mathbf{j} - 400\mathbf{k}\} \text{ N}$

**F2–23.** $\mathbf{F}_B = F_B \mathbf{u}_B$
$= (840 \text{ N})\left(\frac{3}{7}\mathbf{i} - \frac{2}{7}\mathbf{j} - \frac{6}{7}\mathbf{k}\right)$
$= \{360\mathbf{i} - 240\mathbf{j} - 720\mathbf{k}\} \text{ N}$
$\mathbf{F}_C = F_C \mathbf{u}_C$
$= (420 \text{ N})\left(\frac{2}{7}\mathbf{i} + \frac{3}{7}\mathbf{j} - \frac{6}{7}\mathbf{k}\right)$
$= \{120\mathbf{i} + 180\mathbf{j} - 360\mathbf{k}\} \text{ N}$
$F_R = \sqrt{(480 \text{ N})^2 + (-60 \text{ N})^2 + (-1080 \text{ N})^2}$
$= 1.18 \text{ kN}$  *Ans.*

**F2–24.** $\mathbf{F}_B = F_B \mathbf{u}_B$
$= (3 \text{ kN})\left(-\frac{1}{3}\mathbf{i} + \frac{2}{3}\mathbf{j} - \frac{2}{3}\mathbf{k}\right)$
$= \{-1\mathbf{i} + 2\mathbf{j} - 2\mathbf{k}\} \text{ kN}$
$\mathbf{F}_C = F_C \mathbf{u}_C$
$= (2.45 \text{ kN})\left(-\frac{6}{7}\mathbf{i} + \frac{3}{7}\mathbf{j} - \frac{2}{7}\mathbf{k}\right)$
$= \{-2.1\mathbf{i} + 1.05\mathbf{j} - 0.7\mathbf{k}\} \text{ kN}$
$\mathbf{F}_R = \mathbf{F}_B + \mathbf{F}_C = \{-3.1\mathbf{i} + 3.05\mathbf{j} - 2.7\mathbf{k}\} \text{ kN}$  *Ans.*

**F2–25.** $\mathbf{u}_{AO} = -\frac{1}{3}\mathbf{i} + \frac{2}{3}\mathbf{j} - \frac{2}{3}\mathbf{k}$
$\mathbf{u}_F = -0.5345\mathbf{i} + 0.8018\mathbf{j} + 0.2673\mathbf{k}$
$\theta = \cos^{-1}(\mathbf{u}_{AO} \cdot \mathbf{u}_F) = 57.7°$  *Ans.*

**F2–26.** $\mathbf{u}_{AB} = -\frac{3}{5}\mathbf{j} + \frac{4}{5}\mathbf{k}$
$\mathbf{u}_F = \frac{4}{5}\mathbf{i} - \frac{3}{5}\mathbf{j}$
$\theta = \cos^{-1}(\mathbf{u}_{AB} \cdot \mathbf{u}_F) = 68.9°$  *Ans.*

**F2–27.** $\mathbf{u}_{OA} = \frac{12}{13}\mathbf{i} + \frac{5}{13}\mathbf{j}$
$\mathbf{u}_{OA} \cdot \mathbf{j} = u_{OA}(1) \cos \theta$
$\cos \theta = \frac{5}{13}; \quad \theta = 67.4°$  *Ans.*

**F2–28.** $\mathbf{u}_{OA} = \frac{12}{13}\mathbf{i} + \frac{5}{13}\mathbf{j}$
$\mathbf{F} = F\mathbf{u}_F = [650\mathbf{j}] \text{ N}$
$F_{OA} = \mathbf{F} \cdot \mathbf{u}_{OA} = 250 \text{ N}$
$\mathbf{F}_{OA} = F_{OA} \mathbf{u}_{OA} = \{231\mathbf{i} + 96.2\mathbf{j}\} \text{ N}$

**F2–29.**  $\mathbf{F} = (400 \text{ N}) \dfrac{\{4\mathbf{i}+1\mathbf{j}-6\mathbf{k}\}\text{m}}{\sqrt{(4 \text{ m})^2 + (1 \text{ m})^2 + (-6 \text{ m})^2}}$
$= \{219.78\mathbf{i} + 54.94\mathbf{j} - 329.67\mathbf{k}\}$ N

$\mathbf{u}_{AO} = \dfrac{\{-4\mathbf{j}-6\mathbf{k}\}\text{m}}{\sqrt{(-4 \text{ m})^2 + (-6 \text{ m})^2}}$
$= -0.5547\mathbf{j} - 0.8321\mathbf{k}$

$(F_{AO})_{\text{proj}} = \mathbf{F} \cdot \mathbf{u}_{AO} = 244$ N     Ans.

**F2–30.**  $\mathbf{F} = [(-3 \text{ kN}) \cos 60°] \sin 30° \mathbf{i}$
$+ [(3 \text{ kN}) \cos 60°] \cos 30° \mathbf{j}$
$+ [(3 \text{ kN}) \sin 60°] \mathbf{k}$
$= \{-0.75\mathbf{i} + 1.299\mathbf{j} + 2.598\mathbf{k}\}$ kN

$\mathbf{u}_A = -\tfrac{2}{3}\mathbf{i} + \tfrac{2}{3}\mathbf{j} + \tfrac{1}{3}\mathbf{k}$

$(F_A)_{\text{proj}} = \mathbf{F} \cdot \mathbf{u}_A = 2.232$ kN     Ans.

$(F_A)_{\text{per}} = \sqrt{(3 \text{ kN})^2 - (2.232 \text{ kN})^2}$
$= 2.00$ kN     Ans.

## Chapter 3

**F3–1.**  $\xrightarrow{+} \Sigma F_x = 0;\ \tfrac{4}{5} F_{AC} - F_{AB} \cos 30° = 0$
$+\uparrow \Sigma F_y = 0;\ \tfrac{3}{5} F_{AC} + F_{AB} \sin 30° - 2.75 \text{ kN} = 0$
$F_{AB} = 2.39$ kN     Ans.
$F_{AC} = 2.59$ kN     Ans.

**F3–2.**  $+\uparrow \Sigma F_y = 0;\ -2(7.5) \sin \theta + 3.5 = 0$
$\theta = 13.5°$
$L_{ABC} = 2\left(\dfrac{1.5 \text{ m}}{\cos 13.5°}\right) = 3.09$ m     Ans.

**F3–3.**  $\xrightarrow{+} \Sigma F_x = 0;\ \ T \cos \theta - T \cos \phi = 0$
$\phi = \theta$
$+\uparrow \Sigma F_y = 0;\ \ 2T \sin \theta - 49.05 \text{ N} = 0$
$\theta = \tan^{-1}\left(\dfrac{0.15 \text{ m}}{0.2 \text{ m}}\right) = 36.87°$
$T = 40.9$ N     Ans.

**F3–4.**  $+\nearrow \Sigma F_x = 0;\ \tfrac{4}{5}(F_{sp}) - 5(9.81) \sin 45° = 0$
$F_{sp} = 43.35$ N
$F_{sp} = k(l - l_0);\ 43.35 = 200(0.5 - l_0)$
$l_0 = 0.283$ m     Ans.

**F3–5.**  $+\uparrow \Sigma F_y = 0;\ \ (392.4 \text{ N}) \sin 30° - m_A(9.81) = 0$
$m_A = 20$ kg     Ans.

**F3–6.**  $+\uparrow \Sigma F_y = 0;\ \ T_{AB} \sin 15° - 10(9.81) \text{ N} = 0$
$T_{AB} = 379.03$ N $= 379$ N     Ans.
$\xrightarrow{+} \Sigma F_x = 0;\ \ T_{BC} - 379.03 \text{ N} \cos 15° = 0$
$T_{BC} = 366.11$ N $= 366$ N     Ans.
$\xrightarrow{+} \Sigma F_x = 0;\ \ T_{CD} \cos \theta - 366.11 \text{ N} = 0$
$+\uparrow \Sigma F_y = 0;\ \ T_{CD} \sin \theta - 15(9.81) \text{ N} = 0$
$T_{CD} = 395$ N     Ans.
$\theta = 21.9°$     Ans.

**F3–7.**  $\Sigma F_x = 0;\ \ \left[\left(\tfrac{3}{5}\right)F_3\right]\left(\tfrac{3}{5}\right) + 600 \text{ N} - F_2 = 0$   (1)
$\Sigma F_y = 0;\ \ \left(\tfrac{4}{5}\right)F_1 - \left[\left(\tfrac{3}{5}\right)F_3\right]\left(\tfrac{4}{5}\right) = 0$   (2)
$\Sigma F_z = 0;\ \ \left(\tfrac{4}{5}\right)F_3 + \left(\tfrac{3}{5}\right)F_1 - 900 \text{ N} = 0$   (3)
$F_3 = 776$ N     Ans.
$F_1 = 466$ N     Ans.
$F_2 = 879$ N     Ans.

**F3–8.**  $\Sigma F_z = 0;\ \ F_{AD}\left(\tfrac{4}{5}\right) - 900 = 0$
$F_{AD} = 1125$ N $= 1.125$ kN     Ans.
$\Sigma F_y = 0;\ \ F_{AC}\left(\tfrac{4}{5}\right) - 1125\left(\tfrac{3}{5}\right) = 0$
$F_{AC} = 843.75$ N $= 844$ N     Ans.
$\Sigma F_x = 0;\ \ F_{AB} - 843.75\left(\tfrac{3}{5}\right) = 0$
$F_{AB} = 506.25$ N $= 506$ N     Ans.

**F3–9.**  $\mathbf{F}_{AD} = F_{AD}\left(\dfrac{\mathbf{r}_{AD}}{r_{AD}}\right) = \tfrac{1}{3} F_{AD}\mathbf{i} - \tfrac{2}{3} F_{AD}\mathbf{j} + \tfrac{2}{3} F_{AD}\mathbf{k}$
$\Sigma F_z = 0;\ \ \tfrac{2}{3} F_{AD} - 600 = 0$
$F_{AD} = 900$ N     Ans.
$\Sigma F_y = 0;\ \ F_{AB} \cos 30° - \tfrac{2}{3}(900) = 0$
$F_{AB} = 692.82$ N $= 693$ N     Ans.
$\Sigma F_x = 0;\ \ \tfrac{1}{3}(900) + 692.82 \sin 30° - F_{AC} = 0$
$F_{AC} = 646.41$ N $= 646$ N     Ans.

**F3–10.**  $\mathbf{F}_{AC} = F_{AC}\{-\cos 60° \sin 30° \mathbf{i}$
$+ \cos 60° \cos 30° \mathbf{j} + \sin 60° \mathbf{k}\}$
$= -0.25 F_{AC}\mathbf{i} + 0.4330 F_{AC}\mathbf{j} + 0.8660 F_{AC}\mathbf{k}$
$\mathbf{F}_{AD} = F_{AD}\{\cos 120° \mathbf{i} + \cos 120° \mathbf{j} + \cos 45° \mathbf{k}\}$
$= -0.5 F_{AD}\mathbf{i} - 0.5 F_{AD}\mathbf{j} + 0.7071 F_{AD}\mathbf{k}$
$\Sigma F_y = 0;\ \ 0.4330 F_{AC} - 0.5 F_{AD} = 0$
$\Sigma F_z = 0;\ \ 0.8660 F_{AC} + 0.7071 F_{AD} - 300 = 0$
$F_{AD} = 175.74$ N $= 176$ N     Ans.
$F_{AC} = 202.92$ N $= 203$ N     Ans.
$\Sigma F_x = 0;\ \ F_{AB} - 0.25(202.92) - 0.5(175.74) = 0$
$F_{AB} = 138.60$ N $= 139$ N     Ans.

**F3–11.** $\mathbf{F}_B = F_B\left(\dfrac{\mathbf{r}_{AB}}{r_{AB}}\right)$

$= F_B\left[\dfrac{\{-1.8\mathbf{i} + 0.9\mathbf{j} + 0.6\mathbf{k}\}\ \text{m}}{\sqrt{(-1.8\ \text{m})^2 + (0.9\ \text{m})^2 + (0.6\ \text{m})^2}}\right]$

$= -\tfrac{6}{7}F_B\mathbf{i} + \tfrac{3}{7}F_B\mathbf{j} + \tfrac{2}{7}F_B\mathbf{k}$

$\mathbf{F}_C = F_C\left(\dfrac{\mathbf{r}_{AC}}{r_{AC}}\right)$

$= F_C\left[\dfrac{\{-1.8\mathbf{i} - 0.6\mathbf{j} + 0.9\mathbf{k}\}\ \text{m}}{\sqrt{(-1.8\text{m})^2 + (-0.6\text{m})^2 + (0.9\text{m})^2}}\right]$

$= -\tfrac{6}{7}F_C\mathbf{i} - \tfrac{2}{7}F_C\mathbf{j} + \tfrac{3}{7}F_C\mathbf{k}$

$\mathbf{F}_D = F_D\mathbf{i}$

$\mathbf{W} = \{-75(9.81)\mathbf{k}\}\ \text{N}$

$\Sigma F_x = 0;\ \ -\tfrac{6}{7}F_B - \tfrac{6}{7}F_C + F_D = 0$     (1)

$\Sigma F_y = 0;\ \ \tfrac{3}{7}F_B - \tfrac{2}{7}F_C = 0$     (2)

$\Sigma F_z = 0;\ \ \tfrac{2}{7}F_B + \tfrac{3}{7}F_C - 75 \times 9.81 = 0$     (3)

$F_B = 792.3\ \text{N}$     *Ans.*

$F_C = 1.5(792.3\ \text{N}) = 1188.5\ \text{N}$     *Ans.*

$F_D = 1697.8\ \text{N}$     *Ans.*

## Chapter 4

**F4–1.** $\zeta + M_O = 3 \sin 50° (1.5) + 3 \cos 50° (0.15)$

$= 3.74\ \text{kN}$     *Ans.*

**F4–2.** $\zeta + M_O = -\left(\tfrac{4}{5}\right)(100\ \text{N})(2\ \text{m}) - \left(\tfrac{3}{5}\right)(100\ \text{N})(5\ \text{m})$

$= -460\ \text{N} \cdot \text{m} = 460\ \text{N} \cdot \text{m} \downarrow$     *Ans.*

**F4–3.** $\zeta + M_O = [(300\ \text{N}) \sin 30°][0.4\ \text{m} + (0.3\ \text{m}) \cos 45°]$

$\quad - [(300\ \text{N}) \cos 30°][(0.3\ \text{m}) \sin 45°]$

$= 36.7\ \text{N} \cdot \text{m}$     *Ans.*

**F4–4.** $\zeta + M_O = (3\ \text{kN})(1.2\ \text{m} + (0.9\ \text{m})\cos 45° - 0.3\ \text{m})$

$= 4.61\ \text{kN} \cdot \text{m}$     *Ans.*

**F4–5.** $\zeta + M_O = 50 \sin 60° (0.1 + 0.2 \cos 45° + 0.1)$

$\quad - 50 \cos 60°(0.2 \sin 45°)$

$= 11.2\ \text{N} \cdot \text{m}$     *Ans.*

**F4–6.** $\zeta + M_O = 500 \sin 45° (3 + 3 \cos 45°)$

$\quad - 500 \cos 45° (3 \sin 45°)$

$= 1.06\ \text{kN} \cdot \text{m}$     *Ans.*

**F4–7.** $\zeta + (M_R)_O = \Sigma Fd;$

$(M_R)_O = -(600\ \text{N})(1\ \text{m})$

$\quad + (500\ \text{N})[3\ \text{m} + (2.5\ \text{m}) \cos 45°]$

$\quad - (300\text{N})[(2.5\ \text{m}) \sin 45°]$

$= 1254\ \text{N} \cdot \text{m} = 1.25\ \text{kN} \cdot \text{m}$     *Ans.*

**F4–8.** $\zeta + (M_R)_O = \Sigma Fd;$

$(M_R)_O = \left[\left(\tfrac{3}{5}\right)500\ \text{N}\right](0.425\ \text{m})$

$\quad - \left[\left(\tfrac{4}{5}\right)500\ \text{N}\right](0.25\ \text{m})$

$\quad - [(600\ \text{N}) \cos 60°](0.25\ \text{m})$

$\quad - [(600\ \text{N}) \sin 60°](0.425\ \text{m})$

$= -268\ \text{N} \cdot \text{m} = 268\ \text{N} \cdot \text{m} \downarrow$

**F4–9.** $\zeta + (M_R)_O = \Sigma Fd;$

$(M_R)_O = (1500 \cos 30°\ \text{N})(2\ \text{m} + 2 \sin 30°\ \text{m})$

$\quad - (1500 \sin 30°\ \text{N})(2 \cos 30°\ \text{m})$

$\quad + (1000\ \text{N})(2 \cos 30°\ \text{m})$

$= 4.33\ \text{kN} \cdot \text{m}$     *Ans.*

**F4–10.** $\mathbf{F} = F\mathbf{u}_{AB} = 500\ \text{N}\left(\tfrac{4}{5}\mathbf{i} - \tfrac{3}{5}\mathbf{j}\right) = \{400\mathbf{i} - 300\mathbf{j}\}\ \text{N}$

$\mathbf{M}_O = \mathbf{r}_{OA} \times \mathbf{F} = \{3\mathbf{j}\}\ \text{m} \times \{400\mathbf{i} - 300\mathbf{j}\}\ \text{N}$

$= \{-1200\mathbf{k}\}\ \text{N} \cdot \text{m}$     *Ans.*

or

$\mathbf{M}_O = \mathbf{r}_{OB} \times \mathbf{F} = \{4\mathbf{i}\}\ \text{m} \times \{400\mathbf{i} - 300\mathbf{j}\}\ \text{N}$

$= \{-1200\mathbf{k}\}\ \text{N} \cdot \text{m}$     *Ans.*

**F4–11.** $\mathbf{F} = F\mathbf{u}_{BC}$

$= 600\ \text{N}\left[\dfrac{\{1.2\,\mathbf{i} - 1.2\,\mathbf{j} - 0.6\,\mathbf{k}\}\ \text{m}}{\sqrt{(1.2\ \text{m})^2 + (-1.2\ \text{m})^2 + (-0.6\ \text{m})^2}}\right]$

$= \{400\mathbf{i} - 400\mathbf{j} - 200\mathbf{k}\}\ \text{N}$

$\mathbf{M}_O = \mathbf{r}_C \times \mathbf{F} = \begin{vmatrix} \mathbf{i} & \mathbf{j} & \mathbf{k} \\ 1.5 & 0 & 0 \\ 400 & -400 & -200 \end{vmatrix}$

$= \{300\mathbf{j} - 600\mathbf{k}\}\ \text{N} \cdot \text{m}$     *Ans.*

or

$\mathbf{M}_O = \mathbf{r}_B \times \mathbf{F} = \begin{vmatrix} \mathbf{i} & \mathbf{j} & \mathbf{k} \\ 0.3 & 1.2 & 0.6 \\ 400 & -400 & -200 \end{vmatrix}$

$= \{300\mathbf{j} - 600\mathbf{k}\}\ \text{N} \cdot \text{m}$     *Ans.*

**F4–12.** $\mathbf{F}_R = \mathbf{F}_1 + \mathbf{F}_2$

$= \{(100 - 200)\mathbf{i} + (-120 + 250)\mathbf{j}$

$\quad + (75 + 100)\mathbf{k}\}\ \text{N}$

$= \{-100\mathbf{i} + 130\mathbf{j} + 175\mathbf{k}\}\ \text{N}$

$(\mathbf{M}_R)_O = \mathbf{r}_A \times \mathbf{F}_R = \begin{vmatrix} \mathbf{i} & \mathbf{j} & \mathbf{k} \\ 0.8 & 1 & 0.6 \\ -100 & 130 & 175 \end{vmatrix}$

$= \{97\mathbf{i} - 200\mathbf{j} + 204\mathbf{k}\}\ \text{N} \cdot \text{m}$     *Ans.*

**F4–13.** $M_x = \mathbf{i} \cdot (\mathbf{r}_{OB} \times \mathbf{F}) = \begin{vmatrix} 1 & 0 & 0 \\ 0.3 & 0.4 & -0.2 \\ 300 & -200 & 150 \end{vmatrix}$

$= 20 \text{ N} \cdot \text{m}$ *Ans.*

**F4–14.** $\mathbf{u}_{OA} = \dfrac{\mathbf{r}_A}{r_A} = \dfrac{\{0.3\mathbf{i} + 0.4\mathbf{j}\} \text{ m}}{\sqrt{(0.3 \text{ m})^2 + (0.4 \text{ m})^2}}$

$M_{OA} = \mathbf{u}_{OA} \cdot (\mathbf{r}_{AB} \times \mathbf{F}) = \begin{vmatrix} 0.6 & 0.8 & 0 \\ 0 & 0 & -0.2 \\ 300 & -200 & 150 \end{vmatrix}$

$= -72 \text{ N} \cdot \text{m}$ *Ans.*

**F4–15.** $\mathbf{F} = (200 \text{ N}) \cos 120° \, \mathbf{i}$
$+ (200 \text{ N}) \cos 60° \, \mathbf{j} + (200 \text{ N}) \cos 45° \, \mathbf{k}$
$= \{-100\mathbf{i} + 100\mathbf{j} + 141.42\mathbf{k}\} \text{ N}$

$\mathbf{M}_O = \mathbf{i} \cdot (\mathbf{r}_A \times \mathbf{F}) = \begin{vmatrix} 1 & 0 & 0 \\ 0 & 0.3 & 0.25 \\ -100 & 100 & 141.42 \end{vmatrix}$

$= 17.4 \text{ N} \cdot \text{m}$

**F4–16.** $\mathbf{M}_p = \mathbf{j} \cdot (\mathbf{r}_A \times \mathbf{F}) = \begin{vmatrix} 0 & 1 & 0 \\ -3 & -4 & 2 \\ 30 & -20 & 50 \end{vmatrix}$

$= 210 \text{ N} \cdot \text{m}$ *Ans.*

**F4–17.** $\mathbf{u}_{AB} = \dfrac{\mathbf{r}_{AB}}{r_{AB}} = \dfrac{\{-0.4\mathbf{i} + 0.3\mathbf{j}\} \text{ m}}{\sqrt{(-0.4\text{m})^2 + (0.3\text{m})^2}} = -0.8\mathbf{i} + 0.6\mathbf{j}$

$M_{AB} = \mathbf{u}_{AB} \cdot (\mathbf{r}_{AC} \times \mathbf{F})$

$= \begin{vmatrix} \mathbf{i} & \mathbf{j} & \mathbf{k} \\ -0.8 & 0.6 & 0 \\ 0 & 0 & 0.2 \\ 50 & -40 & 20 \end{vmatrix} = -0.4 \text{ N·m}$

$\mathbf{M}_{AB} = M_{AB} \mathbf{u}_{AB} = \{0.32\mathbf{i} - 0.24\mathbf{j}\} \text{ N·m}$ *Ans.*

**F4–18.** $F_x = \left[\left(\tfrac{4}{5}\right)500 \text{ N}\right]\left(\tfrac{3}{5}\right) = 240 \text{ N}$
$F_y = \left[\left(\tfrac{4}{5}\right)500 \text{ N}\right]\left(\tfrac{4}{5}\right) = 320 \text{ N}$
$F_z = (500 \text{ N})\left(\tfrac{3}{5}\right) = 300 \text{ N}$
$M_x = 300 \text{ N}(2 \text{ m}) - 320 \text{ N}(3 \text{ m})$
$= -360 \text{ N} \cdot \text{m}$ *Ans.*
$M_y = 300 \text{ N}(2 \text{ m}) - 240 \text{ N}(3 \text{ m})$
$= -120 \text{ N·m}$ *Ans.*
$M_z = 240 \text{ N}(2 \text{ m}) - 320 \text{ N}(2 \text{ m})$
$= -160 \text{ N·m}$ *Ans.*

**F4–19.** $\zeta + M_{C_R} = \Sigma M_A = 400(3) - 400(5) + 300(5)$
$+ 200(0.2) = 740 \text{ N} \cdot \text{m}$ *Ans.*
Also,
$\zeta + M_{C_R} = 300(5) - 400(2) + 200(0.2)$
$= 740 \text{ N} \cdot \text{m}$ *Ans.*

**F4–20.** $\zeta + M_{C_R} = 300(0.4) + 200(0.4) + 150(0.4)$
$= 260 \text{ N} \cdot \text{m}$ *Ans.*

**F4–21.** $\zeta + (M_B)_R = \Sigma M_B$
$-1.5 \text{ kN·m} = (2 \text{ kN})(0.3 \text{ m}) - F(0.9 \text{ m})$
$F = 2.33 \text{ kN}$ *Ans.*

**F4–22.** $\zeta + M_C = 10\left(\tfrac{3}{5}\right)(2) - 10\left(\tfrac{4}{5}\right)(4) = -20 \text{ kN} \cdot \text{m}$
$= 20 \text{ kN·m} \circlearrowright$

**F4–23.** $\mathbf{u}_1 = \dfrac{\mathbf{r}_1}{r_1} = \dfrac{[-0.2\mathbf{i} + 0.2\mathbf{j} + 0.35\mathbf{k}] \text{ m}}{\sqrt{(-0.2 \text{ m})^2 + (0.2 \text{ m})^2 + (0.35 \text{ m})^2}}$

$= -\tfrac{2}{4.5}\mathbf{i} + \tfrac{2}{4.5}\mathbf{j} + \tfrac{3.5}{4.5}\mathbf{k}$

$\mathbf{u}_2 = -\mathbf{k}$

$\mathbf{u}_3 = \tfrac{1.5}{2.5}\mathbf{i} - \tfrac{2}{2.5}\mathbf{j}$

$(\mathbf{M}_c)_1 = (M_c)_1 \mathbf{u}_1$
$= (450 \text{ N·m})\left(-\tfrac{2}{4.5}\mathbf{i} + \tfrac{2}{45}\mathbf{j} + \tfrac{3.5}{4.5}\mathbf{k}\right)$
$= \{-200\mathbf{i} + 200\mathbf{j} + 350\mathbf{k}\} \text{ N·m}$

$(\mathbf{M}_c)_2 = (M_c)_2 \mathbf{u}_2 = (250 \text{ N·m})(-\mathbf{k})$
$= \{-250\mathbf{k}\} \text{ N·m}$

$(\mathbf{M}_c)_3 = (M_c)_3 \, \mathbf{u}_3 = (300 \text{ N·m})\left(\tfrac{1.5}{2.5}\mathbf{i} - \tfrac{2}{2.5}\mathbf{j}\right)$
$= \{180\mathbf{i} - 240\mathbf{j}\} \text{ N·m}$

$(\mathbf{M}_c)_R = \Sigma \mathbf{M}_c;$
$(\mathbf{M}_c)_R = \{-20\mathbf{i} - 40\mathbf{j} + 100\mathbf{k}\} \text{ N·m}$ *Ans.*

**F4–24.** $\mathbf{F}_B = \left(\tfrac{4}{5}\right)(450 \text{ N})\mathbf{j} - \left(\tfrac{3}{5}\right)(450 \text{ N}) \, \mathbf{k}$
$= \{360\mathbf{j} - 270\mathbf{k}\} \text{ N}$

$\mathbf{M}_c = \mathbf{r}_{AB} \times \mathbf{F}_B = \begin{vmatrix} \mathbf{i} & \mathbf{j} & \mathbf{k} \\ 0.4 & 0 & 0 \\ 0 & 360 & -270 \end{vmatrix}$

$= \{108\mathbf{j} + 144\mathbf{k}\} \text{ N·m}$ *Ans.*

Also,
$\mathbf{M}_c = (\mathbf{r}_A \times \mathbf{F}_A) + (\mathbf{r}_B \times \mathbf{F}_B)$

$= \begin{vmatrix} \mathbf{i} & \mathbf{j} & \mathbf{k} \\ 0 & 0 & 0.3 \\ 0 & -360 & 270 \end{vmatrix} + \begin{vmatrix} \mathbf{i} & \mathbf{j} & \mathbf{k} \\ 0.4 & 0 & 0.3 \\ 0 & 360 & -270 \end{vmatrix}$

$= \{108\mathbf{j} + 144\mathbf{k}\} \text{ N·m}$ *Ans.*

**F4–25.** $\xrightarrow{+} F_{Rx} = \Sigma F_x;\ F_{Rx} = 1000 - \frac{3}{5}(500) = 700$ N
$+\downarrow F_{Ry} = \Sigma F_y;\ F_{Ry} = 750 - \frac{4}{5}(500) = 350$ N
$$F_R = \sqrt{700^2 + 350^2} = 782.6 \text{ N} \qquad Ans.$$
$$\theta = \tan^{-1}\left(\frac{350}{700}\right) = 26.6° \searrow \qquad Ans.$$
$\zeta + M_{A_R} = \Sigma M_A;$
$M_{A_R} = \frac{3}{5}(500)(1.2) - \frac{4}{5}(500)(1.8) + 750(0.9)$
$M_{R_A} = 315$ N·m $\qquad Ans.$

**F4–26.** $\xrightarrow{+} F_{Rx} = \Sigma F_x;\ F_{Rx} = \frac{4}{5}(50) = 40$ N
$+\downarrow F_{Ry} = \Sigma F_y;\ F_{Ry} = 40 + 30 + \frac{3}{5}(50)$
$= 100$ N
$$F_R = \sqrt{(40)^2 + (100)^2} = 108 \text{ N} \qquad Ans.$$
$$\theta = \tan^{-1}\left(\frac{100}{40}\right) = 68.2° \searrow \qquad Ans.$$
$\zeta + M_{A_R} = \Sigma M_A;$
$M_{A_R} = 30(3) + \frac{3}{5}(50)(6) + 200$
$= 470$ N·m $\qquad Ans.$

**F4–27.** $\xrightarrow{+}(F_R)_x = \Sigma F_x;$
$(F_R)_x = 900 \sin 30° = 450$ N $\rightarrow$
$+\uparrow(F_R)_y = \Sigma F_y;$
$(F_R)_y = -900 \cos 30° - 300$
$= -1079.42$ N $= 1079.42$ N $\downarrow$
$F_R = \sqrt{450^2 + 1079.42^2}$
$= 1169.47$ N $= 1.17$ kN $\qquad Ans.$
$\theta = \tan^{-1}\left(\frac{1079.42}{450}\right) = 67.4° \searrow \qquad Ans.$
$\zeta + (M_R)_A = \Sigma M_A;$
$(M_R)_A = 300 - 900 \cos 30°(0.75) - 300(2.25)$
$= -959.57$ N·m
$= 960$ N·m $\circlearrowright \qquad Ans.$

**F4–28.** $\xrightarrow{+}(F_R)_x = \Sigma F_x;$
$(F_R)_x = 750\left(\frac{3}{5}\right) + 250 - 500\left(\frac{4}{5}\right) = 300$ N $\rightarrow$
$+\uparrow(F_R)_y = \Sigma F_y;$
$(F_R)_y = -750\left(\frac{4}{5}\right) - 500\left(\frac{3}{5}\right)$
$= -900$ N $= 900$ N $\downarrow$
$F_R = \sqrt{300^2 + 900^2} = 948.7$ N $\qquad Ans.$
$\theta = \tan^{-1}\left(\frac{900}{300}\right) = 71.6° \searrow \qquad Ans.$
$\zeta + (M_R)_A = \Sigma M_A;$
$(M_R)_A = 500\left(\frac{4}{5}\right)(0.3) - 500\left(\frac{3}{5}\right)(1.8) - 750\left(\frac{4}{5}\right)(0.9)$
$= -960 = 960$ N·m $\circlearrowright \qquad Ans.$

**F4–29.** $\mathbf{F}_R = \Sigma \mathbf{F};$
$\mathbf{F}_R = \mathbf{F}_1 + \mathbf{F}_2$
$= (-300\mathbf{i} + 150\mathbf{j} + 200\mathbf{k}) + (-450\mathbf{k})$
$= \{-300\mathbf{i} + 150\mathbf{j} - 250\mathbf{k}\}$ N $\qquad Ans.$
$\mathbf{r}_{OA} = (2 - 0)\mathbf{j} = \{2\mathbf{j}\}$ m
$\mathbf{r}_{OB} = (-1.5 - 0)\mathbf{i} + (2 - 0)\mathbf{j} + (1 - 0)\mathbf{k}$
$= \{-1.5\mathbf{i} + 2\mathbf{j} + 1\mathbf{k}\}$ m
$(\mathbf{M}_R)_O = \Sigma \mathbf{M};$
$(\mathbf{M}_R)_O = \mathbf{r}_{OB} \times \mathbf{F}_1 + \mathbf{r}_{OA} \times \mathbf{F}_2$
$$= \begin{vmatrix} \mathbf{i} & \mathbf{j} & \mathbf{k} \\ -1.5 & 2 & 1 \\ -300 & 150 & 200 \end{vmatrix} + \begin{vmatrix} \mathbf{i} & \mathbf{j} & \mathbf{k} \\ 0 & 2 & 0 \\ 0 & 0 & -450 \end{vmatrix}$$
$= \{-650\mathbf{i} + 375\mathbf{k}\}$ N·m $\qquad Ans.$

**F4–30.** $\mathbf{F}_1 = \{-100\mathbf{j}\}$ N
$$\mathbf{F}_2 = (200 \text{ N})\left[\frac{\{-0.4\mathbf{i} - 0.3\mathbf{k}\} \text{ m}}{\sqrt{(-0.4 \text{ m})^2 + (-0.3 \text{ m})^2}}\right]$$
$= \{-160\mathbf{i} - 120\mathbf{k}\}$ N
$\mathbf{M}_c = \{-75\mathbf{i}\}$ N·m
$\mathbf{F}_R = \{-160\mathbf{i} - 100\mathbf{j} - 120\mathbf{k}\}$ N $\qquad Ans.$
$(\mathbf{M}_R)_O = (0.3\mathbf{k}) \times (-100\mathbf{j})$
$$+ \begin{vmatrix} \mathbf{i} & \mathbf{j} & \mathbf{k} \\ 0 & 0.5 & 0.3 \\ -160 & 0 & -120 \end{vmatrix} + (-75\mathbf{i})$$
$= \{-105\mathbf{i} - 48\mathbf{j} + 80\mathbf{k}\}$ N·m $\qquad Ans.$

**F4–31.** $+\downarrow F_R = \Sigma F_y;\ F_R = 2.5 + 1.25 + 2.5$
$= 6.25$ kN $\qquad Ans.$
$\zeta + F_R x = \Sigma M_O;$
$6.25(x) = 2.5(1) + 1.25(2) + 2.5(3)$
$x = 2$ m $\qquad Ans.$

**F4–32.** $\xrightarrow{+}(F_R)_x = \Sigma F_x;$
$(F_R)_x = 0.5\left(\frac{3}{5}\right) + 0.25 \sin 30° = 0.425$ kN $\rightarrow$
$+\uparrow(F_R)_y = \Sigma F_y;$
$(F_R)_y = 1 + 0.25 \cos 30° - 0.5\left(\frac{4}{5}\right)$
$= 0.8165$ kN $\uparrow$
$F_R = \sqrt{0.425^2 + 0.8165^2} = 0.917$ N $\qquad Ans.$
$\theta = \tan^{-1}\left(\frac{0.8165}{0.425}\right) = 62.5° \measuredangle \qquad Ans.$
$\zeta + (M_R)_A = \Sigma M_A;$
$0.8165(d) = 1(1) - 0.5\left(\frac{4}{5}\right)(2) + 0.25 \cos 30°(3)$
$d = 1.04$ m $\qquad Ans.$

**F4–33.** $\xrightarrow{+}(F_R)_x = \Sigma F_x;$
$(F_R)_x = 15(\frac{4}{5}) = 12 \text{ kN} \rightarrow$
$+\uparrow(F_R)_y = \Sigma F_y;$
$(F_R)_y = -20 + 15(\frac{3}{5}) = -11 \text{ kN} = 11 \text{ kN} \downarrow$
$F_R = \sqrt{12^2 + 11^2} = 16.3 \text{ kN}$ *Ans.*
$\theta = \tan^{-1}(\frac{11}{12}) = 42.5°$ *Ans.*
$\zeta + (M_R)_A = \Sigma M_A;$
$-11(d) = -20(2) - 15(\frac{4}{5})(2) + 15(\frac{3}{5})(6)$
$d = 0.909 \text{ m}$ *Ans.*

**F4–34.** $\xrightarrow{+}(F_R)_x = \Sigma F_x;$
$(F_R)_x = (\frac{3}{5})5 \text{ kN} - 8 \text{ kN}$
$= -5 \text{ kN} = 5 \text{ kN} \leftarrow$
$+\uparrow(F_R)_y = \Sigma F_y;$
$(F_R)_y = -6 \text{ kN} - (\frac{4}{5})5 \text{ kN}$
$= -10 \text{ kN} = 10 \text{ kN} \downarrow$
$F_R = \sqrt{5^2 + 10^2} = 11.2 \text{ kN}$ *Ans.*
$\theta = \tan^{-1}(\frac{10 \text{ kN}}{5 \text{ kN}}) = 63.4°$ *Ans.*
$\zeta + (M_R)_A = \Sigma M_A;$
$5 \text{ kN}(d) = 8 \text{ kN}(3 \text{ m}) - 6 \text{ kN}(0.5 \text{ m})$
$\quad - [(\frac{4}{5})5 \text{ kN}](2 \text{ m})$
$\quad - [(\frac{3}{5})5 \text{kN}](4 \text{ m})$
$d = 0.2 \text{ m}$ *Ans.*

**F4–35.** $+\downarrow F_R = \Sigma F_z; \quad F_R = 400 + 500 - 100$
$= 800 \text{ N}$ *Ans.*
$M_{Rx} = \Sigma M_x; -800y = -400(4) - 500(4)$
$y = 4.50 \text{ m}$ *Ans.*
$M_{Ry} = \Sigma M_y; \quad 800x = 500(4) - 100(3)$
$x = 2.125 \text{ m}$ *Ans.*

**F4–36.** $+\downarrow F_R = \Sigma F_z;$
$F_R = 200 + 200 + 100 + 100$
$= 600 \text{ N}$ *Ans.*
$M_{Rx} = \Sigma M_x;$
$-600y = 200(1) + 200(1) + 100(3) - 100(3)$
$y = -0.667 \text{ m}$ *Ans.*
$M_{Ry} = \Sigma M_y;$
$600x = 100(3) + 100(3) + 200(2) - 200(3)$
$x = 0.667 \text{ m}$ *Ans.*

**F4–37.** $+\uparrow F_R = \Sigma F_y;$
$-F_R = -6(1.5) - 9(3) - 3(1.5)$
$F_R = 40.5 \text{ kN} \downarrow$ *Ans.*
$\zeta + (M_R)_A = \Sigma M_A;$
$-40.5(d) = 6(1.5)(0.75)$
$\quad - 9(3)(1.5) - 3(1.5)(3.75)$
$d = 1.25 \text{ m}$ *Ans.*

**F4–38.** $F_R = \frac{1}{2}(1.8)(3) + 2.4(3) = 9.9 \text{ kN}$ *Ans.*
$\zeta + M_{A_R} = \Sigma M_A;$
$9.9d = [\frac{1}{2}(1.8)(3)](1.2) + [2.4(3)](3)$
$d = 2.51 \text{ m}$ *Ans.*

**F4–39.** $+\uparrow F_R = \Sigma F_y;$
$-F_R = -\frac{1}{2}(6)(3) - \frac{1}{2}(6)(6)$
$F_R = 27 \text{ kN} \downarrow$ *Ans.*
$\zeta + (M_R)_A = \Sigma M_A;$
$-27(d) = \frac{1}{2}(6)(3)(1) - \frac{1}{2}(6)(6)(2)$
$d = 1 \text{ m}$ *Ans.*

**F4–40.** $+\downarrow F_R = \Sigma F_y;$
$F_R = \frac{1}{2}(1)(2) + 3(2) + 2.5$
$= 9.5 \text{ kN}$ *Ans.*
$\zeta + M_{A_R} = \Sigma M_A;$
$9.5d = [\frac{1}{2}(1)(2)](2)(\frac{2}{3}) + [3(2)](1) + 2.5(3)$
$d = 1.56 \text{ m}$ *Ans.*

**F4–41.** $+\uparrow F_R = \Sigma F_y;$
$-F_R = -\frac{1}{2}(3)(4.5) - 3(6)$
$F_R = 24.75 \text{ kN} \downarrow$ *Ans.*
$\zeta + (M_R)_A = \Sigma M_A;$
$-24.75(d) = -\frac{1}{2}(3)(4.5)(1.5) - 3(6)(3)$
$d = 2.59 \text{ m}$ *Ans.*

**F4–42.** $F_R = \int w(x) \, dx = \int_0^4 2.5x^3 \, dx = 160 \text{ N}$
$\zeta + M_{A_R} = \Sigma M_A;$
$\bar{x} = \dfrac{\int xw(x) \, dx}{\int w(x) \, dx} = \dfrac{\int_0^4 2.5x^4 \, dx}{160} = 3.20 \text{ m}$

# Chapter 5

**F5–1.** $\xrightarrow{+} \Sigma F_x = 0;$ $-A_x + 2500(\frac{3}{5}) = 0$
$A_x = 1500$ N  *Ans.*

$\zeta + \Sigma M_A = 0;$ $B_y(3) - 2500(\frac{4}{5})(1.5) - 900 = 0$
$B_y = 1300$ N  *Ans.*

$+\uparrow \Sigma F_y = 0;$ $A_y + 1300 - 2500(\frac{4}{5}) = 0$
$A_y = 700$ N  *Ans.*

**F5–2.** $\zeta + \Sigma M_A = 0;$
$F_{CD} \sin 45°(1.5 \text{ m}) - 4 \text{ kN}(3 \text{ m}) = 0$
$F_{CD} = 11.31$ kN $= 11.3$ kN  *Ans.*

$\xrightarrow{+} \Sigma F_x = 0;$ $A_x + (11.31 \text{ kN}) \cos 45° = 0$
$A_x = -8$ kN $= 8$ kN $\leftarrow$  *Ans.*

$+\uparrow \Sigma F_y = 0;$
$A_y + (11.31 \text{ kN}) \sin 45° - 4 \text{ kN} = 0$
$A_y = -4$ kN $= 4$ kN $\downarrow$  *Ans.*

**F5–3.** $\zeta + \Sigma M_A = 0;$
$N_B[6 \text{ m} + (6 \text{ m}) \cos 45°]$
$- 10 \text{ kN}[2 \text{ m} + (6 \text{ m}) \cos 45°]$
$- 5 \text{ kN}(4 \text{ m}) = 0$
$N_B = 8.047$ kN $= 8.05$ kN  *Ans.*

$\xrightarrow{+} \Sigma F_x = 0;$
$(5 \text{ kN}) \cos 45° - A_x = 0$
$A_x = 3.54$ kN  *Ans.*

$+\uparrow \Sigma F_y = 0;$
$A_y + 8.047 \text{ kN} - (5 \text{ kN}) \sin 45° - 10 \text{ kN} = 0$
$A_y = 5.49$ kN  *Ans.*

**F5–4.** $\xrightarrow{+} \Sigma F_x = 0;$ $-A_x + 400 \cos 30° = 0$
$A_x = 346$ N  *Ans.*

$+\uparrow \Sigma F_y = 0;$
$A_y - 200 - 200 - 200 - 400 \sin 30° = 0$
$A_y = 800$ N  *Ans.*

$\zeta + \Sigma M_A = 0;$
$M_A - 200(2.5) - 200(3.5) - 200(4.5)$
$- 400 \sin 30°(4.5) - 400 \cos 30°(3 \sin 60°) = 0$
$M_A = 3.90$ kN·m  *Ans.*

**F5–5.** $\zeta + \Sigma M_A = 0;$
$N_C(0.7 \text{ m}) - [25(9.81) \text{ N}] (0.5 \text{ m}) \cos 30° = 0$
$N_C = 151.71$ N $= 152$ N  *Ans.*

$\xrightarrow{+} \Sigma F_x = 0;$
$T_{AB} \cos 15° - (151.71 \text{ N}) \cos 60° = 0$
$T_{AB} = 78.53$ N $= 78.5$ N  *Ans.*

$+\uparrow \Sigma F_y = 0;$
$F_A + (78.53 \text{ N}) \sin 15°$
$+ (151.71 \text{ N}) \sin 60° - 25(9.81) \text{ N} = 0$
$F_A = 93.5$ N  *Ans.*

**F5–6.** $\xrightarrow{+} \Sigma F_x = 0;$
$N_C \sin 30° - (250 \text{ N}) \sin 60° = 0$
$N_C = 433.0$ N $= 433$ N  *Ans.*

$\zeta + \Sigma M_B = 0;$
$-N_A \sin 30°(0.15 \text{ m}) - 433.0 \text{ N}(0.2 \text{ m})$
$+ [(250 \text{ N}) \cos 30°](0.6 \text{ m}) = 0$
$N_A = 577.4$ N $= 577$ N  *Ans.*

$+\uparrow \Sigma F_y = 0;$
$N_B - 577.4 \text{ N} + (433.0 \text{ N}) \cos 30°$
$- (250 \text{ N}) \cos 60° = 0$
$N_B = 327$ N  *Ans.*

**F5–7.** $\Sigma F_z = 0;$
$T_A + T_B + T_C - 1 - 2.5 = 0$
$\Sigma M_x = 0;$
$T_A(0.9) + T_C(0.9) - 2.5(0.45) - 1(0.9) = 0$
$\Sigma M_y = 0;$
$-T_B(1.2) - T_C(1.2) + 2.5(0.6) + 1(0.6) = 0$
$T_A = 1.75$ kN, $T_B = 1.25$ kN, $T_C = 0.5$ kN  *Ans.*

**F5–8.** $\Sigma M_y = 0;$
$600 \text{ N}(0.2 \text{ m}) + 900 \text{ N}(0.6 \text{ m}) - F_A(1 \text{ m}) = 0$
$F_A = 660$ N  *Ans.*

$\Sigma M_x = 0;$
$D_z(0.8 \text{ m}) - 600 \text{ N}(0.5 \text{ m}) - 900 \text{ N}(0.1 \text{ m}) = 0$
$D_z = 487.5$ N  *Ans.*

$\Sigma F_x = 0;$ $D_x = 0$  *Ans.*
$\Sigma F_y = 0;$ $D_y = 0$  *Ans.*
$\Sigma F_z = 0;$
$T_{BC} + 660 \text{ N} + 487.5 \text{ N} - 900 \text{ N} - 600 \text{ N} = 0$
$T_{BC} = 352.5$ N  *Ans.*

**F5–9.** $\Sigma F_y = 0$;  $400\,\text{N} + C_y = 0$;
$C_y = -400\,\text{N}$  *Ans.*
$\Sigma M_y = 0$;  $-C_x(0.4\,\text{m}) - 600\,\text{N}(0.6\,\text{m}) = 0$
$C_x = -900\,\text{N}$  *Ans.*
$\Sigma M_x = 0$;  $B_z(0.6\,\text{m}) + 600\,\text{N}(1.2\,\text{m})$
$+ (-400\,\text{N})(0.4\,\text{m}) = 0$
$B_z = -933.3\,\text{N}$  *Ans.*
$\Sigma M_z = 0$;
$-B_x(0.6\,\text{m}) + -(-900\,\text{N})(1.2\,\text{m})$
$+ (-400\,\text{N})(0.6\,\text{m}) = 0$
$B_x = 1400\,\text{N}$  *Ans.*
$\Sigma F_x = 0$;  $1400\,\text{N} + (-900\,\text{N}) - A_x = 0$
$A_x = 500\,\text{N}$  *Ans.*
$\Sigma F_z = 0$;  $A_z - 933.3\,\text{N} + 600\,\text{N} = 0$
$A_z = 333.3\,\text{N}$  *Ans.*

**F5–10.** $\Sigma F_x = 0$;  $B_x = 0$  *Ans.*
$\Sigma M_z = 0$;
$C_y(0.4\,\text{m} + 0.6\,\text{m}) = 0$  $C_y = 0$  *Ans.*
$\Sigma F_y = 0$;  $A_y + 0 = 0$  $A_y = 0$  *Ans.*
$\Sigma M_x = 0$;  $C_z(0.6\,\text{m} + 0.6\,\text{m}) + B_z(0.6\,\text{m})$
$- 450\,\text{N}(0.6\,\text{m} + 0.6\,\text{m}) = 0$
$1.2C_z + 0.6B_z - 540 = 0$
$\Sigma M_y = 0$;  $-C_z(0.6\,\text{m} + 0.4\,\text{m})$
$- B_z(0.6\,\text{m}) + 450\,\text{N}(0.6\,\text{m}) = 0$
$-C_z - 0.6B_z + 270 = 0$
$C_z = 1350\,\text{N}$  $B_z = -1800\,\text{N}$  *Ans.*
$\Sigma F_z = 0$;
$A_z + 1350\,\text{N} + (-1800\,\text{N}) - 450\,\text{N} = 0$
$A_z = 900\,\text{N}$  *Ans.*

**F5–11.** $\Sigma F_y = 0$;  $A_y = 0$  *Ans.*
$\Sigma M_x = 0$;  $-9(3) + F_{CE}(3) = 0$
$F_{CE} = 9\,\text{kN}$  *Ans.*
$\Sigma M_z = 0$;  $F_{CF}(3) - 6(3) = 0$
$F_{CF} = 6\,\text{kN}$  *Ans.*
$\Sigma M_y = 0$;  $9(4) - A_z(4) - 6(1.5) = 0$
$A_z = 6.75\,\text{kN}$  *Ans.*
$\Sigma F_x = 0$;  $A_x + 6 - 6 = 0$  $A_x = 0$  *Ans.*
$\Sigma F_z = 0$;  $F_{DB} + 9 - 9 + 6.75 = 0$
$F_{DB} = -6.75\,\text{kN}$  *Ans.*

**F5–12.** $\Sigma F_x = 0$;  $A_x = 0$  *Ans.*
$\Sigma F_y = 0$;  $A_y = 0$  *Ans.*
$\Sigma F_z = 0$;  $A_z + F_{BC} - 400 = 0$
$\Sigma M_x = 0$;  $(M_A)_x + 1.8F_{BC} - 400(1.8) = 0$
$\Sigma M_y = 0$;  $9F_{BC} - 400(0.45) = 0$  $F_{BC} = 200\,\text{N}$
*Ans.*
$\Sigma M_z = 0$;  $(M_A)_z = 0$  *Ans.*
$A_z = 200\,\text{N}$  $(M_A)_x = 360\,\text{N} \cdot \text{m}$  *Ans.*

## Chapter 6

**F6–1.** Joint A.
$+\uparrow \Sigma F_y = 0$;  $1\,\text{kN} - F_{AD} \sin 45° = 0$
$F_{AD} = 1.414\,\text{kN (C)}$  *Ans.*
$\xrightarrow{+} \Sigma F_x = 0$;  $F_{AB} - (1.414\,\text{kN}) \cos 45° = 0$
$F_{AB} = 1\,\text{kN (T)}$  *Ans.*
Joint B.
$\xrightarrow{+} \Sigma F_x = 0$;  $F_{BC} - 1\,\text{kN} = 0$
$F_{BC} = 1\,\text{kN (T)}$  *Ans.*
$+\uparrow \Sigma F_y = 0$;  $F_{BD} = 0$  *Ans.*
Joint D.
$\xrightarrow{+} \Sigma F_x = 0$;
$F_{CD} \cos 45° + (1.414\,\text{kN}) \cos 45° - 2\,\text{kN} = 0$
$F_{CD} = 1.414\,\text{kN (T)}$  *Ans.*

**F6–2.** Joint D:
$+\uparrow \Sigma F_y = 0$;  $\tfrac{3}{5} F_{CD} - 1.5\,\text{kN} = 0$
$F_{CD} = 2.5\,\text{kN (T)}$  *Ans.*
$\xrightarrow{+} \Sigma F_x = 0$;  $-F_{AD} + \tfrac{4}{5}(2.5) = 0$
$F_{AD} = 2\,\text{kN (C)}$  *Ans.*
$F_{BC} = 2.5\,\text{kN (T)}$,  $F_{AC} = F_{AB} = 0$  *Ans.*

**F6–3.** $A_x = 0$,  $A_y = C_y = 2\,\text{kN}$
Joint A:
$+\uparrow \Sigma F_y = 0$;  $-\tfrac{3}{5} F_{AE} + 2 = 0$
$F_{AE} = 3.333\,\text{kN (C)}$  *Ans.*
Joint C:
$+\uparrow \Sigma F_y = 0$;  $-F_{DC} + 2 = 0$;
$F_{DC} = 2\,\text{kN (C)}$  *Ans.*

**F6–4.** *Joint C.*
$+\uparrow \Sigma F_y = 0;$   $2F\cos 30° - P = 0$
$F_{AC} = F_{BC} = F = \frac{P}{2\cos 30°} = 0.5774P$ (C)
*Joint B.*
$\pm \Sigma F_x = 0;$   $0.5774P \cos 60° - F_{AB} = 0$
$F_{AB} = 0.2887P$ (T)
$F_{AB} = 0.2887P = 2$ kN
$P = 6.928$ kN
$F_{AC} = F_{BC} = 0.5774P = 1.5$ kN
$P = 2.598$ kN
The *smaller value* of P is chosen,
$P = 2.598$ kN $= 2.60$ kN   Ans.

**F6–5.**   $F_{CB} = 0$   Ans.
$F_{CD} = 0$   Ans.
$F_{AE} = 0$   Ans.
$F_{DE} = 0$   Ans.

**F6–6.** *Joint C.*
$+\uparrow \Sigma F_y = 0;$   $1.299$ kN $- F_{CD} \sin 30° = 0$
$F_{CD} = 2.598$ kN (C)   Ans.
$\pm \Sigma F_x = 0;$   $(2.598$ kN$) \cos 30° - F_{BC} = 0$
$F_{BC} = 2.25$ kN (T)   Ans.
*Joint D.*
$+\nearrow \Sigma F_{y'} = 0;$   $F_{BD} \cos 30° = 0$   $F_{BD} = 0$   Ans.
$+\nwarrow \Sigma F_{x'} = 0;$   $F_{DE} - 2.598$ kN $= 0$
$F_{DE} = 2.598$ kN (C)   Ans.
*Joint B.*
$\uparrow \Sigma F_y = 0;$   $F_{BE} \sin \phi = 0$   $F_{BE} = 0$   Ans.
$\pm \Sigma F_x = 0;$   $2.25$ kN $- F_{AB} = 0$
$F_{AB} = 2.25$ kN (T)   Ans.
*Joint A.*
$+\uparrow \Sigma F_y = 0;$   $1.701$ kN $- F_{AE} = 0$
$F_{AE} = 1.701$ kN (C)   Ans.

**F6–7.**   $+\uparrow \Sigma F_y = 0;$   $F_{CF} \sin 45° - 3 - 4 = 0$
$F_{CF} = 9.899$ kN (T)   Ans.
$\zeta + \Sigma M_C = 0;$   $F_{FE}(1) - 4(1) = 0$
$F_{FE} = 4$ kN (T)   Ans.
$\zeta + \Sigma M_F = 0;$   $F_{BC}(1) - 3(1) - 4(2) = 0$
$F_{BC} = 11$ kN (C)   Ans.

**F6–8.**   $+\uparrow \Sigma F_y = 0;$   $F_{KC} + 33.33$ kN $- 40$ kN $= 0$
$F_{KC} = 6.67$ kN (C)   Ans.
$\zeta + \Sigma M_K = 0;$
$33.33$ kN$(8$ m$) - 40$ kN$(2$ m$) - F_{CD}(3$ m$) = 0$
$F_{CD} = 62.22$ kN $= 62.2$ kN (T)   Ans.
$\pm \Sigma F_x = 0;$   $F_{LK} - 62.22$ kN $= 0$
$F_{LK} = 62.2$ kN (C)   Ans.

**F6–9.**   $\zeta + \Sigma M_A = 0;$   $G_y(12$ m$) - 20$ kN$(2$ m$)$
$- 30$ kN$(4$ m$) - 40$ kN$(6$ m$) = 0$
$G_y = 33.33$ kN
From the geometry of the truss,
$\phi = \tan^{-1}(3$ m$/2$ m$) = 56.31°$.
$\zeta + \Sigma M_K = 0;$
$33.33$ kN$(8$ m$) - 40$ kN$(2$ m$) - F_{CD}(3$ m$) = 0$
$F_{CD} = 62.2$ kN (T)   Ans.
$\zeta + \Sigma M_D = 0;$   $33.33$ kN$(6$ m$) - F_{KJ}(3$ m$) = 0$
$F_{KJ} = 66.7$ kN (C)   Ans.
$+\uparrow \Sigma F_y = 0;$
$33.33$ kN $- 40$ kN $+ F_{KD} \sin 56.31° = 0$
$F_{KD} = 8.01$ kN (T)   Ans.

**F6–10.** From the geometry of the truss,
$\tan \phi = \frac{(3 \text{ m}) \tan 30°}{1 \text{ m}} = 1.732$   $\phi = 60°$
$\zeta + \Sigma M_C = 0;$
$F_{EF} \sin 30°(2$ m$) + 1.5$ kN$(2$ m$) = 0$
$F_{EF} = -3$ kN $= 3$ kN (C)   Ans.
$\zeta + \Sigma M_D = 0;$
$1.5$ kN$(2$ m$) - F_{CF} \sin 60° (2$ m$) = 0$
$F_{CF} = 1.732$ kN (T)   Ans.
$\zeta + \Sigma M_F = 0;$
$1.5$ kN$(3$ m$) - 1.5$ kN$(1$ m$) - F_{BC}(3$ m$)\tan 30° = 0$
$F_{BC} = 1.732$ kN (T)   Ans.

**F6–11.** From the geometry of the truss,
$\theta = \tan^{-1} (1$ m$/2$ m$) = 26.57°$
$\phi = \tan^{-1} (3$ m$/2$ m$) = 56.31°$.
The location of G can be found using similar triangles.
$$\frac{1 \text{ m}}{2 \text{ m}} = \frac{2 \text{ m}}{2 \text{ m} + x}$$
$4$ m $= 2$ m $+ x$
$x = 2$ m

$\zeta + \Sigma M_G = 0;$
26.25 kN(4 m) − 15 kN(2 m) − $F_{CD}$(3 m) = 0
$F_{CD} = 25$ kN (T)   Ans.
$\zeta + \Sigma M_D = 0;$
26.25 kN(2 m) − $F_{GF}$ cos 26.57°(2 m) = 0
$F_{GF} = 29.3$ kN (C)   Ans.
$\zeta + \Sigma M_O = 0;$   15 kN(4 m) − 26.25 kN(2 m)
− $F_{GD}$ sin 56.31°(4 m) = 0
$F_{GD} = 2.253$ kN = 2.25 kN (T)   Ans.

**F6–12.**  $\zeta + \Sigma M_H = 0;$
$F_{DC}$(4 m) + 6 kN(3 m) − 8 kN(7 m) = 0
$F_{DC} = 9.5$ kN (C)   Ans.
$\zeta + \Sigma M_D = 0;$
6 kN(7 m) − 8 kN(3 m) − $F_{HI}$(4 m) = 0
$F_{HI} = 4.5$ kN (C)   Ans.
$\zeta + \Sigma M_C = 0;$  $F_{JI}$ cos 45°(4 m) + 6 kN(7 m)
− 4.5 kN(4 m) − 8 kN(3 m) = 0
$F_{JI} = 0$   Ans.

**F6–13.**  $+\uparrow \Sigma F_y = 0;$   $3P - 300$ N = 0
$P = 100$ N   Ans.

**F6–14.**  $\zeta + \Sigma M_C = 0;$
$-(\frac{4}{5})(F_{AB})(2.7) + 2(1.8) + 2.5(0.9) = 0$
$F_{AB} = 2.708$ kN
$\xrightarrow{+} \Sigma F_x = 0; -C_x + \frac{3}{5}(2.708) = 0$
$C_x = 1.625$ kN   Ans.
$+\uparrow \Sigma F_y = 0; C_y + \frac{4}{5}(2.708) - 2 - 2.5 = 0$
$C_y = 2.334$ kN   Ans.

**F6–15.**  $\zeta + \Sigma M_A = 0; 100$ N(250 mm) − $N_B$(50 mm) = 0
$N_B = 500$ N   Ans.
$\xrightarrow{+} \Sigma F_x = 0;$   (500 N) sin 45° − $A_x$ = 0
$A_x = 353.55$ N
$+\uparrow \Sigma F_y = 0; A_y - 100$ N − (500 N) cos 45° = 0
$A_y = 453.55$ N
$F_A = \sqrt{(353.55 \text{ N})^2 + (453.55 \text{ N})^2}$
$= 575$ N   Ans.

**F6–16.**  $\zeta + \Sigma M_C = 0;$
$F_{AB}$ cos 45°(1) − $F_{AB}$ sin 45°(3)
+ 800 + 400(2) = 0
$F_{AB} = 1131.37$ N
$\xrightarrow{+} \Sigma F_x = 0; -C_x + 1131.37 \cos 45° = 0$
$C_x = 800$ N   Ans.
$+\uparrow \Sigma F_y = 0; -C_y + 1131.37 \sin 45° - 400 = 0$
$C_y = 400$ N   Ans.

**F6–17.** Plate A:
$+\uparrow \Sigma F_y = 0;$ $2T + N_{AB} - 500 = 0$
Plate B:
$+\uparrow \Sigma F_y = 0;$ $2T - N_{AB} - 150 = 0$
$T = 162.5$ N, $N_{AB} = 175$ N   Ans.

**F6–18.** Pulley C:
$+\uparrow \Sigma F_y = 0; T - 2P = 0; T = 2P$
Beam:
$+\uparrow \Sigma F_y = 0;$   $2P + P - 6 = 0$
$P = 2$ kN   Ans.
$\zeta + \Sigma M_A = 0; 2(1) - 6(x) = 0$
$x = 0.333$ m   Ans.

# Chapter 7

**F7–1.**  $+\uparrow \Sigma F_y = 0;$   $N - 50(9.81) - 200(\frac{3}{5}) = 0$
$N = 610.5$ N
$\xrightarrow{+} \Sigma F_x = 0;$   $F - 200(\frac{4}{5}) = 0$
$F = 160$ N
$F < F_{max} = \mu_s N = 0.3(610.5) = 183.15$ N,
therefore $F = 160$ N   Ans.

**F7–2.**  $\zeta + \Sigma M_B = 0;$
$N_A(3) + 0.2 N_A(4) - 30(9.81)(2) = 0$
$N_A = 154.89$ N
$\xrightarrow{+} \Sigma F_x = 0;$   $P - 154.89 = 0$
$P = 154.89$ N = 155 N   Ans.

**F7–3.** Crate A
$+\uparrow \Sigma F_y = 0;$   $N_A - 50(9.81) = 0$
$N_A = 490.5$ N
$\xrightarrow{+} \Sigma F_x = 0;$   $T - 0.25(490.5) = 0$
$T = 122.62$ N

**Crate B**

$+\uparrow \Sigma F_y = 0; \quad N_B + P \sin 30° - 50(9.81) = 0$
$\qquad N_B = 490.5 - 0.5P$

$\xrightarrow{+} \Sigma F_x = 0;$
$P \cos 30° - 0.25(490.5 - 0.5P) - 122.62 = 0$
$\qquad P = 247 \text{ N} \qquad Ans.$

**F7–4.** $\xrightarrow{+} \Sigma F_x = 0; \quad N_A - 0.3 N_B = 0$
$+ \uparrow \Sigma F y = 0;$
$N_B + 0.3 N_A + P - 100(9.81) = 0$
$\zeta + \Sigma M_O = 0;$
$P(0.6) + N_B(0.9) - 0.3 N_B(0.9)$
$\qquad - 0.3 N_A(0.9) = 0$
$N_A = 175.70 \text{ N} \qquad N_B = 585.67 \text{ N}$
$\qquad P = 343 \text{ N} \qquad Ans.$

**F7–5.** If slipping occurs:
$+\uparrow \Sigma F_y = 0; \quad N_C - 125(9.81) \text{ N} = 0 \quad N_C = 1226.25 \text{ N}$
$\xrightarrow{+} \Sigma F_x = 0; \quad P - 0.4(1226.25) = 0 \quad P = 490.5 \text{ N}$
If tipping occurs:
$\zeta + \Sigma M_A = 0; \quad -P(1.35) + 125(9.81)(0.45) = 0$
$\qquad P = 408.75 \text{ N} \qquad Ans.$

# Chapter 8

**F8–1.** $y_G = 0.75 \sin \theta \qquad \delta y_G = 0.75 \cos \theta \, \delta\theta$
$x_C = 2(1.5) \cos \theta \qquad \delta x_C = -3 \sin \theta \, \delta\theta$
$\delta U = 0; \qquad 2W \delta y_G + P \delta x_C = 0$
$\qquad (294.3 \cos \theta - 3P \sin \theta)\delta\theta = 0$
$P = 98.1 \cot \theta|_{\theta=60°} = 56.6 \text{ N} \qquad Ans.$

**F8–2.** $x_A = 5 \cos \theta \qquad \delta x_A = -5 \sin \theta \, \delta\theta$
$y_G = 2.5 \sin \theta \qquad \delta y_G = 2.5 \cos \theta \, \delta\theta$
$\delta U = 0; \qquad -P \delta x_A + (-W \delta y_G) = 0$
$\qquad (5P \sin \theta - 1226.25 \cos \theta)\delta\theta = 0$
$P = 245.25 \cot \theta|_{\theta=60°} = 142 \text{ N} \qquad Ans.$

**F8–3.** $x_B = 0.6 \sin \theta \qquad \delta x_B = 0.6 \cos \theta \, \delta\theta$
$y_C = 0.6 \cos \theta \qquad \delta y_C = -0.6 \sin \theta \, \delta\theta$
$\delta U = 0; \qquad -F_{sp} \delta x_B + (-P \delta y_C) = 0$
$\qquad -9(10^3) \sin \theta \, (0.6 \cos \theta \, \delta\theta)$
$\qquad - 2000(-0.6 \sin \theta \, \delta\theta) = 0$
$\sin \theta = 0 \qquad \theta = 0° \qquad Ans.$
$\qquad -5400 \cos \theta + 1200 = 0$
$\theta = 77.16° = 77.2° \qquad Ans.$

**F8–4.** $x_B = 0.9 \cos \theta \qquad \delta x_B = -0.9 \sin \theta \, \delta\theta$
$x_C = 2(0.9 \cos \theta) \qquad \delta x_C = -1.8 \sin \theta \, \delta\theta$
$\delta U = 0; \qquad P \delta x_B + (-F_{sp} \delta x_C) = 0$
$6(10^3)(-0.9 \sin \theta \, \delta\theta)$
$\qquad -36(10^3)(\cos \theta - 0.5)(-1.8 \sin \theta \, \delta\theta) = 0$

$\sin \theta \, (64\,800 \cos \theta - 37\,800)\delta\theta = 0$
$\sin \theta = 0 \qquad \theta = 0° \qquad Ans.$
$\qquad 64800 \cos \theta - 37\,800 = 0$
$\qquad \theta = 54.31° = 54.3° \qquad Ans.$

**F8–5.** $y_G = 2.5 \sin \theta \qquad \delta y_G = 2.5 \cos \theta \, \delta\theta$
$x_A = 5 \cos \theta \qquad \delta x_C = -5 \sin \theta \, \delta\theta$
$\delta U = 0; \qquad -W \delta y_G + (-F_{sp} \delta x_A) = 0$
$(15\,000 \sin \theta \cos \theta - 7500 \sin \theta$
$\qquad - 1226.25 \cos \theta)\delta\theta = 0$
$\theta = 56.33° = 56.3° \qquad Ans.$

**F8–6.** $F_{sp} = 15\,000 \, (0.6 - 0.6 \cos \theta)$
$x_C = 3[0.3 \sin \theta] \qquad \delta x_C = 0.9 \cos \theta \, \delta\theta$
$y_B = 2[0.3 \cos \theta] \qquad \delta y_B = -0.6 \sin \theta \, \delta\theta$
$\delta U = 0; \qquad P \delta x_C + F_{sp} \delta y_B = 0$
$(135 \cos \theta - 5400 \sin \theta + 5400 \sin \theta \cos \theta)\delta\theta = 0$
$\theta = 20.9° \qquad Ans.$

# Answers to Selected Problems

## Chapter 1
- **1–1.**
  - a. 4.66 m
  - b. 55.6 s
  - c. 4.56 kN
  - d. 2.77 Mg
- **1–2.**
  - a. N
  - b. MN/m
  - c. N/s²
  - d. MN/s
- **1–3.**
  - a. 0.431 g
  - b. 35.3 kN
  - c. 5.32 m
- **1–5.**
  - a. GN/s
  - b. Gg/N
  - c. GN/(kg·s)
- **1–6.**
  - a. 45.3 MN
  - b. 56.8 km
  - c. 5.63 μg
- **1–7.** $W_e = 35.8$ MN
  $W_m = 5.91$ MN
  $m_m = m_e = 3.65$ Gg
- **1–9** 1 ATM = 101 kPa
- **1–10.**
  - a. $W = 98.1$ N
  - b. $W = 4.90$ mN
  - c. $W = 44.1$ kN
- **1–11.**
  - a. 0.447 kg·m/N
  - b. 0.911 kg·s
  - c. 18.8 GN/m
- **1–14.**
  - a. $m = 2.04$ g
  - b. $m = 15.3$ Mg
  - c. $m = 6.12$ Gg
- **1–15.**
  - a. 0.04 MN²
  - b. 25 μm²
  - c. 0.064 km³
- **1–16.** $F = 7.41$ μN
- **1–18.** 26.9 μm·kg/N

## Chapter 2
- **2–1.** $F_R = \sqrt{6^2 + 8^2 - 2(6)(8)\cos 75°} = 8.67$ kN
  $\frac{\sin \alpha}{8} = \frac{\sin 75°}{8.669}$    $\alpha = 63.05°$
  $\phi = 3.05°$
- **2–2.** $F_R = 10.5$ kN
  $\phi = 17.5°$
- **2–3.** $T = 6.57$ kN
  $\theta = 30.6°$
- **2–5.** $\frac{F_u}{\sin 105°} = \frac{200}{\sin 30°}$    $F_u = 386$ N
  $F_v = 283$ N
- **2–6.** $F_u = 150$ N
  $F_v = 260$ N
- **2–7.** $\theta = 78.6°$
  $F_R = 3.92$ kN
- **2–9.** $F_R = \sqrt{8^2 + 6^2 - 2(8)(6)\cos 100°} = 10.8$ kN
  $\frac{\sin \theta'}{6} = \frac{\sin 100°}{10.80}$
  $\theta' = 33.16°$
  $\phi = 3.16°$
- **2–10.** $\theta = 54.9°$
  $F_R = 10.4$ kN
- **2–11.** $F_R = 400$ N
  $\theta = 60°$
- **2–13.** $\frac{-F_{x'}}{\sin 30°} = \frac{360}{\sin 80°}$    $F_{x'} = -183$ N
  $\frac{F_y}{\sin 70°} = \frac{360}{\sin 80°}$    $F_y = 344$ N
- **2–14.** $\theta = 53.5°$
  $F_{AB} = 1242$ N
- **2–15.** $\phi = 38.3°$
- **2–17.** $\frac{F_{2v}}{\sin 30°} = \frac{150}{\sin 75°}$, $F_{2v} = 77.6$ N
  $\frac{F_{2u}}{\sin 75°} = \frac{150}{\sin 75°}$, $F_{2u} = 150$ N
- **2–18.** $F_A = 774$ N
  $F_B = 346$ N
- **2–19.** $F_B = 325$ N
  $F_A = 893$ N
  $\theta = 70.0°$
- **2–21.** $F_1 = 6 \cos 30° = 5.20$ kN
  $F_2 = 6 \sin 30° = 3$ kN
  $\theta = 60°$
- **2–22.** $\theta = 90°$
  $F_2 = 2.50$ kN
  $F_R = 4.33$ kN
- **2–23.** $F_R = 8.09$ kN
  $\phi = 98.5°$

# Answers to Selected Problems

**2–25.** $\dfrac{F}{\sin\phi} = \dfrac{F}{\sin(\theta - \phi)}$
$\phi = \dfrac{\theta}{2}$
$F_R = \sqrt{(F)^2 + (F)^2 - 2(F)(F)\cos(180° - \theta)}$
$F_R = 2F\cos\left(\dfrac{\theta}{2}\right)$

**2–26.** $F_A = 3.66$ kN
$F_B = 7.07$ kN

**2–27.** $F_B = 5.00$ kN
$F_A = 8.66$ kN
$\theta = 60°$

**2–29.** $F_A = 600\cos 30° = 520$ N
$F_B = 600\sin 30° = 300$ N

**2–30.** $\theta = 10.9°$
$F_{min} = 471$ N

**2–31.** $F = 194.7$ N
$\theta = 16.2°$

**2–33.** $F_R = \sqrt{499.62^2 + 493.01^2} = 702$ N
$\theta = 44.6°$

**2–34.** $\phi = 42.4°$    $F_1 = 731$ N

**2–35.** $F_x = 336.5$ N
$F_y = -807.7$ N

**2–37.** $5.196 = -2 + F_2\cos\phi + 4$
$-3 = -3.464 + F_2\sin\phi - 3$
$\phi = 47.3°$    $F_2 = 4.71$ kN

**2–38.** $F_2 = 12.9$ kN
$F_R = 13.2$ kN

**2–39.** $\theta = 29.1°$    $F_1 = 275$ N

**2–41.** $0 = 700\sin 30° - F_B\cos\theta$
$1500 = 700\cos 30° + F_B\sin\theta$
$\theta = 68.6°$    $F_B = 960$ N

**2–42.** $F_R = 839$ N
$\phi = 14.8°$

**2–43.** $F_R = 2.314$ kN
$\theta = 39.6°$

**2–45.** $0 = F_1\sin\phi - 0.9 - 1.2$
$F_R = F_1\cos\phi + 1.2 - 0.5$
$F_1 = 2.1$ kN
$F_R = 0.7$ kN

**2–46.** $\theta = 63.7°$    $F_3 = 1.20F_1$

**2–47.** $\theta = 54.3°$    $F_A = 686$ N

**2–49.** $F_R = \sqrt{(-515.2)^2 + (-212.9)^2} = 557.5$ N
$\theta = 202°$

**2–50.** $1.22$ kN $\leq P \leq 3.17$ kN

**2–51.** $F_R = 391$ N
$\theta = 16.4°$

**2–53.** $F_R = \sqrt{(0.5F_1 + 300)^2 + (0.8660F_1 - 240)^2}$
$F_R^2 = F_1^2 - 115.69F_1 + 147\,600$
$2F_R\dfrac{dF_R}{dF_1} = 2F_1 - 115.69 = 0$
$F_1 = 57.8$ N,    $F_R = 380$ N

**2–54.** $\theta = 103°$
$F_2 = 440.7$ N

**2–55.** $F_R = 803.6$ N
$\theta = 38.3°$

**2–57.** $F_R^2 = (-4.1244 - F\cos 45°)^2 + (7 - F\sin 45°)^2$
$2F_R\dfrac{dF_R}{dF} = 2(-4.1244 - F\cos 45°)(-\cos 45°)$
$\qquad + 2(7 - F\sin 45°)(-\sin 45°) = 0$
$F = 2.03$ kN
$F_R = 7.87$ kN

**2–58.** $\mathbf{F}_1 = \{F_1\cos\theta\,\mathbf{i} + F_1\sin\theta\,\mathbf{j}\}$ N
$\mathbf{F}_2 = \{350\mathbf{i}\}$ N
$\mathbf{F}_3 = \{-100\mathbf{j}\}$ N
$\theta = 67.0°$
$F_1 = 434$ N

**2–59.** $\mathbf{F}_1 = \{-159\mathbf{i} + 276\mathbf{j} + 318\mathbf{k}\}$ N
$\mathbf{F}_2 = \{424\mathbf{i} + 300\mathbf{j} - 300\mathbf{k}\}$ N

**2–61.** $\mathbf{F}_1 = 600\left(\tfrac{4}{5}\right)(+\mathbf{i}) + 0\,\mathbf{j} + 600\left(\tfrac{3}{5}\right)(+\mathbf{k})$
$\qquad = \{480\mathbf{i} + 360\mathbf{k}\}$ N
$\mathbf{F}_2 = 400\cos 60°\,\mathbf{i} + 400\cos 45°\,\mathbf{j}$
$\qquad + 400\cos 120°\,\mathbf{k}$
$\qquad = \{200\mathbf{i} + 283\mathbf{j} - 200\mathbf{k}\}$ N

**2–62.** $F_R = 3.768$ kN
$\alpha = 25.5°$
$\beta = 68.0°$
$\gamma = 77.7°$

**2–63.** $F_x = -200$ N
$F_y = 200$ N
$F_z = 283$ N

**2–65.** $-100\mathbf{k} = \{(F_{2_x} - 33.40)\mathbf{i} + (F_{2_y} + 19.28)\mathbf{j}$
$\qquad + (F_{2_z} - 45.96)\mathbf{k}\}$
$F_2 = 66.4$ N
$\alpha = 59.8°$
$\beta = 107°$
$\gamma = 144°$

**2–66.** $\alpha = 124°$
$\beta = 71.3°$
$\gamma = 140°$

**2–67.** $\mathbf{F}_1 = \{70\mathbf{j} - 140\mathbf{k}\}$ N
$\mathbf{F}_2 = \{450\mathbf{i} - 636.4\mathbf{j} + 450\mathbf{k}\}$ N

**2–69.** $-300\mathbf{i} + 650\mathbf{j} + 250\mathbf{k}$
$= (459.28\mathbf{i} + 265.17\mathbf{j} - 530.33\mathbf{k})$
$+ (F\cos\alpha\mathbf{i} + F\cos\beta\mathbf{j} + F\cos\gamma\mathbf{k})$
$F^2(\cos^2\alpha + \cos^2\beta + \cos^2\gamma) = 1\,333\,518.08$
$F = 1.15$ kN
$\alpha = 131°$ $\quad \beta = 70.5°$ $\quad \gamma = 47.5°$

**2–70.** $F = 882$ N
$\alpha = 121°$ $\quad \beta = 52.7°$ $\quad \gamma = 53.0°$

**2–71.** $F_R = 718$ N
$\alpha_R = 86.8°$
$\beta_R = 13.3°$
$\gamma_R = 103°$

**2–73.** $F_R = \sqrt{(550)^2 + (52.1)^2 + (270)^2} = 615$ N
$\alpha = 26.6°$
$\beta = 85.1°$
$\gamma = 64.0°$

**2–74.** $\alpha_1 = 45.6°$
$\beta_1 = 53.1°$
$\gamma_1 = 66.4°$

**2–75.** $\alpha_1 = 90°$
$\beta_1 = 53.1°$
$\gamma_1 = 66.4°$

**2–77.** $F_2 \cos\alpha_2 = -150.57$
$F_2 \cos\beta_2 = -86.93$
$F_2 \cos\gamma_2 = 46.59$
$F_2 = 180$ N
$\alpha_2 = 147°$
$\beta_2 = 119°$
$\gamma_2 = 75.0°$

**2–78.** $\alpha = 121°$
$\gamma = 53.1°$
$F_R = 754$ N
$\beta = 52.5°$

**2–79.** $F_3 = 9.58$ kN
$\alpha_3 = 15.5°$
$\beta_3 = 98.4°$
$\gamma_3 = 77.0°$

**2–81.** $\alpha = 64.67°$
$F_x = 1.28$ kN
$F_y = 2.60$ kN
$F_z = 0.776$ kN

**2–82.** $F = 2.02$ kN
$F_y = 0.523$ kN

**2–83.** $F_3 = 166$ N
$\alpha = 97.5°$
$\beta = 63.7°$
$\gamma = 27.5°$

**2–85.** $F_2 = \sqrt{(-17.10)^2 + (8.68)^2 + (-26.17)^2}$
$= 32.4$ N
$\alpha_2 = 122°$
$\beta_2 = 74.5°$
$\gamma_2 = 144°$

**2–86.** $\mathbf{r}_{AB} = \{-3\mathbf{i} + 6\mathbf{j} + 2\mathbf{k}\}$ m
$r_{AB} = 7$ m

**2–87.** $z = 5.35$ m

**2–89.** $\mathbf{F}_B = \{2\mathbf{i} - 2\mathbf{j} - 1\mathbf{k}\}$ kN
$\mathbf{F}_C = \{1.25\mathbf{i} + 2.5\mathbf{j} - 2.5\mathbf{k}\}$ kN
$F_R = \sqrt{3.25^2 + 0.5^2 + (-3.5)^2} = 4.80$ kN
$\alpha = 47.4°$
$\beta = 84.0°$
$\gamma = 137°$

**2–90.** $\alpha = 72.8°$
$\beta = 83.3°$
$\gamma = 162°$
$F_R = 822$ N

**2–91.** $F_R = 1.38$ kN
$\alpha = 82.4°$
$\beta = 125°$
$\gamma = 144°$

**2–93.** $\mathbf{F}_A = 300 \dfrac{(1.2\cos 30°\,\mathbf{i} - 1.2\sin 30°\,\mathbf{j} - 1.8\,\mathbf{k})}{\sqrt{(1.2\cos 30°)^2 + (-1.2\sin 30°)^2 + (-1.8)^2}}$
$= \{144.1\,\mathbf{i} - 83.2\,\mathbf{j} - 249.6\,\mathbf{k}\}$ N
$\mathbf{F}_B = \{-144.1\,\mathbf{i} - 83.2\,\mathbf{j} - 249.6\,\mathbf{k}\}$ N
$\mathbf{F}_C = 300 \dfrac{(1.2\,\mathbf{j} - 1.8\,\mathbf{k})}{\sqrt{(1.2)^2 + (-1.8)^2}}$
$= \{166.4\,\mathbf{j} - 249.6\,\mathbf{k}\}$ N
$F_R = 748.8$ N
$\alpha = 90°$
$\beta = 90°$
$\gamma = 180°$

**2–94.** $F = 260.4$ N

**2–95.** $\mathbf{F} = \{297\mathbf{i} - 440.9\mathbf{j} - 415.9\mathbf{k}\}$ N
$\alpha = 63.9°$
$\beta = 131°$
$\gamma = 128°$

**2–97.** $\mathbf{r}_{AB} = \{(0 - 0)\mathbf{i} + [0 - (-2.299)]\,\mathbf{j}$
$+ (0 - 0.750)\,\mathbf{k}\}$ m
$\mathbf{r}_{CD} = \{[-0.5 - (-2.5)]\,\mathbf{i} + [0 - (-2.299)]\,\mathbf{j}$
$+ (0 - 0.750)\,\mathbf{k}\}$ m
$\mathbf{F}_A = \{285\mathbf{j} - 93.0\mathbf{k}\}$ N
$\mathbf{F}_C = \{159\mathbf{i} + 183\mathbf{j} - 59.7\mathbf{k}\}$ N

# Answers to Selected Problems

**2–98.** $\mathbf{F}_A = \{-43.5\mathbf{i} + 174\mathbf{j} - 174\mathbf{k}\}$ N
$\mathbf{F}_B = \{53.2\mathbf{i} - 79.8\mathbf{j} - 146\mathbf{k}\}$ N

**2–99.** $F_C = 1.62$ kN
$F_B = 2.42$ kN
$F_R = 3.46$ kN

**2–101.** $\mathbf{u} = \frac{\mathbf{F}}{F} = -\frac{120}{170}\mathbf{i} - \frac{90}{170}\mathbf{j} - \frac{80}{170}\mathbf{k}$
$x = 0.72$ m
$y = 0.54$ m
$z = 0.48$ m

**2–102.** $F_R = 6.204$ kN
$\alpha = 90°$
$\beta = 90°$
$\gamma = 180°$

**2–103.** $F_A = F_B = F_C = 1.63$ kN

**2–105.** $\mathbf{F}_A = \{150\mathbf{i} - 100\mathbf{j} - 300\mathbf{k}\}$ N
$\mathbf{F}_B = \{150\mathbf{i} + 100\mathbf{j} - 300\mathbf{k}\}$ N
$\mathbf{F}_C = \{-150\mathbf{i} + 100\mathbf{j} - 300\mathbf{k}\}$ N
$\mathbf{F}_D = \{-150\mathbf{i} - 100\mathbf{j} - 300\mathbf{k}\}$ N
$F_R = 1200$ N
$\alpha = 90°$
$\beta = 90°$
$\gamma = 180°$

**2–106.** $F = 0.525$ kN

**2–107.** $\mathbf{F} = \{-33.1\mathbf{i} - 18.6\mathbf{j} + 46.5\mathbf{k}\}$ N

**2–109.** $\mathbf{r}_A = (0 - 0.75)\mathbf{i} + (0 - 0)\mathbf{j} + (3 - 0)\mathbf{k}$
$= \{-0.75\mathbf{i} + 0\mathbf{j} + 3\mathbf{k}\}$ m
$\mathbf{F}_A = \{-1.46\mathbf{i} + 5.82\mathbf{k}\}$ kN
$\mathbf{r}_C = [0 - (-0.75\sin 45°)]\mathbf{i}$
$+ [0 - (-0.75\cos 45°)]\mathbf{j} + (3 - 0)\mathbf{k}$
$\mathbf{F}_C = \{0.857\mathbf{i} + 0.857\mathbf{j} + 4.85\mathbf{k}\}$ kN
$\mathbf{r}_B = [0 - (-0.75\sin 30°)]\mathbf{i}$
$+ (0 - 0.75\cos 30°)\mathbf{j} + (3 - 0)\mathbf{k}$
$\mathbf{F}_B = \{0.970\mathbf{i} - 1.68\mathbf{j} + 7.76\mathbf{k}\}$ kN
$F_R = 18.5$ kN
$\alpha = 88.8°$
$\beta = 92.6°$
$\gamma = 2.81°$

**2–110.** $\mathbf{F} = \{0.717\mathbf{i} + 1.24\mathbf{j} - 1.00\mathbf{k}\}$ kN

**2–113.** $(F_{AO})_\parallel = (24)\left(\frac{3}{7}\right) + (-48)\left(-\frac{6}{7}\right)$
$+ 16\left(-\frac{2}{7}\right) = 46.9$ N
$(F_{AO})_\perp = \sqrt{(56)^2 - (46.86)^2} = 30.7$ N

**2–114.** $r_{BC} = 5.39$ m

**2–115.** $(F_{ED})_\parallel = 334$ N
$(F_{ED})_\perp = 498$ N

**2–117.** $\mathbf{u}_1 = \cos 120°\,\mathbf{i} + \cos 60°\,\mathbf{j} + \cos 45°\,\mathbf{k}$
$|\text{Proj}\,F_2| = 71.6$ N

**2–118.** $F_{BC} = 45.2$ N
$\mathbf{F}_{BC} = \{32\mathbf{i} - 32\mathbf{j}\}$ N

**2–119.** $F_1 = 333$ N
$F_2 = 373$ N

**2–121.** $\mathbf{u}_{AC} = 0.1581\mathbf{i} + 0.2739\mathbf{j} - 0.9487\mathbf{k}$
$(F_{AC})_z = -2.8461$ kN

**2–122.** $F_{AC} = 366$ N
$\mathbf{F}_{AC} = \{293\mathbf{j} + 219\mathbf{k}\}$ N

**2–123.** $(F_{BC})_\parallel = 245$ N
$(F_{BC})_\perp = 316$ N

**2–125.** $\mathbf{u}_{OD} = -\sin 30°\,\mathbf{i} + \cos 30°\,\mathbf{j}$
$\mathbf{u}_{OA} = \frac{1}{3}\mathbf{i} + \frac{2}{3}\mathbf{j} - \frac{2}{3}\mathbf{k}$
$\phi = 65.8°$

**2–126.** $(F_1)_{F_2} = 50.6$ N

**2–127.** $\theta = 97.3°$

**2–129.** $\mathbf{r}_{AB} = \{-4.5\mathbf{i} + 0.9\mathbf{j} + 2.4\mathbf{k}\}$ m
$\mathbf{r}_{AC} = \{-4.5\mathbf{i} - 2.4\mathbf{j} + 3.6\mathbf{k}\}$ m
$\theta = 34.2°$

**2–130.** $F_x = 217.2$ N
$F_{AC} = 206.7$ N

**2–131.** $F_x = -75$ N
$F_y = 260$ N

**2–133.**
$\mathbf{u}_{F_1} = \cos 30°\sin 30°\,\mathbf{i} + \cos 30°\cos 30°\,\mathbf{j} - \sin 30°\,\mathbf{k}$
$\mathbf{u}_{F_2} = \cos 135°\,\mathbf{i} + \cos 60°\,\mathbf{j} + \cos 60°\,\mathbf{k}$
$(F_1)_{F_2} = 5.44$ N

**2–134.** $F_R = 178$ N
$\theta = 100°$

**2–135.** $F_x = -606$ N
$F_y = 350$ N

**2–137.** $\mathbf{r}_{BA} = \{-0.9\mathbf{i}\}$ m
$\mathbf{r}_{BC} = \{1.8\mathbf{i} + 1.2\mathbf{j} - 0.6\mathbf{k}\}$ m
$\theta = 143°$

**2–138.** $F_r = 178$ N
$\theta = 85.2°$

**2–139.** $F_{AB} = 215$ N
$\theta = 52.7°$

**2–141.** $\dfrac{250}{\sin 120°} = \dfrac{F_u}{\sin 40°}$    $F_u = 186$ N
$F_v = 98.7$ N

**2–142.** Proj $F = 48.0$ N

**2–143.** $\mathbf{F}_B = \{-324\mathbf{i} + 130\mathbf{j} + 195\mathbf{k}\}$ N
$\mathbf{F}_C = \{-324\mathbf{i} - 130\mathbf{j} + 195\mathbf{k}\}$ N
$\mathbf{F}_E = \{-194\mathbf{i} + 291\mathbf{k}\}$ N

# Chapter 3

**3–1.** $F_{BA} \sin 30° - 200(9.81) = 0 \quad F_{BA} = 3.92$ kN
$F_{BC} = 3.40$ kN

**3–2.** $F_{BC} = 2.90$ kN, $y = 0.841$ m

**3–3.** $F_{AB} = 29.4$ kN
$F_{BC} = 15.2$ kN, $F_{BD} = 21.5$ kN

**3–5.** $T = 13.3$ kN, $F_2 = 10.2$ kN

**3–6.** $\theta = 36.3°, T = 14.3$ kN

**3–7.** $T_{BC} = 22.3$ kN
$T_{BD} = 32.6$ kN

**3–9.** $F_{AB} \cos 45° - F_{AC}(\frac{3}{5}) = 0$
$F_{AC} = 1473$ N
$W = 2062$ N

**3–10.** $T = 7.20$ kN
$F = 5.40$ kN

**3–11.** $T = 7.66$ kN
$\theta = 70.1°$

**3–13.** $W_C \cos 30° - 1.375 \cos \theta = 0$
$\theta = 40.9° \qquad W_C = 1.20$ kN

**3–14.** $x_{AC} = 0.793$ m
$x_{AB} = 0.467$ m

**3–15.** $m = 8.56$ kg

**3–17.** $F_{CB} \cos \theta - F_{CA} \cos 30° = 0$
$\theta = 64.3° \qquad F_{CB} = 85.2$ N
$F_{CA} = 42.6$ N

**3–18.** $F_{AB} = 98.6$ N $\quad F_{AC} = 267$ N

**3–19.** $d = 2.42$ m

**3–21.** Joint $D, \Sigma F_x = 0$,
$F_{CD} \cos 30° - F_{BD} \cos 45° = 0$
Joint $B, \Sigma F_x = 0$,
$F_{BC} + 8.7954m \cos 45° - 12.4386m \cos 30° = 0$
$m = 48.2$ kg

**3–22.** $\theta = 35.0°$

**3–23.** $200 = 850(\sqrt{1.08} - l'), l' = 0.804$ m

**3–25.** Joint $E, F_{ED} \cos 30° - F_{EB}(\frac{3}{5}) = 0$
Joint $B$,
$1.3957W \cos 30° - 0.8723W(\frac{3}{5}) - F_{BA} = 0$
$W = 0.289$ kN

**3–26.** $F_{BA} = 395.8$ N
$F_{CD} = 323.2$ N
$F_{BC} = 280.2$ N
$\theta = 2.95°$

**3–27.** $W_F = 614.7$ N

**3–29.** $0.5 \cos \theta = W(\frac{5}{13})$
$\theta = 78.7°$
$W = 0.255$ kN

**3–30.** $T = 265.4$ N

**3–31.** $F = 196.4$ N

**3–33.** $2(T \cos 30°) - 50 = 0$
$T = 28.9$ N
$F_R = 14.9$ N, $(A$ and $D)$
$F_R = 40.8$ N, $(B$ and $C)$

**3–34.** $P = 147$ N

**3–35.** $l = 477.3$ mm

**3–37.** $-T_{AC} + F_s \cos \theta = 0$
$d = 176.9$ mm

**3–38.** $k = 1334.5$ N/m

**3–39.** $W_E = 89.9$ N

**3–41.** $-150 + 2T \sin \theta = 0$
$-2(107.1) \cos 44.4° + m(9.81) = 0$
$m = 15.6$ kg

**3–42.** $m = 2.37$ kg

**3–43.** $y = 6.59$ m

**3–45.** $F_{AB} - \frac{2}{3}F_{AD} = 0$
$-F_{AC} + \frac{2}{3}F_{AD} = 0$
$\frac{1}{3}F_{AD} - 981 = 0$
$F_{AD} = 2.94$ kN
$F_{AB} = F_{AC} = 1.96$ kN

**3–46.** $m = 102$ kg

**3–47.** $F_{AB} = 2.52$ kN
$F_{CB} = 2.52$ kN
$F_{BD} = 3.64$ kN

**3–49.** $-\frac{2}{3}F_{AB} - \frac{2}{3}F_{AC} + F_{AD} = 0$
$\frac{1}{3}F_{AB} - \frac{2}{3}F_{AC} = 0$
$\frac{2}{3}F_{AB} + \frac{1}{3}F_{AC} - W = 0$
$F_{AC} = 1125$ N $\quad F_{AD} = 2250$ N
$W = 1875$ N

**3–50.** $F_{AB} = 6.848$ kN
$F_{AC} = 3.721$ kN
$F_{AD} = 8.518$ kN

**3–51.** $F_{AB} = 7.337$ kN
$F_{AC} = 4.568$ kN
$F_{AD} = 7.098$ kN

**3–53.** $0.1330 F_C - 0.2182 F_D = 0$
$0.7682 F_B - 0.8865 F_C - 0.8729 F_D = 0$
$0.6402 F_B - 0.4432 F_C - 0.4364 F_D - 4905 = 0$
$F_B = 19.2$ kN
$F_C = 10.4$ kN
$F_D = 6.32$ kN

**3–54.** $F_{AB} = 1.21$ kN
$F_{AC} = 606$ N
$F_{AD} = 750$ N

3–55. $F_{AB} = 1.31$ kN
$F_{AC} = 763$ N
$F_{AD} = 708.5$ N
3–57. $\frac{4}{14}F_B - \frac{6}{14}F_C - \frac{4}{14}F_D = 0$
$-\frac{6}{14}F_B - \frac{4}{14}F_C + \frac{6}{14}F_D = 0$
$-\frac{12}{14}F_B - \frac{12}{14}F_C - \frac{12}{14}F_D + W = 0$
$m = 2.62$ Mg
3–58. $F_{AB} = 831$ N
$F_{AC} = 35.6$ N
$F_{AD} = 415$ N
3–59. $m = 90.3$ kg
3–61. $(F_{AB})_x - \frac{3}{7}F_{AB} - \frac{3}{7}F_{AB} = 0$
$(F_{AB})_z + \frac{3}{14}F_{AB} + \frac{3}{14}F_{AB} - 490.5 = 0$
$F_{AB} = 520$ N
$F_{AC} = F_{AD} = 260$ N
$d = 3.61$ m
3–62. $y = 0.112$ m
$z = 0.761$ m
3–63. $F = 4157.6$ N
$z = 0.634$ m
3–65. $F_{AD}\left(\dfrac{0.5\cos 30°}{\sqrt{0.5^2+z^2}}\right) - F_{AC}\left(\dfrac{0.5\cos 30°}{\sqrt{0.5^2+z^2}}\right) = 0$

$F_{AB}\left(\dfrac{0.5}{\sqrt{0.5^2+z^2}}\right) - 2\left[F\left(\dfrac{0.5\sin 30°}{\sqrt{0.5^2+z^2}}\right)\right] = 0$

$3F\left(\dfrac{z}{\sqrt{0.5^2+z^2}}\right) - 100(9.81) = 0$

$z = 173$ mm
3–66. $d = 0.502$ m
3–67. $F_{AB} = F_{AC} = F_{AD} = 1839.4$ N
3–69. $-0.3873 F_{OB} + 0.3873 F_{OC} + 500 \sin \theta = 0$
$\theta = 0°$
$-0.4472 F_{OA} - 0.2236 F_{OB}$
$-0.2236 F_{OC} + 500 = 0$
$0.8944 F_{OA} - 0.8944 F_{OB} - 0.8944 F_{OC} = 0$
$F_{OA} = 745.4$ N
$F_{OB} = F_{OC} = 372.7$ N
3–70. $\theta = 11.3°$
3–71. $\theta = 4.69°$
$F_1 = 4.31$ kN
3–73. $1.699(10)^{-3}\cos 60° - F = 0$
$F = 0.850$ mN
3–74. $F_{AB} = 110$ N
$F_{AC} = 85.8$ N $\quad F_{AO} = 319$ N

3–75. $P = 639$ kN
$\alpha_3 = 77.2°$
$\beta_3 = 148°$
$\gamma_3 = 119°$
3–77. $F_2 + F_1 \cos 60° - 800\left(\frac{3}{5}\right) = 0$
$800\left(\frac{4}{5}\right) + F_1 \cos 135° - F_3 = 0$
$F_1 \cos 60° - 200 = 0$
$F_1 = 400$ N
$F_2 = 280$ N
$F_3 = 357$ N
3–78. $F_{CD} = 3.125$ kN
$F_{CA} = F_{CB} = 0.988$ kN
3–79. $F_1 = 0$
$F_2 = 311$ N
$F_3 = 238$ N

## Chapter 4

4–5. $150(\cos 45°)(5.4) = F\left(\frac{4}{5}\right)(3.6)$
$F = 198.9$ N
4–6. $M_A = 7.21$ kN·m $\circlearrowright$
4–7. $\theta = 64.0°$
4–9. $-600 = -F \cos 30°(0.450) - F \sin 30°(0.125)$
$F = 132.7$ N
4–10. $M_O = 120$ N·m $\circlearrowright$
$M_O = 520$ N·m $\circlearrowright$
4–11. $M_A = 38.2$ kN·m $\circlearrowright$
4–13. $M_A = (36 \cos \theta + 18 \sin \theta)$ kN·m
$\dfrac{dM_A}{d\theta} = -36 \sin \theta + 18 \cos \theta = 0$
$\theta = 26.6°, (M_A)_{max} = 40.2$ kN·m
When $M_A = 0$,
$0 = 36 \cos \theta + 18 \sin \theta, \theta = 117°$
4–14. $\circlearrowleft + M_A = 15.4$ N·m $\circlearrowdown$
$F = 118.6$ N
4–15. $(M_R)_A = 2.09$ N·m $\circlearrowright$
4–17. $(M_{F_A})_C = -150\left(\frac{3}{5}\right)(2.7)$
$= -243$ N·m $= 243$ N·m $\circlearrowright$
$(M_{F_B})_C = 389.7$ N·m $\circlearrowleft$
Since $(M_{F_B})_C > (M_{F_A})_C$, the gate will rotate *counterclockwise*.
4–18. $F_A = 144.3$ N
4–19. $M_P = (806.25 \cos \theta + 112.5 \sin \theta)$ N·m
4–21. a. $M_A = 400\sqrt{(3)^2 + (2)^2}$
$M_A = 1.44$ kN·m $\circlearrowleft$
$\theta = 56.3°$

**4–22.** $\circlearrowleft + M_A = 1200 \sin \theta + 800 \cos \theta$
$M_{max} = 1.44 \text{ kN} \cdot \text{m} \circlearrowright$
$\theta_{max} = 56.3°$

**4–23.** $M_{min} = 0$
$\theta_{min} = 146°$

**4–25.** $BC = 7.370 \text{ m}$
$\frac{\sin \theta}{30} = \frac{\sin 105°}{7.370}$ $\quad \theta = 23.15°$
$2250 = F \sin 23.15°(6)$
$F = 953.9 \text{ N}$

**4–26.** $(M_A)_1 = 14.74 \text{ N} \cdot \text{m} \circlearrowright$
$(M_A)_2 = 17.46 \text{ N} \cdot \text{m} \circlearrowright$

**4–27.** $M_A = 73.9 \text{ N} \cdot \text{m} \circlearrowright$
$F_C = 82.2 \text{ N} \leftarrow$

**4–29.** $\circlearrowleft + M_B = 200 \cos 25°(0.750) = 135.9 \text{ N} \cdot \text{m} \circlearrowright$
$\circlearrowleft + M_C = 211.1 \text{ N} \cdot \text{m} \circlearrowright$

**4–30.** $\circlearrowleft + M_A = 291.9 \text{ N} \cdot \text{m} \circlearrowright$

**4–31.** $\circlearrowleft + M_A = 7.71 \text{ N} \cdot \text{m} \circlearrowright$

**4–33.** Maximum moment, $OB \perp BA$
$\circlearrowright +(M_O)_{max} = 80.0 \text{ kN} \cdot \text{m}$
$\theta = 33.6°$

**4–34.** $F = 115 \text{ N}$

**4–35.** $F = 84.3 \text{ N}$

**4–37.** $\mathbf{M}_O = \mathbf{r}_{OA} \times \mathbf{F}_1 = \{22\mathbf{i} - 10\mathbf{j} + 18\mathbf{k}\} \text{ N} \cdot \text{m}$

**4–38.** $\mathbf{M}_O = \{18\mathbf{i} - 26\mathbf{j} - 12\mathbf{k}\} \text{ N} \cdot \text{m}$

**4–39.** $(\mathbf{M}_R)_O = \{40\mathbf{i} - 36\mathbf{j} + 6\mathbf{k}\} \text{ N} \cdot \text{m}$

**4–41.** $\mathbf{M}_O = \mathbf{r}_{OA} \times \mathbf{F}_C = \{1080\mathbf{i} + 720\mathbf{j}\} \text{ N} \cdot \text{m}$
$\mathbf{M}_O = \mathbf{r}_{OC} \times \mathbf{F}_C = \{1080\mathbf{i} + 720\mathbf{j}\} \text{ N} \cdot \text{m}$

**4–42.** $\mathbf{M}_O = \{-720\mathbf{i} + 720\mathbf{j}\} \text{ N} \cdot \text{m}$

**4–43.** $(\mathbf{M}_A)_O = \{-18\mathbf{i} + 9\mathbf{j} - 3\mathbf{k}\} \text{ N} \cdot \text{m}$
$(\mathbf{M}_B)_O = \{18\mathbf{i} + 7.5\mathbf{j} + 30 \mathbf{k}\} \text{ N} \cdot \text{m}$

**4–45.** $\mathbf{M}_A = \mathbf{r}_{AC} \times \mathbf{F}$
$= \{-5.39\mathbf{i} + 13.1\mathbf{j} + 11.4\mathbf{k}\} \text{ N} \cdot \text{m}$

**4–46.** $\mathbf{M}_B = \{10.6\mathbf{i} + 13.1\mathbf{j} + 29.2\mathbf{k}\} \text{ N} \cdot \text{m}$

**4–47.** $y = 1 \text{ m}$
$z = 3 \text{ m}$
$d = 1.15 \text{ m}$

**4–49.** $\mathbf{b} = \mathbf{r}_{CA} \times \mathbf{r}_{CB}$
$\mathbf{u}_F = \frac{\mathbf{b}}{b}$
$\mathbf{M}_B = \mathbf{r}_{BC} \times \mathbf{F} = \{10\mathbf{i} + 0.750\mathbf{j} - 1.56\mathbf{k}\} \text{ kN} \cdot \text{m}$

**4–50.** $M_O = 4.27 \text{ N} \cdot \text{m}$
$\alpha = 95.2°$
$\beta = 110°$
$\gamma = 20.6°$

**4–51.** $\mathbf{M}_{AF} = \{9.33\mathbf{i} + 9.33\mathbf{j} - 4.67\mathbf{k}\} \text{ N} \cdot \text{m}$

**4–53.** $\mathbf{u} = \mathbf{k}$
$\mathbf{r} = 0.25 \sin 30° \mathbf{i} + 0.25 \cos 30° \mathbf{j}$
$M_z = 15.5 \text{ N} \cdot \text{m}$

**4–54.** $M_x = 15.0 \text{ kN} \cdot \text{m}$
$M_y = 4.00 \text{ kN} \cdot \text{m}$
$M_z = 36.0 \text{ kN} \cdot \text{m}$

**4–55.** $\mathbf{M}_{AC} = \{11.5\mathbf{i} + 8.64\mathbf{j}\} \text{ kN} \cdot \text{m}$

**4–57.** $\mathbf{r}_{OB} = \{0.2 \cos 45°\mathbf{i} - 0.2 \sin 45°\mathbf{k}\} \text{ m}$
$M_y = 0.828 \text{ N} \cdot \text{m}$

**4–58.** $M_x = 73.0 \text{ N} \cdot \text{m}$

**4–59.** $F = 771 \text{ N}$

**4–61.** $M_{CD} = \mathbf{u}_{CD} \cdot \mathbf{r}_{CA} \times \mathbf{F}$
$= u_{CD} \cdot \mathbf{r}_{DB} \times \mathbf{F} = -648 \text{ N} \cdot \text{m}$

**4–62.** $F = 810.2 \text{ N}$

**4–63.** $M_{y'} = 772.7 \text{ N} \cdot \text{m}$

**4–65.** $\mathbf{u}_y = -\sin 30° \mathbf{i}' + \cos 30° \mathbf{j}'$
$\mathbf{r}_{AC} = -2 \cos 15° \mathbf{i}' + 1\mathbf{j}' + 2 \sin 15° \mathbf{k}$
$M_y = 469.2 \text{ N} \cdot \text{m}$

**4–66.** $M = 18.92 \text{ N} \cdot \text{m}$

**4–67.** $P = 42.28 \text{ N}$

**4–69.** $M_{OA} = \mathbf{u}_{OA} \cdot \mathbf{r}_{OB} \times \mathbf{W} = \mathbf{u}_{OA} \cdot \mathbf{r}_{OB} \times \mathbf{W}$
$W = 284.2 \text{ N}$

**4–70.** $M_x = 14.8 \text{ N} \cdot \text{m}$

**4–71.** $F = 20.2 \text{ N}$

**4–73.** $M_2 = 424 \text{ N} \cdot \text{m}$
$0 = 424.26 \cos 45° - M_3$
$M_3 = 300 \text{ N} \cdot \text{m}$

**4–74.** $F = 625 \text{ N}$

**4–75.** $(M_c)_R = 260 \text{ N} \cdot \text{m} \circlearrowright$

**4–77.** $F' = 33.3 \text{ N}$
$F = 133 \text{ N}$

**4–78.** $F = 111 \text{ N}$

**4–79.** $\theta = 56.1°$

**4–81.** $\circlearrowright + M_R = 100 \cos 30° (0.3) + 100 \sin 30° (0.3)$
$- P \sin 15° (0.3) - P \cos 15°(0.3) = 15$
$P = 70.7 \text{ N}$

**4–82.** For minimum $P$ require $\theta = 45°$
$P = 49.5 \text{ N}$

**4–83.** $N = 26.0 \text{ N}$

**4–85.** a.
$M_R = 8 \cos 45°(1.8) + 8 \sin 45°(0.3) + 2 \cos 30°(1.8)$
$- 2 \sin 30°(0.3) - 2 \cos 30°(3.3) - 8 \cos 45°(3.3)$
$M_R = 9.69 \text{ kN} \cdot \text{m} \circlearrowright$
b. $M_R = 9.69 \text{ kN} \cdot \text{m} \circlearrowright$

**4–86.** $(M_c)_R = 5.20 \text{ kN} \cdot \text{m} \circlearrowright$

**4–87.** $F = 14.2 \text{ kN} \cdot \text{m}$

**4–89.** a. $\circlearrowleft + M_C = 200 \cos 30°(1.2) - 300(\frac{4}{5})(1.2)$
$= 80.15 \text{ N} \cdot \text{m} \circlearrowright$
b. $\circlearrowleft + M_C = -80.15 \text{ N} \cdot \text{m} = 80.15 \text{ N} \cdot \text{m} \circlearrowright$

**4–90.**  a. $\curvearrowleft +M_C = 80.15$ N·m $\curvearrowright$
  b. $\curvearrowleft +M_C = 80.15$ N·m $\curvearrowright$
**4–91.**  $(M_c)_R = 1.04$ kN·m
  $\alpha = 120°$
  $\beta = 61.3°$
  $\gamma = 136°$
**4–93.**  $\mathbf{M}_c = \mathbf{r}_{AB} \times \mathbf{F} = \mathbf{r}_{BA} \times -\mathbf{F}$
  $M_c = 40.8$ N·m
  $\alpha = 11.3°$
  $\beta = 101°$
  $\gamma = 90°$
**4–94.**  $F = 98.1$ N
**4–95.**  $(M_R)_{x'} = 7.26$ kN·m
  $(M_R)_{y'} = 44.76$ kN·m
**4–97.**  $M_C = F(1.5)$
  $F = 15.4$ N
**4–98.**  $\mathbf{M}_R = \{-12.1\mathbf{i} - 10.0\mathbf{j} - 17.3\mathbf{k}\}$ N·m
**4–99.**  $d = 342$ mm
**4–101.** $0 = -M_2 + \tfrac{2}{3}M_3 + 112.5$
  $0 = M_1 - \tfrac{2}{3}M_3 - 112.5$
  $0 = \tfrac{1}{3}M_3 - 159.1$
  $M_3 = 447.3$ N·m
  $M_1 = M_2 = 430.7$ N·m
**4–102.** $(M_C)_R = 224$ N·m
  $\alpha = 153°$
  $\beta = 63.4°$
  $\gamma = 90°$
**4–103.** $F_1 = 1000$ N
  $F_2 = 750$ N
**4–105.** $F_R = \sqrt{1.25^2 + 5.799^2} = 5.93$ kN
  $\theta = 77.8°$
  $M_{R_A} = 34.8$ kN·m $\curvearrowright$
**4–106.** $F_R = 5.93$ kN
  $\theta = 77.8°$
  $M_{R_B} = 11.6$ kN·m $\curvearrowright$
**4–107.** $F_R = 149.6$ N
  $\theta = 78.4°$
  $M_{R_O} = 214$ lb·in. $\curvearrowright$
**4–109.** $F_R = \sqrt{533.01^2 + 100^2} = 542$ N
  $\theta = 10.6°$
  $(M_R)_A = 441$ N·m $\curvearrowright$
**4–110.** $F_R = 50.2$ kN
  $\theta = 84.3°$
  $(M_R)_A = 239$ kN·m $\curvearrowright$
**4–111.** $F_R = 461$ N
  $\theta = 49.4°$
  $(M_R)_O = 438$ N·m $\curvearrowright$

**4–113.** $\mathbf{F}_R = \{2\mathbf{i} - 10\mathbf{k}\}$ kN
  $(\mathbf{M}_R)_O = \mathbf{r}_{OB} \times \mathbf{F}_B + \mathbf{r}_{OC} \times \mathbf{F}_D$
  $= \{-6\mathbf{i} + 12\mathbf{j}\}$ kN·m
**4–114.** $\mathbf{F}_R = \{-210\mathbf{k}\}$ N
  $\mathbf{M}_{RO} = \{-15\mathbf{i} + 225\mathbf{j}\}$ N·m
**4–115.** $\mathbf{F}_R = \{6\mathbf{i} - 1\mathbf{j} - 14\mathbf{k}\}$ N
  $\mathbf{M}_{RO} = \{1.30\mathbf{i} + 3.30\mathbf{j} - 0.450\mathbf{k}\}$ N·m
**4–117.** $\mathbf{F}_2 = \{-1.768\mathbf{i} + 3.062\mathbf{j} + 3.536\mathbf{k}\}$ kN
  $\mathbf{F}_R = \{0.232\mathbf{i} + 5.06\mathbf{j} + 12.4\mathbf{k}\}$ kN
  $\mathbf{M}_{R_O} = \mathbf{r}_1 \times \mathbf{F}_1 + \mathbf{r}_2 \times \mathbf{F}_2$
  $= \{36.0\mathbf{i} - 26.1\mathbf{j} + 12.2\mathbf{k}\}$ kN·m
**4–118.** $F_R = 53.75$ kN $\downarrow$
  $d = 4.12$ m
**4–119.** $F_R = 53.75$ kN $\downarrow$
  $d = 2.78$ m
**4–121.** $F_R = \sqrt{(0.5)^2 + (4.491)^2} = 4.52$ kN
  $\theta = 6.35°$
  $\phi = 23.6°$
  $d = 1.52$ m
**4–122.** $F_R = 984.8$ N
  $\theta = 42.6°$
  $d = 1.57$ m
**4–123.** $F_R = 984.8$ N
  $\theta = 42.6°$
  $d = 0.247$ m
**4–125.** $F_R = \sqrt{(212.5)^2 + (251.6)^2} = 329.3$ N
  $\theta = 49.8°$
  $d = 0.630$ m
**4–126.** $F_R = 329.3$ N
  $\theta = 49.8°$
  $d = 1.386$ m
**4–127.** $F_R = 542$ N
  $\theta = 10.6°$
  $d = 0.827$ m
**4–129.** $F_R = 140$ kN $\downarrow$
  $-140y = -50(3) - 30(11) - 40(13)$
  $y = 7.14$ m
  $x = 5.71$ m
**4–130.** $F_R = 140$ kN
  $x = 6.43$ m
  $y = 7.29$ m
**4–131.** $F_C = 600$ N    $F_D = 500$ N
**4–133.** $0 = 1000(0.45\cos 45°) - F_B(0.45\cos 30°)$
  $F_B = 816.5$ N
  $F_C = 115.4$ N
**4–134.** $F_R = 215$ kN
  $y = 3.68$ m
  $x = 3.54$ m

**4–135.** $F_A = 30$ kN   $F_B = 20$ kN   $F_R = 190$ kN

**4–137.** $F_R = 26$ kN
$-26(y) = 6(650) + 5(750) - 7(600) - 8(700)$
$y = 82.7$ mm
$x = 3.85$ mm

**4–138.** $F_A = 18.0$ kN   $F_B = 16.7$ kN
$F_R = 48.7$ kN

**4–139.** $F_R = 4.039$ kN
$x = 1.055$ m   $y = 0.0414$ m
$M_W = -1504.2$ N·m

**4–141.** $F_R = 990$ N
$\mathbf{u}_{F_R} = -0.5051\mathbf{i} + 0.3030\mathbf{j} + 0.8081\mathbf{k}$
$M_R = 3.07$ kN·m   $x = 1.16$ m
$y = 2.06$ m

**4–142.** $F_R = 75$ kN $\downarrow$
$\bar{x} = 1.20$ m

**4–143.** $F_R = 30$ kN $\downarrow$
$\bar{x} = 3.4$ m

**4–145.** $F_R = \frac{1}{2}w_0 L \downarrow$
$-\frac{1}{2}w_0 L(\bar{x}) = -\frac{1}{2}w_0\left(\frac{L}{2}\right)\left(\frac{L}{6}\right) - \frac{1}{2}w_0\left(\frac{L}{2}\right)\left(\frac{2}{3}L\right)$
$\bar{x} = \frac{5}{12}L$

**4–146.** $F_R = 23.4$ kN $\uparrow$
$d = 3.38$ m

**4–147.** $w_1 = 3.766$ kN/m
$w_2 = 5.633$ kN/m

**4–149.** $\mathbf{F}_R = \{-108\,\mathbf{i}\}$ N
$\mathbf{M}_{RO} = -\left(1 + \frac{2}{3}(1.2)\right)(108)\mathbf{j}$
$\quad -\left(0.1 + \frac{1}{3}(1.2)\right)(108)\mathbf{k}$
$\mathbf{M}_{RO} = \{-194\mathbf{j} - 54\mathbf{k}\}$ N·m

**4–150.** $b = 1.35$ m
$a = 2.925$ m

**4–151.** $F_R = 30.025$ N
$\bar{x} = 0.0937$ m

**4–153.** $F_R = 107$ kN $\leftarrow$
$$\bar{z} = \frac{\int_0^z zw\,dz}{\int_0^z w\,dz}$$
$$\bar{z} = \frac{\int_0^{4\,\text{m}}\left[\left(20z^{\frac{3}{2}}\right)(10^3)\right]dz}{\int_0^{4\,\text{m}}\left(20z^{\frac{1}{2}}\right)(10^3)dz}$$
$h = 1.60$ m

**4–154.** $F_R = 10.7$ kN $\downarrow$
$\bar{x} = 1$ m

**4–155.** $F_R = 3.460$ kN, $\theta = 47.5°$ ⦨
$M_{RA} = 3.96$ kN·m ↺

**4–157.** $F_R = 448$ kN $\uparrow$
$$448\bar{x} = 192(2) + \int_0^x (x+4)w\,dx$$
$\bar{x} = 4.86$ m

**4–158.** $F_R = 53.3$ kN
$\bar{x} = 1.60$ m

**4–159.** $w_{\max} = 18$ kN/m
$F_R = 53.3$ kN
$x' = 2.40$ m

**4–161.** $(dF_R)_x = 0.4(1 + \cos\theta)\sin\theta\,d\theta$
$F_R = 1.428$ kN $\uparrow$

**4–162.** $F_R = 2.667$ kN $\downarrow$
$M_{R_A} = 0.667$ kN·m ↻

**4–163.** $d = 1.663$ m

**4–165.** $\mathbf{M}_O = \mathbf{r}_{OA} \times \mathbf{F} = \{7.45\mathbf{i} + 0.379\mathbf{j} - 5\mathbf{k}\}$ N·m

**4–166.** $M_A = 4.34$ kN·m ↙

**4–167.** $\mathbf{M}_A = \{-59.7\mathbf{i} - 159\mathbf{k}\}$ N·m

**4–169.** a. $\mathbf{M}_C = \mathbf{r}_{AB} \times (25\,\mathbf{k})$
$\mathbf{M}_C = \{-5\mathbf{i} + 8.75\,\mathbf{j}\}$ N·m
b. $\mathbf{M}_C = \mathbf{r}_{OB} \times (25\,\mathbf{k}) + \mathbf{r}_{OA} \times (-25\,\mathbf{k})$
$\mathbf{M}_C = \{-5\mathbf{i} + 8.75\,\mathbf{j}\}$ N·m

**4–170.** $F = 992$ N

**4–171.** $\mathbf{F}_R = \{-80\mathbf{i} - 80\mathbf{j} + 40\mathbf{k}\}$ kN
$\mathbf{M}_{RP} = \{-240\mathbf{i} + 720\mathbf{j} + 960\mathbf{k}\}$ kN·m

**4–173.**
$M_z = \mathbf{k} \cdot (\mathbf{r}_{BA} \times \mathbf{F}) = \mathbf{k} \cdot (\mathbf{r}_{OA} \times \mathbf{F}) = -4.03$ N·m

## Chapter 5

**5–1.** $W$ is the effect of gravity (weight) on the paper roll.
$N_A$ and $N_B$ are the smooth blade reactions on the paper roll.

**5–2.** $N_A$ force of plane on roller.
$B_x, B_y$ force of pin on member.

**5–3.** $W$ is the effect of gravity (weight) on the dumpster.
$A_y$ and $A_x$ are the reactions of the pin $A$ on the dumpster.
$F_{BC}$ is the reaction of the hydraulic cylinder $BC$ on the dumpster.

**5–5.** $C_y$ and $C_x$ are the reactions of pin $C$ on the truss.
$T_{AB}$ is the tension of cable $AB$ on the truss.
3 kN and 4 kN force are the effect of external applied forces on the truss.

# Answers to Selected Problems

**5–6.** W is the effect of gravity (weight) on the boom. $A_y$ and $A_x$ are the reactions of pin $A$ on the boom.
$T_{BC}$ is the force reaction of cable $BC$ on the boom. The 6.25 kN force is the suspended load reaction on the boom.

**5–7.** $A_x, A_y, N_B$ forces of cylinder on wrench.

**5–9.** $N_A, N_B, N_C$ forces of wood on bar.
50 N forces of hand on bar.

**5–10.** $C_x, C_y$ forces of pin on drum.
$F_{AB}$ forces of pawl on drum gear.
2500 N forces of cable on drum.

**5–11.** $N_B = 245$ N
$N_A = 425$ N

**5–13.**
$T_{AB} \cos 30°(2) + T_{AB} \sin 30°(4) - 3(2) - 4(4) = 0$
$T_{AB} = 5.89$ kN
$C_x = 5.11$ kN
$C_y = 4.05$ kN

**5–14.** $T_{BC} = 55.29$ kN
$A_x = 51.04$ kN
$A_y = 30.77$ kN

**5–15.** $N_B = 700$ N
$A_x = 700$ N
$A_y = 100$ N

**5–17.** $N_C = 28.87$ N
$50 \cos 30°(325 - 43.30) - N_A(125 - 43.30)$
$- 28.87 \left( \dfrac{75}{\cos 30°} \right) = 0$
$N_A = 118.70$ N
$N_B = 60.96$ N

**5–18.** $F_{AB} = 2.003$ kN
$C_x = 1.667$ kN
$C_y = 3.611$ kN

**5–19.** $(N_A)_r = 492.9$ N, $(N_B)_r = 107.1$ N
$(N_A)_s = 500$ N, $(N_B)_s = 100$ N

**5–21.** $T(\tfrac{3}{5})(3) + T(\tfrac{4}{5})(1) - 60(1) - 30 = 0$
$T = 34.62$ kN
$A_x = 20.8$ kN
$A_y = 87.7$ kN

**5–22.** $F_B = 20.94$ kN
$A_x = 16.04$ kN
$A_y = 9.83$ kN

**5–23.** $N_C = 213$ N
$A_x = 105$ N
$A_y = 118$ N

**5–25.** $N_B(0.9) - 1500(0.45) = 0$
$N_B = 750$ N
$A_y = 1500$ N
$A_x = 750$ N

**5–26.** $F_{CD} = 131$ N
$B_x = 34.0$ N
$B_y = 95.4$ N

**5–27.** $F_{AB} = 0.864$ kN
$C_y = 6.56$ kN
$C_x = 2.66$ kN

**5–29.** $F_{BC}(\tfrac{4}{5})(1.5) - 700(9.81)(d) = 0$
$F_{BC} = 5722.5d$
$F_A = \sqrt{(3433.5d)^2 + (4578d - 6867)^2}$

**5–30.** $A_y = 250$ N
$N_B = 8.0$ kN
$A_x = 7.567$ kN

**5–31.** $F = 0.46875$ kN
$A_x = 7.094$ kN
$A_y = 0.234$ kN

**5–33.**
$40\,000(\tfrac{3}{5})(4) + 40\,000(\tfrac{4}{5})(0.2) - 2000(9.81)(x) = 0$
$x = 5.22$ m
$C_x = 32$ kN
$C_y = 4.38$ kN

**5–34.** $N_B = 1.04$ kN
$A_x = 0$
$A_y = 600$ N

**5–35.** $d = 1.8$ m
$w = 1.333$ kN/m

**5–37.** $-490.5\,(3.15) + \tfrac{1}{2} w_B (0.3)(9.25) = 0$
$w_B = 1.11$ kN/m
$w_A = 1.44$ kN/m

**5–38.** $k = 1.33$ kN/m
$A_y = 300$ N
$A_x = 398$ N

**5–39.** $\theta = 23.1°$
$A_y = 300$ N
$A_x = 353$ N

**5–41.** $A_y = 3750$ N
$N_B(1.2 \sin 30°) - 1500(0.3) - 2250(0.9) = 0$
$N_B = 4125$ N
$A_x = 4125$ N

**5–42.** $N_B = 1.27$ kN
$A_x = 900$ N
$M_B = 227$ N·m

**5–43.** $T = 45.38$ N

**5–45.** $12.5(0.42 + 2.52) - 2.5(4.5 \cos 30° - 2.52)$
$- N_A(0.66 + 0.42 + 2.52) = 0$
$N_A = 9.252$ kN
$N_B = 5.748$ kN

**5–46.** $W = 26.69$ kN

**5–47.** $F_A = 2.16$ kN    $F_B = 0$    $F_C = 2.16$ kN

**5–49.** $50(9.81) \sin 20° (0.5) + 50(9.81) \cos 20°(0.3317)$
$- P \cos \theta (0.5) - P \sin \theta (0.3317) = 0$
For $P_{min}$;  $\frac{dP}{d\theta} = 0$
$\theta = 33.6°$
$P_{min} = 395$ N

**5–50.** $F = 5.20$ kN
$N_A = 17.3$ kN
$N_B = 24.9$ kN

**5–51.** $\theta = 63.4°$
$T = 29.2$ kN

**5–53.** $F_C (0.15 \cos \theta) - F_A (0.15 \cos \theta) = 0$
$\theta = 23.6°$

**5–54.** $k = 186.9$ N/m

**5–55.** $\alpha = 1.02°$

**5–57.** For disk E: $-P + N'\left(\frac{\sqrt{24}}{5}\right) = 0$
For disk D: $N_A\left(\frac{4}{5}\right) - N'\left(\frac{\sqrt{24}}{5}\right) = 0$
$N_A = 1.25$ kN
$N_B = 0.046$ kN
$N_C = 0.704$ kN

**5–58.** $P_{max} = 1.048$ kN
$N_A = 1.310$ kN
$N_C = 0.714$ kN

**5–59.** $\alpha = 10.4°$

**5–61.** $95.35 \sin 45°(300) - F(400) = 0$
$F = 50.6$ N
$A_x = 108$ N
$A_y = 48.8$ N

**5–62.** $a = \sqrt{(4r^2 l)^{\frac{2}{3}} - 4r^2}$

**5–63.** $N_C = 289$ N
$N_A = 213$ N
$N_B = 332$ N

**5–65.** $T_{CD}(2) - 6(1) = 0$
$T_{CD} = 3$ kN
$T_{EF} = 2.25$ kN
$T_{AB} = 0.75$ kN

**5–66.** $y = 0.667$ m, $x = 0.667$ m

**5–67.** $R_D = 113.15$ kN
$R_E = 113.15$ kN
$R_F = 68.70$ kN

**5–69.** $C_y = 450$ N
$C_z(0.9 + 0.9) - 900(0.9) + 600(0.6) = 0$
$C_z = 250$ N
$B_z = 1.125$ kN
$A_z = 125$ N
$B_x = 25$ N
$A_x + 25 - 500 = 0$
$A_x = 475$ N

**5–70.** $T_{BD} = T_{CD} = 117$ N
$A_x = 66.7$ N
$A_y = 0$
$A_z = 100$ N

**5–71.** $F_{DC} = 1.875$ kN
$E_x = 0$
$E_z = 2.8125$ kN
$A_x = 0$
$A_y = 0$
$A_z = 0.3125$ kN

**5–73.** $N_B(3) - 200(3) - 200(3 \sin 60°) = 0$
$N_B = 373$ N
$A_z = 333$ N
$T_{CD} + 373.21 + 333.33 - 350 - 200 - 200 = 0$
$T_{CD} = 43.5$ N
$A_x = 0$
$A_y = 0$

**5–74.** $F_{CD} = 0$
$F_{EF} = 100$ kN
$F_{BD} = 150$ kN
$A_x = 0$
$A_y = 0$
$A_z = 100$ kN

**5–75.** $F = 4.5$ kN
$A_x = 0$
$A_y = 0$
$A_z = 3$ kN
$M_{Ax} = 0$
$M_{Az} = 0$

**5–77.** $T_{EF}(L) - W\left(\frac{L}{2}\right) - 0.75W\left(\frac{L}{2} - d \cos 45°\right) = 0$
$d = 0.550L$
$T_{EF} = 0.583W$

**5–78.** $T_{AB} = 1.14W$
$T_{EF} = 0.570W$
$T_{CD} = 0.0398W$

**5–79.** $T_B = 16.7$ kN
$A_x = 0$
$A_y = 5.00$ kN
$A_z = 16.7$ kN

**5–81.** $A_x + \left(\dfrac{0.9}{\sqrt{4.86}}\right)F_{CB} = 0$

$-275(0.9) + \left(\dfrac{1.8}{\sqrt{4.86}}\right)F_{CB}(0.9) = 0$

$F_{CB} = 336.8$ N
$A_x = -137.5$ N
$A_y = -137.5$ N
$A_z = 0$
$M_{Ay} = 247.5$ N·m
$M_{Az} = 0$

**5–82.** $F_{BC} = 17.5$ kN
$A_x = 13.0$ kN
$A_y = -1.0$ kN
$M_{Ax} = -9$ kN·m
$M_{Ay} = 0$
$M_{Az} = -21.6$ kN·m

**5–83.** $F_{BC} = 10.5$ kN

**5–85.** $\Sigma M_{AB} = 0; T_C(r + r\cos 60°) - W(r\cos 60°)$
$\qquad - P(d + r\cos 60°) = 0$
$d = \dfrac{r}{2}\left(1 + \dfrac{W}{P}\right)$

**5–86.** $d = \dfrac{r}{2}$

**5–87.** $P = 0.5\,W$

**5–89.** $600(6) + 600(4) + 600(2) - N_B\cos 45°(2) = 0$
$N_B = 5.09$ kN
$A_x = 3.60$ kN
$A_y = 1.80$ kN

**5–90.** $F = 354$ N

**5–91.** $N_A = 8.00$ kN
$B_x = 5.20$ kN
$B_y = 5.00$ kN

**5–93.**
$25(4.2) + 35(1.8) + 2.5(1.8) - 10(1.8) - A_y(4.2) = 0$
$A_y = 36.8$ kN
$B_x = 2.5$ kN
$B_y = 83.2$ kN

**5–94.** $T = 1.01$ kN
$D_y = -507.66$ N
$F_D = 982$ N

**5–95.** $P = 500$ N
$B_z = 200$ N
$B_x = -178.6$ N
$A_x = 678.6$ N
$B_y = 0$
$A_z = 200$ N

## Chapter 6

**6–1.** Joint $D$: $600 - F_{DC}\sin 26.57° = 0$
$F_{DC} = 1.34$ kN (C)
$F_{DE} = 1.20$ kN (T)
Joint $C$: $-F_{CE}\cos 26.57° = 0$
$F_{CE} = 0$
$F_{CB} = 1.34$ kN (C)
Joint $E$: $900 - F_{EB}\sin 45° = 0$
$F_{EB} = 1.27$ kN (C)
$F_{EA} = 2.10$ kN (T)

**6–2.** $F_{AD} = 4.24$ kN (C)
$F_{AB} = 3$ kN (T)
$F_{BD} = 2$ kN (C)
$F_{BC} = 3$ kN (T)
$F_{DC} = 7.07$ kN (T)
$F_{DE} = 8.00$ kN (C)

**6–3.** $F_{AD} = 5.66$ kN (C)
$F_{AB} = 4.00$ kN (T)
$F_{BD} = 0$
$F_{BC} = 4.00$ kN (T)
$F_{DC} = 5.66$ kN (T)
$F_{DE} = 8.00$ kN (C)

**6–5.** Joint $A$: $F_{AE}\left(\dfrac{1}{\sqrt{5}}\right) - 166.22 = 0$
$F_{AE} = 372$ N (C)
$F_{AB} = 332$ N (T)
Joint $B$: $F_{BC} - 332.45 = 0 \quad F_{BC} = 332$ N (T)
$F_{BE} = 196$ N (C)
Joint $E$: $F_{EC}\cos 36.87°$
$\quad - (196.2 + 302.47)\cos 26.57° = 0$
$F_{EC} = 558$ N (T)
$F_{ED} = 929$ N (C)
$F_{DC} = 582$ N (T)

**6–6.** $F_{CB} = 3.00$ kN (T)
$F_{CD} = 2.60$ kN (C)
$F_{DE} = 2.60$ kN (C)
$F_{DB} = 2.00$ kN (T)
$F_{BE} = 2.00$ kN (C)
$F_{BA} = 5.00$ kN (T)

**6–7.** $F_{CB} = 8.00$ kN (T)
$F_{CD} = 6.93$ kN (C)
$F_{DE} = 6.93$ kN (C)
$F_{DB} = 4.00$ kN (T)
$F_{BE} = 4.00$ kN (C)
$F_{BA} = 12.0$ kN (T)

**6–9.** Joint A: $F_{AF} \sin 45° - P = 0$
Joint F: $F_{FB} \cos 45° - 1.4142 P \cos 45° = 0$
Joint E: $F_{ED} - 2P = 0$
Joint B: $F_{BD} \sin 45° - 1.4142 P \sin 45° = 0$
Joint C: $3P - N_C = 0$
$2P = 4$ kN   $P = 2$ kN
$3P = 3$ kN   $P = 1$ kN (controls)

**6–10.** $F_{BG} = F_{CG} = F_{AG} = F_{DF} = F_{CF} = F_{EF} = 0$
$F_{AB} = F_{DE} = 3.33$ kN (C)
$F_{BC} = F_{CD} = 3.33$ kN (C)

**6–11.** $F_{BG} = F_{GC} = F_{GA} = 0$
$F_{DF} = 2$ kN (C)
$F_{FC} = F_{FE} = 1.67$ kN (T)
$F_{BC} = F_{BA} = 3.54$ kN (C)
$F_{DC} = F_{DE} = 4.13$ kN (C)

**6–13.** Joint A: $F_{AC} \sin \theta = 0$
Joint D: $2.60 P_2 \cos 22.62° - F_{DC} = 0$
Joint B: $F_{BC} - 2.60 P_2 \sin 22.62° = 0$
$P_2 = 0.673$ kN

**6–14.** $F_{BG} = 0$
$F_{BC} = 12.25$ kN (C)
$F_{CG} = 8.84$ kN (T)
$F_{CD} = 6.25$ kN (C)
$F_{GD} = 8.84$ kN (C)
$F_{GF} = 12.5$ kN (T)
$F_{AE} = 12.25$ kN (C)
$F_{AF} = 0$
$F_{ED} = 6.25$ kN (C)
$F_{EF} = 8.84$ kN (T)
$F_{FD} = 8.84$ kN (C)

**6–15.** $P = 10$ kN

**6–17.** Joint A: $0.8333P \cos 73.74° + P \cos 53.13° - F_{AB} = 0$
Joint B: $0.8333P\left(\frac{4}{5}\right) - F_{BC}\left(\frac{4}{5}\right) = 0$
Joint D: $F_{DE} - 0.8333P - P \cos 53.13° - 0.8333P \cos 73.74° = 0$
$P = 1.50$ kN (controls)

**6–18.** $F_{FA} = 7.5$ kN (C)
$F_{FE} = 6.0$ kN (T)
$F_{ED} = 6.0$ kN (T)
$F_{EA} = 0$
$F_{AD} = 6.25$ kN (C)
$F_{AB} = 1.0$ kN (C)
$F_{BC} = 1.0$ kN (C)
$F_{BD} = 0$
$F_{DC} = 1.25$ kN (T)

**6–19.** $F_{FA} = 0.45$ kN (C)
$F_{ED} = 0.36$ kN (T)
$F_{FE} = 0.36$ kN (T)
$F_{EA} = 0.33$ kN (C)
$F_{AD} = 0.925$ kN (C)
$F_{AB} = 0.38$ kN (T)
$F_{BC} = 0.38$ kN (T)
$F_{BD} = 0.33$ kN (T)
$F_{DC} = 0.475$ kN (C)

**6–21.** Joint D: $F_{DC} \sin 45° + F_{DE} \cos 30.25° - W = 0$
Joint A: $F_{AG} - 1.414 W \sin 45° = 0$
$m = 1.80$ Mg

**6–22.** $F_{CD} = 778$ N (C)
$F_{CB} = 550$ N (T)
$F_{DB} = 70.7$ N (C)
$F_{DE} = 500$ N (C)
$F_{EA} = 636$ N (C)
$F_{EB} = 70.7$ N (T)
$F_{BA} = 450$ N (T)

**6–23.** $F_{CD} = 286$ N (C)
$F_{CB} = 202$ N (T)
$F_{DB} = 118$ N (T)
$F_{DE} = 286$ N (C)
$F_{BE} = 118$ N (T)
$F_{BA} = 202$ N (T)
$F_{EA} = 286$ N (C)

**6–25.** Joint A: $1.4142 P \cos 45° - F_{AB} = 0$
Joint D: $F_{DC} - 1.4142 P \cos 45° = 0$
Joint F: $F_{FE} - 1.4142 P \sin 45° = 0$
Joint E: $1.4142 P \sin 45° - P - F_{EB} \sin 45° = 0$
Joint C: $F_{CB} = P$ (C)
$P = 1$ kN (controls)
$1.4142 P = 1.5$
$P = 1.06$ kN

**6–26.** $F_{CD} = 3.90$ kN (C)
$F_{CB} = 3.60$ kN (T)
$F_{DB} = 0$
$F_{DE} = 3.90$ kN (C)
$F_{BE} = 1.485$ kN (T)
$F_{BA} = 3.612$ kN (T)

**6–27.** $F_{FE} = 0.667P$ (T)
$F_{FD} = 1.67P$ (T)
$F_{AB} = 0.471P$ (C)
$F_{AE} = 1.67P$ (T)
$F_{AC} = 1.49P$ (C)
$F_{BF} = 1.41P$ (T)
$F_{BD} = 1.49P$ (C)
$F_{EC} = 1.41P$ (T)
$F_{CD} = 0.471P$ (C)

# Answers to Selected Problems

**6–29.** Joint $A$: $F_{AF} - 2.404P\left(\dfrac{1.5}{\sqrt{3.25}}\right) = 0$

Joint $B$: $2.404P\left(\dfrac{1.5}{\sqrt{3.25}}\right) - P$

$-F_{BF}\left(\dfrac{0.5}{\sqrt{1.25}}\right) - F_{BD}\left(\dfrac{0.5}{\sqrt{1.25}}\right) = 0$

Joint $F$: $F_{FD} + 2\left[1.863P\left(\dfrac{0.5}{\sqrt{1.25}}\right)\right]$

$- 2.00P = 0$
$P = 1.25$ kN

**6–30.** $127° \le \theta \le 196°$
$336° \le \theta \le 347°$

**6–31.** $F_{BH} = 1273$ N (T)
$F_{BC} = 650$ N (T)
$F_{HC} = 900$ N (C)

**6–33.** $A_y = 65.0$ kN
$A_x = 0$
$F_{BC}(4) + 20(4) + 30(8) - 65.0(8) = 0$
$F_{BC} = 50.0$ kN (T)
$F_{HI} = 35.0$ kN (C)
$F_{HB} = 21.2$ kN (C)

**6–34.** $F_{JK} = 11.1$ kN (C)
$F_{CD} = 12$ kN (T)
$F_{CJ} = 1.60$ kN (C)

**6–35.** $F_{EF} = 12.9$ kN (T)
$F_{FI} = 7.21$ kN (T)
$F_{HI} = 21.1$ kN (C)

**6–37.** $E_y = 7.333$ kN
$7.333(4.5) - 8(1.5) - F_{FG}(3 \sin 60°) = 0$
$F_{FG} = 8.08$ kN (T)
$F_{CD} = 8.47$ kN (C)
$F_{CF} = 0.770$ kN (T)

**6–38.** $F_{HI} = 42.5$ kN (T)
$F_{HC} = 100$ kN (T)
$F_{DC} = 125$ kN (C)

**6–39.** $F_{GH} = 76.7$ kN (T)
$F_{ED} = 100$ kN (C)
$F_{EH} = 29.2$ kN (T)

**6–41.** $A_y = 1200$ N
$A_x = 500$ N
$1200(2.4) - F_{BC} \cos 14.04°(1.2) = 0$
$F_{BC} = 2473.9$ N (T)
$F_{HG} = 2100$ N (C)
$F_{BG} = 1000$ N (C)

**6–42.** $AB$, $BC$, $CD$, $DE$, $HI$, and $GI$ are all zero-force members.
$F_{IC} = 5.62$ kN (C)
$F_{CG} = 9.00$ kN (T)

**6–43.** $AB$, $BC$, $CD$, $DE$, $HI$, and $GI$ are all zero-force members.
$F_{JE} = 9.38$ kN (C)
$F_{GF} = 5.625$ kN (T)

**6–45.** $N_A = 6.5$ kN
$F_{KL}(2.5) + 5(2.5) - 4.5(2.5) - 6.5(7.5) = 0$
$F_{KL} = 19$ kN (C)
$F_{CD} = 13$ kN (T)
$F_{LD} = 2.121$ kN (T)

**6–46.** $F_{BC} = 3.25$ kN (C)
$F_{CH} = 1.92$ kN (T)

**6–47.** $F_{CD} = 1.92$ kN (C)
$F_{GF} = 1.53$ kN (T)
$F_{FD} = F_{FC} = 0$

**6–49.** $A_x = 0$
$A_y = 15.5$ kN
$F_{KJ} \sin 33.69°(4) + 5(2) + 3(4) - 15.5(4) = 0$
$F_{KJ} = 18.0$ kN (C)
$F_{KC} = 7.50$ kN (C)
$F_{BC} = 15$ kN (T)

**6–50.** $F_{AB} = 21.9$ kN (C), $F_{AG} = 13.1$ kN (T),
$F_{BC} = 13.1$ kN (C), $F_{BG} = 17.5$ kN (T),
$F_{CG} = 3.12$ kN (T), $F_{FG} = 11.2$ kN (T),
$F_{CF} = 3.12$ kN (C), $F_{CD} = 9.38$ kN (C),
$F_{DE} = 15.6$ kN (C), $F_{DF} = 12.5$ kN (T),
$F_{EF} = 9.38$ kN (T)

**6–51.** $F_{AB} = 43.8$ kN (C), $F_{AG} = 26.2$ kN (T)
$F_{BC} = 26.2$ kN (C), $F_{BG} = 35.0$ kN (T)
$F_{GC} = 6.25$ kN (T), $F_{GF} = 22.5$ kN (T)
$F_{ED} = 31.2$ kN (C), $F_{EF} = 18.8$ kN (T)
$F_{DC} = 18.8$ kN (C), $F_{DF} = 25.0$ kN (T)
$F_{FC} = 6.25$ kN (C)

**6–53.** $G_y = 8$ kN
$8(12) - F_{JI}(9) = 0$
$F_{JI} = 10.67$ kN (C)
$F_{DE} = 10.67$ kN (T)

**6–54.** $F_{CA} = 833$ N (T)
$F_{CB} = 667$ N (C)
$F_{CD} = 333$ N (T)
$F_{AD} = F_{AB} = 354$ N (C)
$F_{DB} = 50$ N (T)

**6–55.** $F_{CA} = 1000$ N (C)
$F_{CD} = 406$ N (T)
$F_{CB} = 344$ N (C)
$F_{AB} = F_{AD} = 424$ N (T)
$F_{DB} = 544$ N (C)

**6–57.** Joint $D$: $-\frac{1}{3}F_{AD} + \frac{5}{\sqrt{31.25}}F_{BD}$
$+ \frac{1}{\sqrt{7.25}}F_{CD} - 200 = 0$
$F_{AD} = 343$ N (T)
$F_{BD} = 186$ N (T)
$F_{CD} = 397$ N (C)
Joint $C$: $F_{BC} - \frac{1}{\sqrt{7.25}}(397.5) = 0$
$F_{BC} = 148$ N (T)
$F_{AC} = 221$ N (T)
$F_{EC} = 295$ N (C)

**6–58.** $F_{BC} = 1.15$ kN (C)
$F_{DF} = 4.16$ kN (C)
$F_{BE} = 4.16$ kN (T)

**6–59.** $F_{CF} = 0$
$F_{CD} = 2.31$ kN (T)
$F_{ED} = 3.46$ kN (T)
$F_{AB} = 3.46$ kN (C)

**6–61.** $D_x = 0.5$ kN
$C_y = 3.25$ kN
$E_x = 2.75$ kN
$F_x = 0.75$ kN
$F_y = 3.25$ kN
$F_z = 3.5$ kN
Joint $C$: $F_{CB} = 0$
$F_{CD} = 3.25$ kN (C)
$F_{CF} = 0$
Joint $F$: $F_{BF} = 1.125$ kN (T)
$F_{DF} = 6.149$ kN (T)
$F_{EF} = 2.625$ kN (C)

**6–62.** $F_{AE} = F_{AC} = 220$ N (T)
$F_{AB} = 583$ N (C)
$F_{BD} = 707$ N (C)
$F_{BE} = F_{BC} = 141$ N (T)

**6–63.** $F = 170$ N

**6–65.** Joint $F$: $F_{FG}$, $F_{FD}$, and $F_{FC}$ are lying in the same plane.
$F_{FE} \cos \theta = 0 \quad F_{FE} = 0$
Joint $E$: $F_{EG}$, $F_{EC}$, and $F_{EB}$ are lying in the same plane.
$F_{ED} \cos \theta = 0 \quad F_{ED} = 0$

**6–66.** $F_{GD} = 157$ N (T)
$F_{GE} = 505$ N (C)
$F_{FD} = 0$

**6–67.** $P = 122.625$ N

**6–69.** Apply the force equation of equilibrium along the $y$ axis of each pulley
$2P + 2R + 2T - 50(9.81) = 0$
$P = 18.9$ N

**6–70.** $P = 24.525$ N

**6–71.** $P = 122.625$ N
$F_A = P = 122.625$ N $\quad F_B = 295.25$ N

**6–73.** $N_B(0.8) - 900 = 0$
$N_B = 1125$ N
$A_x = 795$ N
$A_y = 795$ N
$C_x = 795$ N
$C_y = 1.30$ kN
$M_C = 1.25$ kN · m

**6–74.** $A_y = 300$ N
$C_y = 803.55$ N
$C_x = 450$ N
$A_x = 803.55$ N

**6–75.** $C_y = 5.00$ kN
$B_y = 15.0$ kN
$M_A = 30.0$ kN · m
$A_y = 5.00$ kN
$A_x = 0$

**6–77.** $C_x = 0.5$ kN $\quad B_y = 2.577$ kN
$C_y = 1.711$ kN $\quad A_x = 0.462$ kN
$A_y = 0.597$ kN
$M_A = 1.114$ kN · m

**6–78.** $A_y = 300$ N
$A_x = 300$ N
$C_x = 300$ N
$C_y = 300$ N

**6–79.** $N_D = 333$ N
$A_x = 333$ N
$A_y = 100$ N

**6–81.** Segment $BD$: $B_y = 135$ kN
$D_x = 0$
$D_y = 135$ kN
Segment $ABC$: $C_y = 607.5$ kN
$A_x = 0$
$A_y = 337.5$ kN
Segment $DEF$: $F_y = 607.5$ kN

$E_x = 0$
$E_y = 337.5$ kN

**6–82.** $N_C = 12.7$ kN
$A_x = 12.7$ kN
$A_y = 2.94$ kN
$N_D = 1.05$ kN

**6–83.** $A_x = 167$ N
$A_y = 1.17$ kN
$C_x = 1.33$ kN
$C_y = 833$ N

**6–85.** Member $AB$, $F_{BG} = 264.9$ N
Member $EFG$, $F_{ED} = 158.9$ N
Member $CDI$, $m_s = 1.71$ kg

**6–86.** $m_L = 106$ kg

**6–87.** $F_{FB} = 1.94$ kN
$F_{BD} = 2.60$ kN

**6–89.** Member $AB$: $F_{BD} = 0.8119$ kN
$B_x = 0.4871$ kN
$B_y = 0.6495$ kN
$A_x = 0.2629$ kN
$A_y = 0.6495$ kN

**6–90.** $E_x = 4.7243$ kN
$E_y = 2.5$ kN
$D_x = 4.7243$ kN
$D_y = 5$ kN

**6–91.** $N_A = 4.60$ kN
$C_y = 7.05$ kN
$N_B = 7.05$ kN

**6–93.** Pulley $E$: $T = 1.75$ kN
Member $ABC$: $A_y = 35$ kN
Member $DB$: $D_x = 9.116$ kN
$D_y = 9.216$ kN
$A_x = 9.966$ kN

**6–94.** $W_s = 16.74$ N

**6–95.** $F = 562.5$ N

**6–97.** $80 - N_G \cos 36.03° - N_C \cos 36.03° = 0$
$N_B - N_C = 49.5$ N

**6–98.** $M = 2.43$ kN·m

**6–99.** $F = 5.07$ kN

**6–101.** Member $ABC$
$A_y = 245$ N
Member $CD$
$D_y = 245$ N
$D_x = 695$ N
$A_x = 695$ N

**6–102.** $F_{CD} = 1.01$ kN
$F_{ABC} = 319$ N

**6–103.** $A_y = 183$ N
$E_x = 0$
$E_y = 417$ N
$M_E = 500$ N·m

**6–105.** Member $BC$
$C_y = 1.33$ kN
$B_y = 549$ N
Member $ACD$
$C_x = 2.98$ kN
$A_y = 235$ N
$A_x = 2.98$ kN
$B_x = 2.98$ kN

**6–106.** $F_{AC} = 12.561$ kN
$F_{AB} = 15.384$ kN
$F_{AD} = 17.159$ kN

**6–107.** $F = 875$ N
$N_C = 1750$ N
$F = 437.5$ N
$N_C = 437.5$ N

**6–109.** Clamp
$C_x = 1175$ N
Handle
$F = 370$ N
$F_{BE} = 2719.69$ N

**6–110.** $N_A = 284$ N

**6–111.** $W_C = 0.812W$

**6–113.** $\Sigma M_E = 0;\quad W(x) - N_B\left(3b + \tfrac{3}{4}c\right) = 0$
$\Sigma M_A = 0;\quad F_{CD}(c) - \dfrac{Wx}{\left(3b + \tfrac{3}{4}c\right)}\left(\tfrac{1}{4}c\right) = 0$

$\dfrac{Wx}{12b + 3c}(4b) + W\left(1 - \dfrac{x}{3b + \tfrac{3}{4}c}\right)(b) - W_1(a) = 0$

$W_1 = \tfrac{b}{a}W$

**6–114.** $F_{IJ} = 9.06$ kN (T)
$F_{BC} = 15.4$ kN (C)

**6–115.** $N_E = 187$ N

**6–117.** $l_{AB} = 861.21$ mm, $L_{CAB} = 76.41°$,
$F_{AB} = 9.23$ kN
$C_x = 2.17$ kN
$C_y = 7.01$ kN
$D_x = 0$
$D_y = 1.96$ kN
$M_D = 2.66$ kN·m

6–118. $A_x = 600$ N
$A_y = 0$
$N_C = 75$ N

6–119. $A_x = 400$ N
$A_y = 400$ N
$B_y = 666.67$ N
$B_x = 1666.67$ N
$C_x = 2066.67$ N
$C_y = 266.67$ N

6–121. $N_c = \frac{4P \sin^2 \theta}{\sin \phi}$
$M = \frac{4PL \sin^2 \theta}{\sin \phi}[\cos(\phi - \theta)]$

6–122. $W_1 = 22.5$ N
$W_2 = 157.5$ N
$W_3 = 562.5$ N

6–123. $P = 283$ N
$B_x = D_x = 42.5$ N
$B_y = D_y = 283$ N
$B_z = D_z = 283$ N

6–125. $-\frac{6}{9} F_{DE}(3) + 180(3) = 0$
$F_{DE} = 270$ kN
$B_z + \frac{6}{9}(270) - 180 = 0$
$B_z = 0$
$B_x = -30$ kN
$B_y = -13.3$ kN

6–126. $A_z = 0$
$A_x = 172$ N
$A_y = 115$ N
$C_x = 47.3$ N
$C_y = 61.9$ N
$C_z = 125$ N
$M_{Cy} = -429$ N · m
$M_{Cz} = 0$

6–127. $F_B = 625$ N

6–129. $F_{DB} = F_{BE} = 0$
Joint C: $F_{CB} = 17.9$ kN (C)
$F_{CD} = 8.00$ kN (T)
Joint D: $F_{DE} = 8.00$ kN (T)
Joint B: $F_{BA} = 17.9$ kN (C)
Joint A: $F_{AE} = 8.00$ kN (T)

6–130. $F_{BF} = 0$
$F_{BC} = 0$
$F_{BE} = 500$ kN (T)
$F_{AB} = 300$ kN (C)
$F_{AC} = 583$ kN (T)
$F_{AD} = 333$ kN (T)
$F_{AE} = 667$ kN (C)

$F_{DE} = 0$
$F_{EF} = 300$ kN (C)
$F_{CD} = 300$ kN (C)
$F_{CF} = 300$ kN (C)
$F_{DF} = 424$ kN (T)

6–131. $F_{BF} = 0$
$F_{BC} = 0$
$F_{BE} = 500$ kN (T)
$F_{AB} = 300$ kN (C)
$F_{AC} = 972$ kN (T)
$F_{AD} = 0$
$F_{AE} = 367$ kN (C)
$F_{DE} = 0$
$F_{EF} = 300$ kN (C)
$F_{CD} = 500$ kN (C)
$F_{CF} = 300$ kN (C)
$F_{DF} = 424$ kN (T)

6–133. Member AC: $C_x = 402.6$ N
$C_y = 97.4$ N
Member AC: $A_x = 117$ N
$A_y = 397$ N
Member CB: $B_x = 97.4$ N
$B_y = 97.4$ N

6–134. $P = \frac{kL}{2 \tan \theta \sin \theta}(2 - \csc \theta)$

6–135. $A_x = 62.31$ kN
$A_y = 2.31$ kN
$E_x = 62.31$ kN
$E_y = 42.69$ kN

## Chapter 7

7–1. $P \cos 30° + 0.25N - 50(9.81) \sin 30° = 0$
$P = 140$ N
$N = 494.94$ N

7–2. $P = 474$ N

7–3. $\mu_s = 0.256$

7–5. $90(9.81)(3 \cos \theta) - 0.4(90(9.81))$
$(3 \sin \theta) - 90(9.81)(0.9) = 0$
$\theta = 52.0°$

7–6. $\mu_s = 0.231$

7–7. Yes, the pole will remain stationary.

7–8. $150 (3.9 \cos \theta) - 45 (7.8 \sin \theta) = 0$
$d = 4.013$ m

7–10. $P = 73.6$ N

7–11. $P = 4.905$ N

7–13. $F_B = 280$ N
$N_B = 700$ N
$P = 350$ N

**7–14.** $\mu_s = 0.577$
**7–15.** $F_B = 200$ N
**7–17.** $N_D = 469.32$ N
Boy does not slip.
$F_D = 184.62$ N
$A_y = 2.326$ kN
$B_x = 173.21$ N
$B_y = 1.133$ kN
**7–18.** $\mu_s = 0.595$
**7–19.** $\theta = 10.62°$  $x = 0.046$ m
**7–21.** $N_A = 1000 \cos \theta$
$N_B = 750 \cos \theta$
$\theta = 16.3°$
$F_{CD} = 41.13$ N
**7–22.** $n = 12$
**7–23.** $P = 5.86$ N
**7–25.** Assume $P = 500$ N
$N = 800$ N
$x = 0.431$ m $< 0.45$ m
$P = 500$ N
**7–26.** $P = 225$ N
$\mu_s' = 0.300$
**7–27.** The man is capable of moving the refrigerator.
The refrigerator slips.
**7–29.** $P = 147.6$ N
$N_A = 64.3$ N  $N_B = 361.9$ N
**7–30.** Tractor can move log.
**7–31.** $W = 4.183$ kN
**7–33.** $F_A = 86.60$ N
$N_A = 648.5$ N
The bar will not slip.
**7–34.** $\theta = \tan^{-1}\left(\dfrac{1 - \mu_A\mu_B}{2\mu_A}\right)$
**7–35.** $P = 0.633$ N
**7–37.** $N = wa \cos \theta$
$b = 2a \sin \theta$
**7–38.** $h = 0.48$ m
**7–39.** $\theta = 33.4°$
$\mu_s = 0.3$
**7–41.** $F_A = 0.3714 F_{CA}$
$N_A = 0.9285 F_{CA}$
$\mu_s = 0.4$
**7–42.** He can move the crate.
**7–43.** $\mu_s' = 0.376$
**7–45.** $N_A = 551.8$ N
$B_x = 110.4$ N
$B_y = 110.4$ N
$M = 77.3$ N·m

**7–46.** $F_A = 71.4$ N
**7–47.** $P = 589$ N
**7–49.** $T = 11\,772$ N
$N_l = 9.81 m_l$
$m_l = 1500$ kg
**7–50.** $m_l = 800$ kg
**7–51.** $P = 1.02$ kN
**7–53.** $N = 243$ N
Slipping of board on saw horse $P_x = 121.5$ N.
Slipping at ground $P_x = 95.4$ N.
Tipping $P_x = 106$ N.
The saw horse will start to slip.
**7–54.** The saw horse will start to slip.
**7–55.** $\mu_s = 0.3462$
**7–57.** $P = 294.3$ N
$N' = 735.75$ N
$F' = 294.3$ N
**7–58.** $P = 441.45$ N
**7–59.** $\theta = 16.0°$
**7–61.** $N_C = 377.31$ N
$N_D = 188.65$ N

$N_A = 150.92$ N
$N_B = 679.15$ N
$F_B = 37.73$ N
**7–62.** $P = 225$ N
**7–63.** $P = 49.0$ N
**7–65.** $N_B = 412.8$ N  $N_C = 1376.1$ N
$P = 453.7$ N
**7–66.** $P = 1.98$ N
**7–67.** $P = 863$ N
**7–69.** $N_A = 1212.18$ N
$N_C = 600$ N
$P = 1.29$ kN
**7–70.** All blocks slip at the same time; $P = 3125$ N
**7–71.** $P = 574$ N
**7–73.** $N_A = 0.5240W$
$N_B = 1.1435W$
$F_B = 0.05240W$
$P = 0.0329W$
**7–74.** $P = 1.38W$
**7–75.** $P = 1.80$ kN
**7–77.** $\theta = 7.768°$
$\phi_s = 11.310°$
$F = 620$ N
**7–78.** $M = 0.712$ N·m
**7–79.** $M = 212.0$ N·m

**7–81.** $\theta = 5.455°$
$\phi_s = 14.036°$
$F = 678$ N
**7–82.** $F = 71.4$ N
**7–83.** $F = 49.2$ N
**7–85.** $F_{CA} = F_{CB}$
$F = 1387.34$ N
$F_{BD} = 1387.34$ N
$F_{AB} = 1962$ N
$\theta = 5.455°$
$\phi_s = 14.036°$
$F = 74.0$ N
**7–86.** $F = 174$ N
**7–87.** $N_C = 123$ N
$N_A = 42.6$ N
**7–89.** $T_B = 68.391$ N
$F_C = 68.4$ N
$N_B = 329.0$ N
$F_B = 192.5$ N
**7–90.** $F = 1.31$ kN
$F = 372$ N
**7–91.** $F = 4.60$ kN
$F = 16.2$ kN
**7–93.** $N = 907.425$ N
$F = 671.55$ N
Yes, just barely.
**7–94.** $T_1 = 283.1$ N
**7–95.** $\theta = 24.2°$
**7–97.** $F = 4.75P$
$F' = 19.53P$
$P = 42.3$ N
**7–98.** $M = 187$ N·m
$T_A = 616.67$ N
$T_C = 150.00$ N
**7–101.** $T_2 = 1767.77$ N
$T_1 = 688.83$
$(\mu_s)_{\text{req}} = 0.3$
$M = 216$ N·m
**7–102.** $P = 85.5$ N
**7–103.** Since $F < F_{\max} = 264.9$ N, the man will not slip, and he will successfully restrain the cow.
**7–105.** $T = 486.55$ N $\quad N = 314.82$ N
$\beta = (2n + 0.9167)\pi$ rad
Thus, the required number of full turns is
$n = 2$
**7–106.** The man can hold the crate in equilibrium.

**7–107.** $T_1 = 1.85$ N
$T_2 = 1.59$ N
**7–109.** For motion to occur, block $A$ will have to slip.
$P = 223$ N
$F_B = T = 36.79$ N
**7–110.** $F = 2.49$ kN
**7–111.** $W = 197.6$ N
**7–113.** $T = 20.19$ N
$F_A = 16.2$ N
$N_A = 478.4$ N
$x = 0.00697$ m $< 0.125$ m
No tipping occurs.
**7–114.** $M = 84$ N·m.
**7–115.** $\mu_k = 0.0632$
**7–117.** Apply Eq. 8–7.
$F_{sp} = 8.97$ kN
**7–118.** $M = 270$ N·m
**7–119.** $M = \dfrac{\mu_s PR}{2}$
**7–121.** $N = \dfrac{P}{\cos\theta}$
$A = \dfrac{\pi}{4\cos\theta}(d_2^2 - d_1^2)$
$M = \dfrac{\mu_s P}{3\cos\theta}\left(\dfrac{d_2^3 - d_1^3}{d_2^2 - d_1^2}\right)$
**7–122.** $p_0 = 3.537$ kN/m²
$F = 2.865$ kN
**7–123.** $M = \dfrac{2\mu_s PR}{3\cos\theta}$
**7–125.** $\tan\phi_k = \mu_k$
$\sin\phi_k = \dfrac{\mu_k}{\sqrt{1 + \mu_k^2}}$
$M = \left(\dfrac{\mu_k}{\sqrt{1 + \mu_k^2}}\right)pr$
**7–126.** $P = 215$ N
**7–127.** $P = 179$ N
**7–129.** $\phi_s = 16.699°$
$m_B = 13.1$ kg
**7–130.** $(r_f)_A = 5$ mm
$(r_f)_B = 2$ mm
**7–131.** $(r_f)_A = 7.50$ mm
$(r_f)_B = 3$ mm
**7–133.** $r_f = 2.967$ mm
$R = \sqrt{P^2 + (833.85)^2}$
$P = 814$ N (exact)
$P = 814$ N (approx.)
**7–134.** $P = 209.8$ N
**7–135.** $\mu_s = 0.411$

**7–137.** $\theta = 5.74°$
$P = 96.7$ N
**7–138.** $P = 299$ N
**7–139.** $P = 266$ N
**7–141.** $P = \frac{(1200)(9.81)(0.2 + 0.4)}{2(15)} = 235$ N
**7–142.** $P = 16.4$ N
**7–143.** $s = 0.750$ m
**7–145.** a) $N_A = 5573.86$ N   $T = 2786.93$ N
$W = 6.97$ kN
b) $N_A = 6376.5$ N
$N_B = 5886.0$ N
$T = 6131.25$ N
$W = 15.3$ kN
**7–146.** a) $W = 1.25$ kN
b) $W = 6.89$ kN
**7–147.** $m_B = 1.66$ kg
**7–149.** $N_A = 5$ kN
$N_B = 12.5$ kN
$T = 6.25$ kN
$M = 3.75$ kN·m
**7–150.** $M = 3.31$ kN·m
**7–151.** $\theta = 35.0°$
**7–153.** $N = 39.39$ kN
$F = 6.946$ kN
The wedges do not slip at contact surface $AB$.
$N_C = 40$ kN
$F_C = 0$
The wedges are self-locking.

# Chapter 8

**8–1.** $y_D = 2.4 \sin \theta$
$y_J = 2(2.4 \sin \theta) + b$
$F_{AD} = 3.92$ kN
**8–2.** $\theta = 0°$ and $\theta = 73.1°$
**8–3.** $\theta = 41.2°$
**8–5.** $x_B = 1.8 \cos \theta$
$y_C = 0.9 \sin \theta$
$F_{sp} = 50.2$ N
**8–6.** $F_S = 75$ N
**8–7.** $\theta = 29.1°$
**8–9.** $y_D = 2(0.2 \cos \theta)$
$\delta y_A = 0.5 \, \delta \theta$
$F_E = 177$ N
**8–10.** $F = 60$ N
**8–11.** $P = 2k \tan \theta \, (2l \cos \theta - l_0)$

**8–13.** $y_C = 0.3 \sin \theta$
$y_A = 0.9 \sin \theta$
$\theta = 11.5°$
$\theta = 90°$
**8–14.** $m_l = m\left(\frac{s}{a}\right)$
**8–15.** $\theta = \cos^{-1}\left(\frac{a}{2L}\right)^{\frac{1}{3}}$
**8–17.** $y_{G_b} = 0.25 \sin \theta + b$
$y_{G_t} = 0.25 \sin \theta + a$
$x_C = 0.25 \cos \theta$
$k = 166$ N/m
**8–18.** $F = 200$ N
**8–19.** $\theta = 39.6°$
**8–21.** $y_G = 0.15 \cos \theta$
$y_A = 0.3 \cos \theta$
$x_A = 0.3 \sin \theta$
$F_{sp} = 24.80$ N
$k = 180.0$ N/m
**8–22.** $W_G = 12.5$ N
**8–23.** $x = 400$ mm
**8–25.** $125^2 = y_C^2 + 75^2 - 2(y_C)(75) \cos(90° - \theta)$
$F = 1295$ N
**8–26.** $x = 0.590$ m
$\frac{d^2V}{dx^2} = 12.2 > 0$   stable
$x = -0.424$ m
$\frac{d^2V}{dx^2} = -12.2 < 0$   unstable
**8–27.** $\theta = 90°$
$\frac{d^2V}{d\theta^2} = 16 > 0$   stable
$\theta = 36.9°$
$\frac{d^2V}{d\theta^2} = -25.6 < 0$   unstable
**8–29.** $V = 5886 \cos \theta + 9810 \sin \theta + 39\,240$
$\theta = 59.0°$
**8–30.** $W_D = 1.376$ kN
**8–31.** $h = 0.218$ m
$\frac{d^2V}{dh^2} = 14 > 0$   stable
**8–33.** $V = 6.25 \cos^2 \theta + 7.3575 \sin \theta$
$+ 24.525a + 4.905b$
$\theta = 36.1°$
**8–34.** $x = 1.23$ m

**8–35.** $\theta = 72°$
$\dfrac{d^2V}{d\theta^2} = 346706 > 0$ stable
$\theta = 14.2°$
$\dfrac{d^2V}{d\theta^2} = -3314 < 0$ unstable

**8–37.** $V = -4.415 \, m_E \sin\theta + 202.5 \cos^2\theta - 405 \cos\theta - 9.81 \, m_E b + 202.5$
$m_E = 7.10$ kg

**8–38.** $\theta = 64.81°$
$\dfrac{d^2V}{d\theta^2} = 16.85 > 0$ stable
$\theta = 0°$
$\dfrac{d^2V}{d\theta^2} = -9 < 0$ unstable

**8–39.** $\theta = 20.2°$
$\dfrac{d^2V}{d\theta^2} = 17.0 > 0$ stable

**8–41.** $V = mg(r + a \cos\theta)$
Thus, the cylinder is in unstable equilibrium at $\theta = 0°$ (Q.E.D.)

**8–42.** $h = 0$

**8–43.** $h = \sqrt{3}\, r$

**8–45.** $\bar{y} = \tfrac{1}{4}(h + d)$
$V = \dfrac{W(h - 3d)}{4} \cos\theta$
$d = \tfrac{h}{3}$

**8–46.** $\theta = 0°$, $\dfrac{d^2V}{d\theta^2} = -1575.6 < 0$ unstable

**8–47.** $h = 35.46$ mm

**8–49.** $\bar{y} = \dfrac{6h^2 - d^2}{4(3h - d)}$
$V = W\left[\dfrac{6h^2 - 12hd + 3d^2}{4(3h - d)}\right] \cos\theta$
$d = 0.586\, h$

**8–50.** $F = 512$ N

**8–51.** $\theta = 90°$ and $\theta = \sin^{-1}\left(\dfrac{W}{2kL}\right)$

**8–53.** $V = 73.5 \sin^2\theta - 147 \sin\theta - 70 \cos\theta + 73.5$
$\theta = 38.45°$
$\dfrac{d^2V}{d\theta^2} = 179.5 > 0$ stable

**8–54.** $P = \left(\dfrac{b - a}{2c}\right) mg$

**8–55.** $\theta = 90°$
$\dfrac{d^2V}{d\theta^2} = 918 > 0$ stable
$\theta = 9.47°$
$\dfrac{d^2V}{d\theta^2} = -1368 < 0$ unstable

**8–57.** $V = 45 \sin^2\theta + 22.5 \cos\theta$
$\theta = 0°$
$\dfrac{d^2V}{d\theta^2} = 67.5 > 0$ stable
$\theta = 72.5°$
$\dfrac{d^2V}{d\theta^2} = -80.5 < 0$ unstable

**8–58.** $h = \dfrac{2kl^2}{W}$

## 教学支持申请表

为了确保您及时有效地申请培生整体教学资源，请您务必完整填写如下表格，加盖学院的公章后传真给我们，我们将很快为您处理。

需要申请的资源（请在您需要的项目后划"√"）：

- ☐ 教师手册
- ☐ PPT
- ☐ 题库

请填写所需教辅的开课信息：

| 采用教材 | | | | ☐中文版 ☐英文版 ☐双语版 |
|---|---|---|---|---|
| 作　者 | | 出版社 | | |
| 版　次 | | ISBN | | |
| 课程时间 | 始于　年　月　日 | 学生人数 | | |
| | 止于　年　月　日 | 学生年级 | | ☐专科　☐本科 1/2 年级<br>☐研究生　☐本科 3/4 年级 |

请填写您的个人信息：

| 学　校 | |
|---|---|
| 院系/专业 | |
| 姓　名 | |
| 通信地址/邮编 | |
| 手　机 | |
| 传　真 | |
| Official E-mail（必填）<br>（eg：×××@ruc.edu.cn） | |

| 职　称 | ☐助教　☐讲师　☐副教授　☐教授 |
|---|---|
| | |
| 电　话 | |
| E-mail<br>（eg：×××@163.com） | |

是否愿意接受我们定期的新书讯息通知：　☐是　☐否

系 / 院主任：_____（签字）

（系 / 院办公室章）

___年___月___日

100037　北京市西城区百万庄大街 22 号　机械工业出版社高教分社　姜凤
电话：（010）88379408
传真：（010）68997455

Please send this form to：ajiang2001@sina.com
Website：www.pearson.com